GUIDE TO THE HIGH TECHNOLOGY INDUSTRIES

A SERVICE OF
HARPER & ROW,
PUBLISHERS, INC.

Cambridge		London
Hagerstown		Mexico City
Philadelphia		Sao Paolo
San Francisco		Sydney

1817

GUIDE TO THE HIGH TECHNOLOGY INDUSTRIES

FIRST EDITION

HARFAX • A Service of Harper & Row, Publishers, Inc.

BALLINGER PUBLISHING COMPANY

Cambridge, Massachusetts

A Subsidiary of Harper & Row, Publishers, Inc.

International Standard Book Number: 0-88410-619-5

ISSN 0738-2324

Printed in the United States of America

Table of Contents

Part 2
Data Publishers and Producers

Part 3
Subject Indexes

Part 4
Title Index

Introduction

The first literature on high technology was science fiction. Before computers or microelectronics or satellites, there were fantasies. This directory, alas, must omit science fiction, but the imagery and excitement shines through in much of the marketing and financial analysis that it covers.

We are still adjusting to the speed with which technology has caught up with our imaginations. *Time* magazine's Man-of-the-Year was not a man at all, but rather the personal computer. It was this technology that had the greatest impact on our lives in 1982. The sentimental favorite among the nominees was also not a man, but E.T., the extraterrestrial. He may be alien, but he is still a more comfortable way of embracing the possibilities created by science.

Not only are we psychically unprepared, but also our social and economic institutions are caught without the necessary information for planning and policy making. Mature industries have well established intelligence systems, but high technology is too new and dynamic. Because there is no Bible, there are plenty of prophets. Finding hard data is something else, however. This compendium is a step toward making some sense of it all.

Some market research firms, cited herein, are making reputations as specialists in the field. Their surveys and projections are state-of-the-art analysis by industry experts. They are an excellent source, but they are pioneers. When you find lots of pioneers it is a sure sign that you are in the wilderness. Stock analysts have a more conservative reputation, but they too must guess the winners before the rest of the crowd. In their efforts to plot the course of firms and industries through a very uncertain financial environment, they produce valuable statistics. They too are among the best sources cited here. The biggest compiler of market and industry data is still the government. This directory is a good guide to the many different offices that touch upon high technology.

There are also references to trade associations. Groups of manufacturers and suppliers can simply poll their members for good estimates of production and sales. Some issue statistical bulletins; others publish journals and maintain databases. These sorts of institutions lay the basis for more stable channels of industry data. Of course, there are independent trade journals and database producers as well, and some of these are recognized industry experts. Special issues of these journals are some of the most widely available sources of data, but they can be hard to locate. They are easily accessible using this directory.

More scholarly treatments of the industries are found in the monographs, dissertations, and conference papers. For quick access to addresses of suppliers or manufacturers' annual sales, there are directories and buyer's guides. These can be invaluable for generating lists of clients or employment leads.

All of these follow, but first, what is the question? What is high technology and why is it so important? Suddenly it's here, and science fiction did not prepare us for all the ramifications.

Developments during World War II, the Cold War, and the Space Race laid the groundwork for what we now call high technology. Government sponsored work at Bell Labs, General Electric, IBM, and various universities made major improvements in radar, microwave communications, television, computers, satellites, navigational systems, optoelectronics, instrumentation, and most of all semiconductors and integrated circuits. Between 1950 and 1970 federal contracts for semiconductors alone totaled $900 million.

The rapid post-war recovery of commercial markets enabled labs to turn their equipment, their experienced engineers, and their discoveries to consumer goods. It was like putting a match to tinder. The transistor was invented in 1948 under defense contract work conducted at Bell Labs. By the 1960s transistor radios were a necessity for every teenager in America. Computers were one-of-a-kind, experimental monsters at a few universities in the late 1940s. By 1985 there should be roughly a billion of them worldwide. The first communications satellite was launched in 1958. By 1965 the Intelsat system was ready for regular commercial use.

Rapid growth alone does not define high technology. For our purposes, high technologies are those applied to processing and transmitting information electronically. Semiconductors are the basic units. They can amplify, rectify, transform, transduce, and perform all the other basic electronic functions. In addition, they can sense light, heat, moisture and other environmental qualities; keep time; and light up for displays. Integrated circuits (ICs) combine the functions of many semiconductors on a single chip of silicon. Computers use ICs for logic, memory, and communications. Telecommunications, information handling in the most obvious form, also uses electronics.

Developments in these fields have spawned many new products, but what is really significant, and what enables us now to speak of high technology as one technology, is the convergence of these innovations into the Information Industry.

Just as we are leaving an era of cheap energy, we are entering one of cheap information. Like energy, information is ubiquitous but dependent upon special technologies to collect, process, and distribute it. The information industry not only automates these technologies — it also joins them into a system with potentially unrestricted flows among them. Every element becomes more powerful and more useful with the support of the others.

The synthesis of computers and ICs is illustrative. Computer power has increased a million fold since inception: functions that would have cost $200,000 using tubes can now be performed on a 2¢ chip. From 1975 to the present, memory chips have improved still more, from 1000 bits for 2.3¢ per bit to 256,000 bits for a mere .15¢ per bit! All this and size reductions, energy efficiency, and failure rates thousands of times better do not just make computers better, they are made qualitatively different. Changing them from curiosities to commodities gives them social and economic impact that rearranges values throughout the whole culture.

Telecommunications also have employed ICs to great advantage. Switching, routing, and other functions formerly performed by human or electro-mechanical links, have gone electronic. Satellite systems, packet switching, cellular microwave systems, and fiber optics are hardly conceivable without advanced electronic components. The crux of the information industry, however, is in the synergy of telecommunications and computers. Certainly computers improve telecommunications, but allowing computers to communicate with one another has genuine revolutionary potential. With these technologies working together, aspects of the information industry become possible that are neither computers nor telecommunications but are dependent upon both of them and are, in turn, used by them.

Value-added networks for data communications form an information marketplace. Through them, corporate headquarters can share data with distant subsidiaries or with clients' computers, and database producers and service vendors can deliver their goods.

There is one other aspect of the information industry. Perhaps it would be wrong to call computer software a technology, but it is the glue that holds it all together. The electronics we have been talking about understands little more than off or on. To communicate pertinent information requires a complex architecture of languages and programs. Software is cumulative but far from standardized. Translating between incompatable software remains one of the main tasks. If ever we reach the ideal of a universal information network, it will be thanks to the programmers who ironed out the differences between all the machines and systems.

The saying used to be: the possibilities are limited only by your imagination. It would be more accurate now to say that the only limitation is the market. Futuristic projections make exciting popular literature, but discovery and innovation in electronics have become so institutionalized that the question is not what could be done, but what will sell.

The trick to marketing information-handling equipment is to replace expensive labor with cheap hardware. The most expensive labor is that of programmers, hence the greatest returns come from user-friendly software and machines that run it. It also makes sense to target well-paid executives rather than clerical workers. Unfortunately, investments in total decision support systems are much riskier than piecemeal upgrading of typewriters. One must consider compatability, obsolescence, and vendor viability.

Vendors also must keep one eye on cost and technological trends. The other must watch the industry giants. IBM is the *de facto* leader of the computer market, and AT&T casts its deregulated shadow over telecommunications. Even the giants have their problems. They must plan their strategies with Japan in mind.

In the 1970s the business market surpassed the military as the biggest consumer of electronics and other high tech goods. Households and factories are the other major segments. The problem with trying to describe these markets is not only the proliferation of new products, but also the overlaps among multifunctional machines. Systems are available for automating an entire office around a local area network; for specific applications such as computer-aided design or supermarket point-of-sale systems; or for specific end users such as engineers, pharmacists, or farmers. These systems may use different technologies to perform the same tasks and may distribute functions differently among machines. With these caveats, let us peruse the products in the high tech marketplace.

The usual business computer is what is called a mini, not to be confused with a micro, or personal computer. The largest and most powerful mainframes are used mainly for scientific and engineering applications. The technology of the central processing unit (CPU) is widely shared and well developed. The distinctions among competing products are in the peripheral components affecting communications capability, memory capacity and speed, and numbers and types of input/output devices supported. More than any hardware, however, it is the software that sells machines.

Computers are unique among machines in that they must be told what to do, that is, programmed, before they are of any use at all. The basic software for operating the machine usually comes with the hardware purchase, but each application—accounting tasks, word processing, database management, and so on—requires still more software. This may be written in-house, but programmers are highly skilled, scarce, and hence highly paid. In many cases a software package available from the vendor or a software company will suffice. It may even be more economical to go to an outside service vendor for certain jobs, even if one's own computer could do the job but lacks the software. The cost and difficulty of producing software is, in fact, one of the major factors shaping the information industry. Computer service companies continue to thrive despite the proliferation of computers. Companies are able to carve out a comfortable niche in the market by virtue of the compatability or special applicability of their software.

In their short history, computers have already gone through major changes in the way they are used in the organization. The trend is away from centralized mainframes and toward intelligent terminals or personal computers. For those who still need big computer power, they can gain access through a local network.

Data communication functions are also being built into peripheral equipment like word processors, copiers, and printers. Local area networks (LANs) based on telephone lines, coaxial cable, or fiber optics, can link all the nodes of an integrated information system. Telephone based networks rely on private automatic branch exchanges (PBXs). These offer automatic dialing, lowest cost routing, number storage, and call queuing. They can also be the intelligent link for data communications, performing interface, scheduling, and protocol functions among incompatable machines, while still handling voice communications.

PBXs can also interconnect with telecommunications networks outside the local area, opening access to databases, teleconferencing, subsidiaries' and clients' computers, and other PBXs. Value-added, packet-switched networks, cable and satellite based, also can provide the protocols and interfacing software to connect different equipment designs and brands. The day is coming when universal networks could connect all sorts of equipment no matter what the system architecture, software, or protocol.

Memory innovations have been as important to commercial growth of computers as logic. Capacity and speed of retrieval have increased—with decreasing costs—using tape and disks. Magnetic bubble storage and optical disks are waiting in the wings with the potential of still greater performance. Optical disks are particularly attractive for document storage since the concept of a "document" has expanded to include still and moving images, text, data, and sound, in various combinations. Computer-assisted micrographics systems are already available and offer remarkable convenience, speed, and capacity.

Typewriters are often the first entry point for information in the office. Electronic features have made traditional machines more efficient, and word processors have merged typewriter and computer functions, eliminating intermediary stages and giving data entry personnel the power of the computer to edit, store addresses or form letters, or manipulate data. Xerographic duplicating machines are being made smarter and being interconnected with computer communications networks so that documents can be drawn from storage in a computer in one place, transmitted, and printed at another site.

There are also optical recognition devices that can read text directly from a printed page and store or transmit it in digital form. Voice synthesis and recognition could further smooth the gaps between machines and operators.

High technology affects businesses outside the office too. Remote data entry devices are useful for inventory and shop floor data collection. Point-of-sale systems at supermarkets and discount stores combine bar code recognition sensors with database systems for retrieving prices and storing sales records. Human error is minimized, labor costs reduced, and accounting simplified. Furthermore, the increased detail, timeliness, and manipulability of sales data enables better market monitoring and assessment of marketing strategies.

Automatic teller machines link specialized terminals, via networks, to centralized bank computers, where transactions are recorded, records are updated, and funds electronically transferred. Private computers can make financial transactions from homes or offices via network services offered by some banks and brokers.

The technology is available for other teleshopping from homes plus information, education, entertainment, and electronic mail services. Telemetric systems for monitoring security, fire alarms, heating and air conditioning, and eventually smart appliances are ready to go. The question is which of the many possible routes will information and interactive services take. Homes could receive signals via traditional video broadcasts, direct satellite broadcasts, cable, television lines, microwave, or FM radio signals. Two-systems could use cable, telephone, or hybrids combining, for example, direct broadcast satellite for incoming signals and telephone for outgoing. The bright star of the home market has been video games. The explosion of video game popularity caught most market experts by surprise. It was expected that home entertainment would be popular, but the focus was on video recorders and cable. Arcade games caught the fancy of youth, however, and demand for home games has led the penetration of home computer markets. Once installed, they are ready for other services.

Many of the new products of the information age will be produced in factories that are, themselves, profoundly influenced by electronics, computers and sophisticated new instruments. Computer-aided design (CAD) already is improving the efficiency of some engineers. Systems provide interactive graphic displays, often with color and 3-D features; linkage to a database of technical specifications; and computer storage and retrieval capacity. Eventually CAD systems can be connected with computer-aided manufacturing (CAM) systems. Designs could then be translated directly into instructions for numerically controlled machine tools, photolithographic circuit board printers, or robot assemblers. Not only are labor costs reduced, but the chances for error are minimized. Smaller, more flexible factories will be able to change or upgrade products quickly and outmaneuver older plants.

Test equipment and process controls using new electronic sensors, microprocessors, and optoelectric displays can improve quality control. One hundred percent testing becomes feasible with very high speed devices. Programmability is featured, as well as versatility, efficiency, and compactness. High technology is exacting technology. Semiconductors approach thicknesses of one molecule, state-of-the-art photolithographic techniques require electron beams to cut the minute stencils, and optical fibers are much finer than human hairs. The machinery for producing and testing high tech products is itself highly advanced.

Robots combine sensors with computers in agile mechanical slaves that can assemble, adjust, weld, and more, within fine tolerances, following complex programs, over and over. They promise to take over many of the tasks in the auto industry and elsewhere despite the worries of labor unions. More than any other new product, they personify the threat and the promise of high technology.

Last, we return to military markets where it all began. The requirements of many military tasks are quite demanding: missile guidance, surveillance, avionics, and navigation by satellite, for example. Often the newest techniques and the most exotic materials are tried first in military applications.

The controversial relationship between high technology and the military industrial complex enters into the market and financial literature, as do other political and social issues. Defense budgets argued in the political arena, for example, have major impacts on some firms. Also, technology—or rather the restriction of technical information—is used as an economic weapon against the Soviet Bloc. The sophistication of missiles and aircraft is not just a military issue; the super powers also compete in world armament markets. Military superiority is as important as ever, but high technology changes what is strategic. The vulnerability of suppliers of rare metals needed for semiconductors and found mainly in Southern Africa has become important. The United States is proceeding with plans for the development of laser beam weapons for defense of space because satellites have become indispensible. The latest thinking of Defense Department analysts is that the speed, accuracy, and survivability of command, control, and communications (C3) will determine who has the upper hand in the next round of the Cold War.

Technological competition with the Soviet Union, as serious as it is, has not received as much attention as competition with Japan and Western Europe. Our allies have entered the technology race with careful planning, close cooperation between government and industry, more engineers, and educational policies for maintaining superior human resources. MITI, the Japanese Ministry of Industry and Trade, has convened experts well in advance to target susceptible markets. They have distributed government and industry funds for research and development, staged industrial development, protected domestic markets, and promoted foreign trade. France has launched a concerted, well-publicized effort to build the infrastructure that will make France the model for the information society of the future. Virtually every advanced nation has some policy to promote high technology, except the United States, and a number of bills have been proposed here.

The stakes in the high tech race are higher than they might at first appear. There is money to be made, of course, which also means employment and foreign exchange. High tech is the sector with the highest growth rates and the best returns. Moreover, advanced nations cannot compete with the cheaper labor attracting heavy industry to developing countries. The high tech industries are in flux now, but when they mature, as some commodity semiconductor markets have already, the companies with the best market positions will have the high ground for the foreseeable future.

There is an economic shift underway as fundamental as the transition from agriculture to manufacturing. To see computers and electronic gear only as the hottest selling items in the market is to miss the point. Information is the hottest item; the equipment only handles it. We are leaving an era in which economic power was measured by capital investments. The emerging economy is knowledge based, and knowledge is nothing but the right information at the right time. Capital is invested in machinery while knowledge is invested in people. Most people in advanced nations no longer work in manufacturing, but rather in services. What they do primarily is handle information. There are only two ways to invest them with knowledge: education and better tools for information handling.

Some experts have criticized the current national economic policy for its support of "sunset" industries at the expense of "sunrise" ones. The philosophy that proposes lower taxes and reduced government regulation in order to increase capital investment and productivity is conceived from truths about capital intensive heavy industry, but not high tech. Steel mills and auto factories benefit from accelerated depreciation and investment tax credits. High tech companies, however, do not have massive plants. What they need is liberal credit for high risk R&D investments and lower capital gains taxes. The national policy, unfortunately, maintains high interest rates. Not only is capital diverted to sunset industries, but government revenues that could be supporting high tech, especially through aid to education, are all but dried up. Japan, in contrast, has interest rates on the order of one third those in the United States. Government assumes some risks for R&D investments, and there is much less emphasis on short term returns (see *Global Stakes: The Future of High Technology in America*, Ballinger Publishing Company).

The issues discussed thus far have concerned the importance of high tech in global and national realms. This final note concerns a smaller scale—the community. Economic transitions have regional manifestations too. In the United States high tech has been concentrated in Massachusetts, Southern California, and newly developed areas like North Carolina's Research Triangle. The competition for these new businesses is getting fierce, with local and state governments offering tax breaks, help with financing, infrastructure, and training programs. It is little wonder. They are nonpolluting, they employ skilled labor, and they are growing at a time when factories are closing. Any consideration of a national high tech policy must include the old steel towns of the Northeast. These communities will not support a high tech policy unless it is part of a comprehensive, fairly distributed, reindustrialization plan.

It is not science fiction anymore, it is economic facts. Understanding is the best defense.

How to Use the Guide

The **Harfax** *Guide to High Technology Industries* provides bibliographic citations to sources of financial and marketing data. Statistical and directory information can be found in numerous documents and databases. The Guide organizes these sources industrially, regionally, and by document type. The high technology industries indexed are Communications Equipment, Computers, Computer Services, Electrical and Electronic Equipment, Office Equipment, Scientific and Technical Instruments, and Telecommunications.

Computer Services has been considered a distinct industry since January 1982. Prior to that date, "Computer Services" were categorized under the heading "Computers." Literature searchers requiring Computer Services citations should use both headings.

ORGANIZATION OF GUIDE

Part 1: High Technology Industries
(Industrial/Regional Bibliographic Citations)

Part 2: Section 1: Publisher Listings

Part 2: Section 2: Publisher Index

Part 3: Section 1: SIC Code Index

Part 3: Section 2: Subject Index

Part 4: Title Index

PART 1: HIGH TECHNOLOGY INDUSTRIES
(Industrial/Regional Bibliographic Citations)

The industrial/regional citations are arranged by industry, broad geographic region, document type, and alphabetically by title. Document types covered, and the sequence in which they occur, are: Market Research Reports, Investment Banking Reports, Industry Statistical Reports, Financial and Economic Studies, Forecasts, Directories and Yearbooks, Special Issues and Journal Articles, Numeric Databases, Monographs and Handbooks, Dissertations and Working Papers, Conference Proceedings, Newsletters, Services, Sound Recordings, Maps, Indexes and Abstracts, Bibliographic Databases, and Dictionaries.

Each document is given complete bibliographic coverage and annotations under the industry with which it is concerned. A document containing data on more than one high technology industry is fully cited under the industry that is its prime focus. A numeric "See" reference is made at the end of other relevant industries.

Items Included in Citations in Part 1

1. Industry
2. Region
3. Document type and/or journal title
4. Citation number
5. Document, article, or special issue title
6. Editor or author
7. Publisher
8. Frequency
9. Edition, date, and pages
10. Price, if available
11. Abstract

Part 1: High Technology Industries   **Communications Equipment**[1] **/North America**[2]

Forecasts[3]

227.[4] **THE FUTURE FOR ELECTRONIC MAIL AS A BUSINESS OPPORTUNITY.**[5] *Bengston, R.*[6] *Battelle Memorial Institute.*[7] *October 1979.*[9] *$4000.00.*[10]

Attempts to determine when electronic mail systems will be used commercially in the U.S., who will initially need and purchase systems, what sorts of hardware and software will be required, and other market information of interest to hardware producers.[11]

See 1355, 1364, 1369, 1371, 2463, 2471, 4175, 4176, 4178, 4179, 4180, 4181, 4183, 4184.[4]

Part 1: High Technology Industries   **Computers**[1] **/North America**[2]

Forecasts[3]

1371.[4] **TWO WAY BUSINESS COMMUNICATIONS, SERVICES AND EQUIPMENT.**[5] *Frost and Sullivan, Inc.*[7] *February 1980. 238 pg.*[9] *$950.00.*[10]

Includes information on data communications, communications, word processors, electronic mail, computer reference, facsimile, phone systems, telephone answering, picture phone, video conferencing. User survey results.[11]

Journal articles and special issues appear under the document type "Special Issues and Journal Articles," arranged first by journal title and then by article title.

Special Issues and Journal Articles[3]

Advertising Age[3]

2551.[4] **THE BIG BANK MARKETING OF HOME VIDEO-GAMES.**[5] *Mansfield, Matthew F.*[6] *Crain Communications, Inc.*[7] *v. 53, no. 36. August 30, 1982. pg. M1–M3, M30*[9]

Discusses state of the home video game industry and estimates market size. Provides 1982 estimated advertising budget for each manufacturer. Lists advertising agencies of each manufacturer. Gives 1982 estimated consoles already installed in U.S. homes; 1982 estimated consoles shipped; and 1982 estimated game cartridge shipments.[11]

Appliance[3]

2552.[4] **APPLIANCE: TRADE SHOW IN PRINT 1981.**[5] *Dana Chase Publications, Inc.*[7] *Annual.*[8] *June 1981. pg. 59–74*[9]

Showcases a variety of new hardware, electronic components, paints, and process machinery for manufacturers of home appliances. Entries include photos, descriptions with specifications, and manufacturers' names. Contains readers' service check-off card for product literature.[11]

PART 2: PUBLISHERS

Part 2: Section 1: The "Publisher Listings" section alphabetizes data publishers and producers with their addresses.

Part 2: Section 2: The "Publisher Index" section categorizes publishers by document type.

Part 2: Section 2: Publisher Index   Forecasts

Items Included in Citations in Part 2

1. Bibliographic citation numbers

Forecasts

Administrative Management Society	3298
American Banker, Inc.	4206
Artech House Inc.	4180
Association for Systems Management	592
Bank Administration Institute	362
Battelle–London	3380[1]

PART 3: SUBJECT INDEXES

Part 3: Section 1: SIC Code Index

The United States government's Office of Management and Budget has devised a hierarchical system of coding manufacturing and non-manufacturing industries. Each industry's numerical code and industry category can be found in the *Standard Industrial Classification* (SIC) *Manual*, 1972 edition.

Part 3: Section 1 indexes Part 1's bibliographic citations by SIC code and nomenclature. In numerical order, each code identifies its corresponding industry. The code is followed by the *bibliographic citations' numbers*. All appropriate SIC codes have been applied to each document. Therefore, the same citations may be indexed more than once.

Part 3: Section 1: SIC Code Index

Items Included in Citations in Part 3

1. SIC code

2. SIC industry heading

3. Bibliographic citation numbers

35[1]   MACHINERY, EXCEPT ELECTRICAL[2]
        1077, 1247, 2151, 2525, 2607, 2699, 2762, 2793,
        2804, 2805, 2810, 2811, 3032, 3042, 3067, 3452,
        3682, 3804, 3902, 1805, 2142, 2160, 2202, 2479,
        2590, 2749, 2767, 3004, 3005, 3048, 3754, 3766[3]

351[1]   Engines and turbines[2]
        2118, 2148, 2175, 2388, 2772, 2806, 2954, 2955,
        2999, 2338[3]

3511[1] Turbines and turbine generator sets[2] . . . 2622, 2777,
        3915

3519[1] Internal combustion engines, nec.[2] ....123, 3462[3]

Part 3: Section 2: Subject Index

This section categorizes Part 1's bibliographic citations by industrial subject headings. Under each heading are references to *citation numbers*.

Part 3: Section 2: Subject Index

Items Included in Citations in Part 3

1. Subject heading

2. Bibliographic citation number

Electrical industrial apparatus[1] . . 943, 1695, 2084, 2130, 2131,
  2136, 2175, 2286, 2339, 2505, 2509, 2527, 2531, 2564,
  2590, 2591, 2622, 2629, 2630, 2640, 2644, 2745, 2756,
  2758, 2768, 2777, 2887, 2888, 2889, 2896, 2954, 2955,
  2990, 2999, 3067, 3684, 3800[2]

PART 4: TITLE INDEX

Part 4 lists each document title in alphabetical order with *bibliographic citation numbers.*

Part 4: Title Index

Items Included in Citations in Part 4

1. Document title

2. Bibliographic citation number

K

Kenya Gets Bank Loan for Telephone System.[1] .... 4339[2]
Key Abstracts: Communication Technology.[1] .... 56[2]
Key Abstracts: Electrical Measurement and Instrumentation.[1]
.... 3476[2]
Key Abstracts: Electronic Circuits.[1] .... 2196[2]
Key Abstracts: Industrial Power and Control Systems.[1] ....
.... 2197[2]

Communications Equipment

International

Market Research Reports

1. ASW OUTSIDE THE U.S. *Frost and Sullivan, Inc. August 1980. $1250.00.*

Market analysis and forecast to 1985 for ASW (anti-submarine warfare) platforms, sensors, weapons, trainees and simulators outside the U.S. Considers government purchases and funding, and profiles major manufacturers.

2. CABLE AND INTERACTIVE T.V. *Predicasts, Inc. May 1981. $375.00.*

Report on cable and interactive TV vendors, system operators and equipment manufacturers. Includes government regulation, historical data and forecasts of consumers, worldwide market analysis, and performance capabilities and potentials.

3. COMMERCIAL CONSEQUENCES OF TECHNOLOGY SHARING UNDER NATO ARMS COLLABORATION AGREEMENTS. *Developing World Industry and Technology, Inc. March 1981. 51 pg. $35.00.*

Analysis of commercial impact of sharing production and technology among U.S. and other NATO country manufacturers of aerospace products, electronics and other goods for military markets. Considers management and ownership issues.

4. THE ELECTRONIC WARFARE MARKET OUTSIDE THE U.S. *Frost and Sullivan, Inc. October 1981. $1500.00.*

Gives a 5-year forecast to 1986 for ECM, ECCM, ELINT, ESM systems, chaff and other mechanical systems in West Europe, the Middle East, Latin America, Asia and the Pacific. Gives market trends by system type. Provides company profiles, shares and strategies by region.

5. FIBER OPTICS. *Predicasts, Inc. October 1980. $325.00.*

International market analysis and forecasts for fiber optical communications equipment. Reviews manufacturers and new products. Covers end uses, including telecommunications, CATV, data processing, vehicles, and military markets.

6. FIBER OPTICS. *Creative Strategies International. 1980. $895.00.*

Examines the world market, excluding domestic Japanese market, for fiber optics used in telecommunications and data links. Industry segmented into telecommunications and CATV applications and data links. Shipments in dollars forecast through 1983 by segment. Includes analysis of major competitors.

7. THE FUTURE FOR ELECTRONIC DOCUMENT DISTRIBUTION. *Stamps, George M., editor. Institute for Graphic Communications, Inc. May 24, 1982. $4950.00.*

Analysis of the electronic document distribution industry and the market for new products and services. Examines technologies of plants and equipment including teleprinters, communicating word processors and printers, terminals and communications networks. Examines demand and regulations. Includes forecasts.

8. INTERNATIONAL OPPORTUNITIES IN CABLE AND PAY TV: PROPOSAL. *Kalba Bowen Associates Inc. October 1982.*

Analysis of international market opportunities in cable and pay TV. Examines industry structure, regulation and market size by country. Assesses competition and opportunities for joint ownership arrangements.

9. MOBILE RADIOS AND TELEPHONES. *Predicasts, Inc. August 1980. 325.00 pg.*

Worldwide analysis of the mobile radio and telephone industry. Covers manufacturers, products and services, and end uses such as aviation, ground, marine, business and safety.

10. NEW COMMUNICATIONS SERVICES AND SYSTEMS: THE "MAIL" POTENTIAL FOR ELECTRONIC MAIL. *Probe Research, Inc. September 1979. 150 pg. $180.00.*

This report contains detailed operating and cost specifications for recent domestic and international postal electronic mail proposals, including customer hardware, software and transmission requirements, estimated delivery times, and sample cost comparisons. Similar EMS systems, such as XTEN, SBS, and Mailgram, are described. Market analyses of user industries, common carrier capacity, postal projections, and legal and policy issues are also provided.

11. THE NON-U.S. TRAINER AND SIMULATOR MARKET. *Frost and Sullivan, Inc. June 1979. $900.00.*

Market analysis and five-year forecast for training devices and simulators outside of the U.S. Reviews major manufacturers and demand in various end-use markets.

12. SATELLITE BROADCASTING. *Mackintosh Consultants, Inc. 1982. 600 pg. $9000.00.*

Three-volume report analyzes sales of reception equipment in North America and Western Europe and investment in direct broadcasting satellites to 1990. Includes estimates of revenues generated through pay TV and sales to advertisers, the size of equipment markets, and the extent of market penetration of cable television systems.

13. SATELLITE COMMUNICATIONS. *Creative Strategies International. 3662. January 1979. $1500.00.*

Examines U.S. and world markets for equipment used in satellite communications. Market forecasts through 1983 for sale of communications satellites and growth of communications traffic via satellites. Addresses three major market segments: satellite communications services, spacecraft construction firms, and small earth terminal suppliers. Discusses market strategies of major competitors and details U.S. regulatory environment.

14. THE SATELLITE EARTH STATION MARKET. *International Resource Development, Inc. July 1981. 102 pg. $985.00.*

Analysis of a dozen market segments with current and future shipment levels for each. Examines domestic and overseas use of satellites, needs of users, and trends towards higher-frequency transmission. Reviews earth station technology, direct broadcast satellites, and teleconferencing services. Covers supplier industry structure. Includes ten-year forecasts.

15. SEARCH AND DETECTION EQUIPMENT. *Predicasts, Inc. October 1981. $375.00.*

Worldwide manufacturers and markets for radar, sonar, milli-meter wave and laser search and detection instruments and other end uses.

16. SPACE SATELLITE APPLICATIONS. *Predicasts, Inc. September 1979. $325.00.*

Worldwide analysis of space satellites and end use markets. Covers manufacturers and government actions.

17. THE THIRD WORLD MARKET FOR GENERAL AVIATION AIRCRAFT, ENGINES AND AVIONICS. *Frost and Sullivan, Inc. January 1981. 329 pg. $1300.00.*

Analysis and sales forecast to 1985 of general aviation aircraft, engines, and avionics to North America, Western Europe, Latin America, Near and Middle East, Far East and Pacific and Sub-Saharan Africa. Avionics segments include communications, nav/comm. equipment, VHF nav, area-nav, ADF, transponders, DME, weather radar, radio altimeters, omega and inertial nav, autopilot, flight director / control systems, and emergency locator transmitter. Includes profiles of world manufacturers (product lines, sales, strategies) and technological trends.

18. THE WORLDWIDE COMMERCIAL NAVIGATION AND AIR TRAFFIC CONTROL MARKET: VOLUME 1. *Frost and Sullivan, Inc. December 1979. $975.00.*

First of two-volume report on the worldwide commercial navigation and air traffic control market which analyzes and forecasts, through 1984, sales of 4 major product groups and 10 specific systems types by 7 regional markets and 156 countries. Examines navigation aids and ATC equipment in use, air traffic activity, inventory / planned ships and aircraft, and background economic and trade information, including relevant import-export statistics, for individual countries. The state-of-the-art in navigation and ATC equipment is determined as well as the leading world suppliers and their product lines.

19. THE WORLDWIDE COMMERCIAL NAVIGATION AND AIR TRAFFIC CONTROL MARKET: VOLUME 2. *Frost and Sullivan, Inc. December 1979. $975.00.*

Second of two-volume report on the worldwide commercial navigation and air traffic control market analyzes and forecasts through 1984 sales of 4 major product groups and 10 specific systems types by 7 regional markets and 156 countries. Examines navigation aids and ATC equipment in use, air traffic activity, inventory / planned ships and aircraft and background economic and trade information, including relevant import-export statistics for individual countries. The state-of-the- art in navigation and ATC equipment is determined as well as the leading world suppliers and their product lines.

20. WORLDWIDE GENERAL AVIATION AIRCRAFT AND AVIONICS. *Frost and Sullivan, Inc. January 1979. $775.00.*

Forecast and analysis of worldwide general aviation aircraft and avionics sales to 1985 by region (including North America, Latin America, Western Europe, Eastern Europe, Middle East and Africa, Southeast Asia, Far East). Avionics segments include communications, nav / com, VHF nav, RNAV, ADF, transponders, DEM's, weather radar, altimeters, Omega and

initial nav. Profiles of manufacturers, market strategies, estimated market shares, products, technological capabilities.

21. THE WORLDWIDE MARKET FOR INTRUSION DETECTION AND PHYSICAL SECURITY SYSTEMS. *Frost and Sullivan, Inc. July 1980. 668 pg. $1250.00.*

Worldwide market analysis and forecast to 1985 for electronic security devices. Analyzes the market by market areas, country-by-country; by end uses; and by products. Reviews economic indicators, foreign trade, and product lines of leading manufacturers.

22. WORLDWIDE SATELLITE MARKET FORECAST: 1982-2000. *Western Union. Phillips Publishing Inc. 1982.*

Worldwide satellite market analysis and forecasts. Projects market va

See 769, 773, 783, 784, 2119, 3135, 3443, 3955, 3957, 3964, 3966.

Industry Statistical Reports

See 3969.

Financial and Economic Studies

23. COUNTDOWN: DIRECT - TO - HOME SATELLITE BROADCASTING IN THE EIGHTIES. *Television Digest, Inc. 1981. 70 pg. $75.00.*

Status report on direct-to-home broadcasting by satellite includes forecasts and industry analysis. Covers firms, technology, and issues during the 1980's.

See 3974.

Forecasts

See 3976.

Directories and Yearbooks

24. BSO DIRECTORY 1979 OF BROADCASTING OPERATORS SUPPLIERS AND EQUIPMENT. *Holland, Linda, editor. BSO Publications Ltd. Annual. 1979. 208 pg.*

Directory of broadcasting operators, suppliers, manufacturers and equipment products for 1979. Divided into four sections: international and national (United Kingdom) broadcasting authorities, organizations, and services; manufacturers and suppliers of equipment and services; classified index of systems, equipment and services; and technical descriptions of sound and television broadcasting equipment. Listings are alphabetical by country, by product and by manufacturers and suppliers.

25. EXISTING FIBER OPTIC SYSTEMS WORLDWIDE. *Information Gatekeepers, Inc. Irregular. 1981. 53 pg.*

Arranged geographically by nation, fiber optics systems data includes company name, installation date, location, length, design capability, supplier, and comments. Breakdowns for U.S. for common carriers; operating FT3 systems; CATV broadcasting; process control (industrial control, power, computers, transportation); U.S. government; and U.S. military. Includes detailed field tests listed by CCITT in seven nations; Bell System lightwave applications; and fiber optic systems in the home.

26. FIBER OPTIC OPERATING SYSTEMS. *Information Gatekeepers, Inc. Annual. 1981. $25.00.*

This annual publication describes each of the major fiber optics communication systems currently in operation, throughout the world (there were, as of 1980, 175 telephone, CATV, and other types of systems). Charts list each system's operator, location, installation date, suppliers, length, and design capability.

27. HOME VIDEO YEARBOOK, 1981-1982. *Knowledge Industry Publications. Annual. 1st. ed. 1981. 248 pg. $75.00.*

Overview of home video industry includes information on video discs, VCR's, VCR and disc software, videotape duplication services, videotext, home computers, advertising on cable, interactive cable, pay television services, satellites, and subscription television. Numerous tables detail growth rates, markets, revenues, expenditures, unit sales, imports, manufacturers and brands, ownership by brands, mergers, acquisitions, joint ventures and forecasts.

28. INTERNATIONAL FIBER OPTICS AND COMMUNICATIONS HANDBOOK AND BUYERS' GUIDE. *Information Gatekeepers, Inc. Annual. 1980. $45.00.*

Annual directory includes international manufacturers and suppliers listings, a calendar of fiber optics events, and manufacturers and products matrices. Covers technical characteristics: fibers, cables, sources, detectors, couplers, connectors, splicing equipment, test equipment, and data links.

29. JANE'S MILITARY COMMUNICATIONS: 1979-80. *Raggett, R.J., editor. Franklin Watts Inc. Annual. October 1979. 500 pg. $95.00.*

Details communications equipment and systems designed for military use. Covers radio communication, line communication, microwave and ancillaries. Products are classified by country and further classified by land, sea or air.

30. TELEPHONY'S DIRECTORY AND BUYERS' GUIDE 1982/83. *Telephony Publishing Corporation. Annual. 1982. $49.00.*

Annual address directory of telecommunications products and services includes independent telcos, Bell System companies, telephone administrations outside the U.S., interconnect companies, and specialized common carriers, as well as telecommunications associations, commissions, and regulatory bodies.

31. 1980 VIDEODISC DIRECTORY. *Walker, John, editor. Phillips Publishing Inc. Irregular. 1st ed. 1980. 29 pg. $29.00.*

1980 videodisc directory provides international listing for companies involved in the industry. Sections on capacitance systems, optical systems, mechanical systems, software and programming and an index. Provides names, addresses, telephone numbers, key personnel, prices, products, market, and status for each item.

See 3977, 3979, 3986.

Special Issues and Journal Articles

Air Transport World

32. SPECIAL SURVEY REPORT: SIMULATORS SAVE AIRLINES $138 MILLION YEARLY. *Penton / IPC. Subsidiary of Pittway Corporation. June 1981. pg.40-55*

Examines the world airlines' use of simulators for training pilots. Details airline, aircraft type, dollar value, manufacturer, and planned purchases for next 2 years of simulators. Also details who does what in airline training, by airline.

33. WHAT'S NEW IN GROUND SUPPORT EQUIPMENT. *Henderson, Lisa C. Penton / IPC. Subsidiary of Pittway Corporation. Annual. v.19, no.16. June 1982. pg.61-70*

Discusses new ground support equipment for the international air transport industry. Provides product descriptions and manufacturers' name

Aviation Week and Space Technology

34. AEROSPACE FORECAST AND INVENTORY: PASSENGER MILES, 1962-1980. *McGraw-Hill Book Company. Annual. March 3, 1980. 248 pg.*

International inventory of products in military, space technology, air transport, avionics, and business flying. Includes forecasts of sales to 1981 and industry analyses. Reviews progress of all U.S. and European defense and aerospace programs. Includes growth trends and new products, evaluation of NATO and Soviet military capacity, and management concerns in air transport industry.

CableVision

35. CABLEVISION: RECORD LEVELS: GENERAL INSTRUMENT BOARD CHAIRMAN FRANK HICKEY MAKES PREDICTION TO MORGAN STANLEY ANALYST GROUP. *Panero, Hugh. General Instrument. Titsch Publishing, Inc. November 12, 1981. pg.47-48*

Forecasts total U.S. and world shipments of cable television terminal products in 1981.

Communications News

36. TCA PREVIEW OF PRODUCTS. *Harcourt Brace Jovanovich, Inc. Irregular. v.19, no.9. September 1982. pg.46-63*

Provides names and addresses of manufacturers of communications equipment displaying products at TCA conventions. Provides descriptions and illustrations of products.

37. 1982 TCA CONFERENCE EXHIBITORS. *Harcourt Brace Jovanovich Publications. Annual. v.19, no.11. November 1982. pg.56-62G*

Reviews products exhibited at the 1982 TCA Conference. Includes manufacturers' names, brief descriptions of products, and booth numbers.

Data Communications

38. DATA COMMUNICATIONS 1981 BUYERS' GUIDE. *McGraw-Hill Book Company. Annual. 1981.*

1981 annual guide to worldwide sources of data communications products, services and vendors. Products and services section is divided into 9 categories, with names of businesses. Vendors arranged alphabetically with address, telephone number, contact person, products manufactured or services offered, and sales offices.

Flight International

39. ECM: THE BLACK BOX WEAPON. *Richardson, Doug. IPC Industrial Press Ltd. Annual. v.120, no.3778. October 3, 1981. pg.998-1014*

A worldwide guide to electronic countermeasures (ECM) equipment. Compiled from unclassified sources such as manufacturers' literature, defense electronics, and ECM journals.

Provides a description of the equipment, model numbers and manufacturers' names.

40. TOWARDS THE TOTAL FLIGHT SIMULATOR. *Whitaker, Richard. IPC Industrial Press Ltd. Annual. February 21, 1981. pg.488-499*

Annual listing of the world's airliner flight simulators, in service or on order. Includes manufacturer with address, operator, place of installation, date of service entry, and brief description.

Interavia

41. INTERAVIA: MAJOR EQUIPMENT CONTRACTS. *Interavia SA. Monthly. v.36, no.10. October 1981. pg.1040*

Monthly listing provides information on new major aerospace equipment contracts including communications equipment. Gives supplier's name and address, value of order, customer and type of equipment.

The Office

42. INTERNATIONAL COMMUNICATIONS ASSOCIATION: CONFERENCE AND EXPOSITIONS (MAY 1982) EXHIBITORS. *Office Publications, Inc. Irregular. April 1982. pg.160-166*

A listing of products exhibited at the 1982 International Communicat ons Association Conference. Information includes name of manufacturer, description, and photograph. Further information available through reader card service.

Protector

43. 4 INTERNATIONALE FACHMESSE FUR SICHERHEIT MIT INFORMATIONSTAGUNGEN IN ZURICH. [Fourth International Fair on Security with Information Seminar in Zurich.]. *Dammert, Bruno. Graft Neuhaus AG. German. June 1982. pg.1-4*

Reviews the fourth international fair of security systems (alarm systems, fire detection, lock, etc.) which took place in Zurich, Switzerland. Gives analysis of the security industry.

Satellite Communications

44. 1982-83 SATELLITE INDUSTRY DIRECTORY. *Schaffer, Bruce, comp. Cardiff Publishing Company. Annual. v.6, no.9. 1982. 82 pg.*

Annual, comprehensive listing of international satellite industry firms includes four major sections: agencies, domestic satellite carriers, satellite services, and hardware suppliers. Features name of firm, address, telephone number, contact person, and description of services.

Siemens Review

45. PROGRESS IN COMPONENTS: PROGRESS IN COMMUNICATIONS SYSTEMS. *Baur, Friedrich. Siemens Aktiengesellschaft. January 1980. pg.8-11*

Examines technological advancements in communications equipment and their impact on the communications industry. Data includes production value (1977) of sectors within the telecommunications industry; sales volume of integrated circuits, semi-conductors, and display tables; production costs (1960-1990); and energy requirements (1940-2000). Reviews component applications.

Telecommunication Journal

46. SATELLITE LAUNCHINGS NOTIFIED DURING PERIOD 24 APRIL TO 17 MAY 1982. *International Telecommunication Union. v.49, no.8. August 1982. pg.493*

Provides code name and spacecraft description; international number; country, organization, and launching site; date; initial orbital data; and observations, 1982.

Telephony

47. WORLD TELECOM MARKET: SIZE AND POTENTIAL. *Kammon, Allan B. Telephony Publishing Corporation. Irregular. October 22, 1979. pg.116-121*

Market analysis for the world telecommunications equipment industry. Cover sales, 1976-1977; revenues, 1975, 1980, 1990; equipment costs; value of shipments, 1970-1990; and growth rates, 1965-1990. Data includes forecasts to 1990. From studies by Arthur D. Little, Inc.

See 819, 2171, 2181, 3142, 3143, 3459, 3462, 3988, 3990, 3992.

Monographs and Handbooks

48. INTERNATIONAL COUNTERMEASURES HANDBOOK 1980 - 1981. *Eustace, Harry F., editor. EW Communications Inc. Annual. 1980. 430 pg.*

Annual guide to electronic warfare industry covers U.S. military budgets by branch and Europe's electronic industry. Examines equipment: nomenclatured U.S. equipment; equipment system descriptions; U.S. Army EW, DF, and ESM equipment spectrum; U.S. Army communications interface spectrum; and equipment index.

49. JANE'S MILITARY COMMUNICATIONS 1982. *Raggett, R.J. Jane's Publishing Company Ltd. Biennial. 3rd. ed. 1982. 650 (est.) pg. $140.00.*

Covers underwater and satellite links, global networks, and items an individual can carry. Contents include equipment; electronic warfare; radio, line, and composite systems; glossary; directory; and acronyms and code names.

See 839, 3996, 3997.

Conference Reports

50. INTERNATIONAL BROADCASTING CONVENTION. *Institute of Electrical and Electronics Engineers, Inc. September 18, 1982.*

Conference proceedings cover studio and outside broadcast facilities; sound systems and radio broadcasting; lightweight program equipment; television cameras, electronic graphics, visual effects, and other studio equipment, including recording; propagation and service planning; transmitters, transposers, and antennas; satellite broadcasting equipment and systems; satellite and terrestrial links; cable and optical fiber systems; new services and infomration systems; future possibilities in domestic receivers, display, and recording, including discs; and measurement technology.

51. SEMINAR ON COMMAND, CONTROL, COMMUNICA-TIONS AND INTELLIGENCE. *Harvard University. Information Resources Policy. Order Department. I-81-9. December 1981. $25.00.*

Eight papes cover command, control, communication, and intelligence from the perspective of Congress, combatant command, defense contractors, and technological innovators.

52. SUBMARINE TELECOMMUNICATIONS SYSTEMS. *Institution of Electrical Engineers. 1979. 184 pg. $41.00.*

Forty-three papers cover network planning; system implementation; influence of the submarine environment; present technology; reliability; and future developments.

See 2186, 4000, 4005, 4009.

Newsletters

53. FIBER LASER NEWS: THE BIWEEKLY REPORT ON OPTICAL FIBER AND LASER TECHNOLOGY. *Chaffee, C. David, ed. Phillips Publishing Inc. Bi-weekly. 8 pg. $197.00/yr.*

Gives news summaries of U.S. companies' roles and markets, foreign building of manufacturing plants, vendors' needs and activities, conference news, FCC recommendations, and new patents, contracts, and techniques. Once a month a calendar of upcoming courses, meetings, and conferences is published. A monthly marketing column reports on government procurements awarded or contemplated. Features regular columns on new products including prices and new technology in fiber optics and laser beams. Index included every six months.

Services

54. FIBER OPTICS AND COMMUNICATIONS WEEKLY NEWS SERVICE. *Resources for the Future, Inc. Weekly. 1981. 2 pg. $128.00/yr.*

Weekly service detailing market data; contracts and awards; R and D; new products; and executive hirings in the fiber optics industry.

55. TESS: TELEVISION EQUIPMENT SPECIFICATION SERVICE. *Knowledge Industry Publications. Tri-annually. 1982. 500 pg. $450.00/yr.*

Annual subscription to loose-leaf service includes three or more updates and provides specs for 500 U.S. suppliers and additional foreign suppliers of TV production and post-production equipment. Products are broken down into 400 categories under 80 headings. Contains brand names index and directory of manufacturers.

See 4013.

Indexes and Abstracts

56. KEY ABSTRACTS: COMMUNICATION TECHNOLOGY. *Institute of Electrical and Electronics Engineers, Inc. 1981. $72.00/yr.*

Arranged by topic, abstracts include author, author's address, publication, nation where published, volume and number, pages, publication date, synopsis, and number of references.

See 4014.

North America

Market Research Reports

57. ACCESS CONTROL AND PERSONAL IDENTIFICATION (PI) PRODUCTS MARKET IN THE U.S. *Frost and Sullivan, Inc. July 1981. 157 pg. $975.00.*

Market analysis and forecast for access control and personal identification products and services. Includes market survey to determine utilization of protection systems for facilities and computers. Examines new products, end uses, manufacturers and vendors, market shares and unit sales by product groups.

58. THE ASW MARKET IN THE U.S. *Frost and Sullivan, Inc. October 1981. $1250.00.*

Includes forecast and analysis to 1983 for platform, sensor, sonar, weapon, target, C3 technical support and facility, and countermeasure and R & D programs. Considers basic and applied RDT & E funding. Includes state-of-the-art technology. Profiles leading suppliers.

59. AT&T AIMS AT LONG-HAUL FIBER OPTICS: DETAILED PLANS FOR THE NORTHEAST CORRIDOR TO 1990. *Probe Research, Inc. 1980. 93 pg. $150.00.*

Analysis of effects on the fiber optics market, communications suppliers, users, and competitors from AT&T's plans to 1990 for its Northeast Corridor System. Includes AT&T's economic analysis of fiber optical systems, technical information, demand forecasts and plans to 1990, costs of major components and costs per circuit mile, and comparisons with alternative technologies.

60. AT&T AND COMPETITIVE EQUIPMENT PRICE COMPARISONS. *Probe Research, Inc. March 1979. 67 pg. $90.00.*

This study presents a price comparison between AT&T and its competitors. Includes price comparison data in the areas of transmission, switching, station equipment, wire and cable, and equipment purchasing.

61. AUDIO RECORDERS: U.S. MARKET OPPORTUNITIES. *Strategic Inc. June 1981. 110 pg. $1500.00.*

Analysis of markets for audio recorders for broadcasting, studio production and industrial consumption. Examines competition among formats and technologies. Forecasts annual sales to 1989.

62. BACKPLANE FORECAST. *Gnostic Concepts, Inc. 1981. 666 pg. $9000.00.*

Encompasses all types of backplanes used in electronic equipment including the backplane market 1976-1983, interconnection component market in backplanes, technological trends, competitive environment, and production of backplane-using equipment. Report categorizes backplanes into printed wiring types, metal plates, rack types.

63. BELL SYSTEM EQUIPMENT CROSS-SUBSIDIZATION POTENTIAL: SELLING EQUIPMENT TO THE U.S. TELEPHONE INDUSTRY: PART 1. *Probe Research, Inc. September 1979. 50 pg. $90.00.*

This is an in-depth study on cross-subsidization potential, focusing on how monopoly service rate-payers pay for many services which affect competition to Bell's equipment. Includes research and development, marketing, customer services, public relations, and anti-trust expenses. A presentation of several parties' views on the subject including: AT&T, interconnect and other equipment suppliers and the North American Telephone Association, California and Texas regulatory commissions.

64. THE BELL SYSTEM MARKET FOR TRANSMISSION PRODUCTS. *International Resource Development, Inc. March 1980. 175 pg. $895.00.*

Ten-year outlook for selling cable, microwave gear, fiber optic systems, carrier equipment, etc. to Bell operating companies and Long Lines. Detailed discussion of procurement techniques of each operating company. Analysis of impact of Bell Purchased Products Division. Potential effects of anti-trust suits, pending Federal legislation, FCC actions to open up Bell markets, etc. Opportunities and strategies discussed, with list of vendor contacts.

65. BILL PAYING BY TELEPHONE: AN EMERGING EFTS SUCCESS STORY: GUIDELINES: BUSINESS INTELLIGENCE PROGRAM. *SRI International. Marketing Services Group. June 1979. 13 pg.*

Market report on telephone bill-paying industry in the United States. Banks offering the service are listed, payment systems for 1978-1983 given, and cost structure and hardware and software suppliers are listed. Publication is part of SRI's Business Intelligence Program series.

66. BROADBAND INTRA-COMPANY NETWORKS - THE KEY ISSUES. *Strategic Inc. January 1982. 110 pg. $1200.00.*

Market analysis and forecast, 1981-1990, for broadband intra-company networks. Includes marketing strategies of major vendors, industry structure, and technology of equipment.

67. THE CANADIAN ELECTRONICS MARKET. *Welling, Ernie. Maclean Hunter Ltd. June 1981. 44 pg. $65.00.*

Market analysis for electronics in Canada. Gives market shares by end uses, 1976-1980; domestic production, imports, exports and consumption, 1976-1980; consumers, equipment installations, and capital investments in electronic equipment and power generation, 1978-1979; manufacturers, facilities, number of employees, costs, and value of shipments, 1972-1980; manufacturers'shipment by product groups, 1976-1978; total households and percent of households with ownership of equipment, 1973-1980; breakdown of utilization by end uses; and forecasts of economic indicators to 1985.

68. CELLULAR SYSTEMS AND ALTERNATIVES FOR MOBILE RADIO COMMUNICATIONS (U.S.). *Frost and Sullivan, Inc. December 1980. $900.00.*

Analysis of the U.S. market for central radio stations, cellular stations, and mobile stations. Includes revenue projections, regulations and competition.

69. CENTRAL OFFICE REMOTE TESTING EQUIPMENT. *Frost and Sullivan, Inc. March 1981. $900.00.*

Provides forecast of voice remote testing systems by Bell Telephone Companies and other independent telephone companies. Gives applications and types of testing systems. Provides geographical distribution of market and foreign export prospects. Includes company profiles for manufacturers. Gives review of trends and other factors influencing the market.

70. CHARACTER GENERATORS AND ANIMATION SYSTEMS: U.S. MARKET OPPORTUNITIES. *Strategic Inc. June 1981. 90 pg. $1200.00.*

Analysis of markets for character generators and animation systems for television production. Forecasts annual unit sales and dollar values, 1981-1989. Also includes annual growth rates, 1981-1989. Analysis by end uses and product groups.

71. THE COMMERCIAL AND INDUSTRIAL SECURITY / FIRE ALARM SERVICE AND EQUIPMENT MARKETS IN THE U.S.: VOLUME 1, THE COMMERCIAL MARKET. *Frost and Sullivan, Inc. June 1980. $975.00.*

First of two-volume market analysis and forecast through 1984 of the commercial and industrial security / fire alarm service and equipment industries. Establishes the classes of services and products in use or to be used; describes the types of supplier firms active in the market with profiles and competitive rankings of major participants by product (service area); comments on the problems of commercial and industrial end-users; and assesses the impact of government on the marketplace. Presents the results of surveys of equipment manufacturers, central-station companies, industrial end-users, and commercial end-users. Provides forecasts of sales to the entire non-residential market and separately for the commercial and industrial end-user segments to 1984. Commercial and industrial forecasts are each broken out by 4 major service categories and 14 product segments.

72. THE COMMERCIAL AND INDUSTRIAL SECURITY / FIRE ALARM SERVICE AND EQUIPMENT MARKETS IN THE U.S.: VOLUME 2, THE INDUSTRIAL MARKET. *Frost and Sullivan, Inc. June 1980. $975.00.*

Second of two-volume market analysis and forecast through 1984 of the commercial and industrial security / fire alarm service and equipment industries. Establishes the classes of services and products in use or to be used; describes the types of supplier firms active in the market with profiles and competitive rankings of major participants by product (service area); comments on the problems of commercial and industrial end-users; and assesses the impact of government on the marketplace. Presents the results of surveys' equipment manufacturers, central-station companies, industrial end-users, and commercial end-users. Provides forecasts of sales to the entire non-residential market and separately for the commercial and industrial end-user segments to 1984. The latter broken out by 4 major service categories and 14 product segments. Volume II contains 1977 F&S survey and results of security and fire equipment dealers and distributors.

73. COMMUNICATIONS EQUIPMENT, MEXICO: A COUNTRY MARKET SURVEY. *U.S. Government Printing Office. Irregular. CMS 81-048. January 1981. 14 pg. $.50.*

Report analyzes the current and projected markets for sale of communications equipment in Mexico. Data is provided on the market (local production, imports, exports) by type of equipment; leading suppliers; imports by type by country of origin; the market by product category; the market by end user sector. Describes the competitive environment, provides product market profiles, end user profiles, and information on marketing practices.

74. THE COMPLETE GUIDE TO OPTICAL COUPLERS. *Probe Research, Inc. September 1979. $95.00.*

Outlines detailed operating specifications of the 43 optical couplers currently being marketed by six European and North American manufacturers. Optical couplers - used to divide, combine, or mix optical signals - are becoming more and more important in FOC field installations. Accordingly, Probe analyzes the couplers from both user and manufacturer viewpoints. The guide contains comprehensive tables listing up to 18 coupler parameters for each model of every manufacturer. Tables are broken down into primary characteristics and design and secondary characteristics. The tables are further divided into directional, bidirectional, and star (both transmissive and reflective) couplers. Also included is commentary outlining uses, functions, and designs of various couplers.

75. COMPLETE SERIES ON EDUCATIONAL AND INDUS-TRIAL VIDEODISC MARKETS: VOLUME 2: POTENTIAL EDUCATIONAL MARKETS FOR VIDEODISC PROGRAM-MING. *Kalba, Konrad K. Kalba Bowen Associates Inc. January 1979. 56 pg. $125.00.*

Analyzes the U.S. market for videodisc programming in the educational environment. Focuses on primary and secondary schools, colleges and adult education programs. Discusses AV expenditures, software and hardware markets, and market growth trends. Data compiled from various government agencies and private sources.

76. COMPLETE SERIES ON EDUCATIONAL AND INDUS-TRIAL VIDEODISC MARKETS: VOLUME 3: POTENTIAL EDUCATIONAL MARKETS FOR VIDEODISC PROGRAM-MING: REPORT ON THREE FOCUS GROUPS. *Savage, Maria. Kalba Bowen Associates Inc. September 1979. 64 pg. $125.00.*

Analyzes the U.S. educational market for videodisc programming. Focuses on private and public secondary schools and colleges and computer - assisted instruction (CAI). Discusses current uses of AV media; reactions to videodisc capabilities; and barriers to the use of AV media. Data compiled from the conclusions of 3 focus groups, including directors of AV departments at the secondary school level; professors of college business administration; and program managers of CAI.

77. COMPLETE SERIES ON EDUCATIONAL AND INDUS-TRIAL VIDEODISC MARKETS: VOLUME 4: A SURVEY OF VIDEODISC APPLICATIONS IN INDUSTRIAL TRAINING AND DEVELOPMENT. *Bowen, Carroll B. Kalba Bowen Associates Inc. October 1979. 25 pg. $125.00.*

Analyzes the U.S. market for videodisc technology in the following 7 industry sectors: automobiles; insurance; utilities and related companies; electronics; merchandising; mineral extraction and processing; and medical. Data compiled from interviews with leading industry specialists.

78. CONTRACTING INTELLIGENCE NEWSLETTER. *DMS, Inc. Weekly. $200.00/yr.*

Weekly newsletter on defense contracts and other analysis of military markets.

79. DATA COMMUNICATIONS EQUIPMENT MARKET. *Frost and Sullivan, Inc. August 1981. $1100.00.*

Forecasts market for modems and acoustic couplers; multiplexes; facsimile devices; communications processors including front end processors, remote concentrators, message switches, microprocessor bases processors, communicating word processors; communications minicomputers; and other data communications equipment. Includes opportunities in selected industries and survey results and company profiles.

80. DATA REMOTE TESTING EQUIPMENT MARKET IN THE U.S. *Frost and Sullivan, Inc. September 1981. 350 (est.) pg. $1100.00.*

Analysis of markets for remote diagnostic and test equipment for data communications networks. Derives market value from value of all equipment shipments. Examines historical data and new products; industry structure; regulations; and manufacturers' sales, market shares and product lines.

81. DEFENSE / AEROSPACE CONTRACT QUARTERLY. *DMS, Inc. Quarterly. 1981. $450.00.*

Quarterly guide to defense and aerospace contracts, by manufacturers, government agencies and type of contract. Includes research and development.

82. DEFENSE BUDGET INTELLIGENCE NEWSLETTER. *DMS, Inc. Weekly. $125.00/yr.*

Weekly newsletter covering government budgets for defense contracts and other items of interest in military markets.

83. DEFENSE RESEARCH AND DEVELOPMENT SERIES: VOLUME 1: COMMAND, CONTROL, COMMUNICATIONS. *DMS, Inc. 1980.*

5-part market analysis lists program elements, funding, agencies, contracts and code names for all R&D activities taking place in the U.S. Army, Navy, Air Force and Defense Agencies. Volume 1 covers the command, control and communications area, detailing the past, current, and projected funding (7-year period with 5-year forecast). Also lists, for 3-year period, R&D contracts by discriptive codes, sponsoring agency and recipient company. Includes code name index.

84. DEFENSE RESEARCH AND DEVELOPMENT SERIES: VOLUME 3: FIRE CONTROL SYSTEMS. *DMS, Inc. 1980.*

5-part market analysis for U.S. defense research and development projects lists program elements, funding, agencies, contracts and code names. Volume 3 covers the fire control systems area, detailing past, current and projected funding (7-year period with 5-year forecast). Also lists, for 3-year period, R&D contracts by discriptive codes, sponsoring agency and recipient company. Includes code name index.

85. DEVELOPING AN INTEGRATED COMMUNICATIONS STRATEGY: VOICE COMMUNICATIONS. *Yankee Group. April 1981. $7500.00.*

Report anlayzes new voice technologies and details their potential applications, users, costs and system efficiency. Telephone management, and case studies of leading users provided. Voice communications market includes verbal messaging, store-and-forward voice, voice multiplexing and concentrating, voice synthesis, and voice / data integrating. Publication is one of 12 reports published each year, beginning 1981, by Critical Paths Service. Available to service subscribers.

86. DIGITAL CLASS 5 TELEPHONE SWITCHES. *International Resource Development, Inc. October 1979. 146 pg. $895.00.*

Information on the digital environment; and the technologies, architectures, availability, costs, market size and ten-year projections for class 5 digital switches. Discussion of competition through 1980's.

87. DIGITAL COMMUNICATIONS MARKETS. *Frost and Sullivan, Inc. November 1980. $1100.00.*

Forecast and analysis of digital central office switching, transmission subscriber equipment. Examines existing and new applications emerging from digital environment such as: digital networks, value added and private line networks, facsimile, electronic mail, etc. Analyzes shift from analog to digital. Pricing, market size, trends and types of digital equipment covered. Company profiles. Survey analysis. Market share data.

88. THE DIGITAL IMPACT: TRANSMISSION AND SWITCH-ING. *Yankee Group. 1979. 146 pg. $675.00.*

The report forecasts (1978-1985) and examines market size and share, technology trends, and end user demands in the digital industry in North America. Transmission and switching equipment, PABX, telephone equipment vendors, and major non-Bell competitors are analyzed. Publication is a one-time Yankee Group research report.

89. DIGITAL MICROWAVE RADIO MARKETS. *International Resource Development, Inc. February 1980. 158 pg. $895.00.*

Analysis of U.S. and Canadian markets for digital microwave radio equipment, with assessment of the effects of moves towards digital networks. Market shares of suppliers by frequency band. Review of probable development of 10 GHz market for digital termination systems, including XTEN, GTE / Telenet, etc. Discussion of Bell System captive and non-captive market segments, and ten-year segment-by-segment projections.

90. EDITING SYSTEMS (VIDEOTAPE): U.S. MARKET OPPORTUNITIES. *Strategic Inc: May 1981. 65 pg. $1200.00.*

Analysis of markets for video tape editing systems. Forecasts unit sales, dollar values and average annual growth rates to 1989. Analysis by end uses and product groups.

91. ELECTRONIC BUG KILLERS. *Irwin Broh and Associates. March 1980. $1900.00.*

Consumer survey among 80,000 households showing ownership, 1979 market size brand share trends, distribution structure analysis, retail price segments and user satisfaction of electronic bug killers.

92. ELECTRONIC MAIL: CURRENT AND FUTURE USE. *International Data Group. January 1980. 114 pg. $2000.00.*

Reviews the electronic mail market historically and discusses 1980 market conditions and trends. Objective is an attempt to gain a balanced perspective on this new industry. Scope is limited to end-to-end communication capabilities (Telex / TWX, facsimile, word-processing and computer based message systems). Methodology included vendor and user in-depth telephone interviews. 70 users ranked by Fortune were interviewed. 26 interviews conducted with executives from vendors offering electronic mail products or services.

93. ELECTRONIC TELEPHONE SYSTEMS. *Gnostic Concepts, Inc. 1981. $12000.00.*

Analysis of technologies, applications, markets, and competitive environment in the PABX / KTS area. Report covers key system market analysis, PABX market analysis, data / voice product opportunities, technology, legislative and regulatory issues, and economic outlook. Also covers future system trends, future component technology trends, customer applications, and major suppliers.

94. ELECTRONIC WARFARE. *DMS, Inc. Monthly. 1981. $685.00.*

Monthly survey of over 125 electronic warfare systems installations. Includes research and development, manufacturers, government funding, and forecasts to 1986.

95. ELECTRONIC WARFARE MARKET IN THE U.S. *Frost and Sullivan, Inc. April 1982. 764 pg. $1350.00.*

Analysis of military markets for electronics communications and space vehicle equipment. Examines market size by product groups; technology; government funding and defense contracts by government agencies and manufacturers; export markets; and forecasts to 1986 for government funding of research and development, government purchases and projects, and government budgets.

96. FACSIMILE EQUIPMENT AND SYSTEMS. *Frost and Sullivan, Inc. August 1981. $1100.00.*

Gives analysis and forecast of 4 and 6 minute, 1 and 2 minute, sub-minute, unattended and automatic dial facsimile equipment. Current and potential domestic and foreign competitors are profiled. Includes market share data. Provides effect of new technologies. Facsimile common carriers and networks

are assessed. Discusses standards and compatibility issues and international facsimile.

97. FACSIMILE: PRINTERS IN THE NETWORKS OF THE 80'S. *Strategic Inc. March 1981. 140 pg. $950.00.*

Analysis of markets, technology and utilization of digital facsimile systems, with forecasts to 1985. Examines new products and their manufacturers.

98. FIBER OPTICS AND SEMICONDUCTOR LASERS. *Creative Strategies International. October 1981. $1200.00.*

Analysis of markets for fiber optics and semiconductor lasers. Includes market penetration, market areas, competition and distribution channels.

99. THE FIBER OPTICS EXPERIENCE OF 4 TELCOS. *Telephony Publishing Corporation. 1980. 40 pg. $50.00.*

Interviews telco experts on their current systems. Examines suppliers in the market and details Bell and independent OSP cable acquisitions and mileage added.

100. FIBER OPTICS IN COMPUTER SYSTEMS. *Gnostic Concepts, Inc. 1981. $9800.00.*

Addresses applications, technology, competition, and price trends in fiber optics components and systems for the transmission of information in computer systems. Data includes markets in cables, transmitters, receivers, connectors and couplers. Chapters include fiber optics applications, economic analysis, competitive environment, EDP systems forecasts, systems considerations, light guides, sources and detectors.

101. FIBER OPTICS INDUSTRY SERVICE. *Gnostic Concepts, Inc. 1981. $15000.00.*

Analyzes current status of fiber optic component and system production, applications, technology and competition. Gives strategic information on new plants and facilities; major product directions; plans and programs; investments; and manpower levels. Market data includes demand, production, costs and consumption.

102. FIBER OPTICS INDUSTRY SERVICE - 2. *Gnostic Concepts, Inc. 1981. $18000.00.*

Discusses applications by industry and components, comparison with conventional transmission media, investment opportunities, and forecasts for fiber optics. Total world consumption of systems and components is forecasted to 1990 in terms of total value, quantity and average price. Component data included on cables, modules, repeaters, connectors, splice and couplers.

103. GRADED-INDEX OPTICAL FIBERS: A COMPLETE GUIDE. *Probe Research, Inc. 1980. 60 pg. $95.00.*

Graded-index fiber tables and text compare 78 fiber models using up to 16 operating parameters, including price, loss vs. wavelength, fabrication method, loss, bandwidth, NA, strength, bend radius and core diameter. Recommendations on how to evaluate optical fibers include bandwidth differential, limited phase space, macro- / micro- bending, and differential mode attenuation. Discussion of fiber technology development considers infrared transmission and stress effects on fibers.

104. GROWING MARKETS / SECURITY AND ALARMS. *Business Communications Company. February 1980. 130 (est.) pg. $650.00.*

Fire and theft losses, vs. safety and savings outlined. Methods of protection, installation, remote, electronic and centralized types all clearly discussed. Contents include: security manufacturers by type; government regulations / effects; estimated costs / options; various alarm systems; sales by type of device; future projections by type.

105. THE HOME BURGLAR ALARM PRODUCTS MARKETS. *Frost and Sullivan, Inc. May 1980. 295 pg. $900.00.*

Market analysis and forecast to 1984 for home burglar alarm products in the U.S. Covers distribution channels, sales prospects, industry structure, manufacturers, market penetration strategies and consumers.

106. HOME ELECTRONICS THROUGH 1995: STRATEGIES FOR PROVIDING INFORMATION AND CONTROL. *Venture Development Corporation. November 1981. 175 pg. $2490.00.*

Market survey of consumer attitudes toward new home electronics products and services, with analysis and forecasts to 1995. Includes industry structure, manufacturers and suppliers, demographic statistics, and economic indicators, shipments and revenues, 1980-1995.

107. HOME ENTERTAINMENT IN THE 1980'S. *FIND / SVP. March 1981. 39 pg. $300.00.*

Report considers catalytic effect of pay TV and its influence on cable TV and resulting increased demand for a variety of products and programming services with declining market shares for major networks. Study includes electronics and motion picture industries and evaluates Cox Broadcasting, Storer, North American Philips, Oak Industries, UA Columbia Cablevision, United Cable TV, and Zenith Radio.

108. HOME SATELLITE TERMINAL MARKET. *Frost and Sullivan, Inc. June 1981. $1100.00.*

Forecasts and analyzes home satellite terminals market. Gives history of satellite broadcasting. Gives technological and legal problems. Includes future developments of two-way satellite terminals and marketing approaches. Contains competition and market share analysis. Gives comparison of small satellite terminals and rival hardware developments.

109. HOME SECURITY - THE IMPACT OF CABLE TV. *International Resource Development, Inc. 1982. 177 pg. $1285.00.*

Analysis of markets for two-way cable home security systems. Examines security systems' role in market penetration for other interactive services provided by franchises, forecasts of revenues to 1992, competition with the traditional alarm business, industry structure, and licensing.

110. HOSPITAL VOICE COMMUNICATIONS AND TV MARKET. *Frost and Sullivan, Inc. January 1981. $1000.00.*

Market report gives ten-year forecast and analysis of PBX, station equipment, sound and public address systems, pocket paging and two-way radio systems, nurse call and PAX systems, dictation systems, and TV equipment including cameras, receivers / monitors, video tape recorders, videodisc, and slow-can TV. Gives profile of hospital market, procurement process and sales approach. Includes vendor competition.

111. HOTEL / MOTEL COMMUNICATIONS AND TV MARKET IN THE U.S. *Frost and Sullivan, Inc. April 1982. 235 pg. $1200.00.*

Analyzes the market for communications equipment sales to hotels and motels. Examines the lodging market structure demand for telephone, sound system, television and radio paging product groups; technology; prices; other factors in hotel purchases; market penetration of major manufacturers and regulations. Forecasts unit and dollar sales, 1981-1990. From a market survey of hotels and motels.

112. IMPACT OF COMMUNICATIONS DEVELOPMENTS ON INFORMATION SERVICES VENDORS. *INPUT. March 1982.*

Examines and forecasts the U.S. communications network market to 1985. Covers market shares and competition.

113. INDEPENDENT TELEPHONE COMPANY EQUIPMENT MARKET. *Northern Business Intelligence Ind. March 1980. 120 pg. $475.00.*

Analysis of independent telephone system equipment markets. Includes markets for 15 categories of equipment; market shares; market penetration opportunities by product lines; installed base; forecasts to 1985; and breakdown of manufacturers and telephone services.

114. INDUSTRIAL COMMUNICATIONS. *Industrial Communications. Weekly. $129.60/yr.*

Provides information on two-way radio communication, manufacturers and end-uses, regulations and legislation issues affecting the industry.

115. INDUSTRY STANDARDS FOR MULTISERVICE CATV TERMINALS: PROPOSAL. *Kalba Bowen Associates Inc. 1982. $30000.00.*

Market analysis proposal for multiservice cable TV terminals. Includes rankings of services by return on investment, 1985, 1990; revenues, costs and market penetration forecasts; standards for capacity and performance; industry structure; patents; and demand forecasts.

116. INFORMATION SECURITY SYSTEMS MARKET. *Frost and Sullivan, Inc. December 1981. $1200.00.*

Market report contains 5-year sales forecasts for: standalone data security devices, secured terminals, home entertainment (scramblers for satellite transmissions), and two-way business communications. Includes company and product profiles. Assesses risks. Discusses cryptology and questionnaire survey results.

117. INTEGRATED COMMUNICATIONS. *Gnostic Concepts, Inc. Monthly. 1981. $18000.00.*

Integrated systems information: networks, vendors, markets, competition and technology. Program outlines in 12 monthly reports: data networks - end user requirements; EPABX - integrated voice / data; communicating word processors; conferencing; VLSI circuits; direct satellite communications; store and forward message systems; network services; fiber optics; IBM's strategy in communications; electronic key telephone systems; and voice recognition systems.

118. INTERACTIVE INFORMATION SERVICES IN THE HOME. *Yankee Group. January 1981. $8500.00.*

Statistical data, industry analysis and product opportunities of interactive information services in the home. Includes analysis of costs and examines the potential for banking at home, shopping at home and financial-security services at home. Part of Yankee Group's Home of the Future Service and available only to subscribers.

119. INTRA-SITE COMMUNICATIONS NETWORKS. *Perspective Telecommunications. 1982.*

Multi-client study provides an in-depth analysis of second and third generation PABX's, baseband, broadband and multiple alternatives in terms of performance, compatibility, and cost effectiveness. About 120 interviews conducted with vendors and users.

120. KEY TELEPHONE SYSTEM MARKET. *Northern Business Intelligence Ind. April 1981. 194 pg. $995.00.*

1980 market analysis for key telephone systems, with forecasts to 1985. Includes shipments, dollar value of markets, installed base, prices, and market segmentation by vendors.

121. LOCAL AREA COMMUNICATIONS NETWORKS: TECHNOLOGIES AND VENDORS. *Yankee Group. February 1981. $7500.00.*

Part 1 of Local Communications deals with developments and techniques in the local networking arena as well as specific vendors' strategies and products. Intercompany communications systems including transparent information transfer, electronic messaging and resource sharing discussed. Publication is one of 12 reports published each year, beginning 1981, by Critical Paths Services. Available to service subscribers.

122. THE MEXICAN MARKET FOR COMMUNICATIONS EQUIPMENT. *Batres, Valdes, Wygard y Asociados, S.C., Mexico City, Mexico. U.S. Department of Commerce. International Trade Administration. 1980. 123 pg. $10.00.*

Analyzes current dimensions and likely growth trends of the Mexican communications equipment market, by end-user industry; current and potential market shares accruing to U.S. manufacturers; product groups offering high export potential; and preferred marketing / promotional techniques.

123. MEXICAN MARKET SURVEY FOR COMMERCIAL FISHING EQUIPMENT. *H. L. Wallace y Asociados, Mexico City. U.S. Department of Commerce. International Trade Administration. 1980. 163 pg. $10.00.*

Analyzes current Mexican fishing equipment market and growth trends; current and potential U.S. share of market; product categories offering high export potential to U.S. manufacturers; and preferred marketing and promotional strategies.

124. MICROWAVE INDUSTRY: TRENDS / DEVELOPMENTS. *Averko, Michael. Business Communications Company. February 1981. 132 (est.) pg. $825.00.*

Evaluation of 1981-1985 U.S. and other markets for microwave solid-state devices, tubes, communications systems, burglar alarms, radar and navigational systems, and ovens. Includes discussion of major manufacturers and their market shares during the late 70's; current and potential exports; and likely trends in secondary markets.

125. MICROWAVE TRENDS: 1980-1985. *Telephony Publishing Corporation. 1980. 35 pg. $55.00.*

Forecasts (1980-1985) trends in digital radio and digital switches and fiber optics / microwave. Examines digital analog and fiber optics microwave manufacturers.

126. MILITARY AIRBORNE RADAR MARKET IN THE U.S. *Frost and Sullivan, Inc. April 1982. 385 pg. $1200.00.*

Analysis and forecast to 1987 for military radar market structure and size. Examines manufacturers in competition for defense contracts, research and development of new technology, and funding by end use government agencies.

127. MILITARY AND AEROSPACE DISPLAY MARKET IN THE U.S. *Frost and Sullivan, Inc. September 1981. 229 pg. $1200.00.*

Analysis of military markets for electronic display equipment. Describes government expenditures by defense contracts and specific aircraft. Includes analysis of new products, surveys of suppliers and end users, rankings of manufacturers, exports, government R & D funding and forecasts to 1985.

128. MILITARY AND AEROSPACE DISPLAY SYSTEMS AND COMPONENTS MARKET. *Frost and Sullivan, Inc. July 1978. $750.00.*

Analysis and forecasts to 1982 for electronic display instruments in U.S. aerospace and military markets, by manufacturer and end users. Includes user attitude survey results.

129. THE MILITARY C3I MARKET IN THE U.S. *Frost and Sullivan, Inc. August 1981. 397 pg. $1150.00.*

Market report containing 397 page 5-year analysis and forecast to 1986 of ground communications, command and control centers, intelligence fusion centers, reconnaissance and surveillance systems, communications satellites, airborne C3 and electronic intelligence systems for Army, Navy, USAF, DARPA, and DCA. Includes system descriptions, contractors, agencies, funding, timetables, technological trends, profits of system suppliers, their products, sales, and marketing strategies.

130. THE MILITARY GROUNDBASED AND SHIPBASED RADAR MARKET IN THE U.S. *Frost and Sullivan, Inc. June 1981. $1050.00.*

Market report contains analysis and forecast to 1986 for radar equipment categories including large surface-based, tactical groundbased, and shipbased systems. Applications include surveillance, tracking, search navigation, ATC, target location, bomb scoring, and range instrumentation. Includes market structure, end users and procurement agencies, analysis of all inventory and development systems, modification programs, technology assessment, contractors, and profiles of major industrial market participants.

131. MILITARY LASER AND EO. *DMS, Inc. Monthly. 1981. $685.00.*

Monthly review of all U.S. military projects, research and development, and installations involving lasers and electro-optical devices. Lists manufacturers, government funding, and forecasts to 1986.

132. MILITARY LASER MARKET IN THE U.S. *Frost and Sullivan, Inc. June 1981. 240 pg. $1100.00.*

Analysis of military markets for laser equipment. Covers end uses, installed base, technical characteristics of products, profiles of manufacturers, government funding for research and development, government budgets to 1985, and the market value of lasers.

133. THE MILITARY NAVIGATION MARKET IN THE U.S. *Frost and Sullivan, Inc. March 1979. 383 pg. $850.00.*

Forecast and analysis to 1984 for VLF Omega, Loran, Satellite, Tocan, VOR/DME, Landing Systems, Inertial, Doppler, Terrain Avoidance, Beacons & Direction Finders, Bombing Navigation Systems, CAS, PWI, SKE, Position & Location Systems, Navigation Radar, and Integrated CNI and Celestial Navigation. Includes all procurement & RDT&E work, industrial participants, and technological trends.

134. THE MILITARY PASSIVE NIGHT VISION MARKET IN THE U.S. *Frost and Sullivan, Inc. November 1981. $1100.00.*

Contains a five-year forecast to 1986 for airborne FLIR's, vehicle-mounted imagers and night sights, man-portable and crew-served weapon sights, night-vision goggle and individual night-sights, space borne infrared imaging systems, infrared imagers for missile guidance and weapon delivery, shipboard infrared search and track equipment used by DOD agencies. Includes technological and development trends. Includes competitive environment and profiles of leading suppliers.

135. THE MILITARY RECONNAISSANCE AND SURVEIL-LANCE MARKET IN THE U.S. *Frost and Sullivan, Inc. March 1981. 271 pg. $1100.00.*

Analysis and forecast to 1986 for all tactical and strategic R & S programs, including aircraft, ground ship, and space-based segments. Defines and establishes overall U.S. R & S mission requirements; discusses all acquisition and RDT & E programs including platforms, sensors and processing systems. Examines technology, state of the art, program contractors, agencies, funding; and profiles leading U.S. industrial market participants with estimates of their competitive market standing.

136. MILITARY SATELLITE COMMUNICATIONS MARKET IN THE U.S. *Frost and Sullivan, Inc. April 1982. 296 pg. $1250.00.*

Analysis of military markets for satellite communications systems. Includes market segmentation and size by government agencies and specific defense contracts and programs; manufacturers; new technology trends; government funding for research and development, 1982-1987; competition; and market penetration.

137. MILITARY SIMULATORS. *DMS, Inc. Monthly. 1981. $685.00.*

Monthly report on all military simulator and trainer programs receiving U.S. government funding. Lists manufacturers, costs, and forecasts to 1986.

138. THE MILITARY SPACE DEFENSE MARKET IN THE U.S. *Frost and Sullivan, Inc. November 1981. $1150.00.*

Market report forecasts through 1987 for space surveillance radar, optical surveillance, infrared sensors, DP services, high-energy lasers, precision pointing and tracking systems, miniature homing missiles, particle accelerators, space vehicles, and attack warning sensors. SPADOC, GEODSS, SPA-CETRACK, NAVSPASUR and 10 other space defense programs are discussed in depth. Includes technology programs, future trends, and future markets.

139. MOBILE RADIO MARKETS. *International Resource Development, Inc. August 1980. 165 pg. $985.00.*

Analysis of markets for mobile radios. Examines regulations; present, historical data and forecasts of growth rates; market segmentation; manufacturers; new technology; market size; and forecasts to 1990.

140. MODEM / MULTIPLEXOR / CONCENTRATOR MARKET. *Frost and Sullivan, Inc. December 1981. 160 pg. $1150.00.*

Analyzes and forecasts the modem / multiplexor / concentrator markets. Provides data communications overview which includes history background, products and services, regulatory influences, technological factors, industry trends, and data communications outlook and forecasts. Examines end users, OEM market, tariffed common carriers, systems houses, international market, and competition. Includes market share data.

141. MODEMS AND MULTIPLEXERS. *Creative Strategies International. March 1979. $895.00.*

Examines the world market for modems and multiplexers produced by U.S. manufacturers. Sales in units and dollars through 1983 are detailed for: low-, medium-, and high-speed modems; acoustic couplers; OEMS; FDMs; TDMs; and statistical multiplexers. Analysis of the rapidly changing data communications environment with particular emphasis on digital networks. Includes a thorough discussion of current technology and major competitive factors.

142. THE NASA / DOD SPACE MARKET. *Frost and Sullivan, Inc. May 1979. 380 pg. $800.00.*

Market analysis of the NASA /DOD space market discusses on-going programs, contracts, major industrial participants, and quantitative forecasts to 1984 for 12 key market segments. Market segments include transportation systems, space sciences, space and terrestial applications, research and technology, tracking and data systems, and satellite communications.

143. NATO WEAPONS. *DMS, Inc. Monthly. 1981. $675.00.*

Monthly analysis of NATO military weapons markets.

144. NAVIGATION / GUIDANCE. *DMS, Inc. Annual. 1981. $375.00.*

Analysis of military markets for navigation and guidance systems.

145. THE NEW CATV MARKET. *Frost and Sullivan, Inc. September 1981. $1150.00.*

Market report gives 10-year analysis and forecast for operating revenue and CATV equipment requirements. Gives CATV growth patterns in the U.S. Presents program sources available to CATV. Business and two-way cable use analyzed. Gives penetration of CATV, services offered and franchise requirements. Presents regulatory impact. Includes future of STV including reception equipment. Contains company profiles and survey results.

146. NEW COMMUNICATIONS SERVICES AND SYSTEMS: TRANSACTION NETWORK SERVICES. *Probe Research, Inc. February 1979. 94 pg. $150.00.*

Transaction Network Service (TNS) is AT&T's entry into the financial services, POS (point-of-sale) and fast inquiry / response markets. Report reveals how Bell has quietly entered the field of customer-operated terminals (COTs) with a new banking terminal designed for retail premises. It analyzes the COT marketplace, how AT&T's TNS system has actually been performing in the field, and new user services. The report compares TNS with Bell's ACS and IBM's 3750 system.

147. THE NEW GENERATION OF PBX'S. *Yankee Group. May 1981. $7500.00.*

1981 report focusing on PBX products (including digital, analog, antelope) and vendors (AT&T, IBM). Availability, features, functions and market areas provided. Current and upcoming technology, costs, and user survey provided. Publication is one of 12 reports published each year, beginning 1981, by Critical Paths Service. Available to service subcribers.

148. THE NORTH AMERICAN SATELLITE COMMUNICA-TIONS MARKET. *Frost and Sullivan, Inc. July 1981. 399 pg. $1150.00.*

Market analysis and forecasts for manufacturers of satellite communications equipment, end uses, and service vendors. Shows sales volume, annual production, value of shipments, launch expenditures, revenues, and sales, all for 1981-1991. Also includes prices, market shares, new products and technologies and government regulations.

149. OFFICE TELEPHONE SYSTEMS II. *Creative Strategies International. 1982. $1450.00.*

Provides analysis and forecast to 1990 for office telephone systems. Includes industry structure and manufacturers; end uses, size of businesses and number of employees; new technologies; industry growth rate, 1982-1990; distribution channels and leases; competition; and legislation.

150. OPTICAL COMMUNICATIONS MARKET. *Frost and Sullivan, Inc. September 1980. 255 pg. $975.00.*

Market analysis and forecast of optical communications systems both bound (fiberoptic) and point-to-point atmospheric. Commerical and military applications and activity. Considers lasers and light sources, optical fibers and cables, connectors, detectors and receivers, repeaters and amplifiers, and modulators. Market forecast for telephone, CATV, computer related, and various military functions. Competition and participation.

151. OPTICAL STORAGE MEDIA (VIDEODISCS): NEW DEVELOPMENTS - VOLUME 3. *Strategic Inc. May 1981. 110 pg. $950.00.*

Analysis of new technology for optical storage media (videodiscs), utilization, major vendors, and markets. Includes historical data on development, 1970-1981; performance and cost comparisons; market size and competition among techniques; manufacturers; user attitudes; and applications for businesses and consumers.

152. THE OUTLOOK FOR MOBILE RADIO TELEPHONE COMMUNICATIONS IN THE UNITED STATES. *Lipoff, Stuart J. Arthur D. Little, Inc. September 4, 1980. 13 pg. $600.00.*

Analyzes the current market for Mobile Radio Telephone, matching its use against the Wireline Common Carrier and illustrating its use by business type and job level. Reviews the potentials of a new cellular system MTS and describes current and proposed experimental systems. A synopsis of FCC regulations in regard to MTS with expected future enactments and a listing of leading manufacturers complete the report. Sources from trade journals and ADL estimates.

153. PBX EQUIPMENT MARKET IN THE U.S. *Frost and Sullivan, Inc. no. A1105/A-1. September 1982. 280 pg. $1275.00.*

Analysis of markets for PBX equipment forecasts installed base, sales, and market value to 1990. Includes product group details; industry structure; new technology; manufacturers' market shares, product lines, and sales volume; production; prices; and regulation.

154. PBX'S AND SWITCHING: THE NEW GENERATION OF SUPERCONTROLLERS. *Yankee Group. October 1979. 271 pg. $9500.00.*

A 1979 market analysis of PBX and switching systems. Historical data, market areas, major suppliers and other potential sources; system features (including call waiting, automatic redial, speed dialing, etc), pricing, costs and vendor services. Industry forecast for U.S. through 1988 provided. Volume 1 of yearly service by Yankee Group's CIS service. Available to service subscribers.

155. PRIVATE TELEVISION COMMUNICATIONS: 1980 AND BEYOND. *Brush, Judith M. International Television Association. December 1980. $39.95.*

Market report on non-broadcast television production and equipment, based on a survey of major U.S., Canadian, British and European corporations. Describes users and potential users, and discusses trends in production and equipment expenditures; program formats, applications and costs; networks and distribution formats (including satellite, cable, and video disc systems); and new technologies.

156. PROBLEMS AND CONSEQUENCES OF TECHNOLOGY SHARING UNDER NATO ARMS COLLABORATION. *Developing World Industry and Technology, Inc. April 1980. 51 pg. $50.00.*

Study of effects on U.S. corporations in military aerospace and electronics markets from NATO co-production and research and development sharing.

157. PUBLIC VS. PRIVATE NETWORK ALTERNATIVES. *Yankee Group. August 1981. $7500.00.*

1981 report focuses on current public and private network alternatives in data and voice communications. Major vendors, product offerings (including Tymnet, Telenet, Bell's ACS, SBS, and XTEN), product and service comparisons of costs, expenditures and benefits. Publication is one of 12 reports published each year, beginning 1981, by Critical Paths Service. Available to service subscribers.

158. RADAR AND SONAR. *DMS, Inc. Monthly. 1981. $675.00.*

Monthly analysis of military markets for radar and sonar equipment.

159. RADIO AND TV BROADCAST EQUIPMENT. *Business Communications Company. April 1980.*

Company activities and U.S. markets for audio and video program equipment, transmitters, closed circuit TV equipment, CATV, and other broadcast equipment are summarized. Technology and applications information is reviewed.

160. RAPID DEPLOYMENT FORCE. *DMS, Inc. 1980. 750 pg.*

DMS report covers both short term and long range rapid deployment force plans. Discusses RDF structure: logistical requirements, deployment and support. Details unit equipment requirements, along with a 5-year forecast. Requirements broken down into groups: Aviation, Naval, Army, and Command, Control and Communications. Also reviews budgets with specifics on costly equipment and generalized figures on small budget items.

161. REMOTE MONITORING ALARM AND CONTROL SYSTEMS MARKETS. *Theta Technology. FIND / SVP. March 1980. $850.00.*

Examines market for remote alarm systems and the technological advancement of the microprocessor. Study covers: market competition, technological developments and innovations, market trends, market segments and purchasing industry requirements. Tables summarize future market trends and annual growth rates.

162. THE RESTRUCTURING OF AT & T. *Yankee Group. 1981.*

Analysis of the restructuring of AT & T under pressures of government regulation and competition. Includes an analysis of the communications / computer industry, 1982-1985; the formation of a new Bell subsidiary; new products and services to be offered. Examines distribution channels, potential revenues, sales to 1983, and communications market analysis.

163. RETAIL TELEPHONE MARKETS: 1979-1984. *Boggs, Raymond L. Venture Development Corporation. Irregular. July 1979. 227 pg. $950.00.*

Industry analysis of consumer purchased telephones. Consumer attitudes and behavior toward new and proposed product features examined. Phone shipments by product type, 1979-1984 forecasted. Technological developments and regulatory impact reviewed. Methodology includes: industry forecasts (independent and in-house), industry interviews, surveys of retailers and consumers.

164. SATELLITE COMMUNICATIONS MARKET IN N. AMERICA. *Frost and Sullivan, Inc. July 1981. $1150.00.*

Gives 10-year forecast and analysis of satellite communications market in N. America. Forecasts traffic system revenues, satellites in orbit, launch costs, earth stations and components. Examines international commercial and government systems. Users of satellite systems are identified. Gives policy and regulatory impact, vendors of satellites, launch vehicles and earth stations. Includes market share data. Covers technological developments.

165. SATELLITE PRIVATE BUSINESS TERMINALS MARKET IN THE U.S. *Frost and Sullivan, Inc. April 1982. 302 pg. $1275.00.*

Analyzes markets for private satellite terminals and network services. Profiles manufacturers and consumers. Gives estimated sales, market shares, and installations, 1981; growth rates, 1981-1983; and forecasts of revenues, unit sales and dollar sales, 1981-1991.

166. SECURITY, BURGLARY, AND FIRE ALARM SYSTEMS AND SERVICES MARKETS IN THE UNITED STATES. *International Resource Development, Inc. July 1979. 165 pg. $795.00.*

Report on security, burglar and fire alarms gives analysis of available equipment and technologies, including user requirements and expected future market trends. Examines interrelationships between alarm and other industries, including telephone companies and CATV operators. Evaluates performance of suppliers and analyzes market shares. Discusses influences on market growth and impact of new technology and legislation.

167. SECURITY / MONITORING ALARM SYSTEMS. *Business Communications Company. February 1981. 150 pg. $750.00.*

Report discusses economic and social factors and their effects on the security industry. Outlines fire and theft losses vs. safety and savings. Lists various methods of protection, with costs, companies involved, and forecasts into the 1990's. Contains forty tables.

168. SELLING BUSINESS STATION EQUIPMENT TO THE BELL SYSTEM AND U.S. INDEPENDENTS: 1980-1982 BUSINESS END-USER TRENDS. *Telephony Publishing Corporation. 1980. 107 pg.*

Coverage spans key system and PABX installed base (U.S. independents, Bell System) including rankings, units in operation, and market segmentation; Western Electric (1968-1978) and GTE (1976-1978) automatic electrics and subsidiaries production of telephone sets by product groups; and 1980-1982 business station equipment trends survey.

169. SELLING TO THE BSPPD - BELL SYSTEM PURCHASED PRODUCTS DIVISION: SELLING EQUIPMENT TO THE U.S. TELEPHONE INDUSTRY. *Probe Research, Inc. September 1979. 60 pg. $90.00.*

This report presents the BSPPD, the centralized interface for selling to the Bell System. Procedures for equipment evaluation, price comparisons, technical and economic analysis, contracting and other procurement activities are discussed. Includes interviews with BSPPD personnel and suppliers who have dealt with BSPPD. An analysis of advantages and problems of the BSPPD is presented.

170. SIGNAL PROCESSING MARKETS. *International Resource Development, Inc. July 1980. 159 pg. $985.00.*

Analysis of signal processing equipment and components technology for military markets. Examines utilization and opportunities for market penetration.

171. SMALL SATELLITE EARTH STATIONS: U.S. MARKET OPPORTUNITIES. *Strategic Inc. May 1981. 115 pg. $1500.00.*

Market analysis and forecast for small satellite earth station equipment. Forecasts annual sales by end uses through 1989.

172. SPECIAL EFFECTS GENERATORS. *Strategic Inc. May 1981. 50 pg. $1200.00.*

Market analysis with forecasts to 1989 for television special effects generators. Examines competition among new products and utilization trends. Projects annual sales and prices to 1989, with detailed market segmentation.

173. SPEECH SYNTHESIS AND RECOGNITION EQUIPMENT. *Business Communications Company. November 1981. 125 pg. $975.00.*

Analysis of markets for speech recognition and synthesis equipment. Includes historical data, current utilization and forecasts. Examines new products competition (foreign and domestic), research and development, and military markets.

174. STORE - AND - FORWARD VOICE SWITCHING. *International Resource Development, Inc. January 1980. 161 pg. $895.00.*

Forecasts major growth in systems and services for non-interactive voice message transmission and reception, including voice mail and advance calling services. Discussion of new services from AT&T based on the VSS # 1A switch, and new products from ECS and others. Analysis of pattern of expected market development, with relationship between S&FV, electronic mail and the "office of the future." Ten-year market projections.

175. SWITCHERS: U.S. MARKET OPPORTUNITIES. *Strategic Inc. May 1981. 85 pg. $1200.00.*

Market analysis for television production switches. Examines technologies and new products, and forecasts unit sales, dollar values, and annual growth rates to 1989. Analysis by end uses and product groups.

176. SWITCHING SYSTEMS IN THE 80'S - VOLUME 3. *Strategic Inc. August 1981. 420 pg. $1500.00.*

Report on the changing market structure for CBX / PBX switching systems. Examines technology of integrated voice, data and image services; competition and industry structure; and major manufacturers. Includes market survey results.

177. TELCOS ASK SUPPLIERS FOR ENERGY EFFICIENT EQUIPMENT. *Telephony Publishing Corporation. 1980. 10 pg. $12.00.*

A report based on the telephone companies' internal documents describing needs and wants in the way of energy efficient equipment and why.

178. TELECOM MARKET LETTER. *Northern Business Intelligence Ind. Bi-weekly. $215.00/yr.*

Newsletter for the Canadian telecom equipment market. Covers construction expenditure budgets, new product introductions, manufacturers' strategies, demand for new equipment, regulations, and new technologies.

179. TELECOM POWER EQUIPMENT MARKET IN CANADA. *Northern Business Intelligence Ind. July 1981. 231 pg. $450.00.*

Technology and market analysis for telecom power equipment in Canada. Includes major end users, 1980 installed base, forecasts to 1985, manufacturers, prices, production costs and market penetration opportunities.

180. TELECONFERENCING. *Interactive Data Corporation. July 1979. 58 pg.*

1979 report on teleconferencing includes video, audio and computers. Provides vendor profiles, product costs, growth rates and shipments (1978-1985), and small-medium-large businesses involved in the industry. Teleconferencing types, facilities used and potential market participation of major companies presented. Publication is part of I.D.C.'s Continuous Information Service.

181. TELEPHONE ANSWERING AND RADIO PAGING SERVICES AND ASSOCIATED EQUIPMENT MARKETS. *Frost and Sullivan, Inc. September 1981. 248 pg. $1200.00.*

Analysis of markets for telephone answering services, radio paging, and associated equipment. Examines manufacturers including a survey of product lines, distribution, regulations, and market penetration. Also examines end users, including factors in purchases, utilization of products and services, prices, current installations, and forecasts to 1990.

182. TELEPHONE CONTRACTORS, CONSULTING ENGINEERS AND CONSULTANTS. *Telephony Publishing Corporation. 1980. 12 pg. $10.00.*

Survey of telephone industry contractors, consulting engineers and telephone consultants examining their buying power within telephone industry market segments.

183. TERMINAL EQUIPMENT IN THE BELL SYSTEM. *Probe Research, Inc. 1979. 115 pg. $275.00.*

A statistical study and analysis regarding telephone terminal equipment. The data is derived mainly from Bell documents and includes: PBX equipment; Centrex; key systems; telephone sets; modems; and data terminals. Tables on ancillary equipment include answering devices, dialers, radio equipment, and interconnect penetration in major areas. Analyzes important Bell terminal equipment documents and summarizes Bell's investments, spending, units and revenues for each area of terminal equipment.

184. TRANSPORTATION SECURITY PRODUCTS AND SERVICES MARKETS. *Frost and Sullivan, Inc. July 1981. 265 pg. $975.00.*

Market analysis and forecasts to 1985 for security products and services for truck, rail, air and ocean transport. Examines penetration of end use markets, government and trade anti-crime actions and manufacturers. Includes market survey and estimated sales, 1980-1985.

185. UNITED STATES MARKET FOR FIRE AND SECURITY PRODUCTS, SYSTEMS AND SERVICES. *Industrial Market Research Ltd. 1980.*

Analysis of markets for fire and security products, systems and services.

186. U.S. MILITARY AND AEROSPACE TRAINER AND SIMULATOR MARKET. *Frost and Sullivan, Inc. January 1981. 334 pg. $1000.00.*

Analysis of U.S. military markets for electronic training simulators with forecasts to 1985. Reviews government contracts and purchases by department, competition, and major manufacturers.

187. U.S. MILITARY AUTOMATIC TEST EQUIPMENT (ATE) MARKET. *Frost and Sullivan, Inc. July 1980. $950.00.*

Forecast and analysis to 1985 of military ATE programs for avionics, missiles, vehicles, communications and other electronic systems. Also system engineering services, logistic support and software system support. Covers technology status and trends, profiles, major manufacturers plus user requirements and fundings.

188. THE U.S. MILITARY COMMUNICATIONS MARKET. *Frost and Sullivan, Inc. July 1980. 343 pg. $975.00.*

Analysis of U.S. military markets for communications equipment, with forecasts to 1985. Analyzes market structure;

details government funding for research and development and government contracts for systems; lists executives, by government offices; describes products; and profiles manufacturers.

189. THE U.S. TELECOMMUNICATIONS INDUSTRY, AN OVERVIEW AND ANALYSIS: 1982. *Venture Development Corporation. 1982.*

Analysis of the telecommunications and equipment industries. Includes average annual growth rates for communications equipment product groups, 1981-1984; market segmentation and analysis; vendors' marketing strategies; purchasing patterns; and forecasts of market size and shipments.

190. VOICE COMMUNICATIONS: NEW TECHNOLOGIES AND MARKETS. *Yankee Group. June 1981. 200 pg. $9500.00.*

1981 report is part of Communications / Information Systems Planning Service. Available to service subscribers. Discusses current and potential markets and applications for emerging voice technologies. Reviews suppliers, market shares, user perceptions, innovations and marketing opportunities.

191. WORLDWIDE SATELLITE MARKET DEMAND FORECAST: TASK 11 REPORT PLANNING ASSISTANCE FOR THE 30 / 20 GHZ PROGRAM. *Bowyer, J.M. Phillips Publishing Inc. June 19, 1981. 105 pg.*

Forecast of worldwide satellites and market analysis, 1981-2000. Includes expenditures, rankings of consumer countries, earth station installations, satellite units in operation, technology types and market size, all for 1981-2000.

See 912, 913, 951, 965, 966, 967, 976, 977, 979, 982, 985, 1002, 1003, 1004, 1005, 1022, 1024, 1025, 1027, 1034, 1046, 1063, 1064, 1073, 1074, 1075, 1080, 1110, 1117, 1126, 1144, 1169, 1210, 1212, 1213, 1219, 2230, 2231, 2232, 2235, 2247, 3156, 3159, 3162, 3191, 3518, 3520, 3521, 4018, 4019, 4020, 4021, 4025, 4027, 4029, 4033, 4034, 4035, 4038, 4039, 4040, 4041, 4042, 4044, 4047, 4050, 4053, 4057, 4058, 4059, 4062, 4067, 4071, 4073, 4074, 4075.

Investment Banking Reports

192. DOMINION SECURITIES AMES: INDUSTRY REVIEW: CANADIAN HIGH TECHNOLOGY OPPORTUNITIES FOR THE EARLY 1980'S: AN ENTREPRENEURIAL MANIFESTO: VOLUME 1. *Coupal, C.E. Dominion Securities Ames Ltd. April 16, 1982. 35 pg.*

Analysis of new technology industries in Canada. Examines productivity increases, management and marketing strategies, and finances of selected corporations. Shows telephone units in operation by country, 1980; percent of total world inventory, 1980; telephone growth rates, 1970-1980; telecommunications construction expenditures, 1979-1982; market analysis, 1970-1990; and manufacturers, 1981.

193. DONALDSON, LUFKIN AND JENRETTE: THE COMMUNICATIONS INDUSTRIES: TELEPHONE PBX AND KEY SYSTEMS. *Vitolo, Aristide I. Donaldson, Lufkin and Jenrette Inc. June 11, 1982. 27 pg.*

Analysis of the telephone PBX and key system market and industry. Examines manufacturers, product lines, capacity and marketing strategies; the technology; and new products. Shows installed base, 1978-1984, and annual installations, 1980-1984, for PBX's and key systems.

194. LOCAL AREA NETWORKS. *FIND / SVP. R158. December 1981. 18 pg. $250.00.*

Wall Street analysis of new technology for local area networks and industry trends.

195. MORGAN STANLEY AND CO.: INVESTMENT PERSPECTIVES: AEROSPACE: STRATEGIC ARMS. *Demisch, Wolfgang H. Morgan Stanley and Company, Inc. October 19, 1981. pg.10-11*

Analysis of government budgets for defense programs and products and the impact on corporations in the military markets.

196. MORGAN STANLEY AND CO.: INVESTMENT PERSPECTIVES: RESEARCH COMMENT: DEFENSE: REAGAN EASES OFF. *Demisch, Wolfgang H. Morgan Stanley and Company, Inc. September 8, 1981. pg.13-15*

Analysis of impact on military markets and the aerospace industry of announced reductions in government spending, 1983-1984. Examines stock performance, specific products and prospects for investments.

197. PAINE WEBBER MITCHELL HUTCHINS INC.: STATUS REPORT: THE FISCAL 1983 DEFENSE BUDGET. *Pincavage, John V. Paine Webber Mitchell Hutchins, Inc. April 5, 1982. 10 pg.*

Review of government defense budget, 1981-1984. Shows percent of expenditures for research and development, employees and other items, 1977-1985; government funding by projects, 1982-1983; Department of Defense inflation rates, 1979-1987; and major defense contractors.

198. UNITED BUSINESS AND INVESTMENT REPORT: COMMUNICATIONS, ENTERTAINMENT MEDIA MERGE: A LOOK AT TWO PRIMARY PLAYERS. *United Business Service Company. v.74, no.19. May 10, 1982. pg.182, 191*

Report on communications and entertainment markets and two vendors. Graphs stock prices, dividends, price earnings ratio, and earnings per share, 1970-1982. Includes 1982 estimated sales, revenues from new products, and performance of all product lines.

199. UNITED BUSINESS AND INVESTMENT REPORT: INDUSTRY SURVEY: DEFENSE ELECTRONICS SPENDING HEATING UP. *United Business Service Company. v.73, no.42. October 19, 1981. pg.418*

Report on selected corporations providing electronics for military markets. Includes product lines; sales; earnings per share, 1980-1981; stock prices, 1980-1981; price earnings ratio, 1981; dividend; and yield.

See 1250, 2344, 4097, 4098, 4099, 4102, 4111, 4113, 4115, 4121.

Industry Statistical Reports

200. AUDIOVISUAL WAGES AND SALARIES. *Hope Reports, Inc. August 1980. 20 pg. $30.00.*

Report of 1980 wage and salary figures for 813 audio-visual employees in 33 job classifications of professional and technical media work. Salary data includes jobs and pay by media systems, by type of company, by job tenure, by regions and management salaries related to staff size. The principal functions of AV units are producing and coordinating use of closed-circuit TV, video programs, motion pictures, slides and exhibits.

201. A BASELINE STUDY OF THE TELEPHONE TERMINAL AND SWITCHING EQUIPMENT INDUSTRY: REPORT TO THE SUBCOMMITTEE ON TRADE OF THE COMMITTEE ON WAYS AND MEANS OF THE U.S. HOUSE OF REPRESENTATIVES ON INVESTIGATION NO.332-92 UNDER SECTION 332 OF THE TARIFF ACT OF 1930, AS AMENDED. *Fletcher, William B. U.S. International Trade Commission. USITC Publication 946. February 1979. 108 pg. Free.*

Results of a study undertaken to provide a reliable data base of information on shipments, exports, and imports of telephone terminal and switching equipment. Data was obtained from an ITC questionnaire and from U.S. Department of Commerce figures for the years 1972, 1976 and 1977. Study also analyzes effect of technological changes in the 1970's, and surveys royalty payments, world markets, and international trade barriers. Appendices provide statistical tables, product definitions, Congressional and regulatory actions, and an analysis of the accuracy and comparability of different sets of statistics.

202. CIVILIAN SPACE POLICY AND APPLICATIONS. *Office of Technology Assessment, U.S. Congress. U.S. Government Printing Office. su. doc. no. Y3.T22/2:2C49. June 1982. 391 pg.*

Examines use of communication satellites, land remote sensing, materials processing in space, and space transportation in U.S. civilian sector, 1982. Provides data on government funding, purposes of specific programs, and future uses of technology.

203. COMMUNICATIONS EQUIPMENT MANUFACTURERS, 1979: ANNUAL CENSUS OF MANUFACTURES: FABRICANTS D'EQUIPEMENT DE TELECOMMUNICATION, 1979: RECENSEMENT ANNUEL DES MANUFACTURES. *Canada. Statistics Canada. English/French. Annual. Cat. #43-206. August 1981. 20 pg. 5.40 (Canada).*

For manufacturers of radio and television transmitters, radar equipment, closed circuit TV apparatus, electronic navigational aids, public address equipment, telephone and telegraph apparatus, and electronic signalling equipment, tables give data on establishments, employees and ownership, wages and salaries, fuel and raw materials costs, value of shipments, and value added, 1971-1979, and inventories, 1979, all by province. For 1978-1979, provides tables on materials and supplies used and on manufacturers' shipments, by kind. Includes a list of companies with addresses.

204. CURRENT INDUSTRIAL REPORTS: SELECTED ELECTRONIC AND ASSOCIATED PRODUCTS, INCLUDING TELEPHONE AND TELEGRAPH APPARATUS, 1978. *U.S. Department of Commerce. Bureau of the Census. Customer Service Branch. Annual. MA-36N. January 1980. 30 pg. $.35.*

Report includes statistics on telephone and telegraph apparatus and switching equipment; radio and television communications equipment; search and detection and navigation and guidance systems equipment; tubes, semiconductors, transistors, capacitors, and other electronic devices; X-ray and other electromedical equipment. Data is presented on the quantity and value of shipments and number of companies for selected electronic and associated products, by product, for current and preceding year. Includes statistics on export shipments of electronic products for current and preceding year and imports for consumption for same years.

205. EIS SHARE OF MARKET REPORT: RADIO AND TV COMMUNICATION EQUIPMENT. *Economic Information Systems, Inc. Quarterly. 1982. $350.00.*

Report on radio and TV communication equipment marketplace concentration and ownership structure. Companies in industry are listed and then ranked by production. Other data includes market share (dollars and percent), with groupings of all plants a company operates in the industry, with the parent company's market share broken down on a per-plant basis.

206. EIS SHARE OF MARKET REPORT: TELEPHONE AND TELEGRAPH APPARATUS. *Economic Information Systems, Inc. Quarterly. 1982. $180.00.*

Telephone and telegraph equipment market place concentration and ownership structure. Companies in industry are listed and then ranked by production. Other data includes market share (dollars and percent), with groupings of all plants a company operates in the industry, with the parent company's market share broken down on a per-plant basis.

207. EIS SHIPMENTS REPORT: RADIO AND TV COMMUNICATIONS EQUIPMENT. *Economic Information Systems, Inc. Quarterly. 1982. $350.00/yr.*

Report on radio and TV communications equipment industries. Arranged by state and county, report includes every plant in the industry with annual shipments over $500,000 and/or 20 or more employees. Plant listings detail address, telephone, estimated annual shipments, and percent of market. Similar statistics for each county and state.

208. EIS SHIPMENTS REPORT: TELEPHONE AND TELEGRAPH APPARATUS. *Economic Information Systems, Inc. Quarterly. 1982. $180.00/yr.*

Report on telephone and telegraph apparatus industries. Arranged by state and county, report includes every plant in the industry with annual shipments over $500,000 and/or 20 or more employees. Plant listings detail address, telephone, estimated annual shipments, and percent of market. Similar statistics for each county and state.

209. HOUSEHOLD FACILITIES AND EQUIPMENT, MAY 1981. *Canada. Statistics Canada. English/French. Annual. January 1982. 39 pg. 5.40 (Canada).*

Market survey of Canadian families and household reports May 1981 installations of telephone equipment, bathroom fixtures and other equipment, and ownership of various appliances. Includes size of housing units and fuel consumption by type.

210. INDEPENDENT TELEPHONE STATISTICS: VOLUME 1. *U.S. Independent Telephone Association. 1981. 35 pg. $5.00.*

Volume 1 of a 2-volume statistical report on the U.S. independent telephone industry. Details growth patterns over the past 5 and 10 years. Provides data on operating income, revenue and tax statistics and information regarding plant and types of service. Data is submitted to the USITA by reporting member companies.

211. INTERCONNECT INDUSTRY STATISTICAL REVIEW 1979. *North American Telephone Association. Annual. 1979. 26 pg. Free.*

Annual economic development report and forecast for the U.S. telephone communications equipment industry. Gives census of telephone units with projections; market shares for types of equipment, 1965-1985; revenues, 1979-1982; and growth rates, 1979-1990. Compiled from government and industry sources.

212. INTERCONNECT INDUSTRY STATISTICAL REVIEW 1981. *North American Telephone Association. Annual. September 1981. 28 pg. $25.00.*

Annual statistical survey of the business telephone equipment market. Tables and charts show numbers, dollar amounts, and percentages for instruments and systems arranged by Bell, independent, and interconnect companies. Estimates and projections for the interconnect industry in PBX and key systems through 1990 is charted. Data is derived from government and industry sources.

213. TELEPHONE OPERATING COMPANIES AND SOLAR ENERGY: PRESENT AND FUTURE APPLICATIONS. *Telephony Publishing Corporation. 1980. 23 pg. $12.00.*

A status report on current and future applications of solar energy by telephone operating businesses.

214. U.S. CENTRAL OFFICE SWITCHING TRENDS: 1976-1985. *Telephony Publishing Corporation. 1980. $90.00.*

Offers detailed accounts of switches, by manufacturer and model, in the U.S. and Continental systems. Gives Bell System and GTE 1976-1978 growth, by operating company; Western Electric and GTE actual production, and production for the most often recognized general trade suppliers. Contains forecasts (1980-1985) by the operating companies' own buyer-specifiers. Includes Bell's operating companies investment in switching equipment, by category.

215. 1977 CENSUS OF MANUFACTURES: RADIO AND TV COMMUNICATION EQUIPMENT: PRELIMINARY REPORT: INDUSTRY SERIES. *U.S. Department of Commerce. Bureau of the Census. Customer Service Branch. Irregular. MC77-1-36D-4(P). 1979. 17 pg. $.35.*

This is a preliminary report from the Bureau of the Census 1977 Census of Manufactures. Tables with industry statistics from the U.S. for 1963-1977 and with statistics for selected geographic areas for 1972 and 1977 give data relating to employees, production costs, assets and expenditures, inventories, and ratios. Other tables supply 1972 and 1977 statistics on product classes giving quantity and value of shipments by all producers and statistics on materials consumed by kind.

216. 1977 CENSUS OF MANUFACTURES: TELEPHONE AND TELEGRAPH APPARATUS: PRELIMINARY REPORT: INDUSTRY SERIES. *U.S. Department of Commerce. Bureau of the Census. Customer Service Branch. Irregular. MC77-1-36D-3(P). 1979. 9 pg. $.35.*

This is a preliminary report from the Bureau of the Census 1977 Census of Manufactures. Tables with industry statistics from the U.S. for 1963-1977 and with statistics for selected geographic areas for 1972 and 1977 give data relating to employees, production costs, assets and expenditures, inventories, and ratios. Other tables supply 1972 and 1977 statistics on product classes giving quantity and value of shipments by all producers and statistics on materials consumed by kind.

217. 1981 ANNUAL STATISTICAL VOLUME 2. *U.S. Independent Telephone Association. Annual. 1981. 192 pg. $30.00.*

1981 annual statistical report covering Class A, B, and C independent telephone companies for 1980. Some 765 businesses submitted data which includes operating results and financial statistics.

See 1310, 2383, 2393, 2394, 2425, 2426, 4133, 4144, 4145.

Financial and Economic Studies

219. ABA (AMERICAN BANKERS ASSOCIATION) BANK NETWORK FEASIBILITY STUDY. *American Bankers Association. 1980. $31.25.*

Examines costs of operation of three possible network configurations for communications in the banking industry, based on a study by the AT&T Bank Network Feasibility Study Team. Evaluates volumes, costs, and voice-only calling patterns and other factors of all U.S. commercial banks, and estimates cost-effectiveness of a shared voice switching network for communications.

220. BELL SYSTEM PROCUREMENT - THE NEW RULES. *Probe Research, Inc. February 1981. 61 pg. $130.00.*

Analysis of proposed regulations of Bell System equipment purchases. Includes historical data and analysis, industry structure, and effects on contracts.

221. THE INTERCONNECT INDUSTRY IN CANADA 1982. *ICA Telecommunications Management. August 1982. 164 pg. 150.00 (Canada).*

Analysis of the interconnect industry in Canada with data on the market for privately-owned telephone systems and equipment includes industry size, installation to 1981, costs, revenues to 1981, employees by types, industry structure, market shares, market areas, labor productivity, sales to 1981, sales personnel earnings, financial analysis, and competition.

See 1330, 2430, 4169, 4172.

Forecasts

222. COLOR VIDEO CAMERAS: U.S. MARKET OPPORTUNITIES. *Strategic Inc. June 1981. l500.00.*

Examines the trend toward broadband intraplant communications, with ample channel capacity for video, teleconferencing, manufacturing process monitoring and commmunicating training and marketing messages. Report is based on a survey of broadcast, CATV, institutions and business and industry users, indicating how they perceive the advantages of future solid state equipment. Provides statistical tables.

223. DATA COMMUNICATIONS SYSTEMS AND EQUIPMENT: PREDICASTS' INDUSTRY STUDY NO. E54. *Predicasts, Inc. August 1979. $725.00.*

Analyzes the rapidly growing market for data communications. Competition of leading firms in communications and data processing and computer networks and distributed data processing are the main focus of the study, and historical and projected (1985, 1990) data is provided for computers in use by type; data terminals by type; and equipment such as modems, multiplexers, and processors. The structure of the data communications equipment industry is analyzed, and sales estimates and profiles of leading firms are provided.

224. DATA REMOTE TESTING EQUIPMENT (U.S.). *Frost and Sullivan, Inc. December 1980. $950.00.*

Forecast of equipment by Bell Telephone Companies, other independent telephone companies, computer, terminal and modem companies. Review of applications and types of testing systems. Discussion of end-user market and geographic distribution. Company profiles for manufacturers. Design trends and other factors influencing the market.

225. ELECTRONIC HOME OF THE FUTURE: (ENTERTAINMENT, COMMUNICATIONS, EDUCATION): PREDICASTS' INDUSTRY STUDY NO. E60. *Predicasts, Inc. August 1980. $725.00.*

Analyzes the U.S. market for electronic devices to be found in the home of the future. Historical (1967 to 1979) and projected (1985 to 1995) trends are investigated by device and by market segment. Discussions are included on selected companies, distribution channels, and emerging technologies, including cable-telephone-television interface, speech synthesis and digital audio.

226. FACSIMILE EQUIPMENT AND ELECTRONIC MAIL: PREDICASTS' INDUSTRY STUDY NO. E53. *Predicasts, Inc. March 1979. $725.00.*

Examines the current and future status of facsimile equipment as a major element in the dynamic electronic mail market. Demand for facsimile equipment is analyzed in relation to other message modes. A number of key environmental and industry factors affecting facsimile demand are discussed. The structure of the facsimile industry is reviewed, with special attention given to the role of foreign manufacturers and imports.

227. THE FUTURE FOR ELECTRONIC MAIL AS A BUSINESS OPPORTUNITY. *Bengston, R. Battelle Memorial Institute. October 1979. $4000.00.*

Attempts to determine when electronic mail systems will be used commercially in the U.S., who will initially need and purchase systems, what sorts of hardware and software will be required, and other market information of interest to hardware producers.

228. THE FUTURE OF DIGITAL TECHNOLOGY IN THE PRIVATE RADIO SERVICES. *U.S. Federal Communications Commission. 1981.*

Discusses a variety of digital technologies that could affect the various private radio services. Forecasts the growth of digital communications. Reviews the current state of digital communications including aviation, land mobile, marine, microwave and personal. Analyzes the regulatory, economic and social factors which will influence the rate of adoption of digital technologies.

229. LOCAL NETWORK AND SHORT RANGE COMMUNICATIONS. *International Resource Development, Inc. September 25, 1980. 124 pg.*

1980 report includes: 8-year (1982-1990) forecast of the markets for local networks and short range communication fiber optics, baseband, PABX, and broadband coax; state of the art technology overview; industry structure; review of leading companies; and the results of a user market research survey.

230. MILITARY COMMUNICATIONS: MARKETS, SYSTEMS, OPPORTUNITIES FOR OPTICAL COMMUNICATIONS, 1979-1990. *Probe Research, Inc. September 1979. 230 pg. $545.00.*

In-depth study of optical communications in the military. Contains a detailed analysis of eight key military communications markets: avionics, tactical land, nontactical land, undersea C31, satellite and ground station, shipboard and ashore, intelligence and COMSEC, guided weapons and other. Describes specific programs, comparative economics, alternate technologies. Projections for each of 8 military markets of sales of communications equipment and sales of each fiber optic component.

231. OUTLOOK FOR FIBER OPTIC COMMUNICATIONS SYSTEMS IN NORTH AMERICA. *Lipoff, Stuart J. Arthur D. Little, Inc. September 12, 1979. 11 pg. $500.00.*

1979 report on fiber optic communications systems in 1978 and forecasts for 1985 in North America. Included are transmission equipment (telephone, computer, satellites) growth rates, sales markets, cost distribution, and suppliers. Data derived from Arthur D. Little estimates.

232. PABX, INTERCONNECT AND THE FUTURE OFFICE CONTROLLER: SECOND REPORT. *International Resource Development, Inc. August 1980. 350 pg. $4500.00.*

This report contains detailed analysis of the PABX market in manufacturing, transportation, airlines, utilities, wholesale / retail, banking and finance, insurance, brokerage, service industries, hotels / motels, health care, education and state / local and federal government. Year-by-year, segment-by-segment estimates of shipments and installed base through 1990. Special services using the database available.

233. PABX, INTERCONNECT AND THE FUTURE OFFICE CONTROLLER: VOLUME 1. *International Resource Development, Inc. August 1980. 196 pg. $985.00.*

Analysis of future PABX markets, as the PABX takes on the role of the office controller, handling electronic mail, least-cost-routing and other advanced functions. Review of state-of-the-art in PABX market and product development. Ten-year market projection of shipments (lines and value) segmented into fourteen industry groups. Forecasts are based upon proprietary computer model of U.S. PABX market.

234. THE PBX SWITCHING SYSTEMS MARKET IN THE U.S. *Frost and Sullivan, Inc. August 1979. 270 pg. $900.00.*

Examines the PBX equipment market by system size category (under 100 station lines, 101-350, 351-1000, and over 1000 lines). Market shares for the total market are distributed by Bell Telephone companies, other telephone companies and interconnect suppliers. Includes profiles of the major suppliers with information on capabilities of the product line, marketing approaches and new product possibilities. Market development forecasted through 1989 in dollars and number of units and size of systems.

235. PRIVATE SECURITY SYSTEMS. *Predicasts, Inc. February 1982. $900.00.*

Analysis of markets for security services and equipment, with forecasts to 1995. Includes expenditures, new products, end uses, industry structure and manufacturers.

236. SPACE DEFENSE MARKET IN THE U.S. *Frost and Sullivan, Inc. April 1980. $950.00.*

Five-year analysis and forecast of military space market includes military requirements; current and projected DOD programs, and their funding; and their competitive environment. Program areas include space surveillance and warning; space system survivability; antisatellite systems; space defense operations; and advanced space technology.

237. STEP-INDEX AND SINGLE-MODE FIBERS: A COMPLETE GUIDE. *Probe Research, Inc. 1980. 66 pg. $95.00.*

Analysis covers up to 21 operating specifications of 109 step-index and single-mode optical fibers by 18 manufacturers. Fiber types include pcs multimode, glass-on-glass- multimode, plastic core-multimode, glass bundled and single-mode. Specifications include attenuation, wavelength range, bandwidth, numerical aperture, price, core diameter. Recent R&D develop- ments, new products, and their implication to the industry are discussed in terms of single-mode fiber advances, lower loss plastic fibers, performance data, zero dispersion fibers, mode partition noise and multilevel signaling.

238. TELEPHONE ACCOUNTING AND ROUTING SYSTEMS (TARS) MARKET. *Frost and Sullivan, Inc. July 1979. 277 pg. $950.00.*

Analyzes market for and forecasts sales through 1983 of telephone accounting and routing systems. Focuses on passive accounting systems, active routing and accounting systems, and PBX's with routing and accounting capabilities. Examines TARS sale opportunities by selected end-user industries. Supplies company profiles of a representative list of suppliers, analyzing their product lines and marketing approaches.

239. VENTURECASTS 1981 DATA BOOKS: COMPUTER / COMMUNICATIONS. *Venture Development Corporation. 1981. 300 (est.) pg. $495.00.*

Abstracts of forecasts for computers and communications equipment from electronics and business periodicals, government statistics, and trade associations.

240. WHOLESALE BANKING AUTOMATION MARKET: COMPUTER / COMMUNICATIONS EQUIPMENT, TIME-SHARING SERVICES, SOFTWARE PACKAGES. *Frost and Sullivan, Inc. March 1980. 336 pg. $950.00.*

Examination of the wholesale banking automation market reveals opportunities in the corporate structure for automating complex applications with telecommunications systems. Estimates through 1990 provided in sales volumes (number of units and dollar sales) and installed base for message switching hardware, video terminals and line printers, and for end user systems (funds transfer investments, foreign exchange trading, letters of credit, securities clearance) which include minicomputer and software packages.

See 1355, 1364, 1369, 1371, 2463, 2471, 4175, 4176, 4178, 4179, 4180, 4181, 4183, 4184.

Directories and Yearbooks

241. ALL ABOUT 125 TELEPRINTER TERMINALS. *Datapro Research, Inc. (McGraw-Hill). Annual. 1981. 39 pg. $15.00.*

Provides specifications and prices for 125 teleprinter-style communications terminals from 50 vendors. Includes results of user survey covering 10,600 installed terminals.

242. ALL ABOUT 31 PABX INTERCONNECT SYSTEMS. *Datapro Research, Inc. (McGraw-Hill). Annual. 1981. $15.00.*

Using side-by-side comparison tables, report reviews 31 PABX systems offered by 18 vendors on the U.S. market.

243. ALL ABOUT 84 COMMUNICATIONS PROCESSORS. *Datapro Research, Inc. (McGraw-Hill). Annual. 1981. 31 pg. $15.00.*

Report analyzes benefits and drawbacks of the programmed communications processor concept in general; describes the front-end environment; and presents the experiences of 211 users. Comparison charts summarize the capabilities of 84 communications processors.

244. AMATEUR RADIO EQUIPMENT DIRECTORY. *Kengore Corporation. Annual. 1981. 200 pg. $5.00.*

Directory of 100 manufacturers provides address, telephone, products, prices, and specifications. Citizens band equipment manufacturers not included.

245. AV-USA 1975-80. *Hope Reports, Inc. Annual. 5th. 1980. $165.00.*

Annual compilation of information on AV manufacturers: net profits, capital expenditures, new products, production and distribution costs. Gives dollar and unit sales for 130 products. Has market information and forecasts.

246. BROADCASTING / CABLE YEARBOOK 1982. *Broadcasting Publications, Inc. Annual. 1982. 1300 (est.) pg. $65.00.*

Sections include broadcasting in general (directory of government agencies; group ownership directory; financial data; tabular record of station trading since 1954; industry analysis); television (coverage maps and market rankings of 211 areas of dominant influence according to Arbitron; data on stations in U.S., its territories and Canada, including channel, power, antenna height, address, teletype writer exchange number, licensee, date of acquisition, ownership, print media connections, principal stockholders for independently owned stations, network affiliation, hours of foreign language programming per week, names of rep and Washington attorney, number of cable systems, and other data); radio (similar data as well as frequency, format, minute rate); broadcast advertising, networks, and programming (directories of advertising agencies, station reps, media buying/planning services; directories of producers, radio and TV affiliates, distributors, production services, news services, and other groups; data on audience composition and set usage); equipment and engineering (directories of manufacturers and distributors; directories of satellite owners, operators, resale and common carriers, and programmers and networks using satellites); professional services, associations, broadcast education, and international industry-related groups; and cable (regulations; system directory; directory of U.S. and Canadian multiple system operators).

247. BROADCASTING YEARBOOK. *Broadcasting Publications, Inc. Annual. 1981. $30.00.*

Directory covers TV and radio networks and stations, stations' representatives, program services, and advertising firms as well as industry-related equipment.

248. THE BUSINESS COMMUNICATIONS REVIEW MANUAL OF PBX. *Goeller, Lee. Telecom Library Inc. Irregular. 1980. $127.50.*

Ten-page data sheets on each of 39 PBX systems detail systems design, technology and cost comparisons by line size.

249. CANADIAN INTERCONNECT DIRECTORY. *Northern Business Intelligence. Irregular. 1981. 35.00 (Canada).*

Outlines the interconnect (PBX) suppliers in Canada.

250. CATV BUYER'S GUIDE OF EQUIPMENT, SERVICES AND MANUFACTURERS. *Robert A. Searle. Annual. 1981.*

Annual listing of cable television accessories and devices as accepted by the Cable Television Association.

251. THE COMPLETE GUIDE TO OPTICAL CONNECTORS. *Probe Research, Inc. Irregular. June 1979. $95.00.*

Guide to the 40 optical connectors of 19 manufacturers currently on the market. It contains tables comparing up to 21 different connector operating parameters. In-depth discussions of connector technologies, problems, uses, and feasibility. Sample MilSpec explanations. Manufacturer address lists. There are separate tables for each of the three connector types: single fiber connectors, multi-channel connectors, bundled fiber connectors. Each connector type is further broken down into primary characteristics, design characteristics, and secodary characteristics. Detailed tables on price, insertion loss, ease of field installation, MilSpecs, design, construction, applications, and fiber-compatability.

252. DATAPRO REPORTS ON DATA COMMUNICATIONS: VOLUME 3. *Datapro Research, Inc. (McGraw-Hill). Monthly. 1980. 1000 (est.) pg. $410.00.*

Volume III of a loose-leaf monthly updating service. Discusses transmission facilities; modems; multiplexors, test, monitor and control equipment; remote computing service; telephone systems; and special equipment services. Marketing analysis, system characteristics / features, and costs are provided. Reviews and analyzes systems and information currently available.

253. DIRECTORY OF DEFENSE ELECTRONIC PRODUCTS AND SERVICES: U.S. SUPPLIERS, 1981. *Electronic Industries Association. FIND / SVP. Irregular. 7th rev. ed. March 1981. 229 pg. $60.00.*

10-year military computer forecast features full test of EIA data processing forecast. Examines 40 equipment service categories, including computers, communications, EW, C3, components, and meteorology. Shows leading U.S. supplier capabilities, with technical specifications, illustrations, and supplier addresses.

254. DIRECTORY OF DEFENSE ELECTRONIC PRODUCTS AND SERVICES: U.S. SUPPLIERS, 1982. *Information Clearing House. Annual. 8th ed. February 1982. 225 pg. $85.00.*

Covers defense communications, displays, and naval surface / subsurface electronics; supplier directory for 40 equipment and service categories, with technical specifications and illustrations; and a glossary and military nomenclature guide.

255. FIBER OPTICS HANDBOOK AND BUYERS' GUIDE. *Information Gatekeepers, Inc. Irregular. 3rd ed. 1980. $45.00.*

Handbook, directory, and buyers' guide for fiber optical communications equipment. Lists addresses of manufacturers, end uses, costs, market analysis, and new product characteristics.

256. FIBER OPTICS PATENTS NEWSLETTER: A MONTHLY SURVEY OF U.S. AND INTERNATIONAL PATENTS. *Information Gatekeepers, Inc. Monthly. 1981. 12 pg. $195.00/yr.*

Monthly service listing U.S. fiber optics patents granted during month of coverage (with explanatory data, patent holders' name and address, application date, and patent number); a supplement to previous issues; U.S. government-owned patents and inventories for license; and editor's addendum.

257. FIRE PROTECTION EQUIPMENT DIRECTORY. *Underwriters Laboratories Inc. Annual. January 1982. 285 pg. $3.00.*

Addresses of manufacturers of fire protection equipment and product lines found in compliance with Underwriters Laboratories safety standards.

258. GUIDE TO ELECTRONIC MAIL AND MESSAGE SYSTEMS. *Thorne, Stevenson and Kellogg. Irregular. 1982.*

Guide includes information about electronic mail, store-and-forward message systems, voice messaging and facsimile transmission. Also includes types of systems available, vendors and their products and how to select a system.

259. MARINE PRODUCTS DIRECTORY. *Underwriters Laboratories Inc. Annual. September 1981. 60 pg. $1.10.*

Addresses of manufacturers and marine product lines found in compliance with Underwriters Laboratories safety standards.

260. MOBILE RADIO HANDBOOK: 1981. *Titsch Publishing, Inc. Annual. 1981. $24.95.*

1981 handbook for the land mobile radio industry. Provides

data on manufacturers, services, equipment trade associations, frequency coordinators and FCC personnel.

261. NATA BUSINESS TELEPHONE AND EQUIPMENT DIRECTORY OF THE INTERCONNECT INDUSTRY 1980-1981. *North American Telephone Association. Annual. 1980. 194 pg.*

Directory lists all major suppliers of private telephone systems in the U.S. Contains 2,009 entries from over 1000 cities and towns. Lists contractors by states and alphabetically; PBX and key manufacturers and suppliers alphabetically; ancillary equipment, manufacturers and retailers, alphabetically; and state associations. Provides names, addresses, telephone numbers and key personnel.

262. PBX SYSTEMS. *Telecomsept Services Inc. Irregular. 1981. 195.00 (Canada).*

Covers vendors, characteristics, equipment, and prices of 60 to 4000 line PBX's in Canada.

263. THE SATELLITE SERVICES SOURCEBOOK. *TMS. Irregular. 1st ed. 1982. 300 pg. $75.00.*

Covers electronics, cable TV and related satellite communications services, consultants, brokers, carriers, hardware manufacturers, programmers, videotex, earth stations, government agencies and associations.

264. SSI BUYERS / SELLERS GUIDE. *Surplus Source International. Annual. 1982. $180.00.*

Listings of telecommunications plant and equipment for sale or wanted for purchase, with prices.

265. TELEPHONE ANSWERING DEVICES: BUYERS LABORATORY INC. *Buyers Laboratory, Inc. Irregular. 2nd ed. 1979. 24 pg.*

A directory to telephone answering devices, arranged by type of device. Each entry includes; features, prices, warranty and descriptions. Manufacturers' listing with addresses and phone numbers included.

266. 1979 / 1980 BUSINESS TELEPHONE AND EQUIPMENT DIRECTORY OF THE INTERCONNECT INDUSTRY. *North American Telephone Association. Irregular. 1979. 142 pg.*

1979 directory contains listings of telephone contractors, by states and alphabetically; manufacturers and suppliers, alphabetically; and state associations. Entries provide names, addresses, telephone numbers, key executives, and product lines.

267. 1981-1982 NBFAA ROSTER. *National Burglar and Fire Alarm Association. Annual. July 31, 1981. 130 pg. $25.00.*

Annual directory lists NBFAA (National Burglar and Fire Alarm Association) members and committees. Includes a list of suppliers and dealers.

268. 1981-1982 TELECOMMUNICATIONS SOURCEBOOK. *North American Telephone Association. Annual. 1981. 184 pg. $35.00.*

Lists all major suppliers of telecommunications equipment in the U.S. and Canada. Includes the following sections: contractors by state with addresses and top executives; manufacturers and suppliers; suppliers of add-on equipment and related services listed alphabetically with addresses, telephone numbers, executives, and product lines; telecommunications consultants; state associations, with membership data, and a general alphabetical listing.

See 1375, 1389, 1414, 1421, 1422, 1443, 2506, 2542, 3246, 3260, 3649, 4187, 4191, 4196, 4201, 4202.

Special Issues and Journal Articles

Advanced Technology Libraries

269. 3M, CHECKPOINT DOMINATE SECURITY / SYSTEM MARKET: KNOGO, GAYLORD ARE ALSO FACTORS. *Knowledge Industry Publications. September 1981. pg.5-6*

Discusses size, segmentation and saturation of the U.S. market for library theft detection systems. Identifies the five leading manufacturers of these devices, with number of installations achieved to date, addresses, and comments by executives.

Aerospace Canada

270. CANADA IN SPACE: SPECIAL REPORT. *Maclean Hunter Ltd. v.6, no.1. December 1981. pg.11-25*

Special report on strengths of Canadian space industry. Discusses remote manipulator technology and its potential use in hostile environments; satellites and ground stations; and remote sensing of the earth, its surface features and natural resources. Also discusses size of space industry in Canada, amount of government spending on space programs and the future Canada will play in space technology.

271. CANADA IN THE MILITARY MARKET PLACE. *Maclean Hunter Ltd. v.6, no.2. May 1982. pg.2-39*

Special issue highlighting the accomplishments of Canadian aerospace industry in military market place. Examines an air to ground rocket weapon system and an airborne maritime search radar system. Internal digital communications systems for ships that offer significant cost and space savings over current technology are discussed.

272. HANOVER 1982: THE HANOVER INTERNATIONAL AIR SHOW AND THE INTERNATIONAL DEFENCE ELECTRONICS EXPOSITION. *Maclean Hunter Ltd. Irregular. March 1982. pg.44-48*

An overview of the eleven Canadian firms attending the International Defense Electronics Exposition and the Hanover International Air Show. Gives products and personnel.

AOPA Pilot

273. LORAN C: NAVIGATION'S NEW WAVE. *LaCagnina, Mark M. Aircraft Owners and Pilots Association. Irregular. v.25, no.9. September 1982. pg.58-63*

A technical description of LORAN C air navigation systems includes a short list of manufacturers giving model, storage, capability, functions, price, and address.

Audio-Visual Communications

274. AUDIO-VISUAL COMMUNICATIONS: EQUIPMENT AND MATERIALS. *United Business Publications, Inc. Monthly. v.16, no.5. May 1982. pg.52-66*

A product directory of 55 pieces of audio-visual and photographic equipment. Includes model numbers, descriptions, technical specifications, manufacturer's name, photo, and some prices.

275. WHO'S WHO IN CORPORATE COMMUNICATIONS. *United Business Publications, Inc. Annual. v.16, no.9. September 1982. pg.14-17*

Lists corporate communications executives.

A-V Canada

276. A-V CANADA: HARDWARE. *Maclean Hunter Ltd. Quarterly. v.4, no.4. November 1981. pg.10, 33-36*

Surveys new photographic and video equipment and supplies, including some digital equipment (editing decks, graphics, generators, etc.). Includes descriptions, manufacturers' names, and a check-off card for manufacturers' literature.

277. A-V CANADA: SOURCEBOOK 81: GUIDE TO CANADIAN AUDIO-VISUAL PRODUCTS, DISTRIBUTORS, DEALERS, PRODUCERS AND SERVICES. *Southam Business Publications, Ltd. Division of SouthamCommunications Ltd. Annual. 3662. May 1981. pg.14-48*

Lists Canadian audio-visual equipment manufacturers and dealers, rental and repair services, producers, and production-related services (editing, processing, duplicating, stock and music libraries, etc.). Firms are listed by product / service category, and alphabetically by province and city.

B M/E

278. BROADCAST MANAGEMENT ENGINEERING: THE SOURCE: BUYER'S GUIDE. *Broadband Information Services, Inc. Annual. v.18, no.9. September 1982. pg.43-196*

Lists products for the U.S. broadcasting industry. Presents products in seven major categories with manufacturers listed within each product type. Indicates the specific type of product each company sells. Provides an alphabetical listing of manufacturers including addresses, product lines, and field offices. Additional directory lists representatives and distributors by state and includes addresses.

279. SMPTE EXHIBITORS. *Broadband Information Services, Inc. Annual. v.18, no.10. October 1982. pg.93-96*

Lists exhibitors at the 1982 SMPTE conference. Includes company name and indicates product groups exhibited.

280. 1982 NAB SHOW IN PRINT (CONTINUED). *Broadband Information Services, Inc. Annual. v.18, no.7. July 1982. pg.51-69*

Describes and illustrates new products in following categories: AM, FM, and TV transmitters; ENG microwave advances; fiberoptics; STL's, remote control testing; antennas; and tower arrays. Gives manufacturer.

Bank Systems and Equipment

281. SECURITY EQUIPMENT AND SERVICES: NEW SECURITY PRODUCTS BATTLE BANK ROBBERY HIKE. *Gralla Publications. Irregular. October 1981. pg.65-72*

Part one of a two-part product review covering over 75 products and services for bank security systems. Product specifications and manufacturers are provided. Unique features are highlighted for each product.

Biomedical Communications

282. BIOMEDICAL COMMUNICATIONS 1980 BUYER'S GUIDE. *United Business Publications, Inc. Annual. November 1979. pg.4-58*

A 1980 buyer's guide to the biomedical communications industry. Entries are arranged in the following categories: medical illustration equipment; motion picture equipment; video / CCTV equipment; still cameras and lenses; lighting equipment and accessories; darkroom and processing equipment; audio equipment; projectors / programmers; AV furniture and accessories; special equipment; media sources; and company roster. About 450 manufacturers are covered, with 1800 listings. Company addresses and phone numbers given.

Broadcast Communications

283. NAB "82 DIRECTORY OF EQUIPMENT AND SERVICES. *Globecom Publishing, Ltd. Annual. v.5, no.3. March 1982. pg.83-120*

Directory listing of exhibitors at the annual NAB convention includes 100 categories of products and services for the broadcasting industry. Manufacturers and suppliers are listed alphabetically within each category.

284. NAB "82 EXHIBITORS. *Globecom Publishing, Ltd. Annual. v.5, no.3. March 1982. pg.122-173*

Lists exhibitors at the NAB convention alphabetically and includes a short description of product lines, new products and services. Manufacturers and suppliers of equipment and services for the broadcasting industry are included.

285. NRBA "82: THE BIG WINNER FOR RADIO: EXHIBITORS. *Globecom Publishing, Ltd. Irregular. v.5, no.8. August 1982. pg.48-52*

Lists corporations exhibiting at the 1982 National Radio Broadcasters Association Convention. Includes description of product lines and services.

Broadcast Engineering

286. AUDIO SWITCHING SYSTEMS: PAST, PRESENT AND FUTURE. *Palmer, Mike. Intertec Publishing Corporation. Irregular. v.24, no.3. March 1982. pg.202-218*

Lists 12 manufacturers of audio routing switchers. Includes prices and technical specificaitons of each company's product. Source is 1980 Buyer's Guide issues of Broadcast Engineering.

287. BROADCAST ENGINEERING: NEW PRODUCTS. *Intertec Publishing Corporation. Monthly. v.24, no.3. March 1982. pg.272-276*

Reviews new products used by the broadcasting industry. Includes product description and gives manufacturer's name.

288. GETTING THE PICTURE: VIDEO MONITOR SURVEY. *Bentz, Carl. Intertec Publishing Corporation. Irregular. July 1982. pg.32-48*

A survey of current TV video monitors including color and monochrome composite video models and RGB units. Information includes manufacturer, model number, and description. Further information through reader card service.

**289. NATIONAL ASSOCIATION OF BROADCASTERS -
1982: CONVENTION REPLAY.** *Rhodes, Bill, ed. Intertec Publishing Corporation. Annual. v.24, no.6. May 1982. pg.8-150*

Summarizes issues, trends, new products, and new technology covered at the 1982 Annual Convention of the National Association of Broadcasters. Reviews developments in AM stereo; satellite technology; audio processors; digital video effects; mobile TV vans and trailers; ENG, EFP and studio cameras; recorder-in-camera systems; 3-chip cameras; teletext; and over 80 pages of more new products. Lists manufacturers, product specifications, and descriptions.

**290. TV TEST EQUIPMENT SURVEY: MONITORING VIDEO
PARAMETERS.** *Intertec Publishing Corporation. Irregular. v.24, no.5. May 1982. pg.38-66*

Lists specifications of 1982 products used to monitor quality of television broadcast stations' signals.

**291. 1980 BUYERS' GUIDE: THE BROADCAST INDUS-
TRY'S COMPREHENSIVE PRODUCT DIRECTORY.** *Intertec
Publishing Corporation. Annual. September 1980. pg.113-206*

1980 directory of broadcasting products, manufacturers (with sales officers, executives and addresses), and dealers or distributors (with addresses). Includes advertisers' indexes.

292. 1982 NAB CONVENTION SPECIAL. *Intertec Publishing
Corporation. Annual. v.24, no.3. March 1982. pg.22-200*

Provides a directory to the NAB - 82 / Dallas exhibitor products. Gives for each major equipment heading the exhibitor and booth number.

Broadcaster

293. BROADCASTER: FALL DIRECTORY. *Northern Miner
Press, Ltd. Semi-annual. November 1981. 142 pg.*

2058 listings cover Canadian radio and television stations, cable businesses, equipment manufacturers and suppliers, advertising agencies, and audio and video production houses. Includes names, addresses, chief executives, telephone numbers, unions, research services, government agencies, record companies, and film labs.

294. BROADCASTER: FALL DIRECTORY. *Northern Miner
Press, Ltd. Irregular. November 1980. 141 pg. 7.50 (Canada).*

Directory lists radio stations, television stations, sales representatives, equipment manufacturers and suppliers, and cable television systems, with addresses. Includes audio production businesses, video production businesses, TV and radio program distributors, media consultants, advertising agencies, associations, unions, research services, government agencies, record companies, and film laboratories.

Broadcasting

295. BROADCASTING: STOCK INDEX. *Broadcasting Publications, Inc. Weekly. v.102, no.19. May 10, 1982. pg.103*

Lists stocks of broadcasting and related corporations. Includes broadcasting, broadcasting with other major interests, cable, programming, service and electronics manufacturing companies. Gives for each stock the following: Exchange and company; closing price, May 5, 1982; closing price April 28, 1982; net change in week; percent change in week; P / E ratio; and market capitalization. Sources are Broadcasting and Standard and Poor.

**296. EASTERN SHOW SETS ITS SIGHTS ON THE CABLE
OPERATOR.** *Broadcasting Publications, Inc. Annual. v.103,
no.10. September 6, 1982. pg.46-52*

Directory lists exhibitors at the 1982 Eastern Cable Television Trade Show. Alphabetical listings of manufacturers and suppliers provide company name, address, products and services.

297. EXHIBITION SHOWCASE: BROADCASTING. *Broadcasting Publications, Inc. Irregular. v.102, no.17. April 26, 1982.
pg.46-65*

Lists companies exhibiting at the Las Vegas Convention Center. Listing includes company name and address as well as asterik indicating new products.

298. SCENE SET FOR SMPTE IN NEW YORK. *Broadcasting
Publications, Inc. Annual. v.103, no.18. November 8, 1982.
pg.65, 68*

Outlines annual conference of Society of Motion Picture and Television Engineers, 1982. Gives partial list of exhibitors. Emphasizes broadcast related exhibits.

**299. THE TOP 100 COMPANIES IN ELECTRONIC COMMU-
NICATIONS.** *Broadcasting Publications, Inc. Annual. v.102,
no.1. January 4, 1982. pg.39-73*

Provides rankings of the top 100 corporations in electronic communications. Shows total revenues and net earnings, 1980-1981; earnings per share, 1981; and index of participation in electronic communications. For each corporation, gives chief executive, lines of business, and operating performance.

**300. THE TOP 100 COMPANIES IN ELECTRONIC COMMU-
NICATIONS.** *Broadcasting Publications, Inc. Irregular. January
7, 1980. pg.35-86*

A directory of publicly listed corporations involved with broadcasting and electronics communications. Includes a ranking of the top 100 companies by revenues for 1979, net earnings 1978-79, earnings per share, and growth rates. Also includes an alphabetical listing with brief histories of the corporations, financial statements, and chief executives.

Cable Communications

301. CCTA CONVENTION EXHIBITORS LIST. *Ter-Sat Media
Publications Ltd. Irregular. v.48, no.5. May 1982. pg.38-39*

Lists exhibitors at the 1982 Canadian Cable Television Association convention, Toronto. Gives name, city, and booth number.

302. NEW BROADCAST CABLECAST EQUIPMENT GUIDE.
*Ter-Sat Media Publications Ltd. Irregular. September 1981.
pg.36-40*

A product review covering 23 new products for use in the cable television, broadcasting and telecommunications industries. Products are described with name, address and telephone number of supplier. Includes photographs.

303. 1981 ANNUAL DIRECTORY / BUYER'S GUIDE. *Ter-Sat
Media Publications Ltd. Annual. March 1981. pg.26-73*

Source of business and product listings for the cable television, broadcasting and telecommunications industries compiled in Canada. First section is an alphabetic listing of manufacturers and suppliers serving the industry and includes addresses and phone numbers. The second section lists, by subject (cable television, broadcasting and telecommunications), industrial plants and equipment and services. Notes U.S. suppliers and manufacturers.

304. 1982 DIRECTORY OF COMPANIES AND THEIR KEY PERSONNEL SUPPLYING PRODUCTS AND SERVICES TO THE CABLE TELEVISION, BROADCASTING AND TELE-COMMUNICATIONS INDUSTRIES. *Ter-Sat Media Publications Ltd. Annual. v.47, no.11. November 1981. pg.63-81*

Alphabetical listing of manufacturers, suppliers, and vendors serving the Canadian cable television, broadcasting, and telecommunications industries. Each entry includes company name, address, telephone number, location of sales offices, key sales personnel by name, and description of products and services.

Cable Television Business

305. THE FIRST STEP: 20 EQUIPMENT MANUFACTURERS AND SUPPLIERS. *Cardiff Publishing Company. Irregular. v.19, no.12. June 15, 1982. pg.40-48*

Directory of 20 equipment manufacturers and suppliers includes names, addresses, telephone numbers, and products manufactured.

306. NATIONAL CABLE PROGRAMMING CONFERENCE, 1982. *Cardiff Publishing Company. Annual. v.19, no.21. November 1, 1982. pg.70-73, 78*

Outlines National Cable Television Association's first annual National Cable Programming Conference, 1982. Lists exhibitors.

CableVision

307. CABLE STATS: PAY / BASIC CABLE GROWTH. *Titsch Publishing, Inc. September 28, 1981. pg.79-80 $54.00/yr.*

Ranks U.S. cable television systems by number of consumers at end of preceding month, within various service categories. Lists transponder allocations on major satellites, satellites to be launched through 1984, and new satellite-fed cable services to be introduced through 1982.

308. CABLEVISION: LEADING THE WEEK: PROTECTION STRIPPED. *Evanow, Peter. Titsch Publishing, Inc. v.7, no.31. April 12, 1982. pg.13-16*

Presents a discussion of H.R. 5949, covering copyright regulation pertinant to the broadcasting industry.

309. CABLEVISION: NCTA CONVENTION SCHEDULE AND BOOTH GUIDE. *Titsch Publishing, Inc. v.7, no.33. April 26, 1982. pg.149-206*

Conference calendar includes listings of presentations covering all phases of the broadcasting industry. Listings include moderator and participants profiles.

310. CABLEVISION: STACKING DISHES. *Panero, Hugh. Titsch Publishing, Inc. September 28, 1981. pg.28-34 $54.00/yr.*

Discusses cable television operations' growing tendency to employ multiple dish antennae at ground receiving stations, as relaying satellites proliferate. Covers purchases and installed facilities of major systems, and 1980 sales and market shares of antenna suppliers.

311. NCTA ENHANCED SERVICES CHART. *Titsch Publishing, Inc. Irregular. June 1, 1981. pg.180-201*

National Cable Television Association's listing of businesses in the cable television industry offering enhanced services. Provides business name, start-up date, number of subscribers, and cost to subscribers for installation and monthly fee. Covers the following services: home security including burglar / fire / emergency medical / police; electronic newspapers; energy management / meter reading; traffic light control, home banking; stock reports; information retrieval; opinion polling; videotex; and teletext.

312. SIXTH ANNUAL CONSTRUCTION EXPENDITURES SURVEY: THE $1.5 BILLION SELLERS' MARKET. *Dawson, Fred. Titsch Publishing, Inc. Annual. January 5, 1981. pg.26-28*

Annual review of construction expenditures for the U.S. cable television industry for 1980, with forecasts for 1981. Tables cover miles of new construction, expenditures for new plants and equipment by type, labor costs, and repair costs. Compiled by survey.

313. 1981 NCTA CONVENTION BOOTH GUIDE. *Titsch Publishing, Inc. Irregular. May 18, 1981. pg.158-194*

Lists U.S. businesses that have booths at the 1981 National Cable Television Association convention. Includes those in cable television production and those in communications manufacturing. Provides contact person.

314. 50 OF THE TOP COMPANIES IN CABLE. *Ross, David. Titsch Publishing, Inc. v.7, no.35. May 10, 1982. pg.203-234*

Analysis of the top 50 corporations in the cable television industry gives rankings, business addresses, executives, lines of business, percent of ownership in ventures and mergers. Financial analysis includes cable revenues, 1980-1981; ratios of total revenues to net income and total assets to cable revenues, 1980-1981; and stock prices, 1980-1981.

Canada Commerce

315. CANADIAN ADVANCED TECHNOLOGY ENTERPRISE: THE KEY TO THE FUTURE. *Murray, Paul J. Canada. Department of Industry, Trade and Commerce. 1982. pg.1-5*

Discusses microelectronics, robotics, telecommunications, fiber optics, and biotechnology in the context of Canadian research and development. Includes forecasts and financial support.

Canadian Aviation

316. 1981 AVIONICS BUYERS' GUIDE. *Maclean Hunter Ltd. Annual. v.54, no.7. July 1981. pg.47-93*

1981 annual directory lists avionics equipment under the following categories: VHF communications, VHF navigation, VHF navcoms, HF communications, automatic direction finders, radio magnetic indicators, altimeters, ATC transponders, distance measuring equipment, altitude indicators, flight directors, autopilots, horizontal situation indicators, various navigation systems, glide slope receivers, marker beacon receivers, audio panels, weather radar, ELT / CPI / FDR, and radiotelephone systems. Provides manufacturer's name, model number, description, weight, size and price for each product.

Canadian Secretary

317. TELEPHONE MARKETPLACE. *Maclean Hunter Ltd. Irregular. May 1982. pg.14-16*

Listing of new products available in telephone communications, manufacturers, and addresses.

Canadian Shipping and Marine Engineering

318. CANADIAN SHIPPING AND MARINE ENGINEERING MARINE BUYERS' GUIDE 1982. *Arthurs Publications. Annual. v.53, no.3. December 1981. pg.7-100*

International directory of products and services for the marine industry in Canada. Directory contains an index of products and services with an alphabetical listing of the relevant manufacturers, agents and suppliers. Separate index contains head and branch office addresses and telephone numbers of manufacturers, agents and suppliers.

Coal Age

319. UNDERGROUND COMMUNICATIONS EQUIPMENT GUIDE. *McGraw-Hill Book Company. Annual. October 1980. pg.154-173*

1980 directory of underground communications equipment, manufacturers and product features, listed alphabetically.

Communication Technology Impact

320. UP ON THE ROOF. *International Resource Development, Inc. Elsevier Science Publishers. v.4, no.7. October 1982. pg.10-11*

Summarizes report on U.S. direct broadcast satellite systems. Gives annual unit sales, installed base, and revenues for both backyard and rooftop terminals, 1982, 1984, 1987, 1990. Predicts 1990 prices of home terminals. Names companies involved with DBS systems.

Communications

321. COMMUNICATIONS: 1981 BUSINESS RADIO BUYERS GUIDE EDITION. *Cardiff Publishing Company. Annual. December 1980. 96 pg.*

Classified directory of manufacturers and suppliers of land-based mobile radio components: transceivers, antennas, encoders / decoders, multicouplers, scanners, microphones, crystals, displays, generators, batteries, etc. Also lists manufacturers' representatives, engineers, consultants, FCC officials and relevant trade associations.

Communications News

322. ALL THAT'S NEW IN TWO-WAY RADIO. *Harcourt Brace Jovanovich, Inc. Annual. v.19, no.8. August 1982. pg.41-87*

Outlines the new technologies of the U.S. two-way radio industry. Reports on the number of two-way radio stations by type of service. Provides estimates of demand for two-way communication on 35 major cities. Presents a two-way radio checklist and buyers' guide. List includes suppliers' names, addresses, and products. Additional directory lists suppliers with addresses for each major product group.

323. COMMUNICATIONS NEWS: NEW ITEMS. *Harcourt Brace Jovanovich, Inc. Monthly. v.19, no.9. September 1982. pg.114-125*

Monthly directory lists, describes, and illustrates new communication products giving brand names and manufacturers' names and addresses.

324. COMMUNICATIONS NEWS TWO-WAY BUYERS GUIDE. *Harcourt Brace Jovanovich, Inc. Annual. August 1981. pg.58-72*

Directory of U.S. manufacturers and suppliers of two-way communications equipment. Provides business address and products. Lists new products available for use in two-way radio systems, with descriptions and photographs.

325. DATACOMM UPDATE: LATEST - GENERATION MODEMS PERFORM BETTER AND COST LESS. *Edwards, Morris. Harcourt Brace Jovanovich, Inc. Annual. v.19, no.7. July 1982. pg.91-96*

Reviews new product developments among 18 independent U.S. modem suppliers. Arranged alphabetically by supplier name. Includes technical specifications.

326. DIGITAL MICROWAVE GROWTH. *Harcourt Brace Jovanovich, Inc. v.19, no.10. October 1982. pg.62*

Projects microwave growth by segment, 1982-1990.

327. FORECAST FOR 1981 SEES GREATER PARTICIPATION BY USERS REQUIRED. *Harcourt Brace Jovanovich, Inc. Annual. January 1981. pg.28-39*

Forecasts of and views on economic indicators for 1981, by executives in the communications industry. Comments especially on government regulation.

328. GREAT GROWTH ELECTRONIC MAIL / FAX GEARED TO HIGH-SPEED NEED OF USERS. *Harcourt Brace Jovanovich, Inc. v.19, no.9. September 1982. pg.72-73*

Provides electronic mail systems data and 1981-1990 projected growth by type. Discusses state of industry and CBMS standards.

329. NEW PRODUCT INTRODUCTIONS PACE THE RECORD - SIZE TCA EXPOSITION. *Harcourt Brace Jovanovich, Inc. Irregular. July 1981. pg.72-81*

Overview of the 160 businesses which exhibited their products at the ICA Exhibition (1981). Provides business names and new products.

330. NEW PRODUCTS FOR DATA COMMUNICATIONS SYSTEMS WITH CN READER SERVICE FOR MORE INFORMATION. *Harcourt Brace Jovanovich, Inc. Irregular. December 1981. pg.88-97*

Directory includes photographs, descriptions and manufacturers, withaddresses, for more than 70 new products for data communications systems.

331. NEW PRODUCTS RECENTLY INTRODUCED FOR ANTENNA AND TOWER SYSTEMS. *Harcourt Brace Jovanovich, Inc. Irregular. June 1982. pg.96-99*

A listing of new antennas and tower systems includes manufacturer's address, description, and photograph. Further information through reader card service.

332. OVER 9000 PEOPLE TOUR TCA'S BIGGEST EXHIBIT EVER. *Harcourt Brace Jovanovich, Inc. Irregular. November 1981. pg.66-76*

Provides descriptions and photographs of products displayed at the TCA's exhibit on communications equipment. Gives manufacturers' names.

333. 1981 CN DIRECTORY OF PBX MAKERS / SUPPLIERS. *Harcourt Brace Jovanovich, Inc. Annual. July 1981. pg.46-48*

1981 directory of U.S. manufacturers and suppliers of PBX equipment. Provides business addresses and products.

Computerworld

334. COMMUNICATIONS: LOCAL-NET OFFERINGS NOW NUMBER 40. *Bartik, Joan. CW Communications Inc. v.16, no. 13. March 29, 1982.*

Provides list of local-area network vendors. Includes, for each company, corresponding network; network type (baseband or broadband); access method (contention, token passing, other); transmission speed (to 1Mbps, to 2Mbps, to 10Mbps, over 10Mbps); cable length (to 2000 ft., to 5000 ft., over 5000 ft.); gateways IBM SNA/SDLC X.25 (xerox, ethernet, other); and application area (general business, electronic mail, word processing, industrial, other).

335. VOICE MAIL. *Ulrich, Walter. CW Communications Inc. v.16, no.16. April 19, 1982. pg.13-16*

Briefly discusses the technology involved in voice mail, with a review of the characteristics of voice mail systems. Includes charts listing the features of voice mail compared with telephone answering services and other alternative mail systems; and voice system access modes. Data is presented on system price ranges, and systems currently in use.

Data Communications

336. MULTIPLEXER USERS SHIFT INTO HIGH GEAR. *Datapro. McGraw-Hill Book Company. Irregular. v.10, no.11. November 1981. pg.54-56*

Presents highlights of survey results on the use of multiplexers among data communications users. Contains the responses of 49 users of 393 multiplexers. Ranks the users' ratings of those products included in the questionnaire. Lists some 28 manufacturers of multiplexers, with business address and phone number.

db, the Sound Engineering Magazine

337. DB: NEW PRODUCTS AND SERVICES. *Sagamore Publishing Company, Inc. Monthly. v.16, no.4. April 1982. pg.62-65*

Lists new products for sound engineering. Includes manufacturer's name, product description, and photographs.

Dealerscope

338. WHO'S WHO IN SECURITY EQUIPMENT. *Dealerscope Inc. Semi-annual. v.24, no.8. August 1982. pg.78-80*

Alphabetically lists manufacturers of electronic home and automotive security services. Includes addresses.

Design News

339. FORECASTS PREDICT LARGE GROWTH FOR FIBER OPTICS MARKET. *Cahners Publishing Company. v.38, no.18. September 27, 1982. pg.38*

Analyzes the U.S. fiber optics market, 1982-1990. Forecasts, by component, the size of the U.S. market in dollars for 1982, 1986, and 1990. Discusses end uses for fiber optics. Source is Business Communications Co., Inc.

Educational and Industrial Television

340. DIRECTORY OF CHARACTER GENERATORS. *C.S. Tepfer Publishing Company, Inc. Irregular. v.14, no.9. September 1982. pg.47-61 ff.*

Directory of video character generators lists manufacturers, product lines, technical specifications, and prices.

341. EDUCATIONAL AND INDUSTRIAL TELEVISION: NEW PRODUCTS. *C.S. Tepfer Publishing Company, Inc. Monthly. v.14, no.3. March 1982. pg.88-99*

Lists new products related to the fields of industrial and educational television. Each entry contains brief description of the product, name of manufacturer and price. For some products information on delivery and availability also provided. Typical products listed include oscilloscope kit, 3-camera system, light color changer, and mobile consoles.

342. NEW CAMERAS AND RECORDERS SHOWN AT NAB. *C.S. Tepfer Publishing Company, Inc. v.14, no.5. May 1982. pg.17, 28,ff*

Reviews new communications equipment displayed at the National Association of Broadcasters exhibit, April 1982. Includes manufacturer's name, product name, and detailed product description.

Electrical Consultant

343. CATALOG LISTING OF SECURITY AND LIFE - SAFETY SYSTEMS. *Cleworth Publishing Company, Inc. Irregular. v.62, no.3. May 1982. pg.48-50, ff*

Lists manufacturers of security systems and components including card access systems, detectors, CCTV equipment, panels, security lighting, AV and associated equipment. Gives manufacturer's addresses arranged by product.

344. SECURITY / LIFE SAFETY SYSTEMS. *Cleworth Publishing Company, Inc. Annual. v.61, no.3. May 1981. pg.12-52*

Annual special issue on electronic and electrical apparatus for security and safety systems. Several feature articles are included presenting descriptions of security systems in institutions, along with cost data and technical specifications. Products section lists new products for security and life safety systems. Each entry includes description, photograph, and manufacturer's name and address.

345. SPECIAL ISSUE ON SECURITY AND FIRE ALARM SYSTEMS. *Cleworth Publishing Company, Inc. Annual. v.60, no.3. May 1980. pg.7-43*

Annual special issue on electrical and electronic apparatus for security and safety systems. Covers large systems installed in a variety of institutions including prisons and universities. Includes some information on costs and technical specifications. News and Developments section provides a listing of new products with descriptions, photographs and manufacturer's name and address.

Electronic Business

346. CABLE MARKET IS A HOT ONE BUT MAY BE TOUGH TO CRACK. *Cahners Publishing Company. v.8, no.7. June 1982. pg.62*

Examines demand for addressable CATV converters. Gives number of units shipped, dollar sales, and market share for five manufacturers, 1981, 1982.

347. GI BETS ON NEW CONVERTER TO KEEP IT NO.1 IN CATV. *Cahners Publishing Company. v.8, no.7. June 1982. pg.67-68*

Examines U.S. market for addressable CATV converters. Gives number of units for addressable, pay and plain converters, 1980-1984.

348. PROMISE OF SAVINGS DRIVES FOUR COMMUNICATIONS AREA. *Cahners Publishing Company. v.8, no.7. June 1982. pg.62*

Examines U.S communications market. Gives dollar value, 1980, 1982, 1984, 1986, 1988, 1990. Provides dollar value of production of six types of equipment, 1981, 1986; and average annual compounded growth rate. Mentions influence of AT & T's deregulation.

Electronics

349. MILITARY PURCHASES TO SLOW. *Connolly, Ray. Electronics Industry Association. McGraw-Hill Book Company. October 9, 1980. pg.95-97*

Estimates U.S. military electronics markets. Forecasts 1981-1990 government expenditures and funding for R & D.

Electro-Optical Systems Design

350. VENDOR SELECTION ISSUE 1979-80. *Milton S. Kiver Publications, Inc. Annual. November 1979. 162 pg.*

1979-80 directory of U.S. manufacturers and vendors of products, services, materials and equipment for the manufacture of laser-related instruments. Lists addresses and telephone numbers. Products grouped under radiation detectors, image sensors and detectors, lasers, incoherent radiation sources, image sources and information displays, active and passive optical components and assemblies, electronic support components and equipment, testers and test systems, meters and indicators, support equipment, hardware, materials and chemicals, services, and systems.

FEM (Financieel-Economisch Magazine)

351. AMERIKA WACHT MEDIA - REVOLUTIE: VIDEO - DISC DOET INTREDE. [America Awaits the Media Revolution.]. *Ponsen, Dick. Advisory Group Decision Making - Part of Adviesgroep Strategische Besluitvorming, Amsterdam. B.V. Uitgevers-Maatschappy Bonaventura. Netherlandish. v.12, no. 13. June 25, 1981. pg.59-61*

Examines the media revolution: video disc. Discusses uses and possibilities of the video disc for publishers of specialized magazines, educational institutes and computer manufacturers. Reviews manufacturers, price index, end uses, and capacity.

Home Entertainment Marketing

352. WHO'S WHO IN SATELLITE TV. *Dealerscope Inc. Irregular. v.2, no.7. July 1982. pg.43-53*

Lists manufacturers of consumer equipment required to monitor TV satellites including antennas, receivers, and others. Gives name, address, telephone, and description of products.

Industrial Economics Review

353. INFORMATION EQUIPMENT: AN INTEGRATED APPROACH TO A GLOBAL MARKET. *Sullivan, William J. U.S. Department of Commerce. Bureau of Industrial Economics. Office of Public Affairs and Publications. Irregular. December 1979. pg.1-18*

Analysis of international telecommunications industry emphasizes U.S. market positions and federal regulation. Four tables of 1973-1979 data for U.S., Canada, Japan, and certain European and less developed nations detail information equipment sales; telephone market penetration; 14 major firms' revenues from equipment sales and services; government market for business and communications equipment and computers.

Infosystems

354. MODEM AND MULTIPLEXER UPDATE: COMPACT AND CHEAPER. *Edwards, Morris. Hitchcock Publishing Company. v.28, no.11. November 1981. pg.50-58*

Discusses major manufacturers of modems and multiplexers in t the U.S. Lists the manufacturers of modems and acoustic couplers providing details on short or long haul. Also lists the manufacturers of statistical multiplexers and provides details on bps.

International Business Week

355. THE FASTEST GROWING MARKET FOR DEFENSE CONTRACTORS. *McGraw-Hill Book Company. no.2757-88. September 20, 1982. pg.108, 111*

Report on government funding for defense contracts, 1981-1991, and the industry structure shows 1981 sales of leading corporations and growth rate potential.

International Fiber Optics and Communications

356. TEST AND MEASUREMENT - PRODUCT FUTURE. *Information Gatekeepers, Inc. Irregular. January 1980. 37 pg.*

Report on optical fibers, connectors, cables sources and splices, 1979. Calls for new performance standards, measurement techniques, and instruments. Describes new products and lists addresses of manufacturers.

Laser Focus

357. 1980 LASER FOCUS BUYER'S GUIDE WITH FIBEROPTIC COMMUNICATIONS. *Advanced Technology Publications. Annual. 15th ed. January 1980. 480 pg. $20.00.*

Directory of products and services given manufacturers, suppliers, consultants and representatives with addresses, contacts, and descriptions of products. Covers detectors, fiber optics materials and devices, lasers, modulators, meters, switches, and crystals.

Lloyd's Ship Manager

358. PRODUCT FOCUS: LORAN AND OMEGA. *Lloyd's of London Press Ltd. Irregular. February 1981. pg.19-22*

A directory of marine radio navigation systems and components including plotters, converters, processors, etc. Data includes address of manufacturer / supplier; trade name and technical specifications; price in January, 1980 when available; and service office.

359. PRODUCT FOCUS: SATNAVS. *Lloyd's of London Press Ltd. Irregular. January 1981. pg.19*

A directory of maritime satellite navigation equipment. Data includes address of manufacturer / supplier; trade name, components, and specifications; price in December 1980; and international sales / service representative.

Magazine of Bank Administration

360. A DIRECTORY OF BANK SECURITY PRODUCTS. *Bank Administration Institute. Annual. v.48, no.3. March 1982. pg.52-72*

Lists, by major and minor product category, bank security products offered by over 150 leading U.S. manufacturers and suppliers. List of vendor addresses, alphabetically arranged by company name, follows.

361. DIRECTORY OF BANK SECURITY PRODUCTS. *Bank Administration Institute. Annual. February 1981. pg.29-40*

Directory provides a classified product listing of bank security equipment offered by leading manufacturers and suppliers. Includes an alphabetical list of business addresses.

362. HOME BANKING: WHERE IT STANDS TODAY. *van der Velde, Majolijn. Bank Administration Institute. v.58, no.9. September 1982. pg.16-19*

Examines the trend toward home banking in the U.S. Focuses on telephone banking and video banking systems. Includes discussions of issues affecting the development of these systems and the future role of these systems in the banking industry. Includes tabular data for four types of financial institutions, as follows: number of transactions per call, December 1981; percent of weekly transactions that occur by day; and availability to customers.

Management World

363. TELESOURCES: A DIRECTORY OF TELEPHONE SYSTEM REFERENCES, MANUFACTURERS AND TOOLS. *Administrative Management Society. Annual. v.11, no.3. March 1982. pg.16-17*

Lists sources for further information about telecommunications systems; headquarters of larger manufacturers of telephone switching systems and related equipment in North America, listed in alphabetical order; and new products including a brief description of each product.

Mart

364. MART FOCUS TELEPHONES 1982. *Morgan-Grampian, Inc. Annual. v.28, no.9. May 1982. pg.1S-31S*

Focus on telephones discusses 1982 phone equipment sales. Presents top 10 telephone retailers including phone retailers survey. Catalog shows selection of available equipment from manufacturers and distributors. Lists suppliers along with addresses, key executives, and products.

Merchandising

365. STAFF, STOCK AND PROMOTIONS SELL EARTH STATIONS: SUPPLIERS. *DeSiena, Bill. Gralla Publications. v.7, no.7. July 1982. pg.59, 79*

Retail sales (1982) of earth satellites or television receive-only terminals (TVRD's) are discussed, with suggested strategies for retailers and data on sales and installations, 1981-1982. Market growth is forecast, 1982-1985, with data on retail prices, and estimates of price decreases, 1982-1986. Units in inventory are estimated, 1982. Manufacturers' product lines are described with technical specifications.

Modern Metals

366. CABLE TV: 2.3 BILLION PER YEAR. *Modern Metals Publishing Company. July 1980. pg.12-15*

Examines the expansion of cable television systems in the U.S. and resulting market for production, replacement and maintenance of equipment and components. Covers copper or copper clad aluminum coaxial cable, tubing, aluminum and brass connectors, fabricated microwave antennas, zinc and aluminum diecastings, component extrusions, stampings for equipment, and plated materials. Gives opinions of companies producing equipment, components, and cabling, together with their present and expected output levels to meet demand.

Modern Office Products

367. SHOW PREVIEW: COMMUNICATIONS. *Penton / IPC. Subsidiary of Pittway Corporation. Annual. v.27, no.4. April 1982. pg.112-124*

Previews telecommunications products to be displayed at the 1982 international Communications Association Conference and Telecommunications Exposition.

Motor Boating and Sailing

368. SPORT FISHERMAN'S GUIDE TO MARINE ELECTRONICS. *Hearst Corporation. Irregular. March 1982. pg.65-80*

A directory of electronic equipment used in sport fishing boats, including navigation equipment, depth sounders, sonar, etc. Information includes company name, description, photograph, price, and brief specifications.

Multichannel News

369. WESTERN CABLE SHOW 1982: BOOTH GUIDE. *Fairchild Publications. Annual. v.3, no.45. November 15, 1982. pg.55-92*

Alphabetically lists exhibitors at cable television show, 1982. Includes addresses, representatives, and products or services.

Occupational Hazards

370. NINETEENTH ANNUAL PRODUCT DATAGUIDE: OCCUPATIONAL HAZARDS. *Penton / IPC. Subsidiary of Pittway Corporation. Annual. January 1980. pg.73-200*

19th annual product dataguide for safety equipment and supplies provides a U.S. directory of manufacturers and products. Alphabetically lists firms manufacturing equipment or offering services for industrial safety, hygiene, fire protection, security, and housekeeping programs. Provides names and addresses, alphabetical product listing, and index of products arranged under subject headings.

Progressive Railroading

371. COMMUNICATION AND SIGNAL DIVISION MEETS IN SAINT LOUIS. *Murphy-Ritcher Publishing Company. Annual. v.25, no.10. October 1982. pg.19*

Alphabetically lists exhibitors at annual meeting of AAR's Communication and Signal Division, 1982. Includes their cities.

Radio Communications Report

372. BULLISH ON RCCS. *Titsch Publishing, Inc. November 15, 1982. pg.15-17*

Examines U.S. radio common carrier industry, 1976-1981. Gives 5 companies' annual earnings per share, net income and/or revenues, and number of pagers. Discusses company trends, policies, and acquisitions.

Radio - Electronics

373. BUYER'S GUIDE - CORDLESS TELEPHONES. *McComb, Gordon. Gernsback Publications, Inc. Irregular. v.53, no.11. November 1982. pg.39-42, 104*

Evaluates cordless telephones available in the U.S. Discusses the technology of cordless telephones. Lists manufacturers, addresses, 1982 prices, and technical specifications.

Resort Management

374. PHONE-CALL ACCOUNTING. *Chervenack, Larry. Resort Management, Inc. Annual. v.36, no.5. May 1982. pg.18-21 ff.*

Lists vendor name, equipment base, remote diagnosis option, automatic route selection option, and brief notes for 27 hotel phone-call accounting systems.

Satellite Communications

375. SATELLITE COMMUNICATIONS: DIRECTORY ISSUE. *Cardiff Publishing Company. Annual. September 1980. pg.14-29*

Directory of suppliers of turnkey satellite systems, components, and satellite - related services. Listings include address, phone number, products / services provided, and key contact.

376. SATELLITE COMMUNICATIONS: TECHNOLOGY. *Cardiff Publishing Company. Monthly. v.6, no.5. May 1982. pg.66-70*

Reviews new technology in the satellite communications field. Includes product description (some photographs). Also manufacturer's name, address, and telephone number.

Security World

377. 1981 BUYERS GUIDE TO SECURITY PRODUCTS AND SERVICES: THE SOURCE. *Security World. Division of Cahners Publishing Company. Annual. November 1980. 128 pg. $25.00.*

1981 buyers guide to security products and services lists sources for security equipment, services and accessories. Divided into 6 major categories, the categories are broken down by individual types of equipment and services. Provides names, addresses of manufacturers and suppliers. Includes security-related associations. Lists companies that offer nationwide guard services. Contains master index of products and services.

Telecommunications

378. TELECOMMUNICATIONS: DATACOM PRODUCTS AND PRODUCT SPOTS. *Horizon House. Monthly. v.16, no.4. April 1982. 6 pg.*

Listings of new communications equipment includes brief descriptions, and manufacturers names, addresses and phone numbers.

379. TELECOMMUNICATIONS MAGAZINE 1982 REFERENCE AND BUYERS' GUIDE. *Horizon House. Annual. September 1981.*

Annual reference issue and buyer's guide for 1982.

380. THE 1982 TELECOMMUNICATIONS REFERENCE DATA AND BUYER'S GUIDE. *Horizon House. Annual. 4th ed. September 1982.*

Identifies telecommunications, data communications and computer related equipment to purchase. Provides information on telex, teletex, facsimile, view data, telecommunications in the office, telecom computers, satellite communications and carrier services.

Telephone Engineer and Management

381. MICROWAVE TRENDS AND TECHNOLOGY. *Harcourt Brace Jovanovich Publications. Irregular. July 1, 1981.*

Special issue reviews microwave components trends and technology in the telecommunications industry.

382. OBSERVATIONS: 'A JUNGLE OUT THERE.' *Smith, Roy. Harcourt Brace Jovanovich Publications. v.86, no.16. August 15, 1982. pg.14*

PBX manufacturers' shipments, market shares and growth rates, 1980-1981.

383. SMDA GIVES COMPETITIVE EDGE. *Kopf, Lisa E. Harcourt Brace Jovanovich Publications. v.86, no.22. November 15, 1982. pg.66-67*

Analysis covers 1970-1981 interconnect share of total PBX-key and business telephone instrument markets, with forecasts to 1992.

384. TELEPHONE TERMINAL AND INSTRUMENTS. *Harcourt Brace Jovanovich Publications. Irregular. June 1, 1981.*

Special issue discusses role of telephone terminals and instruments in the communications equipment industry.

Telephony

385. DIGITAL SWITCHING. *Telephony Publishing Corporation. Irregular. June 22, 1981.*

Special issue discusses digital switching equipment industry.

386. EXCLUSIVE INTELEXPO "81 PRODUCT PREVIEW. *Telephony Publishing Corporation. Irregular. July 27, 1981. pg.40-50*

Reviews some of the products which will be on display at INTELEXPO "81 in Los Angeles, Sept. 14-17, 1981. Provides business name and product description with photograph.

387. MICROWAVES. *Telephony Publishing Corporation. Irregular. May 18, 1981.*

Special issue dealing with the microwave market.

388. TELEPHONY: NEW PRODUCTS. *Telephony Publishing Corporation. Weekly. v.202, no.19. May 10, 1982. pg.44-53*

Reviews new products int he telecommunications field. Includes product type, product name, description including photograph, and manufacturer's name.

Television Digest

389. FINANCIAL REPORTS OF TV - ELECTRONICS COMPANIES. *Television Digest, Inc. Weekly. v.22, no.5. February 1, 1982. pg.18*

Weekly financial reports of TV - electronic companies include revenues, net earnings, and per share figures for 16 electronics companies.

390. FINANCIAL REPORTS OF TV - ELECTRONICS COMPANIES. *Television Digest, Inc. Weekly. May 4, 1981. pg.16*

Weekly financial report of T.V. and T.V. - electronics companies. Gives business name, revenues and net earnings for previous week. Number and businesses reviewed each week varies.

TVC

391. TVC 1981 CATV SUPPLIERS PHONE BOOK. *Cardiff Publishing Company. Annual. March 15, 1981. pg.43-69*

1981 annual directory of cable television suppliers and manufacturers. Arranged alphabetically by business name, with address and telephone number. Includes government agencies, associations, and committees of Congress.

392. TWENTY-FIRST ANNUAL TEXAS CABLE TV ASSOCIATION'S EXHIBITORS LIST. *Cardiff Publishing Company. Annual. February 1981. pg.56*

Lists more than 100 businesses specializing in cable television equipment.

Video Systems

393. POST-PRODUCTION FACILITIES SURVEY. *Intertec Publishing Corporation. Annual. v.8, no.4. April 1982. pg.34-49*

Directory of post-production facilities for the video industry lists the following information: facility name, address, phone number, and client contact; editing facilities available; formats; and special services. Gives over 200 entries from a 1982 survey conducted by Video Systems.

Video User

394. NAB PRODUCT REVIEW. *Knowledge Industry Publications. Annual. April 1981. pg.11-17*

Preview, with brief descriptions, of new video product lines to be introduced by major manufacturers at the upcoming National Association of Broadcasters Convention Exhibit. Emphasis is on hardware such as recorders, cameras, switchers, editing decks, lighting equipment, and especially diagnostic instruments.

See 669, 1570, 1600, 1660, 1662, 1694, 1701, 1731, 2566, 2578, 2645, 2656, 2660, 2663, 2687, 2716, 2724, 2726, 2730, 2738, 2765, 2766, 2773, 2794, 3320, 3326, 3653, 3658, 3660, 3681, 3686, 3701,

3721, 3723, 3726, 3745, 4213, 4215, 4216, 4218, 4224, 4225, 4226, 4229, 4234.

Monographs and Handbooks

395. BOOK THEFT AND LIBRARY SECURITY SYSTEMS, 1981-82. *Bahr, Alice Harrison. Knowledge Industry Publications. 2nd ed. 1981. 157 pg. $24.50.*

Coverage includes new products; descriptions and photos of major electronic security systems; and user interviews. 1981 edition features bookstore security chapter that was not present in earlier edition.

396. CELLULAR RADIO: BIRTH OF AN INDUSTRY. *Television Digest, Inc. Irregular. 1983. $125.00.*

Volume includes a primer on cellular workings; how to apply for a license; market analysis and forecasts of cellular mobile telephony; city-by-city status report on applications; and directory of companies in the cellular business.

397. THE CONTINUING REVOLUTION IN COMMUNICATIONS TECHNOLOGY: IMPLICATIONS FOR THE BROADCASTING BUSINESS. *Rosenbloom, Richard S. Harvard University. Information Resources Policy. Order Department. I-81-4. June 1981. 30 pg. $25.00.*

Analyzes the coming challenges to present day broadcasting equipment like the tower and transmitter. New information technologies considered include cable, video discs and cassettes, lightweight personal stereo audio cassette players, and digital records.

398. DEFENSE PROCUREMENT BUDGET. *DMS, Inc. Annual. 1980.*

FY 1981 budget handbook reviews in dollars the defense procurement plan for the present fiscal year, the proposed request for the following fiscal year and two previous year's figures. Data is organized according to budget activity and individual programs within each activity.

399. DEFENSE RDT&E BUDGET HANDBOOK. *DMS, Inc. Annual. 1980.*

FY 1981 budget handbook reviews in dollars all program elements, projects or line items in the Defense Dept.'s annual R&D and Procurement budget request. Years 1979-1982 presented by department: Army, Navy, Air Force defense agencies. Also includes RDT&E funding summary by department, by budget activity and R&D catergory.

400. ELECTRONIC IMAGING: A PROSPECTUS. *Ravich, Leonard E., editor. Institute for Graphic Communications, Inc. Irregular. March 1980. 5000 (est.) pg. $10500.00.*

Discusses recent and projected technological developments in electronic imaging, and their commercial applications. Technologies covered include: image intensifiers, laser holography, fiber optics, charge coupled devices, cathode ray tubes, light-emitting diodes, liquid crystal displays, memory technologies (bubble, magnetic, semiconductor, and optical), television cameras and image processing, videotape and videodiscs, electronic mail, satellites, cable television, printers, medical applications (scanners, ultrasound, nuclear medicine, endoscopy, thermography). Includes 11,000 indexed patents from 16 nations, on microfiche, with narrative summary of patent trends.

401. HANDBOOK OF LOSS PREVENTION AND CRIME PREVENTION. *Fennelly, Lawrence J. Butterworths Publishers, Inc. March 1982. 1000 pg. $60.00.*

Section 1 of loss and crime prevention guide covers closed circuit TV, locks, alarms, guard supervision, fire, and safety. Part

2 (management concept) discusses planning and evaluation, data analysis in crime prevention, budgeting, and public relations. Part 3 addresses security applications in hotels, the cargo and transportation industry, college campuses, retailing, and other sectors.

402. VIDEO IN LIBRARIES: A STATUS REPORT 1979-80. *Bahr, Alice Harrison. Knowledge Industry Publications. 1980. 119 pg. $24.50.*

Issues discussed include: how many libraries have access to video equipment; how widespread is library interest in video; how library administrators decide which machine to buy, what equipment is compatible, which format to adopt; how libraries pay for video; and what services libraries provide with video.

See 3361, 3771, 3772, 4247, 4253, 4257.

Conference Reports

See 4279.

Newsletters

403. VIDEO WEEK. *Television Digest, Inc. Weekly. v.3, no.21. 10 pg.*

Newsletter with news and analysis of the video cassette and disk, pay TV, and other new media industries covers legislation and regulation; activities of vendors and associations; new products; converter and decoder prices and unit sales, 1975-1984; market analysis; executives; retail distribution; home video cassette recorder (VCR) units in o

404. VIDEONEWS. *Phillips Publishing Inc. Bi-weekly. $147.00/yr.*

Biweekly newsletter with analysis of the video, cable and pay TV industries covers technology, management and regulation.

405. WORLD AEROSPACE WEEKLY. *Forecast Associates Inc. Weekly. 10 pg. $295.00/yr.*

Weekly survey of government and private-sector (chiefly airline) awards of aircraft and aerospace equipment contracts and subcontracts, purchases and sales of new and used equipment, and value of contracts and sales. Includes new product developments and activities and financial fortunes of leading airlines.

See 1872, 3773.

Services

406. DATAPRO REPORTS ON DATA COMMUNICATIONS. *Datapro Research, Inc. (McGraw-Hill). Monthly. $595.00/yr.*

Coverage of data communications products, services, and techniques includes product profiles, comparison charts, and user ratings on communications processors, software, terminals, and telephone systems. Annual subscription service includes three loose-leaf volumes, monthly report supplements, monthly newsletter, and a telephone / telex inquiry service.

407. ELECTRONIC TECHNOLOGY 2000 FORECAST. *Gnostic Concepts, Inc. Monthly. 1981. $25000.00.*

Electronic technology forecasting (to 2000) service. Defines state of the art, forecasts technological development, reviews significant research, forecasts market potential and recommends investment strategy for each technology. Program is a continuing service with monthly reports. Subscriber services

include advanced research planning, acquisition candidate analysis, and diversification planning. Examples of potential topics include electronic mail, robotics, automated office, networks, and solar power.

408. LIST OF MATERIALS ACCEPTABLE FOR USE ON TELEPHONE SYSTEMS OF REA BORROWERS. *U.S. Government Printing Office. Irregular. A68.6/5:981. 1981. $18.00/yr.*

Gives names of manufacturers and catalog numbers for materials acceptable for use on telephone systems of REA borrowers. Subscription to looseleaf service includes supplementary material for an indeterminate period.

409. THE OUTLOOK FOR THE U.S. TELEPHONE INTER-CONNECT EQUIPMENT INDUSTRY. *Kay, Phillip M. Arthur D. Little, Inc. June 1980. 55 pg. $2000.00.*

1980 report on U.S. telephone interconnect equipment industry. Product lines, suppliers, applications, market structure, product growth rates, U.S. shipments in U.S., Japan and Europe, market size and distribution from 1977-1985 on equipment provided. Leading suppliers with product lines (PABX, CBX, KTS) included. Publication is part of ADL Impact Service.

See 1885, 2818, 3366.

Indexes and Abstracts

See 1894, 1895.

Dictionaries

See 1897, 4290, 4291.

Central and South America

Market Research Reports

410. COMMUNICATIONS EQUIPMENT IN COLOMBIA: FOREIGN MARKET SURVEY REPORT. *Alarcon, Consuelo. U.S. Embassy - Bogota, Commercial Section. U.S. Department of Commerce. International Trade Administration. NTIS #ITA-82-04-503. November 17, 1981. 95 pg. $15.00.*

Survey performed by the U.S. Embassy on the market in Colombia for communications equipment and services describes market organization and structure, marketing characteristics and trade practices, promising product lines, and end-users. Analyzes statistics on imports by type of equipment and share of market and country of origin and share of market; on exports by product category and share of market and country of destination and share of market; and on local production by product and sales. Describes government and private sector end users, as well as major projects underway or planned. Covers trade regulations and marketing practices. Lists major prospective customers; legally- registered representatives of foreign communications firms; principal trade associations; publications suitable for advertising; and potential agents, representatives, distributors. Includes tables showing imports and exports of specific products.

411. COMMUNICATIONS EQUIPMENT, VENEZUELA: A COUNTRY MARKET SURVEY. *U.S. Government Printing Office. Irregular. CMS 81-047. February 1981. 14 pg. $.50.*

Report analyzes the current and projected markets for sale of communications equipment in Venezuela. Data is provided on the market (local production, imports, exports) by type of equipment; imports by type by country of origin; leading suppliers; market product category; purchases by major user sectors. The text provides product market profiles, end user profiles, and information on marketing practices, and trade restrictions.

412. DEFENSE MARKETS IN LATIN AMERICA. *Frost and Sullivan, Inc. February 1981. $1350.00.*

Five-year forecast and analysis of combat aircraft, support aircraft, tactical missiles, tanks and armored vehicles, naval craft and vessels, major electronic systems, and training, support, and logistical service requirements in Latin American nations. Contains market trends by system categories, national surveys of defense production, major suppliers and their strategies, and economic - political considerations.

413. ELECTRONIC COMPONENTS (BRAZIL). *Lindsey, Richard P. U.S. International Trade Commission. 1980. 34 pg. $10.00.*

Analysis of markets for radios, televisions and other communications equipment in Brazil, 1980. Estimates potential market shares for exports from U.S. manufacturers; analyzes end use industries; specifies promising products; and projects sales.

414. THE MARKET FOR SELECTED COMMUNICATIONS PRODUCTS (ARGENTINA): FOREIGN MARKET SURVEY REPORT. *Webster, Christopher. American Embassy, Buenos Aires (Argentina). U.S. Department of Commerce. National Technical Information Service. DIB-80-05-505. 1979. 93 pg. $10.00.*

The market research was undertaken to study the present and potential U.S. share of the market in Argentina for selected communications products; to examine growth trends in Argentina end-user industries over the next few years; to identify specific product categories that offer the most promising export potential for U.S. companies; and to provide basic data which will assist U.S. suppliers in determining current and potential sales and marketing opportunities. The trade promotional and marketing techniques which are likely to succeed in Argentina were also reviewed.

415. VENEZUELAN MARKET STUDY ON SECURITY DEVICES, EQUIPMENT AND SYSTEMS: FOREIGN MARKET SURVEY REPORT (FINAL). *Ronai y. Asociados S.R.L. Caracas (Venezuela). U.S. Department of Commerce. National Technical Information Service. DIB-80-06-500. 1979. 87 pg. $10.00.*

The market research was undertaken to study the present and potential U.S. share of the market in Venezuela for security devices, equipment and systems; to examine growth trends in Venezuelan end-user industries over the next few years; to identify specific product categories that offer the most promising export potential for U.S. companies; and to provide basic data which will assist U.S. suppliers in determining current and potential sales and marketing opportunities. The trade promotional and marketing techniques which are likely to succeed in Venezuela were also reviewed.

See 4292, 4293.

Special Issues and Journal Articles

Telephone Engineer and Management

416. A TELEPHONE TOUR OF THE CARIBBEAN. *Appelo, Carlton E. Harcourt Brace Jovanovich Publications. v.86, no. 16. August 15, 1982. pg.76-80*

Survey of telephone systems in Caribbean countries. Shows demographic statistics, telephone units in operation, per capita gross national product, and foreign aid.

See 4294.

Western Europe

Market Research Reports

417. AIRBORNE ELECTRONIC EQUIPMENT IN EUROPE TO 1986. *Larsen Sweeney Associates Ltd. 1981. 233 pg. 995.00 (United Kingdom).*

Market research on airborne electronic equipment to 1986 in the following countries: Austria, Belgium, Luxembourg, Denmark, Eire, Finland, France, West Germany, Italy, Netherlands, Norway, Portugal, Spain, Sweden, Switzerland, and U.K. Covers market analysis to 1986 and forecast to 1986; also distribution, production, suppliers, market segmentation, employment and unemployment, profits, sales, and end uses. Airborne electronic equipment include: aircraft instrumentation, communication systems, computers (airborne), navigational systems, and radar.

418. THE ANTI-SUBMARINE WARFARE (ASW) MARKET IN WEST EUROPE. *Frost and Sullivan, Inc. September 1982. 300 pg. $1550.00.*

Analysis of military markets for anti-submarine warfare systems in Western Europe. Includes market size by product groups; defense contracts, manufacturers, and unit costs; rankings of manufacturers; and new technology trends. Provides forecasts for 1983-1987.

419. BRITISH SECURITY COMPANIES. *Jordan and Sons (Surveys) Ltd. 1982. 85.00 (United Kingdom).*

Analysis of British security products markets and corporations. Analysis of financial ratios includes turnover, profits, sales, capital, assets and liabilities. Includes rankings of corporations and, for the top 50, gives business addresses, executives, ownership, employees, and financial statements.

420. BROADCAST EQUIPMENT IN EUROPE TO 1986. *Larsen Sweeney Associates Ltd. 1981. 273 pg. 895.00 (United Kingdom).*

Market research report on broadcast equipment in Europe to 1986 in the following countries: Austria, Belgium, Luxembourg, Denmark, Eire, Finland, France, West Germany, Italy, Netherlands, Portugal, Spain, Sweden, Switzerland and the U.K. Covers market forecast to 1986, distribution, production, suppliers, market segmentation, employment and unemployment, profits, sales and end uses. Product groups covered include: transmitters, television, studio equipment, broadcast equipment, private, defense and governmental.

421. CLOSED CIRCUIT TV IN EUROPE TO 1986. *Larsen Sweeney Associates Ltd. 1981. 264 pg. 895.00 (United Kingdom).*

Market research report on closed circuit TV in the following countries: Austria, Belgium, Luxembourg, Denmark, Eire, Finland, France, West Germany, Italy, Netherlands, Norway, Portugal, Spain, Sweden, Switzerland and the U.K. Covers market forecasts to 1986, distribution, production, suppliers, market segmentation, employment and unemployment, profits, sales and end users. Product groups include: industrial process monitoring, surveillance, security, access control, fire protection, security and military, and low light. 32 end user markets are discussed.

422. THE COMMERCIAL LASER MARKET IN EUROPE. *Frost and Sullivan, Inc. July 1980. 460 pg. $1400.00.*

Market analysis and forecasts to 1990 for commercial applications of lasers in Western Europe. Analyzes end use markets; market shares and products of leading manufacturers; competition from U.S. suppliers; the supply outlook; industry structure and distribution; and prospects for mergers. Forecasts sales and demand to 1990.

423. COMMERCIAL SECURITY PRODUCTS MARKET IN EUROPE. *Frost and Sullivan, Inc. January 1981. 259 pg. $1250.00.*

Market analysis and forecasts to 1990 for commercial security products in Europe. Reviews types of equipment, manufacturers and vendors. Analyses retail distribution, industry structure, demographic statistics, and regulation.

424. COMMUNICATIONS EQUIPMENT AND SERVICES MARKET IN EUROPE. *Frost and Sullivan, Inc. July 1981. 178 pg. $1200.00.*

Market analysis for communications equipment products and services in Western Europe. Covers industry structure, regulations and tariffs, market penetration strategies, major manufacturers, and forecasts of installations, 1979-1987.

425. COMMUNICATIONS EQUIPMENT - SWEDEN: A FOREIGN MARKET SURVEY REPORT. *A B Konsulterna. U.S. Department of Commerce. International Trade Administration. NTIS # ITA-82-03-514. December 1981. 52 pg. $15.00.*

Survey on the market in Sweden for communications equipment and services describes market organization and structure and marketing practices and trade restrictions. Provides data on imports by product category by country of origin. Gives data, by product in each category, on the 1980 market and forecasts for 1986. 3 end user sectors are analyzed: the Swedish Telecommunications Administration, Swedish Radio and Television Authority, and the military. Trade lists include associations and major trade fairs.

426. THE CORDLESS TELEPHONE MARKET IN EUROPE. *Rolco Electronics. January 1981. 45 pg. $700.00.*

Cordless telephone markets in Europe. Includes importers and subsidiaries, new product technologies, utilization, prices, market shares, and regulations.

427. DEFENCE ELECTRONICS: THE FREQUENCY-HOPPING TACTICAL RADIO MARKET. *Simon and Coates. June 1981. 12 pg.*

Electronics research report on defence electronics covers the frequency hopping tactical radio market. Examines General Electric Company and Racal Company, giving an investment analysis including prices, sales, profits, dividends, price earnings ratios and yields.

428. THE E.E.C. MARKET FOR OFFICE COMMUNICATIONS SYSTEMS 1982 / 86. *Tactical Marketing Ltd. March 1, 1982. 15 pg. 280.00 (United Kingdom).*

Report on the European Economic Community market for office communications systems, 1982-1986. Provides forecasts to 1986 and figures concerning market size, demand,

suppliers, products, product groups, and market shares. Also gives opportunities for deeper market penetration.

429. THE EUROPEAN MARKET FOR ACCESS CONTROL AND PERSONAL ID SYSTEMS. *Frost and Sullivan, Inc. July 1981. 315 pg. $1400.00.*

Contains forecasts for plastic financial cards, laminated photo ID cards, access control equipment, entryphones, verification from personal characteristics and others. Factors affecting demand by country are included. Contains supplier profiles.

430. FACSIMILE TRANSCEIVERS IN EUROPE TO 1986. *Larsen Sweeney Associates Ltd. 1981. 215 pg. 595.00 (United Kingdom).*

Market report to 1986 giving forecasts, market analyses, production, marketsegmentation, sales, distribution, employment and unemployment, profits, end users and suppliers. Covers the following countries: Austria, Belgium, Luxembourg, Denmark, Eire, Finland, France, West Germany, Italy, Netherlands, Norway, Portugal, Spain, Sweden, Switzerland, and U.K.

431. GROUND BASED MILITARY ELECTRONICS MARKET IN W. EUROPE. *Frost and Sullivan, Inc. September 1981. 412 pg. $1500.00.*

Analysis of the structure of military markets for ground-based electronics in Western Europe. Forecasts government expenditures, 1980-1986. Examines new products, unit production costs, rankings and competition among manufacturers, and end use markets by country.

432. HOSPITAL TELECOMMUNICATIONS AND TV SYSTEMS MARKET EUROPE. *Frost and Sullivan, Inc. June 1982. 362 pg. $1500.00.*

Analysis of European markets for TV systems and telecommunications in hospitals forecasts unit and dollar sales for eleven product groups to 1990. Includes new construction, size of facilities, medical costs by country, technology, uses and prices of equipment, market penetration, major manufacturers, regulations, and factors in purchases.

433. HOTEL / MOTEL INTEGRATED TELECOMMUNICATIONS SYSTEMS MARKET IN EUROPE. *Frost and Sullivan, Inc. September 1981. 295 pg. $1500.00.*

Analyzes European market for integrated telecommunications systems for hotels / motels. Forecasts and analyses are provided for 13 European countries. Economic constraints are explored. Gives profiles of major manufacturers and suppliers.

434. IBM AND THE EUROPEAN PABX MARKET FIGHT. *Probe Research, Inc. March 1979. 63 pg. $160.00.*

This work discusses IBM's PABX market share in Europe and its growing attacks by North American PABX products (including Rolm, Northern, Telecom, and GTE). The report includes a product analysis of IBM's 3750; user perceptions; IBM marketing and regulatory tactics; IBM's European strategy and implications for the U.S.; European suppliers; the new alliances such as Plessey/Rolm, GEC/Northern Telecom; market outlook for specific companies; United Kingdom market shares; the British P.O.'s role; and plans in the PABX market.

435. INTEGRATED INDUSTRIAL SECURITY, FIRE, AND ENERGY SYSTEMS MARKETS IN EUROPE. *Frost and Sullivan, Inc. January 1982. 305 pg. $1400.00.*

Analyzes and forecasts European markets for integrated, industrial, fire, security and energy systems to 1991. Product sectors isolated are industrial end user security alarm systems, industrial fire protection markets, and industrial end user integrated systems markets. Growth rates in Europe are contrasted to U.S. experience and key factors are isolated.

Includes information on suppliers, users, and demand. Forecasts market size.

436. INTRUDER ALARMS SYSTEMS MARKETS IN EUROPE. *Frost and Sullivan, Inc. September 1982. 324 pg. $1400.00.*

Analysis of markets for intruder alarm systems in Europe includes market segmentation and market size forecasts, 1986, 1992, by product groups; current installations and market shares; market structure; manufacturers and suppliers; marketing strategies; new products; and consumer attitudes. Drawn from market surveys. Lists business addresses, including trade associations.

437. THE MARKET FOR AUTOMATED BUILDING MANAGEMENT SYSTEMS. *Industrial Market Research Ltd. 1981.*

Analysis of markets in the U.K. for building automation and telemetry products and services.

438. THE MILITARY AVIONICS MARKET IN W. EUROPE. *Frost and Sullivan, Inc. July 1981. $1500.00.*

Market report provides analysis and forecast to 1986 for airborne navigation, radar, communications, instrumentation, weapons and fire control, reconnaissance and surveillance, EW and ASW systems in 13 W. European countries. Includes analysis of major end-user programs contributing to the market, evaluation of equipment including unit costs, surveys of state-of-the-art and trends in technology, and major factors influencing avionics procurement. Profiles important European and non-European suppliers to the marketplace.

439. THE MILITARY COMMUNICATIONS MARKET IN EUROPE. *Frost and Sullivan, Inc. 3662. November 1981. $1500.00.*

Market report contains forecast to 1991 for VHF/UHF radars, Los radios, ANALOG, PABXs/PBXs, multiplexes, modems, and 15 other products in Holland, Norway, Denmark, FRG, UK, Italy, Spain, Greece, Turkey, Switzerland and NATO. Includes methods of procurement and state-of-the-art technology. Contains company profiles, and includes political considerations.

440. MILITARY LASER AND EO - EUROPE. *DMS, Inc. Annual. 1981. $375.00.*

Analysis of military markets in Europe for lasers and EO (electro-optics).

441. THE MILITARY MARKET FOR ELECTRO - OPTICAL EQUIPMENT IN EUROPE. *Frost and Sullivan, Inc. October 1981. $1500.00.*

Market report forecasts to 1990 for image intensifying and infrared weapon sights and surveillance devices; thermal imagers, low-light TV; infrared point detectors, missile seekers, perimeter security systems; FLIR; line-scan laser rangefinders, designators, communicators; training devices and security systems; stabilized platforms; and optical sighting systems and sub-systems in U.K., France, FRG, Belgium, the Netherlands, Austria, Switzerland, Sweden, Norway, Denmark, and Italy. Contains company profiles and market shares. Includes future development and applications.

442. MILITARY NAVIGATION MARKET IN WESTERN EUROPE. *Frost and Sullivan, Inc. April 1982. $1500.00.*

Analysis and forecast (to 1990) of demand for military Omega, Loran, TACAN, Decca, Satellite, Landing Systems, Inertial, Dappler, Terrain Avoidance, Beacons and DF, Bombing, Integrated and Navigational Radar in all W. European national markets. Provides inventory levels, new purchases planned, system descriptions, state-of-the-art, and manufacturers profiles and awards.

443. THE MILITARY RADAR MARKET IN W. EUROPE. *Frost and Sullivan, Inc. August 1981. 284 pg. $1300.00.*

Contains 5-year forecast to 1986 for anti-personnel, fire control, navigation, shipdocking, surveillance, target designation, target illumination and tracking radar for the armies, navies and air forces of France, Italy, U.K., FRG and nine other W. European countries. Includes trading patterns between Europe and the rest of the world. State-of-the-art is assessed. Includes 21 company profiles.

444. MOBILE RADIOS MARKET. *Frost and Sullivan, Inc. February 1982. $1500.00.*

Provides analysis and forecast to 1990 for private mobile radio, public telephone, mobile radio, radio paging equipment, and citizen band radio. Includes supplier profiles and market shares by product and by country. Examines regulations and technical trends.

445. NAVAL ELECTRONICS MARKET IN WESTERN EUROPE. *Frost and Sullivan, Inc. January 1981. 740 pg. $1450.00.*

Military market analysis and forecast to 1985 for naval electronics in Western Europe. Notes new construction of ships, end-use markets, exports and government purchases. Includes profiles and rankings of major manufacturers with products, sales, profits, financial statements and addresses.

446. THE NEW PABX - THE 1750: IBM AND TELECOMMUNICATIONS. *Probe Research, Inc. May 1979. 60 pg. $140.00.*

This study analyzes IBM's newest PABX product, the 1750, which was released in the spring of 1979. This report explains the 1750 features, function sets, electronic mail systems, pricing, and how the 1750 fits into European PABX market fight.

447. OPERATIONAL AMPLIFIERS MONOLITHIC IN EUROPE TO 1986. *Larsen Sweeney Associates Ltd. 1981. 222 pg. 595.00 (United Kingdom).*

Market research report with forecasts to 1986 on 8 types of operational amplifiers in Austria, Belgium, Luxembourg, Denmark, Eire, Finland, France, West Germany, Italy, Netherlands, Norway, Portugal, Spain, Sweden, Switzerland and the U.K. Gives market analyses, production, market segmentation, sales, distribution, employment and unemployment, profits, end users and suppliers.

448. PABX AND CALL INFORMATION LOGGING SYSTEMS MARKET IN EUROPE. *Frost and Sullivan, Inc. April 1982. 276 pg. $1600.00.*

Analysis and forecast of Western European markets for PABX and Call Information Logging Systems (CIL), 1982-1986. Includes review of new technology, existing telephone facilities, manufacturers and products including business addresses, installed base, annual shipments and value, investments in equipment, government actions and industry structure.

449. PRIVATE TELEPHONE EXCHANGES (PABX) MARKET IN FRANCE - 1981: THE INSTALLED EQUIPMENT IN 1980 AND MARKET FOR 1978-79-80-81-82. *SEMA METRA. French/English. 1981. 120000.00 (France).*

Market analysis for private telephone exchanges (PABX) in France, 1979-1983. Covers shipments, 1978-1980; current installations; and forecasts of purchases, 1981-1983. Examines patterns of utilization of facilities by types of businesses, number of employees, user attitudes, and capacity of equipment.

450. PUBLIC BROADCAST SYSTEMS IN EUROPE TO 1986. *Larsen Sweeney Associates Ltd. 1981.*

Market report to 1986 on public broadcast systems in Western

Europe. Covers forecasts to 1986, production, sales, distribution, end users, employment and unemployment.

451. RADAR IN EUROPE TO 1986. *Larsen Sweeney Associates Ltd. 1981. 222 pg. 795.00 (United Kingdom).*

Market report to 1986 on radar in Western Europe. Covers forecasts to 1986 on production, sales, distribution, end users, employment and unemployment. The following types of radar are covered: airborne; ground based; shipborne; portable systems and vehicle based; industrial / security applications; phased spray systems; and frequency agile systems.

452. RADIO COMMUNICATIONS, RADAR AND NAVIGATIONAL AIDS SECTOR WORKING PARTY: PROGRESS REPORT 1979. *National Economic Development Office. Annual. 1979. 11 pg.*

Report by the U.K. National Economic Development Council covers radio communications, radar and navigational aids. Shows industrial strategy, market analysis and market forecast.

453. RADIO IN EUROPE TO 1986. *Larsen Sweeney Associates Ltd. 1981. 222 pg. 895.00 (United Kingdom).*

Market report to 1986 on radio equipment in Western Europe. Covers forecasts to 1986 for production, sales, distribution, end users, employment and unemployment. The following products are treated individually for each country: receivers, transmitters; transceivers.

454. SECURITY EQUIPMENT AND SERVICES IN EUROPE: BELGIUM AND THE NETHERLANDS. *Industrial Market Research Ltd. May 1981. 2400.00 (United Kingdom).*

Analysis of markets for fire and security products and services in Belgium and the Netherlands. Examines supply and demand, market structure, competition in each sector, end uses, manufacturers and distribution channels, foreign trade, forecasts to 1986 for legislation, new products, and user attitudes and purchase choice factors including suppliers, performance, and prices.

455. SECURITY EQUIPMENT AND SERVICES IN EUROPE: FRANCE. *Industrial Market Research Ltd. October 1981. 2600.00 (United Kingdom).*

Analysis of the market for fire and security products and services in France. Examines the structure of the market for each product and service, the supply and demand situation, competition, end uses, manufacturers and distribution, foreign trade, forecasts to 1980 considering new products and legislation; and purchase choice factors and user attitudes toward suppliers, performance, and prices.

456. SECURITY EQUIPMENT AND SERVICES IN EUROPE: GERMANY. *Industrial Market Research Ltd. October 1981. 2700.00 (United Kingdom).*

Analysis of market structure for products and services related to fire protection and security in Germany. Examines supply and demand, competition, end uses, distribution, manufacturers, foreign trade, forecasts to 1986 considering new products and legislation, and users' attitudes towards purchases, suppliers, performance, and price.

457. SECURITY EQUIPMENT AND SERVICES IN EUROPE: ITALY. *Industrial Market Research Ltd. October 1981. 2400.00 (United Kingdom).*

Analysis of market structure for products and services related to fire protection and security in Italy. Examines supply and demand, competition, end uses, distribution, manufacturers, foreign trade, forecasts to 1986 for new product trends and

legislation, and users' attitudes toward purchases, suppliers, performance, and prices.

458. SECURITY PRODUCTS IN THE TRANSPORTATION INDUSTRIES IN EUROPE. *Frost and Sullivan, Inc. March 1982. 259 pg. $1350.00.*

Market report gives analysis and forecast (1980-1987) for fencing, barriers and gates, locks, seals, lighting, surveillance, access controls, document equipment, intrusion systems, vehicle equipment, hazardous materials detection, sounder lights, telephone callers, and other equipment. Calculates installation and service cost.

459. SUBSCRIBER APPARATUS IN EUROPE TO 1986. *Larsen Sweeney Associates Ltd. Quarterly. 1981. 246 pg. 995.00 (United Kingdom).*

Market report to 1986 on 21 types of subscriber apparatus in Western Europe. Gives forecasts to 1986 for sales, production, deliveries, distribution costs, marketing costs, capital investments, advertising costs, end users, employment and unemployment. 25 end user sectors are analyzed individually for each country. Report updated quarterly.

460. SURVEY ON THE ITALIAN MARKET FOR COMMUNICATIONS EQUIPMENT: A FOREIGN MARKET SURVEY REPORT. *J.B.A. U.S. Department of Commerce. International Trade Administration. NTIS # ITA-82-05-505. February 1982. 138 pg. $15.00.*

Survey on the market in Italy for communications equipment provides an overview of the total market, imports, and domestic manufacturers. For each major product category, covers sales, market, use, leading suppliers and installations, technology, standards, services, and projected U.S. sales potential. Analyzes each of the following end users: telecommunications authority, broadcasting, public utilities, banking and insurance, transportation. Sections of the report cover marketing practices and trade restrictions. Includes the following lists: sales prospects; potential agents and distributors; trade, industrial, and professional associations; trade journals; major trade fairs and commercial exhibitions.

461. TELECOMMUNICATIONS MARKET STUDY: BELGIUM. *Information Gatekeepers, Inc. March 1980. $75.00.*

Examines the Belgium import market for individual categories of telecommunications equipment, as well as domestic manufacturers. Discusses import regulations and tariffs and identifies potential purchasers.

462. TELECOMMUNICATIONS MARKET STUDY: FRANCE. *Information Gatekeepers, Inc. March 1980. $75.00.*

Examines the French import market for individual categories of telecommunications equipment, as well as domestic manufacturers. Discusses import regulations and tariffs, and identifies potential purchasers.

463. TELEPHONE ANSWERING TERMINAL EQUIPMENT AND SERVICES MARKET IN EUROPE. *Frost and Sullivan, Inc. April 1982. 227 pg. $1700.00.*

Analysis of European markets for telephone answering machines, services and call transferring facilities. Forecasts annual shipments, dollar values, and number of consumers, 1981-1989. Includes industry structure, vendors, products, technology, market penetration, costs, regualtions, installed base, revenues and market structure. Lists business addresses.

464. TRANSMISSION EQUIPMENT IN EUROPE TO 1986. *Larsen Sweeney Associates Ltd. Quarterly. 1981. 251 pg. 895.00 (United Kingdom).*

Market report to 1986 on transmission equipment in Western Europe. Report is updated quarterly and covers: market analysis, forecasts to 1986, sales, production trends, profits, assets, end users, employment and unemployment, products, capital expenditures, investments, and distribution.

465. TV BROADCASTING IN EUROPE - INNOVATION AND FUTURE STRATEGY. *Acumen-System Three. April 1980. 50 pg. 190.00 (United Kingdom).*

Data on Europe's TV broadcasting industry covers direct satellite broadcasting, cable TV, pay TV, viewdata and teletext plus such other developments as hi-fi and digital TV. Provides regional backgrounds for northern European nations.

466. ULTRASONIC CLEANING AND INSPECTION EQUIPMENT IN EUROPE TO 1986. *Larsen Sweeney Associates Ltd. Quarterly. 1981. 231 pg. 595.00 (United Kingdom).*

Market report to 1986 on ultrasonic cleaning and inspection equipment in Western Europe. 23 end user markets are covered individually, and also the following aspects: market analysis, forecasts to 1986, sales, production trends, profits, assets, end users, employment and unemployment, products, capital expenditures, investments, and distribution.

467. ULTRASONIC CLEANING AND INSPECTION EQUIPMENT IN THE U.K. TO 1985. *Larsen Sweeney Associates Ltd. Quarterly. October 1979. 63 pg. 195.00 (United Kingdom).*

Forecast to 1985 for the market in the U.K. for ultrasonic cleaning and inspection equipment. Projects market shares and market values to 1985.

468. ULTRASONIC INSTRUMENTS IN EUROPE TO 1986. *Larsen Sweeney Associates Ltd. Quarterly. 1981. 239 pg. 595.00 (United Kingdom).*

Market report to 1986 on ultrasonic instruments (sensing and switching; intensity; generators; flaw detection; disintegrators; diagnostic; beam counters; and welding instruments) in Western Europe. Report updated quarterly and covers: market analysis, forecasts to 1986, sales, production trends, profits, assets, end users, employment and unemployment, products, capital expenditures, investments, and distribution.

469. ULTRASONIC INSTRUMENTS IN THE U.K. TO 1985. *Larsen Sweeney Associates Ltd. Quarterly. October 1979. 65 pg. 195.00 (United Kingdom).*

Forecast to 1985 for ultrasonic instruments in the U.K. to 1985. Gives projected market growth, market shares, and market values to 1985.

470. THE U.K. MARKET FOR FIRE AND INTRUDER SECURITY PRODUCTS, SYSTEMS AND SERVICES. *Industrial Market Research Ltd. 1979. 220 pg. 1250.00 (United Kingdom).*

Analysis of markets for fire and intruder security products and services. Includes forecasts to 1984, market segmentation and growth rates, market structure, end users, and competition.

471. THE U.K. MARKET FOR TELECOMMUNICATIONS EQUIPMENT: A FOREIGN MARKET SURVEY REPORT. *Systec Consultants Ltd. U.S. Department of Commerce. International Trade Administration. NTIS # ITA-82-05-506. March 1982. 159 pg. $15.00.*

Survey on the market in the United Kingdom for telecommunications equipment provides a market overview and analyzes the position of local manufacturers, U.S., and third country suppliers. Gives market analysis statistics by major product category and covers local production, imports, and exports. Imports are given by major product category by country of origin. A detailed analysis is provided for each major product category, with current and projected markets described for each product. Each end user sector is described. A section of the report covers marketing and distribution practices. In addition, the report includes the following lists: potential purchasers; potential agents and distributors; trade associations and official bodies; trade journals; and national exhibitions.

472. THE U.K. MARKET FOR VIDCOM: DOMESTIC AND COMMERCIAL OPPORTUNITIES IN TV AND VIDEO IN THE 1980'S. *Economist Intelligence Unit Ltd. 1982. 300 pg. 1000.00 (United Kingdom).*

Analysis of video communication technology and the industry in the U.K. Covers cable, satellite, broad and narrowcast, videotext, TV's and video equipment and broadcasting equipment. Includes historical data, policy and regulations, consumer market analysis, foreign trade, distribution, consumer behavior, retail prices, and advertising expenditures.

473. VIDEO RECORDERS: WEST EUROPEAN MARKET OPPORTUNITIES. *Strategic Inc. June 1981. 175 pg. $1500.00.*

Analysis of markets for video recorders in Western Europe. Includes broadcasting, production, cable, and industrial consumption. Forecasts to 1989, by end uses and product groups.

See 726, 1907, 1932, 1952, 1977, 1982, 1983, 2880, 2897, 2898, 2907, 2928, 3840, 4295, 4299, 4304.

Industry Statistical Reports

474. BUSINESS MONITOR: RADIO AND RADAR AND ELECTRONIC CAPITAL GOODS. *United Kingdom. Central Statistics Office. Quarterly. January 1981. 14 pg.*

Quarterly U.K. government statistics (1976-1980) on radio, radar and electronic capital goods.

475. BUSINESS MONITOR: REPORT ON THE CENSUS OF PRODUCTION - RADIO, RADAR AND ELECTRONIC CAPITAL GOODS. *Her Majesty's Stationery Office. Annual. 1981. 10 pg.*

Government of the U.K. report on the census of production on radio, radar and electronic capital goods. Gives production statistics, 1975-1979; capital expenditures, 1975-1979, of all U.K. establishments; stocks, 1975-1979; work in progress, 1975-1979; employment and unemployment, 1975-1979; and operating ratios, 1978-1979.

476. DEFENSE EQUIPMENT MANUFACTURERS. *ICC Business Ratios. Annual. 1981. 80.00 (United Kingdom).*

One in a series of 150 annual industry reports comparing performance between up to 100 individual leading companies, subsectors, and sectors over a three-year period. Fourteen ratio tables by company detail profit margin, profitability, asset utilization, return on capital, liquidity, stock turnover, credit period, exports, profit per employee, average remuneration, sales per employee, capital employed per employee, and other data. Six annual growth rate tables cover sales, total assets, capital employed, average wages, total wages, and exports. Other company data includes names of directors and secretary and the registered office address. Gives analyses of current problems and likely developments within the sector and industry as a whole.

477. SECURITY INDUSTRY. *ICC Business Ratios. Annual. 1981. 80.00 (United Kingdom).*

One in a series of 150 annual industry reports comparing performance between up to 100 individual leading companies, subsectors, and sectors over a three-year period. Fourteen ratio tables by company detail profit margin, profitability, asset utilization, return on capital, liquidity, stock turnover, credit period, exports, profit per employee, average remuneration, sales per employee, capital employed per employee, and other data. Six annual growth rate tables cover sales, total assets, capital employed, average wages, total wages, and exports. Other company data includes names of directors and secretary and the registered office address. Gives analyses of current problems and likely developments within the sector and industry as a whole.

See 2945, 2960.

Financial and Economic Studies

478. FINANCIAL SURVEYS: INTERCOMMUNICATION EQUIPMENT MANUFACTURERS AND DISTRIBUTORS. *I.C.C. Information Group Ltd. Annual. 1981.*

One in a series of annual financial surveys for quoted and unquoted companies in particular fields. Listings for past two years cover turnover, total assets, current liabilities, profit before tax and group relief (excluding extraordinary items), and payment to directors. Listings detail whether firm is part of a group account or a subsidiary of another company. Companies with name changes are noted.

479. FINANCIAL SURVEYS: SECURITY AND FIRE PREVENTION EQUIPMENT MANUFACTURERS AND DISTRIBUTORS. *Inter Company Comparisons Ltd. 1981. 59.80 (United Kingdom).*

Annual financial analysis, with two year comparative figures, for security and fire prevention equipment manufacturers and distributors in the U.K. For all applicable businesses, tabulates turnover of sales; total assets, including fixed assets, current assets, intangible assets and investments; current liabilities; profits before taxes; executive compensation; capital employed; and subsidiaries.

See 2966, 4312.

Forecasts

480. THE FACSIMILE EQUIPMENT AND SERVICES MARKET IN EUROPE. *Frost and Sullivan, Inc. May 1979. 256 pg. $1000.00.*

This report forecasts the European market for facsimile equipment and services, by country and by generation of equipment: six-minute devices, three-minute devices and sub-minute devices. The prospects for both PTT's and private facsimiles are considered. End user requirements are documented and promising areas of application for facsimiles are identified. The facsimile is evaluated in terms of its full market potential as an office product. Company profiles are included.

See 2007, 2011, 2016, 2980, 2981, 4314.

Directories and Yearbooks

481. ASSOCIATION OF SOUND AND COMMUNICATIONS ENGINEERS: ASCE YEARBOOK. *Association of Sound and Communication Engineers. Annual. 1980. 80 pg. 2.50 (United Kingdom).*

Listings of public address system equipment manufacturers include address, telephone, and names of sales and service engineers. Covers P.A. system trading engineers and professional engineers. Contains products index.

482. AUDIO VISUAL DIRECTORY AND BUYERS GUIDE. *Maclaren Publishers Ltd. Annual. 1980. 107 pg.*

Coverage of firms supplying materials and services for audiovisual presentations. Includes address, telephone, products and/or services, director, and date founded.

483. COMMERCIAL TELEVISION AND RADIO YEARBOOK. *Mercury House Business Publications. Annual. 1980. 464 pg. 5.00 (United Kingdom).*

Annual directory for United Kingdom's broadcasting industry covers TV production companies; program contractors; advertisers; advertising agents; directors, producers, and scriptwriters; theater and studio sources; public relations services; schools; research facilities; performers; agents; and equipment manufacturers.

484. PUBLIC ADDRESS ENGINEERS DIRECTORY. *Association of Public Address Engineers. Annual. 1979.*

Annual directory of audiovisual communications industry manufacturers, suppliers, professional engineers, and service firms in Great Britain.

485. SECURITECH: THE INTERNATIONAL GUIDE TO SECURITY EQUIPMENT. *Unisaf Publications Ltd. Annual. 1980. 364 pg.*

Security products directory includes products index in English, French, German, and Spanish.

486. SWITZERLAND - YOUR PARTNER: VOLUME 3: TRANSPORTATION AND COMMUNICATIONS. *Swiss Confederation. Office for the Development of Trade. Irregular. 1980. 156 pg.*

Presents an overall analysis of the transportation and communications industry in Switzerland. Includes indexes to products and services, suppliers, and a list of professional associations and groupings. Company-by-company descriptions present the range of products exported in the following catagories: guided transport systems; road transport; air transport; shipping; and services associated with transport. Lists planning, consulting, engineering, and supply firms.

487. VIDEO YEARBOOK 1980. *Robertson, Angus, editor. Blandford Publications. Annual. 1979. 616 pg. $37.50.*

Covers manufacturers of equipment and suppliers of services for the video industry. Provides addresses, product lines, contacts, and telephone numbers. Includes importers' and distributors' index.

See 3877, 4315.

Special Issues and Journal Articles

Appliance

488. EUROPEAN REPORT: THE VIDEO BOOM IN EUROPE. *Spiro, Richard B. Dana Chase Publications, Inc. v.39, no.4. April 1982. pg.51*

Provides statistics for market shares of the three video recording systems dominating the European market. Includes long-range forecasts for annual production as well as present production time for Philips system. Document also includes data on efforts of various countries to limit Japanese imports and/or to acquire Japanese manufacturing licenses. Results of 1980 market study summarized.

British Business

489. AEROSPACE AND ELECTRONIC COST INDICES. *Her Majesty's Stationery Office. Monthly. v.8, no.5. June 4, 1982. pg.29*

Gives latest monthly cost indices for aerospace, radio, radar, and electronic goods industries in Great Britain, 1981-1982. Breaks down costs into 30 categories.

490. COST INDICES FOR THE AEROSPACE AND RADIO, RADAR AND ELECTRONIC CAPITAL GOODS INDUSTRIES. *Her Majesty's Stationery Office. Monthly. September 4, 1981. pg.31*

Monthly review of the aerospace, radio, radar and electronic capital goods industries in the United Kingdom. Table details 1980-1981 cost indices including bought out finished goods and other materials purchased by the aerospace industry; salaries and general expenses for the radio, radar and capital goods industry; hourly earnings of manual workers in the aerospace industry; and combined costs for all of the above industries.

Business Equipment Digest

491. GETTING TO GRIPS WITH BUZBY. *B.E.D. Business Books Ltd. Irregular. v.22, no.1. January 1982. pg.52*

Lists manufacturers of telephone call logging equipment, their models and system type. Specifies number of lines monitored and method of processing.

Chip, Revista de Informatica

492. ESPANA ENTRA EN LA CARRERA DE LA FIBRA OPTICA. [Spain Enters Fiber Optics Race.]. *Ediciones Arcadia, S.A. Spanish. no.16. August 1982. pg.18*

Examines TRACOF, fiber optic communication system being researched and developed by Spanish Stadard Electrica. New technology and end uses of the new transmission system in electrical stations and substations is described. Capacity utilization of the optic cable are provided.

Economisch Dagblad

493. ONDERZOEK NAAR VIDEO-SOFTWAREMARKT: NOS-MONOPOLIE MOET AFGEBROKEN WORDEN. [Examination of the Video-Software Market.]. *Selles, Gert. Sijthoff Pers B.V. Netherlandish. v.42, no.10704. July 12, 1982. pg.2*

Reviews the development of communications equipment, film vs. video. Examines possibilities for jobs and stimulation for exports. Shows sales, 1981; employees, 1981; forecasts, 1987; and exports, 1981.

FEM (Financieel-Economisch Magazine)

494. MARKT GEZOCHT VOOR NIEUWE TELECOMMUNI-CATIE TECHNOLOGIE. [Looking for a Market for New Telecommunications Technology.]. *Veenis, S. B.V. Uitgevers-Maatschappy Bonaventura. Netherlandish. v.12, no.2. January 22, 1981. pg.24-28*

Reviews telecommunications markets and possibilities in developing countries. Examines competition between American and Japanese manufacturers and software. Gives percent of market size, 1980, and forecasts to 1985.

High-Speed Surface Craft

495. EUROPEAN SPENDING FOR MILITARY RADAR TO REACH $16.8 BILLION. *Kalerghi Publications. January 1982. pg.19*

Reviews a Frost & Sullivan market report, 'The Military Radar Market in Western Europe.' Gives forecast statistics for total spending; growth rates; market shares of airborne systems, seaborne and submarine radar; ASW purchases; and spending by country. Lists major manufacturers.

Interavia

496. AEROSPACE ELECTRONICS: FEARS BELIED BY CLIMBING SALES. *Interavia SA. v.37, no.8. August 1982. pg.798-799*

Developments in the aerospace electronics industry in the United Kingdom are reviewed. Government support of the industry, research and development efforts, and new approaches to electronic design are discussed. Data is presented on sales of radio, radar, and electronic capital equipment (1980-1981) showing export sales, and exports as a percent of total sales.

Investors Chronicle and Financial World

497. DEFENCE STOCKS AFTER FALKLANDS. *Financial Times Business Information Ltd. June 18, 1982. pg.772-774*

Analyzes the possibilities of investment in the British defense industry after the Falklands crisis. Projects domestic and international sales for British aerospace and warship building. Examines 9 corporations for investment opportunities.

Norwegian Shipping News

498. SATCOMS - SMALLER AND SMARTER. *K/S Selvig Publishing A/S. v.38, no.4. September 24, 1982. pg.117-118*

Discusses Inmarsat ship earth station systems. Survey lists new systems scheduled for marketing this year.

ServEx: El Semanario del Comercio Exterior

499. CARTA DE EXPORTADOR INDIVIDUAL: DELIMITA-CION DE SECTORES. [Export License: Product Groups Classifications.]. *Banco de Bilbao. Service de Estudios. Spanish. July 5, 1979. pg.2*

Modifications to export product groups are listed, including chemicals and communications equipment. Product group classifications in Spanish Customs grouping are given. Export value of shipments for the groups are given in Spanish pesetas for 1976-1978. Information relates to export licenses for 1980-1983, and published in the Official Spanish State Bulletin (Boletin Oficial del Estado) of June 22, 1979.

500. LISTA APENDICE DEL ARANCEL DE ADUANAS. [Supplemental List of Spanish Customs Duties.]. *Banco de Bilbao. Service de Estudios. Spanish. No.1352. April 30, 1981. pg.4-7 Free.*

1981 imports information on laser, ultrasonic sounding, and electronic measuring equipment are included in these tables published in the official Spanish state bulletin (Boletin Oficial del Estado) April 22, 1981. Included are additional and continued items and modifications to previous listings of equipment that can be imported to Spain. Percent of Spanish import duties and term of such duties are shown with alpha-numeric classifications of equipment.

501. LISTA APENDICE DEL ARANCEL DE ADUANAS. [Supplement List of Customs Classifications.]. *Banco de Bilbao. Service de Estudios. Spanish. October 12, 1979. pg.2-4*

Spanish royal decree authorizes new inclusions, extensions and modifications of customs product group classifications. Includes machines for the fabrication of communications equipment and electrical electronic equipment. Lists percent of regular customs duties and term of duration of authorization. Document was published in the Official Spanish State Bulletin (Boletin Oficial del Estado) November 3, 1979.

Telephony

502. REMOTE ENERGY SYSTEMS CATCH THE WINTER WIND AND SUMMER SUN. *Forbes-Jamieson, Douglas A. Telephony Publishing Corporation. v.203, no.18. October 25, 1982. pg.52-58, 62*

Assesses the use of wind and solar energy for powering telecommunications equipment in remote areas. Reports 1982 energy costs for diesel, solar, wind, thermoelectric, and energy converter systems. Indicates capital requirements for selected power generation.

Television

503. TELEVISION: EQUIPMENT BRIEFS. *Royal Television Society. Bi-monthly. 1981.*

Gives descriptions of new equipment and products associated with the television industry, such as photographic lenses, cameras, oscilloscopes, etc. Information includes photograph, model number, brief technical specifications, supplier, and addresses of manufacturers.

L'Usine Nouvelle

504. COMMANDES EN REGISTREES PAR LES GRANDS DES TELECOMS. [Orders Received by Leading French Telecommunications Equipment Manufacturers.]. *Fleurot, Olivier. Usine Nouvelle. French. no.42/1982. October 14, 1982. pg.97*

Illustrates the total value of orders received by France's 6 leading telecommunications equipment manufacturers during April 1980 - March 1981 and April 1981 - May 1982. Accompanying article discusses companies' activities.

See 2033, 2042, 2048, 2050, 3034, 3051, 3418, 3903, 4321, 4325.

Monographs and Handbooks

505. VIDEO DISCS - THEIR APPLICATION TO INFORMATION STORAGE AND RETRIEVAL. *Horder, Alan. Hatfield Polytechnic. National Reprographic Centre for Documentation. Irregular. 2nd ed. 1981. 50 pg.*

Presents an overview of video disc development, applications, and implications of their use in education, industry, medicine and the military. Listings of equipment, by manufacturer, describe playbakc principle, equipment configuration, and availability. Also includes extensive reference list.

Services

506. MILITARY COMMUNICATIONS: EUROPE. *DMS, Inc. Annual. 1982. $750.00.*

Analyzes all major receivers, transmitters, transceivers, radio links; plus satellite, underwater, and live communications equipment, multiplexors, modems and others being developed and built in Europe, Canada, and Israel. Includes a world overview and an examination of the European communications industrial base.

Indexes and Abstracts

See 3060, 3061.

Eastern Europe

Indexes and Abstracts

507. REFERTIVNYI ZHURNAL: RADIOTEKHNIKA: ABSTRACTS JOURNAL: RADIOTECHNICS. *Vsesoiuznyi Institut Nouchnoi Informatsii. 1981.*

Monthly journal of abstracts of technical radio material, in Russian, with contents in English.

508. SIGNAL'NAIA INFORMATSIIA PO VOPROSAM AUTOMATIKI I RADIOELEKTRONIKI. [Descriptive Information on Automation and Radioelectronics.]. *Vsesoiuznyi Institut Nouchnoi Informatsii. Semi-monthly. 1981.*

Lists titles of automation and radioelectronics publications in Russian.

Asia

Market Research Reports

509. ASIAN CAMPAIGN MARKET RESEARCH STUDY FOR SINGAPORE, NUMBER 4, SECURITY, SAFETY AND PLANT MAINTENANCE EQUIPMENT: FOREIGN MARKET SURVEY REPORT. *Lim, Albert. U.S. Department of Commerce. National Technical Information Service. DIB-80-02-510. 1979. 49 pg. $10.00.*

The market research was undertaken to study the present and potential U.S. share of the market in Singapore for security, safety and plant maintenance equipment; to examine growth trends in Singaporean end-user industries over the next few years; to identify specific product categories that offer the most promising export potential for U.S. companies; and to provide basic data which will assist U.S. suppliers in determining current and potential sales and marketing opportunities. The trade promotional and marketing techniques which are likely to succeed in Singapore were also reviewed.

510. CHINESE AND RUSSIAN COMMUNICATIONS MARKETS. *International Resource Development, Inc. December 1980. 178 pg. $985.00.*

Review of domestic communications production and technology trends in view of the current market environment in Russia and China. Analyzes marketing activities of major foreign suppliers in each country with the discussion of marketing strategies and tactics. Future market projections, with review of opportunities for Western suppliers. Report covers all aspects of communications including satellite, radio, TV, telephone, data, marine and microwave.

511. COMMUNICATIONS EQUIPMENT, HONG KONG: A COUNTRY MARKET SURVEY. *U.S. Government Printing Office. Irregular. CMS 81-049. January 1981. 14 pg. $.50.*

Report analyzes the communications system in Hong Kong and the current and projected markets for sale of communications equipment in that country. Data is provided on the market (local production, imports, exports) by type of equipment; imports by type by country of origin; the market by product category; leading suppliers (including country of manufacture and type of equipment); purchases by major user sectors; telecommunications statistics. Provides product market profiles, end user profiles, and information on practices, and trade restrictions.

512. COMMUNICATIONS EQUIPMENT, INDIA: A COUNTRY MARKET SURVEY. *U.S. Government Printing Office. Irregular. CMS 81-050. February 1981. 11 pg. $.50.*

Report analyzes the communcations system in Hong Kong and the current and projected markets for sale of communications equipment in that country. Data is provided on the market (local production, imports, exports) by type of equipment; imports by type by country of origin; the market by product category; leading suppliers (including country of manufacture and type of equipment); purchases by major user sectors; telecommunications statistics. Provides product market profiles, end user profiles, and information on practices, and trade restrictions.

513. COMMUNICATIONS EQUIPMENT, KOREA: A COUNTRY MARKET SURVEY. *U.S. Government Printing Office. Irregular. CMS 81-057. September 1981. 13 pg. $.50.*

Report analyzes the current and projected markets for sale of communications equipment in the Republic of Korea. Data is provided on the market (local production, imports, exports) by type of equipment; leading suppliers; imports by type by country of origin; the market by product category; the market by end user sector. Describes the competitive environment, provides product market profiles, end user profiles, and information on marketing practices.

514. COMMUNICATIONS EQUIPMENT, MALAYSIA: A COUNTRY MARKET SURVEY. *U.S. Government Printing Office. Irregular. CMS 81-053. April 1981. 12 pg. $.50.*

Report analyzes the current and projected markets for sale of communications equipment in Malaysia. Data is provided on the market (local production, imports, exports) by type of equipment; imports by type by country of origin; the market by product category; estimated end user budgets for the equipment; leading suppliers. Provides product market profiles, end user profiles, and information on marketing practices and trade restrictions.

515. COMMUNICATIONS EQUIPMENT MARKET RESEARCH IN HONG KONG. *Mobius Research. U.S. Department of Commerce. International Trade Administration. NTIS #DIB-80-04-508. January 1980. 137 pg. $15.00.*

Survey of the market for communications equipment in Hong Kong provides data on the total apparent market and imports 1977-1979 and forecast 1980 and 1985. Profiles each product category and gives the size of the market for that product for 1979 and 1985. Analyzes end user sectors and describes future needs. Marketing practices, trade restrictions, and custom duties are covered. Trade lists include potential purchasers; agents and distributors; trade, industrial, and professional associations; journals; and fairs and exhibitions.

516. COMMUNICATIONS EQUIPMENT, PHILIPPINES: A COUNTRY MARKET SURVEY. *U.S. Government Printing Office. Irregular. CMS 81-051. April 1981. 13 pg. $.50.*

Report analyzes the current and projected markets for sale of communications equipment in the Philippines. Data is provided on the market (local production, imports, exports) by type of equipment; imports by type by country of origin; the market by product category; leading suppliers; purchases by major user sectors; capacity and loads of telephone companies; telephone systems by region; telex traffic volume; television networks. Provides product market profiles, end user profiles, and information on marketing practices and trade restrictions.

517. COMMUNICATIONS EQUIPMENT, THAILAND: A COUNTRY MARKET SURVEY. *U.S. Government Printing Office. Irregular. CMS 81-055. July 1981. 12 pg. $.50.*

Report analyzes the current and projected markets for sale of communications equipment in Thailand. Data is provided on the market (local production, imports, exports) by type of equipment; imports by type of country of origin; market product category; growth in TCT (Telephone Company of Thailand); telephone expansion project; leading suppliers. Provides product market profiles, end user profiles, and information on marketing practices and trade restrictions.

518. THE CONSTRUCTION INDUSTRY IN THE FAR EAST: PRODUCT REPORT: FIRE ALARM SYSTEMS. *Business Management and Marketing Consultants Ltd. Industrial Market Research Ltd. July 1981. $2000.00.*

Analysis of the construction industry in South East Asia and the market for fire alarm systems. Examines imports, end uses, suppliers, distribution, prices, user attitudes, legislation and regulations, and market penetration strategies. Includes a directory of manufacturers.

519. THE CONSTRUCTION INDUSTRY IN THE FAR EAST: PRODUCT REPORT: SECURITY ALARM SYSTEMS. *Business Management and Marketing Consultants Ltd. Industrial Market Research Ltd. July 1981. $2000.00.*

Analysis of the construction industry in South East Asia and the market for security alarm systems. Examines imports, end uses, suppliers, distribution, prices, user attitudes, legislation and regulations, and market penetration strategies. Includes a directory of manufacturers.

520. DEFENSE MARKETS IN THE PEOPLE'S REPUBLIC OF CHINA. *Frost and Sullivan, Inc. January 1982. 285 pg. $1500.00.*

Categorizes and establishes the size of the present day market, and overviews domestic politics, policies, and the defense budget of the PRC. Examines international threat perceptions of the PRC. Describes in detail the indigenous defense production sources. Discusses procurement trends and offers guidance and strategies for the offshore marketing executive. Provides forecast by major groups of major PRC defense equipment procurements to 1986 in dollars and, when possible, unit terms.

521. INDIA MARKET RESEARCH REPORT ON COMMUNICATIONS EQUIPMENT AND SYSTEMS. *Sachdeva, Ram P. U.S. Embassy, New Delhi (India). U.S. Department of Commerce. International Trade Administration. 1980. 93 pg. $10.00.*

Analyzes current dimensions and likely growth trends of the Indian communications equipment market, by end-user industry; current and potential market shares held by U.S. manufacturers; product categories offering high potential for market penetration; and appropriate marketing techniques.

522. THE JAPANESE MARKET FOR SELECTED CONSUMER PRODUCTS: FOREIGN MARKET SURVEY REPORT. *Peat, Marwick, Mitchell and Co. Tokyo (Japan). U.S. Department of Commerce. National Technical Information Service. DIB-80-03-504. 1979. 88 pg. $10.00.*

The market research was undertaken to study the present and potential U.S. share of the market in Japan for selected consumer products; to examine growth trends in Japanese end-user industries over the next few years; to identify specific product categories that offer the most promising export potential for U.S. companies; and to provide basic data which will assist U.S. suppliers in determining current and potential sales and marketing opportunities. The trade promotional and marketing techniques which are likely to succeed in Japan were also reviewed.

523. JAPANESE TELECOM EQUIPMENT MARKET BRIEF. *Northern Business Intelligence Ind. January 1981. 125 pg. $375.00.*

Japanese telecom equipment market analysis. Includes installed base, growth rates, competition, U.S. / Japanese foreign trade, and market penetration opportunities for U.S. manufacturers.

524. KOREAN MARKET FOR COMMUNICATION EQUIPMENT AND SYSTEMS. *Pacific Projects, Ltd. U.S. Department of Commerce. International Trade Administration. NTIS #DIB 80-07-511. February 1980. 179 pg. $15.00.*

Survey of the market for communications equipment systems in Korea provides data on market size and imports by major product category for 1977-1979 with forecasts for 1980 and 1985. Separate tables show market size for individual products in each major category for 1979 and 1985. Provides data on local manufactures, including number of companies by size in 1977 and production and exports by nature of capital. Analyzes each product category, including discussion of leading suppliers, sales potentials for U.S. manufacturers, and levels of technology. Each end user group is profiled and its equipment needs described. Marketing practices and trade restrictions are detailed. Trade lists include potential end users; import agents; associations; trade journals; and Korea electronics show.

525. MARKET RESEARCH SURVEY ON COMMUNICATIONS EQUIPMENT AND SYSTEMS IN THE PHILIPPINES. *SGVE Co. Sycip, Govres, Velayo & Co. U.S. Department of Commerce. International Trade Administration. NTIS #DIB-880-04-506. January 1980. 152 pg. $15.00.*

Survey of the market for communications equipment and systems in the Philippines provides data on market size and imports by major product category for 1977-1979 with forecast for 1980 and 1985. Profiles each major equipment category including data on market size, suppliers (domestic and foreign), end users, market trends and prospects, regulatory aspects, and sales opportunities for U.S. suppliers. Each end user sector is profiled, including data on current equipment, relevant government programs, and sales potential for U.S. suppliers, marketing practices and trade restrictions are detailed. Trade lists include potential purchasers, agents and distributors, associations, and journals.

526. SECURITY, SAFETY AND PLANT MAINTENANCE EQUIPMENT (PHILIPPINES): FOREIGN MARKET SURVEY REPORT. *Del Val, Edgar P. American Embassy, Manila (Philippines). U.S. Department of Commerce. National Technical Information Service. DIB-80-02-521. 1979. 27 pg. $10.00.*

The market research was undertaken to study the present and potential U.S. share of the market in the Philippines for security, safety and plant maintenance equipment; to examine growth trends in Philippinean end-user industries over the next few years; to identify specific product categories that offer the most promising export potential for U.S. companies; and to provide basic data which will assist U.S. suppliers in determining current and potential sales and marketing opportunities. The trade promotional and marketing techniques which are likely to succeed in the Philippines were also reviewed.

527. SURVEY OF THE MARKET FOR COMMUNICATIONS EQUIPMENT AND SYSTEMS IN INDONESIA: A FOREIGN MARKET SURVEY REPORT. *U.S. Department of Commerce. International Trade Administration. NTIS #ITA-82-05-501. February 1982. 98 pg. $15.00.*

Survey on the market in Indonesia for communications equipment and systems describes the market organization and structure, marketing characteristics and trade practices, the total net apparent market, and end users. Includes estimated market size and leading brands and suppliers in the following product categories: telephone and telex equipment, transmission equipment, mobile, radio, video and radio broadcasting equipment, data communications equipment, communications test and measurement equipment and satellite transmission equipment. Covers marketing practices and trade restrictions. Includes lists of potential purchasers of U.S. equipment; firms capable of acting as agents; and trade fairs.

528. A SURVEY OF THE MARKET FOR COMMUNICATIONS EQUIPMENT AND SYSTEMS IN MALAYSIA. *Applied Research Corporation. U.S. Department of Commerce. International Trade Administration. NTIS #DIB 80-02-519. December 1979. 86 pg. $15.00.*

Provides data on the total apparent market by major product category and on imports by major product category and country of origin, for 1977-1979 with forcasts for 1980 and 1985. Each product category is analyzed separately and data is given on market size (1979 and 1985 forecast) and major suppliers. The end user sectors analysis gives budgeted purchases of equipment (1979, 1980, 1985). Examines total capital expenditures, number of posts, and 1979 operating expenses, by sector. Describes marketing practices and trade restrictions. Trade lists include: potential buyers of U.S. manufactured equipment; potential agents or distributors; publication / trade journals; and trade fairs and commercial exhibitions.

529. TECHNOLOGY GROWTH MARKETS AND OPPORTUNITIES: HIGH TECHNOLOGY GROWTH MARKETS IN ASIA. *Creative Strategies International. v.2, no.8. May 1982. 20 pg. $95.00.*

Analysis of markets in Asia for high technology product groups, especially electronics and communications. Includes gross domestic product, national income accounts and five-year growth rates, 1980; manufacturing as percent of gross domestic product, 1974, 1979; imports and exports; foreign trade partners, 1980; TV and telephone units in operation, 1980; energy consumption, 1979; personal consumption expenditures, 1979; market sizes, 1975, 1979; videocassette installed base, 1981; corporations; and expansion plans.

530. TELECOMMUNICATIONS SYSTEM AND EQUIPMENT MARKET IN ASIAN COUNTRIES. *Frost and Sullivan, Inc. February 1981. 584 pg. $1250.00.*

Market analysis and forecasts to 1990 for telecommunications and broadcasting equipment and systems in 21 Asian countries. Examines technology of plants and equipment, services, competition among manufacturers, and political and economic indicators.

See 2077, 2081, 2085, 3079, 3082.

Directories and Yearbooks

531. ASIAN SOURCES: ELECTRONICS. *Pacific Subscription Service. Monthly. 1982. 300 pg. $230.00/yr.*

One in a series of six illustrated monthly guides to export products avaiable from Asian nations. Prices and other trade information for radios, tv's, tape recorders, calculators, audio equipment, telephones, home computers, tv games, and tapes.

Special Issues and Journal Articles

China Business Review

532. TV PRODUCTION: INTERFERENCE IN THE INDUS-TRY. *National Council for U.S. - China Trade. January 1982. pg.25-28*

Reviews the Chinese television industry and examines problems associated with decentralized production and tube technology. Statistics given for imports by country, 1979, 1980, 1981. Lists current assembly plants and capacities. Also lists foreign firms doing business in China's TV industry giving cost, annual capacity, and start-up date, 1981.

IEEE Spectrum

533. A STEP TOWARD "PERFECT" RESOLUTION: PIC-TURES RIVALING 35MM FILM IN QUALITY ARE IN THE OFFING FOR HOME TELEVISION. *Mokhoff, Nicolas. Institute of Electrical and Electronics Engineers, Inc. July 1981. pg.56-58*

Discusses the likely parameters and applications of high-definition television (HDTV) systems currently being developed in Japan, Europe and the U.S. Focuses on the Japanese NHK system, which offers about twice the resolution obtained with current U.S. broadcast standard equipment, and is said to rival the quality of 35mm film. Discusses such prospects as satellite transmission of electronic 'movies' to theaters, and compatibility problems.

See 2100, 3104, 3113, 3114.

Indexes and Abstracts

See 3122.

Oceania

Special Issues and Journal Articles

See 3131.

Middle East

Market Research Reports

534. COMMUNICATIONS EQUIPMENT IN ISRAEL: A MAR-KET BRIEF. *P-E Consulting Group Ltd. U.S. Department of Commerce. International Trade Administration. NTIS #ITA 82-05-509. December 1981. 79 pg.*

Survey of the market in Israel for communications equipment describes the market size, organization, and structure. Analyzes product categories according to current and projected sales, domestic manufacturers, and potential for U.S. suppliers. Provides end-user analysis by industry, with current and future demand for equipment noted. Sections of the report describe marketing practices and trade restrictions, and the appendices include company profiles, government help to industry, and leading suppliers.

535. COMMUNICATIONS EQUIPMENT, SAUDI ARABIA: A COUNTRY MARKET SURVEY. *U.S. Government Printing Office. Irregular. CMS 81-056. August 1981. 8 pg. $.50.*

Report analyzes the current and projected markets for sale of communications equipment in Saudi Arabia. Data is provided on the market for equipment by type (market supplied entirely by imports), and imports by type by country of origin. Provides product market profiles, end user profiles, and information on marketing practices, and trade restrictions.

536. COMMUNICATIONS EQUIPMENT, TURKEY: A COUN-TRY MARKET SURVEY. *U.S. Government Printing Office. Irregular. CMS 81-058. October 1981. 10 pg. $.50.*

Report analyzes the current and projected markets forsale of communications equipment in Turkey. Data is provided on sales of selected equipment; imports by country of origin; projected market by end user sector; total investment during next five years; projects utilizing estimated five-year investments. Provides a competitive assessment of the market, end user profiles, and information on marketing practices, and trade restrictions.

537. MARKET RESEARCH OF TELECOMMUNICATIONS EQUIPMENT AND SYSTEMS IN SAUDI ARABIA. *Inbucon / AIC - Telecommunications Division, London, England. U.S. Department of Commerce. International Trade Administration. NTIS #DIB 80-64-507. February 1980. 87 pg. $15.00.*

This survey of the market for telecommunications equipment and systems in Saudi Arabia graphs the number of telephone, telex, and telegraph lines 1975-1991. Provides tables of data showing the exports of telecommunications equipment to Saudi Arabia by exporting countries, 1977-1978 and imports by product category, 1977-1980 and forecast, 1985. Each product category is discussed and various end user industries are profiled. Marketing practices and trade restrictions are described and several major firms are listed.

Africa

Market Research Reports

538. COMMUNICATIONS EQUIPMENT, EGYPT: A COUN-TRY MARKET SURVEY. *U.S. Government Printing Office. Irregular. CMS 81-054. July 1981. 10 pg. $.50.*

Report analyzes the communications system in Egypt and the current and projected markets for sale of communications equipment in that country. Data is provided on the total market (local production, imports, exports) by type of equipment; imports by type by country of origin; the market by product category; purchases by major user sectors; telecommunications statistics; leading suppliers. Provides product market profiles, end user profiles, information on marketing practices, and trade restrictions.

539. COMMUNICATIONS EQUIPMENT IN THE ARAB REPUBLIC OF EGYPT. *Frith, Kirk and Spinks, Ltd. (London, Eng.). U.S. Department of Commerce. International Trade Administration. 1980. 201 pg. $10.00.*

Analyzes current dimensions and likely growth trends of the Egyptian communications equipment market, by end-user industry; current and potential market shares accruing to U.S. manufacturers; product groups offering high potential for market penetration; and preferred marketing and promotional techniques.

540. COMMUNICATIONS EQUIPMENT, IVORY COAST: A COUNTRY MARKET SURVEY. *U.S. Government Printing Office. Irregular. CMS 81-052. April 1981. 13 pg. $.50.*

Report analyzes the current and projected markets for sale of communications equipment in the Ivory Coast. Data is provided on the total market (local production, imports, exports) by type of equipment; imports by type by country of origin; the market by product category; market by end user sectors; and leading suppliers. Provides product market profiles, end user profiles, and information on marketing practices and trade restrictions.

541. COMMUNICATIONS EQUIPMENT - NIGERIA: A FOREIGN MARKET SURVEY REPORT. *Alan Stratford & Associates. U.S. Department of Commerce. International Trade Administration. NTIS #ITA-82-03-516. January 1982. 96 pg. $15.00.*

Survey on the market in Nigeria for communications equipment analyzes the current situation and future trends for local manufacturer and for imports, by product category and services. Provides an analysis by end user sector, describing equipment currently in use and projected needs. Two sections of the survey report om marketing practices and trade restrictions. Lists potential purchasers; potential agents / distributors; trade and professional associations; trade journals; and trade fairs.

542. COMMUNICATIONS EQUIPMENT - ZIMBABWE: A FOREIGN MARKET SURVEY REPORT. *Alan Stratford & Associates. U.S. Department of Commerce. International Trade Administration. NTIS #ITA-82-03-512. January 1982. 63 pg. $15.00.*

Survey on the market in Zimbabwe for communications equipment analyzes the current situation and future trends for local manufacture, services, and for imports, by product category. Provides analysis by end user sector, describing current equipment in use. Reports on marketing practices and trade restrictions.

543. THE MARKET FOR COMMUNICATIONS EQUIPMENT AND SYSTEMS IN ALGERIA. *Konsulterna. U.S. Department of Commerce. International Trade Administration. NTIS # ITA 81-12-502. August 1981. 50 pg. $15.00.*

Analyzes the current (1977-80) and projected (1985) total apparent market (local production and imports) and provides data on imports by major product categories and country of origin. The size of the market for specific products, and also market size by end user sectors, are give for 1979 and projected 1985. The text analyzes each product category, describing the present situation and projecting future purchases. Covers the end user market, marketing practices, and trade restrictions.

544. MARKET SURVEY OF COMMUNICATIONS EQUIPMENT AND SYSTEMS, THE IVORY COAST: FOREIGN MARKET SURVEY REPORT. *Stratford (Alan) and Associates Ltd. Maidenhead (UK) International Trade Administration. U.S. Department of Commerce. National Technical Information Service. DIB-80-04-505. 1980. 87 pg. $10.00.*

The market research was undertaken to study the present and potential U.S. share of the market in the ivory coast for communication equipment and systems; to examine growth trends in ivory coast end-user industries over the next few years; to identify specific product categories that offer the most promising export potential for U.S. companies; and to provide basic data which will assist U.S. suppliers in determining current and potential sales and marketing opportunities. The trade promotional and marketing techniques which are likely to succeed in Ivory Coast were also reviewed.

545. TELECOMMUNICATIONS SYSTEMS AND EQUIPMENT IN BLACK AFRICAN STATES. *Frost and Sullivan, Inc. May 1979. 513 pg. $1175.000.*

Market analysis and forecasts to 1990 for telecommunications equipment in African nations. Includes high and low demand forecasts and analysis of competition among types of equipment.

Forecasts

See 4338.

Computer Services

International

Market Research Reports

546. INFORMATION RETRIEVAL SYSTEMS. *Predicasts, Inc. April 1982. $375.00.*

Analysis of markets and vendors of databases, data storage and retrieval services, equipment manufacturers and teletext and data communications services and products. Includes effects of government regulations.

See 767.

Directories and Yearbooks

547. ASSOCIATION OF DATA PROCESSING SERVICE ORGANIZATIONS MEMBERSHIP DIRECTORY. *Association of Data Processing Services Organization, Inc. Annual. 1981. 145 pg. $50.00.*

Membership directory of 1000 firms throughout the world details address, telephone, executives, employees, financial information, and products and/or services.

548. ENCYCLOPEDIA OF INFORMATION SYSTEMS AND SERVICES: FOURTH EDITION. *Kruzas, Anthony T., editor. Gale Research Company. Biennial. 4th ed. 1981. 933 pg. $190.00.*

Provides detailed descriptions of some 2000 organizations in the U.S.(and about 60 other countries) that produce, process, store, and use bibliographic and nonbibliographic information. Covers on-line systems, non-computerized services, commercial, governmental, technical, and general sources; videotext / teletext systems, and online vendors.

549. INFORMATION INDUSTRY MARKET PLACE: AN INTERNATIONAL DIRECTORY OF INFORMATION PRODUCTS AND SERVICES. *R.R. Bowker. Annual. 4th ed. October 1982.*

Covers information production; information distribution; information retailing; support services and suppliers; associations and government agencies; conferences and courses; and sources of information. Entries detail address, telephone, key personnel, and descriptions of products, services, or objectives. Subject, service, and geographic indexes. Published in Europe under the title 'Information Trade Directory'.

550. INTERNATIONAL MICROCOMPUTER SOFTWARE DIRECTORY. *Graham, John, ed. Imprint Software. Annual. 1981. 425 pg. $39.95.*

Directory of microcomputer software vendors lists products and services, prices, brand names, and business addresses.

551. INTERNATIONAL MINICOMPUTER SOFTWARE DIRECTORY. *Graham, John, ed. Imprint Software. Annual. 1982. $69.95.*

Directory of minicomputer software vendors lists products and services, prices, brand names, and business addresses.

See 802.

Special Issues and Journal Articles

Datamation

552. SOFTWARE AND SERVICES. *Technical Publishing Company. Irregular. August 25, 1981. 100 pg. $4.00.*

This 100 page special report consists of 15 articles on the growth and future development of the rapidly expanding computer software and services industry.

See 815, 819, 820, 830, 2168, 2171.

Monographs and Handbooks

553. SOFTWARE FOR PRINTED INDEXES: A GUIDE. *Armstrong, C. Association for Information Management. 1981. 98 pg. 10.50 (United Kingdom).*

Guide is designed to acquaint potential index producers with existing possibilities in computerized indexing. Sections cover commercial, ready-for-sale packages with full support facilities; non-commercial, negotiable-sale packages with various nominal levels of support facility; and the range of possibilities offered by a library - oriented software house.

See 837, 839.

Conference Reports

See 843, 845, 848, 850, 858, 859.

Newsletters

See 4010, 4012.

Indexes and Abstracts

See 868, 870, 871, 872, 873.

North America

Market Research Reports

554. COMPUTER - AIDED DESIGN AND MANUFACTURING: MARKETS AND OPPORTUNITIES. *Strategic Inc. October 1981. 150 pg. $1500.00.*

Analysis of markets for equipment and software for computer - based functions in manufacturing. Examines end uses, vendors, and candidates for mergers.

555. COMPUTER GRAPHICS SOFTWARE AND SERVICES MARKET (U.S.). *Frost and Sullivan, Inc. April 1982. 230 pg. $1250.00.*

Analysis of markets for computer graphics software and services. Forecasts growth rates, revenues to 1987 and sales to 1992. Examines industry structure, historical data, and market segmentation. Market surveys of products include technical specifications, prices, utilizations, major end uses, installed base, and user attitudes. Surveys of vendors include size and ownership, product lines, new products, distribution, marketing strategies, market shares and software development costs.

556. COMPUTER MANPOWER - SUPPLY AND DEMAND - BY STATES: FOURTH EDITION 1981. *Hamblen, John W. Information Systems Consultants. 1981. 36 pg. $30.00.*

Survey of computer labor supply by state examines training programs and personnel; computer installations, 1976-1977; and manpower production, 1976-1979. Includes bibliographic information.

557. COMPUTER SERVICES IN THE MEDICAL FIELD. *Channing Weinberg. 1981. $9500.00.*

Analysis of emerging computer service businesses for the medical industry.

558. COMPUTER SERVICES OPPORTUNITIES IN ENERGY MARKETS. *INPUT. 1981.*

Analysis of markets for computer services in energy industries gives expenditures for services by market segments, including exploration and development, production, facilities management and other end uses. Shows market shares and revenues for services; and average annual growth rates, for 1980-1985. Includes technology review, market penetration opportunities, and professional labor supply.

559. COMPUTER SOFTWARE AND SERVICES. *Frost and Sullivan, Inc. June 1981. 18 pg. $375.00.*

Market report on computer software and services. Provides projected growth to 1985 by four areas.

560. DATABASE MANAGEMENT SYSTEM EVOLUTION: FUTURE HARDWARE AND SOFTWARE DEVELOPMENTS. *Strategic Inc. March 1981. $1500.00.*

Analysis of database management systems hardware and software markets and industry trends. Profiles vendors. Projects price and performance of new technologies, 1980, 1985, 1990.

561. THE DATABASE MANAGEMENT SYSTEMS (DBMS) MARKET. *Frost and Sullivan, Inc. January 1982. 36 pg. $340.00.*

Documents size and scope of software industry and DBMS segment. Matrix is prepared on 16 of the major DBMS companies indicating name of package, price, hardware systems used with, database design, number of users, and operating language and software system. Provides market share information. Company profiles include 1979-1982 financial analysis, software sales growth, product introduction, and market direction. Indicates management strategies.

562. EDUCATIONAL SOFTWARE: CONVERGING TECHNOLOGIES AND STRATEGIES. *Strategic Inc. December 1981. 130 pg. $1200.00.*

Analysis of educational computers and software markets. Includes the technology, industry structure and competition, educational consumers' behavior, and profiles of microcomputer and software service vendors and educational publishers. Shows installed base, 1980-1985; prices, and sales by end uses.

563. IBM SOFTWARE ENVIRONMENT. *International Data Group. Annual. February 1982. 137 pg.*

Annual market survey for software services and facilities being used on IBM systems, 1981-1982. Includes user attitudes; vendors; growth rates, 1981-1985; market segmentation, 1980, 1985; installations, 1980-1981; revenues, 1980-1985; distribution of software; and 1982 purchase plans.

564. INFORMATION MARKET INDICATORS. *Infometrics, Inc. Quarterly. 1982. $10000.00.*

Database market survey service includes producers and vendors; total revenues, 1980-1982; market share, 1980-1982; rankings; average costs; utilization patterns; new products, 1980-1982; end uses; and forecasts.

565. INFORMATION SYSTEMS AND RELATED DELIVERY SYSTEMS IN WHOLESALE BANKING. *Frost and Sullivan, Inc. February 1982. $1150.00.*

Market report contains 10-year forecasts of CIF, information processing, deposit accounting systems, commercial loan systems, money transfer systems, service delivery systems, related hardware, time-sharing and OEM market.

566. KEY VERTICAL MARKETS FOR NON-FINANCIAL INFORMATION SERVICES. *International Resource Development, Inc. 1982. 241 pg. $985.00.*

Analysis of non-financial information database services. Examines the market structure, vendors, competition with government agencies, and impacts of new technology.

567. OPPORTUNITIES FOR BUSINESS GRAPHICS SERVICES AND SOFTWARE. *INPUT. May 1981.*

Analysis of markets for business graphics services and software. Forecasts average annual growth rate, market size, and revenues, 1980-1985. Also includes vendors, prices and performance.

568. PRIMER ON SOFTWARE / SERVICES COMPANIES. *FIND / SVP. December 1981. 12 pg. $175.00.*

Report primarily concerns five firms (Automatic Data Processing, Triad, Culinane, EDS, Informatics), with accompanying short software primer and 1982-1986 estimated revenue of U.S. value-added suppliers.

569. PROGRAMMING PRODUCTIVITY: NEW TOOLS AND EMERGING MARKETS. *Strategic Inc. March 1981. 170 pg. $1500.00.*

Analysis of markets for improved productivity in computer software. Gives vendors' marketing strategies and guidelines for performance requirements and utilizations of improved software.

570. SOFTWARE INDUSTRY. *FIND / SVP. June 1981. 12 pg. $175.00.*

Examines software companies. Assesses relations with hardware vendors, marketing strategies, vertical marketing, market growth on turnkey systems, and forecasts computer services to 1985.

571. SOFTWARE REVIEW. *Microform Review Inc. Semi-annual. $38.00.*

Provides the educational and library community with information about pre-written software. Each issue features: articles on software concepts, evaluation, and selection; reports on available software products suitable for education and library applications; and reviews of books and other recent publications pertaining to computer software with direct relevance for library and educational users.

572. STRATEGIC ANALYSIS OF SOFTWARE AND HARDWARE MAINTENANCE. *Strategic Inc. April 1981. 219 pg. $1500.00.*

Analysis of software and hardware maintenance service markets. Profiles vendors and examines the technology, costs and productivity impacts.

573. TEXT PROCESSING SOFTWARE MARKET. *Frost and Sullivan, Inc. January 1982. $1150.00.*

Includes 5-year forecast of text processing packages by DP hardware size. Examines vendor-supplied text packages vs. independents. Covers OEM's and timesharing, hardware / software compatibility, and interfaces. Includes market segmentation by industry. Contains sales approaches for software including marketing, support, and distribution. Gives vendor profiles.

574. TGM TECHNOLOGY GROWTH MARKETS AND OPPORTUNITIES: ARTIFICIAL INTELLIGENCE. *Creative Strategies International. v.2, no.12. October 1982. 19 pg. $95.00.*

Market analysis for artificial intelligence includes research and development of new technology, forecasts, end uses, market values, corporations and products and marketing strategies.

575. VERTICAL MARKET POTENTIAL FOR DISTRIBUTED DATA PROCESSING. *Creative Strategies International. 1982. $1450.00.*

Analyzes distributed data processing markets and vertical integration in the industry structure. Forecasts installations and market penetration to 1985 and 1990. Examines technology, services, prices distribution, competition, and market determinants including 1980 employees by function, size of businesses, and 1980 installed base by end uses.

See 112, 885, 900, 902, 925, 931, 936, 937, 967, 972, 986, 987, 988, 991, 1011, 1013, 1014, 1026, 1041, 1055, 1061, 1084, 1085, 1105, 1111, 1136, 1141, 1154, 1157, 1173, 1192, 1206, 1208, 1220, 2235, 2258, 3519, 4023, 4046, 4054, 4060.

Investment Banking Reports

576. COMPUTER UTILITIES AND THE FUTURE OF THE COMPUTER SERVICES INDUSTRY. *Frost and Sullivan, Inc. June 1982. 11 pg. $150.00.*

Wall Street analysis of the market for computer utility services and the outlook for investments. Includes market forecasts, 1981-1986; profiles of vendors; 1981 sales; and financial analysis of leading corporations including earnings per share, stock valuations, operating ratios, and balance sheets.

577. DATA PROCESSING INDUSTRY. *FIND / SVP. N123. December 1981. 13 pg. $175.00.*

Wall Street analysis of the data processing industry. Includes pricing, gross national product, producer durable expenditures, composite indexes of office equipment, corporations' earnings, currency translations, and new products.

578. DOMINION SECURITIES AMES: RESEARCH COMMENT: SUMMARY HIGHLIGHTS OF F.A.F. HIGH TECHNOLOGY CONFERENCE. *Coupal, C.E. Dominion Securities Ames Ltd. February 2, 1982.*

Report from a conference on high technology companies, especially computers, software and electronics. Includes market segmentation and industry structure, analysis of corporations' financial performance, earnings per share, and product lines.

579. DREXEL BURNHAM LAMBERT INC.: COMPUTER SOFTWARE: AN OASIS OF INVESTMENT OPPORTUNITY IN THE 1980'S. *Kaplan, Janet R. Drexel Burnham Lambert Inc. June 1982. 126 pg. $500.00.*

Analysis of the computer software industry and investment prospects for corporations. Includes composite stock price index performance, 1981-1982; investment ratings; stock prices; earnings per share, 1981-1983; price earnings ratios, 1981-1983; dividends and yields; shares outstanding; return on equity; and market segmentation, growth rates, and market shares. Profiles of corporations include financial analysis, lines of business, management and product lines. Profiles show operating performance, 1979-1983; earnings model, 1980-1983; cash flow, 1980-1983; and balance sheets, 1980-1983.

580. OPPENHEIMER AND CO., INC.: DATA PROCESSING: DATA PROCESSING INDUSTRY: SOFTWARE COMPANY MEETING ROUNDUP. *Dyson, Esther. Oppenheimer and Company. February 11, 1982. 5 pg. $100.00.*

Summaries of management presentations from a conference of computer software companies and investment analysts. Gives corporations' stock prices; earnings per share, 1980-1982; and price earnings ratio, 1982. Also covers new products, performance and marketing strategies.

581. OPPENHEIMER AND CO., INC.: DATA PROCESSING INDUSTRY: SPRING CHECKUP. *Dyson, Esther. Oppenheimer and Company. April 30, 1982. 14 pg.*

Financial analysis of computer and software corporations. Shows earnings per share, 1980-1982; stock prices; price earnings ratio, 1981-1982 and dividends. Includes new accounting guidelines, companies' revenues, sales, operating performance, mergers, and lines of business and services.

582. OPPENHEIMER AND CO., INC.: INDUSTRY REPORT: A PRIMER ON SOFTWARE / SERVICES COMPANIES. *Dyson, Esther. Oppenheimer and Company. December 23, 1981. 12 pg. $120.00.*

Primer for investments in computer software and services vendors. Examines technology, markets and industry structure. Gives worldwide revenues of manufacturers and vendors,

1980-1981, 1986. Also gives selected corporations' stock price range, 1980-1981; earnings per share, 1980-1983; price earnings ratio, 1980-1983; dividends book value and stock valuation, 1981-1983; and return on average equity, 1981-1983.

583. PAINE WEBBER MITCHELL HUTCHINS INC.: STATUS REPORT: COMPUTER SOFTWARE REPORT: DATABASE MANAGEMENT SYSTEMS. *Monarch, Curt A. Paine Webber Mitchell Hutchins, Inc. March 22, 1982. 12 pg.*

Analysis of the computer software industry focusing on database management systems. Includes revenues, 1975-1985; historical data and forecasts for industry, 1968-1983; technology; product lines and services of vendors; prices; income statements, 1980-1981; and analysis of markets and user attitudes.

584. UNITED BUSINESS AND INVESTMENT REPORT: INDUSTRY SURVEY: SOFTWARE COMPANIES PROGRAMMED FOR SUCCESS. *United Business Service Company. February 8, 1982. pg.58*

Reviews computer software industry structure and profiles vendors. Gives lines of business, 1981 revenues, 1981-1982 earnings per share, 1980-1982 stock price range, price earnings ratio, and dividends.

See 1241, 1251, 1261, 1276, 1284, 1289, 4097.

Industry Statistical Reports

585. COMPUTER SERVICE INDUSTRY, 1980: INDUSTRIE DES SERVICES INFORMATIQUES, 1980. *Canada. Statistics Canada. English/French. Annual. Cat # 63-222. April 1982. 21 pg. 5.40 (Canada).*

For companies engaged in the provision of computer services and those engaged in the sale and rental or lease of EDP hardware, tables give number of establishments, employees, wages, salaries and benefits, sources of revenues, sales, operating expenses, and other statistics, for 1980 with historical data to 1976, by province, by class of customer, and by revenue size group.

586. EIS SHIPMENTS REPORT: COMPUTER PROGRAMMING AND SOFTWARE. *Economic Information Systems, Inc. Quarterly. 1982. $75.00/yr.*

One in a series of reports on particular four-digit SIC code industries. Arranged by state and county, report includes every firm in the industry with annual sales over $500,000 and/or 20 or more employees. Listings detail address, telephone, estimated annual sales, and percent of market. Similar statistics for each county and state.

587. EIS SHIPMENTS REPORT: COMPUTER RELATED SERVICES, NOT ELSEWHERE CLASSIFIED. *Economic Information Systems, Inc. Quarterly. 1982. $75.00/yr.*

One in a series of reports on particular four-digit SIC code industries. Arranged by state and county, report includes every firm in the industry with annual sales over $500,000 and/or 20 or more employees. Listings detail address, telephone, estimated annual sales, and percent of market. Similar statistics for each county and state.

588. EIS SHIPMENTS REPORT: DATA PROCESSING SERVICES. *Economic Information Systems, Inc. Quarterly. 1982. $385.00/yr.*

One in a series of reports on particular four-digit SIC code industries. Arranged by state and county, report includes every firm in the industry with annual sales over $500,000 and/or 20 or more employees. Listings detail address, telephone, estimated annual sales, and percent of market. Similar statistics for each county and state.

589. ELECTRONIC DATA PROCESSING SALARY SURVEY, 1981. *M and M Resources Corporation. Subsidiary of Merchants and Manufacturers Association. Annual. 1981. 110 pg.*

1981 survey of wages and salaries for electronic data processing personnel in California gives data by job title, number of employees, and types of installations.

590. FMI DATA PROCESSING SURVEY - 1981. *Food Marketing Institute. September 1981. 23 pg. $8.00.*

Survey of data processing among supermarkets and food wholesalers. Covers utilzation, installations, size of businesses, average number of employees, sales, types of employees, average annual labor turnover, expenses and growth rate, operating budgets, and hours of labor in computing.

591. INTERNATIONAL WORD PROCESSING'S 1982 SALARY SURVEY RESULTS. *International Word Processing Association. Annual. 8th ed. June 1982. $40.00.48.00 (Canada).*

Eighth annual salary survey for word processing and related positions lists salaries and pay scales by job title, industry, and geographic location for the U.S., Canada, the United Kingdom and Australia. Includes job descriptions, along with mean, median, and modal salary figures and weekly rates.

592. PROFILE OF THE SYSTEMS PROFESSIONAL: SALARIES / ACTIVITIES. *Association for Systems Management. 1981. 24 pg.*

Examines the employment (and sectoral distribution), activities / responsibilities, and earnings profile of systems analysts in the U.S. over the period 1977-1981, with forecasts through 1990.

593. USER RATINGS OF PROPRIETARY SOFTWARE. *Datapro Research, Inc. (McGraw-Hill). Annual. 9th ed. 1981. 50 pg. $25.00.*

A summary of findings from Datapro's annual survey on user ratings of software. Presents data on average, median, and total expenditures for software packages for 1980 and 1981, with forecasts for 1982. Respondents using software from their hardware vendor and from independent suppliers are identified by percentage. Over 400 individual software packages from 157 vendors were rated by 3900 respondents. Ratings and rankings are presented in charts for the following features: reliability, efficiency, ease of installation, ease of use, vendor's technical support, vendor's maintenance and overall satisfaction. In addition, respondents rated packages by those which first come to mind, those most likely for next purchase, and those used most frequently in past 12 months.

594. USER RATINGS OF PROPRIETARY SOFTWARE, 1982. *Datapro Research, Inc. (McGraw-Hill). Annual. 1982. 87 pg. $25.00.*

Annual report provides tables showing several thousand users' ratings of 254 individual packages. Other information includes financial data and present and future market shares.

595. 1981 CANADIAN COMPUTER SALARY SURVEY. *Source EDP. Annual. 1981.*

Examines wages and salaries for various groups involved in the Canadian computer industry.

596. 1981 HANSEN SUPPLEMENTAL SURVEY ON PERSONNEL PRACTICES FOR DATA PROCESSING POSITIONS. *A.S. Hansen Inc. Annual. 3rd ed. 1982. 20 pg.*

Surveys data processing positions in 300 varied firms. Covers turnover, shift differentials, exempt status, work week and overtime, hiring specifications, union status, contracted and specialized staffs, cost of living adjustments, and employee referral bonuses, 1981. Lists survey participants.

597. 1981 SALARY SURVEY RESULTS. *International Word Processing Association. Annual. 7th ed. 1982. $26.00.*

Gives average weekly pay rates for eight word processing positions, three administrative support positions, and four management positions including average, high, low figures by city, region, and industry.

See 1300, 1325.

Financial and Economic Studies

598. BANKING AND FINANCE INDUSTRY TRENDS: IMPACT ON COMPUTER SERVICES. *INPUT. 1981.*

Analysis of effects of legislation on banking and finance industry structure and markets for information services. Examines deregulation, costs, mergers, and competition in banking. Gives revenues for information services and average annual growth rates, 1980-1985. Includes marketing strategies.

599. ELECTRONIC MAIL: A REVOLUTION IN BUSINESS COMMUNICATIONS. *Connell, Stephen. Knowledge Industry Publications. July 1982. 144 pg. $32.95.*

Reports on electronic mail, the technology, costs and utilization.

See 1326, 1335, 2440.

Forecasts

600. REVIEW OF PROFESSIONAL MANPOWER 1982. *Technical Service Council. 1982.*

Survey by the council forecasts the demand for professionals for 1982 and beyond in engineering, accounting, and data processing.

Directories and Yearbooks

601. AGRICULTURAL MICROCOMPUTING SOFTWARE CATALOGUE. *Agricultural Microcomputing. Quarterly. 1982. 25.00 (Canada).*

Lists commercially available agricultural software and their manufacturers.

602. THE BLUE BOOK FOR THE APPLE COMPUTER. *Visual Materials Inc. Annual. 2nd ed. 1982. 650 (est.) pg. $24.95.*

Software directory for Apple Computers lists addresses of vendors and descriptions of software and prices including hardware and services.

603. COMPANY ANALYSIS AND MONITORING PROGRAM (CAMP) DIRECTORY. *INPUT. Annual. 1981.*

Provides information in 2500 computer services companies including address, telephone, CEO, recent financials, type of services offered, industry markets served, and geographic markets served.

604. DATAPRO DIRECTORY OF MICROCOMPUTER SOFT-WARE. *Datapro Research, Inc. (McGraw-Hill). Monthly. 1981. $340.00.*

Describes 2000 software products for computer systems under $40,000. Includes data on vendors, system requirements, current prices, maintenance, documentation and training. In 2 loose-leaf volumes with monthly supplements.

605. DATAPRO DIRECTORY OF ON-LINE SERVICES: VOLUME 1. *Datapro Research, Inc. (McGraw-Hill). Annual. 1982. 650 (est.) pg. $390.00.*

Directory of vendors of on-line computer services. Includes addresses, executives, subsidiaries of parent corporations, ownership, histories of corporations, number of employees, annual sales volume, management structure, and other lines of business. Other sections cover services and facilities, prices, contract provisions, user attitudes and performance ratings, and end uses.

606. DATAPRO DIRECTORY OF ON-LINE SERVICES: VOLUME 2. *Datapro Research, Inc. (McGraw-Hill). Annual. 1982. 400 (est.) pg. $390.00.*

Directory of vendors of on-line computer services. Includes addresses, executives, subsidiaries or parent corporations, number of employees, annual sales volume, management structure, other lines of business, services, facilities, contract provisions, and prices. Other sections cover user attitudes, performance ratings and end uses.

607. DIRECTORY OF COMMUNICATIONS MANAGEMENT. *Applied Computer Research. Semi-annual. 1982. $120.00/yr.*

Lists executives with voice and data communications responsibilities in companies with an annual DP budget of $250,000 or more. Includes company name, address, telephone, direct dial number of executive listed, and executive's voice and data responsibilities. Main section is organized by city within state and alphabetically by firm or agency name. Cross-index lists companies or agencies alphabetically by 12 classifications.

608. DIRECTORY OF COMPUTER FACILITIES IN THE SOUTHWEST. *Texas Agricultural and Mechanical University. Industrial Economics Research Division. Biennial. 1981. 300 pg. $20.00.*

Arranged geographically by state and city, firm listings detail address, telephone number, executive, and computer and peripheral equipment and services. Indexed by firm, equipment, and applications.

609. DIRECTORY OF COMPUTER SERVICES. *Edmonton. Business Development Department. Irregular. 1981.*

Lists computer services and products available in Edmonton.

610. DIRECTORY OF COMPUTER SERVICES COMPANIES 1982. *Association of Data Processing Services Organization, Inc. Annual. 1982. $95.00.*

Annual Association of Data Processing Service Organizations membership directory provides profiles of 500 remote processing, software products, professional services, integrated systems, and consulting companies. Lists key contacts, types of services and applications available, and basic company information. Indexes by area, services, and application.

611. DIRECTORY OF PROFESSIONAL AND TECHNICAL CONSULTANTS IN SAN FRANCISCO / SAN JOSE. *Professional Technical Consultants Association. Annual. June 1982. 90 (est) pg. Free.*

Includes approximately 170 resumes of professional and technical consultants located in San Francisco and San Jose, California. A cross-reference index lists consultants including hardware / software applications, manufacturing management, and planning.

612. DIRECTORY OF WORD PROCESSING MANAGEMENT. *Applied Management Services. Annual. 1981. 140 pg. $160.00.*

Directory covers the leading word processing professionals in organizations that provide management positions for this function. Includes over 2000 entries, with company name, address, manager, title, telephone number, industrial classification, Fortune ranking, equipment manufacturers, type and units of word-processing equipment, and number of office employees.

613. INFORMATION SERVICES, 1983-84. *Information Industry Association. Annual. 1983. 400 (est.) pg. $37.50.*

Annual Information Industry Association membership directory covers database suppliers, information consultants, publishers, hardware technologists, communications experts, and others in the information industry. Each entry describes company activities, services, branches, and key executives.

614. LOCATE: A DIRECTORY FOR PURCHASERS OF LAW OFFICE COMPUTER APPLICATIONS AND SOFTWARE. *American Bar Association. Irregular. 1981. $28.00.*

Directory of computer software and services geared to the legal profession. Includes profiles of some 200 vendors, and a guide to systems by application.

615. OFFICIAL DIRECTORY OF DATA PROCESSING: WESTERN DIRECTORY OF EDP SYSTEMS USERS, 1981-1982 ED. *Official Directories of Data Processing. Annual. v.7, no.1. 1982. 264 pg. $175.00.*

Lists firms using electronic data processing equipment, 1982. By state and city gives firm and address, officers, type of business, EDP equipment and personnel, and language. Includes alphabetical company and personnel indexes. Separate editions cover 6 other regions, a system 3/34 directory and a 370/4331 directory.

616. ONLINE MICRO - SOFTWARE GUIDE AND DIRECTORY, 1983-1984. *Gordon, Helen A. Online Inc. Annual. 1st ed. 1982. 346 pg. $40.00.*

Alphabetically lists U.S. microcomputer software producers and publishers, 1982. Gives address; software name, date, and cost; application; operating environment and hardware requirements; and documentation and product description. Indexed by distributor, producer, subject and title. Charts compare word processing software and database management systems. Includes bibliography. Update supplements to be published November 1983, 1984.

617. ONLINE MICRO-SOFTWARE GUIDE AND DIRECTORY, 1983 SUPPLEMENT. *Online Inc. Annual. October 1983. $30.00.*

Annual supplement to triennial includes listings of software packages with a description of each, hardware requirements, source language, operating system compatibilities, media type, and size, with articles on industry-related issues. Other features include listings of distributors and their product lines (with prices) and a resource section covering online databases, newsletters, reports, contests, and directories.

618. SCHOOL MICROWARE DIRECTORY: A DIRECTORY OF EDUCATIONAL SOFTWARE. *Dresden Associates. Semi-annual. 1982. 118 pg. $25.00/yr.*

Biannual directory of educational software for microcomputers lists products, descriptions, prices, and addresses of vendors. Includes software for school administration.

619. SOFTWARE VENDORS DIRECTORY: 1981-1982. *Hitchcock Publishing Company. Annual. 1981.*

Alphabetical nationwide directory of U.S. computer software vendors. Indicates which among 35 categories of applications each firm produces software for.

620. SOURCEBOOK FOR SMALL SYSTEMS SOFTWARE AND SERVICES. *Information Sources, Inc. Irregular. 1982. 500 pg. $135.00.*

Directory identifies 1000 business-, education-, and technically-oriented software packages' program application, program details, hardware, operating system, language, supplier (with address, telephone, and contact), availability date, price / terms, number of installations, documentation, training, and other services.

621. TRANSPORTATION AND TRAFFIC - DATA SYSTEMS SOURCE GUIDE. *J.J. Keller Associates, Inc. Irregular. 1982. 250 pg. $65.00.*

Directory of automated services for fleet operation provides key user data on systems applications, data entry requirements, reports generated, hardware configurations, software elements, and additional aspects of system. Includes systems available for management, financial operations, personnel applications, equipment programs, safety, general transportation, and traffic.

See 1400, 1402, 1404, 1405, 1409, 1414, 1415, 1434, 1437, 1442, 1459, 1473, 1474, 1477, 1480, 3252, 4187, 4203.

Special Issues and Journal Articles

American Banker

622. TIME-SHARING FIRM SEES FUTURE IN PROVIDING SOFTWARE EXPERTISE. *Trigaux, Robert. American Banker, Inc. v.147, no.145. July 8, 1982. pg.12, 14*

The growth of software services is compared with growth and demand for time-sharing service during recent years. Data is presented on U.S. computer services industry growth (1980-1981) for each of the following segments: processing services, professional services, software products, and integrated systems. Segments showing the greatest potential for future growth and companies planning to re-enter the time-sharing field with emphasis on software services are identified.

Bank Systems and Equipment

623. BANKS / THRIFTS LOOK TO VENDORS FOR FUTURE SOFTWARE PACKAGES. *Hyman, Joan Prevete. Gralla Publications. March 1982. pg.57-60*

Survey of financial institutions analyzes preferences for purchasing software packages or developing them in-house. Statistics given for influences of purchasing decisions, opinion trends, current ownership of software packages, and future intentions of developing application software.

Business Computer Systems

624. PROGRAMS TO SPELL WELL. *Cahners Publishing Company. Irregular. v.1, no.4. December 1982. pg.79-85 ff.*

Describes spelling software programs, 1982. Provides producers, location, technical details, and prices.

Byte

625. A GRAPHICS PRIMER. *Williams, Gregg. BYTE Publications Inc. Irregular. v.7, no.11. November 1982. pg.448-470*

Examines computer graphics for microcomputers. Discusses selected products and specifications. Lists U.S. vendors. Includes addresses and 1982 product lines.

626. THE THIRD NCGA AND THE FUTURE OF COMPUTER GRAPHICS. *Pournelle, Alexander. BYTE Publications Inc. Irregular. v.7, no.11. November 1982. pg.30-44*

Surveys computer graphics systems displayed at the 1982 National Computer Graphics Association convention. Describes products and provides vendor's name and address.

Canada Commerce

627. SOFTWARE: THE EDUCATION OF THE COMPUTER AND OPPORTUNITY FOR CANADA. *Davies, J.R. Canada. Department of Industry, Trade and Commerce. January 1982. pg.12-13*

Examines the performance of the Canadian software industry including forecasts of revenues to 1985 and sales of systems and new services.

Canadian Consulting Engineer

628. CAD IN CANADA: 1981 SYSTEMS SURVEY. *MacKay, Richard. Southam Business Publications, Ltd. Division of SouthamCommunications Ltd. Annual. December 1981. pg.23-32*

Analysis of the uses of computer-assisted design (CAD) in Canada since 1979 includes descriptions of interactive CAD systems which have been popularly integrated into proprietary commercial systems for engineering applications. Lists suppliers, basic hardware and options available, primary applications, basic system prices, and sales and technical support offices.

Canadian Datasystems

629. CANADIAN DATASYSTEMS: REVIEW / FORECAST "82. *Maclean Hunter Ltd. Annual. December 1981. pg.32-53*

Canadian industry leaders view the issues and prospects for computer technology in the coming year. Forecasts of new trends, capital utilization, revenues, research and development and new services are noted.

630. COMPUTER STOCKS IN PERSPECTIVE. *Maclean Hunter Ltd. Monthly. October 1981. pg.212*

Review of shares of some 11 computer service businesses. Indicates centers of activity in the industry, but is not designed to provide a guideline for the purchase of these stocks.

631. LOOKING AT CANADA'S SOFTWARE INDUSTRY. *Hodson, Bernard A. Maclean Hunter Ltd. March 1982. pg.78, 80, 82*

Analysis and performance of businesses in the software industry. Covers specialization, government involvement, research and development, new trends for the industry, and forecasts.

632. MINICOMPUTER SOFTWARE MARKET CONTINUES EXPLOSIVE GROWTH. *Maclean Hunter Ltd. v.14, no.10. October 1982. pg.48*

Highlights forecasts for the market for mini-software reported by Frost and Sullivan, Inc. Statistical data provided for U.S.

market (billions), 1981, 1986; sales of standard applications packages (billions), 1981, 1986; sales of system software (billions), 1981, 1986; OEM market share (percent), 1981, 1986; and markets for manufacturing packages, graphics packages, insurance packages, general accounting packages, and CPA packages (millions), 1981, 1986.

633. SOFTWARE UPDATE. *Maclean Hunter Ltd. Monthly. October 1981. pg.13*

News of software and services notes products, suppliers and features.

The Cerebus Report

634. AN ANALYSIS OF THE ACQUISITION AND MERGER ACTIVITY IN THE COMPUTER SERVICES INDUSTRY. *Varga, Charles C. Cerberus Publishing. v.1, no.3. June 1982.*

Analysis of mergers and acquisitions in the computer services industry includes costs of computer services; purchases and mergers, 1969-1981; businesses acquired; prices, 1969-1981; major acquiring vendors; and financial analysis of deals including revenues, net income, profitability, price earnings ratio, and other ratios.

635. HE WHO HAS THE DATA GETS THE GOLD. *Cerberus Publishing. v.1, no.1. January 1982. pg.18-23*

Presents an overview of the on-line data base market. Includes a flow chart of markets; table of revenues of information companies, by service, 1979-1984; and a listing of acquisition, joint venture, and new entry activity in the data base marketplace.

CIPS Review

636. PENETRATING U.S. MARKET DEMANDS CAREFUL STRATEGY. *Potter, Michael. Canadian Information Processing Society. v.6, no.4. August 1982. pg.5-7*

Analysis of strategies for Canadian software businesses in the U.S. market examines market structure, trade deficits, direct sales force, product strategy, tax breaks, and in-house research and development.

637. STRATEGIES AND TRENDS IN DATA PROCESSING. *Murray, Grant. Canadian Information Processing Society. January 1981. pg.18-19*

Forecasts the performance of the Canadian data processing and computer industries. Includes new trends in products, marketing, labor costs, and price / performance comparisons.

Commodities

638. WHERE TO FIND COMMODITY SOFTWARE. *Commodities Magazine, Inc. Irregular. v.11, no.11. November 1982. pg.64, 84-94*

Directory lists names, addresses, and telephone numbers of suppliers of commodities software. Describes products and gives brand names.

Computer Business News

639. COMPUTER BUSINESS NEWS: MPU BOOM CROSS - TOOL GOLD MINE. *O'Connor, Rory J. CW Communications Inc. November 16, 1981. pg.1, 3, 8 $1.00.*

Discusses an upsurge in demand for cross-assemblers and cross-compilers (software which translates assembler code or high level languages written for one computer, into object code for a second machine) as a result of the large number of new

microprocessors on the market. Covers number of major competitors, and hardware for which these programs are commonly purchased.

640. SOFTWARE TOOLS DIRECTORY: PART 1: A THROUGH BROOKVALE ASSOCIATES, INC. *CW Communications Inc. Annual. v.5, no.10. March 8, 1982. pg.19-23*

The first part of an annual directory of software tools which gives name and description of programs, compatability requirements, prices, and address of company.

641. SOFTWARE TOOLS DIRECTORY: PART 2: BUSINESS AND TECHNOLOGICAL SYSTEMS, INC. THROUGH MAGUS SYSTEMS. *CW Communications Inc. Annual. v.5, no. 11. March 15, 1982. pg.23-32*

The second part of an annual software directory giving program name, description, compatibility requirements, pricing, and address of company.

642. SOFTWARE TOOLS DIRECTORY: PART 3: MAMBA, INC. THROUGH TECHNICAL SYSTEMS CONSULTANTS, INC. *CW Communications Inc. Annual. v.5, no.12. March 22, 1982. pg.19-28*

The third section of a directory of software tools for resale. Includes program name, description, price, hardware capability, and address and telephone number of manufacturer.

Computer Decisions

643. COMPUTER DECISIONS: SOFTWARE. *Hayden Book Company, Inc. Monthly. March 1982. pg.222-223*

A monthly listing of new available software packages. Gives function; capabilities; system compatibilities; addresses of companies; and price. Further information available through reader card service.

644. CONVERSION: TRAUMA OR TEA PARTY. *Hayden Book Company, Inc. Irregular. March 1982. pg.35-50*

Examines the difficulties inherent in software conversion. Provides a listing of conversion packages and services, giving company address and telephone, package name, and prices.

645. DATA DICTIONARIES: BONUS BENEFITS. *Snyders, Jan. Hayden Book Company, Inc. Irregular. v.14, no.5. May 1982. pg.30-46, 166*

Lists 13 U.S. vendors of data dictionaries. Includes vendor's name, address, and phone number, type of package, type of equipment, and price.

646. INFORMATION RETRIEVAL: NO TIME LIKE THE PRESENT. *Snyders, Jan. Hayden Book Company, Inc. Irregular. v.14, no.11. November 1982. pg.48-64*

Article on various users' attitudes towards assorted computer software systems and their uses includes a directory of information retrieval systems vendors detailing vendor address, telephone, package, requirements, and price.

647. MANAGING THOSE VALUABLE HUMAN RESOURCES. *Snyders, Jan. Hayden Book Company, Inc. v.14, no.2. February 1982. pg.42-56*

Reviews software products for the management of personnel data. A listing of available products is included, with vendors, addresses, technical specifications and prices. Specific products in use are described with discussion of system implementation, utilization, and benefits realized over previous systems.

648. NEW TRENDS IN DBMS (DATA BASE MANAGEMENT SYSTEMS). *Snyders, Jan. Hayden Book Company, Inc. v.14, no.2. February 1982. pg.100-133*

Review of database management systems lists products by vendor, with brand names, prices, technical specifications, and vendor telephone numbers. Accompanying text discusses new technology, the future of database management systems, and the advantages and disadvantages of various systems.

649. SNYDERS ON SOFTWARE: A WINDOW ON YOUR SYSTEM'S PERFORMANCE. *Snyders, Jan. Hayden Book Company, Inc. v.14, no.10. October 1982. pg.39-42, 47, 52, 56*

Discusses available software for evaluating systems and requirements of 8 organizations. Tables show 33 software packages, including vendors, equipment requirements, and prices.

650. SOFTWARE SHOWCASE "82. *Hayden Book Company, Inc. Annual. March 1982. 86 pg. $198-206.*

A listing and description of available utility software packages giving vendor's name, address, and product. Software arranged by functions including applications development, auditing, conversion, data dictionary, disk management, enhancements, information retrieval, etc. Further information through reader card service. Profiles selected packages giving description and price.

651. 1981'S TOP ONE HUNDRED DP VENDORS. *Gartner Group. Hayden Book Company, Inc. Annual. v.14, no.6. June 1982. pg.77-98*

The top 100 U.S. companies are ranked by 1981 data processing revenues. Includes 1979 and 1980 DP revenues, rankings and the top 20 and the bottom 20 in revenue growth rate with their foreign DP revenue growth for 1980-1081. Ranks top 10 by and including the following: DP opearating margin, DP operating profits, revenue per DP employee, foreign DP revenues, worldwide DP revenues, research and development expenditures, research and development as a percentage of sales, and word-processing revenues. Gives growth by market segment for the top 100, 1980-1981. For the top twelve companies shows rank and DP revenue for 1966, compound growth rate 1966-1981, expected 5-year growth rate, and projected 1985 revenue.

Computer Design

652. NCC "82 PREVIEW. *Computer Design Publishing Corporation. Irregular. v.21, no.5. May 1982. pg.81-198*

Preview of the 1982 National Computer Conference gives registration information, technology topics to be presented and listings of exhibiting manufacturers, with product lines, new product technical specifications, prices, and business addresses.

The Computer Graphics Software News

653. GRAPHIC SOFTWARE INDEX. *Computer Graphics Software News. Irregular. v.1, no.2. November 1981. pg.5-20*

A listing of common computer graphic packages includes package name and type, graphic device, language, cost, and address of vendor.

ComputerData

654. CANADA'S ANNUAL DIRECTORY OF THE DP INDUSTRY 1981 / 1982. *Whitsed Publishing Ltd. Annual. v.6, no.7. July 1981. 184 pg.*

Lists service bureaus in Canada, consultants and services, hardware vendors, and products. Includes addresses, phone numbers, key executives, products / services provided, branches, software products and vendors.

655. CANADA'S FIRST SOFTWARE GUIDE. *Whitsed Publishing Ltd. Annual. August 1982.*

Lists U.S. and Canadian software vendors, as well as U.S. businesses selling their products in Canada. Indicates availability for microcomputer uses.

Computerworld

657. ANALYSTS PROGRAMMERS SEEN TOPPING JOB MART. *CW Communications Inc. v.16, no.23. June 7, 1982. pg.12*

Gives the outlook for systems analysts and programmers in 1990 including number of job openings. Gives 1980 salaries for both professions.

658. COMPUTERWORLD: SYSTEM AIDS IN DEBUGGING COUNTY HEALTH CARE. *CW Communications Inc. v.16, no. 11. March 22, 1982. pg.28*

Discusses the application and advantages of a health care information system in terms of dollars saved and improved public health.

659. DATA ENTRY WAGES UP, BUDGETS DOWN. *CW Communications Inc. v.16, no.11. March 22, 1982. pg.67*

Reviews the annual survey of the Data Entry Management Association with data on labor productivity growth (1980-1981), change in labor turnover (1980-1981), data entry expenditures as a percent of data processing department budgets (1980-1981), and data entry employees' participation in labor unions. Includes a chart detailing high, low and average monthly salaries for five data entry positions (1979-1981).

660. DATAPRO FINDS DECLINE IN "81 SOFTWARE OUTLAYS. *Paul, Lois. CW Communications Inc. v.15, no.51. December 21, 1981. pg.1, 16-33*

Summarizes results of Datapro's ninth annual survey on user ratings of proprietary software. Data on average, median and total expenditures for software packages is presented for 1980 and 1981, with forecasts for 1982. Respondents using software from their hardware vendor, and from independent software suppliers are identified by percentage. Over 400 individual software packages from 157 vendors were rated by 3900 respondents. Ratings and cluster rankings are presented in 14 pages of charts. The following features were rated: reliability, efficiency, ease of installation, ease of use, vendor's technical support, vendor's maintenance, and overall satisfaction. In addition, respondents rated packages by those which first come to mind, those most likely for next purchase and those used most frequently in past 12 months.

661. FEMALE DPERS THE TOP WOMEN EARNERS, BUT STILL GET LESS THAN MEN IN THE SAME JOBS. *Blakeney, Susan. CW Communications Inc. v.16, no.14. April 5, 1982. pg.2*

Results of a U.S. Department of Labor survey on salaries in data processing are presented. A table details male and female median weekly salaries for four job categories, with ratio of female to male earnings, and percent of female workers. Statistical highlights are discussed with data on growth rates of salaries (1981-1982).

662. THE REAL COST OF DP PROFESSIONALS. *Plotkin, Stephen. CW Communications Inc. v.16, no.29. July 19, 1982. pg.10-16*

Reports on the costs of data processing professionals. Discusses methodology of 1982 study. Calculates productivity; employment investment; and costs for labor after three months, after six months, and after one to five years of employment.

663. SURVEY OF DP MANAGERS FINDS PAY, BUDGET FREEZES. *CW Communications Inc. v.16, no.23. June 7, 1982. pg.1, 6*

Gives results of survey of 10 DP managers of U.S. companies. Discusses salaries, hiring policies and budgets in general terms.

664. USERS FIND SOFTWARE, SUPPORT PROBLEM AREAS. *Datapro Research Corp. CW Communications Inc. v.16, no.23. June 7, 1982. pg.59*

Survey shows users in all categories had problems with vendor's software and technical support. Lists models causing difficulty. Discusses systems costs of IBM machines and others; and expansion plans of Amdahl Corp. and IBM systems users.

665. WIP: FEMALE, MALE DPERS GETTING EQUAL PAY. *Women in Information Processing. CW Communications Inc. v.16, no.23. June 7, 1982. pg.4*

Previews survey of differences between salaries of female and male programmer / analysts, senior analysts or marketing support persons, 1982.

DEMA Newsletter

666. THIRD ANNUAL DEMA MEMBER STATISTICAL COMPENSATION SURVEY. *Data Entry Management Association. Annual. February 1982.*

Results of the 1981 annual survey of the Data Entry Management Association are presented. Includes labor productivity growth, 1980-1981; change in labor turnover, 1980-1981; data entry expenditures as a percent of data processing department budgets, 1980-1981; and data entry employee participation in labor unions. High, low, and average salaries are detailed for specific data entry positions, 1979-1981.

Design Engineering

667. DEMAND FOR PROFESSIONALS TO CONTINUE DESPITE LAYOFFS. *Maclean Hunter Ltd. March 1982. pg.5*

The demand for professionals in engineering, accounting, data processing for 1982 is forecast in relation to other professionals. Review also predicts size of labor force growth.

Digital Design

668. ENERGY RELATED COMPUTER SERVICES TO HIT $5 BILLION. *Morgan Grampian Publishing Company. v.12, no.4. April 1982. pg.23*

Summarizes report entitled 'Computer Services Opportunities in Energy Markets,' an analysis of both present and future computer services markets. Chart provides forecast of computer services expenditures in energy markets by subsector, 1980-1985. (Source: INPUT)

669. PABX AND LAN INTEGRATE WITH ENERGY MANAGEMENT SYSTEMS. *Morgan Grampian Publishing Company. v.12, no.9. September 1982. pg.24*

Analysis of the integration of energy management systems into PABX and local area network (LAN) markets includes charts of value forecasts and percent breakdown of system shipments, 1981, 1985; and energy system market segmentation. Examines current vendors.

670. SMALL SYSTEMS PACKAGED SOFTWARE MARKET TO HIT $6.3 BILLION BY 1986. *International Data Corporation; Software and Services Information Program. Morgan Grampian Publishing Company. v.12, no.7. July 1982. pg.18-20*

Forecasts growth of small systems packaged software through 1986. Includes graphs comparing software revenues from hardware vendors, hardware manufacturers and independents, 1980, 1986. Summarizes market growth and future marketing strategies of major hardware manufacturers from a market research report by International Data Corporation.

EDP In-Depth Reports

671. THE CANADIAN PROCESSING SERVICES INDUSTRY. *Evans Research Corporation. v.11, no.4. December 1981. 16 pg.*

Reviews the growth of Canadian interactive processing and service bureaus. Gives historical data, including various threats from data communications, costs and inhouse hardware sales; fast and slow growing markets for processing services. Outlines the top four Canadian bureaus, their areas of business, 1980 revenues, major customers, and forecasts growth to 1985.

672. COMPUTER GRAPHICS. *Evans Research Corporation. v.11, no.1. September 1981. 16 pg.*

Examines technologies involved in computer graphics and their use by businesses. Analysis includes look at U.S. competition and their revenues, markets in Canada, and various businesses involved in Canada.

673. THE COST STRUCTURE OF SELECTED PUBLIC DATA BASES AND VIDEOTEX SERVICES. *Evans Research Corporation. v.11, no.10. June 1982. 20 pg.*

Based on a study completed June 1982, examines prices and pricing structure of selected databases. Analyzes competition within the industry and personnel.

Electronic Business

674. THE TOP U.S. DESKTOP SOFTWARE VENDORS. *Cahners Publishing Company. v.8, no.11. October 1982. pg.128*

Ranks the top 7 U.S. desktop computer software vendors by 1981 revenues (in millions of dollars). Source is International Data Corp.

Electronic Engineering Times

675. SOFTWARE FILE. *CMP Publications, Inc. Weekly. 1982.*

Regular feature lists new software products, with technical descriptions, prices, manufacturer's name and address.

EMMS: Electronic Mail and Message Service

676. COST COMPARISON OF LEADING COMPUTER MAIL-BOX SYSTEMS. *International Resource Development, Inc. v.6, no.22. November 15, 1982. pg.1-6*

Market survey compares operating costs for five leading U.S. computer mailbox systems. Describes services offered by each vendor.

Financial World

677. CASHING IN ON CASH MANAGEMENT. *Kessler, Jeffrey. Macro Communications, Inc. v.151, no.6. March 15, 1982. pg.56-59*

The market for financial computer services and cash management systems is examined, with data on market size and growth rates. Discusses the trend toward computer service companies, instead of banks as the providers of the service; and identifies leading computer service companies in the market. Presents data on revenues from cash management services with recent revenue growth rates and forecast for 1982 performance. Includes table detailing earnings per share 1980-1982; recent stock price; price range; price earnings ratio, dividends and yields for six of the leading companies.

The Fox-Morris Report

678. UNPRECENTED DEMAND FOR SOFTWARE TALENT CONTINUES IN $49 BILLION COMPUTER BUSINESS. *Fox-Morris Personnel Consultants. v.4, no.3. 1981. pg.1-2*

Reports employment and compensation trends in the U.S. data processing industry. Statistics include a ranking (1980-1981) of DP professionals by sector and salary figures (1980-1981) by sector and years of experience.

Handling and Shipping Management

679. DISTRIBUTION SOFTWARE. *Penton / IPC. Subsidiary of Pittway Corporation. Irregular. April 1982. pg.63-64*

Directory of some 18 businesses offering software programs and packages for distribution applications.

Hardcopy

680. HARDWARE QUESTIONS THAT DEMAND SOFT-WARE ANSWERS. *Bismuth, Robert. Seldin Publishing, Inc. Irregular. v.11, no.17. September 1982. pg.72-75*

Points out how software changes can solve problems in computer hardware. Chart presents 1982 software designed to better hardware performance. Information includes vendor's name, functions and features, special features, and operating systems.

681. SOFTWARE VENDOR LISTING FOR THE DEC MARKET. *Seldin Publishing, Inc. Irregular. v.11, no.16. August 1982. pg.78-81, 89*

Alphabetically lists vendors of software for DEC market. Includes addresses.

High Technology

682. INVESTMENTS: MARKET SURGE FOR COMPUTER SOFTWARE. *Technology Publishing Company. v.2, no.5. September 1982. pg.107*

The computer software industry is discussed with data on number of suppliers, market size and growth rates (1982) with forecasts to 1985. Investment opportunities in the software industry are reviewed, and three major suppliers are profiled, with descriptions of product lines, and data on revenues and earnings per share, 1980-1983.

Industrial Marketing

683. SOFTWARE PACKAGES FOR SALES / MARKETING. *Crain Communications, Inc. Irregular. v.67, no.7. July 1982. pg.50-56*

Lists 60 U.S. marketing / sales - strategy analysis programs with cross-industry applicability. Arranged by micro-, mini- and mainframe uses. Includes software product name and description, the vendor / supplier name and city, hardware applications, the current number of users, and the price.

Infosystems

684. NEW PACKAGES SPAN THE SOFTWARE HORIZON: SOFTWARE INFO REVIEW. *Thiel, Carol Tomme. Hitchcock Publishing Company. Semi-annual. v.29, no.2. February 1982. pg.47-74*

Contains information on over 200 packages. Lists products manufacturers and descriptions for 35 categories of software packages.

685. PROGRAMMING TOOLS: IMPACTING DP PRODUC-TIVITY. *Thiel, Carol Tomme. Hitchcock Publishing Company. v.29, no.3. March 1982. pg.56-60*

Reviews software products providing programmer aids and productivity tools for users in the data processing environment. Discusses the advantages of various systems with some technical specifications. Includes a listing of software tools by product group, including vendor names.

686. THIRD ANNUAL SOFTWARE / EXPO. *Hitchcock Publishing Company. Annual. v.29, no.9. September 1982. pg.SE1-SE46*

Directory of exhibitors at the Third Annual Software / Expo, 1982 lists services by application and gives addresses and product lines ofvendors.

Insurance Journal

687. DIRECTORY OF COMPUTER SUPPLIERS FOR THE INDEPENDENT INSURANCE AGENT. *Insurance Journal Inc. Irregular. v.61, no.5. March 8, 1982. pg.12, 22*

Entries for 30 U.S. computer suppliers for the independent insurance agent in California. Arranged alphabetically by company name. Includes corporate address and telephone number, and description of available services.

Library Journal

688. THE AUTOMATED CIRCULATION SYSTEM MARKET-PLACE: ACTIVE AND HEATING UP. *Matthews, Joseph R. R.R. Bowker. v.107, no.3. February 1, 1982. pg.233-235*

An overview of the market for automated circulation systems, with data on the number of libraries with installed system, listed by vendor (1977-1981). Additional charts and graphs detail total cumulative number of installed systems by year (1973-1981); size of installed systems, by selected technical specifications; and functions avaiable from major systems on the market. 1981 data on revenues is presented, with discussion of market growth competition. The place of the circulation software in an integrated library system is reviewed, and major vendors are identified.

Magazine of Bank Administration

689. AUDIT SOFTWARE DIRECTORY. *Bank Administration Institute. Irregular. v.57, no.12. December 1981. pg.32-35*

Lists the latest software products for banking applications. The following data is included for each software package: product name, host computers and operating systems with minimum core storage required; vendor's name, address, and telephone number; and purchase price and annual maintenance fee. Includes other technical specifications on input and output file format and structures, reporting features, field types processed, and database systems supported.

Mini-Micro Systems

690. MINI-MICRO SYSTEMS: SOFTWARE. *Cahners Publishing Company. Monthly. v.15, no.4. April 1982. pg.248-252*

Reviews new software for data processing. Includes software description, price, and vendor's name and address.

691. SOFTWARE: WHICH DBMS IS RIGHT FOR YOU? *Weiss, Harvey M. Cahners Publishing Company. Irregular. v.14, no.10. October 1981. pg.157-159*

Discusses the selection process of a database - management system. Lists and describes 38 systems. Provides business name, DBMS name, type, hardware, features and price.

Mining Engineering

692. AVAILABLE SOFTWARE. *Society of Mining Engineers of AIME. Irregular. v.33, no.11. November 1981. pg.1591-1595*

Directory of software available for the mining and mineral industries. Software suppliers and equipment manufacturers are listed by company name, with address. Software package brand names are listed, with a description of applications and capabilities.

Modern Healthcare

693. HOSPITAL CHAINS BECOME IMPORTANT PROVIDERS. *Dorenfest, Sheldon I. Crain Communications, Inc. v.11, no.11. November 1981. pg.77-79, 82, 84*

Article discusses shared computer services provided by multi-hospital systems to affiliated and unaffiliated hospitals. Findings of a 1979-1980 study on trends in computer use in community hospitals are summarized, and the following data is presented in tables and charts: change in number of shared users in sample hospitals (1979-1980); growth in number of hospitals affiliated with multihospital systems; and number of

multihospital systems using the computer for patient accounting, by type of system. Covers multihospital systems with a central role in computer decisions, by name, with number of beds; and hospitals affiliated with systems using in-house computers, subdivided by computer approach.

694. VENDORS EXPECT GOOD DEMAND FOR UPDATED FINANCIAL SOFTWARE. *Kuntz, Esther Fritz. Crain Communications, Inc. v.11, no.11. November 1981. pg.90*

Reviews the market for data processing shared services, systems in operation and the size of facilities purchasing these systems. The advantages of in-house and time-sharing systems are discussed and future market trends are forecast.

Motor Age

695. A GUIDE TO PICKING SOFTWARE FOR YOUR BUSINESS HARDWARE. *Chilton Book Company. Irregular. November 1982. pg.24-25*

Directory of 21 software manufacturers details address, telephone, examples of business packages available, and examples of computers with which the software will work.

Online

696. SEARCH ASSISTANCE: A CURRENT LIST OF DATA-BASE PUBLISHER TELEPHONE NUMBERS. *Sovner, Judi. Online Inc. Irregular. v.6, no.3. May 1982. pg.62-64*

Directory of database publishers gives name, telephone number, and retrieval system.

Ontario Technologist

697. DEMAND FOR PROFESSIONALS STILL AT RECORD LEVELS. *Ontario Association of Certified Engineering Technicians and Technologists. March 1981. pg.15*

Reviews supply and demand and average monthly wages and salaries for Canadian engineers, data processors, and accountants.

Personal Computing

698. SERVICING YOUR SYSTEM: BE PREPARED. *Hayden Book Company, Inc. Irregular. v.6, no.9. September 1982. pg.50-55, 148-154*

Examines personal computer service contracts. Lists computer manufacturer with address, availability of in-house and/or dealer service, third party service, and contract price and time period.

699. SPECIAL REPORT: EDUCATIONAL SOFTWARE FOR THE HOME. *The, Lee. Hayden Book Company, Inc. Irregular. v.6, no.6. June 1982. pg.48-52, 102-103, 106-114*

Alphabetically lists 63 manufacturers of educational software for the home. Includes manufacturer's name, address, and telephone number.

Plan and Print

700. COMPUTER SYSTEM RAISES PRODUCTIVITY. *International Reprographic Association, Inc. v.n54, no.11. November 1981. pg.51-52, 55*

The impact of computerized design and engineering systems on productivity is discussed. Applications for computer-aided design and record keeping functions are discussed, with examples and descriptions of specific computer products in use.

Plastics Design Forum

701. CAD / CAM: TURNKEY SYSTEM ALTERNATIVES. *Collins, Scott. Predicasts, Inc. Industry Media Inc. v.7, no.4. July 1982. pg.47-55*

Analysis of CAD / CAM technology and the industry gives manufacturers of systems, product lines, and prices. Also shows installation and utilization by design employees, 1977, 1981, 1985, 1990, 1995.

Radio - Electronics

702. YOUR OWN COMPUTER: SOFTWARE. *Friedman, Herb. Gernsback Publications, Inc. Semi-annual. v.53, no.10. October 1982. pg.113-134*

Presents 1982 review of software and services for personal computers. Lists U.S. software vendors and software networks. Discusses services and prices. Provides a listing of modem manufacturers.

RNM Images

703. RNM INFORMATION MATRIX. *W.G. Holdsworth and Associates. v.12, no.2. April 1982. pg.48-58*

Presents brief summaries of new developments in radiology and nuclear medicine. Summaries are coordinated numerically with a reader response card. Areas covered include computed tomography, diagnostic radiology, diagnostic ultrasound, nuclear medicine administrative business, multiple modalities, and therapeutic radiology.

Today's Office

704. DATA BASE SECURITY SYSTEMS: PROTECTING THE SOURCE. *Leveille, Greg R. Hearst Business Communications, UTP Division. v.17, no.5. October 1982. pg.61-65, 104*

Analysis of the market for database security systems. Shows system installations and percent of database management systems with security functions, 1982, 1985.

VideoPrint

705. INVESTING IN THE VIDEOTEX INDUSTRY. *International Resource Development, Inc. Bi-monthly. v.3, no.6. March 22, 1982. pg.1-3 $155.00/yr.*

Lists thirty-six companies in the videotex industry. Listings cover price to earnings ratios, one and five year earnings growth, return on equity, total revenues, and profit margins.

Word Processing Systems

706. SURVEY EMPHASIZES THE IMPORTANCE OF WP SUPERVISORS. *American Management Association, Executive Compensation Service. Geyer-McAllister Publications, Inc. Irregular. v.9, no.7. July 1982. pg.36-38*

Reports on 1981-1982 salaries of word processing supervisors in U.S. and Canada based on a survey by the Executive Compensation Service of the American Management Association. U.S. figures are given by region. All salary figures are detailed by job description, with average salary rates, weighted aver-

age, and average range given, as well as number of employees in the position and average number of employees supervised.

See 334, 1486, 1488, 1490, 1494, 1497, 1498, 1504, 1507, 1512, 1522, 1526, 1528, 1536, 1541, 1544, 1556, 1560, 1563, 1568, 1572, 1576, 1579, 1581, 1591, 1595, 1598, 1600, 1630, 1648, 1659, 1676, 1677, 1695, 1700, 1701, 1705, 1707, 1719, 1720, 1736, 1741, 1743, 1745, 1746, 1751, 1757, 1765, 1774, 1814, 1817, 1818, 1821, 2761, 2780, 2781.

Monographs and Handbooks

707. ADASPO SALES AND USE TAX SURVEY. *Association of Data Processing Services Organization, Inc. 1982. $150.00.*

Loose-leaf survey summarizes each state's sales and use tax laws as they relate to the computer services industry, with citations to relevant laws, regulations, and court cases. Quarterly updates available.

708. COST AND BENEFITS OF DATABASE MANAGEMENT. *Draper, Jesse M. U.S. Government Printing Office. 1981. 101 pg. $5.00.*

Discusses benefits of database management systems: centralized data management, data independence, high-level query facilities, and high-level user languages.

709. GROWTH OF MEDICAL INFORMATION SYSTEMS IN THE UNITED STATES. *Lindberg, Donald A.B. Lexington Books. Division of D.C. Heath and Company. 1979. 208 pg. $22.95.*

Discusses MIS innovations and applications and the effects of changing public policy. Includes definitions and means of comparing MIS's; state of the art and description of MIS's; barriers to development and diffusion of MIS technology; effect of changes in technology on future MIS's; and effect of public policy on developments, adoptions, and diffusion of MIS technology.

710. ISSUES IN CANADIAN / U.S. TRANSBORDER COMPUTER DATA FLOWS. *Cundiff, W.E., ed. Institute for Research on Public Policy. 1979. 89 pg. 6.50 (Canada).*

Discusses factors to be considered in developing transborder data flow policies for the computer communications industry.

711. LIBRARY NETWORKS, 1981-82. *Martin, Susan K. Knowledge Industry Publications. Triennial. 4th ed. 1981. 160 pg. $29.50.*

Volume covers developments in computer-based library networks since previous edition in 1978. Discussion includes competition vs. cooperation among networks; patron access; SOLINET's intention to establish an online facility on a regional basis; and integration of National Periodicals System in networking; and network utilities. Networks listings include membership data, status, and plans.

712. MAPPING THE INFORMATION BUSINESS. *McLaughlin, John F. Harvard University. Information Resources Policy. Order Department. P-80-5. December 1979. 22 pg. $240.00.*

Examines the evolving structure of the information business. Map presents 80 products and services with primary functions ranging from conduit to content. Discusses regulation.

713. SOFTWARE TOOLS DIRECTORY. *Reifer, Don. Reifer Consultants Inc. Irregular. 1981. 400 (est.) pg. $195.00.*

Indexed catalog of specific software tools for computer suppliers of all sorts includes public domain software available free from the U.S. Government. Listings are divided into major software categories: development tools, maintenance tools (including programming language translators), management tools, and tool systems. Indexes provide cross-referencing by tool function, supplier, and system compatibility.

714. SUMMARY OF ONTARIO DP SHOPS. *Ontario. Ministry of Labour. Ontario Manpower CommissioLabour Market Research Group. 1982.*

Examines employment in the Canadian data processing industry from 1981-1985, noting types of businesses with hiring difficulties in DP. Information is collected from 430 businesses.

715. VANLOVE'S BUSINESSMAN'S HANDBOOK AND SOFTWARE DIRECTORY. *Harbor Publishing. Irregular. 1982.*

Directory describes and rates several hundred programs, indexing software by title, subject, and publisher. Accompanying publisher contact list.

See 1831, 1832, 1838, 1840, 1842, 1843, 1845, 1846, 1852, 1853, 1855, 2802, 4245, 4252, 4259, 4262.

Conference Reports

716. NATIONAL CONFERENCE ON INFORMATION SYSTEMS. *Society for Management Information Systems. Irregular. December 1981. 417 pg.*

Twenty-nine papers presented at the second annual conference on information systems cover cognitive processes and information systems design, eliciting information requirements, developing DSS, developing and implementing information systems (I.S.)., measuring in I.S., organizational variables affecting I.S., planning for I.S., and the changing I.S. managerial job.

717. PROCEEDINGS OF THE NBS / IEEE / ACM SOFTWARE TOOL FAIR. *U.S. Government Printing Office. Irregular. 1981. $6.50.*

Summarizes proceedings of the San Diego Tool Fair. Covers current and future software engineering tools. Includes requirements, design, and modeling tools; cost estimation and project management tools; program analysis and testing tools; configuration management systems; formal verification systems; program construction and generation systems; and programming environments.

See 1860, 1865.

Newsletters

718. THE COMPUTER GRAPHICS SOFTWARE NEWS. *Computer Graphics Software News. Semi-monthly. 0000. 12 pg. $50.00/yr.*

Newsletter reviews computer graphics software, and new developments in computer graphics. Describes new software packages and services, with technical specifications. Includes a listing of pertinent conventions and useful hints on software implementation.

719. INFORMATION AND DATA BASE PUBLISHING REPORT. *Knowledge Industry Publications. Bi-monthly. $225.00/yr.*

Informational data publishing analysis covers publishing, marketing, and use of data base information; marketing and investment successes and failures; and user resistance to receiving information from terminals.

720. ONLINE DATABASE REPORT. *Link Resources Corporation. Monthly. 8 pg. $120.00.*

Online database industry analysis and news for executives.

721. PACKAGED SOFTWARE REPORTS. *Management Information Corporation. Monthly. $295.00/yr.*

Evaluates business application packages (i.e., accounting, manufacturing, retail packages) that run on micro- and minicomputers. Covers each program module, interfacing, operation, prompts and menus, data maintained, prices, advantages and disadvantages, and other topics.

See 1867, 1868, 1869, 1870, 1873, 1874, 3365.

Services

722. COMPANY ANALYSIS AND MONITORING PROGRAM (CAMP). *INPUT. Monthly. 1982.*

A subscription program providing information on computer services and turnkey system companies. Monthly reports contain background information on the firm, financial data, acquisition activity, organizational structure, employee data, competitors, revenue, key products / services, industry market data, geographic markets, and location installed hardware and data centers. Includes access to database.

723. DATAPRO DIRECTORY OF SOFTWARE. *Datapro Research, Inc. (McGraw-Hill). Monthly. $420.00/yr.*

Providing product histories, user ratings, time-shared availability, and hardware / system requirements, annual subscription service includes two loose-leaf volumes, monthly report supplements, monthly newsletter, and a telephone / telex inquiry service.

See 1878.

Indexes and Abstracts

See 1888, 1889, 1890, 1896.

Dictionaries

724. GLOSSARY OF DATA PROCESSING TERMS. *Steelcase, Inc. 1981. 12 pg.*

Contains 450 data processing terms intended for the non-data processing executive.

See 1897, 1900.

Western Europe

Market Research Reports

725. BATTELLE AUTOMATED BANKING IN EUROPE (BABE). *Battelle Memorial Institute. Monthly. $100.00.55.00 (United Kingdom).*

Monthly newsletter on EFT products and services in Europe covers credit cards, automatic tellers, home banking, and automated bank clearings. Analyzes markets, technology; reviews costs and profits; and gives bibliographic information.

726. COMPETING AND COMPLEMENTARY SYSTEMS. *Butler Cox and Partners Ltd. March 1981.*

Market assessment of videotext services in the context of the current market growth of cable and pay - TV; interactive TV; satellite broadcasting; video recorders; TV games; home computers; etc.

727. DISTRIBUTION CHANNELS FOR SMALL COMPUTERS 1981-1987. *IDC Europa Ltd. 1982. 150 (est.) pg. 1250.00 (United Kingdom).*

Report on distribution channels for small computers in Western Europe provides market analysis by assessing numbers of systems sold through direct sales, systems houses, retail stores and other methods. Examines ways to distribute systems depending upon size, class and target markets. Analyzes major third parties marketing systems with an emphasis on growth factors.

728. DP OPPORTUNITIES IN THE MANUFACTURING INDUSTRY. *Jordan and Sons (Surveys) Ltd. 1982. 130 pg.*

Market research report provides forecasts and analysis of data processing opportunities in the manufacturing industry. Gives size, share, projection by country and analysis by product category of the data processor market; analysis of user consideration of cost / benefits; work force acceptance and applications; end user profiles, including size and nature of businesses; job titles of decision makers; and examination of current products and how they meet the needs of users.

729. EUROPEAN DATA BASE MARKET. *Frost and Sullivan, Inc. April 1981. $1250.00.*

Gives analysis and forecast by principal European country. Documents available data bases. Gives profiles of on-line, batch, and data base publishing companies. Examines influence of foreign suppliers on the European market. Provides end-user analysis.

730. INDEPENDENT PACKAGE SOFTWARE MARKETS. *IDC Europa Ltd. October 1980. 130 pg. 995.00 (United Kingdom).*

Market researach report provides forecasts and analysis of independent package software markets in Western Europe. Considers current and future market size, as well as the strategies of over 20 major software suppliers.

731. INTERNATIONAL MARKET OPPORTUNITIES FOR ON-LINE DATA BASE SERVICES. *INPUT. 1981.*

Analysis and forecasts of markets for online database services in Western Europe, 1980-1985. Examines government funding, vendors, foreign investments in the European market, industry structure, and market segmentation changes, 1980, 1985.

732. ON-LINE DATABASE SERVICES MARKET, 1981-1987. *IDC Europa Ltd. 1982. 150 (est.) pg. 1250.00 (United Kingdom).*

Report on the on-line database services market in Western Europe, 1981-1987, provides market size and forecast by country. Identifies and analyzes major information providers, including new market entrants. Assesses future direction of industry and the effect of limits to transborder data.

733. PROFESSIONAL SERVICES MARKET, 1981-1987. *IDC Europa Ltd. 1982. 150 (est.) pg. 1250.00 (United Kingdom).*

Report on professional services market in Western Europe, 1981-1987 analyzes independent software, consultancy and facilities management, market size and growth by country, and the attitudes of hardware vendors to in-house and third party consultancy. Indicates market opportunities. Assesses activities of leading professional service suppliers. Examines attri-

butes and intentions of users and their willingness to subscribe to independent services.

734. REMOTE COMPUTER SERVICES MARKET, 1981-1985. *IDC Europa Ltd. October 1981. 130 pg. 1250.00 (United Kingdom).*

Market research report provides forecasts and analysis of the Western European remote computer services market, 1981-1985. Gives size, share, forecast by country, and segmentation by industry and application of the remote computer services market; a profile of all major suppliers and an examination of the influences of PTT monopolies; an assessment of impact of falling hardware prices and emerging viewdata services; and a user survey analysis featuring the roles and attitudes of D.P. departments to current and projected remote computer service usage.

735. SOFTWARE AND SERVICES MARKET REFERENCE BOOK, 1981-1987. *IDC Europa Ltd. 1982. 150 (est.) pg. 1250.00 (United Kingdom).*

Provides overview of state of the market of software and computer services in Western Europe, 1981-1987, with a detailed look at processing services, packaged software, custom software, and facilities management. Forecasts market growth by country. Identifies major participants in the market and assesses strengths and weaknesses of leading contenders. Analyzes end user survey focusing on roles and attitudes of D.P. departments to current and projected usage of third party software and services.

736. TEXT PROCESSING SOFTWARE IN EUROPE. *Frost and Sullivan, Inc. January 1982. 239 pg. $1450.00.*

Market report provides 10-year forecast by country for text processing software applications. Examines dedicated office WP; office documentation packages; ancillary WP packages for professional applications; commercial in-plant photocomposition text production and directory production; and newspaper editorial and advertisement packages. Gives software packages information and associated technology trends. Includes vendor profiles, market share data, and information on distribution channels.

See 1905, 1909, 1910, 1925, 1936, 1942, 1961, 1963, 1976, 2837, 2838.

Industry Statistical Reports

737. BRITISH COMPUTER SERVICES INDUSTRY. *Jordan and Sons (Surveys) Ltd. Irregular. 1981. 120 pg. 85.00 (United Kingdom).*

Survey of 189 computer service businesses operating in the United Kingdom provides fixed assets, current assets, sales, pre-tax profits of businesses, addresses of businesses, stockholders, executives, and ratios for the top 50 businesses under the following headings: turnover, profitability, and sales / net capital employed.

738. COMPUTER SERVICES. *Jordan and Sons (Surveys) Ltd. Annual. 1982. $85.00 (ukd).*

One in a series of 29 annual reports on specific British industries. Ratio section compares performance of top 50 companies using six ratios (turnover, pre-tax profit / sales, pre-int. profit / tangible capital employment, sales / net capital employment, stocks / sales, current assets / current liabilities). Section on top 50 firms (ranked by turnover) includes latest four years of filed accounts; directors and other directorships; major shareholders and registration dates; and address and trading objects. 'Shorter Records' section ranks all firms by sales turnover, with three years of data covering exports, pre-tax profits, employees in the U.K., wages and salaries, current assets, current liabilities, bank overdraft and short-term loans, and net fixed assets.

Financial and Economic Studies

See 1999.

Forecasts

739. SOFTWARE EXPENDITURE ANALYSIS. *IDC Europa Ltd. February 1982. 130 pg. 1250.00 (United Kingdom).*

Report analyzes software expenditure in Western Europe, 1981-1985. Provides examination of expenditure on systems, utility and application software by country; factors influencing decision on third party, hardware vendor or internally produced software; analysis of software spending by industry and consideration of most common applications; and factors which influence spending levels for different processor grops.

740. THE U.K. COMPUTER SERVICES INDUSTRY 1980. *INPUT. 1980. $3000.00.*

Projects the market for computer services in the United Kingdom (1980-1984), and identifies leading processing service companies with market shares. Discusses growth rates by market segment, including data on the software products market.

See 2013, 2018.

Directories and Yearbooks

741. BASES ET BANQUES LE DONNES ACCESSIBLES EN CONVERSATIONNEL EN FRANCE. [Databases Accessible in France.]. *Association Nationale de la Recherche Technique. French. Irregular. December 1981. $53.00.*

A directory of 300 online databases and databanks available in France. Data, on file cards, includes origin, fields covered, nature of information, data recorded, and corresponding publications. Gives address of producers and host systems.

See 2023, 3414.

Special Issues and Journal Articles

British Business

742. COMPUTER SERVICES - ANOTHER PERIOD OF SLOWER GROWTH. *Department of Industry and Trade. Her Majesty's Stationery Office. June 26, 1981. pg.397*

Statistical analysis of the growth rate of the United Kingdom's computer services industry from 1971 to 1980. Gives annual figures for value of services performed under the following headings: computer processing, professional services, data preparation, other billings, and unclassified. Shows the value of services according to several general market areas and provides the number of employees in consultancy, programming / analysis, computer operating, data control, data preparation, administration, selling, and other related fields.

743. COMPUTER SERVICES INDUSTRY'S SLOW GROWTH CONTINUED DURING 1981. *Her Majesty's Stationery Office. August 27, 1982. pg.740-741*

Discusses U.K. computer services industries. Statistics include 1971-1981 billing to clients for work done; 1971-1981 billings by clientele; and 1971-1981 personnel employed. Discusses state of industry.

Chip, Revista de Informatica

744. BANCOS DE DATOS EN EMBRION. [Databases in Embryo.]. *Castro, Maria Antonio. Gesellschaft fur Chemiewirtschaft. Spanish. Irregular. no.11. February 1982. pg.53-57*

Study of Spanish database industry includes statistics of present level of activity and forecasts planned database projects. Charts and diagrams describe functional aspects of information networks in operation in Spain. Lists names and locations of Spanish database facilities with public access. Gives forecasts of additional equipment and improvement of database facilities.

745. EN BUSCA DEL TIEMPO PERDIDO: EL DESAROLLO DE LA INFOMATICA EN ESPANA. [In Search of Time Lost.]. *Segura, Casimiro Alonoso. Ediciones Arcadia, S.A. Spanish. no.16. August 1982. pg.25-26*

Analyzes computer time sharing in Spain. Tables, by Mapter Europe, provide costs statistics of various types of services and installations in Spain, 1980. Costs are given in dollars and in Spanish Pesetas, with percentage of total costs per type of services.

746. ESTIMULO OFICIAL PARA LA INDUSTRIA DE LA INFORMACION. [Official Stimulation for EDP Industry.]. *Ediciones Arcadia, S.A. Spanish. no.16. August 1982. pg.15*

Details program approved by Spanish Cabinet of Ministers to stimulate the EDP industry and the creation, distribution and utilization of databases. Distribution of budgets for 1982 through 1985 for the various facets of the program are listed, including allocation for international actual database industry and statistics of facilities are given.

Economisch Dagblad

747. DE DREIGENDE VERSUKKELING VAN EEN SPEERPUNT - INDUSTRIE. [The Danger of an Ailing Spearhead Industry.]. *Sandee, Berhnard. Sijthoff Pers B.V. Netherlandish. v.42, no.10651. April 22, 1982. pg.2*

Examines Dutch software companies and their products, productivity, growth and place in the European export market. Shows capital investments, 1982-1985; government aid; forecasts; employees, 1981-1983; and volume of sales, 1979, 1981.

L'Expansion

748. SPECIAL INFORMATIQUE 1982 - LES ENTREPRISES JUGENT L'INFORMATIQUE: UNE ENQUETE EXCLUSIVE DE 482 UTILISATEURS. [1982 Data Processing Special - Enterprises Judge Data Processing.]. *Fontaine, Jacques. International Werbegesellschaft m.b.H. French. Annual. no. 198. September 24, 1982. pg.175-189*

Compiles responses from a survey of 482 French businesses of all sizes, on their information-processing needs, attitudes and practices. Covers average number of units employed, applications, staff and management allocation to information processing, supplier loyalty, and reasons for satisfaction or dissatisfaction with suppliers.

Mini-Micro Software

749. MINI-MICRO SOFTWARE: APPLICATION NEWS AND PACKAGES. *A.P. Publications Ltd. Quarterly. v.7, no.1. 1982. pg.21-26*

Gives new software products and applications for mini and micro systems. Includes a brief description of the software package and its application, and vendor's name and address.

Le Nouvel Economiste

750. BUREAUTIQUE: LES CADRES FONT LEURS GAMMES. [Computer Services: Executives Make Their Ratings.]. *Barraux, Jacques. Enterprise les Informations. French. no.355. September 27, 1982. pg.80-84*

Rates the competency in nine service areas of 12 companies (U.S. and French) in the computerized information industry. Source is M. Jean-Paul de Balcet of l'Afcet.

Output Osterreich

751. SOFTWARE PRODUZENTEN IN OSTERRICH. [Software Producers in Austria.]. *Bohmann Druck und Verlag AG. German. Irregular. April 1982. pg.19-25*

Provides an alphabetical list of more than 240 Austrian firms specialized in software services, with full address and specializations, as well as computer system preferences.

Trade and Industry

752. RAPID GROWTH IN COMPUTER SERVICES. *Department of Industry, Trade, Prices and Consumer Protection. Her Majesty's Stationery Office. September 14, 1979. pg.544-545*

Statistical analysis of the growth rate of the United Kingdom computer services industry, 1971-1978. Gives annual figures for value of services performed under the headings computer processing, professional services, data preparation, other billings, and unclassified and value of services according to market areas. Shows number of those employed in consultancy, programming / analysis, computer operating, data control, data preparation, administration, selling, and other related fields.

See 2033, 2035, 2050, 4323.

Monographs and Handbooks

See 2065.

Newsletters

See 2068, 2069.

Dictionaries

See 2070.

Asia

Market Research Reports

See 2082.

Financial and Economic Studies

753. COMPUTER WHITE PAPER 1981. *Fuji Corporation. Annual. 1981. $45.00.*

Annual report covers state of computer utilization in Japan, Japan's computer policy, government utilization of computers, trends in Japan's computer and information processing industries, state of data communication utilization and policies in Japan, computer systems, and research and development.

Special Issues and Journal Articles

Business India

754. SOFTWARE EXPORTS: THE BILLION DOLLAR HOPE. *Alltech Publishing Company. July 6, 1981. pg.66-73*

Reviews software developments in India, discusses the activities of Indian companies developing software products, and assesses their export potential. Data includes annual revenues and exports of software by 9 Indian firms (1980). Lists major world companies by 1979 data processing revenues and total revenues.

See 2089, 2094.

Monographs and Handbooks

755. DATA PROCESSING IN JAPAN. *Welke, H.J., ed. North-Holland Publishing Company. 1982. 200 pg. $46.50.100.00 (Netherlands).*

Report describes the policy of the Japanese government concerning information processing and the information industry; the situation of the computer and software industry; and the merits and disadvantages of Japanese software products and software technology.

Newsletters

See 2104.

Oceania

Directories and Yearbooks

See 2106, 2107, 3436.

Computers

International

Market Research Reports

756. ACCOUNTING SYSTEMS. *Larsen Sweeney Associates Ltd. Annual. 1981. 795.00 (United Kingdom).*

Market analysis for computer accounting systems, with forecasts to 1986. Covers 17 product groups. Includes analysis of marketing and the industry by country and surveys of end users and suppliers.

757. COMMUNICATIONS / INFORMATION SYSTEMS PLANNING SERVICE: APPLICATIONS FOR AUTOMATED INTELLIGENCE - BEYOND ARTIFICIAL INTELLIGENCE. *Yankee Group. November 1981. 200 pg. $9500.00/yr.*

Report focuses on use of drone processors and automated operators for machine and network control; speech, sight, and touch sensing systems; robots for assembly manufacturing; and automated program development. Includes development of friendlier man-machine interfaces through system intelligence. Vendors survey determines what kinds of systems and applications are currently being developed in the laboratories and tested in the field. Users tell what kinds of applications for automated intelligence they foresee, and how much they will pay for these kinds of devices and system tools.

758. COMMUNICATIONS / INFORMATION SYSTEMS PLANNING SERVICE: THE SUPER SERVICE BUREAU - THE CHANGING FOCUS OF DATA SERVICES. *Yankee Group. December 1981. 200 pg. $9500.00/yr.*

Report analyzes evolving market and opportunities for shared service offerings. User survey identifies users' perceived needs and their willingness to purchase and / or to share solutions developed by others. Discusses vendor strategies for integrating hardware / software with shared service offerings, particularly in communications services areas.

759. COMPUTER AND ELECTRONICS MARKETING. *Morgan-Grampian, Inc. Irregular. $25.00.*

Covers the marketplace for microelectronic circuits, printed circuits, hybrid circuits, semiconductor circuits, test equipment and all types of computer peripherals.

760. COMPUTER GRAPHICS. *Predicasts, Inc. August 1979. $325.00.*

Analysis of worldwide markets, manufacturers, and new products and technologies in computer graphics.

761. COMPUTER GRAPHICS: THE BUSINESS APPLICATIONS MARKET. *International Data Group. 1981.*

Analysis of world markets for computer graphics systems, with sales forecasts, 1979-1984. Examines market segmentation and product mix.

762. COMPUTER MANUFACTURERS: 1979. *International Data Group. December 1979. 105 pg. $2500.00.*

Sixth (1979) commercial analysis report analyzes activities of the 6 major U.S.-based computer manufacturers (1974-1978), including revenue data, U.S. and international markets, and rent / lease / purchase distribution, by class size. Data also

provided on smaller U.S. vendors of general-purpose computers.

763. COMPUTER MEMORIES. *Predicasts, Inc. March 1980. $325.00.*

International market analysis for computer memories, including main, add-on, peripheral, semiconductor, optical, bubble and laser. Examines new products and technology, utilization and manufacturers.

764. CRITICAL PATH PLANNING SERVICE. *Yankee Group. 1981. $8500.00/yr.*

Includes monthly industry reports; bimonthly series of topical analyses; at least four seminars per year; an in-house seminar / consulting program; and a call-in / call-out service. 1981 monthly industry reports cover User Experiences with IBM Networking, Local Area Communications Networks: Technologies and Vendors, Developing Plans for Implementing Local Networks and Electronic Messaging Subsystems, and Office Automation and Distributed Data Processing. Includes Developing an Integrated Communications Strategy, The New Generation of PBX's, Public vs. Private Network Alternatives, Network Planning and Implementation, Computer Graphics, Applications for Automated Intelligence Beyond Artificial Intelligence, and The Super Service Bureau - The Changing Focus of Data Services.

765. DATA ENTRY DEVICES. *Predicasts, Inc. May 1981. $375.00.*

Worldwide market analysis for data entry devices including cards, magnetic media, OCR and mark recognition; voice recognition and hand-held terminals. Covers new product developments and vendors' activities.

766. DDP VIA GENERAL PURPOSE SYSTEMS. *Creative Strategies International. 1981. $1200.00.*

An analysis of distributed data processing (DDP) via general purpose systems, including minicomputers, intelligent terminals and small business computers. Covers products, end uses, industry and distribution structure, labor supply, and market analysis, 1980-1985. Examines prices, shipments, 1980-1985, market shares and revenues, 1980, manufacturers and competition.

767. ELECTRONIC BANKING. *Predicasts, Inc. March 1982. $375.00.*

Analysis of world markets for automated banking equipment and services including electronic funds transfer and networks. Includes technology developments.

768. GLOBAL ANALYSIS OF COMPUTER COMMUNICATIONS: CHALLENGES AND OPPORTUNITIES. *Strategic Inc. March 1981. 200 pg.*

Examines videotext products and markets, foreign government - sponsored programs, international legislative and regulatory trends, and the evolution and growth of public data networks. Also examines 'Telematique,' the French program of government-directed and subsidized development of a basic telecommunications infrastructure and numerous advanced related products. Reviews legislation and regulation worldwide and analyzes security and privacy issues.

769. HOME OF THE FUTURE PLANNING SERVICE: THE COMMAND AND CONTROL SYSTEM - SECURITY, ENERGY MANAGEMENT, AND HOME COMPUTING. *Yankee Group. October 1981. 200 pg. $8500.00/yr.*

One of six reports forming a component of " "Home of the Future Planning Service.' Report covers potential for command and control systems, how they will be configured, what they will cost, who will make them and service them, and how they will interact with in-home appliances and information / entertainment utilities. Major issues include the changing demographics of the homemaker; pilot projects in the control field; TOCOM and beyond; security systems: market size and product strategy; energy management: savings to individuals and U.S. economy; utility meter reading via cable or telephone: options; food preparation, maintenance, cleaning and robots; product opportunities in home management areas; and a total appliance management system configuration.

770. HOME OF THE FUTURE PLANNING SERVICE: THE INFORMATION SUPERMARKET - INTERACTIVE INFORMATION AND EDUCATION SERVICES. *Yankee Group. 1981. 200 pg. $8500.00/yr.*

One of a series of bimonthly reports forming part of the " "Home of the Future Planning Service.' Coverage spans costs of providing information, what kinds of information are suited for electronic delivery, what information consumers will pay for, and what information they will use if provided free. Sections include electronic products and perishable information; potential market for viewdata products; strategic choices for newspapers and periodicals; electronic mail and voice-mail; advertising on the new electronic media; computerized classified advertising; demographics of home information utility and cable audience; home education: the interactive univeristy; and what teletext products will be viable.

771. IBM CPU (CENTRAL PROCESSING UNIT) MIGRATION ANALYSIS. *Creative Strategies International. May 1980. $995.00.*

Analyzes results of user migration from IBM 360 and 2nd generation mainframes. Topics include: the impact of the 4300, the migration to non-IBM suppliers, and the anticipated marketing opportunities for IBM competitors. U.S. and worldwide forecasts covering 1979 through 1983 are included. A separate forecast for each model within the IBM 2nd generation and 360 product families is also provided.

772. IBM DATA BOOK. *International Data Group. December 1979. 153 pg. $3500.00.*

Compendium of information on IBM including corporate organization, management structure, operating units, installed base, distribution, markets, systems populations (1977-1981), strategies, and impact on markets.

773. INFORMATION RETRIEVAL SYSTEMS. *Predicasts, Inc. April 1980. $325.00.*

International market analysis for computer-based information services, including data bases; videotext equipment and services; and information storage, retrieval and transmission. Discusses regulations.

774. INFORMATION SYSTEMS ECONOMETRIC SERVICE. *Gnostic Concepts, Inc. Quarterly. 1979. $12000.00.*

This quarterly service analyzes the computer industry in 4 volumes (CPU, storage, input, output, terminals, punched media and controllers) in the U.S. and abroad. Computer equipment, systems, end use industries and U.S. economy are discussed. Gives market analyses and forecasts to 1983 of expenditures, production, costs, shipments and foreign trade (1979-1985).

775. THE INTELLIGENT MEMORY CARD: (EFTS VI). *Heinz, M. Battelle Memorial Institute. August 1981. $9000.00.*

Analyzes the uses for microprocessors in electronic payment systems internationally. Discusses technical, economic, and patent and licensing considerations of the intelligent memory card, and its applications as an identification device in commercial banking, retailing, libraries, transportation, and other sectors.

776. INTERNATIONAL COMPUTER MARKETING: A MULTI-CLIENT STUDY. *Interactive Data Corporation. February 1979. 800 (est.) pg.*

A mulit-client study listing international computer firms alphabetically by country. Revenues, addresses, market shares, rankings, trading, subsidiaries, mergers, facilities and products of suppliers and manufacturers given for 1978.

777. MARKETS FOR LOW PRICE COMPUTERS. *Business Communications Company. 1981. 115 pg. $750.00.*

Studies manufacturers and markets with a look at distribution and presents projections into the 1990's. Evaluates opportunities, requirements and problems. International data, major corporations and favored equipment included.

778. MICROCOMPUTER NEWS INTERNATIONAL. *Mackintosh Publications Ltd. Monthly. $215.00.*

International coverage of market trends, government funding, product applications, market opportunities, new technology, company strategy and developments in the U.S. and European countries.

779. MICROPROCESSORS (INCLUDES MPV'S). *Larsen Sweeney Associates Ltd. Annual. 1981. 795.00 (United Kingdom).*

Analysis of markets for microprocessors including MPV's, with forecasts to 1986. Includes analysis of marketing and the industry by country and surveys of end users and suppliers. Covers 9 product groups.

780. THE OUTLOOK FOR THE MINICOMPUTER INDUSTRY IN THE 1980'S: 32 - BIT SUPER MINIS COME OF AGE. *Martin Simpson and Company, Inc. Technology Note #52. September 8, 1980. 127 pg. $600.00.*

Analysis of the minicomputer industry worldwide, with emphasis on 32-bit 'superminis.' Market analysis and forecasts to 1985 by end uses, utilizations, size of installations and domestic or foreign markets. Also includes vendors, product lines, competition, user attitudes and financial analysis.

781. PERSONAL COMPUTERS. *Predicasts, Inc. 1980. $325.00.*

Worldwide market analysis for personal computers. Reviews new products, regulations and utilization in homes and businesses.

782. PORTABLE TERMINALS. *Creative Strategies International. January 1981. 105 pg. $1195.00.*

Examines current worldwide installed base of portable terminals by U.S. vendors, worldwide shipments by U.S. vendors, and worldwide revenues. Discusses handheld and briefcase portable terminals in terms of market segments. Includes technological changes and effect of competition.

783. VENTURECASTS: THE 1979 COMPUTER / COMMUNICATIONS DATA BOOK. *O'Connor, Mary Jane. Venture Development Corporation. Annual. July 1979. 337 pg. $1950.00.*

Annual compendium of quantitative market facts and forecasts based on published 1978-1979 U.S. and worldwide data. Information sources include electronics and business serials, government and trade association publications, reports and publisher's data. Data arranged by SIC code; additional information includes time period and specific source. Concise format, fully indexed. Special services available from the publisher.

784. VENTURESEARCH. *Venture Development Corporation. 1981.*

Information on demand service for electronics, communications and computers. Analysis of markets and industries. Covers U.S. and foreign vendors and manufacturers.

785. VOICE INPUT / OUTPUT. *Predicasts, Inc. April 1982. $375.00.*

Analysis of markets for voice analysis and synthesis equipment for end uses in industry, banking, retail, offices, consumer products and education. Examines technology and vendors.

See 7, 10, 546, 2119, 2122, 3137, 3138, 3446, 3954, 3955, 3960, 3964.

Industry Statistical Reports

786. IBM SYSTEMS REFERENCE MANUAL. *Creative Strategies International. February 1981. $595.00.*

Reference manual for U.S. and worldwide IBM systems. Shows total units in operation and prices / performance categories. For individual products, shows price, rental rates, performance ratings, installed base, 1969-1982, and shipments and retirements, 1969-1982.

See 3452, 3968.

Financial and Economic Studies

787. COMPUTER AGE: WORLD TRADE REPORT. *Casolaro, Daniel, editor. EDP News Services. Bi-monthly. April 11, 1980. $75.00/yr.*

National and international news about computers and electronic data processing. Forecasts expansion of electronic firms' overseas manufacturing facilities in 1980's. Reports on government regulations and proposals. Discusses international trade including trade shows and conferences.

788. IBM: COLOSSUS IN TRANSITION. *Sobel, Robert. Times Books. Division of the New York Times Company. October 1981. $17.95.*

Presents both a history of IBM and a history of the business machines industry which that company dominates. Discusses the success of IBM's policy of leasing rather than selling equipment, thus ensuring a steady flow of funds and long-term contact with its customers. Reviews IBM's responses to the market and legal challenges it encountered in the 1960's and 1970's; its domestic and world sales, product research and development, and entry into minicomputer production and satellite communications.

789. INVESTMENT IN COMPUTER SERVICES. *Greene and Company. Annual. August 1981. 65 pg. 75.00 (United Kingdom).*

Study of investment in computer services gives a profile of the industry, a statistical intra-firm comparison of 50 of the leading international companies, and a rundown on the quoted sector.

790. MULTINATIONAL COMPUTER NETS: THE CASE OF INTERNATIONAL BANKING. *Veith, Richard H. Lexington Books. Division of D.C. Heath and Company. March 1981.*

Analysis of interorganizational dynamics of transborder data flow covers interorganizational relations; transborder data flow debates; computers, networks and banks; survey of the largest banks; dimensions of interbank relations; and organization of interrelations.

See 2145, 2146, 3139, 3974.

Forecasts

791. CAD/CAM INTERNATIONAL DELPHI FORECAST. *Society of Manufacturing Engineers. 1980. 181 pg. $29.00.*

Volume explores opinions of U.S., U.K., and Japanese engineers on CAD/CAM applications in manufacturing technology coverage spans computer graphics in tooling design; fixture design; product design; computer-assisted process planning; NC, DNC, and CNC; automatic machine tool leading; industrial robots; on-line inspection and other functions. Bar charts at end of each chapter compare findings of the panels from the three nations.

792. MARKETS FOR DESK-TOP COMPUTERS. *Business Communications Company. January 1981. 105 (est.) pg. $750.00.*

Forecasts 1980-1989 markets for minicomputers. Discusses segmentation (by industry group, occupational category, and revenue class), initial and used equipment sales, prices, and true costs (including programming and software). Surveys major manufacturers and their products, retailers by store type and location, and revenues accruing to each.

793. NON-IMPACT MARKET AND TECHNOLOGIES FOR DISTRIBUTED PRINTING. *Strategic Inc. December 1980. 120 pg. $950.00.*

Advantages and disadvantages of the following seven principal technologies, markets and applications are examined: electrophotographic laser printers, thermal printers, ink jet printers, dielectric printers, electrosensitive printers, electrolytic printers and magnetic printers. The needs of end users in large data processing, word processing, graphics, small business, home, and distributed data processing environments are analyzed.

794. THE WORLD COMPUTER INDUSTRY: 1979-1984. *Withington, Frederic C. Arthur D. Little, Inc. March 1980. 35 pg.*

Overview of computer industry. Forecasts U.S. and international shipments. Discusses the positions of the major U.S.-based manufacturers and forecasts their individual shares of future markets. Several tables and charts of statistical data. Discusses IBM, Digital Equipment, NCR, Control Data, Burroughs, and Sperry Univac and plug compatible, mainframe vendors separately.

795. 1979 DISK / TREND REPORT: FLEXIBLE DISK DRIVES. *Porter, James N. James N. Porter. Annual. September 1979. 80 (est.) pg. $385.00.*

Detailed annual review of flexible disk drives. Disk drive specifications arranged alphabetically by manufacturer. Comprehensive listing of flexible disk drives manufacturers includes addresses, telephone numbers, disk sales, total net sales, net income and brief descriptions of corporations.

796. 1979 DISK / TREND REPORT: RIGID DISK DRIVES. *Porter, James N. James N. Porter. Annual. July 1979. 180 (est.) pg. $550.00.*

A detailed annual business review of disk cartridge drives, disk pack drives, storage module drives, and fixed disk drives. Specifications arranged by manufacturers. Comprehensive listing of manufacturers of moving head disk drives includes addresses, telephone numbers, total net sales, net income and brief descriptions of corporations.

See 2149.

Directories and Yearbooks

797. COMPUTER TERMINALS REVIEW. *GML Corporation. Tri-annually. 1980. 750 (est.) pg. $175.00.*

1980 looseleaf directory / guide provides display type, screen type, communications, performance and general characteristics, option prices and marketing data on keyboard display and teleprinter terminals commercially available worldwide, with directory of manufacturers. New equipment updates and price changes are issued three times / year.

798. A DATAPRO FEATURE REPORT: DIRECTORY OF SUPPLIERS. *Datapro Research, Inc. (McGraw-Hill). Annual. July 1980. 202 pg. $25.00.*

Provides profiles of 1,000 companies that offer EDP products and services of all types. Contains guidelines for interpreting the information and selecting qualified suppliers. Entries describe company's location, size, executives, financial status, ownership, European sales and service organization. Includes product line, date founded and gross sales.

799. DIRECTORY OF NON-BIBLIOGRAPHIC DATABASE SERVICES: DRAFT EDITION: FEBRUARY 1979. *Cuadra Associates, Inc. Irregular. February 1979. 149 pg.*

Describes databases currently online in the U.S. and elsewhere in the world through one or more online service organizations. Arrangement is alphabetical by database name. Entries contain type of database, subject, name, producer, availability, content, coverage, and updating. Includes subscription prices when known. Provides list of names and addresses of most of the database producers and online service organizations.

800. DIRECTORY OF ONLINE DATABASES. *Cuadra Associates, Inc. Quarterly. March 1980. 130 pg.*

Coverage of online databases throughout the world spans descriptions of online databases, (name, type, producer, online service, content, coverage updating); addresses and telephone numbers of producers and online services; and indexes to databases by subject, producer, online service, and name.

801. EUSIDIC DATABASE GUIDE 1981. *Learned Information (Europe) Ltd. Irregular. 1981. $42.00.*

Listing of over 1,400 databases available worldwide in machine readable form. Identifies databases produced by a given organization or available through a given operator. Indexed by 49 subject categories with alphabetical listings of relevant databases and organizations.

802. THE FROST AND SULLIVAN COMPUTER GRAPHICS CAD AND CAD / CAM PRODUCT GUIDE AND SUPPLIERS DIRECTORY. *Frost and Sullivan, Inc. Irregular. 3573. January 1982. 800 (est.) pg. $220.00 (2 vols.).*

Covers over three hundred worldwide manufacturers and their representatives for all types of hardware, including input devices, terminals, display monitors, graphics computers, turnkey systems, interface devices, printers, and hard copy devices. Volume 2 is entirely devoted to software, listing the manufacturers and suppliers of software packages for applications in CAD, CAD / CAM, business graphics, and in scientific, medical and military areas. Provides product line information for the manufacturers listed. Gives names, addresses, and telephone numbers of manufacturers. Contains pricing information.

803. GML TELEPRINTER REVIEW SUPPLEMENT. *GML Corporation. Annual. 1980. 137 pg. $45.00.*

Directory of over 250 teleprinter models manufactured by 61 companies worldwide. Contains prices, specifications, features, software, marketing data and an overview of the industry, with graphs and charts.

804. INFORMATION TRADE DIRECTORY 1982. *Learned Information (Europe) Ltd. Annual. October 1981. 250 pg. $35.00.*

Provides contact, personnel and product information worldwide for more than 1,000 database publishers, online vendors, information brokers, telecommunication networks, library networks and consortia, terminal manufacturers, consultants, government agencies, and many other related firms and services. Includes updated lists of information courses, reference works, periodicals and newsletters, a geographical index, and names and numbers index. (Published as 'Information Industry Market Place 1982' in North and South America by R.R. Bowker.)

805. INTERACTIVE COMPUTING DIRECTORIES: VOLUME 1. *Association of Computer Users. Annual. May 1979. 910 pg.*

First volume in three-volume set. Applications directory, company directory, and geographic directory comprise this volume of the annual guide to the interactive computing industry in the U.S. and other nations.

806. INTERACTIVE COMPUTING DIRECTORIES: VOLUME 3. *Association of Computer Users. Annual. May 1979.*

Third book in three-volume annual directory to interactive computing industry (U.S., non-U.S.) covers financial modeling languages, interactive data-base systems, interactive statistical packages, interactive graphics programs, data bases available to users, management sciences programs, interactive accounting systems, engineering programs, specialized application programs, and a company literature section.

807. INTERNATIONAL DIRECTORY OF COMPUTER AND INFORMATION SYSTEM SERVICES. *International Publications Service. Biennial. 1980. 640 pg. $25.00.*

Biennial directory in alpha / geographical order by sector (brokers, consultants, universities, government groups, service bureaus). Coverage includes addresses, principle executives, telephone numbers, types of services available, computer types, storage, number of line printers, and remote processing. Indexed by institution and by computer installation.

808. NEW INFORMATION SYSTEMS AND SERVICES. *Gale Research Company. Irregular. 1981. $125.00.*

Supplement to "Encyclopedia of Information Systems and Services, 4th Edition" includes articles on new systems and services and on major changes affecting those listed in the parent volume.

809. ROBOT SYSTEMS AND PRODUCTS. *Robotics Industry Directory. Monthly. $35.00/yr.*

Monthly report and update newsletter to Robotics Industry Directory. Details new products, new applications, and other developments in industrial robot technology. Includes prices and specifications for new products with name, address and telephone number of contact representative.

810. WORLDWIDE COMPUTER INSTALLATION DATA FILE. *International Data Group. Semi-annual. 1980.*

The semi annual worldwide directory of computer installations is also available in a U.S.-only edition, as are subfiles by geographic region, manufacturer, model, industry (SIC). Summarizing management report included. Files furnished on magnetic tape and/or hard-copy printout.

811. 1980-1981 INTERNATIONAL DIRECTORY OF SOFT-WARE. *CUYB Publications Inc. Irregular. 1980. 1005 pg.*

1980 international directory of software describes 3223 independently-marketed software products under 38 categories of systems software and 69 categories of applications software. Provides an index of products by category, software product descriptions, profiles of suppliers, index of industry specific products and alphabetical index of products. Gives names, addresses, telephone numbers, key personnel, number of employees, product activities and products marketed.

812. 1982 ROBOTICS INDUSTRY DIRECTORY. *Robotics Industry Directory. Annual. 1981. $35.00.*

Annual worldwide directory of robot manufacturers, distributors and research institutes. Includes product prices and specifications, with name, address, and telephone number of knowledgeable representative.

See 31, 2157, 3977, 3978, 3987.

Special Issues and Journal Articles

Apparel World

813. 1982 BOBBIN SHOW FEATURES EXPANDED COMPUTER - AIDED MANUFACTURING UNITS. *National Outerwear and Sportswear Association. Annual. September 27, 1982. pg.51-93*

Provides technical specifications for apparel industry products and services exhibited at the 1982 Bobbin Show in Atlanta, Ga. (October 5-8, 1982). Includes 14 different product categories.

Assembly Engineering

814. JUSTIFYING ASSEMBLY AUTOMATION, PART 2: SOME OTHER CONSIDERATIONS. *Naidish, Norman L. Hitchcock Publishing Company. v.25, no.5. May 1982. pg.48-49*

Considers cost justification of assembly automation. Tables cover world robot population by country, as a percent of manufacturing population, and number of robots required for U.S. equivalency, and typical robot benefits and annual savings.

Business Week

815. THE RETAILING BOOM IN SMALL COMPUTERS. *McGraw-Hill Book Company. no.2725. September 6, 1982. pg.92-97*

Estimates 1980, 1982 computer market size worldwide. Discusses markets, brand names, manufacturers, and prices. Gives 1981 total shipments of personal computers and provides 1981 market shares for manufacturers.

816. WINDOW ON THE WORLD: THE HOME INFORMATION REVOLUTION. *McGraw-Hill Book Company. June 29, 1981. pg.74-83*

Forecasts market penetration to 1990, with attention to certain firms developing videotex, need for government regulation, and competition among firms.

Chemical Times and Trends

817. ROBOTICS: A KEY TO THE PRODUCTIVITY PUZZLE. *Letzt, Alan M. Chemical Specialties Manufacturers Association. v.5, no.3. July 1982. pg.20-24*

Examines the increasing use of robots to increase industrial productivity. Data is presented on the number of units in operation worldwide by country (1981); hourly robot costs compared with hourly human labor costs (1957-1981); sales projections for robots (1980-1990); and manufacturing method unit costs comparing robots, manual labor, and hard automation. Identifies types of robots, the use of robots by U.S. manufacturers and the suitability of robots to particular tasks. Also includes brief case studies.

Chip, Revista de Informatica

818. UNA ANARQUIA TOTAL: INDUSTRIA INFORMATICA NACIONAL. [A Total Anarchy: The National Data Processing Industry.]. *Espinosa, Tomas. Ediciones Arcadia, S.A. Spanish. no.08. November 1981. pg.62-64*

Analyzes several aspects of conditions in the Spanish EDP - computer industry in relation to market penetration by multinational computer businesses. Examines industry activity in international import - export of computer components. Charts show Spanish and international percent of EDP - computer market of worldwide businesses in 6 major countries, including Spain. Shows percent of number of installations, worldwide, and percent of value of equipment.

819. LIBROS: PUBLICACIONES DE INFORMATICA. [Books: Computer Publications.]. *Ediciones Arcadia, S.A. Spanish. Monthly. no.06. September 1981. pg.76-78*

Monthly section in Spanish computer includes book reviews of publications in the computer, computer services, and communications equipment fields, authors, publishers or distributors, and publication dates are provided.

820. RANKING ESPANOL DE EMPRESAS ELECTRONICAS. [Spanish Ranking of Electronic Businesses.]. *Ediciones Arcadia, S.A. Spanish. no.9. December 1981. pg.20*

Ranks 36 electronic computer businesses in Spain for 1980. Shows total income, income per employee, and net income per company, in Spanish pesetas. European rankings of 25 international EDP companies are shown with 1980 European income in American dollars, percent of advances and declines in 1979-1980, and total 1980 world income by each company.

Computer Business News

821. COMPUTER BUSINESS NEWS: INDUSTRY CHECKLIST; FINANCIAL CHECKLIST. *CW Communications Inc. October 19, 1981. pg.20-22 $1.00.*

Reports U.S. and other computer manufacturers' and distributors' recently announced distribution agreements, mergers, relocations, annual revenues and net income, and capital growth.

Datamation

822. THE WORLD'S TOP 50 COMPUTER IMPORT MARKETS. *Technical Publishing Company. January 1981. pg.141-146*

Contains a ranking of the world's top 50 computer import markets for 1977 and 1978, compiled from preliminary U.S. international trade statistics. Also provides tables with data on the

11 fastest growing import markets, and the major computer exporting countries in 1978 with their import to export ratio.

823. THE WORLD'S TOP 50 COMPUTER IMPORT MARKETS. *Szuprowicz, Bohdan. Technical Publishing Company. Annual. December 1979. pg.125-127*

1979 study with comparative trade data on specific commodities (computers, peripherals, spare parts and office equipment) in 113 countries, assessing the size and growth of import markets. Data in charts derived from the National Foreign Assessment Center and United Nations International Trade Statistics (1979). Rankings by country, commodity, growth rates, and imports provided.

Electronic Business

824. THE BATTLE SHAPES UP IN DESKTOP COMPUTERS. *Cahners Publishing Company. v.8, no.9. August 1982. pg.40, 42*

Discusses the competition between established suppliers of small-business systems and younger companies with low-priced, high-performance micro-computer systems for the world desktop-computer market. Evaluates marketing strategies of various manufacturers. Table provides the following data for home, portable, professional, and small-business computers: average unit price; market leader; 1982 retail shipments, compounded annual growth rate; and projected 1985 retail shipments. Source is Future Computing Inc.

825. ELECTRONIC BUSINESS: A MATTER OF DEFINITION. *Cahners Publishing Company. May 1981. pg.48*

Provides diagrams showing 1980 revenues and shipments for minicomputers, desktop computers, small business computers and processing terminals.

826. HIGH-END MICRO-WINCHESTER MARKET BEGINS TO EMERGE. *Furst, Al. Cahners Publishing Company. v.8. no.9. August 1982. pg.78-79, 82-84*

Focuses on the growth of the three-year old 5 1/4-inch Winchester-drive business. Provides data on worldwide fixed-disk drive shipments, worldwide market for under 30 Mbytes fixed-disk drives, and worldwide market for 30 to 200 Mbytes fixed-disk drives, 1980, 1982, and 1984. Also presents a genealogical tree of 5 1/4 inch Winchester disk companies, 1978-1982. Includes names of executives.

827. MARKET FORECASTS: EUROPEAN TERMINAL MARKET TO TOP $4 BILLION BY ''89; MICRO GROWTH; BANKING MARKET EXPANDS; FLAT PANEL DISPLAY SALES. *Cahners Publishing Company. v.8, no.7. June 1982. pg.94*

Gives growth rate and projected revenues for European computer terminal market. 1979, 1983, 1987, 1989. Forecasts worldwide microcomputer sales, 1986, and number of units shipped, 1982-1987. Projects revenue and market share for software and turn key systems used in U.S. banking, 1985; and U.S. market for flat-panel displays, 1985, 1987, 1991.

Industry, Commerce, Development

828. MARCH OF THE INDUSTRIAL ROBOTS. *MCB (Industrial Development) Ltd. v.1. July 1982. pg.26-27*

Summarizes findings of a study conducted by the Japanese Industrial Robot Association (JIRA) on world production of industrial robots through 1990. Chart illustrates annual robot production for 1970, 1975, 1980 and forecasted 1985 and 1990.

Inter Electronique

829. LE TRAITMENT AUTOMATIQUE: QUE DE LA PAROLE. [The Automatic Treatment of Speech.]. *Yugoslavia. Centre for Technical and Scientific Documentation. March 1981. pg.19-30*

Reviews the 1981 international market for speech synthesizers and forecasts 1985-1990 developments. Examines companies, products, and prices.

International Business Week

830. THE WORLDWIDE RACE IN INFORMATION PROCESSING. *McGraw-Hill Book Company. no.2761-92. October 18, 1982. pg.49-62*

Assesses international marketing and technology trends. Reports 1982 efforts in six countries to develop their information processing industries. Discusses research and development goals and government aid.

Mini-Micro Systems

831. WORLDWIDE MANUFACTURERS OF WINCHESTER DISK DRIVES. *Roman, Andrew. Roman Associates International. Cahners Publishing Company. Irregular. February 1981. pg.85-97*

Directory of worldwide manufacturers of Winchester disk drives with markets served: OEM, plug-compatible, or captive. Forecasts worldwide shipments and sales revenues, 1981, 1985.

Online Review

832. ONLINE BIBLIOGRAPHY UPDATE. *Hawkins, Donald, editor. Learned Information (Europe) Ltd. Annual. April 1981. pg.139-182*

Annual update to original '' ''Online Review' bibliography contains 441 references to works published from mid-1979 to mid-1980. Listings include author, publisher, publication date, location of publisher, and length. Sections include books, reviews; conferences; descriptions of online systems, databases, and services; man-machine studies, the user interface, user attitudes, system design and evaluation; profile development, searching techniques, indexing, manual vs. machine searching; usage studies, economics, promotion and impact of online retrieval systems, management; user education and training; and general / miscellaneous.

Robotics Today

833. SURVEYS REVEAL ROBOT POPULATION AND TRENDS. *Robot Institute of America. v.4, no.1. February 1982. pg.79-80*

Surveys current international robot population of 6 types of robots for 16 countries. Gives totals per country and per type of robot. Forecasts robot growth by number of new units installed for 1985 and 1990.

Words

834. OFFICE 2000. *Blackmarr, Brian R. International Information / Word Processing Association. Irregular. v.11, no.1. June 1982. pg.16-20*

Previews the typical office of the future, with descriptions of coming automated office and communications equipment including cellular radio, video disk, fiber optics, pulse signals, videotex, and new communications standards. Also discusses trends in user acceptance of this technology. Contains 2 tables comparing the number of local networks in 1981 with 1990 estimates, and their respective market value; and the number of installed micro-processors (1979-1990) and estimated software sales, with data provided by Frost and Sullivan.

See 38, 2168, 2171, 2177, 2179, 2180, 2181, 3141, 3142, 3143.

Numeric Databases

835. ADP MANAGEMENT INFORMATION SYSTEM (ADP / MIS). *Communication Information Management, ADP / MIS Division. (Available from Direct from Producer).*

System provides a worldwide inventory of the Government's ADP (Automated Data Processing) equipment and annual information concerning utilization, functional use, manpower, cost, and acquisition. Output contains yearly summary of Federal ADP activities and inventory of ADP equipment owned by U.S. government. Generalized retrieval system provides reports tailored to user's requirements.

836. ONLINE HOTLINE. *Information Intelligence Inc. (Available from Direct from Producer). Daily.*

A machine-readable version of Information Intelligence Online Hotline newsletter. Contains information on computer industry news; vendors; suppliers; databases; user aids; microcomputers; hardware and software developments; and industry trends.

Monographs and Handbooks

837. COMPUTING IN THE HUMANITIES. *Patton, Peter C., ed. Lexington Books. Division of D.C. Heath and Company. 1981. 416 pg. $29.95.*

Describes current and prospective computer applications for the humanities focusing on language, literature, history, and Biblical studies. Details uses within these fields for large and small computers, databases, time-sharing programs, and microcomputers for liberal arts faculty, researchers, and students.

838. ENCYCLOPEDIE DES EQUIPEMENTS DE BUREAU ET MATERIELS D'INFORMATIQUE. [Encyclopedia of Office Equipment and Computer Hardware.]. *Centre d'Information sur le Materiel de Bureau. French. Annual. September 1982. 2000 pg. 3424.00 (France).*

Handbook of office equipment and computer hardware gives descriptions and specifications of each product. Lists suppliers, with addresses.

839. INTERNATIONAL IMPLICATIONS OF UNITED STATES COMMUNICATIONS AND INFORMATION RESOURCES. *Ganley, Oswald H. Harvard University. Information Resources Policy. Order Department. P-81-3. April 1981. 263 pg. $240.00.*

Describes new electronic communications and information devices and discusses their effects on U.S. domestic industries as well as the reactions these effects create within other coun-

tries. Discusses complications created by domestic regulatory mechanisms, and the altered role of the U.S. media abroad. Relates communications and information innovations to eight concrete social, cultural, economic, political, military, and security situations of present international importance. Treats the U.S. - Canadian bilateral relationship in depth.

840. THE LOGICA PACKET SWITCHING REPORT. *Logica Inc. 1980. 340 pg. $125.00 (ukd).*

Coverage spans user applications (costs and tariffs; applications; case studies; future developments); and technology: overview; the alternatives; standards and network architectures; interface terminals and host computers. Includes network developments (international survey); and appendices (suppliers and products; glossary; bibliography).

841. TRENDS IN INFORMATION TRANSFER. *Hills, Philip J., ed. Greenwood Press. July 1982. $25.00.*

Eleven essays discuss developing technologies (videotex, fiber optics, computerized laser printing, computer input microfilm); authors' and editors' changing roles in data preparation; redesigning material for online viewing and indexing for computer access; and using a computer as a consultant of artificial intelligence.

842. VIEWDATA: A PUBLIC INFORMATION UTILITY. *Stokes, A.V. Input Two-Nine Ltd. 2nd ed. 1980. 133 pg. 12.50 (United Kingdom).*

Examination of viewdata, teletext, and related services. Includes viewdata history; foreign sales of Prestel systems; technical details; potential residential and business applications; associated services; private viewdata services; international developments; and viewdata and teletext forecasts.

See 3469, 3470, 3996.

Conference Reports

843. COMPUTER APPLICATIONS IN FOOD PRODUCTION AND AGRICULTURAL ENGINEERING. *Kalman, R.E., ed. North-Holland Publishing Company. 1982. 338 pg. $46.50. 100.00 (Netherlands).*

Presents results of a working conference organized by International Federation for Information Processing in Havana, October 26-30, 1981. Twenty-four papers examine whether modern information technology is of any use for developing countries which are short of food, drinking water, and shelter. Headings include computer applications for irrigation; computer control for food production; modelling and simulation;

844. COMPUTER VISION AND SENSOR-BASED ROBOTS. *IFS (Publications) Ltd. 1979. 353 pg. 31.00 (United Kingdom).*

Papers and general discussions of a 1979 symposium held at General Motors Research laboratories concerns current and future applications of computer vision systems. Four sessions covered fundamental issues in vision and robotics; vision and robot systems; future vision systems; and future robot systems.

845. COMPUTERS IN DEVELOPING NATIONS. *North-Holland Publishing Company. 1981. 272 pg. $39.50.85.00 (Netherlands).*

Proceedings of October 13, 1980 seminar in Melbourne, Australia cover strategies and policies for development; computer policies in India (relevance for the Third World); computer futures in the ASEAN / PNG region; computer advances and applications of relevance to developing nations; realities of computer and satellite use in developing nations; and computers and cooperation (aids to development in the 1980's).

846. DATA COMMUNICATION AND COMPUTER NET-WORKS. *Ramani, S., ed. North-Holland Publishing Company. 1981. 310 pg. $46.50.100.00 (Netherlands).*

Twenty-five papers from a February 4-6, 1980 conference in Bombay, India focus on applications, practical problems, and economics of computer networks.

847. ELECTRONIC DOCUMENT DELIVERY 2. *Learned Information, Inc. 1981. 221 pg. $47.00.*

Thirty papers from December 18-19, 1980 conference organized by Commission of the European Communities focus on available and proposed systems and services relevant to electronic document delivery, including technical, operational, management, and economic factors.

848. EURO IFIP 79: PROCEEDINGS OF THE EUROPEAN CONFERENCE ON APPLIED INFORMATION TECHNOLOGY OF THE INTERNATIONAL FEDERATION FOR INFORMATION PROCESSING. *Samet, P.A., ed. North-Holland Publishing Company. 1979. 754 pg. $93.00.200.00 (Netherlands).*

Proceedings of Sept. 25-28, 1979 conference in London include such issues as computer development and production in socialist countries; future needs of European mainstream users; computer-aided control of industrial processors; design problems and guidelines in human-computer communication in office automation systems; a future structure and strategy for IFIP; software in the mass market; and other topics.

849. FIFTH INTERNATIONAL ONLINE INFORMATION MEETING. *Learned Information, Inc. 1982. 502 pg. $72.00.*

Fifty-six papers from December 8-10, 1981 conference in London cover such issues as growth of numeric databases; evolution of private and public videotex systems; maturing of chemical structure search systems; and the sharing of online circulation systems.

850. HOSPITAL INFORMATION SYSTEMS. *North-Holland Publishing Company. 1979. 406 pg. $55.75.120.00 (Netherlands).*

Proceedings of April 2-6, 1979 conference in Capetown, South Africa include perspectives on hospital information systems; implementation considerations of hospital information systems; use of hospital information systems; evaluation of systems; and future of hospital information systems.

851. INFORMATION, COMPUTER, AND COMMUNICATIONS POLICIES FOR THE 80'S. *Elsevier-North Holland Publishing Company. January 1982. 290 pg. $40.50.*

Proceedings of the High Level Conference on Information, Computer and Communications Policies for the 80's (Paris, October 1980) cover such issues as contribution of information technologies to economic growth and employment, management of modern information infrastructures, and freedom and regulation of information.

852. INFORMATION POLICY FOR THE 1980'S. *Learned Information, Inc. 1979. $32.00.*

Proceedings of the 1978 Eusidic Conference cover pricing policies for online services; Euronet; survival among database producers, online services, and custom information services; and centralization versus decentralization trends in the development of brokers of bibliographic retrieval services. Includes online information in the 1980's and its social implications; in-house and outside information; development of national online user groups; restrictions on database use; online user problems in remote countries; and the NIH-EPA chemical information system.

853. MICROPROCESSORS AND THEIR APPLICATIONS. *Tiberghien, J., ed. North-Holland Publishing Company. 1979. 412 pg. $65.00.140.00 (Netherlands).*

Proceedings of fifth Euromicro Symposium in Goteborg (August 28-30, 1979) cover language and language processors; display systems; performance improvements by microcode; microprocessor utilization in database systems; tools for microprogram development; operating systems; date logging; use of microprocessors in communication; special purpose processors; personal computing; societal impacts of computing technology; and other issues.

854. PROCEEDINGS OF THE FIFTH INTERNATIONAL CONFERENCE ON COMPUTER COMMUNICATION. *Salz, J., ed. North-Holland Publishing Company. 1980. 880 pg. $45.00.135.00*

Papers from fifth international conference (Atlanta, Georgia, October 27-30, 1980) cover such topics as computer message systems developments and applications; satellite business systems and new communication applications; electronic funds transfer; systems concepts for videotex services; marketing and business planning issues for packet networks; and distributed network processing.

855. PROCEEDINGS OF THE FIRST INTERNATIONAL CONFERENCE ON ROBOT VISION AND SENSORY CONTROLS. *IFS (Publications) Ltd. 1981. 347 pg. 40.00 (United Kingdom).*

Thirty-four papers cover various aspects of vision sensor technology in advanced manufacturing processes. Topics include OMS-optical measurement system; use of tactile sensing for guidance of a robotic device for welding; and assorted reports on specific firms' activities.

856. PROCEEDINGS OF THE INTERNATIONAL CONFERENCE ON ROBOTS IN THE AUTOMOTIVE INDUSTRY. *IFS (Publications) Ltd. 1982. 300 pg. 40.00 (United Kingdom).*

Proceedings of April 20-22, 1982 conference concern present and future uses of robots in the automotive industry. Sessions cover painting and adhesives; management / I.R. aspects; welding applications; assembly / sensing; and flexible manufacturing.

857. SECOND INTERNATIONAL ONLINE INFORMATION MEETING. *Learned Information, Inc. 1979. 286 pg. $37.00.*

Thirty papers from December 5-7, 1978 conference emphasize user education and training, costs and effects of online searching, and the advent of videotex services.

858. SEMINAR ON COMMAND, CONTROL, COMMUNICATIONS AND INTELLIGENCE, GUEST PRESENTATIONS. *Harvard University. Information Resources Policy. Order Department. I-80-6. December 1980. 183 pg. $25.00.*

Eight papers from a spring 1980 seminar explore technical means for carrying out intelligence, command and control processes in support of the formulation and pursuit of strategic goals; strategic goals of organizations; and processes that decision-makers use to learn about the outside world ('intelligence') and to run and monitor their own organizations ('command and control').

859. SEMINAR ON COMMAND, CONTROL, COMMUNICATIONS AND INTELLIGENCE, STUDENT PAPERS. *Harvard University. Information Resources Policy. Order Department. January 1981. 273 pg. $25.00.*

Seven papers from a spring 1980 seminar explore technical means for carrying out intelligence and command and control processes in support of the formulation and pursuant of strategic goals; strategic goals of organizations; and processes that decision-makers use to learn abo

860. TWELFTH INTERNATIONAL SYMPOSIUM ON INDUS-TRIAL ROBOTS. *IFS (Publications) Ltd. 1982. 49.00 (United Kingdom).*

Papers from June 9-11, 1982 symposium in Paris, France cover robotics research, new industrial applications and economic and social aspects of robot introduction.

861. VIDEOTEX, VIEWDATA AND TELETEXT. *Online Publications, Ltd. 1981. 600 pg. 50.00 (United Kingdom).*

Contains written presentations of 70 contributors at Viewdata 80 conference in London. Coverage includes videotex activity in UK, France, Japan, Canada, USA; videotex developments in Germany, the Netherlands and Nordic countries; viewdata and electronic administrations - electronic mail, the electronic office; design and evaluation of existing services; market projections, both sides of the Atlantic; advertising on viewdata and teletext; international standards; electronic publishing; private systems; Prestel and the travel industry; and telesoftware and end user computing.

See 51, 2189, 4000, 4001, 4006, 4008.

Newsletters

862. FMS UPDATE. *IFS (Publications) Ltd. Monthly. 60.00/yr. (United Kingdom).*

Monthly supplement to 'Robot News International' covers businesses, executives, and technology associated with FMS throughout the world.

863. MICROPROCESSORS AT WORK. *Elsevier Science Publishers. Monthly. 16 (est.) pg. $150.00/yr. 65.00/yr. (United Kingdom).*

Monthly newsletter covering the microprocessor industry at an international level. Focuses primarily on company news, research and development, new products, new areas of employment and training needs and opportunities.

864. PRINTOUT. *Datek of New England. Monthly. $120.00/yr.*

Monthly newsletter covers printer technology, products, industry trends, company profits, and market in the U.S. and elsewhere. Includes support industries such as ribbons, toner, and business forms.

See 4010.

Services

865. DATA DECISIONS COMPUTER SYSTEMS. *Ziff - Davis Publishing Company. Monthly. 1981. $635.00.*

A monthly loose-leaf current awareness service published since November 1980. Lists every computer in alphabetical order. Annual service includes two hardcopy volumes, 124-page monthly supplements and a hot-line consulting service.

866. DATAPRO REPORTS ON MINICOMPUTERS (INTER-NATIONAL EDITION). *Datapro Research, Inc. (McGraw-Hill). Monthly.*

Annual subscription service provides microcomputer, minicomputer, and microprocessor product descriptions, specifications, case histories, user ratings, and evaluations. Includes monthly report supplements, monthly newsletter, and a telephone / telex inquiry service.

Indexes and Abstracts

867. ACCOUNTING AND DATA PROCESSING ABSTRACTS. *Anbar Publications Ltd. Bi-monthly. 1981.*

An abstracting service in the area of accounting and data processing published 8 times a year from 1970 forward. Data elements include title, author, bibliographic data, and abstract. Broad subject access.

868. ACM GUIDE TO COMPUTING LITERATURE 1980. *Association for Computing Machinery. Annual. 1982. $50.00.*

Annual guide to the world's scientific literature on computing lists over 15,000 books, papers, and reports with 6 separate indexes (title, author, keyword, topic, source, reviewer). Lists all items reviewed and abstracted in Association for Computing Machinery's 'Computing Review,' plus all papers from ACM conferences and several thousand other references published by other societies, publishers, government agencies, and other sources.

869. COMPUTER AND CONTROL ABSTRACTS. *Institute of Electrical and Electronics Engineers, Inc. Monthly. 1982. $610.00/yr.*

Monthly service abstracts journals concerned with computer and control engineering completely, with over 2300 worldwide periodicals and serials scanned for inclusion of worthy items. Each issue contains eight indexes: subject, author, patent, report, book, conference, bibliography, and supplementary list of journals. Annual subscription includes semi-annual author and subject indexes and a listing of abstracted journals with publishers' names and addresses.

870. COMPUTER SCIENCE RESOURCES: A GUIDE TO PROFESSIONAL LITERATURE. *Myers, Darlene, editor. Knowledge Industry Publications. 1981. 346 pg. $75.00.*

Covers current books on computer sciences; computer - related journals; technical report literature; indexing and abstracting resources; directories, dictionaries and handbooks; university computer center newsletters; software resources; proceedings of ACM special interest groups; programming languages; publisher's index; bibliography of career and salary trends in the computer industry (1970-1980); society, association and user group acronyms; university computer center libraries; computer industry trade fairs and shows; and proposed expansion of Library of Congress Classification Sections QA 75 and QA 76 (draft).

871. COMPUTERS AND EMPLOYMENT: AN ANNOTATED BIBLIOGRAPHY. *Bessant, J.R. University of Aston in Birmingham Technology Policy Unit. 1980. 61 pg.*

An annotated bibliography identifying sources of information about computers and their effects on future employment. Major headings are impact on manufacturing sector, impact on service sector, key perspectives and key issues. Includes most English-language literature from 1976 to February 1980.

872. COMPUTERS IN DEVELOPING COUNTRIES: A BIBLI-OGRAPHY. *Institute of Electrical and Electronics Engineers, Inc. 1981. $38.00.*

Bibliography on computers in developing countries contains 418 references.

873. CURRENT PAPERS ON COMPUTERS AND CONTROL. *Institute of Electrical and Electronics Engineers, Inc. Monthly. 1982. $130.00/yr.*

Current-awareness journal lists titles of articles and details of source documents for computer and control papers published throughout the world. Entries include title, author, author's address, publication, nation of publication, volume and number, publication date, and number of references.

874. NEW LITERATURE ON AUTOMATION. *Automatic Information Processing Research Centre. Monthly. September 1980. 10 (est) pg.*

A monthly abstracting survey of books, articles, reports, works of reference, proceedings, and standards in the computer industry. Source is divided into subject groups and then alphabetically by publication title. Each entry includes title, author, publisher and address, date of publication, number of pages specific subject area, and comprehensive abstract.

875. PREDI-BRIEFS: COMPUTERS AND BUSINESS EQUIPMENT. *Predicasts, Inc. Monthly. 1981.*

One in a series of 29 industry-intensive abstract periodicals derived from Predicasts' PROMT. Abstracts 200 to 400 articles per month covering such topics as acquisitions, capacities, end uses, government regulation, market shares and statistics, and new technology.

876. QUARTERLY BIBLIOGRAPHY OF COMPUTERS AND DATA PROCESSING. *Applied Computer Research. Quarterly. 1981.*

Quarterly index to computer and data-processing literature contains listings from 123 English-language publications from throughout the world. Publications reviewed are generally computer-related trade publications, general business and management periodicals and publications of computer-and-management-oriented professional societies and organizations, with emphasis on practical rather than academic subjects. Indexing is alphabetical by author, according to subject. Entries include publisher, publication date, price, and an abstract.

See 2201.

Dictionaries

877. INTERNATIONAL MICROCOMPUTER DICTIONARY. *SYBEX. Irregular. 1981. 119 pg.*

Presents definitions of microcomputer - related terms, numbers, and acronyms. Provides a basic computer vocabulary in English and ten other languages. Includes a section on standards and specifications. Also features listings of names and addresses of over 80 U.S. suppliers of microcomputer systems and components and of 11 microcomputer periodicals.

See 2204.

North America

Market Research Reports

878. THE A / D AND D / A CONVERTER INDUSTRY: A STRATEGIC ANALYSIS. *Venture Development Corporation. January 1982.*

Analysis of market demand for analog-to-digital and digital-to-analog converters. Includes forecasts of consumption and growth rates, 1981-1986, by types of devices; plans for expansion of facilities; capacity projections; and manufacturers' strategies.

879. ACCEPTANCE OF THE IBM SERIES/1. *International Data Group. May 1979. 70 pg.*

A follow-up study of a 1977 report on IBM Series/1. Forecasts (through 1983) revenues, shipments and prices for minicomputer systems. Telephone interviews with users and third-party hardware and software suppliers were used to determine present and future user satisfaction. Discusses IBM enhancements (including processors and greater range in memory capacity for the 4593-4 and 4955-5 models) for the series and their effects on the minicomputer industry.

880. ACQUISITION STRATEGIES FOR COMPUTER SERVICE COMPANIES. *INPUT. March 1979. 125 pg. $2500.00.*

1979 report analyzes the acquisition process in the computer services industry and the reasons companies make acquisitions. Data derived from telephone and questionnaire interviews with company executives. Impact of acquisitions on the computer services industry and purchaser's viewpoint given with data on revenues, historical and projection assessments, market shares, and growth rates. Manufacturers rated by product type and costs.

881. ADVANCED LOCAL NETWORKS. *Yankee Group. January 1981. $9500.00.*

Examines developments in local networks for information transfer and resource sharing among various equipment types. Discusses user requirements, vendor selection, and market impact. Part of Yankee Group's Communications Information System Planning Service and available only to subscribers.

882. THE ALPHANUMERIC CRT TERMINAL INDUSTRY: A STRATEGIC ANALYSIS 1979-1984. *Wichard, C.B. Venture Development Corporation. December 1979. 158 pg. $1495.00.*

This report provides information on the market, technology, competition, stategy and distribution of the general purpose CRT terminal. The forecast for five classifications of CRT terminals is given; conversational (dumb); editing (smart); 3270 and non-3270; processing (intelligent); single station and clustered. A directory of over 40 manufacturers is given, with information on past performance, new product announcements and relative position in the terminal industry.

883. ALPHANUMERIC CRT TERMINALS. *Creative Strategies International. December 1981. $1200.00.*

Market analysis for alphanumeric CRT terminals in the U.S. and worldwide. Includes forecasts of shipments and revenues, 1980-1985; vendors' market shares; competition; installed base; foreign trade; and distribution.

884. ANNUAL SURVEY OF MINI COMPUTERS. *Maclean Hunter Ltd. Annual. 1981. $10.00.*

Market survey covering new computer products and uses in Canada. Gives detailed tables on products, including performance and capacity.

885. APPLICATIONS AND THE MARKET RESPONSE. *Butler Cox and Partners Ltd. November 1981.*

Examines applications offered on videotext and discusses consumption trends and future potential.

886. APPLICATION-UNIQUE TERMINALS. *International Data Group. February 1980. 71 pg. $2500.00.*

Contains a summary of data on a market described as a high growth sector of the EDP industry. Historical data is combined with current conditions to show trends and developments in the industry. The scope of the study is to review and forecast the U.S. market for financial, retail and source data collection. Current and future installed base of these markets is defined in number of units and value. Competition in each market segment is examined. Methodology consists of historical, current and future market analyses.

887. ASSOCIATION OF COMPUTER USERS: BENCHMARK REPORT. *Association of Computer Users. Monthly. 1980. $11000.00.*

A monthly service which analyzes a different computer system each month. Summary of Benchmark results, pricing, software, hardware components, support services, and user comments analyzed. All major manufacturers' systems are analyzed.

888. AT&T'S ADVANCED COMMUNICATIONS SERVICE: THE NEXT GENERATION OF DATA TRANSMISSION NETWORKS. *Quantum Science Corporation. February 1979.*

Report describes features and characteristics of AT&T's Advanced Communications Service (ACS), strategies, and potential user benefits. Discusses effects on segments of EDP and communications industries, including terminal suppliers, independent telephone companies, specialized carriers, computer service companies, and computer equipment suppliers.

889. AUERBACH COMPUTER TECHNOLOGY REPORTS: BUSINESS MINICOMPUTERS. *Auerbach Information, Inc. Monthly. 1980. 800 (est.) pg. $365.00.*

A monthly updating looseleaf service on the business minicomputer field discussing peripherals, terminals, and other components. Provides marketing data, analysis, pricing and comparison charts of the various systems available by U.S. vendors.

890. AUERBACH GUIDE TO INTERNATIONAL EDP (ELECTRONIC DATA PROCESSING) MARKETS AND SYSTEMS. *Auerbach Information, Inc. 1979. 92 pg.*

Examines leading mainframe systems manufacturers in the U.S., Canada, Britain, France, West Germany, and Japan; their major product lines; and their home markets. IBM and its products are not included. Contains comparative analysis of product lines and an overview of the international datacommunications market. A brief directory of manufacturers is included.

891. AUTOMATED SERVICE UNITS: FIRST QUARTER RESEARCH PROJECT. *Trans Data Corporation. March 1982. 130 (est.) pg. $3400.00.*

Survey of automated teller machines (ATM's) and cash dispensers, and analysis of effects in banking markets. Examines distribution and productivity of installation by bank asset size, expansion of services, ownership, consumer behavior and utilization, costs and benefits, market share impacts, bank management impact, marketing strategies and prices, and vendor's operating performance. Service includes seminars and consultation.

892. AUTOMATION - CAD / CAM, N / C TOOLS, ROBOTS, FLEXIBLE MANUFACTURING, AUTOMATED MATERIAL HANDLING. *Frost and Sullivan, Inc. January 1982. 138 pg. $750.00.*

Analyzes market prospects in computer integrated manufacturing; automation in design; automation on the factory floor; automation in fabrication; and automation in materials handling. Examines market segmentation, growth rates, and technology.

893. AUTOTRANSACTION INDUSTRY REPORT. *Weiszmann, Carol, editor. International Data Group. Bi-weekly. May 26, 1980. 8 pg. $150.00/yr.*

Provides information and marketing data on the electronic computer industry. Includes reports on mergers, contracts, and programs of individual businesses. Includes two or three charts and tables of marketing data, such as market shares of different products.

894. AUXILIARY MEMORY SYSTEMS. *Yankee Group. May 1981. 200 pg. $9500.00.*

1981 report is part of Communications / Information Systems Planning Service. Available to service subscribers. Examines current and potential markets, technologies, and applications for auxiliary storage devices, and discusses user requirements in the office environment.

895. BANKING AUTOMATION III: COMPUTER AND PERIPHERALS INDUSTRY ANALYSIS SERVICE. *Creative Strategies International. June 1980. 89 pg. $1195.00.*

1980 market study analyzes banking automation, including teller terminals and teller machines. The industry, model features and channels of distribution are examined. Market segmentation, product installations, prices, growth rates and shipments are given. Competitors, revenues, products and market shares are compared.

896. THE BEGINNING OF A MAJOR NEW IBM COMPUTER PRODUCT LINE. *Weizer, Norman. Arthur D. Little, Inc. April 23, 1979. 8 pg. $500.00.*

1979 report discussing IBM's two new mini & small computer mainframe lines; IBM 4331, 4341. Product descriptions, prices, features, performance and competition (vendors & models) are given. Data derived from Arthur D. Little and IBM literature.

897. BOOMING MARKET FOR DESK TOP COMPUTERS - WHEN? *Business Communications Company. February 1981. 130 pg. $750.00.*

Examination of desk-top computer industry includes 60 tables, coverage of economic and social factors contributing to growth; state of the art; product and product accessories; markets, by market segment; industry structure; import / export data; security product manufacturers; and new product and innovations.

898. BOOMING MARKETS FOR COMPUTER STORAGE. *Business Communications Company. Irregular. September 1980. 100 pg. $750.00.*

Survey of the computer storage market covers present and experimental technologies; market and product analyses; sales potential; maintenance; forecasts; product descriptions; and directory of firms providing discs, tapes, media, and services.

899. BRAND PREFERENCE SURVEY: NO.5. *Maclean Hunter Ltd. 1981. $150.00.*

Market survey of consumers' attitudes and brand name preferences for EDP equipment. Includes comparisons with historical data from 1971, 1973, 1976, and 1978.

900. BUSINESS COMPUTERS FOR LARGE COMPANIES. *Frost and Sullivan, Inc. December 1981. $1100.00.*

Gives 10-year sales forecast for microcomputers; minicomputers / super minis; terminals; printers; intermediate storage; data communications; system software; and application software and service in 14 key end use segments. Provides supplier profiles and market shares.

901. THE BUSINESS INFORMATION MARKETS: 1979-1984. *Duke, Judith S. Knowledge Industry Publications. 1979. 266 pg.*

Analysis of market outlook for such business information sources as trade magazines, periodicals and newsletters; data bases, credit information sources and research agencies; and business books. Discusses market size and segmentation, distribution methods and their economics, advertising and other revenues, production costs, apparent consumption, and intermedia competition. Identifies and profiles industry leaders.

902. CAD / CAM. *Predicasts, Inc. April 1982. $900.00.*

Analysis of CAD / CAM systems, components, peripherals and software markets, with forecasts to 1985, 1990, and 1995. Examines technology, industry structure and end uses including robotics and numerical control systems.

903. CAD / CAM IN THE 1980'S: LAYING THE FOUNDATION FOR INDUSTRIAL AUTOMATION. *MSRA, Inc. 1981. 138 pg. $875.00.*

CAD / CAM industry and market analysis. Includes survey of users attitudes, with ratings of performance and vendors. Also includes competition, product lines and financial analysis of vendors. Forecasts annual growth rates, revenues and shipments, 1976-1985, and international sales, earnings and installations, 1983.

904. CAD / CAM SYSTEMS AND SERVICES OPPORTUNITIES. *INPUT. October 1981.*

Market analysis, forecasts, and survey of vendors' and users' attitudes for CAD / CAM systems and services, 1980-1986. Includes installed base and average annual growth rates by end uses, 1980-1986. Comments on increases in industrial productivity.

905. THE CANADIAN COMPUTER INDUSTRY. *Evans Research Corporation. 1982.*

Analysis of Canadian markets for computer products. Includes sales for 1980-1981.

906. CHANGES IN COMPUTER MARKETING. *Business Communications Company. G-068. March 1982. 150 pg. $950.00.*

Market research report covers new directions, trends, and strategies in computer marketing. Contains 60 tables.

907. CHANGING REVENUE AND PROFITS PATTERNS IN THE U.S. DATA PROCESSING INDUSTRY. *Withington, Frederic G. Arthur D. Little, Inc. September 29, 1980. 14 pg. $600.00.*

Updates through the first half of 1980 an earlier study of quarterly revenue and profits data for 28 EDP concentrated companies entitled The Revenue and Profits Squeeze in Data Processing. Charts showing growth of revenue (1978-1980), by company and product, after-tax income trends (1978-1980), running four-quarter profit trends by product and quarterly after-tax margin by company and product are provided. Corporate reports and ADL Access Group provide statistics.

908. THE CICS (CUSTOM INFORMATION CONTROL SYSTEM) MARKETPLACE. *International Data Group. April 1980. 42 pg. $1500.00.*

Review of the IBM Custom Information Control System (CICS). Examines user analysis of the product and marketing opportunities for competition. Methodolgy included telephone survey of 121 users and use of database information on over 900 CICS sites. 1979-1980 Data.

909. CLINICAL LAB - MICROCOMPUTERS. *Theta Technology Corporation. November 1979. 160 pg. $600.00.*

Market information for clinical lab microcomputers includes 1978 figures for market size, growth rate, products, company profits, market share, market segmentation, government regulation, and forecasts to 1985. Provides market outlook for programmable calculators and bibliography.

910. COLOR AND GRAPHICS IN HOME AND BUSINESS. *Strategic Inc. September 1981. 110 pg. $1200.00.*

Analysis of markets for computer graphics and color hardware and software. Examines new products and utilizations and ven-

dors' strategies. Market segmentation includes homes, businesses, education, entertainment, publishing and others.

911. COLOR GRAPHICS IMPACT ON 3270 DISPLAY MARKETS. *Creative Strategies International. June 1980. $1195.00.*

A detailed product and market analysis of the IBM 3270 Information Display segment including 3270-compatible devices. The primary focus of this study is on IBM, their product line, their long-range strategies, and the recently announced 3279 color graphics terminal. Also addresses the product direction and market potential for all types of IBM supplied and non-IBM supplied 3270-compatible devices. The five-year forecasts extend from year-end 1979 through year-end 1984.

912. COMMAND / CONTROL / COMMUNICATIONS. *DMS, Inc. Annual. 1981. $375.00.*

Analysis of military markets for command, control and communications systems.

913. COMMUNICATIONS: FUTURE DIRECTIONS IN NETWORKING, OFFICE SYSTEMS AND REGULATION. *MSRA, Inc. October 1981. 131 pg. $200.00.*

Conference report from corporations in networking, communications, and office systems provides analysis of technology, regulations, and markets.

914. COMMUNICATIONS / INFORMATION SYSTEMS PLANNING SERVICE: COMPUTER GRAPHICS. *Yankee Group. October 1981. 200 pg. $9500.00/yr.*

One in a series of monthly reports as part of Yankee Group's Communications / Information Systems Planning Service. Report examines vendors, technologies, and evolving applications of computer graphics. User survey covers trends in user planning and acquisition of graphics terminals, printers, and image systems.

915. COMMUNICATIONS PROCESSORS. *Creative Strategies International. June 1980. $995.00.*

The U.S. market for user-programmable computers specifically designed and programmed to perform all or part of the control and processing functions of a communications network. Sales in units and dollars forecast through 1984 for front-end processors, message switches, and remote concentrators. Analysis of important trends in the industry, including Bell's ACS and distributed processing.

916. COMPUTER AIDED DESIGN AND MANUFACTURING INDUSTRY, CAD / CAM. *Kurlak, Thomas P. Merrill Lynch, Pierce, Fenner and Smith Inc. 1980.*

Gives growth percentage of the computer-aided design and manufacturing industry since mid-1979. Attributes growth to the productivity enhancement for product design, engineering, mapping and architectural drawing. Forecasts 1981 and 1984 sales.

917. COMPUTER - AIDED DESIGN AND MANUFACTURING: MARKETS WITH OPPORTUNITIES. *Cahners Publishing Company. v.8, no.3. March 1982. pg.80*

Includes statistical data on the projected growth of the factory automation market. Special attention is given to computer-aided design, selected by the authors as the fastest growing segment. Data is included on expected expansion of CAD through 1984; growth of computer-aided manufacturing, planning and control systems; and the combined effect, in terms of productivity improvements for end users, of combining these three technologies.

918. COMPUTER - AIDED PAGE MAKE - UP SYSTEMS. *Strategic Inc. September 1981. 110 pg. $1500.00.*

Analysis of market trends for large computer - aided page make-up systems, utilizations in all kinds of publishing, and new products and technologies.

919. COMPUTER AND COMMUNICATIONS PRINTER MARKET (U.S.). *Frost and Sullivan, Inc. February 1980. $900.00.*

Market outlook by printer technique and end user market - computer systems, terminals, word processing and office automation terminals, communications systems, small business computers. Competitive trends, technology changes. Questionnaire survey results.

920. COMPUTER BASED ELECTRONIC MAIL. *International Resource Development, Inc. November 1981. 141 pg. $985.00.*

Market analysis for computer based electronic mail. Forecasts consumers to 1985 and 1991. Examines cost savings, vendors' revenues, and new products and services.

921. COMPUTER - BASED MANUFACTURING SYSTEMS: EXECUTIVE REPORT. *Inforesearch Institute. 1981. $250.00.*

Analysis of the markets for computer - based manufacturing systems and guidelines for purchases. Includes utilization, vendors' products and services, technology, prices, and industry structure trends.

922. COMPUTER - BASED MANUFACTURING SYSTEMS: INDUSTRY REPORT. *Inforesearch Institute. 1981. 100 pg. $875.00.*

Analysis of markets and growth rates for computer - based manufacturing systems. Includes utilization for inventory, reduced labor costs, productivity and profits; management of systems; vendors' product and service lines and prices, and end users.

923. COMPUTER BASED MEDICAL SYSTEMS. *Theta Technology Corporation. September 1979. 175 pg. $60.00.*

Analyzes present needs and applications for three primary market segments: hospitals, nursing homes, and practicing medical groups. Systems and markets covered include: literature storage and retrieval; patient monitoring; medication and diet control; automatic lab testing equipment; dosimetry; hospital inventory; radiology; record systems; decision analysis; information bank. Financial data includes sales, forecasts, markets, operating costs and demand for the years 1979-1990.

924. COMPUTER BASED NON-MEDICAL SYSTEMS. *Theta Technology Corporation. September 1979. 175 pg. $600.00.*

This report examines current systems and their uses, analyzes the potential 10 year market, and presents an overview of possible problems and companies currently in the market place. Systems and markets covered include: hospital inventory, billing, housekeeping, payroll, admissions, accounting and financial management, procedure schedule, record keeping, etc. Financial data includes usage, forecasts, markets, and sales for the years 1979-1990.

925. COMPUTER EQUIPMENT REVIEW. *Microform Review Inc. Semi-annual.*

Evaluates computer equipment for library applications, emphasizing peripheral equipment suitable for on-line bibliographic searching and conversion of catalog data to machine readable form. Each issue begins with a state-of-the-art report designed to analyze and explain the history, purpose, and distinguishing characteristics of the products reviewed. Technical evaluation features are explained in non-technical terms. The state-of-

the-art report is followed by 6 to 8 reviews of devices pertaining to the theme of the issue.

926. COMPUTER EQUIPMENT STRATEGY PROGRAM: DATA BASE AND INDUSTRY SECTOR FORECAST: VOLUME 3. *Quantum Science Corporation. Annual. 1980. 388 pg. $60.00.*

Volume 3 of 3-volume set provides year-by-year forecasts (1980-85) for the OEM market, giving detailed user analyses by industry sectors. Education, retail, wholesale, health care, transportation, government, utilities, insurance and banking industries computer expenditures are given. Forecasts of expenditures, distribution, applications, new products, vendors, manufacturers, installations, growth rates and employment for equipment and services provided, by industry sector. This publication is part of MAPTEK series.

927. COMPUTER GRAPHICS. *Predicasts, Inc. October 1981. $375.00.*

Manufacturers and markets for computer graphics devices and software. Reviews end uses and technologies.

928. COMPUTER GRAPHICS. *Creative Strategies International. May 1981. 90 pg. $1200.00.*

Forecasts cover market growth to 1985, effects of declining hardware prices and increased availability of software applications packages; new products and end uses; sales; software problems; industry trend toward complete systems; market segmentation (especially in terms of turnkey systems); and competitive structure of the industry.

929. COMPUTER GRAPHICS IN BUSINESS. *International Resource Development, Inc. 1981. 145 pg. $985.00.*

Market analysis for computer graphics in business. Covers shipments of graphics equipment for utilization by executives. Projects market growth rates 1981-1991 and comments on unemployment of graphic artists.

930. COMPUTER GRAPHICS IN BUSINESS APPLICATIONS. *Frost and Sullivan, Inc. December 1980. $1000.00.*

Analysis of computer graphics market by equipment services, software and systems. Analysis of market by end-user. Assessment of end-users and applications. Competition established. Survey results.

931. COMPUTER GRAPHICS IN FACILITIES MANAGEMENT AND PLANNING. *Frost and Sullivan, Inc. August 1981. $1100.00.*

Contains a 5-year forecast of computer-aided facilities management systems including hardware, peripherals and software. Analyzes proprietary software, turnkey software, and service bureaus. Examines specialized graphics digitizing applications, vendor profiles, and survey results.

932. THE COMPUTER INDUSTRY. *Wilkinson, Max. Financial Times Ltd. February 19, 1979. 12 pg.*

Reviews the computer industry in the U.S., the U.K., and Japan. Contains company ratings and data on various end uses. Analyzes markets and projects growth to 1983.

933. COMPUTER INDUSTRY ACQUISITIONS: A MULTI-CLIENT STUDY. *Interactive Data Corporation. April 1979. 500 (est.) pg.*

An updating service of acquisitions and / or mergers in the computer industry. Listed are manufacturers of hardware and vendors and suppliers of software and services. Addresses provided.

934. COMPUTER INDUSTRY ECONOMETRIC SERVICE: 1981. *Gnostic Concepts, Inc. Quarterly. 1981. $12000.00.*

Reviews evolution of computer technology with special emphasis on software interactions and technology of storage. Also reviews emerging trends in distributed computing and multiprocessing. Market outlook includes equipment consumption, production and international trade. Subscriber benefits are annual data base, continuing telephone inquiry privileges, and special seminars and strategy sessions.

935. COMPUTER INDUSTRY FINANCE: A MULTI-CLIENT STUDY. *Interactive Data Corporation. September 1980. 600 (est.) pg.*

1980 study on computer industry finance includes small, medium and large publicly and privately held firms, diversified firms, acquired firms and foreign firms. Addresses, new ventures, revenues, profits, growth rates, stock offerings, mergers, founding dates, loans, products, executives and number of employees given for firm groups, 1970-1980.

936. COMPUTER INDUSTRY OUTLOOK. *Frost and Sullivan, Inc. January 1982. 25 pg. $400.00.*

Reviews critical factors affecting the computer industry. Provides industry analysis in terms of products, vendors, markets, and distribution channels. Gives revenue forecasts for independent developed packaged software, processing services, and professional services.

937. COMPUTER INTEGRATED MANUFACTURING SYSTEMS. *Creative Strategies International. October 1982. $1450.00.*

Analysis of computer integrated manufacturing components includes industry structure, hardware manufacturers and software vendors, demand, labor supply, capital sources, productivity, end uses, foreign markets, technology, and competition.

938. COMPUTER NETWORKING STATUS: TRENDS AND UTILITY FOR THE FEDERAL GOVERNMENT. *Cap Gemini, Inc.; J.G. Van Dyke and Associates, Inc. U.S. Department of Commerce. National Bureau of Standards. 1980. 156 pg.*

Examines current computer networking offerings and projects technological and economic trends for the industry. The products and services of the sixty suppliers included in the survey typify the computer networking resources currently available to users. Summaries and assessments of those products and services surveyed are included in this report. Management and procurement issues, as well as technological impact, are evaluated. The probable penetration of computer networking into the federal government environment is described, and the expected advantages and problems are presented.

939. COMPUTER OUTPUT MARKET: AN OVERVIEW. *International Data Group. December 1979. 39 pg. $995.00.*

Examines in detail the markets of paper, printers and computer output microform (CRT statistics included in summaries only). Industry site analysis (1979-1983), output equipment and output media reviewed in detail.

940. COMPUTER OUTPUT MICROFILM MARKET. *International Data Group. 1980. 81 pg.*

This 1979 study analyzes the computer output microfilm market, including forecasts of number of sites, shipments, market shares, manufacturers, dollar value, market penetration, and distribution of products for 1978-1983. Provides an analysis of vendors, manufacturers, and products with sales, revenue data of competitors, and a user analysis of number of sites, manufacturers, end uses and distribution of systems.

941. COMPUTER OUTPUT PROGRAM: TRENDS IN INTERFACING MICROGRAPHICS: A MARKET STUDY. *IDC & National Micrographics Association. International Data Group. August 1980. 96 pg. $600.00.*

Examines use of new types of micrographics as they integrate with other technologies in 1980. Study arranged by 9 industry categories (insurance, banking, manufacturing, government, publishing, retail / wholesale, service bureaus and transportation communication and utilities) using case studies. Interfacing equipment reviewed includes mini-computers for computer-assisted retrieval (CAR), word processors, electronic mail and computer graphics. Innovative uses for CAR discussed as are problems associated with computer output microforms (COM). COM is studied as it interfaces with other technologies but is not included as an interfacing category. Equipment configurations for implementation of various integrated methods are not given in the report. User analysis presents data from telephone survey. Case studies review in-depth samples.

942. COMPUTER PERIPHERALS IN THE 1980'S. *International Resource Development, Inc. January 1979. 278 pg. $1265.00.*

Technology and market trends for ten types of computer terminals, six types of magnetic disk, drum and tape equipment, eight types of data communications peripherals, data entry devices, ten types of printers and plotters, COM and graphics equipment. Rating of strategic positions of more than 100 peripheral vendors. Ten-year market forecasts.

943. THE COMPUTER / POWER CONDITIONING PROTECTION MARKET. *Frost and Sullivan, Inc. #D-124. June 1982. 23 pg. $250.00.*

Wall Street analysis of computer power and power conditioning equipment and financial analysis of the corporations includes forecasts of growth rates; price and product group details; manufacturers' product lines, market shares, and sales; 1981 market size; technology; marketing strategies; earnings per share; end uses; and new products.

944. COMPUTER PRINTER SUPPLIES. *Creative Strategies International. December 1979. $995.00.*

Discusses and analyzes U.S. market for consumable computer printer supplies through 1984. Market segments discussed include supplies for serial impact printers (both fully formed character and matrix), line impact printers, and non-impact printers (both line and serial). Traces distribution channels for paper, ribbons and other consumables. Analyzes competitive practices of leading companies and indicates markets for over 100 other vendors operating in one or more segments.

945. COMPUTER PRINTERS. *Creative Strategies International. 1982. $1450.00.*

Analysis of computer printer markets worldwide includes market segmentation by technology; forecasts of growth rates, shipments and shipment values, 1982-1986; manufacturers' market shares; prices; and competition.

946. COMPUTER SERVICES MARKETS IN BANKING AND FINANCE: EXECUTIVE SUMMARY. *INPUT. July 1979. 12 pg.*

Brief report on banking markets for computer services divided into commercial banking, savings and loan, credit union, finance, security and commodity, and mortgage banking companies. Size, market forecasts, user expenditures and available computer services (type and mode) are discussed (1978-1984). A summary of a larger report by Input.

947. COMPUTER SERVICES MARKETS IN BANKING AND FINANCE: INDUSTRY REPORT #19. *INPUT. July 1979. 193 pg.*

Analyzes present (1979) and future markets of computers in banking / finance industry. Report is divided into subsectors of commercial banking: savings and loans, credit unions, finance companies, security and commodity firms and mortgage banking houses. Provides an analysis of computer services available based on sizes, market forecasts, and user expenditures.

948. COMPUTER SERVICES MARKETS IN INSURANCE COMPANIES: INDUSTRY REPORT #20. *INPUT. November 1979. 143 pg.*

Analysis of present (1979) and future markets for computer services in insurance companies (life / health, property / casualty, government-funded health insurance, insurance agents and brokers). Company size, (assets, revenues, etc.), user expenditures and market forecasts are determinants in the viability of a computer service and of the type and mode of computer service for insurance companies.

949. COMPUTER SOFTWARE INDUSTRY. *Predicasts, Inc. March 1981. $325.00.*

Covers company activities, technology, applications and markets for the computer software industry. Highlights the development of software packages designed to meet a wide variety of business, manufacturing and commercial needs.

950. COMPUTER SOFTWARE PACKAGES. *International Resource Development, Inc. 1981. 309 pg. $985.00.*

Study discusses market growth due to increased market penetration and price hikes; major segments of the software market; recent innovations and trends; various market characteristics; small business software; market constraints; driving vectors; and the supplier industry structure. Appendix lists names and addresses of companies mentioned in the report.

951. COMPUTER SPEECH COMMUNICATION: BUSINESS INTELLIGENCE PROGRAM. *SRI International. Marketing Services Group. December 1979. 24 pg.*

1979 market study on computer speech communication in the U.S. covering costs and rates for 1979, forecasts through the year 2000, equipment market areas, sales, product end uses, corporations manufacturing equipment, growth rates and market shares through 1988. Publication is part of Business Intelligence Program series.

952. THE COMPUTER SYSTEMS AND SERVICES MARKETS IN THE TRAVEL, LODGING AND ENTERTAINMENT INDUSTRIES (U.S). *Frost and Sullivan, Inc. September 1979. 193 pg. $800.00.*

Market analysis of the computer systems and services markets in the travel, lodging and entertainment industries. An assessment is made on the penetration of automated systems and services into these industry groups with forecasts of sales through 1985. An application analysis is made. Existing systems and installations are discussed with comments on features of the systems. Reasons for earlier failures and economic constraints to market acceptance are documented. Characteristics of the industries are discussed to pinpoint how users justify purchasing. Information system developments are identified. Tables show growth rates (1979-1985) in airline reservation sets and accounting systems in travel agencies. Provides the projected market for small business systems and terminals (1979, 1982, 1985) for the transportation, entertainment and lodging groups.

953. COMPUTERIZED PHYSICIAN'S OFFICE MEDICAL INFORMATION SYSTEMS MARKET. *Creative Strategies International. April 1981. $6500.00.*

Analysis of trends and segmentation of the market for computerized physician's office information systems. Includes five-year forecasts; new products; distribution channels; and manufacturers' management, product lines, growth rates and market position.

954. COMPUTERS AND PERIPHERAL EQUIPMENT, MEXICO: A COUNTRY MARKET SURVEY. *U.S. Government Printing Office. Irregular. CMS 81-308. February 1981. 9 pg. $.50.*

Report analyzes the current projected markets for the sale of computers and peripheral equipment in Mexico. Data is provided on the total market (production, exports, imports, market size); number and value of installed computers by major user sectors; expenditures by major user sectors; total employment and number of establishments by size in major user sectors; percentage breakdown by end-user applications; imports by country of origin; value of installed computers by major suppliers. Describes the domestic market, major market trends, and the competitive environment.

955. COMPUTERS IN THE CLINICAL LAB. *Theta Technology Corporation. no.111. September 1982. $295.00.*

Market analysis and forecast covers increasing use of computers to control clinical instrumentation, computer recording of patient results, technological effects, and corporate positioning.

956. COMPUTERS: KEY TO PRODUCTIVITY IN THE EIGHTIES. *FIND / SVP. October 1980. 200 pg. $500.00.*

Examines growth prospects for market participants in office automation, factory automation, home automation. Analyzes data processing expenditures, Japanese sales growth and barriers to entry, R & D, government role, software, potential for capitalization, IBM direction, 1980-1985 market structure, offices of the future, factories and homes.

957. THE CONSUMER DEDICATED MICROCOMPUTER MARKET IN THE U.S. *Frost and Sullivan, Inc. March 1980. 155 pg. $975.00.*

1980 report on the U.S. consumer-dedicated microcomputer market, analyzing and forecasting microcomputer product and end-user sales through 1985 by U.S. companies to worldwide markets. Probes expected (industry trends, including anticipated changes in technology, product obsolesence, domestic and foreign competition, problems associated with marketing (pricing and distribution), the overseas market, and vertical integration. Analyzes the markets for consumer microcomputers, notably electronic toys and games, personal computers, scientific calculators, home appliances and television receivers, among others.

958. CONSUMER MEDICAL ELECTRONICS. *International Resource Development, Inc. 1981. 159 pg. $985.00.*

Analysis of markets and forecast of expenditures for consumer medical electronics, 1981-1991. Examines new products, regulations, distribution, and manufacturers.

959. THE CRT GRAPHICS TERMINAL INDUSTRY. *Venture Development Corporation. 1982.*

Examines the U.S. CRT graphics terminal market emphasizing IBM's position and competitive factors. Statistics include IBM sales (1981) and 1981 share of dollar shipments by company. Assesses the market outlook for 1982.

960. CRT TERMINAL MARKETS 1980-1985. *Venture Development Corporation. 1982. 150 pg. $1295.00.*

Analysis of markets for computer terminals forecasts installed base and shipments by type, 1980-1985. Includes manufacturers' market shares, leasing, mergers, distribution, marketing strategies, services and a market survey of user attitudes.

961. THE DATA ACQUISITION COMPONENTS MARKET IN THE U.S. *Frost and Sullivan, Inc. January 1981. $1000.00.*

Forecast and analysis to 1985 for monolithic devices, hybrids, and modules used in the industrial, commercial, consumer, and military end-user markets. Examines impact of technology, various configurations of components, designs and functions; profiles leading suppliers and competitive market shares; reviews strategies and marketing channels; assesses the threat of foreign competition and the role of distributors.

962. DATA ACQUISITION SYSTEMS 1979-1983. *Venture Development Corporation. 1981. 152 pg. $950.00.*

Analysis of markets for turnkey and user configured data acquisition systems for industrial process plants and equipment, 1979-1983. Examines installations for manufacturers and power generation.

963. DATA BASE COMPUTERS. *Creative Strategies International. 1981. $1200.00.*

Analysis of the database computer industry, its structure, distribution channels and major manufacturers and vendors. Includes analysis of markets and segmentation, production growth, new products, and competition for market shares. Forecasts worldwide shipments, price trends, revenues, and installed base, 1980-1985.

964. DATA COMMUNICATIONS AND LOCAL AREA NETWORK EQUIPMENT AND RELATED MARKETS. *Anderson, Bud. Marketing Development. 1981. 162 pg. $750.00.*

Market analysis covering data communications and local area network equipment. Includes modems, multiplexers, network control systems, and other products and services, from 1979 to 1986, with some forecasts to 1990. Examines technology of products, distribution channels, market shares of manufacturers and vendors, end uses, market penetration strategies and regulations. Shows world and domestic shipments, 1980-1990.

965. DATA COMMUNICATIONS EQUIPMENT MARKET. *International Data Group. May 1979. 85 pg.*

1979 market study of data communications equipment, including modems, multiplexors and communication processors. Market forecasts, market estimates, product segmentation and shipments are discussed. Discusses the role of AT&T and other U.S. manufacturers.

966. DATA COMMUNICATIONS SYSTEMS AND EQUIPMENT. *Predicasts, Inc. November 1981. $900.00.*

Computer networks and distributed data processing emphasized. Forecasts U.S. market for computers and terminals by type, modems, multiplexers, processors. EFT (automated teller, POS terminals, etc.) facsimile and communicating word processor outlook presented. Transmission facilities and data communications services analyzed.

967. DATA COMPRESSION HARDWARE AND SOFTWARE MARKET. *Frost and Sullivan, Inc. April 1982. 184 pg. $1250.00.*

Analysis of markets for data compression hardware and software, with forecasts to 1987. Examines costs of data storage and transmission, potential savings from data compression, end uses, vendors' marketing strategies and product lines,

prices, technology and performance, and the industry structure.

968. DATA ENTRY EQUIPMENT IN THE 1980'S. *International Resource Development, Inc. May 1980. 181 pg. $985.00.*

1980 report analyzes and reviews the market for data entry equipment for 1980-1990. Markets, shipments, suppliers and vendors are discussed. Leading data entry equipment suppliers are analyzed. Statistics on market shares for 1980 provided. Data entry equipment includes card punching devices and card reader equipment, terminals, OMR,OCR and MICR equipment, and voice data entries equipment.

969. DATA ENTRY: USERS PREPARE FOR THE 1980'S. *Impact Marketing Services. 1979. 108 pg. $450.00.*

Survey of computer industry executives on utilization of data entry products and services. Covers user attitudes and technical aspects of equipment.

970. DATA PROCESSING INDUSTRY. *FIND / SVP. Quarterly. December 1981. 13 pg. $175.00.*

Report looks at producer durable expenditures, S & P Office Equipment Group earnings, and currency translation. Considers the possibility of price-cutting by volume-sensitive minicomputer firms following IBM's recent price cuts in Europe, and includes discussions of large scale processors and new product introductions.

971. DATA PROCESSING INDUSTRY: INVESTMENT OVERVIEW. *FIND / SVP. October 1980. 71 pg. $200.00.*

Study of the computer industry: manufacturing, products, and supplying software. In-depth assessment of Amdahl, Burroughs, Control Data, Data General, Digital Equipment, Honeywell, IBM, NCR, Prime Computer, Sperry.

972. DATA PROCESSING SYSTEMS IN CREDIT UNIONS (U. S.). *Frost and Sullivan, Inc. October 1981. 140 pg. $1050.00.*

Gives credit union trends and DP needs. Includes 5-year sales forecasts for batch and on-line computer services, in-house systems, maintenance and business forms. Provides affect on credit unions of Monetary Control and Financial Institutions Deregulation Act. Gives major competitors with market shares. Provides computer tabulated questionnaire survey results.

973. DATA TEXT AND VOICE ENCRYPTION EQUIPMENT. *International Resource Development, Inc. November 1981. 151 pg. $985.00.*

Analysis of markets for data, text, and voice encryption products and services. Includes market segmentation, market shares, and U.S. exports. Gives forecasts to 1990.

974. DESKTOP BUSINESS COMPUTERS: MARKETS AND STRATEGIES, 1981-1986. *International Resource Development, Inc. 1982. 250 pg. $1695.00.*

Analysis of desktop computer markets and the industry structure, with forecasts to 1986. Includes prices and product groups, manufacturers and vendors, 1981 installed base, market shares, shipments, end uses, factors in purchases, distribution channels, marketing strategies, market surveys, and user attitudes.

975. DESKTOP COMPUTER MARKET. *International Data Group. September 1979. 53 pg.*

Analysis of the desktop computer market includes products (1978), market segmentation (1978), distribution (1978), end-user markets (1978), vendor profiles (1978), market size (1977-1983), value of shipments (1979), and total shipments (1979). Dicusses trends in discrete market areas: business / professional, scientific, educational, and home / hobby.

976. DEVELOPING PLANS FOR IMPLEMENTING LOCAL NETWORKS AND ELECTRONIC MESSAGING SUB-SYSTEMS. *Yankee Group. March 1981. $7500.00.*

1981 report focusing on local networks and electronic messaging implementation. Case studies of leading users, standardization, vendor selection and network management. Products available and costs provided. Publication is one of 12 reports published each year, beginning 1981, by Critical Paths Service. Available to service subscribers.

977. DIGITAL TRANSMISSION TECHNOLOGY: THE WAVE OF THE FUTURE. *Yankee Group. August 1980. 200 pg. $9500.00.*

Examination of services, products, components, systems, and vendors of the digital transmission market. Discusses future developments and trends. Part of Yankee Group's Communication Information System Planning Service and available only to subscribers.

978. DISPLAY TECHNOLOGY AND MARKET FORECAST. *Gnostic Concepts, Inc. 1981. $15000.00.*

Analyzes display technologies: light emitting diode (LED); liquid crystal (LCD); plasma gas discharge (PGD); vacuum flourescent (VFD); cathode ray tubes (CRT); incandescent lamp (ILD); and electromechanical (EMD). Market data for each display technology for period 1979- 1984 includes total value, total units, average price, and the I/O coefficient of value added by the display. Applications categories are analyzed, and competition is identified and described. Includes a special section of flat panel technology.

979. DISTRIBUTED DATA ENTRY MARKET. *Frost and Sullivan, Inc. April 1981. 440 pg. $1100.00.*

Contains 10-year analysis and forecast in dollars and units for key-to-disc; intelligent terminals; non-intelligent terminals; alphanumeric display terminals; optical bar code readers; hand-held readers; optical character readers; push-button telephones; portable data recorders; factory data collection equipment; voice data entry products; and direct data entry service. Contains industry structure, company profiles and user survey results.

980. DISTRIBUTED DATA PROCESSING MARKETS: SURVEY AND ANALYSIS. *Impact Marketing Services. 1980. 150 pg. $550.00.*

Survey of computer using industries. Reports user attitudes and plans for adoption of new distributed data base products and services.

981. DISTRIBUTED MULTIPLE PROCESSOR SYSTEMS FORECAST. *Gnostic Concepts, Inc. 1981. $15000.00.*

Analysis of technological evolution in computer system architecture and quantitative market forecast of distributed computing systems (1976-1983). Market analysis includes the dimensions of the distributed processing trend, user attitudes, market configuration, and key issues affecting product acceptance.

982. DISTRIBUTED PROCESSING: BUSINESS OPPORTUNITY REPORT. *Markwood Research Co. Business Communications Company. April 1980. 110 (est.) pg. $700.00.*

Forecasts 1980-1985 unit and dollar sales of distributed processing computer equipment, by system size. Also examines 1974-1980 sales volume, by manufacturer.

983. DISTRIBUTED TELEPRINTERS FOR THE 1980S. *Creative Strategies International. September 1980. $1195.00.*

The worldwide market for dumb, smart and intelligent communicating teleprinters. This focused analysis covers product trends and user application directions of: no-value added (NVA); some-value added (SVA); and high-value added (HVA) teleprinters (non-CRT) sold into the business, office, network and home environments. Detailed 1980-1985 market forecasts are provided for each of these market segments.

984. DISTRIBUTION STRATEGIES OF PERSONAL COMPUTER MANUFACTURERS. *Strategic Inc. no.1503. September 1982. 110 pg. $2500.00.*

Analysis of distribution strategies in personal computer markets drawn from market surveys of manufacturers, suppliers and retail dealers includes franchises, inventories, competition, prices, profit margins, and marketing strategies.

985. DOD AUTOMATED TELECOMMUNICATIONS MARKET - MILITARY AND COMMERCIAL OPPORTUNITIES. *Frost and Sullivan, Inc. September 1981. 434 pg. $1200.00.*

Analysis of defense contract markets for data communications equipment. Forecasts installations in military markets, and future civilian end uses, demand and expenditures. Profiles manufacturers and vendors.

986. DP AND WP IN PHYSICIANS' OFFICES. *Frost and Sullivan, Inc. August 1981. $1150.00.*

Contains a five-year analysis and forecast of data processing, word processing and office automation products and services used in medical practices. Includes technological and regulatory impact. Selling strategies are examined. Types and operational characteristics of medical practices are defined. Discusses marketing by these distribution channels: direct, OEM / system houses; software and service firms are analyzed. Gives vendor profiles.

987. THE ECONOMIC REALITIES OF VIDEOTEX. *Butler Cox and Partners Ltd. October 1980.*

Examines the commercial costs and revenue flows in videotext public service operations. Discusses profit potentials and popular demand factors.

988. EDP IN HOSPITALS. *Frost and Sullivan, Inc. February 1982. $1150.00.*

Contains 5-year analysis and forecast in units and dollars for EDP product classes by type of product, services, and installation environment. Includes computer systems, services (batch, on-line, interactive), peripherals, and terminals. Examines channels of distribution. Describes operational characteristics of hospitals. Examines types of hospitals and their structure. Includes vendor profiles.

989. EDP INDUSTRY LEADERS: CONTINUOUS INFORMATION SERVICE. *Interactive Data Corporation. April 1979. 55 (est.) pg.*

1979 research report covers small business computers, peripherals, minicomputers, general purpose computers, leasing and used computers, computer services and software, computer suppliers, and accessories markets. Revenues and shipments of equipment are given through 1978. Publication is part of I.D.C.'s Continuous Information Service.

990. EDP INDUSTRY REPORT: REVIEW AND FORECAST PART 1: SMALL BUSINESS AND DESKTOP COMPUTER MARKETS. *International Data Group. Annual. v.17. June 13, 1981.*

Examines markets for desk top and small business computers from 1975-1980 and through 1985. Covers volume and value of shipments, and of units in use. A product survey specifies each unit's manufacturer, price, U.S. and foreign installations, and 1981 orders.

991. EDP INDUSTRY REPORT: REVIEW AND FORECAST, PART 2: GENERAL PURPOSE COMPUTER AND MINICOMPUTERS MARKET. *International Data Group. Annual. v.17. June 26, 1981.*

Reviews volume and value of 1980 shipments of general purpose mainframe and mini computers, by category, with forecasts for 1981 and 1982. Also covers individual market segments, and such secondary markets as software and processing services.

992. EFTRAC - ELECTRONIC FUNDS TRANSFER TRACKING RESEARCH AMONG CONSUMERS: A PLANNING TOOL FOR THE 1980'S. *A.J. Wood Research. 1979. 202 pg. $950.00.*

Market survey examining utilization of electronic funds transfer services and consumers' attitudes toward new products and services. Includes market segmentation by consumer financial behavior categories.

993. EIGHTH ANNUAL STUDY OF CUSTOMER SATISFACTION AMONG LARGE SYSTEM USERS IN THE UNITED STATES. *Customer Satisfaction Research Institute. Annual. 1981. $4300.00.*

Survey of customer satisfaction and end user attitudes for large computer systems in the U.S.

994. EIGHTH ANNUAL STUDY OF CUSTOMER SATISFACTION AMONG MEDIUM SYSTEMS USERS IN THE U.S. *Customer Satisfaction Research Institute. Annual. 1981. $4300.00.*

Survey of customer satisfaction and end users' attitudes toward medium size computer systems in the U.S.

995. EIGHTH ANNUAL STUDY OF CUSTOMER SATISFACTION AMONG SMALL SYSTEMS USERS IN THE U.S. *Customer Satisfaction Research Institute. Annual. 1981. $4300.00.*

Market survey of end users' attitudes and customer satisfaction for small computer systems in the U.S.

996. ELECTRONIC DOCUMENT STORAGE 1980-1990: VOLUME 2: USER NEEDS. *Mackintosh Consultants, Inc. November 1980. $8500.00.*

Presents the results of a survey of about 1000 U.S., British, French and German establishments, regarding their electronic document storage needs and preferences through 1990. Presents twenty-eight case studies from a sample of 100 wide-ranging end users who were subjected to in-depth interviews. Detailed questionnaire responses are reproduced, with breakdowns by industry sector and application.

997. ELECTRONIC FILING. *International Resource Development, Inc. 1982. 179 pg.*

Examines the impact of computers and electronic filing systems on the office environment. Data on prices is presented for selected product groups, with forecasts to the mid-1980's. New technology is described and products destined for the marketplace in the next ten years are reviewed with details on improvements currently under study. Data on market size and percentage of office workers expected to use these systems is also included, with forecasts to 1991.

998. ELECTRONIC FILING AND STORAGE SYSTEMS. *Strategic Inc. December 1981. 150 pg. $1500.00.*

Market analysis for electronic filing and storage systems, and competition among new products. Forecasts for products and end uses, 1981-1986, 1990, 2000. Examines effects on employees and management systems and market penetration opportunities for services and vendors.

999. ELECTRONIC FILING OF THE TOTAL INFORMATION RESOURCE: AUTOMATED STORAGE / RETRIEVAL OF TEXT AND/OR GRAPHICS. *Institute for Graphic Communications, Inc. 1982.*

Analysis of markets for electronic filing. Examines new technologies for automatic storage and retrieval of text and graphics, capacity and cost considerations, new products, vendors, and competition.

1000. ELECTRONIC FILING SYSTEMS - THE KEY ISSUES. *Strategic Inc. January 1982. 150 pg. $1500.00.*

Provides analysis and forecasts of markets for electronic filing 1981-1986, 1990, 2000. Includes technology, end uses, effects on employees and management, costs and competition, vendors, and market structure.

1001. ELECTRONIC INFORMATION SERVICES. *FIND / SVP. July 1981. 35 pg. $350.00.*

Market and industry analyses and forecasts to 1990 cover assorted business and consumer information services in terms of marketing uncertainties, technical issues, politics of information distribution, and i

1002. ELECTRONIC MAIL: WHERE WILL IT BE? *Business Communications Company. March 1979. 93 pg. $650.00.*

Discusses equipment, network and system needs, complicated regulatory, political and economic factors involved. All contents include: equipment; scanners / systems; regulations, USPS, competition; and markets, forecasts.

1003. ELECTRONIC MESSAGING AND FACSIMILE. *Yankee Group. Annual. April 1980. 362 pg. $9500.00.*

1980 report on electronic messaging and facsimiles including computer-based message systems, communicating word processors, Telex / TWX. Forecasts (1978-1983) present and future products and technologies and the facsimile market and vendors. Pricing, suppliers, ranking and revenues, services, market segments and user consumption provided. Volume 7 of series by Yankee Group's Communications and Information Planning Service. Available to service subscribers.

1004. ELECTRONIC SOFTWARE DISTRIBUTION FOR HOME, EDUCATION AND BUSINESS. *Strategic Inc. September 1981. 120 pg. $1200.00.*

Analysis of markets for mass information software products distributed through videodiscs, videotape, videotex / teletext, and other interactive systems. Reviews roles of vendors, information providers, and software producers; examines product development costs; forecasts markets, new products and technologies; and recommends sales and pricing strategies.

1005. THE EL-HI MARKET, 1982-1987. *Knowledge Industry Publications. December 1981. 200 pg. $750.00.*

Analysis of markets for educational books, other published materials, audio-visuals, and microcomputers for elementary and high schools, with forecasts, 1982-1987. Includes market size and structure, 1973, 1980; industry profits; industry structure; demographic statistics and economic indicators; government funding, 1972-1980; impact of legislation; sales, 1972-1980; cost of sales; and foreign market sales. For major corporations, gives expenses, 1980 revenues, 1980 rankings, and 1980 operating performances.

1006. THE EMERGING PERSONAL COMPUTER MARKET. *FIND / SVP. July 1981. 11 pg. $150.00.*

Examines the small business, corporate, educational, and industrial markets. Evaluates three top-of-the-line computers, retail distribution, and direct marketing.

1007. ENERGY EFFICIENT COMPUTER PERIPHERAL EQUIPMENT. *Frost and Sullivan, Inc. April 1981. 190 pg. $1000.00.*

Analysis of energy consumption by computer peripheral equipment and the long-term cost of operation. Discusses energy costs of products, end uses, low energy products markets, and vendor marketing strategies.

1008. ENERGY MANAGEMENT SYSTEMS: OPPORTUNITIES IN THE DISCRETE MANUFACTURING INDUSTRY. *Quantum Science Corporation. 1981.*

Examines the market for multifunction energy management systems in the discrete manufacturing industry to 1985.

1009. EVOLUTION OF MAINFRAME / DDP (DISTRIBUTED DATA PROCESSING) ENVIRONMENT. *Creative Strategies International. December 1980. $1195.00.*

Examines trends and strategies of major vendors in the mainframe / DDP (Distributed Data Processing) environment. Discusses market elasticity, impact of new technologies, electronic mail, intelligent copier / printers, front-end and back-end processors, and IBM's product direction. Also examines enduser expectations and applications trends. Includes U.S. and worldwide market forecasts (1979-1984) and vendor market shares.

1010. THE EXECUTIVE WORKSTATION. *International Resource Development, Inc. February 1981. 170 pg. $985.00.*

Market analysis for the executive workstation. Analyzes utilization of office computers and communications products by executives present and future. Forecasts costs and effects on employees and management structures by end use sectors. Includes revenues of manufacturers, shipments, and expenditures for office communications functions and equipment, 1980-1990.

1011. EXPLODING DATA BASE BUSINESS: TRENDS AND DEVELOPMENTS. *Business Communications Company. September 1982. $1250.00.*

Surveys participants in the field; explores information content of the package and pricing schedules; and considers qualitative and quantitative market prospects for these services by type. Issues include whether the market will be corporate or retail; what place data bases will have in the emerging market for videotex services; what the size and type of market will be for terminal equipment and software generated by data bases; and views of the coming information revolution of the 1980's.

1012. FACSIMILE MARKET FORECAST: CONTINUOUS INFORMATION SERVICES. *Interactive Data Corporation. March 1979. 36 pg.*

1979 report on facsimile equipment market with prices, rent-lease-purchase prices, installations, shipments, and uses of products (1978-1985). Companies' geographical distribution is given by employment. Publication is part of I.D.C.'s Continuous Information Services.

1013. FOURTH ANNUAL STUDY OF CUSTOMER SATISFACTION AMONG MINICOMPUTER USERS - DEDICATED TO APPLICATIONS. *Customer Satisfaction Research Institute. Annual. October 1981. $6400.00.*

Market survey of end users' attitudes toward minicomputer installations for dedicated applications.

1014. FOURTH ANNUAL STUDY OF CUSTOMER SATISFACTION AMONG SMALL BUSINESS SYSTEM USERS - BUSINESS COMMERCIAL APPLICATIONS. *Customer Satisfaction Research Institute. Annual. October 1981. $6400.00.*

Market survey on end users' attitudes toward computers. Covers business commercial installations.

1015. THE FUTURE OF THE MINICOMPUTER INDUSTRY. *Frost and Sullivan, Inc. June 1982. 11 pg. $150.00.*

Analysis by a Wall Street firm of minicomputer market structure and investment outlook. Examines competition; shipment forecasts, 1982-1986; installed base; and corporations' earnings, valuation, operating ratios, and balance sheets.

1016. GROWING MARKETS FOR COMPUTER STORAGE. *Business Communications Company. September 1980. 200 (est.) pg. $750.00.*

Market analysis of the computer storage industry, focusing on disk storage and the competing technologies. Includes suppliers, refinishers and manufacturers, with distribution channels. Outlines OEM situation and opportunities for product programs that support efficient use of disk storage. Includes 60 tables.

1017. THE HARD DISK INDUSTRY: A STRATEGIC ANALYSIS, 1979-1984. *Venture Development Corporation. January 1980. 128 pg.*

Study analyzes the market, the technology and the competition for hard disk drives in the U.S. Chapters include discussion on industry structure; technology (types of disk drives, magnetic heads, disk surfaces, drive motors); competitive technologies including mechanical storage, magnetic storage, semiconductor memories and optical disks; growth trends; shipments; and marketing strategy. Includes a directory of 30 selected independent manufacturers of disk drives with addresses and descriptions. Numerous graphs, diagrams and tables show shipments (1979-1984) and market shares (1979, 1984), by type.

1018. HARVARD BUSINESS SCHOOL FIELD STUDY PROJECT: SMALL BUSINESS COMPUTERS: FINAL REPORT, JUNE, 1980. *Harvard University. Graduate School of Business. June 1980. $150.00.*

Results of Small Systems and Terminals Division and Honeywell Information Systems Corporation sponsored small business computers study, including an industry analysis of 5 competitors. Strategic recommendations provided.

1019. HEALTH CARE EDP. *Predicasts, Inc. October 1981. $900.00.*

Analyzes and projects through 1995 U.S. markets for three categories of health care EDP: hardware, software and services. Covers business, patient management and clinical applications of EDP systems. Investigates factors influencing EDP health markets, and presents detailed industry structure including profiles of major companies.

1020. HIGH SPEED COMPUTER PRINTING MARKET ANALYSIS AND FORECASTS. *Gnostic Concepts, Inc. 1981. 300 pg. $9800.00.*

Discusses user requirements, market forecast, technology outlook, and competition trends of high speed printers. Tables include: aggregate demand forecast, application demand forecast, technology assessment, and paper and forms.

1021. HIGH-SPEED NON-IMPACT PRINTERS AT CENTRAL COMPUTER SITES. *International Data Group. December 1979. 133 pg.*

Analysis of high speed non-impact printers (NIP) at central computer sites includes number of U.S. sites (1978-1983), market share of U.S. computer sites by size (1978-1983), NIP forecast by printer type (1978-1983), distribution (1978-1983) and monthly output costs for selected companies (1979).

1022. HOME BANKING AND ELECTRONIC FUNDS TRANS-FER. *Yankee Group. September 1981. $8500.00.*

Discusses the demand, technology, possible applications of, and impact of home banking and electronic funds transfer. Anticipates the results of Citibank test of 600 home banking terminals and the 1981 Beta test by Chase Manhattan. Examines possible information and financial services banks may offer consumers in light of market considerations. Part of Yankee Group's Home of the Future Service and available only to subscribers.

1023. HOME COMPUTERS. *Irwin Broh and Associates. April 1979. $150.00.*

Consumer survey among 70,000 families. Projections of ownership incidence and 1978 market size. Brand share trends, age of home computer owned, number owned per family, owner demographics and 1979 buying intentions.

1024. THE HOME COMPUTING CENTER: THE COMMAND-AND-CONTROL MODE. *Yankee Group. May 1981. $8500.00.*

The second application of home computers, the home command and control center, controls energy management, security and lighting, and stores home accounts and records. Examines future potential, likely distribution patterns. Predicts who will manufacture and service them. Part of the Yankee Group's Home of the Future Service and available only to subscribers.

1025. HOME ELECTRONIC INFORMATION SYSTEMS: BUSINESS OPPORTUNITIES, NORTH AMERICA AND WESTERN EUROPE TO 1990. *Mackintosh Consultants, Inc. June 1980. 4400.00 (United Kingdom).*

Discusses technological developments in home interactive data services, and market outlook through 1990 for relevant equipment and services. Forecasts are presented in the context of user needs and selection criteria, competing media, and cost factors.

1026. THE HOME ENTERTAINMENT AND INFORMATION MARKET. *Frost and Sullivan, Inc. January 1982. $1000.00.*

Market report gives forecast of the products and services for the computerized home of the 1980's. Provides identification of key industries for home market. Includes current and projected strategies of key corporations. Examines role of the phone system, broadcasting and satellites. Assesses products and services including information sources, interactive cable, videodisc, home banking, videophone. teletext systems, and personal computers. Includes effect of government regulation and foreign competition.

1027. HOME FAX AND OTHER PERIPHERALS FOR THE MASS INFORMATION MARKET. *Strategic Inc. November 1981. 115 pg. $1200.00.*

Market analysis for hard copy devices and other peripherals for mass information systems. Covers utilizations and new products including home fax, audio graphic devices, financial transaction terminals, and 'smart card' handling equipment.

1028. HOME INFORMATION AND CONTROL IN THE ELECTRONIC HOME OF THE FUTURE: A STRATEGIC ANALYSIS 1980-1995. *Venture Development Corporation. 1981.*

Analysis of changing market structure and forecasts, 1980-1995, for consumer information, video, and control equipment. Covers consumer demand trends, new products, manufacturers' market penetration strategies, consumer attitudes, and total market value, 1980-1995.

1029. HOME INFORMATION SYSTEMS. *Booz-Allen and Hamilton, Inc. 1981.*

This study on the market for home information systems is based on a survey of 800 consumers. Services evaluated include shopping, banking, reservations, video games, home finance, and information services. Conclusions are drawn regarding consumer acceptance, and the pace, structure and substance of the marketplace.

1030. HOME INFORMATION SYSTEMS. *Yankee Group. January 1980. 295 pg. $9500.00.*

Discusses current and future developments in the small business/home computer market for 1980-1985. Products available in U.S., (Manitoba) Canada and Japan provided, with marketing data, financial comparisons and analysis, costs, vendors, revenues, and sales and growth rates of the home computer industry. Publication is volume 4 of the CIS yearly service. Available to service subscribers.

1031. AN IBM ALTERNATIVE: HEWLETT PACKARD 3000. *Dietz, Laurence. Strategic Inc. January 1979. 85 pg. $750.00.*

1979 report is based on a survey of HP3000 users. Compares IBM systems and HP3000, including financial analysis, new products, annual sales, features / options, hardware, software, revenues, prices and market analysis. Gives detailed data on configuration analysis printers, CRT, peripherals, installation, lease, rent or buy analysis, software, staffing and future plans.

1032. IBM AND THE PLUG COMPATIBLE MANUFACTURERS IN CANADA. *Evans Research Corporation. v.11, no.7. March 1982. 16 pg.*

Analyzes the Plug Compatible Manufacturers (PCM) market in Canada with specific relation to IBM. Examines IBM's EDP revenues (1979, 1980, 1981) and those of compatible suppliers in Canada, market shares for various products; internal Cpu performance; systems available and present locations, Canadian Disk 33XX populations, and disk drive market shares.

1033. THE IBM H SERIES ANALYSIS. *Creative Strategies International. February 1981. $595.00.*

Statistical report on a survey of end users' impressions and reactions to the IBM 3081 ''H'', the 3033-S Series and the Amdahl 580. Analyzes user ranking of desired features and options, disappointments and areas of concern, order details, status, etc. Includes end-user attitudes toward IBM vs. Amdahl, and IBM vs. plug compatible mainframe (PCM) vendors, with regard to pricing, price / performance, support and convenience. Reviews future of DDP communication trends and anticipated growth rates for end users.

1034. THE IBM IMPACT ON DISTRIBUTED DATA PROCESSING. *Yankee Group. November 1979. 260(est.) pg. $9500.00.*

Focuses on IBM's beginnings in the distributed data processing industry and forecasts the industry through 1983. Overall market analysis of industry, including shipments, growth rates and vendors. New IBM product features, costs, hardware (mainframes, peripherals, memories and terminals), shipments and software (DOS/VSE, VM/370, DPPX/8100). IBM future strategies provided. Volume 11 of yearly CIS service by Yankee Group. Available to service subscribers.

1035. IBM PRODUCT EVOLUTION REFERENCE MANUAL. *Creative Strategies International. December 1980. $495.00.*

This statistical data base reference manual provides a detailed evolutionary history and forecast for IBM's general-purpose systems. The product history begins in 1969 and continues through 1983 for every general-purpose system. Each product is presented individually, and collectively, by price category. Key product parameters include first customer shipments, price, performance ratios, price / performance factors, annualized shipments, retirements, and net installations. The market forecasts are also categorized into separate sections for U.S. and worldwide IBM product installations.

1036. IBM SECOND QUARTER USER SURVEY: 4300 ANALYSIS. *International Data Group. Quarterly. June 1980. 29 pg.*

Evaluation of attitudes toward and use of IBM 4300's, including the 4331-2. Survey conducted to 3 user groups: those with 115, 125, 135, 138, 145 or 148 installed or on order; those who cancelled 4300 order; those with 4331 or 4341 installed. Survey discusses hardware and software issues, giving user attitudes, market penetration, installed base, manufacturers' orders and market forecasts for 1980.

1037. THE IBM SYSTEM / 38. *Creative Strategies International. June 1979. $995.00.*

Analysis of the market impact of the IBM·System / 38 for peripherals manufacturers / other systems and software suppliers. Forecasts to 1983. User attitude survey included.

1038. IBM TAPE AND DISK DRIVE MARKETS: CONTINUOUS INFORMATION SERVICE. *Interactive Data Corporation. October 1979. 45 pg.*

1979 report on IBM disk drive and tape markets covers trends of prices, and rates, 1977-1981; site analysis and distribution of disks and tapes, 1978-1979; vendor market shares, 1978-1980; and forecasts of uses of products, 1978-1982. Publication is part of I.D.C.'s Continuous Information Service.

1039. THE IBM 3270 INFORMATION DISPLAY MARKET AND THE COLOR GRAPHICS IMPACT. *Creative Strategies International. June 1980. $1195.00.*

Analysis of the market impact of the IBM 3270 information display product line, with forecasts to 1984. Surveys user attitudes and competition.

1040. IBM 360 / 370 / 303X MIGRATION: SERVICES AND SOFTWARE INFORMATION PROGRAM. *Interactive Data Corporation. August 1979. 71 pg.*

Research report covering IBM 360 / 370 / 303X Migration in the U.S. Product lines, purchase and lease options and distribution rates and uses are discussed. Leasors of systems listed. This publication is part of I.D.C.'s Services and Software Information Program.

1041. IBM'S BILLION DOLLAR BABY: THE PERSONAL COMPUTER. *Isaacson, Portia. Future Computing Inc. 1981.*

Forecasts the impact of IBM's entry into the personal computer market on such competitors as Apple and Radio Shack, through 1985. Discusses the importance of the IBM system's compatibility with CP/M software, and the distribution networks and product strategies of the various manufacturers.

1042. IBM'S DDP PRODUCTS: PRODUCT ANALYSIS AND USER REACTIONS: SERVICES AND SOFTWARE INFORMATION PROGRAM. *Interactive Data Corporation. April 1979. 63 pg.*

1979 research report on IBM's DDP products. Installations, number of facilities, distribution, price and performance comparisons, applications and competition of products analyzed (1978). User attitudes of IBM products provided. Publication is part of I.D.C.'s Services and Software Information Program.

1043. IBM'S EVOLVING NETWORK STRATEGIES, USERS REACTIONS. *Yankee Group. November 1980. 200 pg. $9500.00.*

1980 report part of Communications / Information Systems Planning Service. Available to service subscribers. Analyzes and positions new IBM products and hardware and software

developments. Examines developments in IBM's antitrust case.

1044. IBM'S LONG RANGE STRATEGIES FOR SMALL BUSINESS COMPUTERS. *Creative Strategies International. August 1979. $995.00.*

A focused analysis examining IBM's anticipated long range market and product objectives for small business computers (SBC). Key discussions include: market / product positioning by price and function; GSD vs. DPD; hardware / software unbundling; firmware packaging; multiple growth path criteria; incremental growth path product packaging; high value added product tools and services; and product life cycles.

1045. IBM'S LONG RANGE STRATEGY FOR MINI-MAINFRAMES. *Creative Strategies International. September 1980. $1195.00.*

A comprehensive analysis of IBM's long-range economic, marketing and product development objectives for the 4300 system and related families (such as the 8100). The competitive implications of recent and anticipated tactical and strategical maneuvers are examined. A detailed categorical review of soft versus hard orders and the resultant five-year forecast is provided. The 1979 through 1984 forecasts include the anticipated 4331 Model II and the 4351.

1046. IMPACT OF PC LOCAL AREA NETWORKS. *Strategic Inc. no.1504. August 1982. 140 pg. $1500.00.*

Analysis of the market for personal computer local area networks includes technology, end uses, products, marketing strategies, performance installations, sales, product mix, costs, user attitudes, competition, and forecasts. Lists business addresses of vendors.

1047. IMPROVING THE PRODUCTIVITY OF ENGINEERING AND MANUFACTURING USING CAD / CAM. *INPUT. 1980.*

A multiclient study analyzing currently available and projected computer aided design and computer aided manufacturing systems (CAD / CAM) worldwide in four application areas: electronics, mechanical engineering, civil engineering and mapping. Research carried out with on-site vendor and user interviews and analysis in the U.S., Europe and Japan.

1048. INDEPENDENT PACKAGED SOFTWARE REFERENCE BOOK: SERVICES AND SOFTWARE INFORMATION PROGRAM. *Interactive Data Corporation. August 1980. 114 pg.*

1980 reference book covers the systems software market, the utility software market and the applications software market. Revenues for 1979 and 1984, growth rates for 1978-1984, market shares of suppliers, and product performance of each given. Vendor and supplier rankings and revenues given through 1979. Brief chart compares independent U.S. suppliers' U.S. and international software revenues. Publication is part of I.D.C.'s Services and Software Information Program.

1049. INDUSTRIAL ROBOTS. *Robertson and Associates, Inc. October 1981. 200 pg. $1500.00.*

Market research report provides information on the industrial robot market in the U.S. Contains data on forecasts for market potentials by market and customers; best applications; prices and estimated sales; market shares of robot manufacturers; and improved robot specifications. Results compiled from some 228 in-depth interviews in the following markets: automotive, aerospace, machinery, domestic appliances, electronics, electrical, plastics, metals, glass, die casting, foundries, forging, welding, and others.

1050. INDUSTRY FORECAST 1979 (COMPUTERS). *International Data Group. May 1979. 50 pg.*

1979 industry forecast discusses forecasts, expenses, revenues, U.S. shipments, budget analyses and trends in the computer industry. Includes hardware, software, supplies, communications expenses and staff-related expenses.

1051. INFORMATION EQUIPMENT STRATEGY PROGRAM: MINICOMPUTER AND MICROCOMPUTER MARKETS: VOLUME 3A. *Quantum Science Corporation. 1980. $60.00.*

Report provides market overview and analysis, industry and applications analysis, and product and vendor information with vendor profiles. Publication is part of MAPTEK series.

1052. INFORMATION EQUIPMENT STRATEGY PROGRAM: PRINTER MARKETS: VOLUME 3B. *Quantum Science Corporation. 1979. $60.00.*

This volume of MAPTEK series provides an overview of the market structure for types of printers and types of vendors. Forecasts market through 1982 including projected changes in OEM / end user, impact / non-impact, line / character, and computer equipment/ office equipment mix. Analyzes printer technology and indicates significant product trends. Lists key vendors and types of products offered.

1053. INFORMATION INDUSTRY SURVEY 1980. *Information Industry Association. Annual. July 1980. $700.00.*

Survey measures the current size and future directions of the U.S. profit-making information industry, in terms of profitability, growth levels, new areas of exploitation, product lines, markets served, types of firms, foreign sales, and number of employees.

1054. INFORMATION PROCESSING AT LARGE ORGANIZATIONS. *International Data Group. November 1979. 51 pg.*

Discusses office automation systems products (1979), such as electronic mail, word processing networks, message handling structures and teleconferencing. Describes social and environmental aspects and vendor opportunities.

1055. INFORMATION SECURITY SYSTEMS MARKET IN THE U.S. *Frost and Sullivan, Inc. December 1981. 360 pg. $1200.00.*

Analysis of markets for computer security equipment, with forecasts to 1986. Examines industry structure, market penetration, the technology, military markets and other end uses, and foreign competition. Profiles vendors, including product lines, market shares and sales volume. Gives market survey data.

1056. INFORMATION SERVICES STRATEGY PROGRAM: DATA BASE AND INDUSTRY SECTOR FORECAST: VOLUME 3. *Quantum Science Corporation. Annual. 1980. 278 pg. $60.00.*

Annual report discusses markets in the information services industry for 1978-1984. Market outlook includes expenditures and distribution; key uses; data base needs of businesses and users; applications; growth including growth rates, demand and end uses; and end user expenditures by industry sector, including manufacturers and sales. The publication is part of MAPTEK series.

1057. INFORMATION SERVICES STRATEGY PROGRAM: MARKETS AND STRATEGIES: VOLUME 2. *Quantum Science Corporation. Annual. 1980. 124 pg. $60.00.*

Annual report analyzes the markets for computer services, software support services and equipment support services. Revenues, expenditures, expenses, and distribution of suppliers and vendors given. Growth rates, costs, features, prices

and installations of product lines given. Analysis are forecasted to 1984. The publication is part of MAPTEK series.

1058. INFORMATION SYSTEMS IN THE AUTOMOTIVE INDUSTRY. *Frost and Sullivan, Inc. March 1981. $1000.00.*

Market survey covering utilization of information systems by automotive dealers. Includes market analysis and forecasts and profiles of vendors.

1059. INK JET PRINTING - THE UNFULFILLED PROMISE. *Strategic Inc. August 1980. 148 pg. $950.00.*

Market analysis for ink jet printing. Includes market segmentation and utilization, user attitudes, major vendors, new products, and market penetration prospects.

1060. INNOVATIONS IN DOCUMENT PROCESSING: THIRD QUARTER RESEARCH PROJECT. *Trans Data Corporation. September 1982. $3400.00.*

Survey and analysis of markets for bank document processing services. Examines effects of new technology, current installations, costs and benefits for banks, marketing strategies and prices. Service includes seminar.

1061. INSURANCE AGENCY AUTOMATION. *Venture Development Corporation. January 1982.*

Analysis of the market for computers and systems for insurance agencies. Includes forecasts of shipment growth rates to 1986, market segmentation, vendors, market shares of installed base in 1981; industry structure, technology, consumption of supplies, market survey of users' attitudes, demographic statistics, factors in purchases, and marketing strategies.

1062. INSURANCE ON-LINE DATA COMMUNICATIONS AND DATA PROCESSING SYSTEMS MARKET (U.S.). *Frost and Sullivan, Inc. April 1979. 185 pg. $700.00.*

Examines insurance industry and factors which will accelerate or retard the use of on-line data communications market. Surveys sent to users to solicit views. Limited on-line networks evaluated.

1063. INTEGRATED WORKSTATION OPPORTUNITIES: WORKSTATION SPECIFICATIONS, MARKET FORECASTS, OCCUPATIONAL ANALYSIS: VOLUME 2: MARKET POTENTIAL, SPECIFICATIONS, AND STRATEGIES. *Quantum Science Corporation. 1979. $9000.00.*

This multiclient study provides a multidisciplinary analysis of the evolving market for integrated workstations. It provides equipment suppliers, systems designers, software companies, network information services companies, communications services suppliers and leading edge users with information and recommendations for product, market, financial and investment planning. Volume 2 focuses on 18 Integrated Workstations: accountants, architects, design engineers, draftsmen, insurance agents, pharmacists, real estate agents, registered nurses, stock brokers, chief executive officers, credit managers, financial managers, lawyers, office managers, personnel managers, purchasing agents, sales managers, and secretaries. For each of the 18, field interviews and analyses were conducted to forecast the specifications, features and potential markets of each integrated workstation.

1064. INTEGRATED WORKSTATION OPPORTUNITIES: WORKSTATION SPECIFICATIONS, MARKET FORECASTS, OCCUPATIONAL ANALYSIS: VOLUME 3: MARKET POTENTIAL FORECASTS AND DATA BASE. *Quantum Science Corporation. 1979. $9000.00.*

This multi-client study provides a multidisciplinary analysis of the evolving market for integrated workstations. It provides equipment suppliers, systems designers, software companies, network information services companies, communications services suppliers and leading edge users with information and recommendations for product, market, financial and investment planning. In volume 3, the key occupational parameters are identified for 60 occupations: total staff forecast for each occupation; industry sector distribution; geographical region; educational level; income level; age; sex; and data base requirements. This data is analyzed to provide potential unit placements for workstation configurations across all industry sectors.

1065. INTELLIGENT COMMUNICATIONS PROCESSOR MARKET. *Frost and Sullivan, Inc. April 1982. 237 pg. $1275.00.*

Analysis of markets for intelligent communications processors (ICP's). Forecasts shipments and sales in units and dollar values, 1982-1987. Includes regulations affecting the market, end uses, 1982 installed base, competition among vendors, market shares in 1987, and market survey results including user attitudes and rankings of ICP's.

1066. INTELLIGENT COPIER / PRINTERS. *Creative Strategies International. November 1979. $995.00.*

IC/Ps using electrophotographic imaging technolgoies (xerography) with intelligent microcircuitry are segmented into three product price classes ranging from $7,000-$15,000; $15,000-$35,000; and $35,000- $100,000. Each IC/P product class is reported by U.S. market penetration for 1979-1983 in units and dollars. Future product features are predicted and compared to current product capabilities. The current suppliers and anticipated competitive entrants into the market are assessed.

1067. INTELLIGENT COPIERS ARRIVE: VOLUME 79. *Quantum Science Corporation. April 1979.*

Describes the basic capabilities of new and forthcoming intelligent copiers. Includes a discussion of the IBM 6670, a comparison with the Wang Image Printer, and the competition from stand-alone copier markets.

1068. INTERACTIVE ENTERTAINMENT AND EDUCATION SERVICES IN THE HOME. *Yankee Group. July 1981. $8500.00.*

Discusses consumer demand for interactive entertainment and education services provided by home computers. Examines market expectations in the context of strong consumer demand for self-improvement. Analyzes the potential for an electronic interactive university and for professional education capabilities of personal computing. Investigates the new markets for games and the evolution of computer game technology over interactive networks. Part of Yankee Group's Home of the Future Service and available only to subscribers.

1069. KEY EDP CONTRACTS. *International Data Group. January 1980. 250 pg.*

Lists known key customers, equipment and services of EDP vendors. Key customers defined as those that account for a high percent of vendor's revenues (to $1M) and/or represent penetration of a new market. Gives tabular data on reported size of customer bases, number of customers and amount of contracts. Includes make vs. buy decision analysis.

1070. KEYBOARD AND KEYSWITCH FORECAST. *Gnostic Concepts, Inc. 1981. $7500.00.*

Report analyzes keyboard and keyswitch markets, end-use trends, technology outlook, competitive environment, economic outlook, and equipment production forecast. Market data includes total U.S. production, noncaptive production, captive production, total U.S. shipments, total U.S. domestic shipments, foreign shipments, total U.S. purchases, total OEM purchases, noncaptive purchases, captive purchases, and

U.S. OEM consumption. Keyboard and keyswitch categories include electromechanical, reed, membrane, hall effect, inductive, and capacitive.

1071. KEYS TO PROFITABILITY IN THE COMPUTER SERVICES AND SOFTWARE MARKETS: SERVICES AND SOFTWARE INFORMATION PROGRAM. *Interactive Data Corporation. August 1980. 87 pg.*

1979 profitability analysis of remote autotransaction services, remote problem-solving services, batch services, packaged software, mixes, and contract services / custom software services. Comparisons of each give product lines, revenues, profit margins, sales of employees, current ratio analysis and costs by company size and by growth rates. Industry structure and overviews give primary suppliers, with average and total revenues. Publication is part of I.D.C.'s Services and Software Information Program.

1072. KEYS TO PROFITABILITY IN THE INDEPENDENT PERIPHERAL MARKET - 1979. *International Data Group. September 1979. 80 pg.*

Examines the financial performance and profitability of various peripheral product areas: keyboard equipment disk / tape drives, memory equipment, printers, interconnect equipment, autotransaction equipment, graphic equipment, remote batch terminals, and card recognition equipment. Market data includes industry structure (1978), revenues (1973-1983), product sector review (1973-1983), leading suppliers (1978), and market shares of high performance drives (1978).

1073. LOCAL AREA NETWORKS 1981 TO 1990: A STRATEGIC ANALYSIS. *Venture Development Corporation. October 1981.*

Examines the potential market for local area computer networking equipment through 1990. Envisions three possible scenarios: standardization of, and market capture by, a single, low-cost system; a 'technological jungle' of competing technologies, followed by a 'shakeout' leading to dominance by the most popular systems; or a segmented market, with computer companies offering systems compatible only with their own hardware. Projects market penetration by end use, and activities and competitive positions of individual manufacturers.

1074. LOCAL NETWORKS AND HOME INFORMATION SYSTEMS. *International Resource Development, Inc. March 1982. 160 pg. $1285.00.*

Market analysis and forecasts to 1992 of the utilization of new technologies in cable and office computing equipment to enable working at home. Projects effects on employees and facilities. Includes analysis of new cable services.

1075. LOW SPEED PRINTER AND TELEPRINTER MARKETS. *Gnostic Concepts, Inc. 1981. 500 pg. $9800.00.*

Analyzes competitive environment, end user analysis, market outlook and technology trends of low speed printers and teleprinters. Market data includes aggregate demand forecast, demand by product and application, competition, product pricing, and market segmentation.

1076. LOW-COST COMPUTER PRINTERS. *Creative Strategies International. December 1980. $1195.00.*

Analyzes markets for low-cost impact and non-impact receive-only and keyboard send / receive computer printers. Analyzes both OEM, PCM, and captive segments as well as vendor distribution segments. Analyzes market trends in units shipments, installed base, and revenues for each product type for period 1980-1985. Analyzes competitive factors and vendor 1980 market shares.

1077. MACHINE CENTERS AND COMPUTER - AIDED MANUFACTURING SYSTEMS. *Frost and Sullivan, Inc. March 1983. $1150.00.*

Market report provides analysis and 10-year forecast for machining centers, vertical, CNC, and microprocessor lathes; mini-computer controls; computer - aided design equipment, part coding systems; engineering services for group technology and manufacturing cell systems; and numerical controls. Includes supplier profiles and market shares.

1078. MAINTENANCE CONSIDERATIONS IN THE INFORMATION HANDLING MARKETPLACE. *Alltech Publishing Company. 1980. $6000.00.*

4-volume study of the current status and future trends in maintenance of information handling systems. Included is a summary of study findings and conclusions; an overview and analysis of personal computing (structure of today's information handling marketplace and implications to maintenance, with extensive analysis of maintenance in personal computing); maintenance in the business automation marketplace (survey and analysis of user installations, maintenance from equipment suppliers and third parties, including directory and profiles of third party firms); and maintenance of ADP equipment in the federal government (analyzes the federal government as regulator and as a user of maintenance products and services).

1079. MAINTENANCE IN COMPUTER INSTALLATIONS. *Alltech Publishing Company. March 1980. $3000.00.*

1980 2-volume study of the market for maintenance vendors of foreign peripherals and CPUs. Study was conducted by questionnaire sent out to hardware maintenance vendors. Volume 1 is an analysis of the market in general, and Volume 2 compares users by CPU vendor(s) and features maintenance profiles of user installations.

1080. MAINTENANCE INFORMATION SERVICE. *Alltech Publishing Company. Annual. 1980. $1500.00.*

1980 report on information handling equipment including communications and computer equipment. Service includes introductory seminar, a working notebook on the marketplace, ongoing reports, newsletters and analyses for 1 year and analysis of specific topics (equipment suppliers, maintenance firms, users, research guides).

1081. MANAGEMENT ACTION REPORT: VOLUME 1. *Quantum Science Corporation. Annual. 1980. 92 pg. $60.00.*

Annual report gives an overview of the information processing industry, covering growth rates, distribution, expenditures, and value of shipments of existing product lines and new products through 1984. Suppliers' and vendors' productivity rates, revenues, employment figures and growth rates provided. Publication is part of MAPTEK series.

1082. MANAGEMENT WORKSTATIONS: MARKETS AND STRATEGIES 1981-1986. *Quantum Science Corporation. 1982. 265 pg. $1695.00.*

Analysis of management workstation markets and the industry structure, with forecasts to 1986. Includes market surveys of user's attitudes and rankings, technology, 1981 installed base, manufacturers and vendors, shipments, and distribution and marketing strategies.

1083. MAPTEK BRIEF: TURNKEY SYSTEM MARKETS: SELLING THE NEXT GENERATION OF INFORMATION SYSTEMS AND SERVICES. *MAPTEK (Management Action Program in Technology). Quantum Science Corporation. June 22, 1979. 40 pg.*

This analysis includes market forecasts and strategic recommendations in turnkey system markets. Compares competi-

tion, ratings, strategies, market values, and sales of turnkey, NIS companies, mainframe manufacturers and software houses. User responses, 6430 vendors and 40 users provide basis for data. Discusses current and future market outlook and provides a market analysis by size of company and industry sector. This information is only available to MAPTEK subscribers.

1084. MARKET FOR COMPUTER ASSISTED INSTRUCTION (CAI) IN U.S. EDUCATION. *Frost and Sullivan, Inc. January 1982. 257 pg. $1200.00.*

Analysis of the U.S. market for computer assisted instruction (CAI), with forecasts to 1987. Includes historical data, 1981 installations, and market segmentation from elementary schools to universities.

1085. MARKET FOR COMPUTERS USED IN SMALL BUSINESSES. *Frost and Sullivan, Inc. December 1981. 250 pg. $1150.00.*

Provides an analysis and forecast of the market for computers applicable to small businesses. Ten-year forecast in units and dollars is furnished for hardware, system software, and application software by 4 computer price ranges. Evaluates industry structure, competition, and changing distribution channels. Current and projected market share is presented for current and potential major small business manufacturers in each price range of computers.

1086. MARKET FOR DATA PROCESSING SYSTEMS IN CREDIT UNIONS. *Frost and Sullivan, Inc. September 1981. 140 pg. $1,050.00.*

Market report of 140 pages provides an analysis and forecast for data processing systems in credit unions. Credit union trends are traced. Five-year forecasts are provided for batch and on-line computer services, in-house systems, maintenance, and for business forms. Major competitors are identified and market shares are determined. The 100 largest credit unions are listed by name, state, assets, and members.

1087. THE MARKET FOR GOVERNMENT DATA BASES SOLD THROUGH COMMERCIAL FIRMS. *Frost and Sullivan, Inc. August 1980. 206 pg. $950.00.*

Analysis and forecast of government data bases sold commercially. Data bases considered include: demographics, consumption economics, finance and investments, housing, etc. Identifies users, providers, and trends in data collection. Discussion of industry structure. Inventory and tabulation of data bases, including characteristics, costs, applicability, timeliness.

1088. MARKET OUTLOOK FOR MAGNETIC TAPE MEDIA AND FLEXIBLE DISK MEDIA. *International Data Group. May 1979. 65 pg.*

Analysis of magnetic tape media and flexible disk media includes examination by market segment (1978), price versus performance comparisons of IBM processors (1978), floppy disk media annual shipments (1978), usage (1978), and product mix (1978).

1089. MARKET STRATEGIES FOR SELLING SMALL BUSINESS COMPUTERS. *Frost and Sullivan, Inc. September 1981. 264 pg. $1,100.00.*

Strategy report contains 264 pages analyzing the present ways in which small business computers are marketed and suggesting which methods have the best chance for success. Covers product planning, distribution, direct sales, retailing, and advertising.

1090. THE MARKETS AND COMPETITIVE ENVIRONMENT FOR ROTATING FLEXIBLE MEDIA. *Creative Strategies International. January 1981. $5000.00.*

Market analysis report on rotating flexible media (floppy disks), with forecasts to 1985. Covers manufacturers, market shares, competition, prices, performance, and capacity. Forecasts manufacturers shipments, revenues, and units in operation, 1985.

1091. MARKETS FOR COMPUTER PRINTERS. *Creative Strategies International. August 1979. $12000.00.*

An in-depth analysis of the computer printer market. Includes market forecasts through 1983 by product segment: line impact, serial impact, non-impact and teleprinters. Details analysis of survey results highlighting end-user needs and requirements. Four volumes available at $4500.00 each.

1092. MEMORIES - PRODUCTS, MARKETS AND OPPORTUNITIES. *Mackintosh Consultants, Inc. 1981.*

Report on new products for computer memories in the U.S., Western Europe, and Japan. Includes market analysis and forecasts. Examines utilizations.

1093. MICROCOMPUTER FORECAST. *Gnostic Concepts, Inc. 1981. 1000 pg. $12000.00.*

Analyzes the microcomputer market and trends at four levels: component level including microprocessor set or chip sets, memory circuits and auxiliary circuits; microcomputer or board level; subsystem level; and system level. Areas covered are market analysis, end-use trends, microprocessor / microcomputer technology, competitive environment, and economic outlook.

1094. MICROCOMPUTER SOFTWARE PACKAGES. *International Resource Development, Inc. no. 168. 1981. $985.00.*

Report examines Japanese share of U.S. microcomputer system market, both hardware and software. Reviews U.S. micro software firms, contracts with Japanese manufacturers, and the total micro hardware and software market. Includes revenues and market potential for 1985 and 1990.

1095. MICROCOMPUTER SOFTWARE STRATEGIES. *Creative Strategies International. 1981. $1200.00.*

Microcomputer software market analysis. Profiles software vendors and hardware manufacturers. Covers marketing strategies, products and prices and distribution channels. Includes small business consumers of minicomputers, 1977-1984; competition, and software costs as a percent of hardware investments.

1096. MICROCOMPUTERS IN EDUCATION. *Creative Strategies International. 1981. $1200.00.*

Market analysis for microcomputers in education. Includes historical data and current industry structure; manufacturer profiles; production costs; and profit margins. Examines distribution channels; government funding for educational computers; market segmentation by type of school, price, and end uses. Covers competition including product features and leases; 1980 market shares, vendor profiles, and 1985 forecasts for shipments, revenues, and retail value of sales.

1097. MICROCOMPUTERS IN LARGE ORGANIZATIONS. *International Resource Development, Inc. March 1982. 134 pg. $1285.00.*

Analysis of microcomputer markets and the impact of volume discounts, offered to large organizations, on the market structure. Includes market survey results concerning user attitudes and plans for purchases. Forecasts shipments and revenues, 1982-1992. Profiles vendors and marketing strategies.

1098. THE MICROPROCESSOR MARKETPLACE. *International Data Group. May 1980. 34 pg. $1500.00.*

Analysis of the microprocessor market and the strategic issues facing the industry in 1980. Types of vendor participants and investments are discussed. Market forecasts expressed in terms of world-wide shipments by U.S. manufacturers; imports of microprocessors and support circuits by foreign manufacturers are not included. Captive production by U.S. manufacturers is considered only to the extent that it reduces demand in the open market.

1099. THE MICROPROCESSOR / MICROCOMPUTER INDUSTRY. *Creative Strategies International. December 1980. $1195.00.*

Examines non-captive market for microprocessor chips, microcomputer chips, microcomputer boards, general-purpose microcomputer systems, and microcomputer development systems. Forecasts given in units and dollars for 1980 base year through 1985 by the following segmentation: architecture (32 segments), end-user industry distribution, geographical producers, and geographical consumers. Report contains unit pricing trend analyses and vendor market share analyses for each product segment. Includes a strategic analysis of the competitive trends of innovators and second sourcers as they move between components and systems.

1100. MID-RANGE COMPUTER PRINTER MARKETS. *Gnostic Concepts, Inc. 1981. 400 pg. $9800.00.*

End user analysis, competitive analysis, technology trends, OEM requirements, and market outlook of mid-range line and page printers. Market data includes: aggregate demand forecast, product forecast by applications, technology assessment, competition, and paper and forms trend.

1101. MILITARY ELECTRONIC MEMORY MARKET IN THE U.S. *Frost and Sullivan, Inc. September 1982. 213 pg. $1250.00.*

Analysis of military markets for computer memory devices. Examines product mix and market size, market segmentation and end use applications, new technology forecasts to 1990, consumption forecasts to 1987, rankings and sales of manufacturers, and profiles of major defense contractors.

1102. THE MILITARY SOFTWARE MARKET IN THE U.S. *Frost and Sullivan, Inc. 1980. 707 pg. $850.00.*

Analysis, and forecast through 1985 of the military market for computer software systems. Categorizes estimated Pentagon expenditures by real-time ("C3I", avionics) and non-real-time (management information systems, logistics and maintenance, and scientific and engineering sytems). Discusses system specifications and projected innovations. Considers market portions not pre-empted and projected employment.

1103. MINI / MICRO PERIPHERAL MARKETS. *Newton, Charles W., editor. International Resource Development, Inc. January 1981. 208 pg. $985.00.*

Study of mini / micro computer markets. Includes analysis of market, industry structure, and end uses. Examines businesses and consumers, market penetration, utilization of specific products, profiles of manufacturers and suppliers. Provides prices, distribution and forecasts of growth rates and revenues, 1980-1990.

1104. MINICOMPUTER HARDWARE MARKET. *Frost and Sullivan, Inc. April 1982. 244 pg. $1250.00.*

Analysis of minicomputer hardware markets. Forecasts unit and dollar sales, 1981-1986, by distribution channels. Examines market size, incustry structure, competition, end uses and new technoloby. Profiles of vendors include market shares, sales, histories of corporations, product lines, marketing strategies and number of employees in sales.

1105. MINICOMPUTER INDUSTRY TRENDS, MARKETS AND OPPORTUNITIES. *Spelman, Le-Ellen. MSRA, Inc. October 1981. 300 pg. $200.00.*

Conference papers by executives from minicomputer corporations analyzing the industry. Includes forecasts of industry revenues and shipments, 1977-1985; new technology and performance; end uses; market size to 1985; product mix; market shares and earnings of vendors; distribution channels, and installed base.

1106. THE MINICOMPUTER MARKETPLACE - 1979. *International Data Group. December 1979. 62 pg.*

Analysis of the minicomputer marketplace includes market value (1975-1983), distribution (1975-1983), shipments (1975-1983), shares (1978), average revenues (1970-1983), and growth rates (1975-1983). Provides information regarding domestic versus inter- national areas, OEM and end-user distribution channels, and application types. Gives typical minicomputer configurations and use of peripherals with selected minicomputers.

1107. MINICOMPUTERS: COMPUTER AND PERIPHERALS INDUSTRY ANALYSIS SERVICE. *Creative Strategies International. February 1980. 113 pg.*

Comprehensive report on minicomputers, including system features, applications, shipments, consumption, costs and prices. Financial analyses include revenues, market shares, market segmentation, purchases and products of competing manufacturers. Data spans 1974 to 1984.

1108. THE MOVING SURFACE STORAGE DEVICE (MSSD) COMPUTER PERIPHERAL MARKETPLACE. *Frost and Sullivan, Inc. April 1980. 244 pg. $1000.00.*

Market survey covering user attitudes toward moving surface storage device (MSSD) computer peripheral equipment such as tapes, disks, bubble memories, and laser storage devices. Shows market segmentation by end use industries, size of corporations, and dollar value of current installations. Forecasts market growth to 1989. Gives ratings of manufacturers and products.

1109. MULTIFUNCTION SYSTEMS IN DISTRIBUTED PROCESSING. *International Data Group. February 1980. 52 pg. $1500.00.*

Broad study of the multifunction systems market broken into 4 major areas of concentration: Training/education; applications development; equipment selection and usage; and the decision-making process. Major trends, developments and recommendations presented. Methodology included multiple personal and telephone interviews of 20 leading vendors and 25 large DDP users. Uses case study approach and emphasizes issues and trends rather than statistical analysis. 1979-1980 data.

1110. NETWORK PLANNING AND IMPLEMENTATION. *Yankee Group. September 1981. $7500.00.*

Report focuses on major organizational impacts of network planning, management and maintenance. Data presented on features (communications equipment) and services users want in their networks. Publication is one of 12 reports published each year, beginning 1981, by Critical Paths Services. Available to service subscribers.

1111. NEW DEVELOPMENTS IN ELECTRONIC BANKING. *Business Communications Company. G-069. May 1982. $1250.00.*

Analysis and forecasts of various markets for electronic funds transfer banking cover automatic teller machines and network developments, electronic point of sales (fund transfers) machines, and network developments.

1112. NEW DIRECTIONS IN ROBOTS FOR MANUFACTURING. *Business Communications Company. September 1981. 100 pg. $850.00.*

Analysis of the robot industry by type, end use industries, and utilization processes. Examines costs, markets, and manufacturers.

1113. A NEW DISTRIBUTION MEDIUM - COMPUTER RETAILING. *Evans Research Corporation. February 1982. 12 pg.*

Synopsis of Evans Research Corporations' study into the Canadian computer retailing marketplace, a study undertaken from both the manufacturers' and store operator's viewpoints. Examines the market, the customer base, products, prices, paid by the customer, and demand and growth in the industry. Includes a brief analysis of major computer store competitors, their 1981 retail sales value and revenues; sales by user groups for micro computers, 1981; and value of shipments of computers 1980, 1981, 1987. Lists the five major competitors for the business computer markets as of January 15, 1982.

1114. THE NEW GENERATION OF DATA TERMINALS. *Yankee Group. April 1981. 200 pg. $9500.00.*

Examines current and emerging data terminal markets, market shares, user requirements and new product opportunities. Part of Yankee Group's Communications Information Systems Planning Service and available only to subscribers.

1115. NEW MARKETS FOR 'DESK - TOP' COMPUTERS. *Business Communications Company. January 1981. 115 pg. $750.00.*

Examines retail distribution of desk top computers in home systems and businesses. Analyzes markets, manufacturers and market penetration strategies.

1116. NEW MARKETS FOR LOW-PRICED COMPUTERS. *Business Communications Company. January 1981. 115 pg. $750.00.*

Analyzes low-priced computer industry in terms of opportunities, requirements, markets and market penetration, company activity and distribution. Includes some 30 tables.

1117. NEW PRODUCTS AND SERVICES THROUGH ELECTRONIC DISTRIBUTION: TELESHOPPING. *Yankee Group. November 1981. $8500.00.*

Home computer systems have the potential to design services for every member of the household. Such social engineering potential is demonstrated by research of interactive technology in the Higasi Ikoma district of Tokyo. Part of Yankee Group's Home of the Future Service and available only to subscribers.

1118. NINTH ANNUAL MINI-MICRO COMPUTER MARKET REPORT: 1980. *Dataquest and Mini-Micro Systems. Cahners Publishing Company. Annual. 9th ed. 1980. $495.00.*

Survey, based on responses received from over 11,000 "Mini-Micro Systems" readers, covers past and forecasted purchases in 22 separate categories, including mini- and micro-computers, tape and disk drives, CRT terminals, printers, modems and related equipment. Gives market trends in each product category. Covers end-users, located at large corporations with volume requirements, at EDP sites, and in scientific and engineering areas, and third-party OEMs such as systems integrators, specialized system OEMs, and software houses and services. Lists OEM and end-user buying plans.

1119. NON-IMPACT PRINTER MARKET. *Frost and Sullivan, Inc. February 1982. $1200.00.*

Forecasts market for non-impact printers by process (xerographic, dielectric, ink, jet and magnetography); application and end user; toner, developer, and paper. Provides future trends.

1120. NON-RETAIL DISTRIBUTION CHANNEL SMALL BUSINESS COMPUTER MARKET. *Frost and Sullivan, Inc. September 1981. 157 pg. $1000.00.*

Market analysis and forecasts for changes in the structure of distribution of small business computers. Profiles manufacturers, suppliers and retail distributors. Shows market shares 1981, 1985.

1121. NON-RETAIL DISTRIBUTION OF SMALL COMPUTERS. *Creative Strategies International. October 1981.*

Analysis of non-retail distribution in small computer markets. Examines trends and market shares for non-retail vendors including industrial suppliers, manufacturers' representatives, and mail order and direct sales distribution.

1122. OEM MINI / MICRO COMPUTER MARKETS. *Allison, Andrew, editor. International Resource Development, Inc. March 1981. 170 pg. $985.00.*

Report on OEM mini / micro computer markets. Includes historical data on computers; and review of products and design trends including peripherals. Covers present market analysis; manufacturers; distribution; and industry structure. Provides forecasts of shipments and revenues for board level CPUs manufactured systems, and OEM systems, 1975-1989; addresses of suppliers; and prices.

1123. OFFICE AUTOMATION AND DISTRIBUTING DATA PROCESSING: PART 1: TWO CONVERGING STRATEGIES. *Yankee Group. April 1981. $7500.00.*

1981 report reviews vendor strategies and offerings in future DDP / Office Automation integration. Network control, standards, software and hardware compatibility and security provided. Publication is one of 12 reports published each year, beginning 1981, by Critical Paths Service. Available to service subscribers.

1124. OFFICE, FACTORY AND HOME AUTOMATION MARKET. *Frost and Sullivan, Inc. October 1980. 175 pg. $800.00.*

Contains financial data; market shares; R & D expenditures; industry outlook and overview. Forecasts market growth in the next 3 years.

1125. OFFICE TECHNOLOGY STRATEGY PROGRAM: DATA BASE AND INDUSTRY SECTOR FORECAST: VOLUME 3. *Quantum Science Corporation. Annual. 1980. $60.00.*

Annual report provides a very detailed analysis of the computer marketplace by vertical industry segmentation. Publication is part of MAPTEK series.

1126. OFFICE TECHNOLOGY STRATEGY PROGRAM: MARKETS AND STRATEGIES: VOLUME 2. *Quantum Science Corporation. Annual. 1980. 214 pg. $60.00.*

Annual report examines major segments of the markets for text preparation equipment, dictation equipment, office communications equipment, facsimile equipment, and PABX equipment. Market segments, competition, product lines, applications, user expenditures, new opportunities, strategies for participants, new technologies, and future actions of leading vendors

and government agencies are analyzed. Part of MAPTEK series.

1127. ONLINE DATABASE SERVICES. *Creative Strategies International. January 1981. 116 pg. $1200.00.*

Forecasts revenues and compound annual growth rates for online database services. Gives four major groups participating in the industry: database producers; online service organizations; integrated services; and users. Over 270 producers and vendors are involved in the U.S., Canada, and Europe.

1128. ON-LINE DATA-BASE SERVICES MARKET. *International Resource Development, Inc. March 1981. 177 pg. $985.00.*

Market analysis for on-line data-base services, with forecasts 1981-1991. Covers industry structure, product groups, businesses involved, distribution systems, and end use industries. Shows revenues, 1981, 1991.

1129. OPPORTUNITIES IN MARKETING SYSTEMS SOFTWARE PRODUCTS. *INPUT. August 1979. 185 pg.*

Market research study considers the U.S. market for systems software. Information was derived from in-person and telephone interviews with top officers of systems software companies. Discusses market analysis, shares, size, segmentation and penetrations; software / hardware sales strategies; prices; revenues; and forecasts for the software and hardware markets. Marketing strategies, personnel factors and vendor types are included.

1130. OPTICAL AND MAGNETIC DISC MEDIA. *International Resource Development, Inc. 1982. 210 pg. $1285.00.*

Analysis of markets for magnetic disc data storage media and competition with new optical disc products. Forecasts shipments and shipment values to 1984 and market shares to 1990. Examines manufacturers, retail distribution and new technology.

1131. OPTICAL DATA STORAGE. *Institute for Graphic Communications, Inc. 1982.*

Analysis of optical data storage markets examines end uses for new technology including military markets, performance and costs, competition with other media, and the corporations involved.

1132. OPTICAL DISCS FOR OFFICE AUTOMATION AND ELECTRONIC PUBLISHING. *International Resource Development, Inc. 1982. 171 pg. $1285.00.*

Analysis of markets for optical discs in office automation and electronic publishing. Covers the technology, performance and capacity; utilizations; competition with other technologies; end uses; manufacturers; revenues; and 1982 shipments of disc drives.

1133. OPTICAL MEMORY MEDIA MARKET. *Frost and Sullivan, Inc. September 1981. 166 pg. $1000.00.*

Market analysis and forecasts for optical memory media, 1981, 1983-1986. Examines demand and capacity trends, product performance, costs, and competition with other media. Includes manufacturer profiles and user attitudes.

1134. ORIGINAL EQUIPMENT MANUFACTURER (OEM) MINICOMPUTER MARKET. *International Data Group. March 1980. 175 pg.*

OEM minicomputer data includes 1980 figures for costs, revenues, prices, suppliers, market size, value added, market structure, and products. Products include automatic test equipment; computer aided design; data acquisition; energy management; factory data collection; industrial process control; materials handling systems; message switching; numerical control; photocomposition; security; and word processing. Discusses selection criteria, hardware requirements, switching potential, technological issues, and maintenance policies.

1135. ORIGINAL EQUIPMENT MANUFACTURER (OEM) PRINTER MARKETPLACE. *International Data Group. February 1980. 60 pg.*

Data for OEM character, serial and line printers includes shipments (1978-1983), product lines (1978-1983), distribution (1979), market shares (1979), revenues (1978-1984), and forecasts to 1984. Information includes configurations, applications, print technology, and financial futures by model.

1136. THE OUTLOOK FOR THE U.S. PERSONAL COMPUTER INDUSTRY. *Meserve, Bill. Arthur D. Little, Inc. December 1981. 40 (est.) pg. $750.00.*

Analysis of markets for small personal computers and the evolving industry structure examines market segmentation, size and growth rates by end uses, retail distribution, and software and information utilities services. Forecasts retail sales, 1978-1985; and gives vendors' market shares, 1979, 1980.

1137. OVER-THE-COUNTER COMPUTER HARDWARE, SOFTWARE AND ASSOCIATED PERIPHERALS MARKET IN THE U.S. *Frost and Sullivan, Inc. September 1981. 332 pg. $1100.00.*

Analyzes and forecasts sales for the market through 1985 for OTC computer products serving hobbyists, consumers, and establishments. Provides company profiles. Examines retail channels of distribution.

1138. PAPER - BASED ELECTRONIC MAIL. *International Resource Development, Inc. 1982. 205 pg. $1285.00.*

Analysis of markets for paper-based electronic mail and competition with other media. Forecasts markets and paper suppliers' revenues, 1982-1992. Profiles vendors including possible mergers.

1139. THE PERSONAL COMPUTER INDUSTRY: AN UPDATE. *Meserve, Everett T. Arthur D. Little, Inc. October 12, 1979. 8 pg. $500.00.*

1979 report on personal computer industry including equipment models, major U.S. vendors, equipment prices, consumption 1978-1983, growth rate and market segments (small business, consumer). An update of 1978 research letter. Data derived from Arthur D. Little, Inc. estimates.

1140. THE PERSONAL COMPUTER INDUSTRY II: A STRATEGIC ANALYSIS. *Venture Development Corporation. 1981. 225 pg. $2950.00.*

Market analysis for personal computers with forecasts, 1981-1986. Examines the competition among manufacturers, industry structure, market shares and distribution channels. Forecasts industry shipments to 1985 from the 1980 installed base. Also examines end uses, foreign markets, user attitudes, purchase factors and demographic statistics.

1141. PERSONAL COMPUTER PRODUCTS AND MARKETS - THE KEY ISSUES. *Strategic Inc. February 1982. 135 pg. $1200.00.*

Analysis of market trends for personal computers and software. Examines the technology, marketing strategies of top vendors, and market penetration opportunities.

1142. PERSONAL COMPUTERS -- STRATEGIES FOR SUCCESS. *Burns, Tom. SRI International. Marketing Services Group. July 1980.*

This program presents an overview of the personal computer market: its size, characteristics, growth and potential; business

implications of projected growth and an evaluation of personal computer product sales, and post-purchase attitudes and buyer preferences. SRI's program includes consumer surveys, extensive datafile integration, consumer modeling and competition review.

1143. PERSONAL WORKSTATIONS - THE IMPACT OF THROWAWAY MEMORY. *International Resource Development, Inc. 1982. 163 pg. $985.00.*

Analysis of markets for data storage and communications equipment and systems for the personal workstation. Forecasts costs and demand to 1992; profiles vendors; and describes new products.

1144. PLANNING FOR THE FUTURE: IBM COMMUNICATION. *Yankee Group. December 1981. $7500.00.*

1981 report discusses facilities, organizational impact, and financing or procurement of EDP and communications. Technology trends, viability studies, IBM strategy analysis, product lines, and pricing given. Publication is one of 12 reports published each year, beginning 1981, by Critical Paths Services. Available to service subscribers.

1145. PLUG COMPATIBLE MAINFRAMES. *Creative Strategies International. February 1981. 145 pg. $1195.00.*

Market study, with forecasts through 1985, covers plug compatible mainframe sales and sales of peripherals and software.

1146. PORTABLE COMPUTING DEVICES. *Creative Strategies International. 1982. $1450.00.*

Analysis of markets for portable computing devices includes industry structure; vendors and product lines; end uses; regulations; distribution channels; marketing strategies; product group market shares, 1981; technology; average prices; competition; and forecasts of shipments and revenues, 1981-1986.

1147. PORTABLE DATA RECORDING SYSTEMS AND ASSOCIATED SOFTWARE MARKET IN THE U.S. *Frost and Sullivan, Inc. April 1981. 286 pg. $1000.00.*

Market analysis for portable data recording devices and associated software, with sales forecasts to 1990. Covers industry structure and competition; manufacturers, products, vendors' market shares, and end uses. Includes a market survey of user attitudes on performance and future requirements.

1148. PORTABLE TERMINALS. *International Resource Development, Inc. March 1980. 168 pg. $895.00.*

Technological and market forecasts through 1990 for portable timesharing terminals, radio data terminals, software-intensive portable consumer terminal products. Expected demand for thermal paper to support increasing installed base of terminals. Impact of cellular radio, teletext and videotex services. Market shares of more than twenty suppliers; expected new products.

1149. PROCESSOR DATA BOOK, 1980. *International Data Group. October 21, 1980. 62 pg. $1500.00.*

1980 marketing and statistical source for IDC clients on each of four processor markets: general purpose, mini, small business and desk-top computers. Market surveys and analyses of products given by vendor, by manufacturer, and by size, world shipments and domestic shipments, by U.S. manufacturers. Including forecasts to 1984 and product prices.

1150. PROSPECTS IN THE VIDEO GAME INDUSTRY. *Frost and Sullivan, Inc. June 1982. 20 pg. $275.00.*

Analysis of video game markets and investment prospects for video game corporations. Includes profiles of vendors and product lines; estimated sales; shipments of games, 1977-1981; rankings of vendors; marketing strategies; forecasts to

1984; market penetration; competition; dollar and unit sales, 1980-1982; and foreign markets.

1151. REMOTE DATA BASE SERVICES. *International Data Group. April 1979. 50 pg.*

Analysis of remote data base services includes market structure (1977-1983), market segmentation (1977-1983), growth rates (1977-1983), and market shares (1978). Provides market size (1977-1983) classified by type: legal, bibliographic retrieval, stock quotation credit / check authorization, financial / economic, marketing / demographic, and scientific / engineering. Defines data base services and their relation to the total market.

1152. REPORT TO MANAGEMENT ON COMPUTER EQUIPMENT SUPPLIERS. *Alltech Publishing Company. June 1980. $25.00.*

1980 report covers industries served by computer equipment suppliers, types of computer systems (printers, plotters, memory, disk, tape and card equipment), supplier data (number of employees, revenues, date established) and sources of maintenance for products. A directory of names, addresses and phone numbers of 91 companies included.

1153. RETAIL AUTOMATION TO 1983. *Creative Strategies International. February 1980. $1195.00.*

U.S. markets for electronic and non-electronic cash registers and electronic point-of-sale terminals. Forecasts to 1983. Shows deliveries by type from 1971-1977 and projections through 1983. Details shipments by U.S. manufacturer and U.S. imports by country of origin. Discusses developments in retailing as related to retail automation and the cash register industry. Includes company profiles and competitive analyses.

1154. RETAIL BANKING INFORMATION SYSTEMS AND SERVICES MARKET IN THE U.S. *Frost and Sullivan, Inc. April 1982. 351 pg. $1150.00.*

Analysis of markets for information systems hardware and software for commercial banking. Forecasts unit and dollar sales for hardware and services, 1982-1991, from calculations of installed base and growth rates. Surveys market demand and banking industry structure, including bank deposits, reserves and assets.

1155. RETAIL DISTRIBUTION OF SMALL COMPUTERS. *Creative Strategies International. September 1980. $1195.00.*

Defines, delineates and analyzes types of computer retail outlets by product line, size, staffing, support services, software, maintenance and repair issues. Examines users by type. Profiles retail stores, based upon a mail survey of all retail stores and personal interviews. Identifies and quantifies revenue size, revenue generating products, customer patterns, repair, store purchasing practices, success and failure factors, store design, size and layout, investment costs, pro-forma P&L, financing, etc.

1156. RETAILING ELECTRONICS PRODUCTS TO THE END USER: A STRATEGIC ANALYSIS. *Venture Development Corporation. January 1982.*

Market survey and analysis of changing patterns of retail distribution of consumer electronics. Includes dealers' sales volume and product lines, relations with manufacturers and consumer shopping behavior.

1157. RETAILING PERSONAL COMPUTERS: NEW DISTRIBUTION TRENDS AND CUSTOMER REQUIREMENTS. *Strategic Inc. March 1981. $950.00.*

Analysis of retail distribution markets for personal computers

examines marketing strategies of manufacturers. Also considers Japanese penetration of the U.S. market.

1158. THE REVENUE AND PROFITS SQUEEZE IN DATA PROCESSING: PROSPECTS AND STRATEGIES FOR RECOVERY. *Withington, Frederic G. Arthur D. Little, Inc. March 28, 1980. 9 pg. $600.00.*

1980 report on leading 50 data processing corporations. Comparisons among IBM and competitors by revenues, profits, profit margin, new products and sales from 1977 to 1979 are given. Data derived from corporate reports and Arthur D. Little estimates.

1159. THE ROBOT MARKET EXPLOSION. *International Resource Development, Inc. 1982. 151 pg. $1285.00.*

Analysis of markets for industrial robots. Includes profiles of manufacturers, production growth, Japanese competition, employment and unemployment effects, end uses, new technologies, and performance. Forecasts shipments, market value, and installed base, 1982-1992.

1160. ROBOTICS MARKET. *Frost and Sullivan, Inc. no.F-146. May 1982. 117 pg. $650.00.*

Wall Street analysis of the robotics industry and market drawn from impressions of the Robotics VI Show and speeches by industry executives includes corporations' sales, licensing agreements, new products, markets, prices, end uses, shipments and backlogs, competition, and marketing strategies.

1161. ROTATING PERIPHERAL MEMORIES 1: FLOPPY DISKS AND LOW-COST WINCHESTERS. *Creative Strategies International. August 1979. $995.00.*

The U.S. and worldwide markets through 1983 for floppy disk and low-cost Winchester drives. Sales in units and dollars by application segment: word processors, personal computers, data entry, microcomputers and small business systems. Discusses the market entry of Asian consumer electronics producers and the competitive effect of backward integration. Includes analysis of market shares by product segment, foreign competition, technological and pricing trends. Details major competitors and market strategies.

1162. ROTATING PERIPHERAL MEMORIES 2: DISK AND TAPE DRIVES. *Creative Strategies International. September 1979. $995.00.*

The U.S. and worldwide markets for mass memories, disk and tape drives. Forecasts through 1983 in units and dollars by segment: market segments (OEM, plug compatible manufacturers, and captive and non-captive) and product segments (disk and tape drives by capacity). Analysis includes various marketing strategies, trends in developing markets, and the structure of marketing channels. Includes competitive analysis and profiles of major manufacturers.

1163. SBC: SMALL BUSINESS COMPUTER SOFTWARE STRATEGIES. *Creative Strategies International. 1981. $1200.00.*

Small business computer software market analysis. Forecasts total expenditures and annual growth rates to 1985. Also includes sales, market shares, distribution, vendors' strategies, production costs and revenues.

1164. SELLING PERSONAL COMPUTERS TO LARGE COMPANIES: VOLUME 1. *INPUT. 1980.*

A market study of small computer usage by major companies in the U.S. Covers attitudes and needs of potential users.

1165. SERIAL AND LINE PRINTER MARKETS. *INPUT. 1981. $995.00.*

Forecasts the outlook to 1985 for computer printers in the U.S. Discusses the fastest growing markets for the printers, shipments, technology, prices, and competition.

1166. THE SMALL BUSINESS COMPUTER INDUSTRY: A STRATEGIC ANALYSIS FOR INDUSTRY PARTICIPANTS 1980-1984. *Rosenfeld, Karen E. Venture Development Corporation. June 1980. 218 pg. $1950.00.*

This report provides information on the structure, market, technology, distribution, competition, trends, and current and potential users of the small business computer. The three price / performance levels of the systems and the five categories of small business computer vendors are described and forecasts are given from 1979-1984. A directory of manufacturers with addresses and brief description is given.

1167. SMALL BUSINESS COMPUTER MARKET IN RETAIL BANKING APPLICATIONS. *Frost and Sullivan, Inc. July 1981. 255 pg. $1000.00.*

Analysis of markets for small business computers for end uses in banking. Includes financial analysis of banking operations; 10-year forecasts; market penetration projections; estimated sales based on installed base, shipments, and growth rates; profiles of vendors; and a survey of user attitudes.

1168. THE SMALL BUSINESS COMPUTER SOFTWARE MARKET. *Frost and Sullivan, Inc. August 1979. 328 pg. $900.00.*

Market analysis of business corporate software includes competitive environment (1979), user analysis (1979), sales (1979), suppliers (1979), manufacturers (1979), and number of employees (1979). Market environment includes customized vs. program packages, program packages, fixed vs. modifiable program packages, fixed program packages, user developed programs via special languages, and programs with high profitability.

1169. SMALL BUSINESS COMPUTER SYSTEMS. *Yankee Group. September 1980. 200 pg. $9500.00.*

Details the technological advances and potential for distribution and use in the small business systems market. Part of Yankee Group's Communication Information System Planning Service and available only to subscribers.

1170. SMALL BUSINESS COMPUTER USERS SURVEY. *Management Information Corporation. Annual. 6th ed. 1981.*

Presents performance ratings of some 600 small-business computer systems, 956 peripheral devices and 245 software packages, based on a survey of 474 corporate users. Parameters include ease of use, time required to train new users, reliability (uptime / downtime), and service and manufacturer support.

1171. SMALL BUSINESS COMPUTERS TO 1985. *Creative Strategies International. April 1981. $1200.00.*

Analysis of U.S. and foreign markets for small business computers. Gives market segmentation by prices, distribution channels, utilizations, performance, vendors' market shares, forecasts of shipments and revenues to 1985, and new product trends.

1172. SMALL BUSINESSES: COMPUTING AND DATA PROCESSING. *Focus Research Systems, Inc. 1980. 160 (est.) pg. $495.00.*

Market survey examining utilization of computers and services by small businesses. Examines vendors and the industry.

1173. SMALL COMPUTER AND ASSOCIATED SOFTWARE MARKET FOR THE RETAIL INDUSTRY. *Frost and Sullivan, Inc. June 1982. 369 pg. $1200.00.*

Analysis of markets for computers and software for small retailers. Includes forecasts of sales and total revenues, 1982-1992; surveys of dealer market structure with size and number of businesses, utilization of computers, technology, and factors in purchases; computer industry analysis; and manufacturers and vendors, market shares, marketing strategies, distribution, product lines, prices, shipments and installations.

1174. SMALL COMPUTER INDUSTRY SERVICE: GENERAL PURPOSE MINICOMPUTERS: VOLUME 1. *Dataquest. Monthly. 1980. 600 (est.) pg.*

This market directory on small business computers is a monthly updated looseleaf series including newsletters. Suppliers and manufacturers are compared and analyzed by product lines. Historical data (1976-1980), competition, and market penetration of manufacturers are analyzed. Product demand is also given.

1175. SMALL COMPUTER INDUSTRY SERVICE: SMALL COMMERCIAL SYSTEMS: VOLUME 2. *Dataquest. Monthly. 1980.*

This monthly updated looseleaf market directory analyzes products in small commercial systems including very small business computers, small business computers, larger business systems, and processor-based terminal and accounting computers. Pricing, sales, comparisons, shipments, installations, and end uses of products provided. Market shares, market segmentation, distribution and historical data of industry provided.

1176. SMALL COMPUTER MARKETPLACE. *International Data Group. June 1980. 110 pg. $3500.00.*

Overview of the desktop computer and small business computer markets including the magnitude, future direction, growth and competitive structure of each market segment. Major manufacturers identified with their products. Users review their experiences, applications and future plans to use small computers. Wordprocessing and personal computer markets examined. Major problems facing the industry discussed. Methodology included telephone interviews with manufacturers and 4000 mail questionnaires sent to known users. Secondary research included publicly accessible src information, annual reports, company / product information, government statistics and previous IDC research reports.

1177. THE SOFTWARE CONNECTION. *Yankee Group. October 1980. 200 pg. $9500.00.*

1980 review of new developments in software for various applications and progress in ease-of-use features. Marketing developments also analyzed. Report is part of Communications / Information Systems Planning Service. Available to service subscribers.

1178. SOFTWARE INTENSIVE PORTABLE PRODUCTS. *International Resource Development, Inc. 1981. 184 pg. $1285.00.*

Analysis of markets for portable computers and software, with sales forecasts to 1985 and new product trends to 1991. Examines effect on labor productivity, user attitudes and prices.

1179. SOFTWARE PACKAGES: AN EMERGING MARKET. *Business Communications Company. November 1980. 122 (est.) pg. $750.00.*

Forecasts 1980-1989 U.S. market for computer software, with breakdowns by consumer type and sector, end use application, and supplier category. Discusses consumer preferences and criteria for software evaluation; emerging software products like plain-English compilers and distributed data base systems; and industry structure.

1180. SPEECH RECOGNITION AND COMPUTER VOICE SYNTHESIS. *International Resource Development, Inc. September 8, 1980. 177 pg. $985.00.*

1980 report discusses development of speech recognition / synthesis technologies by the year 1983. Products projected to cover toys, typewriters, educational devices, home appliances and calculators with speech output capabilities. 1980-1990 forecast includes leading manufacturers, product lines, end uses, U.S. shipments, Japanese market growth; price changes and market shares.

1181. STATISTICAL REFERENCE BOOK 1979: SERVICES AND SOFTWARE INFORMATION PROGRAM. *Interactive Data Corporation. June 1979. 83 pg.*

1979 research report on processing services, remote problem-solving, traditional batch services, remote autotransactions, independent software, custom software and consulting, packaged software and facilities management. Market segmentation, market shares, growth rates, suppliers with rankings, and revenues of each service discussed. Publication is part of I.D. C.'s Continuous Information Service.

1182. STRATEGIC IMPLICATIONS OF 'FORTUNE 1000' COMPANIES ENTERING THE MICROCOMPUTER MARKETPLACE. *Creative Strategies International. July 1981. $1200.00.*

Analysis of microcomputer market penetration by large corporations and impact on the industry structure. Includes mergers and retail distribution.

1183. STRATEGIC OVERVIEW OF THE VALUE-ADDED MARKET. *International Data Group. 1981.*

Analysis of value-added markets. Gives revenue and forecasts for vendors of processor services, packaged software, turnkey systems and professional services.

1184. STRATEGIES OF PERSONAL COMPUTER MANU-FACTURERS. *Strategic Inc. March 1981. $950.00.*

Analysis of marketing strategies of 9 personal computer manufacturers. Includes competition from foreign trade and technology of new products.

1185. SUPERMINICOMPUTERS IN THE 1980'S. *International Resource Development, Inc. May 8, 1981. 144 pg. $985.00.*

Examines increasing user demand for high-performance superminicomputers. Considers influx of new products from Japan and competition due to it.

1186. SURVEY AND MARKET FORECAST OF FLEXIBLE DISCS. *Sutron Corp. (Fairfax, VA). U.S. Department of Commerce. National Bureau of Standards. 1980. 110 pg.*

The report includes the following information: division of a generic flexible disk system into its major addressable components, including the flexible disc cartridge, the drive, and the controller; identification of the system parameters affecting information interchange within each component; a survey of marketed systems to determine the most frequently occurring values within the identified parameters; compilation of parameters and associated values that should be considered in the design of a standard or family of standards for information interchange on the flexible disc cartridge; a survey of a future 5-year period identifying the volume of expected units that would support the proposed standards; a cost / benefit analysis of the federal portion of the 5-year volume that would be influenced by the standards; and a description of positive and

negative effects the proposed standards may have on flexible disk technology and the associated market.

1187. SURVIVABLE SYSTEMS: PITFALLS AND OPPORTU-NITIES. *Strategic Inc. 1981. $1200.00.*

Analysis of the survivable systems market, 1980-1989. Examines technologies and performance, vendors and products, installed base, and market shares. Forecasts sales, 1980-1985.

1188. SYSTEM 38 COMPUTER EQUIPMENT AND SOFT-WARE MARKET. *Frost and Sullivan, Inc. December 1979. $900.00.*

Analysis and forecast of the market impact of the IBM system 38 computers. Covers utilization, pricing policy, and competition.

1189. TARGET MARKETS FOR VALUE ADDED SUPPLI-ERS. *International Data Group. April 1980. 44 pg. $1500.00.*

Overview of value-added suppliers' (processing service companies, turnkey systems houses and software vendors) services in 5 Markets. Methodology included in-depth interviews of vendors and standard research methods using government sources and market research reports. Presents a base of information to guide and generate interest among potential entrants to these markets. Bibliography.

1190. TECHNOLOGY GROWTH MARKETS AND OPPORTU-NITIES: COMPUTER VISION SYSTEMS. *Creative Strategies International. v.2, no.7. May 1982. 24 pg. $95.00.*

Presents analysis of the industry growing out of new computer vision technology. Includes the corporations participating; uses; market analysis: sales, 1981, 1985; U.S. shipments, 1981, 1990; competition with the Japanese; end uses; and research and development.

1191. TECHNOLOGY GROWTH MARKETS AND OPPORTU-NITIES EMERGING SOFTWARE OPERATING SYSTEMS FOR ADVANCED MICROCOMPUTERS. *Creative Strategies International. v.2, no.4. April 1982. 19 pg. $95.00.*

Market report on operating systems software for advanced, 16-bit, microcomputers examines the technology, capital requirements, competition, and regulations. Includes market analysis. Shows 1981 installed base by vendors, and forecasts world shipments and retail value, 1981-1986.

1192. TECHNOLOGY GROWTH MARKETS AND OPPORTU-NITIES: HEALTH INFORMATION SERVICES: A TEN-YEAR PERSPECTIVE. *Creative Strategies International. v.2, no.9. June 1982. 23 pg. $95.00.*

Presents analysis of markets and forecasts to 1992 for information services in health care. Examines utilization of information processing technology; percentages of medical costs spent on the various medical services, 1981, 1990; end uses of information; medical industry structure; opportunities for new businesses; effect of government actions; and the need for skilled employees.

1193. THE TELEPRINTER TERMINAL INDUSTRY: A STRA-TEGIC ANALYSIS: 1980-1985. *Abramowitz, Wendy. Venture Development Corporation. October 1980. 175 pg. $1950.00.*

Study of the general purpose teleprinter terminal market. Industry analysis includes markets, technology, competion, forecasts for shipments and installations through 1985. User preferences and strategic opportunities for manufacturers identified.

1194. TERMINAL USE BY INDUSTRY. *International Data Group. February 1980. 245 pg. $3500.00.*

1980 profile of terminal usage in 21 industries (agriculture, forestry, fishing and mining; construction; process manufacturing; discrete manufacturing; miscellaneous manufacturing; transportation; communications; utilities; wholesale; retail; commercial and mutual savings banks; savings and loan associations and credit unions; investments; insurance; general services; business services; medical; education; membership organizations; miscellaneous services; and government). Each group broken down into industry description; survey sample; data entry terminal usage; future plans and industry overview. Estimated spending levels (EDP and terminal equipment) are given. Methodology included detailed user survey of 1185 businesses and IDC database of information on user spending. Major findings summarized concisely with quantitative information presented in statistical overview.

1195. THIRD ANNUAL STUDY OF CUSTOMER SATISFACTION AMONG MINICOMPUTER USERS - DEDICATED APPLICATIONS. *Customer Satisfaction Research Institute. Annual. May 1980. 275 pg. $3600.00.*

Survey of minicomputer user's attitudes about manufacturers' performance. Examines satisfaction with equipment and services, factors in selection of vendors, and executives' attitudes toward suppliers.

1196. THIRD ANNUAL STUDY OF CUSTOMER SATISFACTION AMONG SMALL BUSINESS COMPUTER USERS. *Customer Satisfaction Research Institute. Annual. 1981. $3600.00.*

Survey of small business computer users' attitudes about the performance of manufacturers. Covers satisfaction with equipment and services, factors in selection of vendors and executives' choice of suppliers.

1197. THIRD PARTY COMPUTER, ASSOCIATED PERIPHERALS, AND DATA TERMINALS MAINTENANCE MARKET. *Frost and Sullivan, Inc. October 1980. 235 pg. $1250.00.*

Market analysis and forecast to 1985 for maintenance services for third party computers, peripherals and terminals in the U.S. Profiles service businesses, covering revenues, sales, market shares, histories and mergers. Surveys users' attitudes including repair costs.

1198. TRADITIONAL DATA ENTRY SYSTEMS. *Creative Strategies International. July 1980. $1195.00.*

The U.S. market for "traditional" equipment used to input data into computerized systems. Projects revenues and unit shipments to 1985. Product segments include keypunch, key-to-tape, key-to-diskette, key-to-disk systems. Discusses industry issues, technological trends, and competitive issues.

1199. TRENDS IN INTERFACING MICROGRAPHICS: A MARKET STUDY. *International Data Group. August 1980. 95 pg.*

Market data for interface micrographics includes current use (1979), distribution (1979), and average number of COM frames produced per year. Provides case studies in various industries: insurance, banking, manufacturing, government, retail / wholesale, service bureaus, and transportation.

1200. THE UNDER $10,000 COMPUTER SYSTEMS MARKET. *Anderson, Bud. Marketing Development. 1980. 206 pg. $750.00.*

Market study for computer systems under $10,000 compiled from interviews with manufacturers and survey of users and owners of personal computers.

1201. U.S. BUSINESS SYSTEMS HOUSES: CONTINUOUS INFORMATION SERVICE. *Interactive Data Corporation. September 1979. 117 pg.*

1980 research report on revenues, growth rates, industry structure, distribution, sales, expenditures, number of shipments and costs of computer products by U.S. business system houses. Comparisons among suppliers and manufacturers given with product lines and revenues. Publication is part of I.D.C's Continuous Information Service.

1202. U.S. COMPUTER INDUSTRY: 1980-1982: A STRATEGIC ANALYSIS. *Boggs, Raymond L. Venture Development Corporation. December 1980. 150 pg. $950.00.*

Forecasts of domestic and foreign markets for mainframe computers, memory and peripherals, including terminals and printers.

1203. U.S. MARKET FOR EXTENDED DBMS AND NATURAL LANGUAGE QUERY OF DATA BASES. *Frost and Sullivan, Inc. December 1980. $1000.00.*

Market outlook for extended DBMS - artificial intelligence. Technological extension of DBMS. Market by application and by product type - minicomputer-based products, firmware microprocessors chip-based products, large scale computer based products.

1204. U.S. MARKET FOR INTELLIGENT TERMINALS. *Frost and Sullivan, Inc. April 1981. 221 pg. $1100.00.*

Analysis of the U.S. market for intelligent terminals, including off line / distributed and interactive terminals. Covers the installed base 1981; shipments, 1976-1980; forecasts for 1981-1985; dollar value of shipments, product mix, industry structure, vendors, products, market shares, 1980, 1984-1985; end uses, new products and technologies, competition, and market penetration. Includes a market survey of user attitudes on future utilization.

1205. U.S. MINICOMPUTER ADD-IN / ADD-ON MEMORIES: CORPORATE PLANNING SERVICE. *Interactive Data Corporation. July 1979. 43 pg.*

1979 report on add-in / add-on memories in computers which analyzes shipments to 1981, products, market shares and competition among vendors, suppliers and manufacturers in the U.S. Publication is part of I.D.C.'s Corporate Planning Service.

1206. UNIX RELATED HARDWARE, SOFTWARE AND SERVICES. *Gnostic Concepts, Inc. 1981.*

Addresses markets within the UNIX industry, by software and hardware manufacturers. Covers software developers, educational institutions, add-on systems to DEC and other existing computer installations, industry and science, government, distributed processing and business applications.

1207. THE VALUE ADDED (INTELLIGENT) NETWORK MARKET IN NORTH AMERICA. *Frost and Sullivan, Inc. January 1980. 268 pg. $950.00.*

1980 assessment of the market potential of existing value-added networks in North America which had revenues of approximately $44 million in 1979, as well as the proposed data communication networks that will or could offer intelligent network services - ACS, SBS, XTEN and others. Major European VAN operations are also reviewed in the study. Evaluates the FCC's actions in current and pending value added network services as well as current and future regulatory developments. Forecasts 1979-1985 traffic range for existing and expected value added carriers.

1208. VIDEODISC - HARDWARE AND PROGRAMMING SERVICES MARKET. *Frost and Sullivan, Inc. May 1981. $1050.00.*

Market analysis and forecasts for videodisc hardware and software in the U.S., Western Europe, and Japan. Includes historical data on video recording, prices, distribution, advertising expenditures, consumer behavior, market structure, competition, industry growth rates, revenues, and demand.

1209. VIEWDATA AND ITS POTENTIAL IMPACT IN DIRECT MARKETING. *Link Resources Corporation. January 1980. 600 pg. $6000.00.*

Analysis of marketing impacts of new products and services of the viewdata industry. Examines changes in distribution, marketing, and consumer shopping behavior.

1210. VIEWDATA AND VIDEOTEXT, 1980-1981: A WORLDWIDE REPORT. *Knowledge Industry Publications. June 1980. 600 pg. $77.00.*

1980 study analyzing the videotext and viewdata market including facsimile equipment and terminals. Projections, costs, forecasts to 1983 and vendors in Japan, France, Netherlands, Finland, Germany and United States provided.

1211. VIEWDATA / VIDEOTEXT REPORT. *Link Resources Corporation. Monthly. 8 pg. $120.00.*

Monthly newsletter monitoring developments in the viewdata / videotext market and industry. Covers related information distribution services.

1212. VLSI (VERY LARGE SCALE INTEGRATION) INFORMATION SERVICE 1980-1981: HIGHLIGHTING IMPLEMENTATION OF THE LOGIC FUNCTION. *Gnostic Concepts, Inc. Monthly. 1981. $15000.00.*

Report analyzes VLSI (very large scale integration) information service, market outlook, and competitive environment. Twelve monthly reports cover gate arrays, microprocessors / microcomputers, custom logic / random logic, memory / logic complementation, insulation processes, conductor processes, silicon processes, product reviews, company technology assessments, wafer fab costs, assembly trends, and data bases. Special services include an annual data base, continuing inquiry privileges, and special seminars and strategy sessions.

1213. VLSI (VERY LARGE SCALE INTEGRATION) TECHNOLOGY AND MARKET FORECAST. *Gnostic Concepts, Inc. 1981. 1000 pg. $12000.00.*

Very large scale integration (VLSI) circuits information: technology analysis, competitive environment, market outlook, and government program summaries. Market outlook discusses VLSI products, processes, applications, and government influence. Competive environment chapter lists major U.S. organizations, U.S. government agencies, and universities, as well as VLSI efforts in Japan, Western Europe and Canada.

1214. THE VOICE DIGITIZER MARKET. *Frost and Sullivan, Inc. January 1982. 170 pg. $1200.00.*

Analyzes and forecasts emerging market for voice digitizers. Two sets of five-year forecasts are presented in units and 1981 dollars for voice digitizers in the multiple voice and voice / data integration segments. Estimates market size, costs, capacity, and market penetration. Examines end-user attitudes.

1215. VOICE INPUT / OUTPUT: MARKETS TECHNOLOGY AND APPLICATION. *Strategic Inc. June 1981. 110 pg. $950.00.*

Analysis of markets for voice recognition and synthesis

devices. Examines technology of new products, and forecasts end uses.

1216. VOICE PROCESSING MARKET. *Frost and Sullivan, Inc. October 1980. $950.00.*

Qualitative and quantitative analysis of voice processing market. Detailed analysis of user requirements for data entry, point of sale, manufacturing, transportation & office. Description of application in present and future systems. Competitive analysis of vendors.

1217. WINCHESTER DISKS IN EMERGING OFFICE SYSTEMS. *Strategic Inc. September 1980. 200 pg. $950.00.*

Analysis of markets and utilizations for Winchester discs. Examines new product technologies and end uses.

1218. THE WIRED HOME: HOME DEVICES AND ELECTRONICS ON AN INTERACTIVE DATA BASE. *Yankee Group. May 1981. 200 pg. $8500.00.*

Examines the impact of chip technology on electronic devices, appliances found in the home and the potential of the home command-and-control center to operate as an electronic housekeeper. Discusses the role of standards and the possibility of a new generation of smart devices and appliances. Predicts who will install home systems, at what costs, and likely distribution in the 80's. Discusses opportunities for networks and electronics industry. Part of Yankee Group's Home of the Future Service and available only to subscribers.

1219. WORKSTATIONS AND LOCAL AREA NETWORKS: PRODUCT / MARKETING STRATEGIES FOR THE 1980'S. *Creative Strategies International. 1982. $12000.00.*

Analysis of markets for workstations and local area networks includes market segmentation and growth rates; new technology; corporations' marketing strategies, market shares, prices and distribution; product lines and performance; user attitudes; and forecasts of shipments, revenues, installations, and demand to 1986.

1220. WORLD MARKETS FOR INFORMATION PROCESSING PRODUCTS TO 1991. *Arthur D. Little of Canada Ltd. 1982.*

Forecasts total markets for information processing products in the free world. Examines sales and world revenues for computers, terminals, peripherals, and software.

1221. WORLDWIDE SHIPMENTS OF RIGID DISK MEDIA: 1978-1981: CONTINUOUS INFORMATION SERVICES. *Interactive Data Corporation. April 1979. 22 pg.*

Volume 1 of 2-volume report on magnetic media markets, shipments, market shares of IBM and other manufacturers, and market segments of the industry for 1978 are discussed. Publication is part of I.D.C.'s Continuous Information Services.

1222. WORLDWIDE VALUE OF U.S. MADE COMPUTER SHIPMENTS. *International Data Group. September 1981. 75 pg. $225.00.*

Analyzes the factors which will contribute to growth rates during the period to 1985 and examines their effect on each equipment segment. Contains projections of shipments through 1985 in each of four market sectors, and censuses showing model by model installation numbers for 1980. Presents information on user spending plans derived from approximately 250 companies surveyed during the budget planning cycle at the end of 1980.

1223. 1981 DISK / TREND REPORT: FLEXIBLE DISK DRIVES. *James N. Porter. September 1981. $510.*

Analysis of 1981 markets for flexible disk drives. Covers shipments, revenues, product groups, market shares, forecasts to 1983, manufacturers, and installed base.

1224. 1981 DISK / TREND REPORT: RIGID DISK DRIVES. *James N. Porter. July 1981. $730.00.*

Analysis of markets for rigid disk drives, 1981. Covers shipments, revenues, product groups, market shares, 1983 forecasts, manufacturers, and installed base.

1225. 1981 FIELD SERVICE ANNUAL REPORT. *INPUT. Annual. November 1981.*

Annual report and analysis of the computer field service industry and market. Includes vendors' profits and forecasts of revenues and growth rates of market shares by product lines, 1981-1986. Also reviews users' attitudes.

See 57, 65, 67, 75, 80, 86, 89, 92, 106, 112, 117, 118, 121, 146, 151, 162, 180, 554, 556, 560, 562, 570, 575, 2209, 2214, 2228, 2240, 2241, 2247, 2252, 2271, 2286, 3144, 3145, 3147, 3148, 3152, 3153, 3155, 3156, 3157, 3158, 3159, 3160, 3162, 3165, 3168, 3171, 3172, 3174, 3176, 3177, 3180, 3183, 3185, 3186, 3190, 3193, 3200, 3203, 3204, 3205, 3206, 3207, 3208, 3210, 3519, 3532, 3565, 3566, 4023, 4035, 4044, 4046, 4052, 4054, 4056, 4058, 4072, 4073, 4074, 4075, 4077.

Investment Banking Reports

1226. ALEX BROWN AND SONS COMPUTER SERVICES MONTHLY: APRIL 1980. *Alex. Brown and Sons. Monthly. April 23, 1980. 22 pg.*

Monthly financial service providing data on the U.S. computer services industry. April 1980 issue focuses on the battle for databases and covers the role of databases in the industry, types of databases, 1980 software expenditures, and comments on 9 major businesses. Tables and charts show earnings estimates, 1977-1979; growth rates and price earnings ratios; and financial data (shares outstanding, book value per share, latest price, annual dividend and yield) for 30 selected companies. Includes computer services price index, 1976-1980, and quarterly earnings digest, 1978-1980.

1227. ALEX BROWN AND SONS COMPUTER SERVICES MONTHLY: AUGUST 1980. *Berkeley, Alfred R. Alex. Brown and Sons. Monthly. August 20, 1980. 27 pg.*

Monthly financial service providing data on the U.S. computer services industry. August 1980 issue comments on earnings reports of 18 businesses and concentrates on the most competitive companies. Charts display revenues versus net income; revenues versus cash flow; revenues versus payout; return on equity versus payout; and revenues versus long-term debt / equity for the major competitors. Tables show data for 30 selected companies: earnings estimates, 1978-1981; growth rates; price earnings ratios; insider transactions (disposed, acquired); quarterly earnings digest, 1979-1981; and financial data (shares outstanding, book value per share, latest stock price as of 8/20/80, yields and annual dividends).

1228. ALEX BROWN AND SONS COMPUTER SERVICES MONTHLY: DECEMBER 1980. *Berkeley, Alfred R. Alex. Brown and Sons. Monthly. December 24, 1980. 29 pg.*

Monthly financial service providing data on the U.S. computer services industry. December 1980 issue focuses on mergers and acquisitions and comments on 7 major businesses. Tables and charts show financial data for 30 selected companies: earnings estimates, 1978-1981; growth rates; price earnings ratios; market performance (percent change in stock price since year-end); shares outstanding, book value per share,

yields, annual dividends and stock price as of December 24, 1980; insider transactions (acquired, disposed); and quarterly earnings, digest, 1979-1981.

1229. ALEX BROWN AND SONS COMPUTER SERVICES MONTHLY: FEBRUARY 1980. *Berkeley, Alfred R. Alex. Brown and Sons. Monthly. February 4, 1980. 20 pg.*

Monthly financial service providing data on the U.S. computer services industry. February 1980 issue focuses on the applications software market and covers competitive trends, market potential and size, and comments on 6 businesses and economic characteristics. Graphs and tables display earnings estimates for 30 companies, 1977-1980; growth rates and price earnings ratios; selected financial data (shares outstanding, book value per share, annual dividend, latest stock price, yield); and insider transactions (acquired, disposed). A listing shows which companies are competing in which niches (time sharing, facilities management, contract software, packaged software, turnkey systems) and which are using delivery systems.

1230. ALEX BROWN AND SONS COMPUTER SERVICES MONTHLY: JANUARY 1980. *Berkeley, Alfred R. Alex. Brown and Sons. Monthly. January 7, 1980. 21 pg.*

Monthly financial service providing data on the U.S. computer services industry. January 1980 issue highlights the software market, covering competition and technology, demand, costs, mergers and acquisitions, company comments and industry issues. Graphs and charts display computer services price index, 1975-1979; earnings estimates of 30 selected businesses, 1977-1980; growth rates and price earnings ratios, 1979; selected financial data (shares outstanding, book value per share, annual dividend, yield, price earnings ratio and price change) as of January 3, 1980; and insider transactions.

1231. ALEX BROWN AND SONS COMPUTER SERVICES MONTHLY: JULY 1980. *Alex. Brown and Sons. Monthly. July 22, 1980. 24 pg.*

Monthly financial service providing data on the computer services industry. July 1980 issue focuses on software distribution channels and covers the market for mainframes, minicomputers and microcomputers. Includes comments and earnings reports for 7 major businesses and news briefs on 7 businesses. Tables and charts show the following: computer service price index, 1976-1980; earnings estimates, 1977-1980; growth rates; price earnings ratios; and financial data (shares outstanding, book value per share, latest stock price, annual dividend, yields) for 30 selected companies. Includes insider transactions (acquired, disposed) and quarterly earnings digest, 1979-1980.

1232. ALEX BROWN AND SONS COMPUTER SERVICES MONTHLY: MARCH 1980. *Alex. Brown and Sons. Monthly. March 3, 1980. 24 pg.*

Monthly financial service providing data on the U.S. computer services industry. March 1980 issue focuses on database management systems and covers prospects, competition, the federal DP budget, and comments on 18 businesses. Tables and charts show computer services price index, 1976-1980; earnings estimates of 30 selected companies, 1977-1980; earnings earnings ratios and growth rates; financial data (shares outstanding, book value per share, latest price, annual dividend, yield); insider transactions (acquired, disposed); and quarterly earnings digest, 1978-1970.

1233. ALEX BROWN AND SONS COMPUTER SERVICES MONTHLY: MAY 1980. *Berkeley, Alfred R. Alex. Brown and Sons. Monthly. May 21, 1980. 26 pg.*

Monthly financial service providing data on the U.S. computer services industry. May 1980 issue focuses on Telecredit, Inc. and the credit card processing method. Tables show

Telecredit balance sheet, 5-year summary of operations, and key financial ratios. Includes tables and charts displaying the computer services price index, 1976-1980; earnings estimates, 1977-1980; growth rates; price earnings ratios and financial data (shares outstanding, book value per share, latest price, annual dividend) for 30 selected companies. Includes rankings, by size for 22 companies; insider transactions (acquired, disposed); and quarterly earnings digest.

1234. ALEX BROWN AND SONS COMPUTER SERVICES MONTHLY: NOVEMBER 1980. *Berkeley, Alfred R. Alex. Brown and Sons. Monthly. December 5, 1980. 82 pg.*

Monthly financial service providing data on the U.S. computer services industry. November 1980 issue focuses on a computer services seminar and reproduces transcripts on 7 major businesses. Includes comments on 9 businesses and computer services price index, 1976-1980. Tables show financial data for 30 selected companies: earnings estimates, 1978-1981; growth rates and price earnings ratios; shares outstanding, book value per share, dividends, yields, stock price as of 12/5/80; return on equity; and insider transactions (acquired, disposed).

1235. ALEX BROWN AND SONS COMPUTER SERVICES MONTHLY: OCTOBER 1980. *Berkeley, Alfred R. Alex. Brown and Sons. Monthly. November 13, 1980. 99 pg.*

Monthly financial service providing data on the U.S. computer services industry. October 1980 issue reproduces transcripts from 8 companies presented at a computer services seminar and comments on another 8 businesss. Tables and charts provide financial statistics on 30 selected businesses: earnings estimates, 1978-1981; growth rates and price earnings ratios; rankings by size; market performance; shares outstanding, book value per share, yields, annual dividends and latest stock price as of 11/13/80; quarterly earnings digest; and insider transactions (disposed, acquired).

1236. ALEX BROWN AND SONS COMPUTER SERVICES MONTHLY: SEPTEMBER 1980. *Berkeley, Alfred R. Alex. Brown and Sons. Monthly. October 7, 1980. 30 pg.*

Monthly financial service providing data on the U.S. computer services industry. September 1980 issue focuses on the markets for packaged software and covers determinants of demand, market performance, life cycle costs, investment dynamics, and comments on 7 major businesses. Charts and tables show data for 30 selected companies: earnings estimates 1978-1981; growth rates; price earnings ratios; shares outstanding, book value per share, latest stock price as of 10/7/80; annual dividends; yields; and insider transactions (acquired, disposed).

1237. BEAR STEARNS AND CO.: COMPUTER INDUSTRY OVERVIEW UPDATE - NO. 2: REMOVAL OF SALE RECOMMENDATION ON 11 COMPUTER COMPANIES. *Elling, George D. Bear Stearns and Company. October 1, 1981. 9 pg.*

Review of performancce of computer stocks. Shows recent closing stock prices, recommended sell prices (for selected corporations), and earnings per share, 1981-1982. Profiles individual manufacturers.

1238. BEAR STEARNS AND CO.: INDUSTRY REPORT: COMPUTER AND OFFICE EQUIPMENT - SPECIAL REPORT: AN ANALYSIS OF THE EFFECT OF THE PROPOSED NEW FASB STANDARD ON FOREIGN EXCHANGE ACCOUNTING ON COMPANIES IN THE COMPUTER AND OFFICE EQUIPMENT INDUSTRY. *Elling, George D. Bear Stearns and Company. November 25, 1981. 15 pg.*

Report on new foreign exchange accounting standards and their impact on the computer and office equipment industry. Shows balance sheet, income statement, inventory, and earn-

ings per share effects, 1979-1981, by corporations. Also includes 1981 exchange rates for major currencies and users' attitudes toward new provisions.

1239. BEAR STEARNS AND CO.: INDUSTRY REPORT: COMPUTER INDUSTRY OVERVIEW UPDATE - NO. 1: SELL RECOMMENDATION OF NINE COMPUTER COMPANIES. *Elling, George D. Bear Stearns and Company. June 22, 1981. 9 pg.*

Investment recommendations in the U.S. computer industry. Gives stock price, 1979-1981; price earnings ratio, 1979-1981; and analysis of corporations' financial prospects.

1240. BEAR STEARNS AND CO.: INDUSTRY REPORT: SEMI-ANNUAL COMPUTER INDUSTRY REVIEW. *Elling, George D. Bear Stearns and Company. Semi-annual. July 28, 1981. 10 pg.*

Analysis of computer markets and financial performance of major corporations. Shows stock price changes, 1980-1981; rankings of best and worst performing companies; and, by product groups, 1980-1982 earnings per share and 1981 price earnings ratio.

1241. BEAR STEARNS AND CO.: INDUSTRY REPORT: YEAR-END COMPUTER HIGHLIGHTS. *Elling, George D. Bear Stearns and Company. January 29, 1982. 13 pg.*

Year-end analysis of computer and office equipment industry and stock performance. Shows stock price by vendors, 1979-1981.

1242. BEAR STEARNS AND CO.: RESEARCH DEPARTMENT HIGHLIGHTS: COMPUTER AND OFFICE EQUIPMENT - SPECIAL REPORT: AN ANALYSIS OF THE EFFECT OF THE PROPOSED NEW FASB STANDARD ON FOREIGN EXCHANGE ACCOUNTING ON COMPANIES IN THE COMPUTER AND OFFICE EQUIPMENT INDUSTRY. *Elling, George D. Bear Stearns and Company. November 27, 1981. pg.22-24*

Report on effects of changes in accounting regulations regarding foreign currencies for computer and office products corporations. Shows results of survey of user attitudes toward proposed changes in financial statements of earnings and losses.

1243. BEAR STEARNS AND CO.: RESEARCH DEPARTMENT HIGHLIGHTS: COMPUTER INDUSTRY OVERVIEW UPDATE - NO. 2: REMOVAL OF SALE RECOMMENDATION OF 11 COMPUTER COMPANIES: SUMMARY OF FORTHCOMING REPORT. *Elling, George D. Bear Stearns and Company. September 25, 1981. pg.6-8*

Removal of sale recommendation for selected computer stocks and analysis of economic indicators. Gives 1981 stock prices and earnings per share, 1981-1982, for selected corporations.

1244. BEAR STEARNS AND CO.: RESEARCH DEPARTMENT HIGHLIGHTS: MAINFRAME COMPUTER COMPANIES: THIRD-QUARTER HIGHLIGHTS. *Elling, George D. Bear Stearns and Company. November 13, 1981. pg.18-23*

Third-quarter performance of U.S. mainframe computer corporations. Shows fiscal 1982 revenues and orders, and 1980-1981 sales, revenues, income, taxes, profit margins, and earnings per share.

1245. DONALDSON, LUFKIN AND JENRETTE: RESEARCH BULLETIN: EDP INDUSTRY: NEW TAX BILL SHOULD INCREASE SHIPMENTS, EARNINGS, AND CASH FLOW. *Geran, Michael I. Donaldson, Lufkin and Jenrette Inc. August 11, 1981. 2 pg.*

Financial analysis of impact of new tax legislation on the U.S.

EDP industry. Examines capital expenditures for research and development, leasing, and new plants and equipment.

1246. DREXEL BURNHAM LAMBERT INC.: COMPUTERS: KEY TO PRODUCTIVITY IN THE EIGHTIES. Kline, Frank R. Drexel Burnham Lambert Inc. October 1980. 200 pg. $495.00.

Analysis of the information processing industry in the U.S., with attention to Japanese competition and forecasts to 1985. Includes commentary, charts, tables and diagrams covering retail prices, 1953-1980; new products; labor productivity, 1967-1977; other productivity factors; industry structure and growth in the 1970's; user attitudes, 1980; market analysis, 1980; annual earnings per share, 1972-1979; stock prices, 1972-1979; earnings per share of Japanese companies, 1972-1980; sales growth rates, 1972-1979; barriers to entry; units in operation, 1956, 1980; market shares 1954-1979; research and development, 1972-1979, including government funding; software development costs, 1965-1980; profit margins, 1972-1979; sales; employees; industry structure, 1950-1985; regulations; end uses; capital investments, 1980-1985; new products; patents; market structure, 1980-1985; installations, 1980-1985; services, 1978-1985; industry analysis; value of shipments, 1980; stock prices, 1980; earnings per share, 1977-1979; price earnings ratios, 1979; dividends and yields, 1979; shares outstanding, 1979; and market segmentation.

1247. DREXEL BURNHAM LAMBERT INC.: PRODUCTIVITY: AN INVESTMENT STRATEGY FOR CAPITAL GOODS. Shapiro, Stephen O. Drexel Burnham Lambert Inc. December 1981. 138 pg.

Analysis of industry structure for capital goods companies concerned with productivity, such as CAD / CAM, robotics, automated machine tools, and automated material handling. Gives 1980 sales for selected corporations and percent of sales by market segments; recommendations for investments; corporations 1980-1981 stock prices; earnings per share, 1980-1982; price earnings ratio, 1981-1982; dividends; economic indicators of productivity; energy consumption; forecasts of demographic statistics, 1950-2000; labor supply, 1950-1990; capital outlook; and depreciation. For individual manufacturers, includes revenues, 1977-1980; 1980 installed base; market shares, 1977-1980; operating performance, 1978-1983; financial statements; and management and product lines.

1248. DREXEL BURNHAM LAMBERT INC.: THE JAPANESE THREAT TO THE U.S. COMPUTER INDUSTRY: REAL OR IMAGINARY? Kline, Frank R. Drexel Burnham Lambert Inc. March 16, 1981. 22 pg. $220.00.

Analysis of impacts on U.S. computer markets and corporations from Japanese competition. Covers market penetration and distribution strategies; recommends investment tactics. Tables and charts show Japanese tariffs, 1980, 1987; U.S. and Japanese sales and worldwide market shares, 1974, 1979, 1980, 1984; new products, 1980-1986; average annual labor productivity, 1971-1979; imports, 1979, 1984, 1989; and profit margins, 1979.

1249. ELECTRONIC PUBLISHING. FIND / SVP. R159. December 1981. 32 pg. $250.00.

Wall Street report on electronic publishing. covers cablenews, teletext, and videotex. Analysis of markets for new technology and services.

1250. KIDDER, PEABODY AND CO.: TECHNOLOGY INDUSTRY PERSPECTIVES. Easterbrook, William D. Kidder, Peabody and Company Inc. Annual. 1982. $1000.00.

Analysis of the effects of new technology on the computer, telecommunications and electronics industries and markets.

Examines manufacturers marketing strategies, operating performance and new products.

1251. KIDDER, PEABODY AND CO.: TECHNOLOGY INDUSTRY PERSPECTIVES: COMPUTERS. Easterbrook, William D. Kidder, Peabody and Company Inc. Annual. 1982. $750.00.

Analysis of the computer industry and market. Includes competition among manufacturers and service vendors, effect of general economic indicators on demand, marketing strategies, prices, orders, and new technology. Profiles of corporations' operating performance include sales profits, product lines and new product development.

1252. MORGAN STANLEY AND CO.: INVESTMENT PERSPECTIVES: EDP: EARNINGS ESTIMATES REDUCED: RESEARCH COMMENT. Weil, Ulric. Morgan Stanley and Company, Inc. November 16, 1981. pg.16

Earnings per share revisions for computer corporations, 1980-1982. Includes current stock prices.

1253. MORGAN STANLEY AND CO.: INVESTMENT PERSPECTIVES: EDP INDUSTRY: CRAY, SPERRY, TANDEM, AND XEROX EARNINGS: RESEARCH COMMENT. Weil, Ulric. Morgan Stanley and Company, Inc. February 1, 1982. pg.13-15

Forecasts of earnings per share for selected EDP corporations, 1981-1982. Also examines revenues, mergers, competition, and performance.

1254. MORGAN STANLEY AND CO.: INVESTMENT PERSPECTIVES: EDP INDUSTRY: RECENT QUARTERLY EARNINGS: OVERVIEW AND OUTLOOK - RESEARCH COMMENT. Weil, Ulric. Morgan Stanley and Company, Inc. January 25, 1982. pg.17-20

Report on performance of computer corporations. Gives orders and earnings per share, 1981-1982.

1255. MORGAN STANLEY AND CO.: INVESTMENT PERSPECTIVES: EDP INDUSTRY: SELECTIVE COVERAGE IN THE MIDST OF RECESSION: RESEARCH COMMENT. Weil, Ulric. Morgan Stanley and Company, Inc. February 22, 1982. pg.12-13

Analysis of computer markets and corporations. Includes orders, shipments to 1982-1983, capacity, and profit margins. Also examines business cycles and outlook for investments.

1256. MORGAN STANLEY AND CO.: INVESTMENT PERSPECTIVES: EDP INDUSTRY: THROUGH THE RECESSION WRINGER. Weil, Ulric. Morgan Stanley and Company, Inc. April 5, 1982. pg.14-15

Financial analysis of computer corporations, including 1982 earnings per share estimates. Comments on business cycles.

1257. MORGAN STANLEY AND CO.: INVESTMENT PERSPECTIVES: EDP: IT'S EASY TO BE ALARMED: RESEARCH COMMENT. Weil, Ulric. Morgan Stanley and Company, Inc. March 29, 1982. pg.17-19

First quarter operating performance review for computer vendors. Covers earnings, new products, and executive changes.

1258. MORGAN STANLEY AND CO.: INVESTMENT PERSPECTIVES: EDP: PROSPECTS IN A RECESSIONARY ENVIRONMENT: RESEARCH COMMENT. Weil, Ulric. Morgan Stanley and Company, Inc. November 9, 1981. pg.7-9

Prospects for 1982 performance by U.S. computer companies. Reviews economic indicators and projects earnings and orders for corporations.

1259. MORGAN STANLEY AND CO.: INVESTMENT PER-SPECTIVES: EDP: SOMEWHAT SLOWER GROWTH BECAUSE OF THE WORLDWIDE ECONOMIC SLOW-DOWN: RESEARCH COMMENT. Weil, Ulric. Morgan Stanley and Company, Inc. May 17, 1982. pg.24

Analysis of gross national product growth rates, inflation and the impact on EDP markets. Recommends investments in specific corporations.

1260. MORGAN STANLEY AND CO.: INVESTMENT PER-SPECTIVES: EDP: THE BUNCH COMPANIES. Weil, Ulric. Morgan Stanley and Company, Inc. January 4, 1982. pg.14-15

Report on the EDP industry including new order trends, activity in foreign markets and other economic indicators. Includes analysis of financial position for individual corporations. Shows earnings per share, 1980-1982.

1261. MORGAN STANLEY AND CO.: INVESTMENT PER-SPECTIVES: EDP: VISITS WITH AMDAHL, APPLE, TAN-DEM, AND TYMSHARE: RESEARCH COMMENT. Weil, Ulric. Morgan Stanley and Company, Inc. March 1, 1982. pg.13-15

Financial analysis of several computer and software corporations. Includes products, performance, competition, and earnings per share estimates.

1262. MORGAN STANLEY AND CO.: INVESTMENT PER-SPECTIVES: RESEARCH COMMENT: EDP: TAX BENEFIT TO LIFT 1982 EARNINGS. Weil, Ulric. Morgan Stanley and Company, Inc. August 12, 1981. pg.5-7

Analysis of impact of new legislation governing corporate income taxes, depreciation, and research and development expenditures on U.S. computer companies. Shows 1982 stock price and earnings per share estimate revisions for major corporations.

1263. MORGAN STANLEY AND CO. RESEARCH BROAD-CAST: EDP: INDUSTRY ORDER TRENDS. Weil, Ulric. Morgan Stanley and Company, Inc. April 3, 1981.

Brief on new orders for computers reported by U.S. corporations.

1264. MORGAN STANLEY AND CO.: RESEARCH COM-MENT: EDP: APPLIED TECHNOLOGY; ENCROACHMENT FROM BELOW. Weil, Ulric. Morgan Stanley and Company, Inc. June 22, 1981.

Personal computer market analysis. Discusses new products, prices, and competition among manufacturers.

1265. MORGAN STANLEY AND CO.: RESEARCH COM-MENT: EDP: HAS THE CURRENCY MONSTER STRUCK AGAIN? Weil, Ulric. Morgan Stanley and Company, Inc. June 4, 1981.

Analysis of impact of foreign currency exchange rates on earnings of U.S. computer companies.

1266. MORGAN STANLEY AND CO.: RESEARCH COM-MENT: EDP: HIGHLIGHTS OF THE ROSEN RESEARCH PERSONAL COMPUTER FORUM. Weil, Ulric. Morgan Stanley and Company, Inc. May 18, 1981.

News from a conference covering new products, prices, other activities of corporations, and competition with the Japanese.

1267. MORGAN STANLEY AND CO.: RESEARCH COM-MENT: EDP INDUSTRY: 1982 EARNINGS ESTIMATES SUB-JECT TO DOWNWARD REVISION. Weil, Ulric. Morgan Stanley and Company, Inc. July 10, 1980.

Revision of earnings per share estimates, 1981-1982, for U.S.

computer companies in light of weakness in foreign markets, interest rates and other economic indicators.

1268. MORGAN STANLEY AND CO.: RESEARCH MEETING COMMENT: EDP: A GUARDED PROGNOSIS FOR EURO-PEAN MARKET IN 1981. Weil, Ulric. Morgan Stanley and Company, Inc. January 30, 1981.

Analysis of the European market for computers. Covers market shares of corporations and competition.

1269. MORGAN STANLEY AND CO.: RESEARCH MEETING COMMENT: EDP: A QUICK VIEW FROM EUROPE: JAPA-NESE ENTER WORD PROCESSING MARKET. Weil, Ulric. Morgan Stanley and Company, Inc. December 18, 1980.

European computer market analysis showing orders of U.S. manufacturers. Includes new products from Japan. Analyzes competition among all manufacturers.

1270. MORGAN STANLEY AND CO.: RESEARCH MEETING COMMENT: EDP: INFLECTION POINT FOR INDUSTRY ORDER RATES? Weil, Ulric. Morgan Stanley and Company, Inc. December 12, 1980.

Computer market analysis. Projects order rates through 1981.

1271. OPPENHEIMER AND CO., INC.: DATA PROCESSING: A LAFFER CURVE FOR SOFTWARE: MARKET SHARE AND MARKET SATURATION. Dyson, Esther. Oppenheimer and Company. April 16, 1981. 10 pg. $100.00.

Market analysis for computer software and services using Laffer curves to illustrate strategies in light of different market share and saturation situations. Examines selected corporations using this analysis and showing 1980-1981 stock price; 1981 shares outstanding, dividend, and yield; 1979-1982 earnings per share, 1979-1982; and 1980-1982 price earnings ratio.

1272. OPPENHEIMER AND CO., INC.: DATA PROCESSING: COMPUTER INDUSTRY REVIEW: MAINFRAMES / MINI-COMPUTERS: THE NEW COMPUTER WARS - PART II. Stevens, Jay P. Oppenheimer and Company. June 1, 1982. 14 pg. $140.00.

Analysis of computer markets, including foreign markets, and the major vendors covers orders and shipments, new products, marketing strategies, and financial analysis. Includes comparisons of expenditures for equipment in the U.S., Europe, and Japan, 1970-1982; producer earnings and stock prices, 1970-1982; U.S. gross national product, 1981-1983; revenues, operating profits, and assets in the U.S., Japan and Europe, 1979-1981; exchange rates, 1976-1982; gross domestic product and unemployment in Japan and Western Europe, 1973-1983; earnings per share, 1981-1983; price earnings ratio, 1981-1982; book values, 1981-1982; returns on equity, 1981-1982; and other valuation measures, 1981-1982.

1273. OPPENHEIMER AND CO., INC.: DATA PROCESSING: COMPUTER SOFTWARE / SERVICES INDUSTRY: AUTUMN THOUGHTS. Dupon, Esther. Oppenheimer and Company. November 2, 1981. 7 pg. $100.00.

Analysis of the evolving structure of the computer software and services industry. Reviews management and market penetration strategies. For major corporations, shows stock price, 1980-1981; shares; earnings per share, 1980-1982; price earnings ratio, 1980-1982; and dividends and yield.

1274. OPPENHEIMER AND CO., INC.: DATA PROCESSING: DATA PROCESSING INDUSTRY - MAINFRAMES / MINI-COMPUTERS QUARTERLY REVIEW. Stevens, Jay P. Oppenheimer and Company. Quarterly. June 6, 1981. 8 pg. $100.00.

Quarterly review of economic indicators for the computer industry and news of corporations. Shows earnings per share,

1980-1982; stock price, 1981; price earnings ratio, 1981-1982; 1981 dividend and yield. Includes exchange rate of major foreign currencies, 1976-1981; and valuation of computer stocks, showing book value, price, and return on equity, 1981-1982.

1275. OPPENHEIMER AND CO., INC.: DATA PROCESSING: DATA PROCESSING INDUSTRY - MAINFRAMES / MINI-COMPUTERS QUARTERLY REVIEW. Stevens, Jay P. Oppenheimer and Company. Quarterly. March 18, 1981. 5 pg.

Financial analysis of U.S. computer corporations and the industry. Shows stock prices, 1981; earnings per share, 1980-1982; price earnings ratio, 1981-1982; and dividends and yields, 1981. Covers shipments, 1979-1981; new products; and government purchases, showing units in operation by department and manufacturers. Includes investment recommendations.

1276. OPPENHEIMER AND CO., INC.: DATA PROCESSING INDUSTRY: IBM AMONG THE MIDGETS: WHAT'S THE MEANING OF THIS. Dyon, Esther. Oppenheimer and Company. December 29, 1981. 5 pg. $100.00.

Report on IBM's marketing strategy and the data processing industry structure. Examines IBM's penetration of new markets for products and services and effects on other vendors.

1277. OPPENHEIMER AND CO., INC.: DATA PROCESSING INDUSTRY: IBM'S NEW PERSONAL COMPUTER. Dyson, Esther. Oppenheimer and Company. September 1, 1981. 3 pg. $100.00.

Brief report on IBM's new personal computer. Comments on other vendors' software and hardware, prices, and competition with other manufacturers. Shows 1981 stock prices of major companies; earnings per share and price earnings ratio, 1980-1982; and 1981 dividend and yield.

1278. OPPENHEIMER AND CO., INC.: DATA PROCESSING INDUSTRY: MAINFRAMES / MINICOMPUTERS QUARTERLY REVIEW. Stevens, Jay P. Oppenheimer and Company. Quarterly. December 21, 1981. 13 pg. $130.00.

Quarterly review of new orders, prices and other economic indicators for mainframes and minicomputers, and investment recommendations for corporations. Includes stock prices, 1980-1981; earnings per share, 1980-1982; price earnings ratio, 1981-1982; dividend and yield; stock valuation; book value, 1981-1982; price to book value ratio, 1981-1982; return on equity, 1981-1982; dollar exchange rates, 1976-1982; graph of expenditures for new plants and equipment, 1970-1981; and charts of computer product lines.

1279. OPPENHEIMER AND CO., INC.: DATA PROCESSING INDUSTRY: MAINFRAMES / MINICOMPUTERS QUARTERLY REVIEW: LOWERING EARNINGS ESTIMATES TO REFLECT EFFECT OF HIGH INTEREST RATES ON THE ECONOMY. Stevens, Jay P. Oppenheimer and Company. Quarterly. September 24, 1981. 7 pg. $100.00.

Quarterly analysis of financial performance of U.S. computer corporations and economic indicators for the industry. Shows revised earnings per share estimates, 1980-1982; stock price, 1980-1981; changes in dollar exchange rates versus major foreign currencies; price earnings ratio, 1981-1982; current dividend and yield; stock valuation, 1981-1982; return on equity, 1981-1982; ratio of stock price and book value, 1981-1982; and orders and other news of specific corporations.

1280. OPPENHEIMER AND CO., INC.: DATA PROCESSING INDUSTRY: ROSEN RESEARCH RECAP: THE IMPORTANCE OF DISTRIBUTION. Dyson, Esther. Oppenheimer and Company. May 24, 1982. 6 pg. $100.00.

Report on retail distribution and marketing strategies of microcomput

1281. OPPENHEIMER AND CO., INC.: DATA PROCESSING: MAINFRAMES / MINICOMPUTERS: QUARTERLY REVIEW: THE NEW COMPUTER WARS. Stevens, Jay P. Oppenheimer and Company. Quarterly. March 31, 1982. 7 pg. $100.00.

Quarterly report on mainframes and minicomputers. Examines effects of order rates, earnings, and other economic indicators on major vendors and competition among them. Financial analysis includes earnings per share, stock price range, and price earnings ratio, 1981-1982; dividend and yield; and stock valuation, including book value and returns on equity, 1981-1982.

1282. OPPENHEIMER AND CO., INC.: DATA PROCESSING: MONTEREY DEBRIEFING. Dyson, Esther. Oppenheimer and Company. November 10, 1981. 8 pg. $100.00.

Report from a conference gives capsule financial analysis of vendors and new products. Lists stock prices, shares outstanding, 1980-1982 earnings per share, 1980-1982 price earnings ratio, dividend and yield.

1283. OPPENHEIMER AND CO., INC.: DATA PROCESSING: SOFTWARE INDUSTRY: MARKETING HAY WHILE THE SUN SHINES. Dyson, Esther. Oppenheimer and Company. June 3, 1981. 12 pg.

Analysis of the U.S. computer software industry and market. Shows market value of turnkey systems by types of vendors, 1980-1984; value of shipments, 1979, 1984; lines of software business by corporations; end uses, 1979-1985; stock prices, 1980-1981; shares, 1981; price earnings ratio, 1980-1982; earnings per share, 1979-1982; and 1981 dividends and yields.

1284. OPPENHEIMER AND CO., INC.: DATA PROCESSING: SOFTWARE / SERVICES AND MICRO INDUSTRIES: CURRENT EVENTS. Dyson, Esther. Oppenheimer and Company. July 8, 1982. 10 pg. $100.00.

Report from a recent computer conference. Examines response of software market structure to downward business cycles, new products and the vendors, retail distribution, mergers and financial analysis of corporations. Shows stock prices; earnings per share, 1980-1983; price earnings ratio, 1981-1983; dividends; and yields.

1285. OPPENHEIMER AND CO., INC.: DATA PROCESSING: SOFTWARE, SERVICES, AND SMALL COMPUTERS: COMPANY ROUNDUP. Dyson, Esther. Oppenheimer and Company. August 24, 1981. 11 pg. $110.00.

Review of earnings, mergers, and new products and services of U.S. computer corporations. Covers impact of credits on taxes for research and development. Table shows stock price, 1980-1981; shares outstanding; earnings per share, 1979-1982; price earnings ratios, 1980-1982; dividend; and yield.

1286. OPPENHEIMER AND CO., INC.: DATA PROCESSING: THE COMPUTER INDUSTRY: MARKETING SMALL COMPUTERS IN THE EIGHTIES: NEW DOGS, AND OLD DOGS LEARNING NEW TRICKS. Dyson, Esther. Oppenheimer and Company. January 28, 1981. 14 pg. $140.00.

Analysis of the U.S. small computer industry, and market in the 1980's. Includes financial analysis of selected companies showing stock prices, 1981; earnings per share, 1980-1981; dividends, 1981; and yields, 1981. Gives installations, shipments, utilization and dollar value of units in use, 1974-1984. Market analysis covers users, distribution channels and market shares.

1287. OPPENHEIMER AND CO., INC.: DATA PROCESSING: THE SMALL COMPUTER, II: NOTES ON THE LAS VEGAS CONSUMER ELECTRONICS SHOW: THE JAPANESE, NEXT CHAPTER. *Dyson, Esther. Oppenheimer and Company. February 11, 1981. 4 pg. $100.00.*

Report from a computer show. Discusses competition from Japanese market penetration. Reviews manufacturers, products, and distribution channels. Discusses new products for voice recognition and synthesis.

1288. OPPENHEIMER AND CO., INC.: DATA PROCESSING: THE SOFTWARE SHORTAGE IS DEAD! LONG LIVE THE DISTRIBUTION SHORTAGE! *Dyson, Esther. Oppenheimer and Company. July 20, 1981. 6 pg. $100.00.*

Analysis of markets, vendors' strategies, and products in the U.S. computer software industry. Shows stock price, 1980-1981; shares outstanding; earnings per share, 1979-1982; price earnings ratio, 1980-1982, and current dividend and yield.

1289. OPPENHEIMER AND CO. INC.: INDUSTRY REPORT: DATA PROCESSING INDUSTRY: NEWSLINE: COMPANY UPDATES AND WINTER QUARTER ESTIMATES. *Dyson, Esther. Oppenheimer and Company. January 26, 1982. 10 pg.*

Report on data processing vendors. Examines regulations and industry structure and performance of individual vendors. Shows earnings per share, 1981-1982.

1290. PAINE WEBBER MITCHELL HUTCHINS INC.: STATUS REPORT: MAINFRAME / OFFICE EQUIPMENT MONTHLY. *Garrett, Sandy. Paine Webber Mitchell Hutchins, Inc. Monthly. June 12, 1981. 13 pg.*

Monthly mainframe and office equipment industry analysis. Contains news of mergers, new products, and prices. Tables cover average exchange rates and indexes for major currencies, 1980-1981; 1981 stock price; and 1981 returns. Gives revenues, 1977-1982; earnings from foreign investments, 1980; and performance of corporations.

1291. PAINE WEBBER MITCHELL HUTCHINS INC.: STATUS REPORT: PERSONAL TECHNOLOGY: IMPRESSIONS FROM 1980 WINTER CONSUMER ELECTRONICS SHOW. *Isgur, Barbara. Paine Webber Mitchell Hutchins, Inc. January 27, 1981. 5 pg.*

Financial analysis of the leading U.S. companies in the personal technology industry, including market analysis, 1981. Reviews company performance, sales, prices, and new products.

1292. SMITH BARNEY HARRIS UPHAM AND CO.: DATA PROCESSING: CONDITIONS IN THE HIGH - PERFORMANCE DISK SYSTEMS MARKET. *Labe, Peter. Smith Barney, Harris Upham and Company. November 12, 1981. 6 pg. $200.00.*

Market analysis and update on new high-performance disk systems. Covers market shares, manufacturers' shipments, demand, new products, domestic shipments, capacity of disks, and revenues.

1293. SMITH BARNEY HARRIS UPHAM AND CO.: DATA PROCESSING: MAINFRAME INDUSTRY COMMENTS. *Labe, Peter. Smith Barney, Harris Upham and Company. October 21, 1981. 9 pg. $200.00.*

Analysis of financial performance of computer mainframe corporations. Shows earnings per share, revenues, income, profit margins, equity and shares, 1980-1982.

1294. SMITH BARNEY HARRIS UPHAM AND CO.: DATA PROCESSING: MINICOMPUTER INDUSTRY COMMENTS: THIRD QUARTER REPORTS. *Labe, Peter. Smith Barney, Harris Upham and Company. November 13, 1981. 7 pg. $200.00.*

Report on minicomputer vendors. Gives earnings per share estimates, 1980-1982; revenues; new products; profits; operating performance; sales; income statements, 1981-1982; and forecasts to 1982.

1295. SMITH BARNEY HARRIS UPHAM AND CO.: ELECTRONIC INFORMATION SERVICES: STRUCTURE OF AN EMERGING INDUSTRY. *Atorino, Edward J. Smith Barney, Harris Upham and Company. July 29, 1981. 33 pg. $200.00.*

Market analysis for electronic information services in the U.S. Covers corporations operating as vendors of services, the market for services, industry structure, and competition. Shows industry revenues, 1979, 1980; percent of total company revenues from electronic services, 1980; forecasts of annual growth rates, 1980-1990; and breakdown by size of businesses.

1296. UNITED BUSINESS AND INVESTMENT REPORT: INFORMATION PROCESSORS - A TRANSITION YEAR. *United Business Service Company. v.74, no.21. May 24, 1982. pg.206*

Analysis of information processing markets and corporations. Financial analysis includes earnings per share, 1981-1982; stock prices, 1981-1982; price earnings ratio, 1982; dividends; and yield. Includes orders and new products.

1297. UNITED BUSINESS AND INVESTMENT REPORT: SPECIAL STUDY: 'HOME' COMPUTERS STEPPING OUT. *United Business Service Company. v.74, no.19. May 10, 1982. pg.186*

Analysis of markets for home computers. Financial analysis of selected vendors includes earnings per share, 1980-1982; stock price range, 1981-1982; price earnings ratio, 1982; and dividends. Also includes marketing strategies, sales, and new products.

1298. UNITED BUSINESS AND INVESTMENT REPORT: THE WONDER OF CAD / CAM SYSTEMS. *United Business Service Company. v.73, no.45. November 9, 1981. pg.442*

Brief analysis of the CAD / CAM industry and profiles of major corporations. Gives earnings per share, 1979-1981; stock prices, 1980-1981; price earnings ratio, 1981; dividend; product lines; sales; and orders.

See 194, 577, 578, 581, 2324, 2341, 3211, 4097, 4101, 4124, 4126.

Industry Statistical Reports

1299. ANNUAL SURVEY OF THE COMPUTER SERVICES INDUSTRY. *INPUT. Annual. January 1981. $695.00.*

Annual review of industry composition (by service type), industry financial trends, public company financial data, and revenues by type of company and service.

1300. AUTOMATIC DATA PROCESSING ACTIVITIES SUMMARY IN THE U.S. GOVERNMENT, 1981. *U.S. Government Printing Office. Annual. GS 12.9:980. April 1982. 69 pg.*

Summarizes U.S. government automatic data processing activities as of September 30, 1981. Contains inventory, cost, and utilization data supplied by federal agencies and their cost-reimbursement contractors. Gives number of manufacturers' models in each department; and numbers owned and leased.

1301. AUTOMATIC DATA PROCESSING EQUIPMENT INVENTORY IN THE U.S. GOVERNMENT AS OF END OF FY 79. *U.S. General Services Administration. Annual. GS12. 10:979. June 1980. 905 pg. $15.75.*

Yearly publication with 14 charts and tables detailing inventory by cost groupings, number of computers, and equipment (FY 1965-FY 1979). Six inventory tables give broad summary and detail (by bureau, location, and type of system) by department or agency, by location, and by manufacturer.

1302. BALANCE OF TRADE: U.S. COMPUTER AND BUSINESS EQUIPMENT INDUSTRY. *Computer and Business Equipment Manufacturers Association. Monthly. November 24, 1981. 6 (est.) pg.*

Tabulates previous month's and year-to-date statistics on value of U.S. exports and imports of computers and other categories of office / business equipment (including office forms and supplies); and on respective export / import ratios and trade balances.

1303. CANADA'S DATA PROCESSING MARKET. *Maclean Hunter Ltd. Annual. May 1981. 100.00 (Canada).*

Reports market background for computers, trends and forecasts. Statistical data includes computers installed as of May, 1980, by supplier and industry, and imports of computers and related equipment.

1304. COMPUTER DECISIONS: *DMS / DBMS MARKET STUDY.* *Hayden Publishing Company, Inc. Computer Decisions Division. 1980. 60 pg.*

Fifty-nine tables and graphs detailing purchase plans for assorted products; operating systems usage; distribution of uses and data processing; proposed applications; and consumer attitudes.

1305. COMPUTER SERVICE INDUSTRY. *Canada. Statistics Canada. French/English. Annual. 1981. 15 pg. 4.50 (Canada).*

Includes statistics on operating revenue and expenses of companies engaged in providing computer services as a major activity. Contains notes on methodology of survey, including objectives and questionnaire content. Data issued since 1972.

1306. EIS SHARE OF MARKET REPORT: ELECTRONIC COMPUTING EQUIPMENT. *Economic Information Systems, Inc. Quarterly. 1982. $350.00.*

Shows computer market concentration and ownership structure. Companies in industry are listed and then ranked by production. Other data includes market share (dollars and percent), with groupings of all plants a company operates in the industry, with the parent company's market share broken down on a per-plant basis.

1307. EIS SHIPMENTS REPORT: ELECTRONIC COMPUTING EQUIPMENT. *Economic Information Systems, Inc. Quarterly. 1982. $350.00.*

Report on the electronic computing industry. Arranged by state and county, report includes every plant in the industry with annual shipments of over $500,000 and/or 20 or more employees. Plant listings detail address, telephone, estimated annual shipments, and percent of market. Similar statistics for each county and state.

1308. ELECTRONIC MARKET DATA BOOK: 1979. *Electronic Industries Association. Annual. 1979. 130 pg. $50.00.*

Statistics of the U.S. electronic industries include manufacturers' shipments, 1970-1978; annual factory sales by type of equipment, 1959-1978; annual factory production, 1959-1978; annual imports and exports, 1974-1978, by type; and equipment in use, 1977-1978. Equipment covered includes television, home video tape equipment, videodisc systems, elec-

tronic games, personal computers, calculators, electronic watches, radios, phonographs, audio components, audio tape equipment, autosound, CB, telephone devices and other products.

1309. EMPLOYMENT TRENDS IN COMPUTER OCCUPATIONS. *Bureau of Labor Statistics, U.S. Department of Labor. U.S. Government Printing Office. su. doc. no. L2.3:2101. October 1981. 43 pg. $3.50.*

Presents results of BLS study of employment of workers in 5 computer-related occupations in U.S. Statistics cover employment in computer occupations, employment of computer workers by industry division, 1970, 1980, 1990; and data on college programs and degrees, 1966-1979. Provides value of computer systems produced by U.S. manufacturers by type of computer, 1974-1983.

1310. ENERGY EFFICIENT DATA COMMUNICATIONS NETWORK. *Frost and Sullivan, Inc. September 1980. $975.00.*

Twenty-five distinct components of data communications networks examined in terms of energy efficiency, with computations for energy operational costs. Tabulation of components' energy consumption, by device and vendor. Case study shows contrast of energy-efficient and energy-inefficient network. Operational cost of Bell System modems and digital service units covered. Survey results.

1311. FINANCIAL AND DATA PROCESSING PREVAILING STARTING SALARIES 1980. *Robert Half of New York. Annual. 1980. Free.*

Annual release details starting salaries for numerous varieties of financial and data processing personnel at assorted levels of professional status. Data broken down by employer type (public accounting, banking) and by size of firm or data processing installation.

1312. HANSEN'S 1980 WEBER SALARY SURVEY ON DATA PROCESSING POSITIONS. *A.S. Hansen Inc. Annual. September 1980. 143 pg. $140.00.*

Annual salary survey presents June, 1980 data for 84 data processing positions in almost 1700 U.S. companies. Data is broken down by size of installations, by region (72 cities), and by industry: communications, finance, government, insurance / real estate, services, transportation, utilities, wholesale trade, and total. Gives highest and lowest, average, and first and third quartile figures.

1313. INDUSTRY WAGE SURVEY: COMPUTER AND DATA PROCESSING SERVICES: MARCH 1978. *U.S. Department of Labor. Bureau of Labor Statistics. Irregular. Bulletin 2028. June 1979. 51 pg.*

Report provides data on weekly earnings and employee benefits for professional, technical, and office clerical workers in the computer and data processing services industry. Data is tabulated separately for professional and technical employees and for clerical workers, for 18 selected areas.

1314. MID-YEAR NATIONWIDE SURVEY: U.S. "79 DATA PROCESSING JOB / SALARY MARKET. *Fox-Morris Personnel Consultants. Annual. 1979. Free.*

Annual report analyzes data processing job and salary statistics for applications programmers, computer science B.S. graduates, EDP auditors, MIS directors, software / systems programmers, systems analysts, systems / software / programming managers, and telecommunications specialists.

1315. NATIONAL SALARY SURVEY - COMPUTER PRO-GRAMMERS AND COMPUTER SYSTEMS ANALYSTS. *Organization Resource Counselors Inc. Annual. 1979. $425.00.*

Annual report gives salary breakdowns by geographic area; by job title (programmer 1-4; programmer analyst 1-3; systems analyst 1 and 2); by industry (consulting, financial, industrial, research, other nonmanufacturing). Statistics include median, mean, first and third quarterly, first and ninth decile, and frequency distribution figures.

1316. POS TRENDS IN THE 80'S: AN EVALUATION OF THE CURRENT STATUS OF POINT-OF-SALE TERMINALS AND RELATED SYSTEMS THROUGHOUT THE RETAILING INDUSTRY. *Allen, Randy L. National Retail Merchants Association. 1982. 185 pg.*

Gives results of survey on point of sale terminals in retail industry in U.S. and Canada, 1981. Gives technical and statistical profile of participants; uses of POS; vendor ranking; costs; staffing; and technical data.

1317. SALARY SURVEY OF INFORMATION SYSTEMS ASSOCIATES. *Western Electric Corporate Administration. Annual. September 1979. Free.*

Annual release covers salaries of information systems associates in the U.S.

1318. STATISTICAL REFERENCE BOOK--DATA ENTRY / TERMINAL MARKET, 1980. *International Data Group. Annual. June 1980. 92 pg. $3500.00.*

Annual review and forecast. Contains detailed analysis of the data entry and terminal products end of the computer industry. Discusses growth, projections, technology, problems, competition, and economic indicators which contribute to major trends and developments in the industry. Special section is devoted to the role of IBM. Market strategies discussed. Methodologies employed were historical, current and future market analyses; specific information is supplied on all types.

1319. SURVEY OF DATA PROCESSING SALARIES IN NEW YORK, 1979. *New York Chamber of Commerce and Industry. Annual. September 1979. $140.00.*

Annual report provides New York City and vicinity salary data for 23 job titles.

1320. A SURVEY OF ON-FARM COMPUTER USE IN ALBERTA. *Alberta. Department of Agriculture. 1981. 30 pg.*

Examines uses for computers on farms in Alberta.

1321. 1977 CENSUS OF MANUFACTURES: ELECTRONIC COMPUTING EQUIPMENT: PRELIMINARY REPORT: INDUSTRY SERIES. *U.S. Department of Commerce. Bureau of the Census. Customer Service Branch. Irregular. MC77-1-35F-2(P). August 1979. 10 pg. $.35.*

This is a preliminary report from the Bureau of the Census 1977 Census of Manufactures. Tables with industry statistics from the U.S. for 1963-1977 and with statistics for selected geographic areas for 1972 and 1977 give data relating to employees, production costs, assets and expenditures, inventories, and ratios. Other tables supply 1972 and 1977 statistics on product classes giving quantity and value of shipments by all producers and statistics on materials consumed by kind.

1322. 1980 COMPUTER SALARY SURVEY AND CAREER PLANNING GUIDE. *Source EDP. Annual. 1980. 24 pg.*

Gives 1980 annual wages for positions in the computer industry based on a survey of 30,000 professionals and 14,000 organizations. Data is arranged according to years of experience for non-management positions, and size of system for man-

agement positions. Discusses responsibilities forpositions listed.

1323. 1980 EXECUTIVE COMPENSATION ANALYSIS OF PROFESSIONAL SERVICES FIRMS. *D. Dietrich Associates, Inc. Annual. 4th ed. June 1980. 74 pg. $75.00.*

Report provides information on executive compensation and benefits for consulting engineers, management consultants, computer systems and soft-ware executives, architects, and scientific and economic researchers employed in 227 U.S. firms.

1324. 1981 LOCAL METROPOLITAN COMPUTER SALARY SURVEY. *Source EDP. Annual. 1981. Free.*

Lists average salaries paid for each of 48 data processing job categories in 35 U.S. metropolitan areas and subareas, and in Toronto and Don Mills, Ontario.

1325. 1982 CANADIAN COMPUTER CENSUS. *Canadian Information Processing Society. Annual. April 1982. 125.00 (Canada).*

Contains information as of December 31, 1981, examining the growth of the 'midget' population of computers in the U.S. and Canada.

See 595, 2383, 2384, 2385, 2390, 2393, 3214, 3218, 3221.

Financial and Economic Studies

1326. COMPUTER-BASED NATIONAL INFORMATION SYSTEMS: TECHNOLOGY AND PUBLIC POLICY ISSUES. *U.S. Congress, Office of Technology Assessment. U.S. Government Printing Office. September 1981. 166 pg. $6.50.*

Presents results of a study on the use of computer technology in national information systems and related public policy issues. Discusses the effect of computer systems on labor productivity and employment; the structure of the industry by segment, with rankings of leading manufacturers and 1979 revenues; product groups and description of each industry segemnt; world shipments of major manufacturers; and number of companies in four sales categories. Data on the new technology is also presented in terms of cost per computed calculation or unit of memory.

1327. THE EXPANDING ROLE OF ELECTRONIC DATA PROCESSING IN THE BANKING INDUSTRY: SURVEY 2. *Laventhol and Horwath. September 1979. 30 pg.*

Presents the responses of 189 U.S. commercial banks to a questionnaire in 1979 on data processing's role in the banking industry, and compares it with the 1976 survey. Includes data in current and future uses of electronic data processing, use of outside services, and electronic funds transfer systems.

1328. FINANCIAL PERFORMANCE OF LEADING U.S. DATA PROCESSING COMPANIES IN 1978 AND EARLY 1979. *Cassidy, Donald L. Arthur D. Little, Inc. September 26, 1979. 11 pg. $500.00.*

1979 financial report on 50 data processing industry vendors including peripherals, terminals and mini computer vendors. Rankings by percent of sale, revenues, return on equity and by employee. Comparisons by revenues between IBM and other data processing companies. Data derived from corporate reports, Standard and Poor's Compustat Services, Inc., and Arthur D. Little.

1329. THE MEDIUM-TERM EMPLOYMENT OUTLOOK: SELECTED SUBSECTORS OF THE ELECTRONICS INDUSTRY. *Schwartz, Harvey. Canada. Task Force on Labour Market Development; Employment and Immigration Canada. Canada. Department of Supply and Services. Publishing Centre. Cat. # L41-19/1981E. July 1981. 122 pg. Free (Canada).*

1981 technical study for the Canadian Task Force on Labour Market Development examines the electronics industry in terms of the supply of skilled labor in Canada and the constraints that might be placed on expansion on account of labor shortages. The subsectors studied are consumer products (video-cassette recorders and cameras, videodiscs, home information systems, TV receivers and projection TV, microwave ovens, and other products), and computers, peripherals, and office equipment. The analysis is primarily qualitative rather than quantitative, and the discussion addresses competition from the United States, Western Europe, and Japan.

1330. MICROELECTRONICS, PRODUCTIVITY AND EMPLOYMENT. *Organization for Economic Cooperation and Development. Publications and Information Center. Working Party on Information, Computer and Communications Policy Publication #5. June 1981. 290 pg. $18.00.*

Proceedings of the OECD working party on Information, Computers, and Communications Policy Special Session on Microelectronics, Productivity and Employment. Examines the impact of microelectronics for industries and economies in the member OECD countries. Topics covered include the contribution of information and related technologies to productivity growth; the impact of electronic digital technology on traditional job profiles and employment; market structures, firms' strategies, and organizational structures; and industry location. Includes selected employment and market data. Tables detail employment and sales of main firms in the telecommunications industry (1970-1978); R & D expenditures in the telecommunications industry (1970-1977); market concentration of computers by type in the U.S. market; investment in industry in Germany (1956-1976); white collar vs. blue collar employment in Germany (1900-2000); employment and production changes in the printing industry (1960-1975); and forecasts of the changing job market for software engineers and other microelectronics industry personnel.

1331. NEWSPAPERS AND COMPUTERS: AN INDUSTRY IN TRANSITION. *Desbarats, Peter. Canada. Royal Commission on Newspapers. Canada. Department of Supply and Services. Publishing Centre. Cat. #Z1-1980/1-41-8E. 1981. 116 pg. 7.15 (Canada).*

Examines the role of computers in the production of the newspapers of the future. Discusses new technology such as videotex and Telidon, the role of the newspaper in society, freedom of the press, concentration of ownership, and other matters.

1332. SMALL COMPUTER INDUSTRY SERVICE: COMPANIES: VOLUME 3: A-L. *Dataquest. Monthly. 1980.*

Volumes 3 and 4 of monthly looseleaf series are analyses of the major companies participating in the small computer industry. Companies' historical financial performance, company activities, balance sheets, operating performance, assets, revenues, liabilities, flow of funds, expenditures, dividends, working capital and equity given.

1333. SMALL COMPUTER INDUSTRY SERVICE: COMPANIES: VOLUME 4: M-Z. *Dataquest. Monthly. 1980.*

Volumes 3 and 4 of monthly looseleaf series are analyses of the major companies participating in the small computer industry. Companies' historical financial performance (1975-79), company activities, balance sheets, operating performance,

assets, revenues, liabilities, flow of funds, expenditures, dividends, working capital and equity given.

1334. THIRTEENTH ANNUAL SURVEY OF THE COMPUTER SERVICES INDUSTRY. *INPUT. Association of Data Processing Services Organization, Inc. Annual. July 1979. 194 pg. $495.00.*

Annual on Computer Services Industry (includes operating performances of computer services companies, processing services companies, software product companies, professional services companies and vendors) and 1979 revenues, profit margins, financial analyses and overall performance. Figures based on questionnaire sent to computer services companies by ADAPSO.

1335. THE WORLD OF COMPUTERS AND INFORMATION PROCESSING. *Kelley, Rob. John Wiley and Sons Canada Ltd. May 1982. 15.95 (Canada).*

General overview of computers includes programming, applications, historical data, performance.

See 599, 2427, 2440, 2441, 2442, 3231, 3234, 3235.

Forecasts

1336. APPLICATIONS AND BENEFITS OF INDUSTRIAL ROBOTS. *Battelle Memorial Institute. 1983. $9200.00.*

Forecasts of utilization of industrial robots, 1982-1992, and guidelines for taking advantage of the new technology. Projects demand of major end uses.

1337. CAD/CAM. *FIND / SVP. May 1981. $100.00.*

New York Society of Security Analysts seminar transcript features nine computer graphics industry leaders' overviews of trends and prospects for the CAD/CAM (computer-aided design and computer-aided manufacturing) industry.

1338. CODEC MARKETS. *International Resource Development, Inc. April 1979. 103 pg. $895.00.*

Reviews technologies, markets, usage and prospects for codecs, with a ten-year forecast for the requirement for codecs in each of twelve specific market areas. Discusses the possible emergence of semiconductor vendors as suppliers of certain types of tele- communications equipment and components.

1339. COMPUTER ARCHITECTURE STRATEGY PLANNING FOR THE 80'S. *Strategic Inc. 1981. $1500.00.*

Discusses the apparent strategies for computer architecture development during 1980-1990 employed by the three leading contenders: the U.S. and Japanese governments and I.B.M.. Examines R & D programs and expenditures.

1340. COMPUTER EQUIPMENT STRATEGY PROGRAM: DATABASE AND INDUSTRY SECTOR FORECAST: VOLUME 3. *Management Action Program in Technology (MAP-TEK). Quantum Science Corporation. 1979. 418 pg.*

Volume 3 of a 1978-1982 forecast on computer equipment. Discusses the end-user EDP expenditures forecast (including hardware, software development, computer services, and data communications) and end-user EDP products forecast (including small business, general purpose, telecommunication, and data preparation / entry systems) across the 14 major U.S. industry sectors. Forecasts of expenditures, shipments, growth rates, and competition of computer equipment provided.

1341. COMPUTER EQUIPMENT STRATEGY PROGRAM: VOLUME 2: MARKETS AND STRATEGIES. *Management Action Program in Technology. Quantum Science Corporation. 1979. 238 pg.*

This volume examines major segments of the computer equipment industry. Market segments, competition, market shares, applications, user expenditures, opportunities, strategies, new technologies and scenarios of future actions by suppliers and government agencies are analyzed. Top manufacturers are discussed. The publication is only available to MAPTEK subscribers.

1342. THE COMPUTER GRAPHICS SOFTWARE AND SERVICES MARKET. *Frost and Sullivan, Inc. February 1979. 234 pg. $825.00.*

A study of the current computer graphics and software services market includes structural overview of industry and market segments by product (independent, software, turnkey systems and services). Reviews major proprietary and independent packages with estimates on number of packages sold and installed and the principal users of the packages including specific companies. Descriptions of 19 graphic software and turnkey system vendors and 5 major graphics time-sharing vendors. 10-year market forecasts included.

1343. COMPUTER MAINFRAME TRENDS. *Martin Simpson & Co., Inc. Martin Simpson and Company, Inc. April 1980. 15 pg. $150.00.*

Three reports discuss mainframe trends and prospects for IBM, Burroughs, Honeywell, NCR and Sperry.

1344. THE COMPUTER SOFTWARE PACKAGES AND PACKAGED SERVICES MARKET. *Frost and Sullivan, Inc. 1980. 225 pg. $700.00.*

Analyzes the computer software and package services market potential, with 10-year estimates for 8 basic categories (hardware dependent / dedicated, operating systems software packages, data-base management systems, systems software packages, applications software packages--general and industry, general-industry packaged services, and special-industry and data-base packaged services). Profiles provided for 10 best known independent suppliers. Discusses the impact of the mini-computer on the industry and the question of state sales tax on software.

1345. CONSUMER TIME-SHARING SERVICES. *International Resource Development, Inc. August 1980. 148 pg. $985.00.*

Analysis of current and expected future markets for home time-sharing services. Potential strategies for market entry by remote computer services companies, publishers, software houses, etc. Current and future user demographics; current positions of The Source and CompuServe (MicroNet) and future expected plans of other RCS vendors. Results of survey of leading timesharing companies. Ten-year market growth forecasts.

1346. DATA BASE MANAGEMENT SYSTEMS. *Creative Strategies International. January 1980. $1195.00.*

1980 market analysis of end-user migration trends to data base management systems (DBMS) and the yet-to-come data resource management systems (DRMS). Market forecasts through 1983 are presented. Market projections for DBMS products include segmentation by standalone versus distributed; by mainframe versus small business computers; by IBM versus non-IBM installations; and by captive vendor versus third party suppliers. Key issues discussed include the emergence of relational models, the rising demand for enhanced software tools, the impact of inflation, and the rapid decline in hardware pricing. Vendor strategies, shares and potential future opportunities are presented.

1347. DATA COMMUNICATIONS EQUIPMENT IN FINANCIAL INSTITUTION BRANCH NETWORKS (U.S.). *Frost and Sullivan, Inc. January 1979. $800.00.*

Market forecast for branch terminals - teller, platform / back office, customer operated; distributed processors - minicomputers / CPU's, peripherals, check handling / processing equipment; terminal systems - controllers, concentrators, etc.; front-end telecommunication processors and associated hardware, application software and services. Market profile of U.S. banks and their offices and front office data communications. User survey.

1348. DATA PROCESSING: RECENT DEVELOPMENTS IN THE LARGE-SCALE SYSTEMS MARKET. *FIND / SVP. January 1981. 18 pg. $175.00.*

Provides state-of-the-art information about the 3081 from IBM and 580 series from Amdahl. Study anticipates that value of shipments will exceed twice the level of 1981 by 1984.

1349. EDP INDUSTRY REPORT: REVIEW AND FORECAST. *Upton, Molly, editor. International Data Group. Bi-weekly. May 28, 1980. 28 pg. $60.00.*

1980 report discusses different computer systems (mini / micro, general purpose, small business, desktop) manufactured in the U.S. but shipped internationally. Provides rankings by manufacturer, shipments by computer system and by year (1974-79), and forecasts of growth rates and revenues to 1985. Gives census of computer systems listed by U.S. manufacturers and model, with prices and number of installations.

1350. FINANCIAL SERVICES AND ASSOCIATED FINANCIAL TERMINALS / SYSTEMS MARKETS IN THE RETAIL ENVIRONMENT. *Frost and Sullivan, Inc. September 1979. 298 pg. $900.00.*

Provides forecasts through 1989 for installed base, unit shipments and dollar sales for three terminal and associated equipment categories broken out further by retail outlet type. Presents an operational and marketing overview on current banking services. Major suppliers, their products, prices and applications reviewed.

1351. INDUSTRIAL ROBOTS: A DELPHI FORECAST OF MARKETS AND TECHNOLOGY. *Society of Manufacturing Engineers. 1982. 230 pg. $65.00.*

Technology section covers equipment, capabilities, design, and applications including programming methods, control types, grippers, sensing devices, and scene analysis, with forecasts of robots' sociological impact. Appendix provides data in tabular form.

1352. INK JET PRINTING: ITS ROLE IN WORD PROCESSING AND FUTURE PRINTER MARKETS. *Martin Simpson & Co., Inc. Datek of New England. Martin Simpson and Company, Inc. March 1980. 247 pg. $575.00.*

A 1980 analysis of ink jet printing and its relationship to the word processing market. The report includes an end-user survey of 100 customers of the IBM 6640 ink jet printer who cite their views of the product and the industry it represents. Detailed tabulations and review of test results are provided. Comments are included from 36 firms engaged in ink jet printing. Current product analysis and competitive strategies of each are reviewed.

1353. INTELLIGENT TERMINALS. *Creative Strategies International. September 1980. $1195.00.*

Analyzes role of the markets for intelligent terminals including standalone and clustered terminal systems. Market forecasts and trends are given for standalone and clustered as well as for storage categories of cassette-based, diskette-based, hard disk systems, and others. Forecasts presented in unit shipments, installed base, and revenues for time period 1980-1985. Anatomy, features and characteristics of intelligent terminal products discussed along with relationships to very small business computers which have emerged as a competitive alternative approach to intelligent data entry. Assesses 1980 market shares for major vendors.

1354. KEYBOARD MARKETS. *International Resource Development, Inc. July 1980. 131 pg. $985.00.*

Analysis of end-user equipment (office, home, industrial) using keyboards. Discussion of impact of membrane and other new technologies, and analysis of move away from captive to open markets. Future expected configurations of keyswitches and impact of non-keyboard data entry technologies. Market positions of more than 25 suppliers of keyboards. Ten-year market projections.

1355. THE MAIL POTENTIAL FOR ELECTRONIC MAIL. *Probe Research, Inc. November 1979. 150 pg. $180.00.*

Report focuses on the development of electronic mail systems. It contains detailed operating, and cost specifications for recent domestic (U.S.) and international postal electronic mail proposals, including customer hardware, software, and transmission requirements, estimated delivery times, and sample cost comparisons. Market analyses include identification of user industries, common carrier capacity, and postal projections. Legal and policy issues are analyzed. Brief descriptions of other "EMS" systems currently proposed or in existence, such as XTEN, SBS, mailgram and history of postal and Western Union electronic mail activities followed. The major legal battle which involves Congress, the President, the FCC, the Postal Rate Commission, the U.S. Postal Service, major mailers and trade groups, labor unions, AT&T, Xerox, IBM, and many common carriers is discussed.

1356. MARKET ANALYSIS SERVICE (COMPUTER SERVICES): 1979 ANNUAL REPORT. *INPUT. Annual. December 1979. 303 pg.*

1979-1984 computer services market annual forecast and analysis. The forecast is divided by mode of service delivery: processing (remote computing, batch processing and facilities management); software products (applications and systems products); and professional services. Demographic data, market forecasts, and EDP managers' perspectives are analyzed for 14 major industry sectors. Information regarding mode of collection of data is provided in one of the 7 appendices.

1357. MEDICAL INFORMATION SYSTEMS. *Creative Strategies International. May 1980. $1200.00.*

1980 study analyzes the markets for computerized medical information systems (MIS). Forecasts through 1984 for major application segments: hospitals, physician office-based practices (small medical groups), nursing homes, and dentist office-based practices. Includes discussion of the industry, future markets and products, technology trends, competitive dynamics, and the future of in-house microcomputer systems and shared service systems. Profiles of major manufacturers are provided.

1358. THE MILITARY AND AEROSPACE COMPUTER MARKET. *Frost and Sullivan, Inc. January 1979. 227 pg. $850.00.*

Surveys the market for special purpose digital computers in U.S. military aircraft, U.S. Navy shipboard, U.S. battlefield and missile systems, U.S. military and aerospace communications and U.S. civilian aircraft. Analyzes over 40 specific projects (NTDS, FFG, Harpoon, etc.) for computer content and attendant funding. Supplies program forecasts in dollars and quanti-

ties through 1984. Lists suppliers in each segment with their market shares.

1359. NON-IMPACT PRINTERS. *International Resource Development, Inc. June 1981. 186 pg. $985.00.*

Examines current market position of manufacturers of non-impact printers, and installed base. Gives shipment estimates and ten-year market forecasts. Reviews technologies and equipment by application. Discusses specialty papers, toners and other consumables. Covers U.S., European and Japanese markets.

1360. OFFICE TECHNOLOGY STRATEGY PROGRAM: DATABASE AND INDUSTRY SECTOR FORECAST: VOLUME 3. *Management Action Program in Technology. Quantum Science Corporation. 1979. 315 pg.*

This forecast for 1978-1983 describes the end users' expenditures for office equipment (text preparation and copy duplicating). Forecasts are for each of the 14 major industry sectors of the U.S. economy. Options for systems are based on needs of firms (employees, distribution, price, etc.).

1361. THE OUTLOOK FOR HANDHELD DATA ENTRY TERMINALS. *Meserve, Everett T. Arthur D. Little, Inc. April 17, 1980. 6 pg. $600.00.*

1980 report on data entry terminals including keyboards, alphanumeric displays, two-way communication and user programmability. Market areas discussed are growth rate, shipments, new product features, new applications, suppliers and potential competition. Data derived from Arthur D. Little estimates.

1362. OVERVIEW REPORT: TECHNOLOGY PLACES NEW FOCUS ON TERMINALS. *Hersch, A. Maclean Hunter Ltd. August 1981. pg.36-42 3.00 (Canada).*

Review and forecasts of current and emerging trends in 'user-friendly' terminals. Includes a look at research and development, new techniques, self-help facilities, integrated office systems, prices and voice terminals.

1363. THE PERSONAL BUSINESS TERMINAL. *International Resource Development, Inc. Irregular. September 1979. 168 pg. $895.00.*

This report analyzes the structure and size of more than twenty sectors of the U.S. economy considered significant current or potential users of special-purpose intelligent terminals. Discusses the relationship with PABX and "office of the future." Ten-year projections, with review of effects on paper and business forms are included.

1364. THE PRINTER MARKET (U.S.): COMPUTER, COMMUNICATIONS, OFFICE EQUIPMENT. *Frost and Sullivan, Inc. December 1980. 222 pg. $900.00.*

Computer industry and related equipment report provides analysis and sales forecasts to 1988 by end use equipment market and by impact and non-impact printing technique. Competitive trends are evaluated by type of supplier (system, OEM, and independent) and vendors. Company profiles included for major printer suppliers and for selected suppliers in specific printer technique categories. Technological trends are traced with indications of product improvements to be made in end user segments and various printer techniques. Discusses new application opportunities. User survey results and directory included.

1365. REMOTE COMPUTING SERVICES. *Creative Strategies International. September 1980. $1195.00.*

Analysis of the structural forces of the 1980s which will impact the remote computing services (RCS) industry. Key discussions include: the re-entrance of IBM, the emergence of public data networks, the proliferation of low-cost DDP and micro-computing equipment, office automation, the declining cost of hardware, and the creation of new marketing opportunities for RCS vendors. Market projections are included for 1980 through 1985.

1366. SALES PRODUCTIVITY AND ALTERNATE DISTRI-BUTION CHANNELS. *INPUT. 1980.*

An international multiclient study of trends in distribution chan-nels in the computer industry and their effects on the cost of sales.

1367. THE SMALL BUSINESS COMPUTER MARKET. *Frost and Sullivan, Inc. March 1979. 338 pg. $850.00.*

Market analysis and forecast to 1987 for number of units, unit price and dollar sales of three major types of small business computer: microprocesser based microcomputers (average price $10,000), single terminal minicomputers (average price $65,000), and multiple terminal and distributed data processing machines (average price $140,000). Provides user analysis survey and market share data and company profiles for 11 major suppliers.

1368. SOON: THE COMPUTER 'UTILITY': MARKETS, OPPORTUNITIES. *Business Communications Company. December 1981. 100 pg. $850.00.*

Forecasts of changes in computer industry and market struc-ture. Projects extension of computer utilization to families and households.

1369. STRATEGIC IMPACT OF INTELLIGENT ELECTRON-ICS IN U.S. AND WESTERN EUROPE: 1977-1987: VOLUME 6: CONSUMER MARKETS. *Seabury, Frank. Arthur D. Little, Inc. April 1979. 113 pg. $35000.00.*

Volume 6 of 11-volume multi-client study discusses the micro-processor in the consumer field for 1974-1987. Main areas for strategic impact discussed are home terminals, telephone, communication systems, and toys and games. Consumers grouped by country (U.S., France, West Germany, United King-dom) & by household income to determine market segments, areas, and penetration; sales; and consumption of micropro-cessors. Price listed is for complete set. Contact publisher for individual prices. Data derived from MacKintosh Ltd., Merchan-dising and Arthur D. Little, Inc. estimates.

1370. STRATEGIES FOR MEETING THE INFORMATION NEEDS OF SOCIETY IN THE YEAR 2000. *Boaz, Martha, ed. Libraries Unlimited Inc. September 30, 1981. 197 pg. $22.50.*

Essays forecasting changes in the information industry struc-ture due to effects of new telecommunications and computer technology through the year 2000. Examines effects on end users especially businesses, productivity, information distribu-tion networks, legislation and regulation, ownership issues, and finances of information networks.

1371. TWO WAY BUSINESS COMMUNICATIONS, SER-VICES AND EQUIPMENT. *Frost and Sullivan, Inc. February 1980. 238 pg. $950.00.*

Includes information on data communications, communica-tions, word processors, electronic mail, computer reference, facsimile, phone systems, telephone answering, picture phone, video conferencing. User survey results.

1372. THE USED COMPUTER MARKETPLACE. *International Data Group. June 1979. 75 (est.) pg.*

Analyzes the development of the third-party market for com-puters. Gives market structure, market growth, financial analy-sis, revenues, and U.S. shipments by IBM secondary market (with mention of other mainframe vendors policies concerning secondary equipment market). Telephone interviews con-ducted with EDP managers using IBM/370 for past 4 years provided IDC's information.

1373. VERY SMALL BUSINESS COMPUTERS. *Creative Strategies International. April 1980. $1195.00.*

The commercial markets for microcomputer systems priced under $15,000. Discusses and analyzes U.S. markets and manufacturers including international sales by U.S. companies. Market segments covered include home and hobby, very small businesses and professional services, large businesses, edu-cational institutions, government and non-profit organizations, and international. Traces history of VSBCs, analyzes current market size and forecasts market to 1984. Reviews status of hardware and software technologies impacting industry, trends in marketing and distribution, and leading companies competi-tive practices.

See 223, 224, 226, 227, 232, 235, 239, 240, 2458, 2463, 2465, 2466, 2467, 2468, 2469, 2470, 2471, 3236, 3239, 3620, 3622.

Directories and Yearbooks

1374. ADMINISTRATIVE DIRECTORY OF COLLEGE AND UNIVERSITY SCIENCE DEPARTMENTS AND COMPUTER CENTERS. *Association for Computing Machinery. Annual. 1981. 175 pg. $10.00.*

Directory of college and university computer and data process-ing centers and academic programs and related associations. Listings detail address, chairperson, and academic degrees offered.

1375. ALL ABOUT DATA COMMUNICATIONS FACILITIES. *Datapro Research, Inc. (McGraw-Hill). Annual. 1981. 32 pg. $15.00.*

Report summarizes the communications facilities and rate structures for interstate data transmission over the public tele-phone network, private lines, TWX / Telex satellite channels, and packet switching networks. Survey of user experience with over 10,000 communications links includes user ratings.

1376. ALL ABOUT PERSONAL COMPUTERS 1981. *Datapro Research, Inc. (McGraw-Hill). Irregular. 1981. $29.00.*

Provides comparative data on 18 of the more popular models. Covers capabilities, confirmations and applications, options, warranties, prices, retail availability, and discounts; and char-acteristics of each major peripheral and software product the vendor provides with the system. Contains directories of nearly 200 software vendors, 300 peripheral equipment vendors, 34 system vendors, and 19 periodicals geared to personal com-puter users. Tracks development of personal computer indus-try, with current and projected markets and trends.

1377. ALL ABOUT 114 MICROPROCESSORS. *Datapro Research, Inc. (McGraw-Hill). Annual. 1981. 28 pg. $15.00.*

Details characteristics and prices of 114 current microproces-sors from 21 manufacturers.

1378. ALL ABOUT 117 TIME - SHARING AND REMOTE COMPUTING SERVICES. *Datapro Research, Inc. (McGraw-Hill). Annual. 1981. 45 pg. $15.00.*

Report explains interactive time-sharing and remote batch pro-cessing; discusses their advantages and drawbacks; summa-rizes services offered by 106 remote computing companies; and details user ratings of major companies. Selected from 'Datapro 70.'

1379. ALL ABOUT 136 USER - PROGRAMMABLE TERMINALS. *Datapro Research, Inc. (McGraw-Hill). Annual. 1981. 42 pg. $15.00.*

Report defines a user-programmable terminal and discusses the features and functions of the range of equipment from portable teleprinters to multistation shared processors, which this category includes. Provides specifications of 136 user-programmable terminals from 61 vendors and presents the results of a survey of Datapro subscribers with an installed base of 1,782 terminals. Includes user ratings.

1380. ALL ABOUT 149 MICROCOMPUTERS. *Datapro Research, Inc. (McGraw-Hill). Annual. 1981. 37 pg. $15.00.*

Comparison charts describe characteristics and pricing of microcomputers from 49 vendors. Includes information on development systems and kits.

1381. ALL ABOUT 210 ALPHANUMERIC DISPLAY TERMINALS. *Datapro Research, Inc. (McGraw-Hill). Annual. 1981. 57 pg. $15.00.*

Report provides specifications and prices of 210 alphanumeric display terminals from 68 vendors. Presents the results of a user survey covering over 19,000 installed terminals.

1382. ALL ABOUT 246 MINICOMPUTERS. *Datapro Research, Inc. (McGraw-Hill). Annual. 1981. $15.00.*

Comparison charts cover characteristics and prices of 246 minicomputers from 63 manufacturers. Includes user ratings.

1383. ALL ABOUT 247 MINICOMPUTER DISK STORAGE UNITS. *Datapro Research, Inc. (McGraw-Hill). Annual. 1981. 57 pg. $15.00.*

Survey of 247 disk storage units. Chart describes product lines of 39 suppliers.

1384. ALL ABOUT 278 SMALL BUSINESS COMPUTERS. *Datapro Research, Inc. (McGraw-Hill). Annual. 1981. 68 pg. $15.00.*

Reports characteristics and prices of 278 small business computer systems from approximately 70 vendors on comparison charts. Includes user ratings.

1385. ALL ABOUT 30 PHOTOCOMPOSITION UNITS. *Datapro Research, Inc. (McGraw-Hill). Annual. 1981. 21 pg. $15.00.*

Reviews 30 models of photocomposers from 14 manufacturers. Models range from smaller, keyboard-operated lines to large, computer-operated systems.

1386. ALL ABOUT 346 MINICOMPUTER PRINTERS. *Datapro Research, Inc. (McGraw-Hill). Annual. 1981. 78 pg. $15.00.*

Survey looks at the products of 71 vendors via comparison charts detailing the characteristics and prices of 346 printers and printer families.

1387. ALL ABOUT 400 MODEMS. *Datapro Research, Inc. (McGraw-Hill). Annual. 1981. 53 pg. $15.00.*

Report summarizes characteristics and prices of over 400 modems from 48 suppliers. Presents results of users' survey covering experiences with 27,000 modems.

1388. ALL ABOUT 71 GRAPHIC DISPLAY DEVICES. *Datapro Research, Inc. (McGraw-Hill). Annual. 1981. 23 pg. $15.00.*

Presents characteristics of 71 devices from 22 vendors. Includes user ratings and experiences of 28 users.

1389. ALL ABOUT 76 DATA COMMUNICATIONS MULTIPLEXERS. *Datapro Research, Inc. (McGraw-Hill). Annual. 1981. 28 pg. $15.00.*

Report explains the techniques and applications of multiplexing, and summarizes the characteristics of 76 current units from 21 vendors. Analyzes the experience of multiplexer users. Includes user ratings.

1390. ALL ABOUT 78 FACSIMILE MODELS. *Datapro Research, Inc. (McGraw-Hill). Annual. 1981. $15.00.*

Report compares and rates the features and characteristics of 78 facsimile computer models from 18 facsimile computer equipment suppliers.

1391. ALL ABOUT 84 MICROCOMPUTER CASSETTE AND CARTRIDGE MAGNETIC TAPE UNITS. *Datapro Research, Inc. (McGraw-Hill). Annual. 1981. 23 pg. $15.00.*

Surveys and summarizes the specifications of 84 cartridge and cassette tape subsystems available from 21 suppliers. Comparative charts for minicomputers interfaced, primary market, type, format, capacity, characteristics, software support, packaging, pricing, and availability.

1392. ALL ABOUT 84 OPTICAL READERS. *Datapro Research, Inc. (McGraw-Hill). Annual. 1981. 31 pg. $15.00.*

Analyzes place of optical readers in the data processing world. Tabulates experiences of 50 optical reader users and details characteristics of 84 current optical character, mark, and bar readers from 31 manufacturers.

1393. AMERICAN NATIONAL STANDARDS COMMITTEE X3 - INFORMATION SYSTEMS, MEMBERSHIP AND OFFICERS. *Computer and Business Equipment Manufacturers Association. Irregular. January 1982. 17 pg. Free.*

Directory of member companies of the American National Standards Committee X3. Includes name, mailing address and telephone number of the principal and alternate representatives. Lists membership of standing committees and officers of technical committees.

1394. ASSOCIATION FOR COMPUTING MACHINERY 1981 ROSTER OF MEMBERS. *Association for Computing Machinery. Triennial. 1981. 488 pg. $30.00.*

Provides alphabetic and geographic cross-listing of names and addresses of 50,000 regular, associate, and student members of the Association for Computing Machinery as of January 1, 1981.

1395. AUERBACH COMPUTER TECHNOLOGY REPORTS: PLUG COMPATIBLE PERIPHERALS. *Auerbach Information, Inc. Monthly. 1980. 600 (est.) pg. $235.00.*

Monthly updating looseleaf reference service for data processing includes: computer support equipment, storage and retrieval systems, readers / printers / plotters, magnetic tapes, drives and phototypesetters. Marketing information is presented in analytical reports on individual products and on comparison and pricing charts.

1396. AUERBACH COMPUTER TECHNOLOGY REPORTS: SYSTEMS SOFTWARE. *Auerbach Information, Inc. Bimonthly. 1980. 1000 (est.) pg. $495.00.*

Bimonthly updating looseleaf service on proprietary software packages, including operating systems, data base management, information storage and retrieval, system development and analysis. Gives marketing analysis of reports and packages with comparison charts detailing features and pricing information.

1397. AUERBACH DATA COMMUNICATIONS EQUIPMENT. *Auerbach Information, Inc. Quarterly. 1980. 800 (est.) pg. $650.00.*

Quarterly looseleaf updating service on digital data communications equipment and techniques. Gives analytical reports on terminals and processing equipment and data communication systems and facilities. Marketing comparisons and pricing information in text and charts are provided.

1398. AUERBACH TIME SHARING REPORTS. *Auerbach Information, Inc. Quarterly. 1979. 50 (est.) pg. $305.00.*

A two volume quarterly, looseleaf updating service covering all aspects of commercial time sharing. Includes reports on the state of the art, languages, applications and equipment. Individual reports on commercial services, directory of companies, and a listing of data banks.

1399. AUERBACH'S ELECTRONIC OFFICE: MANAGEMENT AND TECHNOLOGY: VOLUME 1. *Auerbach Information, Inc. Quarterly. 1980. 1000 (est.) pg.*

Volume 1 of 2-volume set on the electronic office covers applications, implementation, and technologies of office equipment. Business minicomputers, office computers, input-output terminals, intelligent computers, printers and teleprinters, OCR readers, and clustered systems are covered with product reports. Comparisons among competitors, user reactions, specifications, performance, price data, and maintenance rates are given for each.

1400. AUTOMATIC DATA PROCESSING EQUIPMENT INVENTORY IN THE U.S. GOVERNMENT, FISCAL YEAR 1981. *General Services Administration. U.S. Government Printing Office. Annual. su.doc.no. GS 12.10:981. February 1982. 708 pg.*

Provides inventory of computers, computer systems, and related equipment in U.S. Government, September 30, 1981. Tables arranged by agency, location, and manufacturer. Covers responsible office; system name, type and management classification; and manufacturer, model, number owned and number leased.

1401. A BUYERS' GUIDE TO SMALL BUSINESS COMPUTERS. *Canadian Information Processing Society. Irregular. 1981. 2.50 (Canada).*

Guide on how to buy a small business computer with step-by-step approaches. Reviews products and manufacturers.

1402. BUYING COMPUTER HARDWARE. *Alberta. Department of Agriculture. Monthly.*

Contains buying guides and a hardware checklist with comparison tables.

1403. CBEMA: CHALLENGES FOR THE 80'S. *Computer and Business Equipment Manufacturers Association. Irregular. June 1981. 15 pg. Free.*

Presents an overview of the computer and business equipment manufacturing industry. Contains a listing of CBEMA member companies and their products.

1404. COMPUTER COMPATIBLE DIRECTORY AND TECHNOLOGY REVIEW 1982/83. *Morgan Grampian Publishing Company. Annual. October 1982.*

Directory covers new products, manufacturers, technology and marketing trends in the computer industry.

1405. COMPUTER DIRECTORY AND BUYERS' GUIDE 1979-1980. *Berkeley Enterprises, Inc. Annual. v.28, no.10B. October 31, 1980. 60 pg. $27.90.*

Roster of organizations in computer and data processing industries provides names, addresses, telephone numbers, and products and services of 2400 organizations.

1406. COMPUTER DISPLAY REVIEW: ALPHANUMERIC DISPLAY TERMINALS: VOLUME 1. *GML Corporation. Tri-annually. 1979. 1200 pg. $600.00 (3 vol.).*

Three-volume compilation, updated quarterly, provides an overview of all major display and graphic systems manufactured in the U.S. and abroad. Features include: available system hardware; display type and deflection; display design; controller design; software; operational characteristics; character specifications; transmission type; keyboard layout; cursor characteristics; editing functions; memory expansion; interfacing; transmission rates; marketing data; general applications; and item prices. Also provides: photograph of unit; summary of highlights; general features; purchase and lease prices; and manufacturer's address.

1407. COMPUTER DISPLAY REVIEW: ALPHANUMERIC DISPLAY TERMINALS: VOLUME 2. *GML Corporation. Quarterly. 1979. $600.00 (3 vol.).*

Three-volume compilation, updated quarterly, provides an overview of all major display and graphic systems manufactured in the U.S and abroad. Features include available system hardware; display type and deflection; display design; controller design; software; operational characteristics; character specifications; transmission type; keyboard layout; cursor characteristics; editing functions; memory expansion; interfacing; transmission rates; marketing data; general applications; and item prices. Also provides: photograph of unit; summary of highlights; general features; purchase and lease prices; and manufacturer's address.

1408. COMPUTER DISPLAY REVIEW: GRAPHIC DISPLAY SYSTEMS: VOLUME 3. *GML Corporation. Quarterly. 1979. $600.00 (3 vol.).*

Three-volume compilation, updated quarterly, provides an overview of all major display and graphic systems manufactured in the U.S. and abroad. Features include: available system hardware; display type and deflection; display design; controller design; operational characteristics; display generator design; vector generation; character generation; display timing and performance; display instruction words; interfacing; transmission rates; marketing data; item prices; configuration diagrams; and operating software. Also provides: photograph of unit; summary of highlights; general features; purchase and lease prices; and manufacturer's address.

1409. COMPUTER GRAPHICS MARKETPLACE. *Cosentino, John, editor. Oryx Press. Irregular. 1981. 53 pg. $18.50.*

A six-part directory lists manufacturers and suppliers of computer graphics equipment, along with addresses, telephone numbers, executives' names, product lines and number of employees. Additional sections of the directory feature consultants and services, professional organizations, educational programs, conferences and conventions, and books and periodicals for the computer graphics industry.

1410. COMPUTER PERIPHERALS REVIEW. *GML Corporation. Annual. 1980. 350 pg. $195.00.*

Annual directory (updated tri-annually) provides product evaluations of more than 4200 computer peripherals manufactured in the U.S. and abroad. Includes technology profile, device type and prices for disks and drums, flexible diskettes, magnetic tapes, tape cassettes, line printers, serial printers, card equipment and paper-tape equipment. Directory of manufacturers also presented.

1411. COMPUTER - READABLE DATA BASES: A DIRECTORY AND DATA SOURCEBOOK: 1979. *American Society for Information Science. Biennial. October 1979. 1367 pg. $95.00.*

1979 directory and data sourcebook for 528 data bases available in the U.S. Data bases included are bibliographic, factographic and numeric. An alphabetical list of data base contents is followed by data base descriptions. All listings are in computer-readable form, are publicly available, and are used for information retrieval purposes or are available through major online vendors of information retrieval services. Provides names, addresses, key personnel, time span covered, frequency of updates; subject matter and scope, and services offered, for each data base.

1412. COMPUTER REVIEW. *GML Corporation. Tri-annually. 1980. 550 pg. $175.00.*

Annual directory (updated tri-annually) provides product evaluations of more than 350 U.S. and foreign computer systems and company profiles including annual financial data of more than 50 computer firms. Reviews computer categories and model summaries as well as peripherals (disk & drum storage, magnetic tape, line printers and card equipment). Also includes discussion of the industry in review and minicomputers on the market.

1413. COMPUTERS: INVESTMENT DATA SERVICE. *Luning Prak Associates, Inc. Irregular. December 1980. $2400.00.*

For each of the 15,000 entries in the data bank on computers in the health care industry, information includes location, brand, model number, year acquired, application, tests / month acquisition mode, reagent source and service provider.

1414. DATA SOURCES. *Ziff - Davis Publishing Company. Quarterly. v.1, no.4. August 1982. 2000 pg. $20.00.*

Directory of computer hardware and software vendors, manufacturers of data communications equipment, telecommunications systems, consultants and services. Lists products and services, business addresses, executives, lines of business, and sources for leasing.

1415. DATA SOURCES. *Data Sources. Quarterly. 1981. 1460 pg. $60.00.*

Quarterly directory lists over 6000 computer systems and peripheral hardware products. Over 7000 software products are offered under data management, communications, utilities, compilers, aids, system software, and applications software. Cross-indesed to company profiles, dir

1416. DATAGUIDE: THE MASTER CATALOG AND DIRECTORY OF OEM COMPUTER PRODUCTS: SPRING 1980. *Sentry Publishing Company. Semi-annual. March 1980. 150 (est.) pg. $50.00/yr.*

1980 semi-annual directory (spring / fall) containing lists of products, advertisers, and manufacturers (sales, number of employees and addresses) of OEM computer products. Products include processors, memory boards, CRT, disk drives, software packages, and printers.

1417. DATAPRO DIRECTORY OF SMALL COMPUTERS: VOLUME 1. *Datapro Research, Inc. (McGraw-Hill). Quarterly. 1980. 1000 (est.) pg.*

Volume 1 of 2-volume marketing directory on small computers includes applications index, user guide and guidelines and then reports on specific small computer systems available. Addresses of manufacturers, installations, rates, prices, perfor-

mance, specifications and number of shipments are given for each product.

1418. DATAPRO DIRECTORY OF SMALL COMPUTERS; VOLUME 2. *Datapro Research, Inc. (McGraw-Hill). Quarterly. 1980. 800 (est.) pg.*

Volume 2 on small computers includes company profiles with addresses, products, presidents, sales revenues, sales concentration, dates founded, number of employees and primary customers. Listings are then divided into sections on manufacturers, maintenance companies, computer services, peripheral and software vendors, user ratings and user groups.

1419. DATAPRO DIRECTORY OF SOFTWARE: VOLUME 1. *Datapro Research, Inc. (McGraw-Hill). Semi-annual. 1980. 1000 (est.) pg. $270.00.*

Semi-annual publication includes a looseleaf monthly updating service. Data on all varieties of proprietary software designed for making decisions on purchasing software products. Two volume set is divided into 26 sections. This volume includes banking / finance, data communication, data management, education, and computer management aids. Demographic data, system features/ options and pricing data are provided. Review and analysis is of already existing data.

1420. DATAPRO DIRECTORY OF SOFTWARE: VOLUME 2. *Datapro Research, Inc. (McGraw-Hill). Semi-annual. 1980. 1000 (est.) pg. $270.00.*

Volume II is semi-annual looseleaf publication with a monthly updating service. Provides a review and analysis of marketing data on all varieties of proprietary software designed for making decisions on purchasing software products. This volume includes engineering, insurance, language processors, manufacturing, math and statistical, and medical and health care through utility programs. Software options, features, marketing analysis and pricing information provided.

1421. DATAPRO REPORTS ON DATA COMMUNICATIONS: VOLUME 1. *Datapro Research, Inc. (McGraw-Hill). Monthly. 1980. 1000 (est.) pg. $410.00.*

Monthly looseleaf updating service contains detailed coverage of communications equipment and services including standards, management / system guides, networks, processors and software. Characteristics / features / pricing / availability of systems are provided in text (management summaries) and comparison charts / tables. This is Volume 1 of 3-volume set. Reviews and analyzes systems and information currently available.

1422. DATAPRO REPORTS ON DATA COMMUNICATIONS: VOLUME 2. *Datapro Research, Inc. (McGraw-Hill). Monthly. 1980. 1000 (est.) pg. $410.00.*

Volume 2 of monthly looseleaf updating service includes programmable terminals, batch terminals, display terminals, teleprinters and terminal attachments. Analyses, comparison charts (price, cost of operation and availability) and vendor systems available (including management summary and characteristics) are provided. Reviews and analyzes systems and information currently available.

1423. DATAPRO REPORTS ON MINICOMPUTERS: VOLUME 1. *Datapro Research, Inc. (McGraw-Hill). Annual. 1980. 800 (est.) pg.*

Vol. 1 (of 3) of 1980 service provides detailed coverage of all varieties of minicomputers, microcomputers and small business computers. Contains sections including: computers, peripherals, software, vendors, mininews, feature reports, user's guide, inquiry service, micros, and an index. "U.S. Vendor Product" section (part of Volume 1 and all of Volume 2) contains a management summary and a listing of computers' characteristics.

1424. DATAPRO REPORTS ON MINICOMPUTERS: VOLUME 3. *Datapro Research, Inc. (McGraw-Hill). Monthly. 1980. 800 (est.) pg.*

Volume 3 (of 3-volume report) deals with software (user ratings, aids to system selection, available software systems with management summaries, and list of characteristics); peripherals (charts / tables of memories, disk storage, floppy disk storage, reel to reel magnetic tape, cassette & cartridge, printers, punchcards and alphanumeric display terminals); and a directory of vendors. Charts providing system features include price, availability and options. This volume is a monthly updating looseleaf service reviewing and analyzing already published market data.

1425. DATAPRO REPORTS ON WORD PROCESSING: VOLUME 1. *Datapro Research, Inc. (McGraw-Hill). Monthly. 1980.*

Monthly 2-volume publication is a comprehensive information service devoted to word processing systems. Includes shared resource systems, cluster systems, and intelligent and electronic typewriters. Product features, manufacturers, applications and pricing are analyzed and compared among competitors.

1426. DATAPRO REPORTS ON WORD PROCESSING: VOLUME 2. *Datapro Research, Inc. (McGraw-Hill). Monthly. 1980.*

Volume 2 of monthly marketing directory on word processing. Software services, including pricing, installations, features, functions, hardware, operating performance and number of users, are discussed. Systems discussed are dictation, composition and auxiliary equipment. Word processing supplies (magnetic media, ribbons and printwheels), company directories and profiles and newsletters are given.

1427. DATAPRO 70: THE EDP BUYER'S BIBLE: COMPUTERS: VOLUME 1. *Datapro Research, Inc. (McGraw-Hill). Monthly. 1980. 1000(est.) pg. $270.00/yr.*

Volume I: comprehensive hardware / software reference service for EDP management. Discusses computers, broken into two sections; characteristics and management summary; general data and analyses on minicomputers (pricing, manufacturers, features); small business computers; and user ratings of computer systems, plug-compatible mainframes, and the top vendors' computer systems. Data is presented in text and charts or tables. A looseleaf, monthly updating service regarding U.S. computer industry analyzes and reviews currently available data.

1428. DATAPRO 70: THE EDP BUYER'S BIBLE: PERIPHERALS AND TERMINALS: VOLUME 2. *Datapro Research, Inc. (McGraw-Hill). Monthly. 1980. 1000 pg. $270.00/yr.*

Volume II: discusses distribution processing and intelligent terminals, display terminals, teleprinters, data entry, graphics, input-output memory and storage, printers and COM and peripherals. Purchase pricing, availability, monthly maintenance charges, operating performance, new product announcements, management summary and characteristics provided for each system mentioned. Monthly updating, looseleaf service which reviews and analyzes currently available systems.

1429. DATAPRO 70: THE EDP BUYER'S BIBLE: SOFTWARE, COMMUNICATIONS, SUPPLIERS, MEDIA AND SUPPLIES: VOLUME 3. *Datapro Research, Inc. (McGraw-Hill). Monthly. 1980. 1000 (est.) pg. $270.00/yr.*

Volume III: discusses software, communication facilities, suppliers and other peripherals. Buyer's guide gives a listing of existing systems, with management summaries and a list of

characteristics. Comparisons of systems and/or features and/or manufacturers is provided in text and charts.User experience information is provided throughout. Monthly looseleaf updating service reviews and analyzes currently available information in the computer market.

1430. DATEK PRINTER DATABASE SERVICE. *Datek of New England. Tri-annually. 1982.*

Published three times yearly, service provides definitions; an alphabetic listing of printer manufacturers and their printer products with OEM users of each; alphabetical OEM listing showing printer identifications and the printer manufacturer's name and model identification; and printer company and OEM company address listings. Fall 1981 edition typically contains over 400 OEM / printer records representing 125 printer models referenced to approximately 130 OEM users. Each printer record includes data on technology, character type, printer type, line length, date of introduction, speed, ribbon information, and OEM users including their designation for that printer. Publisher anticipates that future editions will include new printer introductions, captive printers, other printers for which OEM users have not been identified, and specialized printers.

1431. DIRECTORY OF COMPUTER DEALERS: 1981 ISSUE. *Computer Equipment Information Bureau. Annual. 1981. 54 pg. $12.00.*

Provides a listing of companies which regularly engage in buying, selling, brokering, or leasing second-market data processing equipment (i.e. new or used, not marketed directly by manufacturer). Also includes data on companies which provide market-related services including maintenance, refurb / repair, appraisals and transportation. Gives business address, telephone number, key executive and specialty. Arranged alphabetically by company name. Covers U.S., with some international firms included.

1432. DIRECTORY OF COMPUTER SERVICE COMPANIES. *Association of Data Processing Services Organization, Inc. Annual. 1981. 113 pg. $95.00.*

Profiles each of the 440 member firms of the Association of Data Processing Service Organizations. Lists key personnel, areas of expertise, hardware owned, and branch office locations for 1980.

1433. DIRECTORY OF EDP SUPPLIERS. *Datapro Research, Inc. (McGraw-Hill). Annual. 1981. 214 pg.*

Contains profiles of 1000 companies supplying all types of products and services of interest to computer users, vendors and operating staff. Each entry includes company's location, size, management, financial status, product lines, and sales and service organization.

1434. THE DIRECTORY OF INFORMATION / WORD PROCESSING EQUIPMENT AND SERVICES. *Information Clearing House. Annual. March 1982. $60.00.*

This directory lists information and word processing equipment and services, with product pages for each item providing technical specifications in a simple text format.

1435. DIRECTORY OF ON-LINE DATA BASES. *Cuadra Associates, Inc. Annual. 1st ed. September 1979. 129 pg.*

Directory of bibliographic and non-bibliographic on-line data bases available in the U.S., 1979. Lists type of product, producer, subject, vendor, acquisition procedures, coverage, and updating frequency. Addresses of producers, vendors and on-line services included.

1436. DIRECTORY OF ONLINE INFORMATION RESOUR-CES. *Capital Systems Group, Inc. Semi-annual. 6th. September 1980. 65 pg. $12.00.*

This semi-annual directory lists 254 bibliographic and non-bibliographic databases which are publicly available. These products are listed alphabetically giving database description, file size, time period covered and supplier, and are indexed by subject and supplier. Includes a list of database vendors and producers with addresses and telephone numbers.

1437. DIRECTORY OF TOP COMPUTER EXECUTIVES. *Applied Computer Research. Semi-annual. April 1982. $150.00/yr.*

Listings of 8400 computer equipment and service buyers at 7750 computer sites is organized geographically with cross-referencing by industry. Entries detail company name and address, subsidiary or division names, type of industry, telephone, computer systems installed, and names and titles of top EDP executives.

1438. DIRECTORY OF TOP COMPUTER EXECUTIVES. *Applied Computer Research. Semi-annual. September 1979. 309 pg. $75.00.*

Semi-annual directory lists organizations in industry and government in geographical order, including company name, address, telephone, top computer executives and major computer systems in U.S.A. Includes an index listing companies and their locations by the following categories: manufacturing and service, commercial banking, diversified financial, insurance, retail, transportation, utilities, education, health services, and federal, state or local government.

1439. DIRECTORY OF TOP COMPUTER EXECUTIVES: SPRING 1980. *Applied Computer Research. Semi-annual. March 1980. 330 pg. $150.00.*

1980 directory listing of top executives in data processing companies. Divided into two sections: geographic (complete listing by city and state) and cross reference by industry section (financial, retail). Addresses of corporations and executives given.

1440. DIRECTORY OF USERS GROUPS AND TRADE ASSOCIATIONS. *Datapro Research, Inc. (McGraw-Hill). Annual. 1981. 22 pg. $15.00.*

Report on 155 groups includes data on vendor and product - oriented users' groups, EDP - oriented professional societies, and trade associations. Listings include headquarters address, telephone, objectives, number of members, and membership data.

1441. ELECTRONIC CALCULATORS: BUYERS LABORA-TORY SURVEY REPORT. *Buyers Laboratory, Inc. Irregular. 1979. 72 pg.*

A survey of the electronic calculators market with directory information on 252 machines from 17 manufacturers. Each machine listed has detailed information on suggested retail price, description, dimensions, components, warranty and convenience features.

1442. ELECTRONIC MAIL EXECUTIVES DIRECTORY 1982. *Knowledge Industry Publications. Annual. 1982. 200 pg. $120.00.*

Directory lists 2800 managers responsible for voice and message services, word processing, and data processing. Each entry details company name, address, sales, number of employees, manager responsible for voice communications, data communications, office automation, facsimile transmission, word processing, in-house printing, purchasing, and tele-

communications including satellites, current equipment and plans for future use. Geographic index.

1443. ELECTRONIC NEWS FINANCIAL FACT BOOK AND DIRECTORY: 1980. *Fairchild Publications. Annual. 19th. 1980. 610 pg.*

Directory of electronics industry includes: company addresses, officers, directors, areas of work, subsidiaries, transfer agent(s), stock exchange(s), ticket symbol, number of employees, and plant footage. Annual dollar figures, generally from 1974-1978, given for: sales and earnings, revenues by line of business, common stock, and common stock equity. Further categories, for years 1977-1978, include income accounts, assets, liabilities, and statistical summaries. Statistical chart, "Leaders in the Electronics Industry - 1979", lists 51 companies in order of electronic sales and provides percentage of gross sales in electronics and gross sales in latest available four quarters. Accompanied by alphabetical arrangement of identical information.

1444. GUIDE TO SMALL BUSINESS SYSTEMS, 1980. *Hugo, Ian St. J. Computer Guides Ltd. Annual. 2nd ed. 1980.*

Handbook and directory for small business computer systems. Includes papers on characteristics, specifications, selection of equipment and consultants. Guides to contracts, software, upgrading and acquisition. Lists consultants, services and information sources.

1445. ICP DIRECTORY "80: BUSINESS MANAGEMENT: CROSS INDUSTRY APPLICATIONS. *Hamilton, Dennis, editor. International Computer Programs, Inc. Semi-annual. 1980. 292 pg. $85.00.*

This 1980, semi-annual (Feb / Aug) directory is part of a 6 volume series. This volume discusses application products (software) and services for the administrative, accounting and financial fields capable of medium and/or large computer systems. Systems listed, with contact data (vendor address); number of installations, narrative, maintenance, operating environment (hardware required) and pricing.

1446. ICP DIRECTORY "80: BUSINESS MANAGEMENT: INDUSTRY SPECIFIC APPLICATIONS. *Hamilton, Dennis, editor. International Computer Programs, Inc. Semi-annual. 1980. 411 pg. $85.00.*

This 1980, semi-annual (Mar / Sep) directory is part of a 6 volume series. This volume provides software products and services available for medium and large computers developed for a specific industry. Alpabetically by industry is a listing of systems including contact data, number of installations, narrative on systems; maintenance, pricing and operating environment (hardware). Vendor / Product Index & Hardware Index also.

1447. ICP DIRECTORY "80: DATA PROCESSING MANAGE-MENT. *Hamilton, Dennis, editor. International Computer Programs, Inc. Semi-annual. 1980. 475 pg. $85.00.*

This 1980, semi-annual (Jan / Jul) directory is part of 6 volume series. This volume provides a listing of systems software for mainframe computers. Either by function (the kind of software service desired), by vendor or by product (advertisers) one can locate system names, addresses, narrative, operating environment, maintenance, cost and number of installations.

1448. THE ICP DIRECTORY "80: INFORMATION PRODUCT AND SERVICE SUPPLIERS. *Hamilton, Dennis, editor. International Computer Programs, Inc. Semi-annual. 1980. 371 pg. $125.00.*

1980 semi-annual directory (June / Dec.) is part of a 6-volume ICP directory series. Top 100 U.S. vendors in information processing industry are ranked by revenues and information on their products (software suppliers, mini-OEM distributors, turnkey vendors) processing services (processing by batch, remote, on-site) professional services (custom programming, consulting, education, industry group), addresses, number of employees, and geographic area served is provided. Firms offering products services in Canada, Europe and other countries listed.

1449. ICP DIRECTORY ''80: MINI-SMALL BUSINESS SYSTEMS: CROSS INDUSTRY. *Hamilton, Dennis, editor. International Computer Programs, Inc. Semi-annual. 1980. 435 pg. $85.00.*

This (1980) semi-annual (Feb / Aug) directory is part of a 6 volume ICP Directory Series. This volume discusses software products and services available in accounting, business planning and administration, plus systems software for mini and small business computer environments. Marketing information including product title, number of current installations, narrative listing purpose, function, and features of the system, maintenance arrangements (cost), operation environment needed to support the system (hardware, operating system) & contact data.

1450. ICP DIRECTORY ''80: MINI-SMALL BUSINESS SYSTEMS: INDUSTRY SPECIFIC. *Hamilton, Dennis, editor. International Computer Programs, Inc. Semi-annual. 1980. 315 pg. $85.00.*

This 1980, semi-annual (May / Nov) directory is part of a 6 volume series. This volume provides products (software) and services available for mini or small business computer environments. Systems available, arranged alphabetically by industry; gives system name, vendor address, price, maintenance, operating environment (hardware) and a narrative of the system.

1451. INTERACTIVE COMPUTING DIRECTORIES: VOLUME 2. *Association of Computer Users. Annual. May 1979. 1565 pg.*

Volume 2 of three-volume annual directory to America's interactive computing industry. Includes impact printing terminals, thermal printing terminals, video display terminals, printers, remote batch terminals, intelligent terminals and remote batch services, with listings of major corporations' addresses and executives.

1452. LIBRARY COMPUTER EQUIPMENT REVIEW. *Microform Review Inc. Semi-annual. 1981. 60 pg. $150.00/yr.*

Semiannual publication provides detailed coverage of new products (vendor, specification summary, discussion, basic design, input / output, secondary storage, system software, application software, communications, prices, summary).

1453. LISTING OF COMPUTER SERVICE COMPANIES. *Association of Data Processing Services Organization, Inc. Monthly. 1981. $750.00.*

A print-out of over 10,000 computer software, data processing services, and OEM system houses. Updated monthly; also available in label form, or in partial listings.

1454. LOCALNETTER DESIGNERS HANDBOOK. *Architecture Technology Corporation. Irregular. April 1982.*

A guide containing a complete overview of local network equipment ranging from cables, to attachable local network word processors, with features on gateways and local network performance.

1455. MEMBERS OF THE COMPUTER AND BUSINESS EQUIPMENT MANUFACTURERS ASSOCIATION. *Computer and Business Equipment Manufacturers Association. Irregular. April 1981. 1 pg. Free.*

Lists about three dozen member manufacturers.

1456. MICROCOMPUTER BUYER'S GUIDE. *QED Informational Services. Irregular. 1981. $25.00.*

Provides specifications of 200 microcomputer systems available from approximately 60 vendors.

1457. MINICOMPUTER REVIEW. *GML Corporation. Tri-annually. 1980. 800 pg. $175.00.*

Annual directory (updated tri-annually) provides product evaluations of more than 1,000 minicomputers and microcomputers and company profiles of about 270 domestic and foreign manufacturers. Tabulates information on the applications, features, hardware characteristics, systems software, software languages, prices and marketing data of each system. Also includes alphabetical listing of microcomputers on the market and a discussion of the industries in review.

1458. MULTIPLE-UNIT BUYERS FILE FOR MINI AND MICRO COMPUTERS. *International Data Group. Irregular. 1980.*

Identifies and profiles firms that buy micro and mini computers and peripherals for resale. Covers four categories: traditional OEM buyers, systems houses, dealers or distributors, and internal systems houses. U.S. - only edition also available, as are subfiles by geographic region, manufacturer, model, industry (SIC). Summarizing management report included. Files furnished on magnetic tape and/or hard-copy printout.

1459. ONLINE TERMINAL / MICROCOMPUTER GUIDE AND DIRECTORY 1982-83. *McAllister, Shirley A., ed. Online Inc. Biennial. 3rd ed. December 1981. 286 pg. $40.00.*

Directory of U.S. terminal and micro-computer vendors and suppliers. Gives business addresses of all sales and service firms, inlcuding those in foreign markets. Includes handbook information on technology and factors in purchases.

1460. PERSONAL COMPUTER BUYERS GUIDE. *Grimes, Dennis, ed. Ballinger Publishing Company. Semi-annual. January 1983. $14.95.*

Semiannual buyers guide of personal computers provides product references with specifications, a suppliers directory, and a glossary, with index arranged by software application.

1461. ROBOTICS TODAY 82. *Society of Manufacturing Engineers. Annual. 1982. 368 pg. $42.00.*

Annual compilation of 'Robotics Today' articles and features includes chapters on basics, human factors, machine loading / material handling, welding assembly, casting and forging, painting, other applications, implementation, and vision and sensing. Directory of 30 robot and accessory manufacturers and distributors with specifications for 50 robot models. Technical digest section provides titles and abstracts of all papers on robotics published by the Society of Manufacturing Engineers from 1974 to 1979.

1462. SERVICES AND SOFTWARE DATA FILE. *International Data Group. Irregular. 1980.*

A profile of services and software vendors and the markets in which they participate. Subfiles available by geographic region, U.S. as a whole, manufacturer, model or industry (SIC). Files furnished on magnetic tape and/or hard-copy printout. Summarizing management report included.

1463. SKARBEKS SOFTWARE DIRECTORY. *Skarbeks Computers Inc. Irregular. 1980. 364 pg. $11.95.*

Directory of over 800 Apple software programs, from over 100 vendors. Each listing includes program title, publisher, memory requirements, program description, and prices. Listed alphabetically by title, with a broad subject index.

1464. SMALL BUSINESS COMPUTERS: A GUIDE TO EVALUATION AND SELECTION. *Isshiki, Koichiro R. Prentice-Hall, Inc. Irregular. 1982. 478 pg.*

General handbook for evaluation and selection of small computers describes hardware and software systems, uses, and characteristics. Appendix contains directory of small business computer system vendors, listing available models and specifications of CPU's, disks, and printers.

1465. SOFTWARE VENDOR DIRECTORY. *Micro-Serve Inc. Semi-annual. 1981. $59.95.*

A reference directory for the microcomputer industry. Provides data on 800 software vendors, arranged in 105 categories. Lists 32 hardware vendors.

1466. TOWERS' INTERNATIONAL MICROPROCESSOR SELECTOR. *Towers, T.D. Tab Books, Inc. Annual. June 1982. 244 pg. $19.95.*

Directory of 7,000 U.S., European, and Japanese microprocessors, memories, interfaces, and LSI circuits includes electrical and mechanical specifications, sources, and substitution data for each circuit. Listings in alpha-numeric order by type number. Accompanying data on manufacturer house codes, glossary of terms.

1467. TURNKEY CAD / CAM: PART 3: U. S. DIRECTORY OF VENDORS. *Anderson Publishing Company. Irregular. 1981. 156 pg. $99.00.*

Directory of companies selling turnkey CAD / CAM systems for mechanical applications: company profiles, product data, sales figures, list of key executives, and a breakdown of existing installations by application - mechanical, architectural, electrical, PCB, IC, technical publishing, mapping. Gives customer service profile: details field service, customer support, education, user group resources, R & D and manufacturing capacity.

1468. TYMNET: COMPUTER AND DATA SERVICES AVAILABLE THROUGH TYMNET. *Tymnet Inc. Irregular. June 1979. 34 pg.*

1979 directory describes computer and data services offered by Tymnet Inc. customers in the U.S.A., Mexico and Italy, through the Tymnet data communications network. Provides addresses, contact people and telephone numbers of subscribers as well as brief descriptions of company services. Lists computers through which services are accessed by a local telephone call in over 200 metropolitan areas in the U.S., Canada and Mexico, and other countries via interconnect arrangements.

1469. USER RATINGS OF COMPUTER SYSTEMS. *Datapro Research, Inc. (McGraw-Hill). Annual. 1981. 55 pg. $25.00.*

Provides selection guidelines based on the experiences of 4218 users of desk-top, personal, micro- and mini-computers, small business computers, and general purpose systems. Covers 191 models from 55 vendors, pinpointing strengths and weaknesses of each manufacturers' equipment, software, and support systems.

1470. WEBSTER'S MICROCOMPUTER BUYER'S GUIDE. *Computer Reference Guide. Annual. April 1981. 326 pg. $25.00.*

Examines specifications and applications of some 150 microcomputer systems from about 50 U.S. and Japanese suppliers. Also surveys display and printing terminals, and software packages.

1471. WEBSTER'S REVISED AND UPDATED MICROCOMPUTER BUYER'S GUIDE. *Computer Reference Guide. Irregular. 1982. 424 pg. $29.95.*

Microcomputer buyer's guide lists new products, including microcomputers, peripherals, and software. Includes evaluative details and technical specifications for 16-bit systems and 32-bit micros for all major manufacturers.

1472. WHAT TO BUY FOR BUSINESS 1982 EDITION. *Oppenheim Derrick, Inc. Annual. 1982. $20.00.*

Directory evaluates 60 low-end business computer systems. Includes charts showing price and key specifications by model. Discusses printers and software.

1473. WHO'S WHO IN DATA PROCESSING. *Whitsed Publishing Ltd. Annual. 1979.*

Lists data processing service centers and Canadian producers, manufacturers and users of services and products.

1474. WHO'S WHO IN DATA PROCESSING 1980 / 1981. *Whitsed Publishing Ltd. Annual. 1981. 195.00 (Canada).*

Directory of Canadian users and equipment in data processing. Also lists equipment / products in use, number, makes and model numbers, location of head offices, key executives. Data is broken down by province and classified by industry.

1475. 1980 DIRECTORY OF COMPUTER GRAPHICS SUPPLIERS: HARDWARE, SOFTWARE, SYSTEMS, AND SERVICES. *Harvard University. Laboratory for Computer Graphics. Annual. 1980. $17.00.*

Lists 140 U.S. suppliers of computer graphics hardware, software, systems and services.

1476. 1981 PRINTOUT ANNUAL. *Datek of New England. Annual. 1981. 80 pg. $50.00.*

Covers market trends, company profiles, and product listings, with an industry directory of 125 U.S. companies manufacturing printers and terminals. Articles examine trends in the printer market with projections for that market in terms of market segment, product application, and technology; changing face of impact technology; and historical overview and forecast of the non-impact printer market. Industry profiles provide one-page reviews of 32 printer manufacturers including company history, sales figures, and marketing and sales strategies.

1477. 1982 COMPUTER SALARY SURVEY. *Source EDP. Annual. 1982. 28 pg. Free.*

Lists salary data for 48 position categories at various levels of computer experience and installation sizes, at both technical and managerial levels. In addition a seven-step career management and planning model is presented.

1478. 1982 PRINTOUT ANNUAL. *Dower, Jonathan, ed. Datek of New England. Annual. 2nd ed. May 1982. 100 pg. $25.00.*

Examines printer industry, 1975-1982. Gives annual sales of 12 manufacturers, 1975-1982; market and distribution data; technical specifications on ink and print method characterization. Profiles 39 printer manufacturers. Gives names of officers, address, products with technical data, and sales and production statistics. Lists with addresses other manufacturers. Includes analysis of word processing market, 1982.

1479. 1982 ROBOTICS INDUSTRY DIRECTORY. *IFS (Publications) Ltd. Annual. 1982. 234 pg. 33.00 (United Kingdom).*

Directory provides specifications for a range of robots available in the U.S. with details of firms supplying them) with basic information on robotics available in Japan, France, and Great Britain. Directory includes names and addresses of robot suppliers as well as makers of vision systems, sensors, controllers and electronics, end effectors, and hydraulic and pneumatic equipment. Contains accompanying listings of consultants and research institutes.

1480. 1983 CLASSROOM COMPUTER NEWS DIRECTORY OF EDUCATIONAL COMPUTING RESOURCES. *Intentional Educations, Inc. Annual. September 1982. 150 pg. $9.95.*

Directory covers computer stores, mail order suppliers, hardware manufacturers, hardware suppliers, consultants, programmers, computer camps, services and repair, computer books, book publishers, and software catalogs. Other sections detail computer science courses and degree programs at 2000 colleges and universities and a one-year calendar of national and regional educational computing events.

See 241, 243, 252, 602, 608, 612, 615, 2476, 2496, 2516, 2542, 3244, 3245, 3246, 3252, 3253, 3254, 3260, 4188, 4203.

Special Issues and Journal Articles

American Banker

1481. WHAT TO LOOK FOR WHEN BUYING A PERSONAL COMPUTER: A REVIEW OF BANKING'S POPULAR MICRO-COMPUTERS. *American Banker, Inc. Annual. v.147, no.116. June 16, 1982. pg.9-11*

Provides a brief history of microcomputers and suggests various criteria when purchasing them. Contains profiles of 9 specific models, including manufacturer's name and address; technical specifications; retail price; and software support.

American Machinist

1482. AMERICAN MACHINIST: NEW PLANT EQUIPMENT. *McGraw-Hill Book Company. Monthly. v.125, no.9. September 1981. pg.225-262*

Monthly product directory lists and describes new plant equipment under 6 categories: cutting, forming, computers and control, tooling, robots and plant operations. Provides name of manufacturer.

1483. AMERICAN MACHINIST'S NEW PLANT EQUIPMENT. *McGraw-Hill Book Company. Monthly. v.126, no.2. February 1982. pg.173-202*

Monthly directory lists and describes new plant equipment including cutting and forming tools, computers and controls, tooling, welding and quality assurance equipment. Provides name of U.S. manufacturer and a photograph of the product.

1484. THE TOOLS: DISTRIBUTED MICROPROCESSORS PUT THE COMPUTER WHERE THE ACTION IS. *McGraw-Hill Book Company. v.126, no.6. June 1982. pg.142-148*

Summarizes micro-computer use in the U.S. tool industry. Reviews systems and components currently available. Data (1970-1985) on semiconductors includes active devices per chip, cost of equivalent CPU and memory, and speed-power product.

Apparel World

1485. THE TRIUMPH OF AUTOMATION IN APPAREL. *Hertz, Eric, ed. National Outerwear and Sportswear Association. v.3, no.6. June 28, 1982. pg.25-39*

Analysis of impact of new computer technology on the sewing

machinery industry. Reviews new product lines of manufacturers.

Architectural Record

1486. COMPUTERS: THE EVOLUTION IS OVER; THE REVOLUTION IS ON. *Mileaf, Harry. McGraw-Hill Book Company. June 1982. pg.19-25*

Report on the utilization of computers by architecture and engineering firms shows percent of businesses using computers, by size, 1975-1986; utilization by engineering versus architectural professionals, 1981-1986; use of services, 1980-1981; ownership of specific devices, 1980-1981; and percent of utilization by applications, 1981-1982.

Automotive Age

1487. COMPUTER UPDATE: PRODUCT OVERVIEW FOR "82. *Deitz, Lawrence D. Freed-Crown Publishing Company. v.31, no.4. April 1982. pg.45-58*

Report on new computer products for auto dealers, including utilizations, capacity, and vendors.

Bank Systems and Equipment

1488. BANK SYSTEMS AND EQUIPMENT: ANNUAL SOFTWARE REVIEW. *Gralla Publications. Annual. v.18, no.8. August 1981. pg.72-79*

Lists, in tabular form (alphabetically by manufacturer), software packages and services available. Briefly states their bank / thrift applications, such as mortgage loan, market / branch analysis, report writer, etc.

1489. BANK SYSTEMS AND EQUIPMENT: PRODUCTS. *Gralla Publications. Monthly. v.19, no.3. March 1982. pg.117, 120*

A monthly listing on data processing and other equipment used in banking information systems. Information includes name of manufacturer, description, and photograph. Further information available through reader card service.

1490. HARDWARE, SOFTWARE COMBINE IN READY-TO-USE TURNKEY SYSTEMS. *Goldberg, Joan. Gralla Publications. v.18, no.8. August 1981. pg.88-89*

Reviews turnkey systems: ready-to-use hardware / software packages. Discusses advantages compared to in-house systems and specific banking applications.

1491. SOFTWARE NEWS. *Gralla Publications. Monthly. 1981.*

Reviews new products and developments of interest to banking operations and systems management. Provides product specifications, information on manufacturers; and services.

Building Design and Construction

1492. COMPUTERS AID IN STEEL ANALYSIS AND FABRICATION. *American Institute of Steel Construction. Cahners Publishing Company. v.23, no.6. June 1982. pg.70-71, 74*

Discusses the role of computers and two computer programs in the design of steel framed structures. By eight building types, gives market share of steel, concrete, and pre-engineered metals, 1981; and growth rate for steel and concrete since 1981. Provides total regional growth rates for steel and concrete, 1981.

Business Computer Systems

1493. BRED FOR WORK. *Cahners Publishing Company. Irregular. v.1, no.4. December 1982. pg.105-121*

Describes dot-matrix printers, 1982. Lists manufacturers, with addresses. Gives technical descriptions and prices.

1494. THE ECONOMY OF MULTIUSER SYSTEMS. *Cahners Publishing Company. Irregular. v.1, no.4. December 1982. pg.63-75 ff.*

Explains microcomputer timesharing, 1982. Gives microcomputer manufacturers of multiuser systems, with addresses. Includes technical descriptions and prices.

1495. IT'S EVERY MAKER FOR HIMSELF. *Cahners Publishing Company. Irregular. v.1, no.4. December 1982. pg.87-102*

Discusses computer networking, 1982. Gives price data. Lists network manufacturers, with addresses. Gives technical descriptions.

Business Week

1496. THE SPEEDUP IN AUTOMATION. *McGraw-Hill Book Company. Irregular. August 3, 1981. pg.58-67*

The first section describes the transformation taking place on factory floors and evaluates the consequences for the U.S. economy. The second section examines the impact that automation will have on American workers.

1497. SUPERMARKET SCANNERS GET SMARTER. *McGraw-Hill Book Company. no.2701. August 17, 1981. pg.88-92*

Describes competition among the industry's suppliers of both hardware and software. Covers new software programs and new applications.

Byte

1498. USER'S COLUMN: TERMINALS, KEYBOARDS, AND HOW SOFTWARE PIRACY WILL BRING PROFITS TO ITS VICTIMS. *Pournelle, Alexander. BYTE Publications Inc. Irregular. v.7, no.11. November 1982. pg.394-415*

Discusses selected keyboards, terminals, and software available in 1982. Includes vendor's name and address, and product name and price.

Canadian Datasystems

1499. ANNUAL REVIEW: MINI / MICROCOMPUTERS AND SMALL BUSINESS COMPUTERS. *Maclean Hunter Ltd. Annual. July 1979. pg.32-94*

Annual survey issue includes a directory of manufacturers of mini / micro-computers and small business systems. Gives addresses, phone numbers and products, with a description of capabilities; a discussion of data networks; effective maintenance; choosing software packages; software physics; and industry briefs.

1500. ANNUAL SURVEY OF MINICOMPUTERS AND SMALL BUSINESS COMPUTERS PART 1. *Maclean Hunter Ltd. Annual. July 1980. pg.32-36*

Annual survey of mini and small business computers. Compiled from questionnaire to North American manufacturers and vendors. Model features software support, system capabilities, pricing and availability given by vendor or manufacturers name. Part I (of two) discusses only small business systems.

1501. THE BOOM IN BUSINESS COMPUTER GRAPHICS. *Kolodziej, Stan. Maclean Hunter Ltd. Irregular. March 1982. 4 pg.*

Analysis of the performance of the Canadian computer graphics market and the capabilities of various products. Chart notes manufacturers and their models, host computers, operator interaction, display characteristics, addressable matrix, zoom, color, software support, prices, and additional features.

1502. CANADA'S DATA PROCESSING MARKET 1980. *Maclean Hunter Research Bureau. Maclean Hunter Ltd. Annual. May 1980. 42 pg. $50.00.*

Market report on Canada's data processing market includes: installations, 1976-1979; market shares, 1976- 1979; imports, 1970-1979; exports, 1975-1979; facilities, 1979; and forecasts to 1985. Sections include: U.S. exports to Canada, 1975-1978; firms primarily engaged in computer services, 1975-1978; firms primarily engaged in sales, lease or rental of EDP hardware, 1975-1978; source of operating revenue, 1977-1978; and number of major establishments or firms.

1503. CANADA'S PCM MARKET - WHAT'S HAPPENING? *Maclean Hunter Ltd. April 1982. pg.64, 66, 68*

Examines plug compatible manufacturers. Covers six major PCMs marketing IBM lookalike processors and storage subsystems, their revenues for 1979, 1980, 1981, and forecasts for the industry.

1504. CANADIAN DATASYSTEMS ANNUAL REVIEW AND FORECAST. *Maclean Hunter Ltd. Annual. December 1981.*

Features analyses by top government officials and executives in the EDP field for EDP equipment and products with general economic forecasts for the information industry.

1505. CANADIAN DATASYSTEMS: EDP EQUIPMENT SUPPLIES AND SERVICES: USER PREFERENCE SURVEY NO. 4. *Maclean Hunter Research Bureau (for Canadian Datasystems). Maclean Hunter Ltd. Annual. October 1979.*

Market survey to determine brand preferences for a wide variety of equipment, supplies and services associated with the EDP market in Canada. Comparisons of brand standings with previous studies. Based on replies to questionnaires.

1506. CANADIAN DATASYSTEMS INDUSTRY REVIEW AND FORECAST 1981. *Maclean Hunter Ltd. Annual. December 1980. pg.52-63*

Canada's computer industry leaders view the issues and prospects for 1981. Forecasts include new computer technology, tariffs and duties, new products, sales, and the economy in general.

1507. CANADIAN DATASYSTEMS USER PREFERENCE SURVEY, NO.5 (1981). *Maclean Hunter Ltd. Annual. May 1981.*

Report of a survey among EDP personnel to determine brand preference for a wide variety of EDP products and services. Results reported by geographic area and compared with previous surveys conducted in 1971, 1973, 1976, and 1978.

1508. CANADIAN DATASYSTEMS: WHAT'S NEW. *Maclean Hunter Ltd. Monthly. December 1981. pg.5-9*

Monthly analysis of new products and trends in computers, terminals, printers / plotters, data entry, data communications, components, storage and accessories. Notes products, suppliers and features.

1509. CHOOSING A PRINTER INVOLVES MORE THAN FIRST IMPRESSIONS. *Kolodziej, Stan. Maclean Hunter Ltd. December 1981. pg.64-67*

A look at a representative section of inexpensive printers on the Canadian market and an assessment of trends. Summary includes manufacturers of 25 different printers, model numbers, character generation, speed, character format, features and approximate prices.

1510. COMPUTER STOCKS IN PERSPECTIVE. *Maclean Hunter Ltd. May 1980. pg.108*

Review of shares of 10 computer service businesses. Monthly trading summary indicates centers of activity in the industry. Statistics are for March, 1980.

1511. COMPUTER STOCKS IN PERSPECTIVE. *Maclean Hunter Ltd. October 1979. pg.132 5.00(Canada).*

Review of shares of 12 computer service businesses. The monthly trading summary indicates centres of activity in the business. Not a guideline for purchase.

1512. COST OF INDUSTRIAL SELLING SOARS. *Maclean Hunter Ltd. Annual. August 1981. pg.25*

Reports findings from the sixth nationwide survey in Canada of the average cost of an industrial sales call. Includes 368 businesses which provided data.

1513. ENGINEERS SELECTED COMMADORE 'BEST OVER-ALL SYSTEM.' *Maclean Hunter Ltd. April 1982. pg.100*

A recent survey for the Society of Professional Engineers and Associates (SPEA), examines computer users at the games, student, computer scientist, and business levels. Rankings include second choices.

1514. INTERNATIONAL COMPUTER SHOW / SALON INTERNATIONAL DE L'ORDINATEUR: PREVIEW REPORT. *Maclean Hunter Ltd. Annual. May 1980. pg.59-84 5.00 (Canada).*

Preview of the computer show to take place from June 4-6, 1980 at Place Bonaventure, Montreal, Quebec. Notes more than 100 of the computer industry's leading suppliers of systems and related equipment. Provides addresses, phone numbers, products shown, and key employees.

1515. LATEST CENSUS OF COMPUTERS SHOWS NEW GROWTH. *Shackleton, L.A. Maclean Hunter Ltd. May 1980. pg.97-99 5.00 (Canada).*

Brief article on the annual computer census conducted by Canadian Information Processing Society. Statistical tables note market shares of 12 leading suppliers, ranked by number of computers 1976-1979, the annual rental value of the computers and computer installations by industry for 13 industries 1976-1979, including their annual rental value. Discusses sales for 1979.

1516. PRODUCTS ON REVIEW: 1982 CANADIAN COMPUTER SHOW PREVIEW. *Maclean Hunter Ltd. Annual. v.14, no.10. October 1982. pg.144-199 ff.*

Lists exhibitors and briefly describes products to be shown at 1982 Canadian Computer Show. Provides names, addresses and telephone numbers of exhibitors.

1517. REFERENCE MANUAL 1979. *Maclean Hunter Ltd. Annual. January 1979. 142 pg.*

Guide for buyers and users of computer hardware, software, services and supplies. Includes a calendar of major conferences and shows during 1979 listed with dates, places, topics and contacts (mainly North America). Lists EDP organizations with names, addresses and executives of associations serving

the Canadian data processing community; and suppliers and businesses involved in the Canadian EDP market, including addresses, phone numbers, branches and representatives. Provides a cross-referenced index to product categories of hardware, software, services and supplies. The products section notes hardware and software products, computing services, consulting services, accessories, discs, tapes, and has an alphabetical trade names listing.

1518. REFERENCE MANUAL 1980. *Maclean Hunter Ltd. Annual. January 1980. 162 pg.*

Guide for buyers and users of computer hardware, software, services and supplies. Covers major conferences and shows during 1980; EDP organizations, including executives of associations serving the Canadian DP community; suppliers and other businesses. Lists products and trade names.

1519. REVIEW / FORECAST "80. *Maclean Hunter Ltd. Annual. December 1979. pg.30-50 5.00 (Canada).*

Annual review and forecast for 1980 of the EDP business in Canada. Lists approximately 20 Canadians in the EDP business. Covers costs of services and employees, new technologies, inflation, revenues, growth, networking, marketing and the Canadian dollar value.

1520. SALARY SURVEY. *Peat, Marwick, and Partners, Toronto. Maclean Hunter Ltd. Annual. November 1979. pg.66-67 5.00 (Canada).*

Thirteenth Annual EDP wages and salaries survey contains data from more than 500 EDP installations throughout Canada on nearly 11,000 employees within the computer environment. Includes table on salary trends for comparable positions in management, systems and operations for 1978 and 1979.

1521. SALARY SURVEY: END OF AIB YIELDS 9% SALARY BOOST. *Maclean Hunter Ltd. March 1979. pg.28-29,33-34,39,41-42 1.50(Canada).*

Survey report on the wages and salaries of DP staff, reviewing policies and turnover in Canada. Statistical data includes 6 levels of management and percent annual increases 1975-1978; percentage of organizations reporting staff turnover in 5 ranges of percentage of staff leaving organization or moving to another job. Covers average salaries in approximately 20 Canadian cities; average salaries by industry; and average salaries by size of installation.

1522. SOURCES OF SUPPLY: 1982 CANADIAN DATASYSTEMS REFERENCE MANUAL. *Maclean Hunter Ltd. Annual. v.14, no.1. January 1982. 166 pg.*

Guide for buyers and users of computer hardware, software, services and supplies. Lists names and addresses of suppliers, with representatives, branches and products. Contains a listing of data processing service businesses, EDP organizations and associations, consulting services, accessories, discs, tapes, and listing of Canada's DP service firms and their capabilities.

1523. SURVEY OF MINICOMPUTERS AND SMALL BUSINESS SYSTEMS: PART 2. *Maclean Hunter Ltd. Annual. August 1980. pg.54-75*

Part II of annual survey (1980) on mini and small business computers deals only with minicomputers. Data compiled by questionnaires sent to North American manufacturers and vendors. Model features, software support, system capabilities, pricing and availability by vendor or manufacturer's name given.

1524. 1979 CANADIAN COMPUTER SHOW AND CONFERENCE PREVIEW. *Maclean Hunter Ltd. Annual. October 1979. pg.69-129*

Preview lists trends and expectations in the Canadian EDP business over the next decade. Notes more than 200 registrants in the show, along with their products, services, addresses, phone numbers and key employees. Listing is alphabetical and includes a description of the key products being shown.

1525. 1980 DIRECTORY: CANADA'S DATA PROCESSING SERVICE FIRMS. *Maclean Hunter Ltd. Annual. June 1980. pg.59-86*

1980 annual directory listing of data processing firms in US and Canada. Addresses of manufacturers / corporations, services of corporation (systems and programming, OCR, COM, MICR), specialization and packages (financial or insurance), branch offices, software products, professional services and equipment given.

1526. 1981 CANADIAN COMPUTER SHOW AND CONFERENCE: A PREVIEW. *Maclean Hunter Ltd. Annual. October 1981. pg.150-200*

Notes speakers, topics and themes of the conference held in Toronto Nov. 16-19, 1981. Includes a floor plan of exhibits and businesses as well as a listing of exhibitors and their products, addresses, phone numbers and personnel present at show.

1527. 1981 REFERENCE MANUAL. *Maclean Hunter Ltd. Annual. January 1981. $15.00.*

Directory listing addresses of manufacturers and vendors providing products and services to the Canadian computer hardware and software industries.

Canadian Doctor

1528. COMPUTER SURVEY: SURVEY SHOWS THE TIMES THEY ARE A'CHANGING. *Cara, William J. Southam Business Publications, Ltd. Division of SouthamCommunications Ltd. v.48, no.5. May 1982. pg.41-46*

Results of a 1981 questionnaire establishes the link between vendors and physicians and demonstrates the trend toward increased computer usage in Canadian medical offices. Tables give 28 vendors' names, addresses, equipment costs, activities, and available software. Tabulated user survey results include computerized office functions, factors influencing vendor choice, user satisfaction, and reasons for computer use.

Canadian Electronics Engineering

1529. MAXI GROWTH FOR MINI / MICRO PERIPHERALS. *Maclean Hunter Ltd. February 1981. pg.13*

Statistical data on shipments (1980-1990) of peripherals for mini and micro computer systems. Chart divided by 6 peripheral categories, with number of shipments (1980) and forecasts for 1982, 1985, 1990.

Canadian Machinery and Metalworking

1530. CANADIAN MACHINERY AND METALWORKING'S 1982 SURVEY OF INDUSTRIAL ROBOTS: A SELECTION GUIDE FOR THE POTENTIAL ROBOT USER. *Maclean Hunter Ltd. Annual. v.77, no.1. January 1982. pg.23-39*

Annual 1982 selection guide provides overview of industrial robots available in North America with a comparative list of detailed specifications. Each entry includes manufacturers' names, model number, and specs. Addresses and telephone numbers are provided for each manufacturer.

1531. ROBOT SUPPLIERS. *Maclean Hunter Ltd. Irregular. February 1981. pg.33*

Directory listing 28 manufacturers of industrial robots in the U.S. and Canada. Gives addresses and telephone numbers.

Canadian Transportation and Distribution Management

1532. SURVEY REVEALS COMPUTER USES IN PHYSICAL DISTRIBUTION. *Southam Business Publications Ltd. Division of Southam Communications Ltd. Irregular. January 1979. pg.35-38*

Examines current computer usage for transportation and physical distribution functions and usages being developed or planned. Covers types of computers used, and how they are used. Five tables show: types of computer-related equipment used or planned for use in transportation and distribution-related activities; computer support for the transportation / distribution function; computer communication capabilities; computer applications to transportation-related activity; and computer applications to other distribution-related activity.

Carbide and Tool Journal

1533. ROBOT INVASION COMING TO MID-AMERICA. *Edgerton, Michael. Society of Carbide and Tool Engineers. v.14, no.4. September 1982. pg.4-6*

Market analysis (1982) examines the use of industrial robots in Mid-America. Discusses market penetration and end uses. Reports annual sales, 1980; estimated sales, 1981; and forecasted sales, 1990.

CBEMA Comment

1534. JAPAN DP IMPORTS BRING U.S. TRADE SURPLUS DIP. *Computer and Business Equipment Manufacturers Association. v.4, no.4. June 1982. pg.2*

Examines U.S. trade surplus in computers and business equipment 1982. Gives export and import data, 1981, 1982. Specifies Japan, Taiwan, and Western Europe.

Chilton's Automotive Industries

1535. INDUSTRIAL ROBOT GROWTH. *McElroy, John. Chilton Book Company. v.162, no.9. September 1982. pg.43-45*

Report on forecasts for robotics, 1985, 1990, 1995. Examines employment and unemployment effects, foreign markets and competition.

CIPS Review

1536. CANADIAN INFORMATION PROCESSING SOCIETY 1981-1982 NATIONAL BOARD OF DIRECTORS. *Canadian Information Processing Society. Annual. July 1981. pg.40*

Lists executives of the society, their titles and job positions.

Circuits Manufacturing

1537. CIRCUITS MANUFACTURING: SELECTION AND IMPLEMENTATION OF CAD SYSTEMS. *Votapka, R.E. Morgan Grampian Publishing Company. November 1980. 5 pg.*

Discusses applications of computer-aided design systems in the manufacture of printed electronic circuit boards. Forecasts market value through 1984, with growth rates. Discusses major manufacturers' activities and market shares (1980 and earlier years). Presents results of a survey of user attitudes, including quality rankings of manufacturers.

Computer Business News

1538. CBN / COMDEX ''81 DEAL-MAKING DIRECTORY: PART 2. *CW Communications Inc. Semi-annual. November 9, 1981. pg.31-37*

Lists manufacturers of small computer systems and relevant OEM suppliers, in search of multiple unit buyers, distributors / dealers, representatives, and/or software houses. Also lists a few software vendors seeking hardware affiliates, and distributors seeking suppliers. Each entry specifies company's address, telephone numbers, products offered, types of affiliates and services sought, executive contact, and discounts or special services offered.

1539. COMDEX INTROS SPUR COLOR GRAPHICS COMPETITION. *CW Communications Inc. v.4, no.48. November 30, 1981. pg.14*

Reviews the market for color graphics terminals, with emphasis on new products and lower prices. Major manufacturers are discussed and new products are described, giving specifications and prices.

1540. COMPUTER BUSINESS NEWS: CBN / COMDEX ''81 DEAL - MAKING DIRECTORY: PART 1. *CW Communications Inc. Irregular. November 2, 1981. pg.27-34 $1.00.*

Listing of potential computer business deals worth more than a billion dollars. Covers small computer systems manufacturers, OEM suppliers and software companies looking for distributors, dealers, representatives, multiple-unit buyers (MUB's), software affiliates, and/or new products (as applicable). Alphabetic entries specify company address, telephone number, contact person, product lines, and what sorts of distribution channels or new products are sought. Identifies companies planning to attend the Comdex ''81 trade show.

1541. COMPUTER BUSINESS NEWS: CHECKLIST. *CW Communications Inc. Weekly. v.5, no.6. February 8, 1982. pg.25-38*

Weekly checklists list new products and developments in software, I/O, and memory systems. Data listed includes manufacturers' addresses and prices.

1542. COMPUTER BUSINESS NEWS: IRON MAKERS ATTACK SOFTWARE NEEDS. *CW Communications Inc. October 19, 1981. pg.1, 4 $25.00/yr.*

Discusses a trend among software - pressed major computer manufacturers toward the contracting of programming work to independent software firms. Surveys deals made by individual corporations, with attention to varying stipulations regarding exclusivity and marketing arrangements.

1543. COMPUTER BUSINESS NEWS SPECIAL REPORT: THE WORLD OF GRAPHICS. *Allen, John. CW Communications Inc. Irregular. February 9, 1981. pg.12-22*

Report on the U.S. computer graphics market. Tables show market segments for computer graphics terminals, with sales dollars; market share summary (1975-1982) for industry sales and Newco sales; and a listing of 70 vendors of computer graphics VDT's, with business address and phone number. Discussion provides data on prices, new products, performance of selected terminals and new technology.

1544. COMPUTER BUSINESS NEWS STOCK INDEX. *CW Communications Inc. Weekly. February 15, 1982. pg.23*

A weekly report of stock performance of U.S. computer companies classified by equipment: computer systems; peripherals and subsystems; software and EDP services; semiconductors and components; systems houses; leasing companies; and supplies and accessories. Includes business name; market, dividends; price to earnings ratio; volume of trading for the last five days; high, low, and closing prices; and net change. Graphs depict yearly trends in overall trading by hardware category.

1545. COMPUTER BUSINESS NEWS STOCK INDEX. *CW Communications Inc. Weekly. 1981. 1 pg. $25.00/yr.*

Weekly statistical table providing stock prices for computer systems, peripherals and subsystems, software and EDP services, semiconductors and components, systems houses, leasing companies and supplies and accessories. Gives company name, exchange, price earnings ratio, dividend and high, low and closing prices for the previous Monday.

1546. CUSTOM ENGINEERING SERVICES DIRECTORY. *CW Communications Inc. Irregular. January 12, 1981. pg.9-10.*

Alphabetical listing of 40 U.S. companies specializing in custom engineering services for the computer industry. Includes business address, phone number, key executive and specialty.

1547. HIGH-SPEED PRINTERS: SPECIAL REPORT. *CW Communications Inc. Irregular. v.4, no.50. December 14, 1981. pg.1, 11-19*

Analyzes the market for high-speed digital printers through 1991, and of sales mix in 1980 and 1984. Discusses printing speeds and other technical characteristics of various technologies and specific models, as well as prices. Lists vendors, with addresses and telephone numbers.

1548. IMPORTS POSE CHALLENGE TO U.S. PRINTER INDUSTRY. *CW Communications Inc. v.4, no.49. December 7, 1981. 29 pg.*

Report of a study examining four segments of the computer printer industry, and the effect of Japanese imports on each. The four product groups discussed include fully-formed character printers, dot matrix printers, ink-jet printers and page printers. Market shares by type of printer are forecast for 1985. Discusses Japanese manufacturers and new products.

1549. MICROS, MINIS GAIN IN PAPER CHASE. *CW Communications Inc. February 15, 1982. pg.1, 6*

Reviews trends in the increased substitution of micro and mini computers for dedicated terminals in banking, insurance, and financial institutions. Forecasts data given for mini / micro computer shipments for use in retail banking (1980-1990) by market segment. Examines use by sector; highlights automated teller machine trends.

1550. PLOTTER MART TO DOUBLE BY 1986, STUDY PREDICTS. *CW Communications Inc. v.5, no.19. May 10, 1982. pg.1, 21*

Briefly summarizes results of a recent study conducted by Venture Development Corporation, 'The Hard Copy Graphics Industry.' Provides statistical data for plotter market shares by type of plotter in 1986 (percent of dollars).

1551. SPECIAL REPORT: THE PRACTICAL USE OF GRAPHICS. *Greitzer, John. CW Communications Inc. Irregular. v.5, no.2. January 11, 1982. pg.1, 5-8*

Examinesthe market for graphics terminals, with data on graphics usage growth rates. Describes market segmentation and reviews leading manufacturers, suppliers and new products, with prices for selected items. Includes a directory of graphics terminal vendors, listing about 70 vendors with addresses and telephone numbers.

1552. TWO STUDIES SHOW IBM MISSING GRAPHICS BOOM. *CW Communications Inc. March 8, 1982. pg.1, 23*

Reviews two studies on IBM's role in graphics display terminal and hard copy graphics markets. Statistics include 1981 share by 3 companies of dollar shipments of CRT graphics terminals.

1553. THE WORLD OF FLOPPIES. *CW Communications Inc. Irregular. v.5, no.8. February 22, 1982. pg.1-23*

Special report discusses floppy disk market. Gives data on storage capacity, and corresponding track density. Graphs 1982 OEM shipments and lists manufacturers.

1554. 32-BIT MACHINES AMONG TRENDS CITED IN STUDY. *CW Communications Inc. v.5, no.22. May 31, 1982. pg.11*

Estimates 1979-1982 CAD/CAM revenues by market share for 7 U.S. companies. Source is MRSA, Inc.

1555. 68000 - BASED SYSTEMS PREVALENT AT COMDEX. *CW Communications Inc. v.4, no.48. November 30, 1981. pg.10*

A review of new systems based on 16 - bit 68000 microprocessor. Describes new products and manufacturers' developments, with product specifications and retail prices.

Computer Decisions

1556. APPLYING MICROSYSTEMS: DATA-MANAGEMENT ILLS CURED AT NBC. *Good, Philip I. Hayden Book Company, Inc. v.14, no.10. October 1982. pg.66-68*

Discusses applications of microcomputers to solve NBC Enterprises data management problems. Table shows 5 data management systems, including vendors' addresses and product prices.

1557. COMPETITION INTENSIFIES IN BURGEONING WP INDUSTRY. *Bashford, Eric R. Hayden Book Company, Inc. v.14, no.2. February 1982. pg.76-77*

Examines the market for word processors, with discussion of competition in the market place and current industry trends. Includes a table detailing worldwide word processing shipments in units and value (1976-1981) with estimates (1981-1985), showing yearly growth rates and average value per unit. Leading manufacturers are identified and marketing strategies and prices are discussed.

1558. COMPUTER DECISIONS: ACCESSORIES. *Hayden Book Company, Inc. Monthly. March 1982. pg.227*

Monthly listing of computer hardware accessories including disks, disk drives and storage aids includes description, price range, and address of manufacturer. Further information available through reader card service.

1559. COMPUTER DECISIONS: PERIPHERALS. *Hayden Book Company, Inc. Monthly. March 1982. pg.231*

Monthly listing of new computer peripherals and equipment gives description, brief specifications, price, and address of company. Further information available through reader card service.

1560. DRAWING ON DESK-TOP GRAPHICS FOR PRESENTATIONS. *Good, Phillip I. Hayden Book Company, Inc. Irregular. July 1982. pg.70-72*

Discussion of computer-aided graphics used as aids in presentations lists available products giving vendor, address, telephone, description, and price. Further information available through reader card service.

1561. FOLLOWING THE LEADERS: WILL COMPETITION GIVE IBM THE BLUES? *Gartner, Gideon I. Hayden Book Company, Inc. v.14, no.10. October 1982. pg.70, 72*

Discusses mainframe market and competition. Charts compare 1977, 1980, 1984 markets for mainframe and fault-tolerant systems of IBM, PCM, and Bunch, including percentage of total market.

1562. MICROCOMPUTERS INVADE THE EXECUTIVE SUITE. *Seaman, John. Hayden Book Company, Inc. Irregular. February 1981. pg.68-74, 158-174*

Narrative on the U.S. microcomputer industry. Tables display sampling of popular microsystems, with models, standard features, prices, options and addresses and phone numbers of vendors; and a guide to software for microcomputers, with vendor address and phone number, package, function and price.

1563. MIS/DP PAY: ARE YOU KEEPING UP? *Whieldon, David. Hayden Book Company, Inc. v.14, no.2. February 1982. pg.160-175*

Results of a survey on wages and salaries (1980-1981) for Management Information Systems and Data Processing managers are presented. Graphs detail the percentage of respondents in salary ranges by $5000 increment; median salaries by type of organization; amount of past increase in salary; and increase amount required for respondent to change jobs. The relationship of salaries to data processing department budget is also illustrated. Accompanying text discusses employee attitudes, general industry trends, career growth opportunities, and employee benefits.

1564. PCM'S: ROLLING WITH THE PUNCHES. *Gebremedhin, Elinor. Hayden Book Company, Inc. Irregular. v.14, no.11. November 1982. pg.96-112*

Analysis of plug-compatible mainframe industry and market includes market segmentation, advantages of PCM's, IBM's leadership, pricing, and market expansion. Includes directory of eight manufacturers, a directory of actively marketed IBM mainframes, and a directory of actively marketed plug-compatibles.

1565. PORTABLES: AT YOUR SERVICE. *Brown, Julia G. Hayden Book Company, Inc. Annual. v.14, no.5. May 1982. pg.86*

Lists 16 manufacturers of computers, 6 manufacturers of display terminals, 8 manufacturers of hand-held and mid-size terminals, and 10 manufacturers of printing terminals. Includes manufacturer's name, address and telephone number, model name and/or number, model features, and price.

1566. PUTTING DATA-COMM HARDWARE TO WORK. *Seaman, John. Hayden Book Company, Inc. Irregular. v.14, no.11. November 1982. pg.114-151*

Article covers activities of specific firms; hardware manufacturers directory including address, telephone, and products (modems, multiplexers, concentrators); discussion of modems, multiplexers, and other product groups; and a glossary.

1567. SALARY SURVEY: HOW DO YOU STACK UP? PART 1. *Snyders, J. Hayden Book Company, Inc. Irregular. December 1979. pg.12-18*

Salary survey of data processing managers in the U.S., 1979. Reports on average annual salaries by geographical area, number of personnel supervised, size of business and length of tenure.

1568. SAY IT WITH PICTURES. *Synders, Jan. Hayden Book Company, Inc. Irregular. July 1982. pg.36-52*

Discussion of current developments in graphics software is accompanied by listing of available equipment giving vendor, address, telephone, package name, compatible equipment, and price. Further information on software through reader card service.

1569. TESTING YOUR METTLE. *Lasden, Martin. Hayden Book Company, Inc. v.14, no.10. October 1982. pg.162-174, 190*

Discusses and gives examples of cost-saving potential of measuring performance of data processing and telecommunications systems. Lists selected hardware monitors including vendors, models, prices, and model descriptions.

1570. THIRD ANNUAL CRT AND TELEPRINTER COMPARISON CHARTS. *David Jamison Carlyle Corp. Hayden Book Company, Inc. Annual. v.14, no.6. June 1982. pg.230*

Video terminal comparison chart gives manufacturer, model number and equipment characteristics in the following categories; display, keyboard, edit, communications, general and physical, 1982. The teleprinter comparison chart provides manufacturer, model number, type, and equipment characteristics in the following categories; printer, keyboard, operator controls, indicators, communications, and special features, 1982.

1571. WHEN POWER PROBLEMS STRIKE. *Hayden Book Company, Inc. Irregular. March 1982. pg.102-131*

Examines the difficulties surrounding interruption of electrical power to computer systems. Lists companies manufacturing equipment that provide electrical-power protection including uninterruptible power generators. Further information available through reader card service. Contains other technical information.

Computer Design

1572. SYSTEM COMPONENTS. *Computer Design Publishing Corporation. Monthly. v.21, no.5. May 1982. pg.291-342*

Description of new products for computers, peripherals and software. Includes technology, prices, manufacturers and addresses.

Computer Graphics and Applications

1573. COMPUTER GRAPHICS AND APPLICATIONS: MONTHLY NEW PRODUCTS. *IEEE Computer Society. Monthly. October 1981. pg.127-146*

Monthly listing of new products introduced in the computer industry. Also covers some office equipment, such as copiers. Provides manufacturer's name, product description, and price.

1574. A GUIDE TO RECENT COMPUTER - ANIMATED FILMS AND VIDEOTAPES. *Speer, L.R. IEEE Computer Society. Irregular. October 1981. pg.118-124*

Listing of 37 computer - animated films and videotapes available from individuals or film distribution firms. The listing includes films presented at the Siggraph ''80 and ''81 and from private correspondence with film-makers. Gives film name; production date; description and supplier. Following is a list of suppliers, with addresses and phone numbers.

ComputerData

1575. CANADIAN DATA PROCESSING DIRECTORY, 1980. *Whitsed Publishing Ltd. Annual. 6th ed. 1980. 166 pg. 2.50 (Canada).*

Directory gives addresses of businesses in the data processing field in Canada: services, installations, and products. Includes vendors. Shows number of employees and sales for 1979 and estimated sales for 1980. Tables list end uses of software provided by vendors. Correlates vendors' software packages and the hardware on which they may be used.

1576. COMPUTER DATA: WHAT'S NEW! *Page Publications. Monthly. v.7, no.9. September 1982. pg.31-38*

Product guide presents new computer hardware and software available in Canada as of September 1982. Provides product description and vendor's name.

1577. HARDWARE REPORT 1981. *Whitsed Publishing Ltd. March 1981. pg.28-37 25.00/yr. (Canada).*

Discusses trends in 1981 for the computer hardware industry. Notes several small hardware manufacturers in Canada and provides discussion of their products. Contains few statistics.

1578. PCMERS SAY THEY'RE WINNING ON PRICE AND PERFORMANCE. *White, Ted. Whitsed Publishing Ltd. March 1982. pg.33*

Reviews the market performance of Canadian compatible suppliers. Discusses investment trends and competitive conditions. Graphs depict market share (1977, 1979, 1981) of IBM, noncompatibles, and plug compatibles. Also gives the market share by individual plug compatible companies.

1579. WESTERN COMPUTER SHOW. *Whitsed Publishing Ltd. Annual. September 1981. pg.77-83 2.50 (Canada).*

Lists exhibitors at the Western Computer Show in Calgary October 28-29, 1981, their addresses and telephone numbers, and products.

1580. WHAT'S NEW IN HARDWARE. *Whitsed Publishing Ltd. Monthly. March 1982. pg.7*

Monthly listing of computer hardware includes model name, manufacturer, description, and photograph. Further information available through reader card service.

1581. WHO'S WHO IN DATA PROCESSING IN CANADA. *Mann, Richard I. Whitsed Publishing Ltd. Annual. May 1982. 203 pg. $395.00.*

Directory of data processing personnel in Canada, includes end users, suppliers, services, and government agencies. Lists business addresses, executives and top management, affiliates and parent corporations, sites of computer installations, number of EDP employees and total employees, lines of business, manufacturers and product lines of equipment. Includes 1982 expenditures forecast which tabulates companies; EDP employees; annual revenues; 1981 EDP operating budget; 1982-1983 budget expansion; 1981-1982 planned purchases or leases; and uses.

1582. THE 1981 EXPENDITURE FORECAST SURVEY. *Whitsed Publishing Ltd. Annual. December 1980. pg.38-45 25.00 (Canada).*

Results of the 1981 expenditures forecast survey were tabulated from a survey sent to top 500 businesses in Canada (according to the Financial Post's 1980 ranking of the top 500 firms). Findings based on 100 companies with an EDP budget. Charts are alphabetical according to business name and list EDP employees, annual revenues, EDP budget (1980), forecasted increase of EDP budget (1981 or 1981-82), and planned acquisition of equipment (buy, rent, lease). Includes forecasted use of services (consulting, service bureaus, executive search firms, etc.) and notes if the business can / well develop its own hardware.

Computers and People

1583. COMPUTER DIRECTORY AND BUYER'S GUIDE. *Berkeley Enterprises, Inc. Annual. October 31, 1979. 52 pg.*

Annual directory of over 1,500 organizations involved with computers and data processing provides addresses, brand names, and number of employees. Also included is guide to products and services in computers and data processing, including minicomputers, microcomputers, computer dealers and geographic proliferation. Categories include databases, computer systems, hardware, software, networks, peripherals, timesharing and turnkey systems. Characteristics of digital computers included.

Computerworld

1584. ANALYST SEES BIG WINNERS, LOSERS IN "80S. *Kirchner, Jake. CW Communications Inc. v.16, no.14. April 5, 1982. pg.73, 75*

Discusses growth and changes in the computer industry with forecasts of the top ten computer firms in 1990. Reviews the demand for software and services with data on the growth of selected manufacturers (1979-1981). Competition and changing market shares are discussed.

1585. ARRAY PROCESSORS SPEED TO POPULARITY. *Rochester, Jack B. CW Communications Inc. v.16, no.16. April 19, 1982. pg.63-64*

Discusses array processors as an alternative to supercomputers. Describes technology and presents data on prices, costs of performing specific operations, and original market size. Specific products on the market are reviewed.

1586. BEWARE: BENCHMARK TESTS NOT FINAL WORD. *Segal, Hillel. CW Communications Inc. Irregular. v.16, no.26. June 28, 1982. pg.30*

Reviews methodology used in the Association of Computer Users benchmark tests. Discusses considerations other than test results. Lists 24 small computers and 12 multiuser systems. Includes product names and prices.

1587. COMPUTERWORLD: CURRENT LINEUP OF TOP-OF-THE-LINE MINICOMPUTERS. *CW Communications Inc. September 21, 1981. $1.25.*

Table presents specifications, estimated performance, and prices of six top-of-the-line minicomputers made by four manufacturers. Includes assessments of costs per 1 million bytes of main memory.

1588. COMPUTERWORLD FORECAST "81. *CW Communications Inc. Annual. January 5, 1981. pg.1-80*

Forecast (1981) for the U.S. computer industry covering data base management systems, telecommunications, computer graphics, employment, venture capital, network architecture, user versus vendor, defensive software development, software litigation trends; and technology for the Bell System Network.

1589. COMPUTERWORLD: NATIONAL SEMI BURSTS BUBBLE OPERATION: LEAVES INTEL TOP SUPPLIER. *Blumenthal, Marcia. CW Communications Inc. September 7, 1981. pg.69 $1.25.*

Reports on the withdrawal of National Semiconductor Corp. from the manufacture of magnetic 'bubble' memory devices -

the third exit from that market in 1981, by what had been its largest supplier. Discusses corporations' reasons for abandonment of the 'bubble' technology, and remaining manufacturers' continuing activities. Estimates total market value through 1983.

1590. COMPUTERWORLD: TRADE SURPLUS TOPS $3.5 BILLION. *Computer and Business Equipment Manufacturers Association. CW Communications Inc. September 7, 1981. pg.73 $1.25.*

Presents figures compiled by the Computer and Business Equipment Manufacturers Association regarding total U.S. exports, imports and net trade surplus in computers during the first half of 1981. Estimates indicated total 1981 trade surplus.

1591. CONTINUED RISE IN MVS USE AMONG FIVE TRENDS SPOTTED AT IBM MAINFRAME SITES. *Paul, Lois. CW Communications Inc. v.16, no.14. April 5, 1982. pg.10*

Reviews results of an Interactive Data Corporation survey of IBM mainframe users. Data is presented on the percent of respondents using selected operating systems, database management systems, inhouse time-sharing systems, online editors and checkout systems (1981); and growth rates of utilization since the late 1970's. A graph details percent of IBM mainframe users running the MVS operating systems (1977-1981).

1592. IPL, SINGER RATED TOPS IN MAINFRAME SURVEY: 1982 USER RATINGS OF COMPUTER SYSTEMS. *Datapro Research Corp. CW Communications Inc. Annual. v.16, no.23. June 7, 1982. pg.1, 56-63*

Gives user ratings of mainframes, 1982. Gives manufacturer and model number and rates their principle applications, source of applications programs, database management system, communications monitor and planned acquisitions / implementation programs, 1982.

1593. MAJOR FIRMS' EARNING SLUMP. *Blumenthal, Marcia. CW Communications Inc. v.16, no.13. March 29, 1982. pg.57-58*

Reports first quarter earnings for major computer companies. Gives composite price to earnings ratio for 72 high technology companies. Sou

1594. MINI USERS STAY WITH PRESENT UNITS - SURVEY. *Scannell, Tim. CW Communications Inc. v.16, no.24. June 14, 1982. pg.60-65*

Presents detailed results of survey of U.S. minicomputer and small business users. Includes ratings of a wide variety of minicomputers and their manufacturers. Survey items include significant advantages, significant disadvantages, system ratings, technical support, manufactu

1595. SPECIAL REPORT DATA COMMUNICATIONS TERMINALS: DISCUSSING THE ALTERNATIVES. *Hoard, Bruce, ed. CW Communications Inc. v.16, no.13. March 29, 1982. pg.47*

Provides a wide variety of information on communications terminals, including data on terminals offering IBM compatibility; variety available in intelligent terminals; vendors offering single, multifunction terminal systmes; forecasts on the nature of the future terminal system; summary of study indicating relationship between increased use of videotex and videophone equipment and increased alphanumeric sales; use of terminals in private homes; trends in the graphics market; protection of terminals; interface standardization; software problems posed by terminal variety; the future of CRT and microfiche technologies; and new developments in the use of automatic bank teller machines.

1596. VECTOR GRAPHIC 3005 OFFERS QUALITY, LOW PRICE. *Segal, Hillel. CW Communications Inc. v.16, no.14. April 5, 1982. pg.32*

Fiftieth in a series of articles giving the highlights of benchmark tests conducted on small computer systems. Vector Graphics' 3005 model is reviewed with data on performance and discussion of technical specifications. A table lists 22 computer systems under $25,000 with results of benchmark tests for scientific and business applications.

1597. WOMEN OFFICE WORKERS IN A RACE AGAINST TIME AS AUTOMATION IMPACTS THE WORK PLACE. *Nussbaum, Karen. Working Women Education Fund (Cleveland, OH). CW Communications Inc. September 28, 1981. pg.40-41 $1.25.*

Reviews Bureau of Labor Statistics figures, and conclusions of several studies, on the likely impact of office automation on female clerical workers. Presents statistics on the concentration of women in clerical positions.

Country Guide

1598. A CONSUMER'S GUIDE TO COMPUTERS. *Robertson, Lois. Public Press. Irregular. July 1982. pg.18-19, 22*

Analysis of farm software available in Canada and how to choose a system appropriate for one's farming needs. Lists sources of agricultural software information.

Data Communications

1599. DATA COMMUNICATIONS: BUYER'S GUIDE 1980. *Davis, George R. McGraw-Hill Book Company. Electronics. Annual. November 1979. 466 pg. $15.00.*

1980 Annual Buyer's Guide to products and services associated with data communications listing company names and addresses. Products include: communication carriers, transmission equipment, storage devices and terminals. This publication is a separate issue. Vendor directory of equipment suppliers and services with complete company information included.

1600. DATA COMMUNICATIONS: 1982 BUYERS' GUIDE. *McGraw-Hill Book Company. Annual. November 1981. 518 pg.*

Product and services directory includes communications carriers, data concentration equipment, data transmission equipment, diagnostic and test equipment, distributed data processing, storage devices and media, terminals, software, and support equipment and services. Includes vendor directory and addresses. Discusses current technological trends.

Data Management

1601. DATA MANAGEMENT: ACCESS LINE. *Data Processing Management Association. Monthly. May 1981. pg.36-47 $16.00/yr.*

Monthly listing of new computer products with data-processing applications, chiefly peripherals, terminals, and software, plus some processing units. Access to manufacturers' literature is provided through a check-off information card.

1602. FOR THE BIGGIES, MICROS ARE THE RIGHT FIT. *Etelman, Katherine. Data Processing Management Association. v.20, no.10. October 1982. pg.27K-26M*

Assesses the use of personal computers in large U.S. organizations. Discusses software applications. Chart shows esti-

mated microcomputer units in operation in 1982 and 1992 by end use.

1603. PERSONAL COMPUTERS: THE TWELVE MOST ASKED QUESTIONS. *Blundell, Greggory S. Data Processing Management Association. v.20, no.10. October 1982. pg.26E-26I*

Analyzes the personal computer market from 1981 to 1991. Reports 1981 market shares for 9 manufacturers. Forecasts annual worldwide personal computer shipments and the U.S. installed base by market segment, 1981-1991.

Datamation

1604. ALIVE AND WELL: THE TOP 50 DATA COMMUNICATION INDUSTRY MANUFACTURERS. *Technical Publishing Company. Annual. June 1980. pg.112-142*

Analysis of the data communications equipment market in the U.S. Includes commentary on the utilization of the available technologies and market segmentation; a table ranking the top 50 manufacturers by revenues, 1978-79; and a directory listing manufacturers, addresses, products, and brief histories of the corporations.

1605. BRAND PREFERENCE STUDY OF THE DATA PROCESSING INDUSTRY: OEM AND END USER MARKETS. *Technical Publishing Company. Annual. January 1979. 171 pg.*

Annual study measuring brand preference for 94 types of EDP products, services, and software. Scores were determined by individuals in the OEM and end-user markets listing manufacturers they would consider in planning acquisition of specific types of EDP products and services. Includes results of the last three studies.

1606. THE DATAMATION 100: THE TOP 100 U.S. COMPANIES IN THE DP INDUSTRY. *Technical Publishing Company. Annual. June 1981. pg.91-192 $4.00.*

Ranks the leading 100 U.S. companies in the data processing industry by 1980 DP revenue. Provides their 1979 ranking, percent growth, number of employees for 1980, and their total revenue in 1980. Includes an analysis of each company's performance in 1980.

1607. THE DATAMATION 100: THE TOP 100 U.S. COMPANIES IN THE DP INDUSTRY. *Technical Publishing Company. Annual. July 1980. pg.87-182*

1980 annual survey of the top 100 U.S. companies in the data processing industry. Rankings are derived from a survey of over 250 businesses and are based on calendar year 1979 revenues. Includes net income (1979) with percent change from 1978 and number of employees per company. The top 50 companies are detailed according to revenues from mainframes, mines and micros, peripherals and terminals and software and services. Provides a summary of each company's activities. Rankings are broken down by mainframe, minicomputer, peripheral, terminal, software and services, and personal computers / computer media companies.

1608. DATAMATION'S ANNUAL SOFTWARE PACKAGE EVALUATIONS: EVALUATING OFF-THE-SHELF PROGRAMS. *Data Decisions, Beta Research. Technical Publishing Company. Annual. December 1980. pg.85-122*

1980 user attitude survey report, presenting ratings of computer software program packages, by type and manufacturer. Packages rated by feature, function, and performance. Results arranged by category: data management, communications, applications, and other systems packages. Entries include package name, business name, address, and phone number.

1609. MAINFRAME INDUSTRY SURVEY. *Technical Publishing Company. Irregular. July 1979. pg.88-92*

U.S. mainframe industry survey (1979) conducted by Datamation and G.S. Grunman / Cowan & Co. Data derived from responses of mainframe system users in the U.S. Comparison charts given (1977 vs 1979) of users' intentions to spend more on systems, to increase system capacity, and to purchase IBM or competitor's software.

1610. MINI / MICROCOMPUTER USER SURVEY. *Chamberlin, Dorothy. Technical Publishing Company. Annual. 1979. $445.00.*

Market survey of 35,000 U.S. mini / microcomputer installations by manufacturer and model, rankings of top ten systems, average prices, rankings by unit sales and dollar value of manufacturers' shipments, forecasts of 1979-1980 purchases and industry growth rates, trends of competition and effects on main frames.

1611. THE SOFTWARE SCENE: USERS' RATINGS OF SOFTWARE PACKAGES. *Severino, Elizabeth F., editor. Datapro Research Corp. Technical Publishing Company. Annual. December 1979. pg.138-165*

Annual (1979) Datapro / Datamation questionnaire survey of 35,000 proprietary software users (Datamation readers). Survey does not employ statistical sampling methods due to small sample size. Rankings of top software packages in U.S. for 1979, brief descriptions of packages, and users' ratings given.

1612. SPECIAL EDITION: DATAMATION. *Technical Publishing Company. Annual. January 1981. 76 pg.*

Annual digest of market and industry analyses of the U.S. computer industry. Covers vendors' strategies, new markets, new products, foreign trade, competition and technical issues.

1613. A SURVEY: THE SMALL SYSTEMS MARKET. *Technical Publishing Company. Annual. November 25, 1979. pg.95-98*

Annual survey of the mini / microcomputer market in the U.S. and Western Europe. Reports 1979 sales and forecasts growth rates for 1980. Covers consumers' attitudes about products, plans to buy systems or switch vendors, reasons for dissatisfaction, and trends in utilization. Includes a ranking of major manufacturers by sales and some information on market shares.

1614. SURVEYING THE SOFTWARE: A COMPILATION OF AVAILABLE MRP SYSTEMS AND THEIR VENDORS. *Technical Publishing Company. Irregular. October 1980. pg.101-120*

A directory of vendors of Management Resource Planning (MRP) software packages, including vendor names, addresses, product descriptions, prices, and utilization in terms of the number of users.

1615. THAT OLD BUGABOO, TURNOVER. *McLaughlin, R. A. Technical Publishing Company. Annual. October 1979. pg.97-101*

Study (1979) of percentage of data processing personnel turnover. Data available by Datamation's Computer Executive Panel (700 Data Processing managers), according to data processing positions, annual personnel loss, growth / shrinkage of staff and jobs waiting to be filled were determined. Methods of hiring by effectiveness presented.

1616. 1980 DATA PROCESSING BUDGET SURVEY. *Shaw, Louise C. Technical Publishing Company. Annual. January 1980. pg.126-130*

1980 survey based on user sample of data processing managers on "Datamation" mailing list. Gives data processing budget (hardware, software and salaries) expenditures in industry (by size), in education and in government. Lists corporate revenues for small, medium and large industries in the areas of education, metal products and health care.

1617. 1980 SALARY SURVEY. *Shaw, Louise C. Technical Publishing Company. Annual. April 1980. pg. 110-118*

1980 Annual salary survey for U.S. and Canadian data processing personnel, obtained by questionnaire to Data Processing managers. Fifty-five job descriptions, employee position rankings, average salaries by installation size, by selected industry and by major metropolitan area provided.

1618. 1981 DP SALARY SURVEY. *Technical Publishing Company. Annual. May 1981. pg.98-111 $4.00.*

Annual survey based on the information in 300 completed questionnaires. Provides data on data processing salaries by installation size, by type of industry, and by geographic region.

Dealerscope II

1619. THE NEW KIDS ON THE RETAIL BLOCK. *Dealerscope Inc. v.2, no.6. June 1982. pg.16*

Discusses the relationship between manufacturers and computer and video specialty stores. Gives percentage of manufacturers carried by retailers for more than three years.

Design Engineering

1620. CANADIAN CAD/CAM EXPOSITION AND CONFERENCE PREVIEW. *Maclean Hunter Ltd. Annual. May 1982. pg.46-50*

A listing of design and manufacturing products and systems using the latest computer technology is previewed for the Canadian CAD/CAM show in Toronto June 1-3, 1982. Includes names of businesses and products to be shown.

Design News

1621. PERSONAL HOME COMPUTERS TAKE OFF. *Cahners Publishing Company. v.38, no.19. October 11, 1982. pg.28-29*

Summarizes results of a recent study done by the Yankee Group on projected U.S. sales of personal computers for home use. Figures list data for growth of selected consumer electronics products by installed base (units millions), 1977-1990; and changing segmentation of home personal computer market by units sold, 1980-1983.

Digital Design

1622. ALPHANUMERIC CRT TERMINAL SHIPMENTS TO EXCEED 3 MILLION UNITS IN 1986. *Morgan Grampian Publishing Company. v.12, no.10. October 1982. pg.22*

Summarizes a study of the U.S. alphanumeric CRT terminal industry. Forecasts 1986 shipments and market shares by market segments. Source is Venture Development Corporation.

1623. KEYBOARD INPUT FAVORED OVER TOUCH AND VOICE. *Morgan Grampian Publishing Company. v.12, no.9. September 1982. pg.25*

Analysis of market survey of executives, testing user attitudes toward vendors, new touch and voice sensitive terminal products. Charts forecast of percent distribution of work station product shipments, 1981, 1986.

1624. MARKET TRENDS: COMPUTERS PENETRATE SMALL BUSINESS MARKET. *Morgan Grampian Publishing Company. v.12, no.6. June 1982. pg.24*

Discusses application software, hardware, and system software costing less than $100. Gives statistics for growth rate in 1980's and number of computers projected to be sold. Forecasts sales of business computers during 1980's.

1625. MASSIVE GROWTH OF DATA CONVERTER CONSUMPTION. *Morgan Grampian Publishing Company. v.12, no. 11. November 1982. pg.22-24*

Examines the expanding market for A / D and D / A converters. Figure provides statistical data for shipments of sample / hold amplifiers, and analog multipliers (millions of dollars), 1975-1986. (Data based on Venture Development Corporation study: 'The A / D and D / A Converter Industry: A Strategic Analysis).

1626. SUPERMINIS GIVE MAINFRAMES A RUN FOR THEIR MONEY. *Mokhoff, Nicolas. Morgan Grampian Publishing Company. Irregular. v.12, no.10. October 1982. pg.44-51*

Compares superminicomputers from 7 U.S. manufacturers. Provides product name, technical specifications, and 1982 U.S. prices. Forecasts the value of shipments, 1981, 1983, 1985, and 1990.

1627. 5 1/4' WINCHESTER MANUFACTURERS. *Morgan Grampian Publishing Company. Irregular. v.12, no.10. October 1982. pg.34-39*

Lists U.S. manufacturers of 5 1/4' Winchester disk drives. Includes model name, technical specifications, and 1982 prices.

Discount Store News

1628. CE BUYERS: DECISIONS, DECISIONS. *Taylor, Betty J. Lebhar-Friedman, Inc. v.21, no.14. June 28, 1982. pg.1, 29*

Report on retail distribution of video games and personal computers shows unit sales to dealers, 1979-1982. Discusses manufacturers' price cuts.

1629. RADIO SHACK PLANS TO UNWRAP 10 COMPUTER CENTERS A MONTH. *Lebhar-Friedman, Inc. v.21, no.12. June 14, 1982. pg.65*

Ranks top nine home computer stores in dollar sales in last fiscal year. Gives number of units May 1981 and May and December 1982.

Distribution

1630. GUIDE TO COMPUTER SYSTEMS, SERVICES, HARDWARE AND SOFTWARE. *Chilton Book Company. Monthly. v.81, no.5. May 1982. pg.74-80*

Guide to computer systmes, services, hardware and software covers costing models, fleet management, freight bill auditing, freight claims, graphics and distribution data, inventory control, logistics, materials requirements, order entry, parcel systems, productivity measurement, purchasing systems, rail care tracking, rate maintenance, rating, routing, vehicle maintenance and warehousing. Lists suppliers along with addresses and profiles of products.

Drug Store News

1631. PHARMACY COMPUTER MANUAL. *Lebhar-Friedman, Inc. September 7, 1981. pg.27-34, 39-42 $10.00/yr.*

Examines capabilities and prices of several time-sharing computer systems and minicomputer / software packages for drugstore and drug wholesaler applications. Specifies manufacturers / vendors, and describes activities of, and number of installations by, major suppliers and purchasing chains. Discusses total operating costs, product reliability questions, and overall market penetration and outlook.

Dun's Business Month

1632. SMALL COMPUTER PRODUCT SHOWCASE. *Dun and Bradstreet Publication Corporation. Annual. v.120, no.2. April 1982. pg.73-82*

Lists the following directory information for 33 U.S. small computer manufacturers: manufacturer's name, model name and primary applications, CPU memory capacity, additional outside memory, terminals and peripherals available, communications network capabilities; price category, typical system configuration and cost, and lead time for delivery.

Dun's Review

1633. SPECIAL COMPUTER REPORT. *Dun and Bradstreet, Inc. National Credit Office Division. Annual. August 1980. pg.67-86*

An analysis of the U.S. computer industry with comments on domestic competition, prices, market structure, investment prospects, and Japanese investments in the industry. Also includes forecasts and histories of corporations.

EDN Magazine

1634. ALTERABLE - MEMORY INNOVATIONS ENHANCE NEW PRODUCT DESIGNS. *Cahners Publishing Company. Irregular. January 20, 1982. pg.88-98*

Lists manufacturers of PROM's, UV PROM's, EAROMS's, or EEPROM's, giving addresses and telephone number. Charts listing these products give technical specifications. Further information available through reader card service.

1635. EDN: LEADTIME INDEX. *Cahners Publishing Company. Monthly. 1981.*

Monthly index of leadtime to fill new orders for products in the computer industry. Products are arranged by product group.

1636. EDN: NEW PRODUCTS. *Cahners Publishing Company. Bi-monthly. November 1981.*

Bi-monthly product review featuring new products for electronic data processing. Each entry includes manufacturers' name, address, telephone number, product description and specifications, brand names, prices, and photographs.

1637. EDN'S NINTH ANNUAL UP / UC CHIP DIRECTORY. *Cahners Publishing Company. Annual. v.27, no.21. October 27, 1982. pg.99, 105*

Discusses uP / uC chip state of the art, 1982. Gives suppliers and manufacturers with addresses.

1638. EDN'S THIRD ANNUAL MICROCOMPUTER OPERATING SYSTEMS DIRECTORY. *Kotelly, George. Cahners Publishing Company. Annual. v.27, no.18. September 15, 1982. pg.80-160*

Examines microcomputer operating systems introduced in the U.S. from September 1981 through September 1982. Focuses on new products technologies. Compares technical specifications of selected manufacturer's products.

1639. INDEPENDENTS' ADD-IN MEMORIES IMPROVE ON COMPUTER MAKERS UNITS. *Cahners Publishing Company. Irregular. v.27, no.21. October 27, 1982. pg.43-54*

Lists manufacturers, board numbers, storage elements and capacities, timing, power requirements, price and special features of 8/16 bit and 16/18 bit add-in memory boards, 1982. Gives price and technical data on boards with large bytes.

1640. MICROCOMPUTER OPERATING SYSTEMS DIRECTORY. *Kotelly, George. Cahners Publishing Company. Annual. September 16, 1981. pg.101-167 $2.00.*

Surveys 80 operating systems available in 1981 for 8- and 16-bit microcomputers. Lists each system's hardware and language compatibility, processor multitasking and/or multiuser allocation capabilities, memory and file management features, peripheral management capabilities, price, and manufacturer's address.

1641. NCC "82: PRODUCTS. *Cahners Publishing Company. Irregular. May 26, 1982. pg.109-148*

Lists computer products exhibited at the National Computer Conference 1982 (Houston). Gives name and address of manufacturer, description, and photograph. Further information through reader card service.

1642. PERSONAL SCIENTIFIC CALCULATORS: SPECIAL REPORTS. *McDermott, Jim. Cahners Publishing Company. Irregular. September 16, 1981. pg.71-81*

Surveys new hand-held calculators offering alphanumeric displays and keyboards, and programmability in BASIC. Discusses peripherals available, prices and battery life, and provides a list of manufacturers' addresses.

1643. SPECIAL REPORT: PRODUCT SHOWCASE 14. *Cahners Publishing Company. Semi-annual. v.26, no.24. December 16, 1981. pg.48-358*

Product showcase in six sections features computers and peripherals, IC's and semiconductors, components, hardware and interconnect devices, power sources, and instruments. Each section begins with a review of products featuring the latest technology, and a description of manufacturers' new product lines. New product listing follows introductory section and features a description with technical specifications, price, photograph, manufacturers' address, and phone number for each entry.

EDP In-Depth Reports

1644. ANALYSIS OF THE CANADIAN COMPUTER INDUSTRY. *Evans Research Corporation. v.8, no.10. June 1979. 16 pg.*

Analysis of industry revenues for 1977 and performance of the industry in general examines customer base; profitability of various systems; number of businesses and their revenues, 1975-1977; employees in EDP hardware and their wages and salaries.

1645. THE CANADIAN COMPUTER INDUSTRY - TRENDS AND FORECASTS. *Evans Research Corporation. Annual. v.9, no.11. July 1980. pg.1-16*

Covers computer revenues (1970-1990) and computer population forecasts to 1990. Examines factors affecting the computer industry in future. Includes areas of highest growth (1970-1990) and the size of the industry including number of businesses.

1646. THE CANADIAN COMPUTER INDUSTRY - TRENDS AND FORECASTS. *Evans Research Corporation. Annual. v.8, no.11. July 1979. 16 pg.*

Analysis of computer revenues, with population forecasts to 1990 examines factors affecting the future of the computer industry. Includes areas of growth and number of businesses.

1647. COMPUTER COMMUNICATIONS TECHNOLOGY - ARE ADVANCES LEADING DECLINES? *Evans Research Corporation. v.9, no.8. April 1980. 16 pg.*

Reviews research and technology and worldwide computer-industry markets for semiconductors, 1979, 1984.

1648. THE COMPUTER SERVICES INDUSTRY IN NORTH AMERICA. *Evans Research Corporation. v.9, no.4. December 1979. 16 pg.*

Presents analysis and performance of the computer services industry in Canada and the U.S., 1975-1978; revenues; exports; regulations; effects of new technology; profitability; growth rates of various services and market shares in Canada.

1649. THE LEADING COMPANIES IN THE UNITED STATES COMPUTER INDUSTRY. *Evans Research Corporation. Annual. v.9, no.12. August 1980. pg.1-16*

Analysis of the performance of the major U.S. businesses in the computer industry, including shipments; revenues; areas of growth; research and development budgets; market shares, 1977, 1979; rankings of the 52 top businesses and their revenues and rankings in Canada.

1650. THE LEADING COMPUTER COMPANIES IN THE U.S.: AN ANALYSIS. *Evans Research Corporation. Annual. v.8, no. 12. August 1979. 16 pg.*

Analysis and performance of the U.S. computer business examines market shares of several businesses, their revenues and sales and the Canadian presence in the U.S. marketplace. Lists top 50 businesses.

1651. A NEW DISTRIBUTION MEDIUM - COMPUTER RETAILING. *Evans Research Corporation. v.11, no.6. February 1982. 12 pg.*

Highlights results of ERC's study on the Canadian computer retailing marketplace. Examines various computer stores and bases, average spending at stores, market segmentation, shipments and growth. Includes abrief analysis of the major competition (retail sales value, number of Canadian dealers / outlets and average revenues), with value of shipments 1980, 1981, 1987.

1652. PERSONAL COMPUTERS. *Evans Research Corporation. v.9, no.3. November 1979. 16 pg.*

Presents analysis and performance of the personal computer market including sales, markets, various types available, and distribution of computer store sales by application.

1653. SELECTED MAJOR VENDORS OF MULTI-TERMINAL WORD PROCESSING SYSTEMS. *Evans Research Corporation. v.11, no.8. April 1982. 20 pg.*

Study analyzes the performance of six major businesses in the word processing market which have proven multi-term or cluster systems in their product lines. Includes value of shipments of dedicated word processing systems 1981; market shares; products and prices, production facilities, and revenues of the six businesses.

1654. SERVICE BUREAU STRATEGIES FOR SELLING SMALL COMPUTERS. *Evans Research Corporation. v.8, no.6. February 1979. 16 pg.*

Analyzes small computer strategies of several major U.S. processing services bureaus. Examines competition; expenditures for small computer systems; number of installations on customer premises, 1977-1979; revenue growth; and revenues of Canadian Suppliers.

1655. TERMINALS: PART ONE. *Evans Research Corporation. v.10, no.2. October 1980. 16 pg.*

First of a two-part series examines the performance of Canadian markets for computer terminals; various models available; growth rates for the market; the terminal population by province, 1976-1979; and by major cities.

1656. TERMINALS: PART TWO. *Evans Research Corporation. v.10, no.3. November 1980. 16 pg.*

Second of a two-part series on terminals examines directions in which terminal technology appears to be leading and how this technology will fit into offices of the future. Includes product trends and various products presently available.

1657. THE TOP COMPUTER COMPANIES IN CANADA. *Evans Research Corporation. Annual. v.9, no.9. May 1980. 16 pg.*

Surveys the revenues of the 100 major Canadian businesses in Canada. Includes areas of greatest revenues; ownership and reliability codes; the top 42 EDP hardware suppliers; 34 service bureaus, IBM's revenues, 1975-1979; and profitability. Forecasts growth.

1658. THE TOP COMPUTER COMPANIES IN CANADA. *Evans Research Corporation. Annual. v.8, no.9. May 1979. 16 pg.*

Analysis of the performance and revenues (1976-1978) of the top 83 businesses in the Canadian computer idustry. Includes market shares and ownership.

1659. THE U.S. COMPUTER INDUSTRY - 20% GROWTH RATE IN 1980. *Evans Research Corporation. v.10, no.11. July 1981. 16 pg.*

Analyzes the performance of the U.S. information processing industry. Includes major leaders, their growth rates; ranks the top 66 businesses and their 1978-1980 revenues; examines mainframe suppliers including revenues, research and development expenditures, and working capital.

1660. VIDEOTEX SYSTEMS. *Evans Research Corporation. v.9, no.7. March 1980. 16 pg.*

Analysis of videotex as a communications system includes historical data, research and development areas, and a description of various systems presently available.

1661. WORD PROCESSING AND OFFICE AUTOMATION. *Evans Research Corporation. v.9, no.5. January 1980. 16 pg.*

First of a two-part series reviews the performance of various word processing systmes available; top suppliers in Canada, their sales, and products; systme costs; and estimated Canadian WP revenues by market segments.

Educational and Industrial Television

1662. DIRECTORY OF SPECIAL EFFECTS GENERATORS AND COMPUTER GRAPHIC SYSTEMS. *C.S. Tepfer Publishing Company, Inc. Annual. v.14, no.3. March 1982. pg.43-52*

Provides list, alphabetized according to manufacturer, of special effects generators and computer graphic systems. Each entry contains description of product, available options and price. Directory compiled from manufacturer specification sheets and/or press releases.

Electronic Business

1663. BATTLE FOR PRINTER MARKET RAGES AT HIGH AND LOW ENDS. *Cahners Publishing Company. v.8, no.7. June 1982. pg.78-82*

Gives 1980 unit shipments for line, page, and serial-character printers. Breaks down data by manufacturer. Discusses marketing, sales, and prices of specific models, 1982.

1664. COMPUTER: CHANGES FAIL TO DAMPEN MARKET OUTLOOK. *Furst, Al. Cahners Publishing Company. v.8, no.6. May 15, 1982. pg.6-10*

Article on computer industry estimates market size for desktop computers. Gives graphic illustration of 1981-1986 projected growth rate and market segmentation. Provides 1981 market shares.

1665. DESKTOPS LED THE WAY, NOW HANDHELDS FOLLOW. *Cahners Publishing Company. v.8, no.7. June 1982. pg.84, 86*

Analyzes the marketing of handheld and portable computers, 1981-1987. Gives handheld distribution channels, 1981 and 1987; and 1982 price ranges.

1666. ELECTRONIC BUSINESS: TECHNOLOGY: IN THE WAKE OF THE STING. *Cahners Publishing Company. v.8, no.9. August 1982. pg.28-29, 32*

Focuses on competition and interdependency in the world computer industry. Explores the effects of a June 1982 indictment of 22 persons and a leading Japanese computer manufacturer for conspiring to transport material stolen from IBM to Japan. Presents data on 1981 worldwide computer sales (in billions of dollars) for IBM and for six Japanese companies. Sources include Quantum Science Corp. and the Yankee Group Inc.

1667. FACTORY-OF-THE-FUTURE DEMAND. *Cahners Publishing Company. v.8, no.11. October 1982. pg.164*

Forecasts U.S. shipments of programmable machines (industrial robots, CAD / CAM equipment, and numerical controllers) for production of discrete parts, 1981-1986. Figure illustrates shipments (in billions of dollars) of factory automation systems for discrete manufacturing, 1981-1986. Source is Venture Development Corp.

1668. A GOOD TIME FOR BARGAINS IN 'DUMB' CRT TERMINALS. *Shaw, Stephen. Cahners Publishing Company. July 1981. pg.68-72*

Compares prices of low-cost computer terminals including editing terminals and dumb terminals. Discusses competition, profit margins, price cuts, and manufacturing overseas. Provides market shares (1979, 1980) for the top 11 U.S. vendors of dumb CRT terminals. Graph depicts price plunge (1972-1980) for all terminals, dumb terminals and editing terminals.

1669. GOVERNMENT SPENDING FOR CPU'S SURVIVES BUDGET CUTS. *Cahners Publishing Company. v.8, no.12. November 1982. pg.38*

Discusses U.S. government spending for minicomputer based data processing equipment during fiscal 1981. Figure illustrates market shares of top 15 U.S. government suppliers of data processing equipment, fiscal 1981.

1670. HOME - COMPUTER MAKERS WHEEL AND DEAL ON PRICE. *Cahners Publishing Company. v.8, no.11. October 1982. pg.41-42*

Reports on U.S. vendors' lists prices for personal / home computers, 1978-1983. Graph charts advances and declines in list prices of 7 home computers and indicates discounts, 1978-1983. Source is the Yankee Group, Inc.

1671. INVESTORS' 'DISCOVERY' OF ELECTRONICS DISTRIBUTION. *Rausch, Howard. Cahners Publishing Company. July 1981. pg.50-55*

Discusses the electronics distribution market in the U.S., including computer and electronic components, and it's investment outlook. Forecasts 1985 computer components share of electronics distributors' sales. Details financial data for the top 20 electronics distributors. Data provides 1980 industrial sales; net income as percent of sales; sales per employee; 1979 / 1980 sales growth rate; number of stocking locations; and size of field sales force.

1672. LARGE ORGANIZATIONS PLUS SMALL COMPUTERS EQUAL $3 BILLION. *Cahners Publishing Company. v.8, no.10. September 1982. pg.144-145*

Forecasts 1992 shipments and sales of personal computers to large U.S. organizations. Charts illustrate 1981 and 1985 supplier market shares of total micorcomputer sales to large U.S. organizations. Source is International Resource Development Inc.

1673. LEADTIMES FOR 8-BIT AND 16-BIT CHIPS EXPECTED TO LENGTHEN. *Rozak, Laura. Cahners Publishing Company. v.8, no.10. September 1982. pg.168-171*

Forecasts leadtimes for microprocessors, 1982-1984. Surveys 1982 leadtimes of representative U.S. microprocessor vendors and manufacturers. Graph shows microprocessor leadtimes in weeks, 1980-1984.

1674. MARKET LOOKS BRIGHT FOR GRAPHICS HARD-COPY DEVICES. *International Data Corp. Cahners Publishing Company. Irregular. March 1981. pg.118-120*

Summarizes U.S. market for graphics hard-copy devices including major manufacturers, shipments, and growth rates. Graph displays 1979-1984 revenues of U.S. manufacturers, from worldwide shipments.

1675. MICROCOMPUTERS SCORE AN 'A' IN THE EDUCATION FIELD. *Gerstel, Michele. Creative Strategies International. Cahners Publishing Company. May 1981. pg.58-60*

Discusses microcomputers share of the U.S. educational field. Bar graphs show retail sales and shipments (1980) with forecasts to 1985 for the educational computer market. Provides 1980 market shares for 13 microcomputer manufacturers, with their retail value, units share and dollars share.

1676. OFFICE AUTOMATION: THE PROFITABLE SYNERGY OF OFFICE AUTOMATION. *Cahners Publishing Company. v.8, no.3. March 1982. pg.84-85*

Examines U.S. growth of office automation equipment and services. Lists equipment and services along with 1981 actual and 1986 forecasted installed base plus forecasted cumulative of shipments (1982-1986).

1677. OFFICE AUTOMATION: STRONG WINDS OF CHANGE BUFFET WORD PROCESSING. *Mead, Tim. Cahners Publishing Company. v.8, no.3. March 1982. pg.98-99*

Gives graphic illustration of forecasted shipments and revenues (1981-1986) of word processing equipment. Includes 1981 leaders in word-processing workstations. Includes company name, number of units and market share.

1678. OPTICAL STORAGE EXPECTED TO RESHAPE MEMORY MARKET. *Cahners Publishing Company. v.8, no.11. October 1982. pg.163*

Assesses the impact of optical-disk throwaway memory systems on multifunction digital workstations in the U.S. Provides data on digital workstations shipped annually and number (percent) of workstations with optical memory, 1982, 1984, 1987 and 1992. Source is International Resource Development Inc.

1679. PERSONAL COMPUTERS - STRATEGIES FOR SUCCESS. *Cahners Publishing Company. v.8, no.3. March 1982. pg.80*

Provides statistical data on emerging personal computer market, including present status and forecasts for 1985. Figuires included for percent of business and individual sales.

1680. RASTER REFRESH GAINING A ROLE IN COLOR GRAPHICS. *Cahners Publishing Company. v.8, no.8. July 1982. pg.92-94*

Reviews the market for graphics-display-terminals and discusses the effect of raster refresh technology. Tables give world CRT shipments by U.S. manufacturers by type of terminal (1981); and world graphic terminal shipments (1981, 1985) by sub-market. Lists 1980 market leaders by value of shipments.

1681. SMALL - COMPUTER MAKERS BRACE FOR CHANGES. *Shaw, Stephen. Cahners Publishing Company. May 1981. pg.38-46*

Examines the small-computer industry in the U.S. in relation to the changing market. Diagrams and tables provide statistics on the top 10 share of the market; the top 10 ranked by 1979 and 1980 sales; and financial data on 14 leading businesses including 1980 sales, net income, revenues, research and development and cost of sales.

1682. SMALL COMPUTERS TO GROW 25 PERCENT IN 1980. *Mini-Micro Systems. Cahners Publishing Company. Irregular. August 1980. pg.88*

Reviews and forecasts the 1980 U.S. market for mini- and micro- computers. Tables show actual 1979 percent of total purchases from 10 computer manufacturers, with forecasts for 1980 purchases; and acquisitions of selected peripherals (1979, 1980), with expected growth rates.

1683. STICKING TO A TARGET DID THE TRICK FOR ATLAS. *Cahners Publishing Company. v.8, no.7. June 1982. pg.102, 104*

Analyzes the sales and marketing strategies of a small-business computer manufacturer, 1977-1985. Gives number of units shipped and market value of personal computers, 1981, 1987.

1684. THE TOP 10 SMALL COMPUTER MAKERS CHALLENGE IBM AND COME OUT WINNERS. *Swartz, R. B. Cahners Publishing Company. Irregular. May 1980. pg.53-60*

Examines U.S. small computer market. Ranks top 10 U.S. small computer manufacturers by 1979 sales. Graphs display revenues (1978-1979) of suppliers and 1979 market shares of top 10. Financial data is provided for top 10 small computer suppliers including 1979 net income, revenues and sales.

1685. UNIVAC GOES AFTER SALES OF NON - PCM MAIN-FRAMES. *Cahners Publishing Company. v.8, no.10. September 1982. pg.30-31*

Provides data on 1981 and 1986 U.S. shipments, in billions of dollars, of mainframe computers. Also reports 1981 market shares of 10 mainframe manufacturers. Source is International Data Corp.

Electronic Design

1686. MICROCOMPUTER DATA MANUAL AND DESIGN GUIDE. *Hayden Book Company, Inc. Irregular. March 15, 1980. pg.83-225*

Manual of technical information on new products in the microcomputer industry. Includes performance specifications and manufacturers' addresses.

1687. MICROPROCESSOR DATA MANUAL. *Bursky, Dave, editor. Hayden Book Company, Inc. Annual. November 22, 1979. pg.49-112*

Manual of microprocessor information including performance specifications and addresses and telephone numbers of manufacturers.

1688. NCC PRODUCT HIGHLIGHTS: SPECIAL REPORT. *Chester, Michael. Hayden Book Company, Inc. Irregular. v.30, no.11. May 27, 1982. pg.85-96*

Contains product information, prices, and delivery time for impact printers, disk-drive controllers, floppy-disk drives, storage systems, tape transporters, and other computer peripherals.

Electronic Engineering Times

1689. DEVELOPMENT SYSTEMS MARKET TO BREAK $1 BILLION MARK BY 1986, STUDY CLAIMS. *Frost and Sullivan. CMP Publications, Inc. no.245. October 25, 1982. pg.70*

Summarizes Frost and Sullivan study of microprocessor development systems market, 1981-1986. Gives estimated dollar value of sales of dedicated and universal systems, and terminals, 1981, 1986; and annual growth.

1690. ELECTRONIC ENGINEERING TIMES: PRODUCTS OF THE TIMES. *CMP Publications, Inc. Weekly. 1982. pg.79-82*

Weekly feature lists new electronics and computer products, including microcomputers, components, and accessories. Each entry includes product description with technical specifications, photograph, manufacturer's name, address, and telephone number, and prices for selected products.

1691. MANUFACTURERS' LOCAL OFFICE LISTINGS. *CMP Publications, Inc. Weekly. 1982.*

Regular feature lists manufacturers of electronic and computer components and accessories, with their suppliers and representatives throughout the U.S. Sales offices and representatives are listed by state with telephone numbers for each manufacturer. Manufacturer addresses are also included.

1692. SPECIAL REPORT: BOARD-LEVEL COMPUTERS. *CMP Publications, Inc. Annual. no.240. August 16, 1982. pg.48-62*

Describes new products. Includes listing of U.S. and Canadian makers of bus-compatible single-board microcomputers. Listing includes manufacturer, address, and telephone number.

1693. SPECIAL REPORT: DISK DRIVES. *Koopman, Phil. CMP Publications, Inc. Annual. no.244. October 11, 1982. pg.53-77*

Reports on new technology in the field of disk drives. Discusses proposed industry standards. Highlights selected products of U.S. disk drive manufacturers. Includes an alphabetical listing of U.S. makers of floppy and small Winchester disks. Includes address, telephone number, and type of drives manufactured.

1694. TECHNOLOGY STRATEGIES: SPEECH SYNTHESIS AND RECOGNITION. *Lubkin, Yale Jay. CMP Publications, Inc. no.241. September 1, 1982. pg.33-45*

Advances in the technology and development of speech synthesis and recognition equipment are discussed. Products under development are reviewed, with discussion of technical specifications. Includes data on market size (1981, 1989) and a listing of speech synthesis and recognition equipment manufacturers and vendors with addresses and telephone numbers.

Electronic News

1695. CAD / CAM DEMAND SPURS COMPUTER GRAPHICS GROWTH. *Fairchild Publications. Annual. v.28, no.1400. May 7, 1982. pg.17*

Reports results of survey of CAD / CAM computer graphics. Presents revenues of five major U.S. software vendors. Discusses survey strategies. Source is CAD / CAM Technical Survey 1982.

Electronic Products Magazine

1696. BAR CODES SCAN IN DATA, PROGRAMS, AND COMMANDS. *Hearst Business Communications, UTP Division. Irregular. March 26, 1982. pg.50-54*

Reviews trends in the use of computer programmable / readable bar codes over the traditional keyboard entry. A list provides addresses and telephone numbers of producers of bar code readers, printers, verifiers, and label applicators.

1697. ELECTRONIC PRODUCTS MAGAZINE: EP / IC UPDATE. *Hearst Business Communications, UTP Division. Monthly. v.25, no.3. July 23, 1982. pg.93-97*

Lists complete and up-to-date data on new IC offerings, including digital, interface, linear, memory and microprocessor devices. Listing includes type, function, price (ea/100), and source and contact.

1698. SIMPLIFIED SOFTWARE INTERFACE BLENDS PERIPHERALS, PROCESSORS, AND PROTOCOLS. *Yencharis, Len. Hearst Business Communications, UTP Division. v.25, no.6. September 30, 1982. pg.48-54*

Assesses 1982 development of software interfaces to blend peripherals, processors, and protocols. Discusses current and new technologies used in intelligent I/O processor products.

Electronics

1699. ROUNDUP: PRINTERS UNDER $1,000 THRIVE. *Bishop, Ana. McGraw-Hill Book Company. Irregular. January 13, 1981. pg.224-229*

Summarizes the market for low-cost matrix printers. Table provides data on 8 recently introduced printers. Gives manufacturer and model, unit price, dimensions, speed and other features.

Engineering News-Record

1700. CONTRACTOR COMPUTER USE UP. *McGraw-Hill Book Company. v.208, no.17. April 29, 1982. pg.11-12*

Market survey and analysis covers utilization of computers and software by construction contractors and user attitudes.

Financial World

1701. THE ALLURE OF HIGH TECH. *Kessler, Jeffrey. Macro Communications, Inc. v.151, no.11. June 1, 1982. pg.16-21*

Examines the recent stock market performance of 9 high-tech industries. Industries include: semi-conductor equipemnt and instrumentation, computers, peripherals, computer software and services, telecommunications, defense electronics, and biotechnology. Gives data as follows: company; earnings per share, 1981-1983; recent price; 52-week price range; current P / E; projected P / E 1982; indicated dividend; and yield.

1702. HERE COME THE SUPER MINIS. *Macro Communications, Inc. v.151, no.14. August 1982. pg.16-21*

Examines sales and growth rates of super mini computer companies. Gives total sales and growth, 1981, 1982, 1986. Provides sales, profits, earnings per share, dividends, and yields for 5 companies, 1981-1983. Gives super mini market share for 5 companies. Gives projected annual growth rates for mainframes, all minis, super minis and micros, 1981-1986.

1703. PLAYING THE PERSONAL COMPUTER SWEEP-STAKES. *Harris, Diane. Macro Communications, Inc. v.151, no.5. March 1, 1982. pg.48-62*

The personal computer market is discussed with emphasis on investment opportunities. Total 1981 sales for the industry are detailed with forecasts of the market for 1985. Manufacturers are reviewed with data on stock performance and discussion of marketing strategies. Includes a table listing mainframe and peripheral manufacturers with earnings per share, 1980-1982; recent stock prices and price ranges; price earnings ratios; dividends and yields.

1704. TERMINALS: WHO'S GOT THE SMARTS? *Taub, Steve. Macro Communications, Inc. v.151, no.18. October 1, 1982. pg.18-20*

Assesses the U.S. computer terminal market, 1982-1985, discussing the technology of computer terminals. Examines the investment potential of four terminal manufacturers. Data includes annual sales, 1981-1982; earnings per share, 1981-1983; stock prices, 52-week range; recent stock price; price earnings ratio, 1982-1983; dividends, 1982; and yields, 1982. Estimates 1982-1985 growth rates and forecasts the size of the 1985 market.

Florida Banker

1705. COMPUTER BANKING. *Florida Bankers Association. v.9, no.3. March 1982. pg.27-30*

Presents short summaries of development in computer banking; credit card technology and discusses computer losses.

Forbes

1706. THE GROWTH INDUSTRY'S GROWTH INDUSTRY: COMPUTERS. *Seneker, Harold. Forbes, Inc. July 6, 1981. pg.142-161*

Rankings of the top 20 computer services firms detail main service area, revenues, computer services revenues as a percent of the firm's total revenues, and revenue growth rate (1979-1980). Industry-wide narrative analysis is accompanied by brief articles on assorted leading firms.

1707. TOMORROW HAS ARRIVED. *Pearl, Jayne A. Forbes, Inc. v.129, no.4. February 15, 1982. pg.111-119*

Reviews the home computer industry, from historical development to future growth speculations. Includes specific references to manufacturers and their products, as well as prices and 1981 market share information.

1708. TOP 50 BUSINESS EQUIPMENT COMPANIES: FORBES SPECIAL ADVERTISING SUPPLEMENT. *Wright, Peter A., editor. The Gartner Group, Inc. Forbes, Inc. Irregular. May 25, 1981. 18 (est.) pg.*

Text / advertising insert, covering leading digital office equipment and software producers. Ranks the top 50 firms by revenues, providing 1980 data-processing growth rates. Identifies leaders in individual product groups, with some discussion of market shares. Examines sectoral market segmentation of the data-processing industry as a whole, and of individual segments. Includes several product advertisements.

Foreign Investment Review

1709. CANADA'S ADVANCED - TECHNOLOGY INDUSTRY. *Plante, Marie. Foreign Investment Review Agency. March 1982. pg.4-7 Free (Canada).*

Analysis and performance of Canada's advanced technology industry. Looks at computer applications, microelectronics, fiber optics, lasers and biotechnology, as well as the structure of the industry, research and development and foreign control. Includes a listing of 13 major Canadian R & D businesses, sales and principal fields of activity for 1979, and a listing of the largest review producers in the Canadian computer industry, their total and EDP revenues for 1980 and their ownership.

Fortune

1710. INFORMATION SYSTEMS FOR TOMORROW'S OFFICE. *Time, Inc. v.106, no.8. October 18, 1982. pg.17-82*

Analyses and forecasts cover number of vendors, models and 1981 and 1986 installations in the U.S. market, with breakdowns by product group; 1971, 1981 information processing market breakdown; 1980, 1986 office system shipments by product group; market shares of installed word processors, desktop computers, and electronic typewriters for 11 firms; desktop computer end uses; shares of desktop computer sales (mail order, direct sales, stores, other); where desktop computers are used; computing expertixe of buyers; and other data.

Foundry Management and Technology

1711. FOUNDRY MANAGEMENT AND TECHNOLOGY'S PRODUCT SPOTLIGHT: ELECTRONIC DATA PROCESSING SYSTEMS AND EQUIPMENT. *William L. Teweles and Company. Irregular. v.109, no.9. September 1981. pg.86-88*

Provides a brief description of data processing hardware and software available to the foundry industry. Vendors' and manufacturers' names are supplied.

High Technology

1712. HIGH TECHNOLOGY STOCK INDEX. *Technology Publishing Company. Bi-monthly. v.2. 1982.*

Details high technology stock price index for the previous 12 months. This is compared to the Dow Jones and Standard and Poors Indexes. An additional graph details the change over the previous two months by industry sector.

1713. MARKET OUTLOOK: 16-BIT MICROCOMPUTERS. *Technology Publishing Company. v.2, no.5. September 1982. pg.52*

Trends in the market for 16-bit microcomputers are reviewed. Data is presented on market size in dollars, 1982-1986; growth rates, 1982-1986; manufacturers' shipments, 1982-1986; and retail price range, 1982. Major 16-bit micro manufacturers are identified and technical developments are discussed.

1714. MICROCOMPUTERS: THE SECOND WAVE. *Lu, Cary. Technology Publishing Company. Irregular. v.2, no.5. September 1982. pg.36-52*

Presents an overview of 16-bit microcomputers presently on the market, with discussion of product features and comparisons of design, memory, and operating system characteristics across product lines. Includes a glossary of frequently used terms, a table detailing technical specifications for products currently available, and a list of manufacturers with addresses.

Highway and Heavy Construction

1715. MINI-COMPUTER BUYER'S GUIDE. *Technical Publishing Company. Irregular. v.125, no.3. March 1982. pg.40-42*

Presents a chart of detailed technical specifications comparing 57 models of mini-computers from 15 major manufacturers. 23 categories are covered under the main headings of central processing unit, language compilers, storage and retrieval capacity, and typical system cost. Information was provided by the manufacturers in response to a publisher survey. Includes a glossary of terms and brief explanations of selected concepts.

Home Entertainment Marketing

1716. WHO'S WHO IN COMPUTERS AND GAMES. *Dealerscope Inc. Annual. v.2, no.9. September 1982. pg.34-31*

Assesses the U.S. market for home computers and computer games. Directory lists vendors and their addresses.

IDP Report

1717. TOP TEN DATA BASE DISTRIBUTORS TOOK IN $250 MILLION IN 1980. *Knowledge Industry Publications. August 14, 1981. pg.1, 3-4*

Ranks the ten leading U.S. electronic database distributors by estimated 1980 revenues. Also lists number of customers reported by 22 distributors, to demonstrate an asserted lack of correlation between customer volume and revenues. Forecasts value of overall database services market, and of specific services / market segments, in 1981 and 1991.

IEEE Computer Graphics and Applications

1718. TOOLS FOR AUTOMATED STATISTICAL GRAPHICS. *Caporal, P.M. IEEE Computer Society. Irregular. October 1981. pg.72-82*

Summarizes automated statistical graphics; types of software available for statistical graphics; and technical features of the software. Includes a listing of 19 software packages offered for the graphics. Gives software name; business and address of the developer; prices; and hardware requirements.

Industrial Distribution

1719. COMPUTERS: 1982. *Morgan-Grampian, Inc. Annual. v.72, no.11. October 1982. pg.43-50*

Annual review includes summary of 22 minicomputers (number of work stations, RAM, floppy disc, data disc pack, and 7 other feature listings); survey data on where purchasers get their software; and how owners see their computers' most significant advantages and most significant problems.

Industrial Engineering

1720. PERSONAL COMPUTERS: WHAT THEY CAN DO FOR IES AND HOW TO CHOOSE ONE: COMPUTERS AND THE INDUSTRIAL ENGINEER. *Dossett, Royal J. American Institute of Industrial Engineers. v.14, no.10. October 1982. pg.70-82*

Discusses how to shop for and specifications to consider in purchasing computer systems. Gives computer applications including word processing, work measurement, scheduling, financial calculations, inventory, computer aided design, and games. Lists types of hardware and software, including suppliers' business addresses.

Industry Week

1721. 2000 FUTURE FOCUS: WILL COMPUTERS OUTNUMBER THEIR MASTERS?: PART 7 OF A SERIES. *Much, Marilyn. Penton / IPC. Subsidiary of Pittway Corporation. Irregular. June 12, 1978. pg.66-78*

Industry analysis and forecast to 2000 for the U.S. computer industry. Includes views of executives, technological developments and effects on the management of business, industry and homes.

Infosystems

1722. INFOSYSTEMS: ANNUAL PRODUCT REVIEW. *Hitchcock Publishing Company. Annual. v.29, no.1. January 1982. pg.88-106*

Lists products and manufacturers with brief descriptions.

1723. INFOSYSTEMS ANNUAL PRODUCT REVIEW. *Hitchcock Publishing Company. Annual. January 1981. pg.76-96*

Annual directory of new products released by vendors during 1980, listed in the following categories: computer systems; minis and small systems; personal and desk-top systems; computer peripherals; communications devices; terminals, data entry devices and facsimile; micrographics; word processing; and data supplies; equipment and office products. Entries include description and manufacturer.

1724. INFOSYSTEMS: ANNUAL PRODUCT REVIEW. *Hitchcock Publishing Company. Annual. January 1980. pg.72-85*

Annual review of products released by vendors during 1979, grouped into the following categories: computer systems; mini and small systems; computer peripherals; and network devices. Covers terminals and data entry devices; data supplies and services; micrographics; word processing; and office products. Each entry provides vendor name and description of product.

1725. INFOSYSTEMS: NEW PRODUCT INFO. *Hitchcock Publishing Company. Monthly. v.29, no.2. February 1982. pg.110-118*

Infosystems monthly new products directory contains product listing, manufacturers, and descriptions.

1726. INFOSYSTEMS: SEMIANNUAL REVIEW OF NEW SOFTWARE PACKAGES. *Hitchcock Publishing Company. Semi-annual. February 1981. 154 pg.*

Describes 250 software packages introduced or enhanced since July 1980, by function / purpose. Examines some new printers, data processing devices, micrographics instruments, and database systems.

1727. INFOSYSTEMS: 1979 DP SALARY REPORT - NET GAIN OF 10% FOR POSITIONS SURVEYED. *Hitchcock Publishing Company. Annual. June 1979. $5.00.*

Annual data processing salary survey with breakdowns for 51 urban areas, 18 industries, and approximately 30 job titles. Statistics cover averages, medians, modes, and first and third quartiles.

1728. INFOSYSTEMS: 1981 ANNUAL DP SALARY REPORT: DP SALARIES LOOKING UP! *Lusa, John M. Hitchcock Publishing Company. Annual. 23rd annual survey. June 1981. pg.38-50*

Reports representative national salaries (highest and lowest reported, mean, median, mode, 25th and 75th percentile) for a variety of data processing / programming job categories. For the same job titles, breaks down average weekly earnings by industry group, city / region, and data processing budget (monthly equipment rental range). Analyzes percent change since 1980. Also examines relationship between total operating budget and DP budget, and between total employment and DP employment.

1729. INFOSYSTEMS: 1981 SOFTWARE INFO EXHIBITORS. *Hitchcock Publishing Company. Irregular. September 1981. pg.78-84*

Lists software vendors exhibiting their products at the September 1981 'Software Info' exhibition in Chicago. Gives names and uses of the systems they were to present (where available).

1730. LIST OF INDEPENDENT DBMS VENDORS FOR MAINFRAME COMPUTERS. *Hitchcock Publishing Company. Irregular. March 1981. pg.76*

Lists some 30 independent data base management system vendors in the U.S. Gives business addresses, software packages and types of computer they run on.

1731. MODEM AND MULTIPLEXER UPDATE: COMPACT AND CHEAPER. *Edwards, Morris. Hitchcock Publishing Company. Irregular. November 1981. pg.50-58*

Identifies manufacturers of modems, acoustic couplers and statistical multiplexers for long-distance data transmission over telephone lines. Discusses capabilities and prices of state-of-the-art systems. Indicates capacity ranges (in bits per second) of multiplexers produced by each company.

1732. SOFTWARE INFO REVIEW. *Thiel, Carol Tomme. Hitchcock Publishing Company. Semi-annual. July 1981. pg.64-96*

Tenth semi-annual review of the U.S. computer software industry. Lists vendors and their software packages under 33 subject headings.

1733. SOFTWARE INFO REVIEW: PART 1. *Hitchcock Publishing Company. Semi-annual. July 1980. pg.52-78*

A selected list of nearly 200 new or enhanced software packages made available since December 1, 1979. Includes vendors' names and short descriptions of features and compatability of programs.

1734. USER SURVEY: DBMS IS STILL A NEW TECHNOLOGY. *Schanstra, Carla R. Hitchcock Publishing Company. Irregular. September 1979. pg.70-74*

1979 market survey of users of data base management systems (DBMS) conducted by "Infosystems" readers' response cards. Comments on users' utilization, in-house verses packaged software, and reasons for dissatisfaction. Forecasts of future usage.

1735. WORD PROCESSING SALARY AND EQUIPMENT SURVEY. *Miller, Frederick W., editor. Hitchcock Publishing Company. Annual. June 1980. pg.56-57*

Infosystems' first U.S. (1980) salary and equipment survey. Full survey results are available for $75.00 by writing to Infosystems. Given are salaries of six job categories, (Manager of Data Processing to Proof Reader), rankings of equipment used, by type, and rankings of corporations utilizing the equipment, and breakdown areas of data processing budget with percentages. Lacks documentation of methodology and sources.

1736. 1980 USAGE SURVEY: COMPUTERS MOVE INTO MICROGRAPHICS. *Lusa, John M. Hitchcock Publishing Company. Annual. v.28, no.4 / 11th ed. April 1981. pg.34-38*

Market survey on the utilization of micrographics and the interface with computers. Examines management of micrographics operations; average expenditures, 1976-1981; percent of utilization by industry; employment of outside services; and end uses of micrographics in the firm.

Interface Age

1737. MICRO SYSTEMS SURVEY: LET THE BUYER COMPARE. *Fox, Tom. McPheters, Wolfe and Jones. Irregular. January 1981.*

Comparison of 34 microcomputer systems, presented in five different tables. The first table carries general information about hardware. It lists manufacturer, model number, price, type of enclosure, CPU, and standard peripherals. Table 2 examines memory data storage areas. Manufacturer, model number, random access memory and type of disk drive are presented. The third table, systems software, shows programs normally provided by the manufacturer. Table 4 lists the major applications packages. The last table lists manufacturer's name and address, gross sales, date established, number of U.S. and foreign dealers, number of service centers, date of first delivery, and number of total deliveries.

International Business Week

1738. THE COMING SHAKEOUT IN PERSONAL COMPUTERS. *McGraw-Hill Book Company. no.2766-97. November 22, 1982. pg.48-53*

Examines future market for personal computers. Lists nine computer manufacturing companies expected to survive (listing based on results of interviews with more than 40 leading hardware, software and peripheral equipment makers, industry consultants, and analysts). Also listed are top four in personal computers according to 1982 sales (millions of dollars.)

Iron Age

1739. BUYERS' GUIDE TO COMPUTERS IN MANUFACTUR-ING. *Chilton Book Company. Irregular. June 22, 1981. pg.A-18-A-19*

Lists names and locations of businesses which have indicated an interest in serving the data processing needs of the manufacturing industries. Includes mainframes; mini- and microcomputers; software suppliers; personal computers; word processors / office automation; components / terminals; time sharing services; CAD / CAM; numerical control; and programmable controllers.

1740. ROBOTS SWING INTO THE INDUSTRIAL 'ARMS' RACE. *Frost & Sullivan. Chilton Book Company. July 21, 1980. pg.48-57*

Reviews use of industrial robots in manufacturing, technology, manufacturers, and products. Shows purchases of robots by metalworking industries.

1741. WILL TURNKEY CAD / CAM SYSTEMS REMAIN VIABLE? *Greene, Alice M. Chilton Book Company. v.225, no.33. November 19, 1982. pg.49-53, 56*

Discusses state of CAD / CAM systems and 1979-1982 CAD / CAM market shares.

Lloyd's Ship Manager

1742. PRODUCT FOCUS: LOADING CALCULATORS. *Lloyd's of London Press Ltd. Irregular. July 1981. pg.10-13*

A directory of maritime loading calculators including complete loading systems, microcomputers, loadmeters, and components. Data includes address of manufacturer / supplier; trade name, model number, components, and brief specifications; prices in June 1981; and internatonal sales and service representative.

Machine Design

1743. MACHINE DESIGN: HARDWARE AND SOFTWARE. *Penton / IPC. Subsidiary of Pittway Corporation. Bi-monthly. v.54. 1982.*

Bimonthly feature lists new computer hardware and software products including product descriptions with technical specifications and prices. Photographs are included for some items.

The Magazine of Bank Administration

1744. THE MAGAZINE OF BANK ADMINISTRATION: BANK SOFTWARE DIRECTORY; DIRECTORY OF SELECTED VENDORS AND TELEPHONE BANKING SYSTEMS. *Bank Administration Institute. Annual. January 1980. pg.28-29, 36-47*

Business addresses and product descriptions are provided in the directory of selected vendors and telephone banking systems and the bank software directory. Sections of the software directory include demand deposits, savings, credit card / authorization, audit, single statement, installment loans, mortgage loans, commercial loans, trust / investment / share-

holder records, general ledger, proof-transit / MICR and other programs.

Management World

1745. COMMITTEE OF 500 REPORT: MANAGERS TAP INTO SMALL BUSINESS COMPUTERS. *McKendrick, Joseph. Administrative Management Society. Irregular. v.11, no.7. July 1982. pg.16-17*

Reports results of AMS survey on managers' use of small business computers as compared to processing personnel. Tables show end uses of computers, number of small business computers per company, personnel with access to small business computer and popular manufacturers. Provides directory of products, addresses and telephone numbers of manufacturers, and prices.

1746. SECOND ANNUAL WORD PROCESSING GUIDE. *Administrative Management Society. Annual. v.11, no.4. April 1982.*

Guide to word processing presents overview of latest trends in word processing equipment. AMS committee of 500 analyzes use of word processing systems and products. Lists companies and WP products they use. Provides compilation of word processing references, manufacturers, and equipment.

Merchandising

1747. ELECTRONIC CHESS SETS CHALLENGE ADVANCED PLAYERS' ABILITY. *Gralla Publications. v.7, no.8. August 1982. pg.33,76*

Examines new electronic chess games, 1982. Gives manufacturer, price,and ability rating. Includes some illustrations.

1748. HOME COMPUTER PROFITABILITY DEPENDS ON RETAIL FORESIGHT. *Silverman, William. Gralla Publications. v.7, no.7. July 1982. pg.6*

Retail marketing strategies and the potential market for home computers are discussed. Data is presented on average retail prices, shipments (1982) of home computers under $1000, and suggested inventories of home computer programs.

Minicomputer News

1749. BUYERS' GUIDE - SPEECH SYNTHESIS / SPEECH RECOGNITION. *Hayden Book Company, Inc. Irregular. October 23, 1980. pg.19-21*

Buyers' guide to speech synthesis and speech recognition machines. Lists manufacturers with addresses, telephone numbers, and detailed descriptions of products and prices.

1750. HARD-COPY PLOTTER BUYER'S GUIDE LIST. *Hayden Book Company, Inc. Irregular. November 20, 1980. pg.16-21*

Hard-copy plotter buyer's guide lists manufacturers, with addresses and telephone numbers. Includes specifications list with company names and models.

Mini-Micro Systems

1751. BOX SCORE OF EARNINGS. *Cahners Publishing Company. Monthly. v.15, no.10. October 1982. pg.69*

Lists revenues, net earnings and earnings per share for 3 months and previous 3 months for 19 computer and computer-related businesses.

1752. HP'S RESOURCES COULD GIVE HAND-HELD CPU IMPETUS. *Cahners Publishing Company. v.15, no.10. October 1982. pg.22, 29-41*

Presents sources of hand-held computers, 1982. For each company and model, gives data on dimensions, processor, memory, keyboard, display, interface, peripherals, software, distribution and retail price.

1753. MEMORIES: 64K DYNAMIC RAM MARKET WILL SOAR. *Klesken, Daniel L. Cahners Publishing Company. v.14, no.11. November 1981. pg.179-184*

Forecasts the market for 64K dynamic RAM's in the 1980's. Discusses prices, major businesses and market growth. Graphs show worldwide shipments of 16K MOS DRAM's and DRAM bits (1971-1981) with forecasts for 1986; and average selling prices per bit for 16K and 64K RAM's (1980-1983).

1754. MEMORY MARKET TRENDS: COMPETING IN THE DEC - COMPATIBLE ADD-IN MEMORY MARKET. *Furst, Al. Cahners Publishing Company. v.14, no.11. November 1981. pg.165-168*

Discusses competition with Digital in the add-in memory market, including prices and repricing, technology and major businesses. Diagrams show 1981-1985 worldwide add-in / add-on memory market forecast; and market shares for minicomputers, microcomputers and mainframes in 1981 and 1985.

1755. MINI-MICRO SYSTEMS: COMPONENTS. *Cahners Publishing Company. Monthly. v.15, no.4. April 1982. pg.254-256*

Reviews new components for the data processing field. Includes product description, price, and gives manufacturer's name and address.

1756. MINI-MICRO SYSTEMS: DATACOMM. *Cahners Publishing Company. Monthly. v.15, no.4. April 1982. pg.258-260*

Reviews new products for data communication. Includes products' technical specifications and price and gives manufacturer's name and address.

1757. MINI-MICRO SYSTEMS: NEW PRODUCTS. *Cahners Publishing Company. Monthly. v.14, no.11. November 1981. pg.267-300*

Monthly listing of new computer products (hardware and software) includes supplier or manufacturer's business address and a description of their product. Prices are given where available.

1758. MINI-MICRO SYSTEMS: PERIPHERALS. *Cahners Publishing Company. Monthly. v.15, no.4. April 1982. pg.268-294*

Reviews new peripherals for the data processing industry. Includes technical specifications and prices. Gives manufacturer's name and address.

1759. MINI-MICRO SYSTEMS: SYSTEMS. *Cahners Publishing Company. Monthly. v.15, no.4. April 1982. pg.235-246*

Reviews new products for data processing systems. Includes peripherals and complete systems. Gives product description, price, and manufacturer's name and address.

1760. MINI-MICRO WORLD: BOX SCORE OF EARNINGS. *Cahners Publishing Company. Monthly. January 1981. pg.30*

Monthly table listing revenues, net earnings and earnings per share for 28 businesses in the computer and computer-related industries. Time periods vary for each company, but generally cover past 2 years.

1761. OPERATING SYSTEMS COST MORE - BUT ALSO DO MORE. *Kotelly, George. Cahners Publishing Company. Irregular. v.14, no.10. October 1981. pg.113-122*

Survey of operating systems classifies and lists each system in terms of user function. Includes business name, languages supported, hardware configuration, primary applications and prices for 54 systems.

1762. PERSONAL COMPUTERS: A NEW GENERATION EMERGES. *Kenealy, Patrick. Cahners Publishing Company. Annual. v.15, no.8. August 1982. pg.161-185*

The market for personal computers is discussed, with data on growth rates (1982), retail price distribution, market shares of leading manufacturers (1981) with forecasts to 1985, and a review of current technology. Includes a directory of personal computers including manufacturers and brand or model, technical specifications, retail price, and photograph.

1763. PRINTERS: FIFTH ANNUAL SURVEY ISSUE. *Roman, Andrew. Roman Associates International. Cahners Publishing Company. Annual. January 1981. pg.61-112*

1981 annual survey of line printers includes a collection of articles, written by industry leaders, focusing on technology, industry structure, market growth, considerations in choosing a printer, and competition. Tables show forecasts of shipments (1980-1984) of impact line printers and listings of manufacturers of impact line printers and serial printers, with descriptive data and prices.

1764. PRODUCT PROFILE: ADD-IN MEMORY SUPPLIERS OFFER VARIED MENU. *Stiefel, Malcolm L. Cahners Publishing Company. Irregular. v.14, no.11. November 1981. pg.141-158*

Examines the outlook for the add-in memory board market for mini computers. Describes several different types of products. Table summarizes the major characteristics of minicomputer add-on memories and add-in memory boards manufactured by respondents to a survey. Data includes manufacturer; type of product; capacity (bytes and bits); model number and access time; and prices and delivery time.

1765. PRODUCT PROFILE: SMALL BUSINESS COMPUTERS. *Simpson, David. Cahners Publishing Company. Irregular. v.15, no.6. June 1982. pg.201-216*

Provides the following information for over 65 small business computer systems on the U.S. market: manufacturer; model; processor, memory (bytes); number of users; storage capacity (bytes), diskette and disk; software supported, price, and notes.

1766. SBC'S (SINGLE-BOARD COMPUTERS): SMALL BUT CAPABLE. *Kenealy, Patrick. Cahners Publishing Company. Annual. v.15, no.8. August 1982. pg.147-160*

Reviews the current technology and market for single-board computers. Identifies manufacturers and discusses popular systems. Lists single-board microcomputers including manufacturer and model number, technical specifications, and price.

1767. SINGLE - BOARD COMPUTERS OFFER GREATER CHOICE AND POWER. *Stiefel, Malcolm L. Cahners Publishing Company. Irregular. September 1981. pg.121-132*

Discusses single - board computers including 16-bit, 32-bit and 8-bit. Table details a representation of some 45 single - board computers showing manufacturer, model, CPU, memory, input / output, software support, board size and price.

1768. SIXTH ANNUAL SURVEY ON PRINTERS. *Cahners Publishing Company. Annual. v.15, no.1. January 1982. pg.143-306*

Annual printer survey gives profiles of products plus details of line-printer industry and its price / performance. Lists current serial printer and line printer manufacturers and their products including model(s), printing method, speed, and price. Provides directory of teleprinter suppliers. Discusses new technologies.

1769. SYSTEMS IN MANUFACTURING: VOICE - INPUT SYSTEMS MAKE INROADS INTO INDUSTRIAL APPLICA-TIONS. *Cahners Publishing Company. v.15, no.10. October 1982. pg.165-171*

Examines industrial applications of voice recognition by computers, 1982. Gives specific U.S. companies and instances. Shows, by type of system, annual dollar value of U.S. shipments of speech processing equipment, 1980-1989. Lists 10 producers, with locations, of voice recognition systems. Gives product name, maximum vocabulary, compatibility, and price.

1770. TABLE OF FLOPPY-DISK DRIVE MANUFACTURERS. *Cahners Publishing Company. Irregular. February 1981. pg.156-159*

Provides manufacturer, model number, descriptive data and price for 40 vendors of floppy-disk drives.

Mining Engineering

1771. MINICOMPUTERS IN THE MINERALS INDUSTRY. *Society of Mining Engineers of AIME. v.33, no.11. November 1981. pg.1579-1610*

Contains twelve articles on minicomputers for use in the minerals industry. Available systems are reviewed with prices and applications. Discusses attitudes toward software, development of future applications, and growing markets. Presents selection considerations and criteria for product evaluation, along with case studies of systems in use. Includes a directory of available software and glossary of computer terms.

Modern Machine Shop

1772. MODERN MACHINE SHOP: BETTER PRODUCTION. *Gardner Publications Inc. Monthly. v.54, no.11. April 1982. pg.110-124*

Presents one page summaries of new equipment, parts, procedures and computer technology. Technical specifications and potential to increase production are discussed.

Modern Materials Handling

1773. MODERN MATERIALS HANDLING: NEW PRODUCT DATA. *Cahners Publishing Company. Monthly. v.37, no.6. April 6, 1982. pg.138-175*

Reviews new products in the materials handling field. Includes manufacturer's name and product description with photograph.

1774. TALKING WITH COMPUTERS: A DISPATCHER AT YOUR SIDE! *Cahners Publishing Company. v.37, no.5. March 19, 1982. pg.54-57*

Discusses the advantages of using rider trucks with on-board computer terminals.

Modern Office Procedures

1775. MODERN OFFICE PROCEDURES' 1982 PERSONAL COMPUTER SPECIFICATIONS CHART. *Penton / IPC. Subsidiary of Pittway Corporation. Annual. v.27, no.5. May 1982. pg.142-158*

Lists the personal computer systems available for under $10,000 in 1982. Data catagories are manufacturer, model, purchase price, memory, technical specifications, and storage system. Survey results based on manufacturer-provided information.

The Office

1776. BUYERS' GUIDE TO BUSINESS (MINI) COMPUTERS. *Office Publications, Inc. Irregular. v.94, no.3. September 1981. pg.182, 187-198*

Directory lists about 66 manufacturers of business or mini-computers. List includes model numbers, price, monthly lease or rental fee, and descriptions.

1777. DATA PROCESSING PRODUCTS: NEW EQUIPMENT IN AUTOMATION. *Office Publications, Inc. Monthly. v.94, no.2. August 1981. pg.136-148*

Monthly directory lists and describes some 25 data processing products available from U.S. manufacturers. Includes addresses of manufacturers and, where available, prices for the equipment.

1778. 1981 ANNUAL READER'S FORUM. *Office Publications, Inc. Annual. January 1981. pg.24-90*

Market survey for office machines and equipment in the U.S. Gives executives' forecasts of expenditures, 1981. Reports users' attitudes and plans.

Office Equipment and Methods

1779. COMPUTER SECURITY COSTS. *Unger, Harlow G. Maclean Hunter Ltd. July 1981. pg.40-43*

Examines the costs of computer crime in Canada and the U.S. Losses are noted in comparison to average walk-in thefts. Includes suggestions for security.

Office Products Dealer

1780. COMPUTER SYSTEMS DEALER: SYSTEMS HARD-WARE. *Hitchcock Publishing Company. Monthly. 1982.*

Monthly directory of products available for the computer systems dealer. Computer systems hardware products for the automated office are listed, with brand names, brief descriptions, and technical specifications. Includes photographs and manufacturers' names.

1781. PROGRAM: SELLING ELECTRONIC OFFICE SUP-PLIES AND ACCESSORIES. *Plant, Janet. Hitchcock Publishing Company. v.110, no.2. February 1982. pg.40-45*

Examines the market for computer and word processing office supplies and accessories, with forecasts of market growth to 1985. Leading product groups are reviewed, and strategies for office products dealers to benefit from the demand for these products are examined. Includes a product guide listing common supplies and accessories with technical specifications; and a statistical profile of the market, which forecasts market growth rates for selected product groups. Factors contributing to dealer success in the high technology supplies market are identified in terms of sales and distribution strategy.

Office Products News

1782. BUSINESS MICROCOMPUTERS. *Hearst Business Communications, UTP Division. Irregular. November 1981. pg.14-19*

Directory lists 37 U.S. manufacturers of business microcomputers. Gives model number or numbers; word size; RAM capacity; language used; software packages; features; options; and price of the hardware.

1783. OCR EQUIPMENT. *Hearst Business Communications, UTP Division. Irregular. v.16, no.7. December 1981. pg.16-17*

Product review lists optical character recognition and scanning equipment. Each entry includes manufacturer's or vendor's name, product brand name and model number, price, applications and intended end uses, and technical specifications including type faces read, scanning speed, accuracy rate, and additional features and options.

Office World News

1784. CBEMA REPORTS TRADE SURPLUS DECLINE. *Hearst Business Communications, UTP Division. v.10, no.17. September 1, 1982. pg.12*

Summarizes report of the Computer and Business Equipment Manufacturers Association on the U.S. trade surplus in computers and business equipment. Statistical data provided for imports of DP equipment, 1981; exports of DP equipment, 1981; imports and exports of business equipment, 1981; and imports and exports of business forms and suppliers, 1981.

1785. COMDEX: LAS VEGAS, NOVEMBER 19-22, 1981. *Hearst Business Communications, UTP Division. Irregular. November 15, 1981. pg.7-18*

Lists computers and computer-related products exhibited at the COMDEX show in Las Vegas, November 19-22, 1981. Gives trade name, manufacturer, brief description, booth number, and photographs. Further information available through reader card service.

1786. 1981 WAS YEAR OF DEALER IN COMPUTER RETAILING FIELD. *Hearst Business Communications, UTP Division. v.9, no.22. December 15, 1981. pg.1, 8*

Reviews the trend for computer suppliers and manufacturers to distribute through dealers. Projects world shipments for 1982 and the current number of dealers, with increases over the past year. Presents comments on effective market penetration, consumer attitudes, and distribution channels from a cross-section of computer vendors.

Ontario Business

1787. IMPACT OF COMPUTERS - WHEN? *Ontario Chamber of Commerce. March 1982. pg.27*

Forecasts the availability of the following to the year 2020: banks interconnected by computer networks, 'paperless' administration, and TV-interactive information systems.

Pension World

1788. COMPUTER APPLICATIONS FOR EMPLOYEE BENEFIT PLANS. *Deverman, Jerone N. Communication Channels, Inc. v.18, no.3. March 1982. pg.37-40*

Presents an overview of computer applications to employee benefit programs. Discusses databases, computer security, agent level responsibilities and administrator level functions. Also gives 9 step outline on choosing a system.

Personal Computing

1789. BIG FOUR ACCOUNTING SOFTWARE ROUNDUP. *Hayden Book Company, Inc. Irregular. v.6, no.9. September 1982. pg.82-112, 146 ff.*

Examines general ledger, accounts receivable, accounts payable, and payroll software packages, giving alphabetical listing and addresses of vendors. Identifies appropriate equipment. Gives price range including features of each vendor's products in four areas.

1790. COMPUTER GAMES: THE FUTURE PRESENT. *DeKoven, Bernie. Hayden Book Company, Inc. Irregular. v.6, no.6. June 1982. pg.68-72, 122-123*

Lists the 28 best-selling computer games. Includes name of game, computer system, and vendor's name, address, and telephone number.

1791. EVERYTHING YOU WANTED TO KNOW ABOUT PRINTERS. *Hayden Book Company, Inc. Irregular. March 1981. pg.58-69*

Evaluates personal computer printers and compares 80 models available from 33 manufacturers, in terms of prices, character form, columns, speed, interface, graphics, character set and KSR / RO. Separate listing of vendors with addresses and telephone numbers.

1792. GUIDE THROUGH THE PRINTER JUNGLE. *Hayden Book Company, Inc. Irregular. October 1981. pg.68-71*

Special report on printers lists products, model numbers, prices, character formats, and number of columns. Covers print speed, paper transport, interface, character setting, graphics, ability, and options or features.

1793. IF IT'S WORTH ITS WEIGHT IN PAPER: WHAT DO YOU NEED IN A PRINTER? *Hayden Book Company, Inc. Irregular. v.6, no.7. July 1982. pg.66-71, 76 ff.*

Gives factors to consider when purchasing a printer for a computer. Buyers' guide gives manufacturer with address, model number, price, type, speed, matrix size, paper width and graphics capability.

1794. MODEMS: YOUR LINE TO THE WORLD. *Hayden Book Company, Inc. Irregular. September 1981. pg.96-97*

Directory of popular personal computer modems lists product names, addresses of the businesses which produce the modem, the type of modem, the application, prices and features.

1795. OUTLOOK: MICROCOMPUTERS IN LARGE ORGANIZATIONS. *Hayden Book Company, Inc. v.6, no.6. June 1982. pg.25-26*

Figure illustrates 1981 market shares and projected 1985 market shares of 5 leading vendors in the organization microcomputer market. Data based on annual shipments. Source is International Resource Development Inc.

1796. PERSONAL COMPUTING DEALERS. *Hayden Book Company, Inc. Monthly. v.6, no.2. February 1982. pg.158-159*

Directory includes lists of computer dealers by state and their respective telephone numbers.

1797. A USER'S GUIDE TO OPERATING SYSTEMS. *Boyd, Alan. Hayden Book Company, Inc. Irregular. May 1981. pg.27-32*

Directory of operating systems lists names and addresses of businesses alphabetically who sell the operating system, the computer / microprocessor used with the operating system, the prices, and the uses for the operating system.

1798. WORD PROCESSING FOR PERSONAL COMPUTERS. *Hayden Book Company, Inc. Irregular. v.6, no.8. August 1982. pg.82-106*

Discusses features of word processors generally. Lists producers of software packages. Gives addresses, product name and price, and acceptable hardware. An alphabetical listing of companies and packages gives word-processing editing features.

1799. WORD PROCESSING SOFTWARE ROUNDUP. *Hayden Book Company, Inc. Irregular. January 1981. pg.23-33*

Examines word processing software packages, including editing, formation, printer and formatting program features. Includes a vendor guide with names, addresses, software packages, systems required and prices.

Plan and Print

1800. PLAN AND PRINT: NEW PRODUCTS. *International Reprographic Association, Inc. Monthly. October 1981. pg.36-45 $10.00/yr.*

Monthly survey of new drafting, engineering design, and reproduction equipment. Includes such objects as drafting tables, microformat readers, slide storage units, printers and plotters, platemaking machines, computer graphics systems, and accessories like adhesive spray and electric erasers. Includes product descriptions and manufacturers' names.

Plastics and Rubber Weekly

1801. U.S. SET FOR ROBOT BOOM. *Maclaren Publishers Ltd. no.953. September 4, 1982. pg.11*

Summarizes data on robot technologies, applications, and markets in the U.S. through 1990. Table reports number of robots in use (units) for the following years: 1972, 1977, 1980, 1985, 1990, and 1995. Data source is Predicasts, Inc.

Plastics Design Forum

1802. CAD / CAM: ITS IMPACT ON PLASTICS PART DESIGNERS. *Collins, Scott. Industry Media Inc. v.7, no.3. May 1982. pg.61-64*

Discusses the use of computer-aided design (CAD) and computer aided manufacturing (CAM) systems in plastics part design and manufacture. Current technology and system features are reviewed, and major system producers are identified. End uses are discussed and data on the price range of CAD / CAM systems is presented.

Production

1803. COMPUTERS IN MANUFACTURING. *Production Publishing Company. Annual. 4th ed. December 1981. pg.64-96*

Lists manufacturers and software companies associated with application of computers to manufacturing systems. Provides information on numerical and computer numerical controls, programmable computers, CAD/CAM, interactive graphics and voice recognition systems. Further information available from reader card service.

1804. DIRECTORY OF DOMESTIC ROBOT AND ACCESSORY BUILDERS / DISTRIBUTORS. *Production Publishing Company. Annual. December 1981. pg.107*

Lists manufacturer's products, and distributors associated with robotics and equipment used in manufacturing systems. Information includes company, and address.

1805. NEW EQUIPMENT PRODUCTS AND PROCESSES. *Production Publishing Company. Monthly. December 1981.*

A monthly feature which lists equipment and processes associated with manufacturing systems including heavy machinery, robots, computer controllers, lathes, etc. Data includes company names and brief specifications. Further information available from reader card service.

1806. ROBOTS AT IMTS - 82. *Production Publishing Company. Biennial. v.90, no.5. November 1982. pg.37-39*

Surveys robots exhibited at IMTS - 82. Includes manufacturers' names and provides brief product descriptions.

Purchasing Magazine

1807. DESKTOP COMPUTERS: THE NEW GENERATION. *Cahners Publishing Company. v.93, no.8. October 21, 1982. pg.85-89*

Reviews desktop computers of 6 large manufacturers, 1982. Compares 9 models in terms of price, bits, memory, and floppy and hard disk storage.

Radio - Electronics

1808. RADIO ELECTRONICS BUYERS GUIDE TO HOME COMPUTERS: 1980. *Gilder, Jules H. Gernsback Publications, Inc. Annual. October 1980. pg.45-86*

1980 buyers guide to home computers covers peripherals and accessories, dial-up networks, and useful programming languages. Each entry provides descriptions, illustrations, applications, prices and special features.

1809. YOUR OWN COMPUTER: HARDWARE. *Stern, Marc. Gernsback Publications, Inc. Semi-annual. v.53, no.10. October 1982. pg.43-112*

Presents 1982 review of personal computers. Lists manufacturers and their addresses. Includes a product description and retail price.

Robotics Today

1810. ROBOTICS TODAY: NEW PRODUCTS. *Robot Institute of America. Bi-monthly. v.4, no.1. February 1982. 5 pg.*

Reviews new products for the robot manufacturing industry. Includes product description with photograph and manufacturer's name.

1811. THIRTEENTH ISIR / ROBOTS 7 EXHIBITORS. *Society of Manufacturing Engineers. Annual. v.4, no.5. October 1982. pg.48*

Lists exhibitors at the 1982 ISIR / Robots 7 conference.

Savings and Loan News

1812. MORE ASSOCIATIONS INSTALL MORE ATMS. *United States League of Savings Associations. v.103, no.2. February 1982. pg.104-105*

Examines automatic teller machine market. Presents tables on quantity of associations with ATMS; states with largest ATM programs; installation patterns; ATM locations; ATM cardholders; ATM overall performance measures; and ATM transaction volume.

Skylines

1813. 1982 BOMA TRADE SHOW EXHIBITORS. *Building Owners and Managers Association International. Irregular. v.7, no.6. June 1982. pg.3*

Provides an alphabetical listing of 120 exhibitors that will display state-of-the-art technology, and new services, products, and equipment at the BOMA International 75th Anniversary Convention (June 20-24, 1982, Washington, DC).

Small Business Computer News

1814. SIXTH ANNUAL SMALL BUSINESS COMPUTER USERS SURVEY. *Management Information Corporation. Annual. v.8. August 1981. pg.6-23*

Compiles users' rankings of small business computers and peripherals, and of 245 software packages, according to criteria of performance, reliability, ease of use, service, and manufacturer support. Includes overall manufacturer ratings and an analysis of software applications.

Solid State Technology

1815. SOLID STATE TECHNOLOGY: PATENT REVIEW. *Marshall, Sidney. Technical Publishing Company. Monthly. v.25. 1982.*

Monthly patent review lists patents filed with the U.S. Patent Office. Each entry includes patent number, title, author / inventory, patent assignee (corporation), and filing date, with a short description of product.

The Southern Banker

1816. RIDING HIGH IN THE ATMOSPHERE. *Zoller, Harriet. McFadden Business Publications. v.157, no.4. April 1982. pg.14-16, 48*

The use of automatic transaction machines (ATM) in banking is discussed with data on the number of units in operation, 1981-1982; number of installations, 1981-1982; and growth rates of installations, 1982. The advantages of using ATM's are reviewed, with discussion of equipment costs, consumer behavior, and future trends.

Telecommunications

1817. NCC PREVIEW: MEETING / PREVIEW. *Horizon House. Irregular. v.16, no.5. May 1982. pg.84-86*

Includes profiles of 15 computer technology companies that will display products at the 1982 National Computer Conference (Houston, Texas, June 7-10, 1982). Includes company name, address, telephone number, telex number, booth number, product(s) description, and key contact personnel.

Television / Radio Age

1818. COMPUTERIZATION OF NEWSROOMS: PROGRESS IS SLOW DUE TO ECONOMIC AND PSYCHOLOGICAL BARRIERS. *Television Editorial Corporation. v.30, no.7. November 1, 1982. pg.30-31, 64 ff.*

Examines use and cost of specific computer hardware and software in U.S. newsrooms, 1982. Lists 7 manufacturers of broadcast and cable newsroom computer systems. Includes stations purchasing equipment and price range.

Today's Office

1819. TODAY'S OFFICE: PRODUCTS. *Hearst Corporation. Monthly. March 1982. pg.69-82*

Lists new automated equipment including word processors, calculators, records management equipment, etc. Information includes company name, description, photograph, price, and telephone number. Further information available through reader card service.

1820. TRACKING OA IN 1983. *Pomerantz, David. Hearst Business Communications, UTP Division. v.17, no.7. December 1982. pg.21-30*

Analysis and forecast of computer office equipment markets. Shows market size and shipment growth rates, 1975-1990.

Toronto Business Magazine

1821. TORONTO BUSINESS COMPUTER GUIDE. *Zanny Ltd. April 1982. pg.21-51*

Analysis of trends in Canada's computer industry includes new research and development, new products and systems.

Venture

1822. YOUR PICK OF PORTABLES. *Venture Magazine, Inc. Irregular. v.4, no.12. December 1982. pg.20*

Examines 7 portable computers, 1982. Gives manufacturer and location, model, weight and basic price.

Videodisc / Videotex

1823. ELECTRONIC RETAILERS BECOMING 'INFORMATION GATEKEEPERS.' *Meckler Publishing. v.2, no.3. June 1982. pg.168-169*

Summarizes a 1982 market survey on retailing electronics products to U.S. consumers. Graph illustrates the influence of sales staff on user purchase decisions for six product groups. Source is Venture Development Corporation.

What to Buy for Business

1824. COMPUTERS - A WORKING MAN'S GUIDE. *Oppenheim Derrick, Inc. Irregular. 1st ed. 1982. pg.3-37*

Explains the technical aspects of computer hardware and software in an introductory section which provides background material on the development of computers and current technology. Discusses factors important to the purchase decision. Presents results of a survey of small business computers users along with a list of computer suppliers with addresses and telephone numbers. A table lists min and microcomputers on the market with manufacturers, prices, technical specifications, and remarks from the user survey.

Wood and Wood Products

1825. WOOD AND WOOD PRODUCTS: PRODUCT REVIEW OF ELECTRONIC EQUIPMENT. *Vance Publishing Corporation. Irregular. v.87, no.3. March 1982. pg.56-58*

Presents an overview of new technological developments in wood processing equipment. All products are electronic: many are programmable. Products are listed by manufacturer, and perform a wide range of funtions, including: veneering, photoelectric sensors, and computerized numerical control.

Woodworking and Furniture Digest

1826. ROUNDUP OF NC AND CNC ROUTERS. *Hitchcock Publishing Company. Irregular. v.84, no.6. June 1982. pg.16-26*

Lists, for 11 alphabetically-arranged company and brand names, a technical description of numerically controlled and computer numerically controlled routers. Represents 6 domestic lines and 5 import lines.

Word Processing World

1827. SOURCE GUIDE TO WP (WORD PROCESSING) PRODUCTS AND SERVICES. *Geyer-McAllister Publications, Inc. Annual. August 1977. pg.57-79*

1977 directory lists firms with addresses offering word processing products and services and a classified section of specific items or services. Listing includes general office needs and industry specific items (collating machines, filing equipment and photocomposers). Three product catagories detailed in charts: dictation equipment, media (dictation and text editing equipment) and WP typing equipment.

Words

1828. WORDS: EQUIPMENT UPDATE. *International Information / Word Processing Association. Bi-monthly. v.10, no.5. February 1982. pg.6-8*

A bi-monthly list of word processing and electric office equipment, including software. Gives manufacturer's name, description, brief compatibility requirements, and prices.

See 272, 330, 334, 335, 351, 353, 361, 378, 380, 625, 628, 629, 632, 637, 639, 645, 651, 652, 654, 656, 658, 664, 669, 671, 678, 685, 688, 690, 692, 698, 700, 701, 705, 2596, 2599, 2601, 2603, 2609, 2612, 2613, 2615, 2616, 2642, 2645, 2647, 2656, 2659, 2660, 2661, 2662, 2663, 2664, 2672, 2687, 2690, 2693, 2694, 2695, 2697, 2698, 2699, 2718, 2723, 2726, 2728, 2730, 2736, 2738, 2763, 2773, 2780, 2781, 2783, 2787, 2795, 3263, 3265, 3266, 3267, 3269, 3270, 3280, 3281, 3282, 3283, 3284, 3286, 3289, 3290, 3291, 3298, 3309, 3310, 3311, 3317, 3319, 3320, 3321, 3330, 3334, 3340, 3343, 3349, 3350, 3352, 3355, 3358, 3359, 3666, 3670, 3686, 3750, 4222, 4223, 4237.

Numeric Databases

1829. AUERBACH COMPAR. *Auerbach Information, Inc. (Available from Tymshare, Inc.). Irregular.*

Contains information on computer systems and components. Data includes prices; vendors, addresses of vendors; products, product line; and model number. Provides terms of purchase, lease, rental, license, and maintenance; equipment specifications and characteristics. Covers microframes, minicomputers, software, word processors, terminals peripherals, and data communication equipment.

1830. CENSUS OF COMPUTER SITES USA. *Hayden Book Company, Inc. (Available from Direct from Producer). Monthly.*

Contains information on installed computers throughout the United States. Data includes company name, address, telephone number, organization size, and types of peripheral equipment. Provides information on whether the equipment is owned, rented, or leased, and what plans companies may have about future purchases. Information is retrievable by manufacturer, model number, operating systems, programming languages, job title / job function, business / industry, and geographic. Available on magnetic tape, computer printout, or sales cards.

Monographs and Handbooks

1831. ADVANCES IN COMPUTER COMMUNICATIONS AND NETWORKING. *Chu, Wesley, ed. QED Informational Services. November 1979. 653 pg. $38.00.*

Presents 61 papers which explore the improved resource sharing capabilities as well as the problems born of rapidly evolving 'comunications' technology. The following is a partial list of contents: Computer traffic characteristics, error control techniques, statistical multiplexing, local networks, communications processors for store and forward networks, and computer communication security.

1832. ADVANCES IN COMPUTER SYSTEM SECURITY. *Turn, Rein, ed. QED Informational Services. 1981. 375 pg. $40.00.*

Presents more than 30 referenced papers selected from such sources as IEEE Proceedings and Transactions, AFIPS Conference Proceedings and ACM Transactions. Contents include computer security framework, model of secure systems, operating system security, database security, computer communication security, public key cryptosystems, computer network security, legal requirements for computer security, computer security auditing.

1833. BEFORE YOU DECIDE TO COMPUTERIZE. *Educational Institute of the American Hotel and Motel Association. 1979. 60 pg. $10.00.*

Report is an aid to hotel industry in decisions on computerized systems. Examines literature, computer systems, hidden costs, cost justifications, maintenance options, purchase and lease options and installations.

1834. THE BUSINESS SYSTEM BUYER'S GUIDE. *Osborne, Adam. McGraw-Hill Book Company. April 1981. $15.*

Geared for readers with little knowledge of computers or computer terminology, this book is oriented towards defining computer needs and choosing appropriate systems.

1835. BUYING COMPUTER HARDWARE. *Alberta. Department of Agriculture. Agdex 818-11. April 1982. 8 pg.*

Examines microcomputer technology as a tool in agriculture. Includes tips on purchasing systems as well as an alphabetical listing of available hardware, list and system prices, MPU used, RAM minimum / maximum, disk capacity, display, text characters, graphics, sound and keyboard.

1836. THE CANADIAN OFFICE TODAY. *Tarnai, Lois. John Wiley and Sons Canada Ltd. April 1982. 14.95 (Canada).*

Combines theories on office practice with information on technological offices. Chapters on computer filing and micrographics, telecommuni

1837. THE COMPUTER ESTABLISHMENT. *Fishman, Katherine Davis. Harper and Row Publishers, Inc. October 1981. $20.00.*

Discusses the development of computers from early electromechanical devices to today's miniaturized solid-state equipment, and the concurrent development of the computer industry. Presents a picture of a segmented market, with IBM's large share largely unchallenged, and the so-called 'Seven Dwarves' competing among themselves for the remainder. Asserts that at least 3200 firms are engaged in software development, facilities management, and processing services in the U.S. Analyzes the conflict between the computer companies and the AT&T network over remote computing linkage. Discusses the potential dangers of the current wave of automation.

1838. COMPUTER REVIEW. *GML Corporation. Tri-annually. 1982. 500 (est.) pg. $175.00.*

Product reports on large computers give current data on processor features, peripherals, communications, software, pricing, and applications. Gives details on configurations, upgrades, and special features. Includes series of lists, directories, and indexes.

1839. ELECTRONIC DOCUMENT STORAGE 1980-1990: VOLUME 3: TECHNOLOGIES AND PRODUCTS. *Mackintosh Consultants, Inc. November 1980. $7500.00.*

Discusses current and emerging technologies in (digital and other) electronic document storage systems and input-output devices. Systems covered include magnetic tape and disc, optical / video disc, and automated micro fiche retrieval. Price trends, and major manufacturers' activities, through 1990 are forecast.

1840. GUIDE TO MINICOMPUTERS. *Frederick, Franz J. Syracuse University Printing Services. 1980. 160 pg. $11.50.*

Guide covers computer languages, operating systems, compatible systems, special accessories, time sharing, service and maintenance, and potential instructional and media center applications, with a listing of resources.

1841. HANDBOOK OF WORD PROCESSING SYSTEMS. *Buyers Laboratory, Inc. 1981. 104 pg. $35.00.*

Survey of word processing systems covers 66 systems produced by 38 vendors, with one-page write-ups on each standalone test edition and shared logic / shared resources system.

1842. HOW TO BUY A PERSONAL COMPUTER. *Shrum, Carlton. Alfred Publishing Company. Irregular. 1981. 64 pg. $2.95.*

Guide to buying microcomputer hardware and software includes chart comparing various systems.

1843. HOW TO BUY A WORD PROCESSOR. *Seriven, Michael. Alfred Publishing Company. Irregular. September 1982. $2.95.*

Buying guide covers electronic typewriters, personal computers, and specialized word processing systems with a comparison chart of the most popular systems and software.

1844. MICROCOMPUTER REFERENCE HANDBOOK. *Computer Reference Guide. 1981. 250 pg. $25.00.*

Reviews more than 130 microcomputer systems from over 50 major microcomputer suppliers, including some of the latest Japanese manufacturers. Part I covers general data; Part II covers a range of microcomputer software from independent vendors; Part III summarizes more than 130 different microcomputers and microcomputer systems from over 50 suppliers; and Part IV includes a summary on a selection of terminals and printers.

1845. MICROCOMPUTERS: A TECHNOLOGY FORECAST AND ASSESSMENT TO THE YEAR 2000. *Wise, Kenneth. John Wiley and Sons, Inc. 1980. 251 pg. $19.95.*

Forecasts microcomputer technology and its social impact to 2000. Coverage spans integrated electronics (state of the art); technology forecasts of microcomputer peripheral hardware and microcomputer architecture; and software and systems. Examines microcomputers' impact on the national aviation system and policy implications. Contains a glossary.

1846. MICROCOMPUTERS AND LIBRARIES: A GUIDE TO TECHNOLOGY, PRODUCTS AND APPLICATIONS. *Roruig, Mark E. Knowledge Industry Publications. 1981. 134 pg.*

Discusses library applications of microcomputers, with case studies. Provides technical and feature / capability comparisons of eleven common microprocessors, in tabular format. Also discusses computer technology at a general level. Includes a list of microcomputer manufacturers and their addresses.

1847. MICROCOMPUTERS IN PATIENT CARE. *Eden, Henry S. Noyes Data Corporation. 1981. 191 pg. $28.00.*

Analysis of microcomputer use in medical care covers technology and applications. Examines microcomputer technology advantages and applications; the medical milieu; the technology's effects on the health care system; computers in medicine; and providing intelligent medical advice. Includes improving physician productivity; arrhythmia monitoring; anesthesia delivery; portable microcomputer-based instrumentation; and limb control and power braces.

1848. MINICOMPUTERS IN LIBRARIES, 1981-82. *Grosch, Audrey. Knowledge Industry Publications. Triennial. 2nd ed. 1981. 263 pg. $34.50.*

Covers issues, approaches, and experiences of libraries involved with minicomputer applications, with particular emphasis on distributive data processing. Directory of installed systems covers 186 U.S. and non-U.S. libraries.

1849. POTENTIAL HEALTH HAZARDS OF VIDEO DISPLAY TERMINALS. *Public Health Services, U.S. Department of Health and Human Services. U.S. Government Printing Office. SuDoc # HE 20.7111:v.66. June 1981. 75 pg. $4.00.*

Reports on National Institute for Occupational Safety and Health's study of video display terminal users, 1979. Covers radiation, industrial hygiene, health complaints, and ergonomics. Includes a bibliography.

1850. SHOULD THE POSTAL SERVICE OFFER ELEC-TRONIC MAIL? *House Committee on Government Operations, 97th U.S. Congress, 1st Session. U.S. Government Printing Office. SuDoc # Y4.G74/7:P84/7. 1982. 308 pg.*

Gives proceedings of U.S. House subcommittee hearing on postal service use of electronic mail, October 5, 1981. Cost statistics interspersed with narrative text. Gives excerpts from electronic computer originated mail market study for U.S. Postal Service by Opinion Research Corporation.

1851. SYSTEMS DEVELOPMENT PRODUCTIVITY. *International Data Group. 1981.*

Study of motivation and labor productivity of systems development employees.

1852. USING MICROCOMPUTERS IN BUSINESS: A GUIDE FOR THE PERPLEXED. *Veit, Stanley. Hayden Book Company, Inc. 1981. 142 pg. $9.95.*

This guide to purchasing microcomputers and software discusses the needs of different types of businesses, the advantages and disadvantages of computerization, specific computer applications, how to determine the scale of hardware required, package versus custom software, different sorts of vendors and consultants, and contracts. Also explains computer and disk drive operation, ASCII and other codes, and typical system specifications.

1853. WORD PROCESSING SYSTEMS USER RATINGS. *Datapro Research, Inc. (McGraw-Hill). 1982. $25.00.*

Analyzes 4000 word processing users' ratings of over 40 makes and models of equipment from 25 vendors of standalone and multiterminal word processors, with accompanying vendor summaries.

1854. WORD PROCESSORS AND INFORMATION PROCESSING: A BASIC MANUAL ON WHAT THEY ARE AND HOW TO BUY. *Poynter, Dan. Para Publishing. Semi-annual. 2nd ed. 1982. 172 pg. $11.95.*

Introductory text outlines benefits of word processing and the problems of choosing the appropriate machine. Includes directories of manufacturers, supply houses, books, pamphlets, magazines, newsletters and report services, courses and seminars, media conversion services / printers and associations. Glossary also provided.

1855. 16 BIT MICROPROCESSORS. *Howard W. Sams and Company, Inc. Irregular. 1981. $14.95.*

Compares specifications of 'the most popular' 16-bit microprocessors, including degree of conformity to key software standards.

See 400, 709, 2799, 2802, 2803, 3361, 4246, 4247, 4253, 4254, 4265, 4266, 4267, 4268.

Conference Reports

1856. AUTOFACT 3 CONFERENCE PROCEEDINGS. *Society of Manufacturing Engineers. 1981. 688 pg. $45.00.*

Conference proceedings cover productivity growth leadership, implementation of CAD / CAM in industry, economic considerations of CAD / CAM, producibility by design, applications of group technology, quality planning, human factors, robots, test and inspection, and other topics. Charts, diagrams, tables.

1857. CAN VIDEOTEX BE PROFITABLE: A LEGAL, TECHNOLOGICAL AND USER ASSESSMENT. *Kelly, Bruce, ed. Hoke Communications. 1982. $22.00.*

Cassette recording of research and policy implications panel discussion at a 1982 information utilities conference addresses questions pertaining to legal barriers, technological acceptance, perceived user needs, and the viability of videotex as a mass market delivery system. Topics include Evaluation Results and Policy Implications of the Green Thumb Videotex Experiment; and Videotex and the Potential Market.

1858. COMMUNICATING INFORMATION: PROCEEDINGS OF THE AMERICAN SOCIETY FOR INFORMATION SCIENCE ANNUAL MEETING, VOLUME 17, 1980. *Benenfeld, Alan R., editor. Knowledge Industry Publications. 1980. 417 pg. $19.50.*

Conference proceedings on information science include electronic messaging and office automation; video conferencing; user characteristics; high speed text searching; organizational communication for corporate decision-making; microcomputer and video disc library applications; online reference services; online library catalogs and data bases; and information management and productivity.

1859. COMPUTER MANUFACTURER'S ROLE IN THE INFORMATION INDUSTRY. *McDonald, Allan, ed. Hoke Communications. 1982. $7.00.*

Cassette recording of information utilities 1982 conference discussion focuses on new technologies such as teletext and videotex in terms of evolution of hybrid systems, and how computer companies in general and Digital in particular are embracing this evolutionary process.

1860. MASS MARKET FOR THE SMALL COMPUTER. *FIND / SVP. January 1981. 168 pg. $100.00.*

Seminar transcript from New York Society of Security Analysts considers new methods of distribution; software and databases; and the selling environment. 24 companies made presentations.

1861. MICROCOMPUTERS: BIG BUCKS FROM LITTLE BOXES. *Hoke Communications. 1982. $15.00.*

Information utilities 1982 panel discussion on cassette recording addresses growing array of business-oriented microcomputer applications, personal applications, and peripheral industries that have been spawned by the microcomputing industry. Covers the information revolution as it effects small business, technological options for coping with it, and an overview of typical systems and how to choose one. Other topics include Microcomputing: Adding Profits to the Intelligent and Informed Business User; and Impact of New Informational Utilities on the Consumer.

1862. THE PRINT PUBLISHER IN AN ELECTRONIC WORLD: DECEMBER 1980 PROCEEDINGS. *Knowledge Industry Publications. 1980. $95.00.*

Conference proceedings include the traditional communications company and the new information technologies; public policy issues; how major publishers are preparing for electronic publishing; the educational publisher and electronic distribution; emerging education markets for information technologies; consumer markets for new technologies; and the electronic newspaper, 24-hour news service.

1863. PROCEEDINGS OF THE SECOND NATIONAL ONLINE MEETING. *Learned Information, Inc. 1981. 554 pg. $42.00.*

Seventy-seven papers from March 1981 conference cover such topics as how to introduce and run online services, online industry growth, and forecasts.

1864. PROCEEDINGS OF THE THIRD NATIONAL ONLINE MEETING. *Learned Information, Inc. 1982. $52.00.*

Approximately 70 papers from March 30 - April 1, 1982 conference cover chemical structure research systems; teletext and videotex systems; techniques for online system management; and education and training of online searches.

1865. TRENDS AND OUTLOOK FOR LARGE COMPUTER SYSTEMS, PERIPHERALS AND SOFTWARE MARKETS. *MSRA, Inc. October 1981. 257 pg. $200.00.*

Conference reports from computer corporations covers Japanese competition, marketing strategies of major vendors, analysis of markets and technology, and outlook for software services.

See 2812, 2813, 4273, 4274, 4277.

Newsletters

1866. COMPUTER MARKETING NEWSLETTER. *MV Publishing Inc. Monthly. 12 pg. $96.00/yr.*

Covers such topics as top executives' compensation by firm; market share strategies; monthly and annual growth rate leader rankings; listing of vendors recently introducing micros with portable OS's; industry events round-up; evaluation of industry directories; industry's independent sales organizations; job and business opportunities; and specific industry people.

1867. COMPUTING NEWSLETTER FOR SCHOOLS OF BUSINESS. *Couger, J. Daniel, ed. University of Colorado. Business Research Division. Monthly. v.15, no.7. 9 pg.*

Reviews new publications on computers and associated areas (e.g. databases). Also covers meetings and conferences.

1868. DATA ENTRY AWARENESS REPORT. *Management Information Corporation. Monthly. $295.00/yr.*

Evaluates a data entry system, including key-to-disc, intelligent terminals, direct data entry, optical character readers, portable data recorders, voice data entry, and application data entry. Descriptions include hardware, software, prices, and advantages and disadvantages.

1869. DATACOMM AND DISTILLED PROCESSING REPORT. *Management Information Corporation. Monthly. $295.00/yr.*

Concentrates on larger business minicomputers and mid-size computer systems. Describes and evaluates distributed processing systems and communications networks. Descriptions cover hardware, software, communications, prices, and advantages and disadvantages.

1870. DATAPRO NEWSCOM. *Datapro Research, Inc. (McGraw-Hill). Monthly. 4 pg. $65.00/yr.*

Newsletter covering the computer and software industry. Includes new products of vendors and changes in prices.

1871. EDP INDUSTRY REPORT. *International Data Group. Bi-weekly. $125.00/yr.*

News of electronic data processing industry and markets, proprietary statistics and forecasts. Covers new products and legal decisions.

1872. THE HIRSCH REPORT. *Hirsch, Phil. Hirsch Organization, Inc. Monthly. 13 (est.) pg. $80.00.*

Monthly narrative survey of new developments in data telecommunications technology, services and regulation, in North America and Western Europe. Discusses activities of major vendors, and technical aspects and applications of specific systems such as videotext / teletext.

1873. IMPACT: INFORMATION TECHNOLOGY. *Administrative Management Society. Monthly. 12 pg. $90.00/yr.*

Discusses human resources, word processing including local network designs, videotext, office automation economies, and digital data switching, data processing including small business computers and elimination of 'garbage-in, garbage-out.' Covers nonpaper systems including records problems and micrographic correspondence management, and work-flow - workspace. Other departments examine trends, telecommunications, and education. Lists new literature including abstracts.

1874. TRADE-A-COMPUTER. *Trade-A-Computer. Monthly. 8 pg.*

Primarily contains classified advertisments by individuals wanting to buy or sell used computer hardware or software.

See 3365, 4283.

Services

1875. ADL IMPACT SERVICES: ELECTRONIC DISPLAY DEVICES: MARKETS, APPLICATIONS AND TECHNOLOGY. *Alam, Mahbub U. Arthur D. Little, Inc. May 1979. 25 pg. $1800.00.*

1979 report discussing digital and alphanumeric displays (calculators, digital watches & automobiles) and large screen displays (cathode ray tubes as terminals and planar gas discharge displays). Market areas (products), estimated demand by U.S., Japan and Western Europe, applications, growth rate and major manufacturers of displays presented. Data derived from Arthur D. Little, Inc. Publication part of ADL Impact Services.

1876. DATAPRO APPLICATIONS SOFTWARE SOLUTIONS. *Datapro Research, Inc. (McGraw-Hill). Monthly. 1981. $365.00.*

Monthly service includes guide to software products, trade directories, applications directories, user ratings, vendors and suppliers, user's guide, inquiry service, software development concepts, industry applications, planning and cost justification, make or buy tradeoffs, software design, software production, selection / acquisitions, reliability / vendor support, installation / testing, maintenance, performance measurement, future systems and a glossary, plus a monthly newsletter.

1877. DATAPRO CUSTOM TELEPHONE CONSULTING SERVICE. *Datapro Research, Inc. (McGraw-Hill). 1981.*

Direct-access telephone consulting service includes new product announcements, supplementary information to reports contained in three-volume 'Datapro 70', and personal consultation on EDP and office product related problems.

1878. DATAPRO MANAGEMENT OF SMALL COMPUTER SYSTEMS. *Datapro Research, Inc. (McGraw-Hill). Monthly. $360.00/yr.*

Service includes basic computer system and data processing concepts, applications, vendor directory (arranged alphabetically by product, with addresses and telephone numbers), and user ratings. Also includes two loose-leaf volumes, monthly report supplements, monthly newsletter, and a telephone / telex inquiry service.

1879. DATAPRO REPORTS ON BANKING AUTOMATION. *Datapro Research, Inc. (McGraw-Hill). Bi-monthly. 1981. $390.00.*

Bimonthly service on banking automation industry provides data on product features, prices, and performance.

1880. DATAPRO REPORTS ON EDP SOLUTIONS. *Datapro Research, Inc. (McGraw-Hill). Monthly. 1981. $350.00.*

Monthly service includes EDP directory along with sections covering user's guide, inquiry service, industry briefs, management and administration, systems development, operations, technical support, selection and acquisition, user ratings, and glossary.

1881. DATAPRO REPORTS ON MINICOMPUTERS. *Datapro Research, Inc. (McGraw-Hill). Monthly. 2000 pg. $645.00/yr.*

Providing product descriptions, specifications, case histories, user ratings, and evaluations of microcomputers, minicomputers, and microprocessors, annual subscription service includes three loose-leaf volumes, monthly report supplements, monthly newsletter, and a telephone / telex inquiry service.

1882. DATAPRO REPORTS ON OFFICE SYSTEMS. *Datapro Research, Inc. (McGraw-Hill). Monthly. 1981. $590.00.*

Monthly service covering office products, systems, and techniques, including word processing equipment, copiers, data processing services, microform systems, telephone and voice communication systems, addressing and labeling machines, facsimile devices, and calculators. Coverage spans product / price comparison charts, user ratings, management summaries, "how to" pieces, and case histories.

1883. DATAPRO REPORTS ON RETAIL AUTOMATION. *Datapro Research, Inc. (McGraw-Hill). Bi-monthly. 1981. 400 pg. $390.00.*

Bimonthly service on automation in the retail trade industry includes comparision tables on integrated POS systems, electronic cash registers, EFTS, credit and payment systems, vendors, applications, specialized equipment, and software.

1884. IBM USER SURVEY: FIRST QUARTER 1980. *International Data Group. Quarterly. April 1980. 20 pg.*

Results of telephone survey to IBM users gives user reactions to the announcement of the Newport 3033N system, "H" and "Whitney-Coronado" products. Records and forecasts IBM market penetration.

1885. INFORMATION SYSTEMS ECONOMETRIC SERVICE: A NEW APPROACH TO FORECASTING THE COMPUTER INDUSTRY. *Gnostic Concepts, Inc. Quarterly. 1981.*

Service provides data to analyze current and future product opportunities, market prospects, impact of technology trends, implications of changing economic conditions, and potential competition in the computer industry. Features a computerized model of the industry covering equipment markets, EDP systems markets, industry-by-industry EDP expenditures, systems installed base, and EDP supplier revenues. Services include quarterly reports, data base, and consultation.

1886. PLANNING, SELECTING AND MANAGING REMOTE COMPUTING SERVICES. *Newton Evans Research Company. Monthly. 737. 1981.*

Two-volume subscription service with monthly updates offers information on the remote computing services industry. Service includes guide to planning, selecting, and managing remote computing services.

1887. VIDEOTEXT MONITORING SERVICE: CONTINUOUS INFORMATION SERVICE. *Link Resources Corporation. Monthly. 1981. $15000.00.*

A continuous information service covering the U.S. videotext and teletext industries from 1979 forward. Examines hardware trends, design factors, applications, carriers, and marketing.

Includes bi-monthly reports; monthly memoranda; news briefs; viewdata, videotext reports; and ongoing consultations.

See 406, 722, 3368, 3369, 4287, 4288.

Indexes and Abstracts

1888. A BIBLIOGRAPHY OF DOCUMENTS ISSUED BY THE GENERAL ACCOUNTING OFFICE ON MATTERS RELATED TO ADP, 1976-1980. *U.S. General Accounting Office. U.S. Government Printing Office. GA1.16:Au2/2. September 1981. 453 pg. $3.25.*

Abstracts and gives bibliographic information on G.A.O. audit reports, staff studies, speeches, testimonies, Comptroller General decisions, and other ADP-related documents, 1976-1980. Indexed by subject, agency or organization, and document number.

1889. A BIBLIOGRAPHY OF DOCUMENTS ISSUED BY THE GENERAL ACCOUNTING OFFICE ON MATTERS RELATED TO AUTOMATIC DATA PROCESSING, 1981. *U.S. General Accounting Office. U.S. Government Printing Office. SuDoc # GA1.16:D65. April 1982. 180 pg. $3.25.*

Supplements earlier version of same title, 1976-1980.

1890. COMPUTER LITERATURE INDEX. *Applied Computer Research. Quarterly. 1982. $75.00/yr.*

Bibliographic service covers over 100 periodicals, books, and special reports classified under 300 subject categories. Subscription includes annual cumulation.

1891. COMPUTER LITERATURE INDEX. *Applied Computer Research. Quarterly. 1981. $75.*

The quarterly index to computer literature covers over 100 periodicals, books and special reports. Coverage is generally limited to computer-related trade publications, general business and management periodicals, and publications of computer and management oriented professional groups; highly academic computer literature is not included. Entries are classified into 300 reference categories and are listed alphabetically within those categories. Author and publisher indexes are included.

1892. COMPUTING REVIEWS. *Association for Computing Machinery. Monthly. 1981. $40.00/yr.*

Abstracts are arranged by category (general topics and education; computing milieu; applications; software; mathematics of computation; hardware; analog computers; and functions). Coverage includes books, periodicals, Ph.D and Master's theses, tape recordings, movies and exhibits. Includes an author index and a list of periodicals received.

1893. MICRO COMPUTER INDEX. *Microcomputer Information Services. Quarterly. v.3, no.1. January 1982. 127 pg.*

Index of abstracts of articles on microcomputers includes index by subject and bibliographic information for the quarter by issue. Includes business addressess of periodicals.

1894. VENTURECASTS: THE 1982 COMPUTER / COMMUNICATION DATA BOOK. *Venture Development Corporation. 1982. $595.00.*

Provides abstracts of market analysis and forecasts from 1981 electronics and business periodicals, and association's and government statistics covering computers and communications equipment in the Americas. Includes sales, shipments, revenues, installed base, annual growth rates, consumption, expenditures, market size, production, and imports and exports by product groups.

1895. VENTURECASTS: THE 1982 CONSUMER ELEC-TRONICS DATA BOOK. *Venture Development Corporation. 1982. $595.00.*

Provides abstracts of market analysis and forecasts from 1981 electronics and business periodicals, and associations' and government statistics covering consumer electronics in the Americas. Includes sales, shipments, revenues, installed base, annual growth rates, consumption, expenditures, market size, production, and imports and exports by product groups.

1896. 1980 ACM GUIDE TO COMPUTING LITERATURE. *ACM. Annual. 1982. 800 (est.) pg. $50.00.*

Annual index covers 15,439 proceedings, papers, and reports from 18,000 authors. Indexed by titles, authors, key words, topics, sources, a

See 3774.

Dictionaries

1897. DICTIONARY OF NEW INFORMATION TECHNOLO-GY. *Meadows, A.J. Kogan Page Ltd. 1982. 206 pg. $27.50.*

Dictionary provides definitions of information processing technology terminology.

1898. GLOSSARY OF 1400 EDP TERMS. *Datapro Research, Inc. (McGraw-Hill). 1981. 70 pg. $19.00.*

A reference to more than 1400 terms that are relevant either to data processing products and systems or to the management of data processing activities. The definitions reflect current usage, as well as conforming to 'official' definitions published by ANSI, NBS and other recognized sources. Selected from Datapro EDP Solutions.

1899. MICROCOMPUTER DICTIONARY. *Sippl, Charles J., ed. Bobbs Merrill Educational Publishing. 2nd ed. August 1981. $15.95.*

Defines terms related to microelectronics, microcomputer hardware and software, and the processes of electronic computing.

1900. WORD PROCESSING COMMUNICATIONS GLOSSA-RY. *Teleprocessing Products, Inc. 1981.*

Designed to help computer personnel and management effectively communicate needs, possibilities and options to one another. Provides non-technical definitions of 'word processing' technology.

See 724, 3373, 3374, 4290.

Central and South America

Market Research Reports

1901. COMPUTERS AND PERIPHERAL EQUIPMENT, BRA-ZIL: A COUNTRY MARKET SURVEY. *U.S. Government Printing Office. Irregular. CMS 80-304. February 1980. 11 pg. $.50.*

Report analyzes the current and projected markets for the sale of computers and peripheral equipment in Brazil. Data is provided on the total market (production, imports, exports, market size); number and value of installed computers by major user sectors; total employment and number of establishments, by size, in major user sectors; percentage breakdown by end-user applications; imports by country of origin; value of installed

computers by major suppliers. Describes the domestic market, major market trends, and the competitive environment.

1902. COMPUTERS AND PERIPHERAL EQUIPMENT, VEN-EZUELA: A COUNTRY MARKET SURVEY. *U.S. Government Printing Office. Irregular. CMS 81-312. March 1981. 10 pg. $.50.*

Report analyzes the current and projected markets for the sale of computers and peripheral equipment in Venezuela. Data is provided on the market by product category; number and value of installed computers by major user sectors; expenditures by major user sectors; total employment and number of establishments, by size, in major user sectors; percentage breakdown by end-user applications; imports by country of origin; value of installed computers by major suppliers. Describes the domestic market, major market trends, and the competitive environment.

See 410, 3782.

Western Europe

Market Research Reports

1903. APPLICATION UNIQUE TERMINAL MARKETS 1981-1987. *IDC Europa Ltd. 1982. 150 (est.) pg. 1250.00 (United Kingdom).*

Report on the present and future state of the application unique terminal markets in Western Europe, 1981-1987 analyzes market size, market share, and market projections on point-of-sale and banking terminals. Provides profiles of major vendors focusing on their strengths, weaknesses and application specialization. Examines new product development and industry opportunities. Includes analyses of user survey focusing on use of and level of satisfaction with current products and provisions for new and extended usage.

1904. AUTOMATED OFFICE OVERVIEW. *IDC Europa Ltd. March 1981. 119 pg. $3000.00.*

Market analysis for the automated office in Europe includes management summary; vendor classifications; forecasts to 1990; the general situation in individual countries; and specific techniques.

1905. AUTOMATION IN RETAILING (EPOS) IN EUROPE. *Frost and Sullivan, Inc. June 1982. 378 pg. $1500.00.*

Analysis of markets for electronic point of sale (EPOS) systems among retailers in Europe includes market penetration; installed base, 1982; forecasts of shipments, values of shipments and unit sales, 1982-1987; retail distribution; industry structure; prices; factors in purchases; consumer attitudes and market surveys; and manufacturers' product lines and software services.

1906. COMMUNICATION SERVICES WESTERN EUROPE. *IDC Europa Ltd. January 1981. 143 pg. 1250.00 (United Kingdom).*

Market analysis (1980) of communication services in Western Europe includes product segmentation; PTT provisions and plans; vendor network offerings; users' current and future network offerings; data transmission facilities; and user survey questionnaire.

1907. COMMUNICATIONS NETWORKS AND NETWORK COMPONENTS: A MULTI-CLIENT STUDY. *Creative Strategies International. 1981. $9500.00.*

Multi-client study of likely trends in the design and operation of public and corporate communication networks in the 1980's. Aims to develop five-year market forecasts, in units and dollars, for modems, multiplexors, concentrators, front-end processors, terminals, and other components.

1908. COMPUTER AND WORD PROCESSING PRINTER MARKET IN EUROPE. *Frost and Sullivan, Inc. December 1981. 385 pg. $1550.00.*

Analyzes and forecasts markets through 1985 in sales volume, number of units and unit price for specific types of computer and data processing printers in three major categories: impact serial dot matrix printers, fully-formed character serial printers, impact line printers, and high-speed non-impact page printers. Evaluates major printer suppliers, their products, market position, and marketing strategies.

1909. COMPUTER GRAPHICS BUSINESS EQUIPMENT IN EEC. *Frost and Sullivan, Inc. August 1981. $1500.00.*

Contains a 10-year forecast and analysis by country for graphic displays, input and output devices and turnkey systems by these end user markets: banking and insurance, publishing, advertising, accounting, and data processing. Software requirements are addressed; vendor profiles are given; and market shares are forecasted. Gives applications analysis.

1910. COMPUTER GRAPHICS IN EEC INDUSTRIAL MARKETS. *Frost and Sullivan, Inc. May 1981. 324 pg. $1400.*

Contains a 10-year forecast and analysis bycountry for input, operator, output devices and displays including digitizers, optical readers, keyboards, light pens, joysticks, graphic tablets, plotters and computer output microfilm. Forecasts vendor profiles and market shares. Identifies end-user industries and addresses software requirements.

1911. THE COMPUTER MARKET 1980 IN THE U.K. AND THE CONTINENT OF EUROPE. *G.G. Baker and Associates. July 1980. 55 pg. 125.00 (United Kingdom).*

Market analysis for 1980 for the COM recorders, COM camera film, and duplicating films.

1912. COMPUTER PERIPHERALS IN WESTERN EUROPE: VOLUME 1: EUROPEAN MARKETS AND MARKET TRENDS. *Mackintosh Consultants, Inc. October 1980. 110 pg. $2400.00.*

Evaluates Western European (principally French, British and German) markets for computer peripherals through 1989, in terms of volume of sales, value and installed base. Statistics are broken down by product type. Discusses market structure and shares and market segmentation.

1913. COMPUTER PERIPHERALS IN WESTERN EUROPE: VOLUME 2: WORLDWIDE TECHNOLOGY AND PRODUCT TRENDS. *Mackintosh Consultants, Inc. October 1980. 238 pg. $2400.00.*

Survey of current technology in computer peripherals. Covers terminals, printers, displays, and memory systems. Discusses major manufacturers and their activities, and price trends. Complements European and global market research reports in same series.

1914. COMPUTER PERIPHERALS IN WESTERN EUROPE: VOLUME 3: PROFILE OF THE WORLDWIDE PERIPHERALS INDUSTRY. *Mackintosh Consultants, Inc. October 1980. 187 pg. $2400.00.*

Completed multi-client study profiles major European manufacturers of computer peripherals and their activities, as of

1980. Focuses on the French, German and British industries. Discusses market segmentation and outlook, and computer industry trade flows and balances.

1915. COMPUTER TERMINALS IN EUROPE. *Frost and Sullivan, Inc. January 1981. 907 pg. $1,400.00.*

Market analysis through 1989 covers hardware and software shipments; new products on the market; country-by-country analysis; and breakdowns by terminal type.

1916. COMPUTERS AND PERIPHERAL EQUIPMENT, BELGIUM: A COUNTRY MARKET SURVEY. *U.S. Government Printing Office. Irregular. CMS 81-314. September 1981. 10 pg. $.50.*

Report analyzes the current and projected markets for the sale of computers and peripheral equipment in Belgium. Data is provided on the total market (production, imports, exports, market size); number and value of installed computers by major user sectors; expenditures by major user sectors; total employment and number of establishments by size in major user sectors; percentage breakdown by end-user applications; imports by country of origin; value of installed computers by major suppliers. Describes the domestic market, major market trends, and the competitive environment.

1917. COMPUTERS AND PERIPHERAL EQUIPMENT, FRANCE: A COUNTRY MARKET SURVEY. *U.S. Government Printing Office. Irregular. CMS 81-313. July 1981. 11 pg. $.50.*

Report analyzes the current and projected markets for the sale of computers and peripheral equipment in France. Data is provided on the total market (production, imports, exports, market size); number and value of installed computers by major user sectors; expenditures by major user sectors; total employment and number of establishments by size in major user sectors; percentage breakdown by end-user applications; imports by country of origin; value of installed computers by major suppliers. Describes the domestic market, major market trends, and the competitive environment.

1918. COMPUTERS AND PERIPHERAL EQUIPMENT, SPAIN: A COUNTRY MARKET SURVEY. *U.S. Government Printing Office. Irregular. CMS 81-315. November 1981. 11 pg. $.50.*

Report analyzes the current and projected markets for the sale of computers and peripheral equipment in Spain. Data is provided on the total market (production, imports, exports, market size): imports by country of origin; total employment and number of establishments by size in major user sectors; expenditures by major user sectors; number and value of installed computers by major user sectors; percentage breakdown by end-user applications; value of installed computers by major supplier. Describes the domestic market, major market trends, and the competitive environment.

1919. COMPUTERS AND PERIPHERAL EQUIPMENT, SWEDEN: A COUNTRY MARKET SURVEY. *U.S. Government Printing Office. Irregular. CMS 80-307. February 1980. 12 pg. $.50.*

Report analyzes the current and projected markets for the sale of computers and peripheral equipment in Sweden. Data is provided on the total market (production, imports, exports, market size); imports by country of origin; number and value of installed units by major user sectors; total employment and number of establishments by size in major user sectors; expenditures by major user sectors; percentage breakdown by end-user applications; value of installed computers by major suppliers. Describes the domestic market, major market trends, and the competitive environment.

1920. COMPUTERS AND PERIPHERAL EQUIPMENT, UNITED KINGDOM: A COUNTRY MARKET SURVEY. *U.S. Government Printing Office. Irregular. CMS 79-301. March 1979. 11 pg. $.50.*

Report analyzes the current and projected markets for the sale of computers and peripheral equipment in the United Kingdom. Data is provided on the total market (production, imports, exports, market size); number and value of installed computers by major user sectors; expenditures by major user sectors; total employment and number of establishments by size in major user sectors; percentage breakdown by end-user applications; value of installed computers by major suppliers. Describes the domestic market, major market trends, and the competitive environment.

1921. COMPUTERS AND RELATED EQUIPMEMT - ITALY: A FOREIGN MARKET SURVEY REPORT. *Directa, Ricerche e Marketing Internazionale. U.S. Department of Commerce. International Trade Administration. NTIS #ITA-82-05-513. November 1981. 106 pg. $15.00.*

Survey on the market in Italy for computers and related equipment describes the market size and special market influences. Provides data and analysis on market size (production, imports, exports) by major equipment subcategories. Analyzes each specific product market according to current size and future potential. Provides competitive assessment of U.S. and third country suppliers, with data on imports by major subcategory and country of origin, both by value and market share. Describes principal end user sectors with current and future needs detailed. A section of the survey describes marketing practices, trade restrictions, and technical requirements. Trade lists include principal potential agents and distributors; trade, industrial, and professional associations; publications; and major national and international trade fairs and commercial exhibitions.

1922. COMPUTERS AND RELATED EQUIPMENT - NORWAY. *Konsulterna AB. U.S. Department of Commerce. International Trade Administration. NTIS #ITA-82-04-502. October 1981. 55 pg. $15.00.*

Analyzes the Norwegian market for computers and related equipment. Describes the current and potential markets for minicomputers, small, medium, and large computer systems, peripherals, and data communications equipment as groups and by specific products. Provides a competitive assessment of U.S., domestic, and third country suppliers, describing specific manufacturers and their products. Analyzes specific end user sectors, and gives data by sector on number and value of installed equipment. Section on market access data provides information on business practices, trade restrictions, and technical requirements. Lists principal trade associations and potential agents and distributors.

1923. COMPUTERS IN EUROPE TO 1986: ELECTRONICS ELECTRICAL SERIES. *Larsen Sweeney Associates Ltd. Irregular. 1981. 261 pg.*

Analyzes and forecasts market conditions for the computer industry in Austria, Belgium, Luxembourg, Denmark, Ireland, Finland, France, West Germany, Italy, the Netherlands, Norway, Portugal, Spain, Sweden, Switzerland and the United Kingdom to 1986. Gives market areas, market analysis, consumption, growth rates, distribution of customer base, channels of distribution, trade balance, product groups, and research and development. Includes market segmentation, suppliers, ratios, cost of sales, sales, production, investments, employment, profits, assets, capital, employees, end users, and dealers. Covers eighteen product groups and is based on OECD Econometric Model, PIMS type industry model, medium term forecasts and surveys of 432 computer manufacturing

companies, 5,332 end user companies, and 452 importers, distributors or dealers.

1924. THE DATA COMMUNICATIONS MARKET IN WESTERN EUROPE, 1981-1987. *Online Publications, Ltd. 1981. 195.00 (United Kingdom).*

Presentations and 1981-1987 forecasts made by Logica team at March 1981 Online seminar to identify characteristics of European data communications market in light of Eurodata 79 study by the 17-country Eurodata Foundation. Based on findings of a survey of nearly 3,000 users. Contents include market characteristics, size, and growth; data communications development by country and industry sector; development of the data communications equipment market; developments in user terminals; distribution over terminal classes; local and remote terminals; developments in the communications processor market; integration trends; communications processors markets in the 1980's; user applications; forecasting; user choices of networks and services; the future of wideband data transmission in western Europe; market by country; developing applications and technology; and competitive aspects.

1925. DATA PROCESSING OPPORTUNITIES IN MANUFACTURING: WESTERN EUROPE. *IDC Europa Ltd. May 1982. 92 pg. 1250.00 (United Kingdom).*

Forecast report showing, for each Western European region, the market size and structure and potential promised by manufacturing industries for the benefit of the computer industry. Principal vendors are profiled, revealing strategies and market strength. Emphasis is placed on determining user attitudes and relating this to their particualr requirements and purchasing intentions. Focuses on the 205,500 manufacturing establishments in Western Europe that employ more than 20 people.

1926. DESKTOP MICROCOMPUTER MARKETPLACE WESTERN EUROPE 1981-1987. *IDC Europa Ltd. July 1982. pg.162 1250.00 (United Kingdom).*

Report on the desktop / personal computer markets in Western Europe, 1981-1987 analyzes market share, market size, and projected shipments by country, as well as the activities of leading suppliers. Examines the growth of the desktop computer in the commercial environment, and assesses the effect of desktop computers on traditional small business system and minicomputer markets.

1927. DESKTOP / PERSONAL COMPUTER MARKETS IN WESTERN EUROPE, 1979-1983. *IDC Europa Ltd. November 1980. 100 pg. 995.00 (United Kingdom).*

Analysis of desktop / personal computer markets in Western Europe for 1979-1983 includes product and market segmentation; market factors of price, software and end user awareness; market sectors; distribution methods; vendor profiles and strategies; country profiles; and shipments for each end use sector.

1928. DISTRIBUTED DATA PROCESSING MARKETS, WESTERN EUROPE, 1979 - 1983. *IDC Europa Ltd. January 1980. 90 pg. 995.00 (United Kingdom).*

Market analysis of distributed data processing markets in Western Europe for 1979-1983 includes European overview; market forecasts, growth rates, segmentation by country; European market profile; strategies of suppliers and vendors; and user survey analysis.

1929. EDP USER EXPENDITURE ANALYSIS: WESTERN EUROPE, 1979-1984. *IDC Europa Ltd. January 1981. 169 pg. 2500.00 (United Kingdom).*

EDP user expenditure analysis for Western Europe for 1979-1984 includes current user expenditures (1979); market movements (1978-1983); country and regional overview (1976-

1979); forecasts (1979-1984); budgets for 1979-1980 aggregated by country, processor size, and industry sector.

1930. ELECTRONIC DOCUMENT STORAGE 1980-1990: VOLUME 1: EXECUTIVE REPORT. *Mackintosh Consultants, Inc. November 1980. $15000.00.*

Presents results of a multi-client market study of British, French, German, other European, and U.S. markets for digital and other electronic document storage systems, through 1990. Surveys user needs by end-use application, state-of-the-art and upcoming technolgies, major manufacturers, demand by market sector, supply and price trends, and installed base and value of shipments.

1931. ELECTRONIC DOCUMENT STORAGE 1980-1990: VOLUME 4: MARKET ENVIRONMENT AND FORECASTS. *Mackintosh Consultants, Inc. November 1980. $13500.00.*

Analyzes markets for electronic document storage systems as of 1985 and 1990, by detailed product, consumer and end-use categories. Gives projections (in terms ofshipment value and installed base) for West Germany, France, the U.K. and the U.S. Profiles major manufacturers and their likely roles.

1932. ELECTRONIC FUNDS TRANSFER MARKET IN EUROPE. *Frost and Sullivan, Inc. April 1981. $1400.00.*

Market analysis and forecasts for electronic funds transfer in Western European countries. Examines effects on banking industry structure from teleprocessed payment and exchange services. Examines equipment.

1933. EUROPEAN COMPUTER MARKET DATABOOK. *IDC Europa Ltd. January 1980. 468 pg. 2350.00 (United Kingdom).*

Market analysis of the Western European computer market gives 1978 data for market overview by individual country; installed base (1978) for general purpose computers, minicomputers, SBS computers; market shares; growth rates; peripherals; and systems software. Country market data includes national economic indicators.

1934. EUROPEAN COMPUTER MARKET FORECAST AND SPENDING PATTERNS, 1981-1987. *IDC Europa Ltd. 1982. 150 (est.) pg. 1250.00 (United Kingdom).*

Report on European computer markets with an emphasis on forecast and spending patterns to 1987 analyzes spending by category, supplier, country / region and by industry. Forecasts both short-term and long-term planned expenditures. Identifies reasons contributing to changes in user expenditure patterns. Determines structure of processor installed base ranging from personal computers to very large mainframes.

1935. EUROPEAN MARKET FOR COMMUNICATING TEXT TERMINALS. *Frost and Sullivan, Inc. July 1981. 289 pg. $1450.00.*

Forecast of European markets for communicating text terminals, 1981-1986. Shows dollar value, annual shipments, installations, and market shares; examines end use demand for services; profiles manufacturers; surveys user attitudes; and lists addresses of suppliers. Reviews economic indicators and regulations.

1936. EUROPEAN MARKET FOR DISTRIBUTED DATA PROCESSING SYSTEMS. *Frost and Sullivan, Inc. April 1981. $1450.00.*

Market report includes forecast and analysis of small,medium and large distributed data processing systems and applicable software products for 12 countries. Contains supplier profiles. Includes application and user analysis and technological changes. Contains impact for major general purpose computer suppliers.

1937. EUROPEAN MARKET FOR FACTORY DATA COLLECTION SYSTEMS. *Frost and Sullivan, Inc. March 1981. 219 pg. $1350.00.*

European market analysis and forecasts to 1990 for factory data collection terminals, peripherals, systems, software and production monitoring instruments. Gives unit sales and dollar value projections, market shares, vendors, and market survey analysis.

1938. THE FUTURE OF THE U.K. MARKET FOR DATA PROCESSING EQUIPMENT AND SERVICES: WHAT WILL USERS BE SPENDING? *Pedder Associates Ltd. May 1980. 2850.00 (United Kingdom).*

Four volume tabulation of planned expenditures for computer equipment, services, and employees in the U.K., with forecasts to 1983. Analyzes market growth areas and surveys user attitudes.

1939. GLOBAL INFORMATION NETWORKS. *Strategic Inc. June 1981. 360 pg. $1500.00.*

Analysis of markets for computer communications networks, regulations and utilization. Examines import controls and national legislation in the U.S., Japan and Western Europe; public versus private ownership; home and business services; and prospects for foreign trade.

1940. HARDWARE VENDORS' PACKAGED SOFTWARE MARKET IN WESTERN EUROPE, 1979 - 1984. *IDC Europa Ltd. April 1981. 151 pg. 1250.00 (United Kingdom).*

Market analysis for packaged software market in Western Europe gives revenues (1979-1984) by country, machine category, and manufacturer; vendor (1979-1984) profiles and strategies; user survey; market environment; and key economic indicators per region (1979-1984).

1941. IBM EUROPEAN MARKET REFERENCE BOOK. *IDC Europa Ltd. November 1981. 1250.00 (United Kingdom).*

Market analysis of IBM market in Western Europe (1981-1985) includes review of processor, software and software services. Includes shipments, market penetration, user survey, census of major models, installed peripherals, and reviews by country.

1942. IBM SOFTWARE ENVIRONMENT, 1981-1987. *IDC Europa Ltd. 1982. 150 (est.) pg. 1250.00 (United Kingdom).*

Report on IBM software environment in Western Europe 1981-1987. Analyzes major IBM models by country showing forecast six-year penetration, IBM revenues by country, and attitude of IBM to third party software, products and suppliers. Examines major independent suppliers and their standing within various software categories. Provides user survey including purchasing intentions and attitudes towards IBM and third party software suppliers.

1943. INDEPENDENT PACKAGED SOFTWARE MARKETS IN WESTERN EUROPE, 1979 - 1983. *IDC Europa Ltd. August 1980. 995.00 (United Kingdom).*

Market analysis of Western European packaged software markets provides software overview; vendors (1979-1983); user survey (1979-1983); customization data (1979-1983); product trends (1979-1983); forecasts (1979-1983); marketing methods (1979-1983); European vendors (1979-1983); and market size (1979-1983) by individual country.

1944. JAPANESE VENDORS IN THE EUROPEAN MARKETPLACE. *IDC Europa Ltd. August 1981. 1250.00 (United Kingdom).*

Market analysis for Japanese vendors in Western Europe for 1981-1985 includes market penetration, current markets, and marketing arrangements by country; supplier profiles and strategies; sales, support and software strategies; reaction to Japanese competition; attitudes of governments; andJapanese products in Europe.

1945. KEY NOTE REPORTS: MICROCOMPUTERS. *Keynote Publications Ltd. 1981. 20.00 (United Kingdom).*

A report on microcomputers in the U.K. covering the structure of the industry, distribution networks, major manufacturers, markets and finances. Analysis includes products by sales volume and value market shares, foreign trade, new products, mergers, legislation, financial analysis, operating ratios, and forecasts of trends.

1946. KEYBOARDS AND KEYSWITCHES IN EUROPE TO 1986. *Larsen Sweeney Associates Ltd. Semi-annual. 1981. $1850.00.*

Market analysis and forecasts to 1986 for computer keyboard and keyswitch product groups. Gives total consumption in terms of market value.

1947. MAGNETIC STORAGE PERIPHERAL EQUIPMENT (MSPE) MARKET IN EUROPE. *Frost and Sullivan, Inc. September 1981. 387 pg. $1550.00.*

Analysis of markets in Western Europe for magnetic storage peripheral equipment, with forecasts to 1985. Examines product groups and prices; distribution channels; market segmentation; national economic indicators; vendors and new products. Shows dollar sales, unit sales, and average unit price.

1948. MARKET OPPORTUNITIES IN SYSTEM NETWORKING 1981-1987. *IDC Europa Ltd. 1982. 150 (est.) pg. 1250.00 (United Kingdom).*

Report on market opportunities in system networking in Western Europe 1981-1987, analyzes current and projected market size and vendor share by country. Examines suppliers and products with an emphasis on hardware vendor network architectures and local area networks. Assesses user satisfaction with current offerings, and present and future net

1949. MARKET POTENTIAL FOR PRODUCTION TECHNOLOGIES IN THE MANUFACTURING AND CAPITAL GOODS INDUSTRY TO 1990. *Gewiplan GmbH. German/English. April 1981.*

Market analysis and forecasts to 1990 for new production technology products. Includes sensors, microprocessors, lasers, automatic test equipment, computer control systems, robots, and others.

1950. METALWORKING EQUIPMENT INCLUDING ROBOTS: A FOREIGN MARKET SURVEY REPORT. *Boss Enterprise AB. U.S. Department of Commerce. International Trade Administration. NTIS #ITA 82-05-510. March 1982. 30 pg. $15.00.*

Survey of the market in Sweden for metalworking equipment describes the current market situation and forecasts markets for the future. Analyzes the competitive environment for domestic, U.S., and other suppliers. Gives market size (domestic production, imports, exports) for various products, 1976-1980. Lists major Swedish manufacturers of machine tools; major end users; principal distributors; trade; journals; and trade, industrial, professional associations.

1951. MINICOMPUTER SYSTEMS MARKET IN EUROPE. *Frost and Sullivan, Inc. 1981. 950 pg. $1950.00.*

Examines minicomputer systems market in Europe. Forecasts hardware shipments to 1989. Covers growth rates; applications in word processing, industrial control, small business systems, and data communications; OEM and systems houses; and software and peripheral markets.

1952. MODEMS, MULTIPLEXERS, PROGRAMMABLE CONCENTRATORS AND NETWORK CONTROL MARKET IN EUROPE. *Frost and Sullivan, Inc. April 1982. 415 pg. $1750.00.*

Analysis of European markets for modems, multiplexers, programmable concentrators and network controls. Forecasts dollar and unit sales, unit prices, shipments and shipment values, 1981-1986, by product groups and countries. Examines industry structure; manufacturers, with financial analysis, mergers, distribution and marketing strategies; competition; market structure, new technologies; and regulations an ddata communications facilities in Europe.

1953. MULTIFUNCTION WP / DP TERMINAL MARKET IN EUROPE. *Frost and Sullivan, Inc. June 1982. 258 pg. $1250.00.*

Analysis of European markets for multifunction, word and data processing terminals. Examines market penetration, utilization, user attitudes, manufacturers, new technology, regulations, and economic indicators by country. Forecasts shipments, value of shipments, prices, and installed base, 1981-1986.

1954. OFFICE AUTOMATION AT THE TOP 500 EUROPEAN COMPANIES. *IDC Europa Ltd. 1982. 150 (est.) pg. 1250.00 (United Kingdom).*

Report on office automation at the top 500 companies in Western Europe examines current equipment, related applications, planned hardware implementation, and factors that influence pattern of usage. Assesses user selection process, the role of inter- and intra-office communications, and planned changes to communications systems.

1955. ON-LINE BRANCH BANKING SYSTEMS IN WESTERN EUROPE: VOLUME 1. *Frost and Sullivan, Inc. April 1980. 235 (est.) pg. $1350.00.*

Volume one of a two volume market analysis for on-line branch banking systems equipment in France, West Germany, Italy, the U.K., and Spain. Includes forecasts to 1989 for revenues and sales of products. Covers the installed base; market structure by end uses; and banking industry analysis by country including rankings of banks by assets, the number of banking businesses, and planned and existing installations.

1956. ON-LINE BRANCH BANKING SYSTEMS IN WESTERN EUROPE: VOLUME 2. *Frost and Sullivan, Inc. April 1980. 235 (est.) pg. $1350.00.*

Volume two of a two volume market analysis for on-line branch banking systems equipment in Western Europe. Covers banking industry structure, computer installations, and rankings and distribution by assets of banks for selected countries. Includes market survey reporting users' attitudes on banking facilities, 1982, and computer installations, 1979, 1982. Sales forecasts include products, installed base, shipments and sales volume to 1989. Lists manufacturers and products.

1957. THE OPTICAL MEMORY MARKET IN EUROPE. *Frost and Sullivan, Inc. January 1981. 252 pg. $1250.00.*

Analysis of Western European markets for new optical memory products. Includes sales forecasts to 1990, competition, utilization, manufacturers and product lines.

1958. THE OVER THE COUNTER (OTC) SMALL COMPUTER MARKET IN EUROPE. *Frost and Sullivan, Inc. April 1980. 700 pg. $1400.00.*

Analysis of Western European markets for over the counter (OTC) small computers, with forecasts to 1989. Review products, distribution channels, economic indicators, manufacturers, sales, and market shares.

1959. PORTABLE DATA RECORDERS MARKET IN EUROPE. *Frost and Sullivan, Inc. September 1981. 255 pg. $1500.00.*

Market analysis for portable data recorders in Western Europe. Examines economic indicators, installed base, end use markets, individual product prospects, market penetration potential, forecasts to 1990 for shipments and dollar values, manufacturers profiles and market shares, and suppliers' and users' attitudes.

1960. PRINTER MARKETPLACE, 1981 - 1985. *IDC Europa Ltd. August 1981. 1250.00 (United Kingdom).*

Market analysis for printers in Western Europe for 1981-1985 includes market size, share and forecast for character, line, and page printers; supplier profits and strategies; new technologies; channels of distribution; and printer usage by processor type.

1961. PROFESSIONAL DESK TOP COMPUTERS, PERIPHERALS AND SOFTWARE. *Frost and Sullivan, Inc. February 1982. $1500.00.*

Gives analysis and sales forecast to 1989 for desk top computers, mini and micro computers, electronic desk-top programmable calculators, and associated products, by country. Provides market shares by product and by country.

1962. RETAIL AUTOMATION IN WESTERN EUROPE. *Creative Strategies International. April 1979. $995.00.*

The markets for electronic cash registers and point-of-sale terminals in 13 West European countries. Unit forecasts through 1983 by equipment type and country-by-country penetration. Includes vendor market shares for systems registers. Discusses European retailing environment and its impact upon retail automation equipment configuration. Includes competitive analyses and profiles major vendors.

1963. SCIENTIFIC DESK TOP COMPUTER AND ASSOCIATED PERIPHERAL AND TERMINAL EQUIPMENT AND SOFTWARE MARKETS IN EUROPE. *Frost and Sullivan, Inc. April 1982. 625 pg. $2000.00.*

Analyzes European markets for desk top computers, terminals peripherals, and software. Examines market size and growth rates by country and by product groups. Forecasts sales, 1982-1987, and unit and dollar sales, 1989, 1992, based on the value of manufacturers' shipments. Includes market segmentation and integration, impacts of new technology, end uses, distribution and national economic indicators. Industry structure analysis covers vendors. market penetration, shares, and marketing strategies.

1964. SERVICES AND SOFTWARE: STATISTICAL REFERENCE BOOK, WESTERN EUROPE, 1978 - 1982. *IDC Europa Ltd. June 1980. 155 pg. 995.00 (United Kingdom).*

Market analysis of Western European market for services and software includes 1978-1982 data for growth rates, revenues, distribution channels, vendor strategies, suppliers, market share, products by country, and forecasts.

1965. SMALL BUSINESS COMPUTER SYSTEMS AND ASSOCIATED SOFTWARE MARKET IN EUROPE. *Frost and Sullivan, Inc. July 1979. 472 pg. $1250.00.*

Analyzes and forecasts the market for small computer systems and associated software products in Europe through 1988 in number of units, unit price and sales value. Analyzes major small business system applications, technological developments, and the supplier industry structure, by products and marketing strategies.

1966. SMALL BUSINESS SYSTEMS MARKETS, 1981 - 1985. *IDC Europa Ltd. December 1981. 1250.00 (United Kingdom).*

Market analysis of small business systems in the Western European market for 1981-1985 includes market size; market share; projected shipments; forecasts by country; supplier profiles and strategies; impact of the desktop computer; and user survey analysis.

1967. SMALL SYSTEMS MARKET REFERENCE BOOK 1981-1987. *IDC Europa Ltd. 1982. 150 (est.) pg. 1250.00 (United Kingdom).*

Report on small systems market in Western Europe, 1981-1987, examines leading suppliers with market shares by shipments and revenues, and aspects of individual country markets with an emphasis on local factors having a bearing on market development. Analyzes market showing shipment forecasts by sector for all categories of small system and competitive products. Reviews the way in which systems will be intertwined and the effect of falling hardware prices.

1968. SOFTWARE EXPENDITURE ANALYSIS, 1981 - 1985. *IDC Europa Ltd. August 1981. 1250.00 (United Kingdom).*

Market analysis of software expenditures in Western Europe for 1981-1985 includes spending analysis on systems, utility, and application by country; software sources and factors affecting software sources; industry specific software by industry; and software spending by machine type.

1969. SOLID STATE MEMORY MARKET IN WEST EUROPE: VOLUME 1. *Frost and Sullivan, Inc. September 1979. 334 (est.) pg. $1650.00.*

Market analysis for solid state memory devices in Western Europe, with forecasts to 1990. Covers products; decline or growth rate projections; prices to 1990; and price factors including production costs and product mix. Examines national political and economic indicators, 1977-1990; government regulations; consumption by end uses to 1990; production to 1990; manufacturers, market shares, 1977-1985, 1990; and major users.

1970. SOLID STATE MEMORY MARKET IN WEST EUROPE: VOLUME 2. *Frost and Sullivan, Inc. September 1979. 334 (est.) pg. $1650.00.*

Second volume of a market analysis for solid state memory devices in Western Europe. Gives country-by-country economic indicators, 1977-1980 with forecasts to 1990; market structure; and markets by products, 1977-1985, 1990. Includes major consumers to 1990; major manufacturers to 1985; production to 1990; consumption by end uses to 1990; average prices to 1990; distribution to 1990; factors influencing purchases; and competition with imports.

1971. STRATEGIES FOR COMPUTER SERVICES COMPANIES IN WESTERN EUROPE. *INPUT. 1980.*

A market analysis of computer services in Western Europe. Covers competition, financial needs, and recommendations for revitalization of the industry.

1972. SURVEY OF THE MARKET FOR COMPUTERS AND RELATED EQUIPMENT IN GERMANY: A FOREIGN MARKET SURVEY REPORT. *Kindel, Karl-Wilhelm. U.S. Department of Commerce. International Trade Administration. NTIS #ITA 81-08-502. January 1981. 124 pg. $15.00.*

Survey of the market in Germany for computers and related equipment provides information on market size, special influences, specific product markets, competitive assessment, end users, and market access data.

1973. SURVEY OF THE MARKET FOR COMPUTERS AND RELATED EQUIPMENT IN SWITZERLAND: A FOREIGN MARKET SURVEY REPORT. *Kindel, Karl-Wilhelm. U.S. Department of Commerce. International Trade Administration. NTIS #ITA 81-08-503. April 1981. 97 pg. $15.00.*

Survey of the market in Switzerland for computers and related equipment, provides information on the market size, special influences, specific product markets, competitive assessment, end users, and market access data. Provides data and text analyses by major subcategory on market size and imports (including supplying country's share of the market) for current and future years. Lists U.S. products with good sales prospects. Gives data on number and value of computers in use, and projected uses and expenditures, by end user sector. Information is provided on marketing practices, trade restrictions, and technical requirements. Trade lists include principal agents and distributors; trade associations; publications; and trade events.

1974. SURVEY OF THE MARKET FOR COMPUTERS AND RELATED EQUIPMENT IN THE NETHERLANDS: A FOREIGN MARKET SURVEY REPORT. *International Trade Administration. U.S. Department of Commerce. National Technical Information Service. DIB-80-09-502. 1980. 84 pg. $10.00.*

The market research was undertaken to study the present and potential U.S. share of the market in the Netherlands for computers and related equipment; to examine growth trends in Dutch end-user industries over the next few years; to identify specific product categories that offer the most promising export potential for U.S. companies; and to provide basic data which will assist U.S. suppliers in determining current and potential sales and marketing opportunities. The trade promotional and marketing techniques which are likely to succeed in the Netherlands were also reviewed.

1975. THIRD ANNUAL STUDY OF CUSTOMER SATISFACTION AMONG LARGE SYSTEMS USERS IN EUROPE. *Customer Satisfaction Research Institute. Annual. 1981. $7465.00.*

Market survey covering end users' attitudes and customer satisfaction with large computer systems in Europe.

1976. TRENDS IN TECHNOLOGY, COMPETITION AND MARKET NEEDS: THE FUTURE OF VIDEOTEX. *Butler Cox and Partners Ltd. December 1981.*

Market forecast of the videotext market examines future trends, technlogy, and competition.

1977. U.K. INFORMATION PROCESSING CENSUS SERIES: VOLUME I: ELECTRONIC OFFICE SYSTEMS. *Bis-Peddar. Benn Publications Ltd. 1982. $1800.00.*

Survey data is broken down by supplier and by model for the electronic office systems installed base in the U.K.

1978. U.K. INFORMATION PROCESSING CENSUS SERIES: VOLUME II: SMALL PURPOSE GENERAL SYSTEMS. *Bis-Peddar. Benn Publications Ltd. 1982. $1800.00.*

Survey data broken down by supplier and by model for the small purpose general computer system installed base in the U.K.

1979. U.K. INFORMATION PROCESSING CENSUS SERIES: VOLUME III: LARGE GENERAL PURPOSE SYSTEMS. *Bis-Peddar. Benn Publications Ltd. 1982. $1800.00.*

Market research report data on large general purpose computer systems is broken down by supplier and by model for the installed base in the U.K.

1980. U.K. INFORMATION PROCESSING CENSUS SERIES: VOLUME IV: SUNDRY SYSTEMS SUPPLEMENT. *Bis-Peddar. Benn Publications Ltd. 1982. $500.00.*

Research report on information processing in the U.K. provides survey data broken down by supplier and by model for the installed base.

1981. U.K. INFORMATION PROCESSING CENSUS SERIES: VOLUME V: MARKET STATISTICS SUMMARY. *Bis-Peddar. Benn Publications Ltd. 1982. $700.00.*

Provides summary of survey data broken down by supplier and by model for computer installed base in the U.K.

1982. THE VALUE ADDED NETWORK SERVICES MARKET IN WESTERN EUROPE. *Frost and Sullivan, Inc. April 1980. 239 pg. $1250.00.*

Analysis of Western European markets for value-added networks: enhanced data communications systems and equipment for electronic mail, reservation systems, on-line information retrieval systems, and the like. Discusses regulations and end use applications in the U.S. and Europe. Forecasts revenues, 1980-1985, and shows annual shipments and current installations.

1983. VIEWDATA / VIDEOTEX MARKET IN EUROPE. *Frost and Sullivan, Inc. January 1981. 204 pg. $1375.00.*

Market analysis with forecasts to 1989 for viewdata / videotext systems in Western Europe. Includes extensive consideration of the U.S., Canada and Japan. Covers competition, demand, market penetration, regulations, shipments, installations, and revenues.

1984. WESTERN EUROPEAN MARKET FOR MEDICAL INFORMATION SYSTEMS. *Creative Strategies International. March 1982. $1450.00.*

Assesses market for computer-based medical information systems (MIS), covering hospital and general practice environments. Discusses MIS growth rate compared with other sectors of the European computer industry.

1985. WESTERN EUROPEAN ROBOTICS. *Creative Strategies International. December 1982. $1200.00.*

Presents analysis of the Western European market for industrial robots. Forecasts growth rates 1983-1986. Examines labor contracts and regulations, end uses, and competition among manufacturers.

1986. 1982 OUTLOOK FOR THE EUROPEAN COMPUTER AND RELATED MARKETS, JANUARY 1982. *MSRA, Inc. January 1982. 101 pg. $495.00.*

Analysis of European markets for mainframes, minicomputers, office automation, communications systems and semiconductors. Gives marketing strategies and performance evaluations for major multinational vendors. Also provides economic indicators for Europe, including gross national product growth rates by country to 1986.

See 425, 428, 437, 460, 471, 725, 727, 728, 736, 2837, 2838, 2855, 2867, 2891, 3378, 3383, 3387, 3388, 3390, 3393, 3394, 3395, 3397, 3398, 3399, 3400, 3401, 3402, 3789, 3810, 3814, 3836, 3852, 3862, 4296, 4304.

Investment Banking Reports

1987. MORGAN STANLEY AND CO.: INVESTMENT PER-SPECTIVES: EDP: EUROPE REVISITED: RESEARCH COM-MENT. *Weil, Ulric. Morgan Stanley and Company, Inc. November 30, 1981. pg.15-16*

Analysis of computer markets in Canada and Western Europe. Includes new orders, manufacturers and subsidiaries, currency problems, and competition.

1988. WARBURG PARIBAS BECKER - A.G. BECKER: INDUSTRY COMMENT: THE VIEW FROM EUROPE: EXPECTED FIRMNESS OF EUROPEAN ECONOMIES IN 1983 BODES WELL FOR FIVE COMPUTER COMPANIES. *Orlansky, Aharon. Warburg Paribas Becker - A.G. Becker, Inc. July 12, 1982. 15 pg.*

Analysis of indicators for the European economies and prospects for computer markets shows exchange rates, 1980-1982, and gross national products, 1980-1983. For major U.S. computer corporations, shows stock prices; earnings per share, 1981-1983; price earnings ratio, 1981-1982; dividends and yield; revenues from European foreign markets and percent of total revenues, 1978-1981; percent contribution to earnings per share, 1980-1981; and orders, 1980-1982.

1989. WARBURG PARIBAS BECKER - A.G. BECKER: INDUSTRY REVIEW: THE VIEW FROM EUROPE: MINICOM-PUTER VENDORS. *Orlansky, Aharon. Warburg Paribas Becker - A.G. Becker, Inc. August 16, 1982. pg.8*

Financial analysis of European minicomputer corporations includes income statements, 1981-1982; sales; orders; earnings per share, 1982-1983; breakdown by lines of business; stock prices; and expected growth rates.

Industry Statistical Reports

1990. BUSINESS MONITOR: COMPUTER SERVICES: QUARTERLY STATISTICS. *Government Statistical Service. Her Majesty's Stationery Office. Quarterly. 1981. 14 pg. 5.10 (United Kingdom).*

Presents quarterly statistics, including five previous quarters, representing the turnover (billings) in pounds sterling and employment figures for over 150 United Kingdom computer service companies. Billings tables give breakdown for custom-built programs, package programs, time hire, professional services, total systems supply, data preparation, other billings and totals. A summary of tables provides comments on industry growth rates and employment trends.

1991. BUSINESS MONITOR: ELECTRONIC COMPUTERS: QUARTERLY STATISTICS. *Government Statistical Service. Her Majesty's Stationery Office. Quarterly. 1981. 5 pg. $5.10 (ukd).*

Tables give quarterly manufacturers' sales statistics, including the four previous quarters, in pounds sterling based on United Kingdom computer businesses employing 25 or more people and computer export and import figures for the United Kingdom based on data provided by H.M. Custom and Excise. Quarterly statistics for employment levels in the U.K. computer industry are also supplied.

1992. BUSINESS MONITOR: ELECTRONIC COMPUTERS: REPORT ON THE CENSUS OF PRODUCTION 1976. *Government Statistical Service. Her Majesty's Stationery Office. Annual. 1979. 7 pg. .60 (United Kingdom).*

Provides 1973-1976 production, production costs, capital expenditure, and stocks statistics, and 1976 business size, employment, wage and salary, sales, value and employee sta-tistics for the United Kingdom computer industry. Figures are based on census of United Kingdom computer businesses employing twenty or more people.

1993. BUSINESS MONITOR: ELECTRONIC COMPUTERS: REPORT ON THE CENSUS OF PRODUCTION 1977. *Government Statistical Service. Her Majesty's Stationery Office. Annual. 1980. 8 pg. .95 (United Kingdom).*

Provides 1973-1977 production, production costs, capital expenditures, stocks statistics and 1977 business size, employment, wage and salary, sales, value, employer and operating ratio statistics for the United Kingdom computer industry. Figures are based on census of United Kingdom computer businesses employing twenty or more people.

1994. BUSINESS MONITOR: ELECTRONIC COMPUTERS: REPORT ON THE CENSUS OF PRODUCTION 1979. *Government Statistical Service. Her Majesty's Stationery Office. Annual. 1981. 8 pg. 2.50 (United Kingdom).*

Provides 1975-1979 production, production costs, capital expenditure, and stocks statistics, 1979 business size, employment, wage and salaries, sales, value and employee statistics and 1978-1979 operating ratio statistics for the United Kingdom computer industry. Figures are based on census of United Kingdom computer businesses employing twenty or more people.

1995. BUSINESS RATIO REPORT: COMPUTER SERVICES. *ICC Business Ratios. 1980. 80.00 (United Kingdom).*

1980 statistical report gives fourteen key ratios involving profits, assets, sales, capital, current assets, current liabilities, liquidity, stock turnover, debt, credit, employees, wages and salaries, and exports. Includes average annual growth rates concerning sales, total assets, capital employed, average wages, total wages and exports. 60 United Kingdom computer service businesses are compared, for which addresses and principal activities are given. Three years financial data and ratios are shown for each business.

1996. COMPUTER EQUIPMENT. *ICC Business Ratios. Annual. 1981. 95.00 (United Kingdom).*

One in a series of 150 annual industry reports comparing performance between up to 100 individual leading companies, subsectors, and sectors over a three-year period. Fourteen ratio tables by company detail profit margin, profitability, asset utilization, return on capital, liquidity, stock turnover, credit period, exports, profit per employee, average remuneration, sales per employee, capital employed per employee, and other data. Six annual growth rate tables cover sales, total assets, capital employed, average wages, total wages, and exports. Other company data includes names of directors and secretary and the registered office address. Gives analyses of current problems and likely developments within the sector and industry as a whole.

1997. COMPUTER SERVICES. *ICC Business Ratios. Annual. 1981. 80.00 (United Kingdom).*

One in a series of 150 annual industry reports comparing performance between up to 100 individual leading companies, subsectors, and sectors over a three-year period. Fourteen ratio tables by company detail profit margin, profitability, asset utilization, return on capital, liquidity, stock turnover, credit period, exports, profit per employee, average remuneration, sales per employee, capital employed per employee, and other data. Six annual growth rate tables cover sales, total assets, capital employed, average wages, total wages, and exports. Other company data includes names of directors and secretary and the registered office address. Gives analyses of current problems and likely developments within the sector and industry as a whole.

See 2945, 2956, 2960, 3403, 3405, 3406, 3407, 3410.

Financial and Economic Studies

1998. COMPUTER EQUIPMENT. *Inter Company Comparisons Ltd. May 1980. 80.00 (United Kingdom).*

Report analyzes the performance of 99 companies (1976-1979), giving market analysis, profits, imports, suppliers, and return on capital employed.

1999. COMPUTER INDUSTRY FINANCE. *IDC Europa Ltd. October 1980. 300 pg. 650.00 (United Kingdom).*

Multiclient study examines 400 Western European computer companies.Focuses on computer industry finances, giving information on revenues, growth rates, profits, and capital.

2000. ELECTRONIC COMPUTERS SWP: INDUSTRIAL STRATEGY PROGRESS REPORT 1979. *Electronic Computers Sector Working Party. National Economic Development Office. Annual. January 1979. 6 pg.*

Outlines opportunities and constraints in the United Kingdom computer industry regarding exports, imports, government procurement policy, research and development, and employment. Includes a table giving the United Kingdom computer industry's turnover and balance of trade figures for 1976-1977 in pounds sterling, and 1976-1977 employment level statistics.

2001. ELECTRONIC COMPUTERS SWP: PROGRESS REPORT 1980. *Electronic Computers Sector Working Party. National Economic Development Office. Annual. 1980. 9 pg.*

Reports on the competition between the United Kingdom computer industry and foreign competitors and outlines opportunities and constraints in the U.K. computer industry regarding exports, imports, government procurement policy, and labor. Includes a table giving the U.K. industry's turnover and balance of trade figures in pounds sterling for 1977-1978, in addition to 1977-1978 employment level figures.

2002. FINANCIAL SURVEYS: COMPUTER AND DATA - PROCESSING EQUIPMENT. *Inter Company Comparisons Ltd. 1981. 59.80 (United Kingdom).*

Annual financial analysis, with two year comparative figures, for computer and data processing equipment in the U.K. For all applicable businesses, tabulates turnover of sales; total assets, including fixed assets, current assets, intangible assets, and investments; current liabilities; profits before taxes; executive compensation; capital employed; and subsidiaries.

2003. INDUSTRIAL ROBOTS IN JAPAN, U.K. AND U.S.A.: ADVANCED TECHNOLOGY SERIES FEATURING U.K. APPLICATIONS. *Inbucon Management Consultants Ltd., Advanced Technology Group. 1982. 152 pg.*

Explains robot anatomy, limitations on use, and general applications. Reviews the development, demand, installed base, productivity, effects on labor, and manufacturers of robots in Japan and in the U.S.A. Presents and analyzes nine case studies of robots in the U.K. Includes sections on the future of robotics, and advanced manufacturing technology a

2004. INVESTING IN SMALL COMPUTER COMPANIES. *Simon and Coates. August 1981. 54 pg.*

A financial study of the United Kingdom computer industry to serve as a guide to investing in small computer businesses. The report's first section provides forecasts for the following general market areas: software, central processing units and systems, peripherals, computer bureau and CAD/CAM. The second part provides forecasts based upon turnover, pretax profit, earnings, dividend, yield, share capital, asset value and sales figures for twelve small British computer businesses.

2005. A STRATEGY FOR INFORMATION TECHNOLOGY: A SUMMARY OF A REPORT PREPARED FOR THE N.E.B. *PA International Group. Pactel. 1981. 44 pg. 3.50 (United Kingdom).*

Economic report examining the state of the United Kingdom computer and telecommunications industries, and providing forecasts regarding future development. Comprises the competitive position of the U.K.; industry goals and key issues; opportunities for the U.K.; and conclusions and recommendations. Growth rates, value of shipments, markets, turnover, competition, imports, exports, costs, and sales are considered.

See 2961, 3412.

Forecasts

2006. DATA BASE MARKETS IN EUROPE. *Frost and Sullivan, Inc. March 1980. 274 pg. $1250.00.*

Analyzes and forecasts the European market for data bases in the following areas: science and technology; economic and econometric; business and finance; credit reference; legal; special marketing information; patents and trademarks; and others. Estimates markets by year to 1988 on the basis of conservative GDP growth rates. Analyzes the relationship between on-line and batch processing and sectors and major data base suppliers. Profiles principal data base originators and marketing organizations, and attitudes and reactions.

2007. DATA COMMUNICATIONS EQUIPMENT MARKETS, 1981-1985. *IDC Europa Ltd. February 1982. 130 pg. 1250.00 (United Kingdom).*

Market research report provides forecasts and data analysis of Western European data communications equipment markets, 1981-1985. Gives size, share, projection by country, and analysis by product category ofthe data communications equipment market; analysis of user consideration of cost / benefits, workforce acceptance and applications; end user survey; and vendor profiles and strategies. Focuses on market sizeand projection for modems, multiplexors, and communications systems.

2008. DATA ENTRY / TERMINAL MARKETS. *IDC Europa Ltd. December 1981. 130 pg. 1250.00 (United Kingdom).*

Market research report provides forecasts and analysis of Western European data entry / terminal markets. Provides consumers, vendors, market analysis, market structure, installed base, value of shipments, unit sales, forecasts, sales, market areas, demand, user attitudes, supply, market size, market survey, and market shares.

2009. ELECTRONIC COMPUTERS SWP: MANPOWER SUB-COMMITTEE FINAL REPORT. *Electronic Computers Sector Working Party. National Economic Development Office. June 1980. 11 pg.*

The final report of the Manpower Sub-Committee of the Electronic Computers Sector Working Party, identifies the United Kingdom's computer industry's labor requirements to the mid-1980's, and examines the possibilities, constraints and social consequences of technical change, the current stocks of computer skilled manpower, the supply of skilled manpower, and manpower imbalances. Recommendations are made to the government, unions and employers.

2010. ELECTRONIC COMPUTERS SWP: MANPOWER SUB-COMMITTEE SECOND INTERIM REPORT. *Electronic Computers Sector Working Party. National Economic Development Office. January 1979. 4 pg.*

An interim report by the Manpower Sub-Committee of the Electronic Computers Sector Working Party, examining the future labor requirements in the United Kingdom computer industry. General forecasts for labor requirements are given and recommendations are made to the government, employees, and trade unions.

2011. THE ELECTRONIC MAIL MARKET IN EUROPE. *Frost and Sullivan, Inc. July 1979. 267 pg. $1000.00.*

Assesses the potential for electronic mail systems in European markets. Includes prospects for effective standards, and trends in technology and constraints (human, organizational and political). Analyzes prospects of two market segments: external (inter-company) and mainly internal (intra-company). Forecasts markets by products used and by country.

2012. THE EUROPEAN COMPUTER AND OFFICE EQUIPMENT MARKET: TRENDS AND OUTLOOK FOR THE 1980'S. *Martin Simpson & Co., Inc. Martin Simpson and Company, Inc. December 26, 1979. 34 pg. $395.00.*

1980 market analysis of developments in the European computer and office equipment market, with emphasis on the data processing, copier and word processing industries. The report highlights the strategies, recent business performance and outlook for the following leading multi-national manufacturers: Sperry Univac, CII Honeywell Bull, IBM, Amdahl, Nixdorf Computer, Wang Laboratories, Tandem Computers, Texas Instruments, Intel, Nashua and Rank Xerox.

2013. MICRO AND MINI MARKETS AT BUSINESS SYSTEMS HOUSES - WESTERN EUROPE 1980-1984. *IDC Europa Ltd. July 1981. 130 pg. 1250.00 (United Kingdom).*

Market research report providing forecasts and analysis of Western European micro and mini computer markets at business systems houses, 1980-1984. Aimed at mainframe vendors and OEM's (systems houses) distributors. Provides suppliers' strengths and weaknesses, as well as hardware and software policy.

2014. MINICOMPUTER MARKETPLACE, 1981-1985. *IDC Europa Ltd. December 1981. 130 pg. 1250.00 (United Kingdom).*

Market research report providing forecasts and analysis of the WestEuropean minicomputer markets, 1981-1985. Gives size, share and forecast of minicomputer markets by size, class, distribution methods and country region; profile of major vendors with attitudes to micros, and of target market segments; analysis of the effect of IBM 4300 on the supermini markets and an evaluation of IBM Series 1 and its low profile; and OEM / Systems House survey concerning organization and value added elements, target markets and leading applications.

2015. OEM PERIPHERAL MARKETS IN WESTERN EUROPE, 1979-1983. *IDC Europa Ltd. December 1980. 130 pg. 995.00 (United Kingdom).*

Market researach report provides forecasts and analysis of the OEM peripheral markets in Western Europe, 1979-1983. Major sections are tape, disk and floppy drives, serial / line printers, and terminals.

2016. STABILIZED POWER SUPPLIES IN EUROPE. *Frost and Sullivan, Inc. September 1980. 455 pg. $1250.00.*

This report analyzes and forecasts the stabilized power supplies markets in Western Europe from 1977-1986 by annual increment estimates. Major market-sector breakdown by coun-

try in constant U.S. dollars covers electronic data processing, control and instrumentation, and communications.

2017. TAPE AND DISK DRIVE MARKETPLACE, 1981-1985. *IDC Europa Ltd. October 1981. 130 pg. 1250.00 (United Kingdom).*

Market research report provides forecasts and analysis of tape and disk drive markets of Western Europe, 1981-1985. Gives size, share, projection by country, and analysis by product category of the tape anddisk drive markets; vendor profiles and strategies with an emphasis on technological developments; the extent to which OEM's distributors and direct sales methods are used for distribution; and tape and disk drive usage by machine type with consideration given to all levels of processor and projected device usage.

2018. TERMINAL MARKETS IN THE DISTRIBUTED DATA PROCESSING ENVIRONMENT, 1981-1985. *IDC Europa Ltd. December 1981. 130 pg. 1250.00 (United Kingdom).*

Market research report provides forecasts and analysis of the West European terminal markets in the distributed data processing environment, 1981-1985. Gives market size, share, and projection by country of DDP market, terminal systems; supplier profiles and strategies; exten

See 2980, 3413.

Directories and Yearbooks

2019. THE COMPUTER USERS' HANDBOOK. *Computer Users' Year Book. 1980. 1,018 pg. $75.00.*

Annual directory covering computer industry in Great Britain and Ireland (Eire), including 25,000 computers in 9,000 locations. . Gives user's address, telephone and executives; type of computer, and name of manufacturer. Other sections offer guides to computer supplies and equipment, consultants, system houses, batch and computing service bureaus, and data preparation services.

2020. COMPUTER USERS' YEAR BOOK DIRECTORY OF SOFTWARE. *Computer Users' Year Book. Irregular. 1980. 478 pg. 24.95 (United Kingdom).*

Listings of software suppliers include address, telephone, executives, parent company, number of employees, date founded, activities, and branches outside the U.K. Software product listings include supplier, end use, and descriptions with indexes by application and by industry of application.

2021. COMPUTER WEEKLY YEARBOOK. *IPC Business Press Ltd. Annual. 1979. 240 pg. 2.00 (United Kingdom).*

Covering the United Kingdom, directory lists computer software suppliers, auxilliary services, bureau services, consulting, and time - sharing.

2022. EUROPEAN USER RATINGS OF COMPUTER SYSTEMS. *Datapro Research, Inc. (McGraw-Hill). Annual. 1981. $25.00.*

Report recaps how more than 3,500 European users from England, Germany and France rate their systems (mainframes to personal computers) on ease of operation, reliability, technical support, software, maintenance and much more. Covers systems built by CII Honeywell Bull, International Computers Ltd., Siemens, other European manufacturers and U.S. manufacturers.

2023. GUIA CHIP "82: DIRECTORIO DE INFORMATICA. [1982 Chip Computer Industry Directory.]. *Chip, Revista de Informatica. Ediciones Arcadia, S.A. Annual. 1982. 322 pg. 3000.00 (Spain).*

Lists 700 businesses in Spain in the computer and computer services industry alphabetically with a numeric product line classification cross indexed to the product listings. A list of the businesses by Spanish province is provided. Sections include businesses not Spanish but represented in Spain, and listings of associations and other public and private organizations allied with the EDP-Computer field.

2024. GUIDE TO PRODUCTION CONTROL SYSTEMS 1981 / 82. *Knight, Barry, editor. Computer Guides Ltd. Annual. 2nd ed. 1981. 233 pg.*

Coverage spans U.K. supplier guide (address, contact person, telephone, systems). Contains system guide with an analysis by function (supplier, product, stock control, order processing, requirements planning, forecasting, shop order control, shop loading, cost, computer, U.K. users, bureau). Includes analysis by computer system (product; supplier); extensive data (59 questions) on individual systems (background, hardware considerations, software considerations, costs, production control functions, industry match).

2025. A GUIDE TO PROGRAMMABLE LOGIC CONTROL-LERS AND SYSTEMS BUILDERS IN THE U.K. *Systec Consultants Ltd. Irregular. 1981. 150.00 (United Kingdom).*

Market research report describing in detail the programmable logic controllers on the market in the U.K. and the builders who configure them for specific applications. Lists details of over 60 PLC's from 30 companies and also 90 systems builders. Gives information on company size, experience, type of work undertaken and other details. Also gives technology trends and a wide market analysis.

See 2990, 2996, 2998, 3414, 3415, 3887.

Special Issues and Journal Articles

British Foundryman

2026. MICROCOMPUTERS IN FOUNDRY MANAGEMENT. *IBF Publications. Irregular. v.75, part 3. March 1982. pg.ix-xi*

Includes listing of products exhibited at event organized by the Loughborough Metallurgical Engineers' Association on December 17, 1981 at the Loughborough University of Technology. Names and addresses of manufacturers also included.

La Champagne Economique

2027. LE MICRO ORDINATEUR: GUIDE PRATIQUE. [The Microcomputer: A Practical Guide.]. *Chambre de Commerce et d'Industrie de Reims. French. November 1981. pg.13-25*

Provides specifications of microcomputers available on the French market. Shows distribution of users, estimated budgets, and annual costs.

Chip, Revista de Informatica

2028. ESPANA PUEDE CONVERTIRSE EN 'PAIS-TALLER' DE LA INFORMATICA. [Spain Can Be Converted Into 'Factory Country' of Computer Industry.]. *Ediciones Arcadia, S.A. Spanish. no.15. June 1982. pg.12-13*

5th Congress of computers and automation in Madrid (1981), by the AIA, analyzes the level of the industry in Spain. Gives

aspects of market penetration by international businesses and provides plans by Spain to overcome industry deficiencies. Percentages of industry imports and foreign market penetration are given.

2029. IBM EL MAYOR PROVEEDOR EUROPEO. [IBM The Major European Supplier.]. *Quantum Science Corp. Ediciones Arcadia, S.A. Spanish. no.03. May 1981. pg.22*

Study provides 1980 statistics of the European computer industry. Gives equipment expenditures in U.S. dollars and shows estimates of percent of growth, 1979-1984, by product groups. Gives number of employees and gross median income per employee in the industry by country, including Spain. Figures are given in dollars for major investing countries. Study by Quantum Science Corp.

2030. ICL O EL CHAUVINISMO BRITANICO: QUE OPINAN LOS USUARIOS ESPANOLES? [ICL or British Chauvinism.]. *Datapro Research Co. Ediciones Arcadia, S.A. Spanish. no.05. August 1981. pg.61-64*

Results of 1980 study by Datapro Research Co., of Spanish buyers of ECL computer systems. Tables show grading by users of equipment and systems in many categories of reliability, operational and service area. ICL is economically analyzed generally, and specifically in Spain. Charts indicate comparative gross and net income of 10 large computer businesses.

2031. INFORMATICA PARA LOS SIETE GRANDES BANCOS. *Gesellschaft fur Chemiewirtschaft. Spanish. no.11. February 1982. pg.82*

Gives statistics of expenditures by Spain's 7 largest banks for computer and EDP equipment. Shows computer firms' market shares. Includes forecasts of expenditures by the largest banks for 1982-1985.

2032. EL MERCADO COMUN DE LA INFORMATICA. [Computer - Telecommunications Market in the EEC.]. *Garcia-Borreguero, Carlos. Ediciones Arcadia, S.A. Spanish. no.04. June 1981. pg.40-44*

Analyzes the computer and telecommunications systems in the EEC and the effect of Spain's entry into the EEC on the implementation of a homologous equipment system. Discusses present EURONET system. Lists 25 European database information centers. Contains basic points of plan of action through 1985, and statistically analyzes impact of U.S. and Japanese penetration of the Spanish computer market.

2033. NUEVO: LAS ULTIMAS NOVEDADES DE SOFTWARE Y HARDWARE SALIDAS AL MERCADO. [New: The Latest in Software and Hardware on the Market.]. *Ediciones Arcadia, S.A. Spanish. Monthly. no.06. September 1981. pg.68-70*

Includes the latest product lines from national and international suppliers of computers, computer services and communications equipment. Gives technical and functional specifications of products.

2034. EL PLAN INFORMATICO DE AVIACION CIVIL. [Civil Aviation Computer Plan.]. *Ediciones Arcadia, S.A. Spanish. no.16. August 1982. pg.12-14*

Describes plan to install civil aviation airport computer communication system in all Spanish airports. The system approved by The Ministry of Transportation gives number of installations for various size airports, and charts provide central control details. Number of employees and equipment and software costs are given. Includes computer businesses involved in bidding for contracts.

2035. PRESUPUESTO INFORMATICO DE LA ADMINISTRA-CION. [Administration Computer - Telecommunications Budget.]. *Gesellschaft fur Chemiewirtschaft. no.14. May 1982. pg.14*

Gives Spanish government administration 1982 budget statistics for computer hardware and software purchases and rentals. Shows allotments for communications services. Gives amounts per ministries in the government for equipment and services, in Spanish pesetas.

2036. RESULTADOS FINANCIEROS DE IBM ESPANA EN 1980. [1980 Financial Results of IBM Spain.]. *Ediciones Arcadia, S.A. Spanish. no.06. September 1981. pg.18*

1980 financial report for IBM Spain gives total income, gross and net profits, investments, exports, and number of employees at year-end. Monetary amounts are shown in Spanish pesetas.

2037. LA SERIEDAD DE UN RETO NACIONAL: SECOINSA, EN MARCHA. [The Seriousness of a National Challenge.]. *Ediciones Arcadia, S.A. Spanish. no.08. November 1981. pg.76-77*

Discusses the growth of the government-backed computer manufacturer, Secoinsa, from 1977 to 1981. Gives annual income, percent of advances and declines, number of employees, and income per employee, 1977 to 1981. Income is given in Spanish pesetas.

2038. SIFD AMBICIOSO PLAN NACIONAL DE INFORMA-TICA FISCAL. [SIFD Ambitious Plan for National Fiscal Computer System.]. *Ediciones Arcadia, S.A. Spanish. no.05. August 1981. pg.15-16*

Describes computer and data processing network, SIFD (Distributed fiscal Information Integraded System), for service of Spanish Ministry of the Treasury. Chart shows data processing capacity of system, 1977-1982, and forecasts capacity and utilization of system through 1985. Gives technical description of current system equipment and number and planned installations. Discusses selection of equipment, capacity and personnel to be employed in the system, through 1985.

Dispo

2039. DRUCKER FUR KLEINCOMPUTER: DAS MARK-TANGEBOT IN OSTERREICH. [Printers for Small Computers: The Market in Austria.]. *Johann L. Bondi and Sons. German. Irregular. September 1981. pg.43-45*

Table lists most of the printers (more than 80) for small computers available on the Austrian market, classified under 24 manufacturers (domestic and foreign). Gives the following specifications: type, speed, paper roll width, cutting intervals, matrix size, graphical abilities, number of possible copies, buffer memory size, and prices.

Distribution

2040. ITALY: A BIG MARKET FOR SMALL COMPUTERS. *Chilton Book Company. June 1981. pg.57-60*

Examines the demand and market for small computers in Italy. Discusses the role the U.S. can play in servicing this demand. Summarizes air and shipping routes between the U.S. and Italy. One table lists Atlantic, Gulf, Pacific and Great Lakes ports and the ocean lines they service.

Documentaliste

2041. BASES DE DONNEES ET CENTRES SERVEURS. [Data Bases and Vendors.]. *Association Francaise des Documentalistes et des Bibliothecaires Specialises. Irregular. November 1980. 64 pg.*

Special issue profiling data base services available in France, and major systems operating throughout Western Europe. Discusses fields covered, number of references online, user costs, and years of operation of most systems. Also provides addresses of some of the larger French systems and operators' and users' organizations, and discusses trends in user needs and operator responses.

Electronic Business

2042. ELECTRONIC BUSINESS: EUROPE'S TOP 100. *Cahners Publishing Company. Annual. July 1981. pg.34-42*

Examines the W. European electronics market including Japanese and U.S. businesses selling in W. Europe. Details the top 100 businesses including sales, gross profits, investment as percent of sales, and sales per employee (1980). Covers growth over 5 years, percent change over prior year for capital expenditures and return on gross capital employed. Lists U.S. suppliers in Europe by product.

Electronic News

2043. SURVEY SAYS U.S. FIRMS DID 33% OF 1981 WEST EUROPEAN SALES. *Fallon, James. Fairchild Publications. v.28, no.1405. August 9, 1982. pg.D*

Evaluates the Western European computer market in 1981. Includes information on manufacturers' sales, profits, and capital expenditures. Source is Bennelectronics Publications.

Electronic Publishing Review

2044. EUROPEAN DATABASE DEVELOPMENTS AND THE EUROPEAN COMMUNITY. *Martyn, John. Learned Information (Europe) Ltd. June 1981. pg.123-129*

Analyzes types, subject scopes, and national origins of 266 information system development proposals received by the Commission of the European Communities. Tables detail countries of origin correlated with types of organization; distribution of proposals by subject field and by type of information product offered; and distribution of selected proposals according to subject field by country.

Engineer

2045. COMPUTERS: INDUSTRY'S CHANCE FOR SURVIVAL. *Davis, Brian. Morgan Grampian Book Publishing Company Ltd. v.255, no.6599. September 16, 1982. pg.33-37*

Examines computer sales to engineering companies in the U.K. Table lists figures for computer expenditures in manufacturing sector U.K., 1981 (millions of pounds Sterling); provides figures for data processing manufacturer market shares in U.K., 1980.

European Plastics News

2046. ROBOT NUMBERS GROW IN THE U.K. *British Robot Association. IPC Industrial Press Ltd. v.9, no.3. March 1982. pg.5*

Examines sales of robot systems to British industry in 1981. Gives figures for British and Japanese built systems, 1980, 1981; and U.S. share of British import market, 1980, 1981.

L'Expansion

2047. LA COURSE FOLLE AU MICROPROCESSEUR. [The Mad Race for Microprocessors.]. *Fontaine, Jacques. International Werbegesellschaft m.b.H. French. September 19, 1980. pg.148-153*

Studies the French market for microprocessors. One graph shows the size of the market and the length of life of a microprocessor.

2048. L'ENCYCLOPEDIE DE L'INFORMATIQUE, TELEMATIQUE, BUREAUTIQUE. [Encyclopedia of Computers, Commuinications, Office Equipment.]. *Clairvois, Marc. International Werbegesellschaft m.b.H. French. Irregular. September 19, 1980. pg.177-215*

Reviews computer hardware and software and communications and office equipment giving prices, applications, and specifications.

2049. OU EN EST L'INFORMATISATION DE LA SOCIETE FRANCAISE. [Where Does the Computer Stand in French Society.]. *De Witt, Francois. International Werbegesellschaft m.b.H. French. September 19, 1980. pg.133-139*

Studies the French society's attitudes toward computers. Tables show the growth of computer installations in French companies, schools, and families, from 1980 to 1985.

2050. LE WHO'S WHO DE L'INFORMATIQUE EN FRANCE: SPECIAL INFORMATIQUE 1981. [The Who's Who in the Computer Industry in France.]. *Clairvois, Marc. International Werbegesellschaft m.b.H. French. Irregular. September 18, 1981. pg.159-195*

Directory gives names, addresses of executives, number of employees, companies, turnover, and main activity for all computer-related industries in the French market. Includes suppliers to these industries.

Die Geschaftsidee

2051. UNTERNEHMENSKONZEPT COMPUTER POR-TRAITS. [Business Conception Computer Portraits.]. *Verlag Norman Rentrop. German. March 1981. pg.1-12*

Offers a conception for newcomers in the computer branch, who want ot do business with computer portraits. Provides calculations, market study, and income possibilities for the branch.

The Industrial Robot

2052. AUTOMAN "81: ROBOTS STEAL THE LIMELIGHT AND POINT THE WAY FORWARD. *IFS (Publications) Ltd. Annual. V.8, No.1. March 1981. pg.16-24*

Lists exhibitors at the 1981 European Automated Manufacturing Exhibition and Conference in England in May 1981. Provides names and addresses for U.K. manufacturers of indus-trial robots. Provides a description of each product and in some instances a photograph.

2053. THE STATE OF ROBOTICS IN ITALY AT THE START OF THE 80'S. *IFS (Publications) Ltd. V.8, No.3. September 1981. pg.176-177*

Examines the industrial robot industry in Italy. Discusses research, labor costs, costs of robots, major manufacturers, and reasons for adapting robots in Italy. One table shows the number of robots in operation in Italy from 1973-1980.

Management Totaal

2054. COMGE VECHT VOOR COMPUTERGEBRUIKER. [COMGE Fights for Computer User.]. *Steenks, Oele. Samson Uitgeverij BV. Netherlandish. March 1980. pg.46-49 12.50 (Netherlands).*

Opinions of the chairman of the COMGE (organization for computer users) in the Netherlands, about standardization in computer appliances and programs from organization line. Examines problems of starting automatization plans and function of government and large industries.

Materials Handling News

2055. U.K. - FASTEST ROBOT GROWTH. *British Robot Association. IPC Industrial Press Ltd. no.293. April 1982. pg.14*

Gives number of industrial robots in U.K. , 1981, 1983; market share of British, American and Japanese made robots in U.K., 1980, 1981; robot production for Japan, USA, Sweden and Germany, 1981; and government spending on robots, 1981, 1982. Describes industrial robots in Japan, USA and U.K.

Mini-Micro Systems

2056. MINI - MICRO WORLD: U.S. COMPUTER MAKERS DOMINATE EUROPEAN MARKET. *Jones, Keith. Cahners Publishing Company. v.15, no.4. April 1982. pg.89*

Graphs 1980 and 1986 sales of general-purpose minicomputers and smallbusiness systems in the Western European market.

MINTEL Market Intelligence Reports

2057. CALCULATORS. *Mitel Publications Ltd. June 1981. pg.21-27*

Original consumer research on electronic calculators other than desk models in the U.K. by the British Market Research Bureau. Gives market penetration, 1973-1981; market size, 1975-1980; products and companies; and distribution, including source of purchase, and future prospects.

Output Osterreich

2058. COMPUTER AUS OSTERREICH. [Computer Made in Austria.]. *Rozsenich, Norbert. Bohmann Druck und Verlag AG. German. June 1979.*

Article surveys the small computer industry in Austria, as well as the production of some on-line apparatus. Examines proportion between production of central units and peripheral apparatus to the mid-eighties.

*OVD / Offentliche Verwahung und
Datenverurbeitung*

2059. DIE WICHTIGSTEN MESSE - THEMEN SIND PER-SPEKTIVEN FUR'S GANZE JAHRZEHNT. [The Most Important Fair Discussions are the Perspectives for the Whole Decade.]. *Schramm, Herbert. Verlagsges. Rudolf Muller GmbH. April 1982. pg.41-45*

Analyzes trends within the computer industry. Examines the microcomputer and the peripheral equipments and software systems, forecasting market segmentation.

ServEx: El Semanario del Comercio Exterior

2060. ARANCEL DE ADUANAS: MAQUINAS CONTABLES Y DE INFORMATICA. [Customs Duties: Electronic Accounting Machines and Computers.]. *Banco de Bilbao. Service de Estudios. Spanish. February 8, 1979. pg.3*

Modifications of Spanish Customs import duties on electronic accounting machines and computers are given with corresponding descriptions and customs code classifications. Duties are given in percent of cost of imports. The royal decree for the changes was published in the Official Spanish State Bulletin (Boletin Oficial del Estado) January 26, 1979.

2061. ARANCEL DE ADUANAS: MAQUINAS DE CALCU-LAR. [Customs Duties: Electronic Calculators.]. *Banco de Bilbao. Service de Estudios. Spanish. no.1366. September 3, 1981. pg.4 Free (Spain).*

Spanish customs duties are modified to change the percent of duties on types of electronic calculators. Decree was published August 13, 1981.

2062. CONSTRUCCION MIXTA DE MAQUINAS DE INFOR-MATICA. [Manufacturing of Computers.]. *Banco de Bilbao. Service de Estudios. Spanish. February 8, 1979. pg.3-4*

Spanish royal decree published in Official Spanish State Bulletin (Boletin Oficial del Estado) January 22, 1979 approves the manufacture in Spain of digital computers and authorizes imports of components necessary for the fabrication of the machines. Percent of decrease in customsimports duties on components is detailed. Regulations and norms for the import of computer parts and for the manufacture of the finished product are provided.

2063. VALORACION DE IMPORTACIONES EN ADUANA: VALOR EN ADUANAS DE SISTEMAS AUTOMATICOS DE MANDO Y DE CONTROL. [Valuation of Imports in Customs: Computers.]. *Banco de Bilbao. Service de Estudios. Spanish. no.1382. April 22, 1982. pg.11-16*

Establishes norms and instruction in the valuation of Spanish imports of computers, industrial automation components, and other control and data processing systems, including software. Discusses foreign exchange related to import valuation determination and gives guidelines. Regulation was published in the official Spanish state bulletin, March 30, 1982.

See 491, 746, 748, 750, 3010, 3020, 3021, 3024, 3027, 3028, 3029, 3039, 3051, 3418, 3420, 3421, 3422, 4317, 4319, 4321, 4326.

Monographs and Handbooks

2064. COMPUTERS: THE ESSENTIAL FOR SENIOR MAN-AGEMENT. *Bedell-Pearce, Keith. Financial Times Business Information Ltd. 1979. $190.00.85.00 (United Kingdom).*

Handbook for purchases and utilization of computers by businesses covers management, employees, choices of products, technology brief, guidelines for contracts, and financial analysis, including investments and returns, cash flow, taxes, leases, and services.

2065. EDV - EUISATZ IN SCHWEIZER BETRIEBEN 1981. [Application of EDP in Swiss Companies 1981.]. *Kuehn, R. Institut fur Automation und Operations Research. German. 1982. 174 pg.*

Shows representative data of installation and application of electronic data processing in business and administration. The investigation is restricted to hardware and software, which dominate in the fields of business and administration. Excludes process computers. Because of operational reasons, systems used for word processing were not taken into consideration.

2066. L'INFORMATIQUE DANS LES ADMINISTRATIONS ET LES ENTREPRISES PUBLIQUES EN FRANCE: AU 1ER JANVIER 1980. [Computers in Administrations and Public Companies in France: First of January 1980.]. *Mission a l'Informatique. Documentation Francaise. French. September 1980. 91 pg.*

Describes use of computers in France: which departments use computers, which kind of computers, types of system used, expenditures and purchases. Provides comparisons between forecasts and carrying out of plans.

2067. PURCHASING COMPUTERS: A GUIDE FOR BUYERS OF COMPUTERS AND COMPUTING EQUIPMENT. *Sambridge, Edward R. Gower Publishing Company Ltd. 1979. 172 pg. $31.00.*

Handbook for computer and computing equipment buyers based on model form.of conditions of contract prepared by the Institute of Purchasing and Supply.

Conference Reports

See 4332, 4333.

Newsletters

2068. INFORMATIONS TECHNOLOGIE UND - MEDIEN. [Information Technology and Information Media.]. *Lange, Wilhem, ed. Wilhem Lange. German. Monthly. 4 pg.*

Newsletter offers information for the German computer and related industries, such as communications. Includes market studies.

2069. MIC BUSINESS MICRO / MINI REPORTS. *Management Information Corporation. Monthly. $175.00/yr.*

Report on micro- and minicomputers developed and sold in Western Europe covers hardware, software, and communications, with prices given in dollars, pounds sterling, and the currency of the nation where the system was developed. Includes system advantages and disadvantages.

Indexes and Abstracts

See 3061.

Dictionaries

2070. THE CONCISE ENCYCLOPEDIA OF COMPUTER TERMINOLOGY. *Stokes, Adrian V. Gower Publishing Company Ltd. 1980. $37.00.*

Defines terms and acronyms from all branches of computer science, including information management and technology, telecommunications, word processing, and microprocessors.

2071. THE PENGUIN DICTIONARY OF MICROPROCESSORS. *Chandor, Anthony. Penguin Books Ltd. 1981. 182 pg. $5.95.*

A comprehensive English-language dictionary of words and phrases used in the computer industry, with an emphasis on new vocabulary arising from advances in microprocessor technology.

Asia

Market Research Reports

2072. CHINESE AND RUSSIAN COMPUTER MARKETS. *International Resource Development, Inc. Irregular. February 1979. 178 pg. $945.00.*

Report contains information on Chinese computer markets and technologies; Russian computer markets and technologies; American computer suppliers' efforts in China and Russia; other Western computer manufacturers' efforts in China and Russia; and marketing strategies and the problem of export control.

2073. COMPUTERS AND PERIPHERAL EQUIPMENT, JAPAN: A COUNTRY MARKET SURVEY. *U.S. Government Printing Office. Irregular. CMS 79-302. March 1979. 11 pg. $.50.*

Report analyzes the current and projected markets for the sale of computers and peripheral equipment in Japan. Data is provided on the total market (production, imports, exports, market size); total employment and number of establishments by size in major user sectors; expenditures by major user sectors; number and value of installed computers by major user sectors; percentage breakdown by end-user applications; value of installed computers by major suppliers; imports by country of origin. Describes the domestic market, major market trends, and the competitive environment.

2074. COMPUTERS AND PERIPHERAL EQUIPMENT, KOREA: A COUNTRY MARKET SURVEY. *U.S. Government Printing Office. Irregular. CMS 81-311. March 1981. 9 pg. $.50.*

Report analyzes the current and projected markets for the sale of computers and peripheral equipment in Korea. Data is provided on the market by product category; number and value of installed computers by major user sectors; expenditures by major user sectors; total employment and number of establishments by size in major user sectors; percentage breakdown by end-user applications; imports by country of origin; value of installed computers by major suppliers. Describes the domestic market, majormarket trends, and the competitive environment.

2075. COMPUTERS AND PERIPHERAL EQUIPMENT, SINGAPORE: A COUNTRY MARKET SURVEY. *U.S. Government Printing Office. Irregular. CMS 80-306. February 1980. 9 pg. $.50.*

Report analyzes the current and projected markets for the sale of computers and peripheral equipment in Singapore. Data is provided on the total market (imports, exports, market size); number and value of installed computers by major user sectors; total employment and number of establishments by size in major user sectors; percentage breakdown by end-user applications; imports by country of origin; value of installed computers by major suppliers. Describes the domestic market, major market trends, and the competitive environment.

2076. COMPUTERS AND PERIPHERAL EQUIPMENT, TAIWAN: A COUNTRY MARKET SURVEY. *U.S. Government Printing Office. Irregular. CMS 81-310. March 1981. 9 pg. $.50.*

Report analyzes the current and projected markets for the sale of computers and peripheral equipment in Taiwan. Data is provided on the market by product category; number and value of installed computers by major user sectors; expenditures by major user sectors; total employment and number of establishments by size in major user sectors; percentage breakdown by end-user applications; imports by country of origin; value of installed computers by major suppliers. Describes the domestic market, major market trends, and the competitive environment.

2077. FUTURE OF THE JAPANESE ELECTRONIC INDUSTRY. *Strategic Inc. October 1980. 626 pg. $425.00.*

Forecasts to 1990 for Japanese electronic products including computers, video discs, lasers, optical fibers, telephone equipment, and home electronics. Reports imports and exports.

2078. INDIA - MARKET RESEARCH REPORT ON COMPUTERS AND RELATED EQUIPMENT. *Sachdeva, Ram P. U.S. Department of Commerce. International Trade Administration. NTIS #ITA 81-11-500. June 1981. 88 pg. $15.00.*

Survey of the market for computers and related equipment in India describes market dimension and potential, sources of supply, end users, and marketing practices. Gives the current and projected total apparent market by major product categories. Gives data on imports by major product category and country of origin. Lists specific U.S. products with good sales potential and provides an assessment of the competitive position of U.S. and third country suppliers. A brief analysis is given of each major end user sector. Trade lists cover principal potential agents and distributors; principal trade, industrial, and professional associations; trade journals; trade fairs and commercial exhibitions; Indian manufacturers, assemblers, licensees of minicomputers, microprocessors, and peripherals; and potential purchasers.

2079. INDUSTRIAL ROBOTS INDUSTRY IN JAPAN. *Benn Publications Ltd. 1982. 230 pg. $595.00.*

Report profiles 50 Japanese robot manufacturers, with market analysis by robot application, suppliers' shares, and evaluation of future technological trends.

2080. THE JAPANESE CHALLENGE IN THE COMPUTER, COPIER AND ELECTRONIC INDUSTRIES. *MSRA, Inc. January 15, 1981. 145 (est.) pg. $850.00.*

Survey of the Japanese mainframe and minicomputer, semiconductor / integrated circuit, office copier, still camera, and related industries (calculators, office equipment, tape recorders, etc.) Discusses major vendors' domestic and export market shares, unit and dollar sales, revenues (by product line), exports, and capital and R & D expenditures during the late 1970's, with projections through 1985.

2081. JAPANESE COMPUTER THRUST: THE NEXT TEN YEARS. *Strategic Inc. July 1982. 125 pg. $1500.00.*

Analysis of the Japanese computer, telecommunications, office automation and electronics industries and impact on markets in the 1980's and 1990's. Examines new products, research and development, market penetration and marketing strategies.

2082. JAPANESE THREAT TO THE U.S. COMPUTER INDUSTRY. *Frost and Sullivan, Inc. March 1981. 22 pg. $400.00.*

Looks at penetration by Japan into worldwide markets, 1974-1984. Provides computer sales in Japan by top six companies. Includes MITI's role, Japanese vendors, and their products.

2083. MARKET RESEARCH SURVEY ON COMPUTERS AND RELATED EQUIPMENT IN THE PHILIPPINES: A FOREIGN MARKET SURVEY REPORT. *SGU and Co. U.S. Department of Commerce. International Trade Administration. NTIS #ITA 81-08-504. April 1981. 221 pg. $15.00.*

Survey of the market in the Philippines for computers and related equipment analyzes the market by major product subcategory, domestic manufacturers, U.S. and third country suppliers, and end user industry sectors. Provides data by major subcategory on market size and imports (including supplying country's share of the market) for current and future years. Lists U.S. products with good sales prospects. Gives data on number and value of computers in use, and projected uses and expenditures, by end user sector. Information is provided on marketing practices, trade restrictions, and technical requirements. Trade lists include principal potential agents and distributors; trade associations; publications; and trade events.

2084. THE ROBOTICS INDUSTRY OF JAPAN: TODAY AND TOMORROW. *The Japan Industrial Robot Association. Fuji Corporation. 1982. 500 (est.) pg. $525.00.*

Report on the robotics industry in Japan based on market surveys. Includes data on production, orders and shipments; size of manufacturers; research and development; installed base, structure and supply of labor; cost comparisons; demand forecasts; and analysis of end use markets.

2085. TECHNOLOGY GROWTH MARKETS AND OPPORTUNITIES: JAPAN'S DRIVE FOR DOMINANCE IN HIGH TECHNOLOGY MARKETS. *Creative Strategies International. v.2, no.2. March 1982. 16 pg. $95.00.*

Analysis of Japan's economy and its drive for dominance in new technologies covers economic indicators, employment and unemployment, government aid to business, financial analysis and debt structure, 1981 foreign investments, research and development, and changes in gross national product, 1980, 1990.

See 515, 524, 525, 527, 528, 529, 3064, 3081, 3429, 3926.

Industry Statistical Reports

2086. THE COMPUTER INDUSTRY IN JAPAN. *Lucas, Rod. Nihon Keizai Shimbun. Vickers da Costa Ltd. January 23, 1980. 5 pg.*

1979 survey of computer usage in Japan provides statistical information regarding market shares held by computer manufacturers in different industries. Covers 503 computer - using businesses and notes contemporary trends in Japanese computer usage.

Directories and Yearbooks

2087. APPLICATIONS AND SPECIFICATIONS OF INDUSTRIAL ROBOTS IN JAPAN. *IFS (Publications) Ltd. Annual. 1982. 700 (est.) pg. 44.00 (United Kingdom).*

Volume details over 150 robots produced by 46 Japanese firms, with firms' addresses, robot specifications, and diagrams outlining general configurations. Includes details of grippers and associated equipment. Covers machining, plastics moulding, assembly, press feeding, spot welding, arc welding, diecasting, transfer handling, painting, forging, inspection and measurement, heat treatment, and miscellaneous applications. Each application includes name of robot supplier and user with diagrams and information about the application.

See 531, 3090.

Special Issues and Journal Articles

Asian Finance

2088. EDP IN ASIA. *Asian Finance Publications Ltd. July 15, 1981. pg.38-40*

Examines the potential for growth in the Asian computer industry. Provides a financial profile for 13 of the world's top computer manufacturers in Asia. Includes 1980 sales, assets and net income. Gives the number of computer installations and computers in 7 Southeast Asian countries as of June 30, 1981.

2089. JAPAN'S COMPUTER VISION SURVIVES SPYING CHARGES. *Asian Finance Publications Ltd. v.8, no.7. July 1982. pg.94*

Discusses U.S. versus Japanese computers and includes figures for manufacturer shares of Japanese market giving Japanese and foreign producers, number of systems, percentage share of total, and value.

Business Japan

2090. AUTOMATION REVOLUTIONIZES OFFICE PROCEDURES. *Furusawa, Akira. Nihon Kogyo Shimbun Publications. December 1982. pg.101-106*

Examines trends in Japan's office automation equipment industry including facsimile transmission systems. Annual data includes office computer shipments, 1976-1980, with a forecast of shipments, 1981-1985, 1990, and value and volume of facsimile production, 1973-1980.

Business Week

2091. JAPAN: THE BOTTOM DROPS OUT OF HIGH-TECH STOCKS. *McGraw-Hill Book Company. Industrial Edition, no. 2738. May 10, 1982. pg.49-52*

Analyzes changing investment trends from Japanese blue-chip high technology export companies toward domestically oriented Japanese companies. Provides graphical data for 1981 and 1982 second quarter percentage of shares owned by foreigners for Hitachi, Fujitsu, Nippon Electric, and Pioneer Electric as compared with domestically oriented companies including Kashiyama and Co., Tokyo Style Co., Renown Inc., and Taisei Corp.

China Business Review

2092. COMPUTER SALES TO CHINA. *Berney, Karen. National Council for U.S. - China Trade. September 1980. pg.25-32*

Reviews exports of computers to China and Chinese technology. Shows 1974-1980 exports and value of licenses for sales.

Computer Business News

2093. STUDY SEES JAPAN LOSING IN BID FOR U.S. MARKET. *Strategic Inc. CW Communications Inc. v.5, no.27. July 5, 1982.*

Examines Japanese penetration of U.S. computer market, 1980-1992, based on Strategic Inc.'s 'Japanese Computer Thrust: The Next 10 Years (No. 809).' Discusses marketing strategies.

Datamation

2094. COMPUTING IN THE NEW INDIA. *Technical Publishing Company. June 1980. pg.146-150*

Market analysis for computers and data processing services in India, 1978. Shows installations by vendor and end use sector.

Diamond's Industria

2095. FEATURES: INDUSTRIAL ROBOTS IN JAPAN. *Diamond Lead Company, Ltd. v.12, no.7. July 1982. pg.8-20*

Analysis of the Japanese industrial robot industry, including the corporations, product lines, technology and government aid. Shows production and market value, 1970-1981; product group breakdown, 1979-1981; shipments by end use industries, 1979-1980; rankings of manufacturers, 1981; manufacturers' sales, 1980-1982; and production forecasts for 1986.

Electronics

2096. U.S. BEACHHEAD FOR JAPANESE COMPUTERS IS ONLY THE START. *Durniak, Anthony. McGraw-Hill Book Company. March 27, 1980. pg.113-136*

Review of Japanese computer industry examines growth and competitive efforts in U.S. and other world markets. Shows production, imports and exports (1961-1985), and Japanese market shares by computer size.

FEM (Financieel-Economisch Magazine)

2097. HITACHI: FLEXIBEL EN HARD. [Hitachi: Flexible and Hard.]. *B.V. Uitgevers-Maatschappy Bonaventura. Netherlandish. May 1, 1980. pg.11-18 5.30 (Netherlands).*

Examines present situation of and provides forecasts for the Japanese firm Hitachi. Reviews competition from I.B.M. diversification, research and development, and micro-computer developments and innovations.

Fuji Bank Bulletin

2098. INDUSTRY REPORT: INDUSTRIAL ROBOTS. *Fuji Bank Ltd. Research Division. v.32, no.3. March 1981. pg.55-57*

Article examines the industrial robot industry in Japan. Discusses what they are; classifies them according to type; pro-

vides data on 1979 production; and summarizes the advantages, product growth, and industry structure. One table shows value of shipments (1975, 1978), by demand sector.

Japanese Research

2099. IBM VERSUS JAPAN. *Vickers da Costa Ltd. May 1981. pg.9-28*

Report provides sales, product group and export information regarding the ten mainframe computer businesses operating in Japan. Highlights the relative strength and weaknesses of each company in the Japanese and international markets.

Siemens Review

2100. JAPAN AND EUROPE: DOES THE MARKET HAVE ROOM FOR BOTH? *Siemens Aktiengesellschaft. March 1980. pg.2-6*

Examines the competition between Japan and Europe in the high technology products industry. Data includes market share of the EEC, Japan, and U.S. (1960-1978); number of U.K., French, German, and U.S. contracts (1971, 1975); and market size of U.S., Japanese, and European market (1950-1985). Reviews Japan's export policy, trade in components, overcapacity, and the telecommunications market.

The Wheel Extended

2101. MECHATRONICS SWEEPS JAPAN. *Higuchi, Takeshi. Toyota Motor Sales. v.12, no.1. January 1982. pg.20-29*

Analyzes increasing use of industrial robots in Japan. Summarizes in the following tables: industrial robot production in units and in value, from 1968-1980; percentage breakdown by type of robot produced in units and in value, 1977-1980; value of machine tool production including NC machines from 1971-1981; industrial robot sales broken down by industrial sector, 1976-1980; and purchases broken down by use in units and in value, 1978-1980. Source is Ministry of International Trade and Industry.

See 3097, 3112, 3430.

Monographs and Handbooks

2102. ROBOTS IN THE JAPANESE ECONOMY. *IFS (Publications) Ltd. Irregular. 1982. 256 pg.*

Discusses industrial robots in Japan; research and development; robots and their applications; technological development and small and medium-sized firms; future for industrial robots; and problems involving introduction of industrial robots. Appendix lists major Japanese robot-building firms.

See 755.

Newsletters

2103. EDP JAPAN REPORT. *International Data Group. Biweekly. $98.00/yr.*

Covers all phases of the Japanese computer industry including analyses, forecasts, legal decisions, new products, and annual reviews.

2104. JAPAN COMPUTER NEWS. *Fuji Corporation. Monthly. 8 (est.) pg. $80.00/yr.*

Monthly newsletter covers Japanese computer industry events, new technology, research and development, business transactions, company profiles, market trends, and statistics.

Indexes and Abstracts

See 3122.

Oceania

Market Research Reports

2105. COMPUTERS AND PERIPHERAL EQUIPMENT, AUSTRALIA: A COUNTRY MARKET SURVEY. *U.S. Government Printing Office. Irregular. CMS 81-309. March 1981. 11 pg. $.50.*

Report analyzes the current and projected markets for the sale of computers and peripheral equipment in Australia. Data is provided on the total market (production, imports, exports, market size); expenditures by major user sectors; total employment and number of establishments by size in major user sectors; number and value of installed computers by major user sectors; percentage breakdown by end-user applications; imports by country of origin; value of installed computers by major suppliers. Describes the domestic market, major market trends, and the competitive environment.

See 3434, 3435, 3946.

Directories and Yearbooks

2106. DP INDEX. *Peter Isaacson Publications Pty. Ltd. Annual. 1980. 223 pg. 25.00 (Australia).*

Directory of data processing companies arranged by product and/or service group. Covers DP and personnel consulting; ancillary equipment and services; data communications; engineering; construction / maintenance / transportation; hardware systems; microfilm / microfiche; peripherals; processing; software; and word processors.

2107. RYDGE'S EDP MANUAL. *Rydge Publications Pty. Ltd. Annual. 1981. 292 pg. 30.00 (Australia).*

Computer manufacturer and supplier listings provide address, telephone, branches, managers, and models available. Other listings cover software, ancillary equipment, and services suppliers.

See 3436.

Middle East

Market Research Reports

See 534.

Special Issues and Journal Articles

The Israel Economist

2108. INPUT - OUTPUT. *Kolleck and Sons Ltd. January 1982. pg.23*

A brief article describing increased use of computers in Israel contains tables giving number of installations at the end of 1979 and 1980, by size; rental fees per month; and installations by sector.

Africa

Market Research Reports

See 541, 542, 543.

Electrical and Electronic Equipment

International

Market Research Reports

2109. AUTOMATIC DISCRETE WIRING: FROM PROTOTYPE TO PRODUCTION. *BPA Management Services. September 1981. 2800.00 (United Kingdom).*

Report on automatic discrete wiring worldwide. Examines utilization of new products, production costs, industrial plants and equipment, performance, competition with traditional PCB's, and raw material consumption. Market analysis, 1981-1986, covers end uses, franchises, patents, and licenses.

2110. THE BUSINESS OF THICK FILM. *BPA Management Services. November 1981. $1100.00.*

Analysis of changes in the structure of the electronic interconnection industry and the impact on the thick film business.

2111. CARBON AND GRAPHITE FIBERS. *Predicasts, Inc. January 1982. $375.00.*

Examination of new technology for production of new products from carbon and graphite fibers. Includes industrial capacity by manufacturers and countries, consumption and price trends, and end uses.

2112. THE CHANGING BATTERY MARKET: DEVELOPMENT / TRENDS. *Business Communications Company. January 1981. 160 (est.) pg. $825.00.*

Evaluates 1979-1980 U.S. and world markets for primary and secondary batteries, with market value projected through 1985 (by detailed product group and end use). Discusses manufacturers' shipments, sales, imports / exports, growth rates, and such major secondary markets as photovoltaics and electrical and conventional motor vehicles.

2113. ELECTRIC VEHICLE: INEVITABLE? *Business Communications Company. Irregular. E-016R. October 1980. 204 pg. $800.00.*

International market, industry, and technology analyses and forecasts cover manufacturers; battery developments by type; electric vehicles currently in use; companies involved; costs; and U.S. Dept. of Energy activities. Contains 50 tables.

2114. ELECTRONIC CONTROL SYSTEMS. *Larsen Sweeney Associates Ltd. Annual. 1981. 695.00 (United Kingdom).*

Analysis of farm markets for electronic control systems, with forecasts to 1986. Includes marketing and industry analysis by country and surveys of farm users' attitudes and farm machinery suppliers.

2115. ELECTRONIC GAMES AND TOYS. *Predicasts, Inc. December 1980. $325.00.*

International market analysis for electronic games and toys, educational aids, and translators. Examines industry structure and new products.

2116. ELECTRONIC MARKETS AND INDUSTRIES. *Larsen Sweeney Associates Ltd. Annual. 1981.*

Market analysis for electronics with forecasts to 1986. Includes analysis of the industry and marketing by country.

2117. ELECTRONICS IN THE DEVELOPING WORLD. *Predicasts, Inc. January 1980.*

Reviews company activities, markets, and trade for electronics in the developing world. Historical and projected regional data on components, consumer electronics, communications, and computer products.

2118. INDUSTRIAL ENGINES. *Larsen Sweeney Associates Ltd. Annual. 1981. 995.00 (United Kingdom).*

Analysis of markets for industrial engines, with forecasts to 1986. Covers 11 product groups. Includes analysis of marketing and the industry by country and surveys of end users and suppliers.

2119. INTERNATIONAL MARKETING OPPORTUNITIES / ELECTRONICS. *Ness, E. McGraw-Hill Book Company. Electronics. 1980. $154.00.*

International market opportunities for computers and peripherals, production and test equipment, communications systems and equipment, electronic components, office and business machines, and security systems. Extracted from U.S. government intelligence reports.

2120. LSI IN ELECTRONICS. *Predicasts, Inc. March 1980.*

Discusses technological developments and applications for large-scale integration (LSI) and very-large-scale integration in various nations. Includes company activities and current and potential LSI markets.

2121. THE MARKET FOR ELECTRONIC SWITCHES. *Marketdata Enterprises. FIND / SVP. 1980. 161 pg. $225.00.*

Study covers component manufacturers in 10 foreign nations who represent competition abroad for U.S. switch producers. Includes an analysis of commercial, industrial, military, and consumer markets for electronic switches. Complete examination of the industry structure and a helpful section containing a computer-searched survey of published magazines and trade journal articles are included.

2122. MICROPROCESSORS. *Predicasts, Inc. August 1980. $325.00.*

International market analysis report for microprocessors. Covers industry structure, prices, end use industries, manufacturers and market areas, by country.

2123. OPTOELECTRONICS. *Predicasts, Inc. December 1981. $375.00.*

Worldwide market analysis for optoelectronic devices. Includes technology and manufacturers.

2124. PROFILE OF THE WORLD SEMICONDUCTOR INDUSTRY: COMPLETE EDITION. *Benn Publications Ltd. June 1982. 300 pg. $595.00.*

Analysis of the semiconductor market and industry worldwide includes production and market forecasts to 1986, manufacturers' market shares and facilities, expansion and foreign investments, and new technology. Profiles facilities including addresses, industrial plant and equipment, number of employees, size of facilities and sales in 1981.

2125. RESISTORS INDUSTRY. *Marketdata Enterprises. FIND / SVP. 1981. 205 pg. $295.00.*

Based on Commerce Dept. and private sector information, report analyzes U.S. and selected non-U.S. markets; competition by state and country; and nature and composition of major end uses. Sections include selling to and collecting information from the U.S. government; export markets and import competition; industry economic structure; and bibliographic abstracts of technological developments and sales forecasts to 1990.

2126. SEMICONDUCTOR ASSEMBLY EQUIPMENT AND MATERIALS OUTLOOK - 1985. *Mackintosh Consultants, Inc. October 1980. $975.00.*

Forecasts Western European, U.S., Japanese, and rest of world markets through 1985 for wafer dicing, die attach, wire bonding, and packaging (injection molding / sealing, marking, etc.) equipment. Covers packaging materials and lead frames.

2127. SEMICONDUCTOR MICROLITHOGRAPHY EQUIPMENT AND MATERIALS OUTLOOK - 1985. *Benn Publications Ltd. 1981. $975.00.*

Market forecasts to 1985 for Europe, U.S., Japan, and rest of the world cover wafer exposure equipment (including optical, electron beam, and x-ray) and reticle mask making equipment (including pattern generation, mask making, and inspection). Evaluates technology trends and suppliers' market shares.

2128. WELDING CONSUMABLES. *Larsen Sweeney Associates Ltd. Annual. 1981. $3950.00.*

Analysis of markets for welding materials, with forecasts to 1986. Covers 17 product groups. Includes analysis of marketing and the industry by country and surveys of end users and suppliers.

2129. WELDING MACHINES. *Larsen Sweeney Associates Ltd. Annual. 1981. 995.00 (United Kingdom).*

Analysis of markets for welding machines, with forecasts to 1986. Covers 14 products groups. Includes analysis of marketing and the industry by country and surveys of end users and suppliers.

2130. THE WORLD MARKET FOR ELECTRICAL PRODUCTS: VOLUME 1: IMPORTS INTO 126 THIRD WORLD COUNTRIES. *Goulden, O.A. O.A. Goulden and Partners. 1981. 129 pg. 150.00 (United Kingdom).*

Statistics for imports of electrical products into each of 126 Third World countries from 15 leading industrial nations, 1974-1979. Broken down by product groups, including motors and alternators; wires and cables; and transformers, switchgear, controls, and accessories. Also includes demographic statistics, growth rates, foreign trade, installed base capacity for power generation, total sales of electricity and average production for each country. Updates 1978 publication.

2131. THE WORLD MARKET FOR ELECTRICAL PRODUC-TS: VOLUME 2: STATISTICAL REPORT FROM 15 INDUS-TRIAL COUNTRIES. *Goulden, O.A. O.A. Goulden and Partners. 1981. 150.00 (United Kingdom).*

Export statistics for electrical products from fifteen leading industrial nations, 1974-1979. Lists foreign trade partners, and market value of exports. Product groups covered include motors and alternators, wires and cables, transformers, switchgear, controls, and accessories. Updates 1978 publication.

2132. THE WORLD MARKET FOR ELECTRICAL PROD-UCTS: VOLUME 9: THE WORLD MARKET FOR MCB'S, MCCB'S AND MOTOR CONTROL COMPONENTS. *Goulden, O.A. O.A. Goulden and Partners. July 1981. 300 (est.) pg. 150.00 (United Kingdom).*

World market analysis for MCB's, MCCB's and motor control components. Contains tables for major countries showing installed base capacity, growth rates, total electricity sales, average power generation per capita, demographic statistics, market value, and market shares of manufacturers.

2133. THE WORLD MARKET FOR ELECTRICAL PROD-UCTS: VOLUME 10: THE WORLD MANUFACTURERS OF MCB'S, MCCB'S AND MOTOR CONTROL COMPONENTS. *Goulden, O.A. O.A. Goulden and Partners. July 1981. 300 (est.) pg. 150.00 (United Kingdom).*

World market analysis of MCB's, MCCB's and motor control components. Lists manufacturers' addresses, factories, products, and affiliates.

2134. THE WORLD MARKET FOR ELECTRICAL PROD-UCTS: VOLUME 13: THE WORLD MARKET FOR ELECTRI-CAL ACCESSORIES. *Goulden, O.A. O.A. Goulden and Partners. January 1980. 200 (est.) pg. 150.00 (United Kingdom).*

World market analysis for electrical accessories including switches, sockets, and plugs. For each of 154 countries, gives domestic market, imports, market shares, manufacturers, and demographic statistics.

2135. THE WORLD MARKET FOR ELECTRICAL PROD-UCTS: VOLUME 14: THE WORLD MANUFACTURERS OF ELECTRICAL ACCESSORIES. *Goulden, O.A. O.A. Goulden and Partners. January 1980. 200 (est.) pg. 150.00 (United Kingdom).*

Analysis of markets and industrial structure for electrical accessories, including switches, sockets and plugs, in 154 countries. Gives addresses and profiles of manufacturers.

2136. THE WORLD MARKET FOR ELECTRICAL PROD-UCTS: VOLUME 15: FACTS AND FORECAST 1981-2000: A CORPORATE PLANNING SUMMARY FOR THE ELECTRI-CAL INDUSTRY. *Goulden, O.A. O.A. Goulden and Partners. June 1981. 150.00 (United Kingdom).*

Summary facts from world market analysis covering electrical products such as motors and alternators, switchgear, transformers, wires and cables, controls, and accessories. Includes forecasts covering foreign trade and manufacturers market shares, 1980-1990.

2137. THE WORLD MARKET FOR H.T. SWITCHGEAR. *O.A. Goulden and Partners. 1982. 284 pg. 250.00 (United Kingdom).*

Summarizes the supply authority and its system for each country. Indicates market size and market shares for the following three sectors: 1-36KV, 37-110KV; and over 110 KV.

2138. WORLDWIDE DISPLAY DATABASE. *Gnostic Concepts, Inc. April 1981. $7500.00.*

Worldwide market analysis and database for display devices for electronic equipment. Gives 1980 production of electronic equipment and forecasts, 1981-1990. Marketing database gives complete technical specifications for display products.

2139. WORLDWIDE PHOTOVOLTAICS MARKET. *Frost and Sullivan, Inc. July 1981. 298 pg. $1250.00.*

Market analysis and forecasts for 1990 and 2000 for photovoltaic cells and systems worldwide. Examines end uses, current and projected installations by use, sales and prices, power generation capacity, capacity of industrial plants and equipment, industry structure, competition, market penetration opportunities, market shares, major manufacturers, market areas, and government funding for research and development.

2140. WORLDWIDE SEMICONDUCTOR INDUSTRY OUT-LOOK - 1985. *Mackintosh Consultants, Inc. 1980.*

Forecast of the market for semiconductor manufacturing, materials and equipment to 1985, including share for U.S. companies. Capital expenditures forecast including effect on over-capacity and breakdown of expected purchases for major equipment categories.

See 17, 759, 775, 783, 784, 785.

Investment Banking Reports

2141. DREXEL BURNHAM LAMBERT INC.: SEMICONDUC-TOR INDUSTRY TREND ANALYSIS AND SHIPMENT FORE-CAST. *Glinski, Vincent J. Drexel Burnham Lambert Inc. Monthly. April 5, 1982. 22 pg.*

Monthly analysis of the world semiconductor market. Includes trade association data for orders and shipments, 1979-1982; dollar and unit sales and prices, 1980, by product groups; foreign trade; forecasts of market value, 1980-1983; investment ratings of corporations; stock prices, 1981-1982; earnings per share, 1980-1983; price earnings ratio, 1980-1983; dividends; and yield.

Industry Statistical Reports

2142. ENGINEERING INDUSTRIES IN OECD MEMBER COUNTRIES, 1976-1979. *Organization for Economic Cooperation and Development. Publications and Information Center. February 1982. 93 pg. $10.00.*

1976-1979 delivery statistics cover 100 products and 28 product groups, including tools, power generating machinery, agricultural machinery, computers, office equipment, metal working machinery, electrical equipment, telecommunications equipment, motor vehicles, scientific and medical equipment, aircraft and related equipment, ships, boats, clocks, and watches.

2143. SEMICONDUCTOR INDUSTRY ASSOCIATION: 1979 YEARBOOK AND DIRECTORY. *Semiconductor Industry Association. Annual. 1979. 39 pg. $35.00.*

Provides statistical data of the semiconductor industry: world and domestic shipments by U.S. based companies (1954-1978); semiconductor shipments by quarters (1966-1979);

average selling prices of transistors (1960-1978); world shipments by product lines (1973-1978); by distribution share (1973-1978); by geographic market (1973-1978), and forecasts by products (1979-1981). Discusses trade policy and lists producers and SIA members.

2144. SIA MONTHLY TRADE STATISTICS PROGRAM. *Semiconductor Industry Association. Monthly. 1980. $900.00.*

Worldwide sales and order data by 34 product lines and 13 geographic regions.

See 3452.

Financial and Economic Studies

2145. THE IMPACT OF MICRO-ELECTRONICS. *United Nations Publications. Sales Section. 1980. 109 pg. $11.50.*

Coverage spans the nature of the new information technology; characteristics and direct cost; a world industry in a world economy; applications and sectoral employment effects; spread of the new technology; international competition and employment; and effects on developing countries.

2146. IMPLICATIONS OF MICRO-ELECTRONICS FOR DEVELOPING COUNTRIES: A PRELIMINARY OVERVIEW OF THE ISSUES. *United Nations. Industrial Development Organization Industrial Information Section. Industrial and Technological Information Bank. UNIDO-UNIDO/IS-246/Cors.1. 1981. 45 pg. Free.*

Report covers microelectronics for the following: the electronics industry, information activities (electronic data processing, computers, and telecommunications industries), industrial applications, and special applications to developing countries. Includes national actions by developing countries (government policy, technological capabilities); and the role of international organizations.

2147. LONG-RUN SUPPLY OF GALLIUM FOR PHOTOVOL-TAICS. *Charles River Associates. 1980. 161 pg. $30.00.*

Study of the supply of gallium available for photovoltaics. Forecasts demand to 2020. Examines resources, technology for extraction, cost of recovery as a by-product of aluminum production, capacity and required investments.

2148. WORLD ENGINE STUDY, 1979. *John Martin Publishing Ltd. October 1979. 250.00 (United Kingdom).*

Study of world production of industrial engines, 1979. Analysis by fuel, capacity, and area of utilization. Includes captive and non-captive manufacturers.

Forecasts

2149. MARKET MONITOR RESEARCH PROGRAMME: CONSUMER ELECTRONICS. *E.R.C. Statistics International Ltd. April 1980. 3000.00 (United Kingdom).*

Analysis of industrial market trends for the world-wide consumer electronics industry from 1974 to 1979, with 1980 estimates. Includes compilation of consumer data analyzing market penetration, household consumption and expenditures, consumer purchases, product usage, retail prices, consumer buying behavior, retail distribution, brand shares, volume / value and profile of consumption. Computer-based forecasts included.

2150. MICROELECTRONICS INTO THE EIGHTIES. *Mackintosh Publications Ltd. 1st ed. 1979. 30.00 (United Kingdom).*

A digest of international scenarios and government strategies for the microelectronics industry in the 1980's.

2151. PREVISIONS GLISSANTES DETAILLEES: HORIZON 1986: VOLUME 3: INDUSTRIES D'EQUIPEMENT. [Detailed Rolling Forecasts: Horizons 1986.]. *Bureau d'Informations et de Previsions Economiques. French. Annual. June 1981. 220 (est.) pg. 1999.20 (France).*

Forecasts world demand, through 1986, for individual segments of the industrial machinery, electrical and electronic equipment, and transportation equipment sectors. Forecasts are presented by consuming nation.

See 795, 796.

Directories and Yearbooks

2152. ELECTRONIC DESIGN'S GOLD BOOK 1979-1980. *Hayden Book Company, Inc. Annual. 1979. $35.00.*

Provides a catalog and directory of suppliers of electronic components / systems. Product directory lists products and services alphabetically, with suppliers divided into 3 subgroups. Contains a trade name directory with over 9,000 entries; manufacturers directory, with over 10,000 U.S. and non-U.S. listings; and 1600 suppliers, organized alphabetically and geographically. Entries include names, addresses, telephone numbers.

2153. ELECTRONIC INDUSTRY TELEPHONE DIRECTORY 1982. *Harris Publishing Company. Annual. 1982.*

Provide 60,000 listings of manufacturers, representatives, and distributors throughout the world. Includes company name, address, and telephone. Yellow pages section lists firms alphabetically by 3500 product headings.

2154. FISHING ELECTRONICS 1980. *IPC Business Press Ltd. Annual. 1979. 5.00 (United Kingdom).*

A guide to the worldwide availability of electronic equipment for the fishing vessels installation.

2155. INFORMATION SOURCES ON THE ELECTRONICS INDUSTRY. *United Nations Publications. Sales Section. Irregular. 1979. 103 pg. $4.00.*

Covers professional, trade, and research organizations, learned societies, and special information services; directories; sources of statistics, marketing, and other economic data; basic handbooks, textbooks, and manuals; monograph series; current periodicals; current abstracting and indexing periodicals; proceedings, papers, and reports; specialized dictionaries and encyclopedias; bibliographies; and other potential sources of information.

2156. TOWERS' INTERNATIONAL TRANSISTOR SELECTOR. *Towers, T.D. Tab Books, Inc. Irregular. 3rd ed. February 1982. 280 pg. $19.95.*

Listing of transistor products with technical specifications, performance ratings, utilizations, manufacturers and addresses of manufacturers.

2157. WORLDWIDE SURVEY AND DIRECTORY ON ROBOTICS. *Robot Institute of America. Annual. 1982. $26.00.*

Annual directory provides detailed information on the worldwide robot population, applications and distribution of equipment. Includes projected forecasts and a directory listing the names and addresses of each country's leading robotic organizations. All information supplied by the national coordinators of the 18 countries surveyed.

2158. 1982 SOLID STATE PROCESSING AND PRODUC-TION WORLDWIDE BUYERS GUIDE AND DIRECTORY. *Technical Publishing Company. Annual. 1982. 500 pg. $50.00.*

Lists industry's suppliers and includes 20,000 company products and services listings.

See 27.

Special Issues and Journal Articles

American Metal Market

2159. GALLIUM ARSENIDE GROWS IN SPEEDY ELEC-TRONIC USES. *Brown, Stuart F. Fairchild Publications. v.90, no.65. April 5, 1982. pg.34-35*

Reviews growing use for integrated circuits. Consumption and prices for 1981 are compared to 1990 forecasts. End uses and market value are detailed by other applications and company activities are highlighted.

American Printer

2160. DRUPA 82: A TRIP TO THE INDUSTRY'S LARGEST TECHNOLOGICAL SUPERMARKET. *Berglund, Elizabeth G. Maclean Hunter Publishing Corporation. Irregular. v.189, no.5. August 1982. pg.34-42*

Describes products shown by exhibitors at DRUPA 82.

The Battery Man

2161. THE ELECTRIC BOAT: A GROWING MARKET FOR THE BATTERY INDUSTRY. *Palumbo, Donna J. Independent Battery Manufacturers Association, Inc. v.24, no.3. March 1982. pg.5-11*

Provides data on a new market for batteries, resulting from increased popularity of electrically powered boats. Data includes statistics documenting sales of electric boats. Much of the information included in the document provided by the follwoing organizations: Lead Development Association, Lead Industries Association, and Gould Inc.

2162. INTERNATIONAL SLIG BUYERS' GUIDE: STARTING, LIGHTING, IGNITION AND GENERATING SYSTEMS: 1979. *Independent Battery Manufacturers Association, Inc. Biennial. 6th ed. 1979. 95 pg. $5.00.*

1979 buyers guide for starting, lighting, ignition and generating systems lists battery manufacturers in the U.S. and 94 other countries. Lists companies who supply equipment, components, and services used by battery manufacturers, as well as manufacturers of parts, tools and equipment used by retailers, wholesalers and fleets. Provides names and addresses.

Electrical Review

2163. CHIPS CHANGE DRIVES AND CUT CONSUMPTION. *IPC Electrical-Electronic Press Ltd. Division of IPC Industrial Press. v.211, no.13. October 22, 1982. pg.24*

Article discusses future of electric drives and power electronics. Illustration shows 1982-1985 forecasts for sales of stepper motors. Table shows 1985-2000 U.S. electric energy saving from power conversion and controls systems.

2164. INTERNATIONAL BUSINESS: MARKET RESEARCH IN ELECTRICAL PRODUCTS. *IPC Electrical-Electronic Press Ltd. Division of IPC Industrial Press. v.210, no.20. May 21, 1982. pg.20-21*

Presents tables showing electricity demand, 1980 production by country, 1975-1980 growth by country, and 1980 installed capacity by country. Examines Algerian imports of electrical products from 20 leading industrial countries, exports from France and West Germany by country, and world market for electrical accessories in terms of standard plug and socket systems.

Electronic Business

2165. ELECTRONICS MARKETS TEN-YEAR WORLDWIDE FORECAST FOR ELECTRONIC EQUIPMENT AND COMPO-NENTS. *Gnostic Concepts, Inc. Cahners Publishing Company. Irregular. February 1981. pg.92-96*

Forecast (1980-1990) provides outlook for the electronic equipment market worldwide. Graphs and charts display worldwide production of equipment (1980-1990), by type and by geographic region; and worldwide purchases (1980-1990) of electronic components, arranged by active, passive, electromechanical interconnection and mechanical components.

2166. FOR SEMICONDUCTOR MAKERS, THE UPTURN PROVES ELUSIVE. *Arnold, William F. Cahners Publishing Company. v.8, no.4. April 1982. pg.40-52*

Provides statistical data relative to future market trends for the semiconductor industry. Five tables provide data for free world semiconductor production, including percent of worldwide production of United States, Japan, Western Europe and rest of world for 1980-1984; free world semiconductor consumption, 1980-1984; ranking of semiconductor companies; U.S. manufacturers of semiconductors ($ million), including figures for MOS, bipolar and IC total for 1981-1982 and percent of yearly change; and production figures for 1981 ($ billion), including producers, consuming area and total production.

2167. WORLDWIDE SEMICONDUCTOR USE WILL TRIPLE BETWEEN 1979 AND 1985. *Mackintosh Consultants, Inc. Cahners Publishing Company. Irregular. February 1981. pg.116-120*

Discusses worldwide semiconductor market. Forecasts (1979-1985) semiconductor consumption by region and shipments by type.

Electronic News

2168. ELECTRONICS NEWS: LOOKING AT THE LEADERS 1981. *Fairchild Publications. Annual. July 20, 1981. 63 pg.*

Review of leaders in the electronics industry (1981) ranks U.S. and foreign companies giving electronic sales. Gives stock price earnings, price earnings ratios, and projected 1981 sales. Provides company profile; officers; facilities; directors; net sales and net profit, 1976-1980; number of common shares; and earnings.

Electronics

2169. BUSINESS: AUTOMATION MAY ERASE ADVAN-TAGES OF OFFSHORE ASSEMBLY OPERATIONS. *Lineback, J. Robert. McGraw-Hill Book Company. v.55, no.8. April 21, 1982. pg.92, 95*

Covers the automation of U.S. integrated circuit chip industry. Data includes hourly pay rates in dollars for seven foreign countries. Data is for 1980 and from Integrated Circuit Engineering Corp.

2170. ELECTRONICS ABROAD. *McGraw-Hill Book Company. 1981. pg.2-3*

Analyzes markets for electronic equipment in selected foreign countries. Shows retail sales.

2171. ELECTRONICS: ANNUAL TECHNOLOGY UPDATE. *McGraw-Hill Book Company. Annual. v.55, no.21. October 20, 1982. 328 pg.*

Reviews major technological developments in international electronics fields in 1982. Features a section on new products in seven major categories of electronics. Includes technical specifications and manufacturers' names, addresses, and telephone numbers.

2172. RECESSION'S BITE IS SHALLOW, SO FAR. *McGraw-Hill Book Company. June 19, 1980. pg.100-101*

Reviews supply and demand in the world electronics industry. Forecasts semiconductor sales, 1979-1982.

Elektronik Report

2173. ELEKTRONIK - HERBSTMESSEN 1981. [Electronics - Autumn Fairs 1981.]. *Verlag Technik Report Ges. m.b.H. German. Annual. September 1981. 122 pg.*

Special supplement section devoted to the Autumn 1981 Fairs on Electronics: INELTEC (Basel, Switzerland, 8-12 Sep.); IE "81 (Vienna, 14-17 Oct.); systems (Munich, 19-23 Sep.); productronica (Munich, 10-14 Nov.); and Laser "81 / Opto Elektronik (Munich, June 1981). Brief reports on all the new products exhibited in these fairs, with illustrations. Gives manufacturers and addresses.

Elektronikschau

2174. IE "81: BESUCHERREKORD. [ie 1981: A Record of Visitors.]. *Technischer Verlag Erb Ges. m.b.H. German. November 1981. pg.56-66*

Reviews the sixth international ie (industrial electronics) Exhibit / Fair in Vienna, October 11-17, 1981, where 244 participant firms from 14 countries attracted more than 25,000 visitors. Stresses the new products shown: 16 measuring instruments, 6 soldering techniques, and 6 electronic circuit elements, together with names and addresses of manufacturing firms.

Energy International

2175. ENERGY INTERNATIONAL MANUFACTURERS INDEX AND BUYERS' GUIDE 1980. *Miller Freeman Publications Inc. Annual. December 1980. pg.60-92*

1980 annual directory provides alphabetical listing of equipment and products used by energy supply and generating industries. Gives principal manufacturer and code to indicate country of head office. Includes list of manufacturers, with addresses, alphabetically and geographically.

FEM (Financieel - Economisch Magazine)

2176. HALFGELEIDER: PUNTJE VAN DE ELEKTRONIKA-YSBERG. [Semi-Conductors: Top of the Electronics Iceberg.]. *Veenis, S. B.V. Uitgevers-Maatschappy Bonaventura. Netherlandish. June 12, 1980. pg.42-45 5.30 (Netherlands).*

Projections of current needs and future possibilities in the semiconductor market world wide. Examines present level of semiconductor sales world wide. Projects domination of the future electronics industry by a small number of companies. Examines manufacturing large frame systems from chips and development of software capabilities. Only electronic industries with sales of $2 billion or more, can afford to develop own semiconductor manufacturing.

IBM Journal of Research and Development

2177. RECENT IBM PATENTS. *International Business Machines Corporation. IBM. Bi-monthly. v.26, no.2. March 1982. pg.266-267*

Lists patent number, inventor(s), and patent title of recently issued IBM patents. Arranged by name of country in which the inventions were developed.

Microwaves

2178. 1982 INTERNATIONAL MICROWAVE SYMPOSIUM: JUNE 15-17, 1982. *Hayden Book Company, Inc. Irregular. v.21, no.5. May 1982. pg.57-116*

Contains abstracts, technical report numbers, and author information for 1982 symposium papers. Divides the material into 21 categories introduced by field experts discussing trends. Provides product listings with symposium booth numbers and an exhibitor key with corporation names and booth numbers.

Mini-Micro Systems

2179. INTERPRETER: POTENTIALLY HUGE MARKETS DEVELOPING FOR ROBOT SYSTEMS. *Cahners Publishing Company. v.15, no.4. April 1982. pg.105-112*

Tables give information on market potential for robotic systems. Data includes: market growth forecasts, 1980-1990; typical robot system breakdown; and technical improvement needs.

Siemens Review

2180. AUTOMATION KEEPS COMPETITIVENESS UP. *Siemens Aktiengesellschaft. v.49, no.5. September 1982. pg.2-6*

Analysis covers international productivity and labor costs; process and production automation; automation hierarchy drives, controls, on-site programming, simplified programming, ancillary controls); robots with strong growth rates; and factories of the future. Figures cover 1960-1980 development of job productivity in manufacturing; 1980-1981 value of production per unit of labor cost for Japan, West Germany, U.S., France, and Great Britain; 1980 utilization of job resources in industry in Japan, West Germany and U.S.; 1960-1985 price / performance development for numerical controls; 1960-1990 development of workpiece programming; and technological data.

Toshiba Review

2181. TOSHIBA REVIEW: PATENTS. *Toshiba Corporation. Bi-monthly. no.134. July 1981.*

Gives patent information on medical and electric equipment including broadcast and communications, computers, semiconductors, etc. Data includes patent number, invention title, and a 'Toshiba' reference number.

See 45, 814, 829, 833.

Numeric Databases

2182. SIMCOST II. *Arthur D. Little, Inc. Annual.*

Provides forecasts of manufacturing costs of electronics, electrical, and light mechanical industries in 14 countries and Hong Kong. Countries include Brazil, Taiwan, Singapore, Germany, France, Indonesia, Japan, Korea, Mexico, Malaysia, Philippines, Thailand, U.K., and the U.S. Forecast data includes employment costs, transportation, price indexes, inflation, exchange rates, and wages. Access requires full subscription to Arthur D. Little Simcost II services.

Monographs and Handbooks

2183. ELECTRICAL INSTALLATIONS HANDBOOK: VOLUME 1. *Seip, Gunter G., ed. Siemens Aktiengesellschaft. Heyden and Son Ltd. 1979. pg.1-759*

Provides a comprehensive guide to the installation of electricity supply systems and electrical equipment in commercial, public and domestic applications. Contains graphics, tables, and definitions.

2184. ELECTRICAL INSTALLATIONS HANDBOOK: VOLUME 2. *Seip, Gunter G., ed. Siemens Aktiengesellschaft. Heyden and Son Ltd. 1979. pg.760-1184*

Handbook of special power installation engineering. Describes electrical installations in hotels, hospitals, theatres, film and television studios, industrial buildings, exhibition halls, airports, and others. Reviews various applications of electrical goods.

2185. ELECTRONIC SWITCHING: DIGITAL CENTRAL OFFICE SYSTEMS OF THE WORLD. *Joel, Amos E., ed. IEEE Computer Society. 1982. 280 pg. $30.95.*

Collection of papers on 26 digital time-division switching systems. Includes diagrams, tables, and references arranged by 11 nations.

See 842.

Conference Reports

2186. FROM ELECTRONICS TO MICROELECTRONICS. *Kaiser, W.A., ed. North-Holland Publishing Company. 1980. 792 pg. $67.50.145.00 (Netherlands).*

Proceedings of the EUROCON 80 Conference (Stuttgart, March 24-28, 1980) include 165 papers concerning technology of microelectronics; telecommunications and data processing; electrical pawn systems and controls; medical and other applications of microelectronics; and tutorial program.

2187. INTERCONNECTION IN THE EIGHTIES. *BPA Management Services. November 1979. 126 pg. 50.00 (United Kingdom).*

Papers from a conference held in October 1979. Participants included international executives from the electronics industry. Covers industry structure, market analysis, end uses including military markets, and new products.

2188. SPACE TRACKING AND DATA SYSTEMS. *Grey, Jerry,, editor. American Institute of Aeronautics and Astronautics. AIAA Aerospace Assessment Series, Vol.8. 1982. $27.00.*

Proceedings of June 1981 conference sponsored by American Institute of Aeronautics and Astronauts and NASA. Includes 23 papers concerning international activities in the civil use of space tracking and data systems for the years 1980-2000. The assessment addresses worldwide activities in data acquisition and tracking, data distribution, and mission control. It outlines current and future developments, identifies associated issues, and provides recommendations on steps to take to resolve these issues and foster increased international cooperation in the support of future civilian space missions. It concludes that major advances appear possible through cooperative international efforts without impairing the basic autonomy and defense needs of participating nations.

2189. WORLD ELECTRONICS 1981. *Benn Publications Ltd. 1981. $120.00.*

Proceedings of 1981 Mackintosh / Financial Times Conference held in London covers such topics as international battle for electronics markets, financing the electronics industry, electronics as an economic driving force, and sociological implications of technological developments.

See 860, 4007.

Newsletters

2190. APA MIKROELEKTRONIK. [APA Microelectronics.]. *Austria Presse Agentur (APA) reg. Gen. m.b.H. German. Weekly. 0000. 450.00 (Austria).*

Contains news about international microelectronics markets which affect the Austrian market. Covers technical developments, exhibits and fairs, sales trends, and forecasts.

2191. ELECTRONIC DIGEST: NACHRICHTEN UND TRENDS AUS DER INTERNATIONALEN FACHPRESSE. [Electronic Digest.]. *Presse - und Brancheninformationsdienst. German. Monthly. 95.00/yr. (West Germany).*

The service offers news and trends out of the international electronics trade press for German interests. A wide range of international publications are analyzed.

2192. INTERNATIONAL PRESS CUTTING SERVICE: ELECTRONICS AND ELECTRICAL INDUSTRY. *International Press Cutting Service. Weekly.*

Weekly service offers processes and selected topical information on the electronics industry.

2193. WORLDWIDE INVESTMENT NOTES. *Worldwide Investment Research. Bi-weekly. 0000. $145.00.*

A bi-weekly report on strategic materials, their supply and demand plus information on international investment opportunities in high technology companies in pharmaceuticals, electronics and heavy industry.

See 862.

Services

See 866.

Indexes and Abstracts

2194. EKSPRESS - INFORMACIIA ELEKTRONIKA. [Rapid Information on Electronics.]. *Mezhdunarodnaia Kniga. Weekly. 1981.*

Loose-leaf service provides abstracts of recent scientific and technical developments in electronics.

2195. ELECTRICAL AND ELECTRONIC ABSTRACTS. *Institute of Electrical and Electronics Engineers, Inc. Monthly. 1982. $1000.00/yr.*

Monthly service abstracts 89 electrical and electronic engineering journals completely and over 2000 worldwide periodicals and serials scanned for inclusion of worthy items. Each issue contains eight indexes: subject, author, patent, report, book, conference, bibliography, and supplementary list of journals. Yearly subscription includes a semi-annual author and subject index and a listing of abstracted journals with publishers' names and addresses.

2196. KEY ABSTRACTS: ELECTRONIC CIRCUITS. *Institute of Electrical and Electronics Engineers, Inc. 1981. $72.00/yr.*

Arranged by topic, abstracts include author, author's address, publication, nation where published, volume and number, pages, publication date, synopsis, and number of references.

2197. KEY ABSTRACTS: INDUSTRIAL POWER AND CONTROL SYSTEMS. *Institute of Electrical and Electronics Engineers, Inc. 1981. $72.00/yr.*

Arranged by topic, abstracts include author, author's address, publication, nation where published, volume and number, pages, publication date, synopsis, and number of references.

2198. KEY ABSTRACTS: SOLID STATE DEVICES. *Institute of Electrical and Electronics Engineers, Inc. 1981. $72.00/yr.*

Arranged by topic, abstracts include author, author's address, publication, nation where published, volume and number, pages, publication date, synopsis, and number of references.

2199. KEY ABSTRACTS: SYSTEMS THEORY. *Institute of Electrical and Electronics Engineers, Inc. 1981. $72.00/yr.*

Arranged by topic, abstracts include author, author's address, publication, nation where published, volume and number, pages, publication date, synopsis, and number of references.

2200. PREDI-BRIEFS: ELECTRONIC COMPONENTS. *Predicasts, Inc. Monthly. 1981.*

One in a series of 29 industry-intensive abstract periodicals derived from Predicasts' PROMT. Abstracts 200 to 400 articles per month covering such topics as acquisitions, capacities, end uses, government regulation, market shares and statistics, and new technology.

2201. ROBOMATICS REPORTER. *EIC / Intelligence. Monthly. no.1. January 1983. $250.00/yr.*

Provides abstracts from worldwide journal articles, conference proceedings, government reports, research studies, market reports and forecasts, corporate reports, and monographs. Covers robotics research, applications in science and industry, policy and regulatory issues, and economics. Indexed by subject, source, author, and SIC code.

2202. WORLDCASTS (PRODUCT EDITION): PRIMARY METALS, MACHINERY, ELECTRONICS, TRANSPORTATION EQUIPMENT. *Predicasts, Inc. November 1982. $400.00.*

Abstracts published forecasts from government agencies, journals, bank letters, and special studies. Each one-line summary contains subject description, base period data, short- and long-term forecasts, and source information. Specific topics include capacities, production, shipments, and sales distribution. Accompanying outlook summaries contain short- and long-term composite forecasts as well as historical data. Presents statistical data for all major companies.

See 869.

Bibliographic Databases

2203. ELECTRICAL PATENTS INDEX. *Derwent Publications Ltd. Weekly.*

Provides patent information from 1980 to date including: patent number, first disclosure, patentee, priorities, and title. Available on microform.

Dictionaries

2204. DICTIONARY OF ELECTRONICS. *Oppermann, Alfred, ed. K.G. Saur Verlag KG. English/German. 1980. 692 pg. $120.00.*

English - German electronic dictionary covers 100,000 concepts with 300,000 translations in alphabetical order, including official translations of American and British terminology by the Bundessprachenanstalt. Technologies covered include cybernetics, data processing and computer science, semiconductor technology, navigation and radar, electrical power, atomic physics, aeronautical power, atomic physics, aeronautics and aviation, and motor and machine construction.

North America

Market Research Reports

2205. AUTO ELECTRONICS: OEM. *Predicasts, Inc. June 1981.*

Analyzes U.S. original equipment markets and impact of safety regulations, ecological considerations and technological innovations on electronic systems in autos. Market data presented by type of system with a technological description and projections to 1985, 1990 and 1995. Detailed industry structure provided.

2206. AUTOMATED WELDING EQUIPMENT MARKET (U.S.). *Frost and Sullivan, Inc. January 1981. $1050.00.*

Sales forecast for welding equipment to 1986 for more than 20 automated items in key end user industries. Manufacturer and user questionnaire survey results; company and product profiles; and market potential for subcontractors.

2207. AUTOMOTIVE CONVENIENCE ELECTRONICS FORECAST. *Gnostic Concepts, Inc. November 3, 1980. $15000.00.*

Three-volume report contains 120 figures and tables focusing on evolving applications, technological developments and price trends in controls of instrumentation and convenience systems for production of passenger cars and light-duty trucks. Chapters include motor vehicle production, electronic conveniences production, component use analysis, aftermarket use analysis, technology outlook, and competitive environments.

2208. AUTOMOTIVE ELECTRONICS AND AUTOMOTIVE ELECTRONIC TEST EQUIPMENT IN THE U.S. *Frost and Sullivan, Inc. February 1979. 194 pg. $800.00.*

Forecast and analysis to 1987 of electronic alternators, voltage regulators, ignitions, spark advance modules, exhaust recirculation modules, automatic transmission modules, antiskid devices, auto microprocessors, clocks, displays, fuel injection systems, speed control system modules, automatic temperature control modules, windshield wiper control system modules. Separate sections on electronic test equipment, major participants, government impact and trends.

2209. BUBBLE DOMAIN MEMORIES 2: 1980-1985: A STRA-TEGIC ANALYSIS. *Ross, Edward A. Venture Development Corporation. October 1980. 150 pg. $1950.00.*

Industry analysis of bubble domain memory devices and subsystems. Market forecasts through 1985. Technological trends and user attitudes studied in depth.

2210. CANADA'S ELECTRONICS MARKET. *Maclean Hunter Ltd. June 1981. 65.00 (Canada).*

Market report covers Canadian production, net production, imports, and exports. Product by product analysis of Canadian imports from the U.S. Includes coverage of related market areas such as broadcasting and avionics.

2211. CONNECTOR INDUSTRY FORECAST. *Gnostic Concepts, Inc. 1981. $15000.00.*

Examines impact of new technology on demand for electronic connector products. Gives market analysis and forecasts, 1979-1986; end uses; manufacturers and importers; market shares; industrial plants; equipment and processes; production and labor cost comparisons; and economic indicators.

2212. THE CONNECTOR INDUSTRY IN TRANSITION: AN ANALYSIS OF HOW MAJOR CONNECTOR COMPANIES ARE ADAPTING FOR THE 1980'S. *MSRA, Inc. November 17, 1980. 124 pg. $825.00.*

Worldwide connector industry analysis. Includes market forecasts, suppliers' and users' attitudes toward manufacturers, inventories, factors in purchases, prices and demand trends.

2213. CONSUMER ELECTRONICS: PRODUCT AND MARKET TRENDS. *Dambrot, Stuart M. Business Communications Company. January 1980. 160 (est.) pg. $725.00.*

Market forecast through 1985 for a wide variety of consumer electronics products, including home audio and video equipment, video games, television cable services, microwave ovens, burglar alarms, CB radios and radar detectors, digital watches, calculators and home computers, and fancy telephones. Forecasts are generally in terms of manufacturers' shipments. Examines major manufacturers and their market shares, and relevant software sales trends.

2214. CRT TERMINAL APPLICATIONS AND MARKETS. *Gnostic Concepts, Inc. 1981. $12000.00.*

Market structure and trends, applications and features, competition and market shares, distribution and service trends, impact of communications, systems technology, component technology impact, and forecasts through 1985 for CRT's. Forecast includes: U.S. production and market, export-import analysis, shipments, market value, price trends and product segmentation.

2215. DESIGN TRADEOFF STUDIES AND SENSITIVITY ANALYSIS (HYBRID ELECTRIC VEHICLE): APPENDICES B1-B4. *South Coast Technology Inc., Santa Barbara National Aeronautics and Space Administration, Washington. U.S. Department of Commerce. National Technical Information Service. May 1979. 190 pg.*

Documentation is presented for a program which separately computes fuel and energy consumption for the two modes of operation of a hybrid electric vehicle. The distribution of daily travel is specified as input data as well as the weights which the component driving cycles are given in each of the composite cycles. The possibility of weight reduction through the substitution of various materials is considered as well as the market potential for hybrid vehicles. Data relating to battery compartment weight distribution and vehicle handling analysis is tabulated. Sponsored by NASA. Prepared for JPL.

2216. DIRECTIONS IN PHOTOVOLTAICS. *Business Communications Company. February 1980. 100 (est.) pg. $750.00.*

Technical review and market projections for solar photovoltaic generating equipment. Discusses current technical state-of-the-art, manufacturing techniques, leading manufacturers and their products. Reviews "worldwide" activities and provides market forecasts.

2217. DISCRETE ELECTRONIC COMPONENTS. *Creative Strategies International. 1981. $1200.00.*

Analysis of the discrete electronic components industry including prices, end uses and employees. Also includes analysis of markets; shipments, 1980-1985, and market shares, 1980, both by product groups; technology of industrial plants and equipment; and competition factors.

2218. ELECTRICAL / ELECTRONIC APPLICATIONS FOR POLYMERS, 2. *Skeist Laboratories, Inc. June 1982. 600 (est.) pg. $7500.00.*

Market analysis report on the use of polymers in electrical / electronic industries in 1981, with forecasts to 1986. Based upon research including surveys of user attitudes. Covers industry structure including manufacturers, product mix, market shares, distribution channels, and mergers. Examines end uses and specific polymer products. Reviews government regulations and legislation. Shows consumption, 1981 and 1986. Contains growth rates and new products.

2219. ELECTRICAL EQUIPMENT IN THE U.S. *Larsen Sweeney Associates Ltd. Annual. 1981. 395.00 (United Kingdom).*

Market analysis reports, by product, for electrical equipment in the U.S. Includes analysis of manufacturers' installed base of plants and equipment, marketing and industry issues, production and new product prospects and opportunities, and end users.

2220. ELECTRONIC COMPONENT AND CIRCUIT BOARD ATE MARKET (U.S.). *Frost and Sullivan, Inc. May 1980. $975.00.*

Analysis and forecast to 1985 of ATE for components (IC logic, linear devices, memory devices, discretes); systems (circuit boards, function boards); and microprocessor development systems. End user analysis for computer, communications, instrumentation, consumer, government segments. Includes economic considerations, technological trends, major supplier profiles and barriers to market entry.

2221. ELECTRONIC COMPONENT MARKETING FORECAST. *Gnostic Concepts, Inc. 1981. 475 pg. $12000.00.*

Report covers electronic equipment production analysis and forecast, electronic component demand analysis and forecast, OEM production, and regional component demand. Chapters include sales channel forecast, equipment production trends, technology analysis, component market forecast, competitive environment, and OEM profiles.

2222. THE ELECTRONIC COMPONENTS DISTRIBUTION INDUSTRY. *Frost and Sullivan, Inc. September 1982. 34 pg. $285.00.*

Analysis by a Wall Street firm of electronics components distribution industry structure, investment valuations, and rankings of corporations. Profiles of vendors include relations with manufacturers, performance, 1983 sales, product mix, marketing strategies, market shares and financial analysis. Includes business cycles, franchises and services provided by suppliers.

2223. ELECTRONIC GAMES AND AMUSEMENTS INDUSTRY FORECAST. *Gnostic Concepts, Inc. 1980. 400 pg. $9800.00.*

Analyzes and forecasts aspects of the electronics game industry including production, growth rate, product mix, imports, consumption, trends, and shipments for the years 1976-1982. Discusses changes in the field in light of continuing microprocessor technology. Chapters include component production trends, component applications, technological analysis, economic outlook, and competitive environment.

2224. THE ELECTRONIC GAMES MARKET. *International Resource Development, Inc. 1981. 166 pg. $985.00.*

Market analysis for electronic games in the U.S. Covers market growth rates; prices; opportunities for market penetration; sales, 1980-1990; retail distribution outlets favored; and manufacturers' strategies.

2225. ELECTRONIC POWER SUPPLIES. *Creative Strategies International. 3679. 1982. $1200.00.*

Analysis of the electronic power supplies industry structure and markets, 1980-1985. Covers manufacturers; regulations; market segmentation, including consumers, industrial consumption, and military markets; technology; and competition. Shows sales and markets by product and end uses, 1980-1985.

2226. ELECTRONIC SYSTEMS: DMS MARKET INTELLIGENCE REPORT. *DMS, Inc. Monthly. 1980. $500.00.*

Reports on more than 100 major U.S. electronic programs for military and civil applications, including weapons systems and funding, contractors, systems make-up, and sponsoring agencies. Updated monthly.

2227. ELECTRONIC TOYS AND GAMES, VIDEO GAMES, PERSONAL COMPUTERS. *Frost and Sullivan, Inc. August 1979. 48 pg. $350.00.*

Examines the market for electronic toys and games, video games and personal computers in terms of product offerings, volume, and strategies of major participants. Risks and additional opportunities are evaluated.

2228. EQUIPMENT AND MATERIALS FOR ELECTRONIC COMPONENTS PRODUCTION, MEXICO: COUNTRY MARKET SURVEY CMS 79-111. *Industry and Trade Administration U.S. Department of Commerce. U.S. Government Printing Office. February 1979. 8 pg. $.50.*

A summary of a detailed market research report made in 1977 for the Office of International Marketing. To aid U.S. exporters. Text summarizes current economic situation in Mexico, the production equipment and materials market, the electronic components industry, and business practices and tariffs. Tables and graphs show market size , imports, end- user profiles for 1973-1976, with projections to 1980, by product, in U.S. dollars. Chart lists best sales opportunities for U.S. exporters, by product. Data is derived from U.S. government sources.

2229. FIRE CONTROL SYSTEMS. *DMS, Inc. Annual. 1981. $375.00.*

Analysis of military markets for fire control systems.

2230. FOREIGN MILITARY MARKETS: VOLUME 1: NATO / EUROPE. *DMS, Inc. Monthly. 1981. $575.00.*

Monthly service detailing military markets in NATO nations of Western Europe. Examines organization of government agencies; exports and imports; dollar values; foreign trade partners; government purchase procedures; present and future orders; U.S. manufacturers with licenses to sell to other countries; U.S. government funding of foreign military expenditures; and regulations governing sales in foreign markets.

2231. FOREIGN MILITARY MARKETS: VOLUME 2: MIDDLE EAST / AFRICA. *DMS, Inc. Monthly. 1981. $575.00.*

Monthly service detailing military markets in the Middle East and Africa. Examines organization of government agencies; exports and imports; dollar values; foreign trade partners; government purchase procedures; present and future orders; U.S. manufacturers with licenses to sell to other countries; U.S. government funding of foreign military expenditures; and regulations governing sales to foreign markets.

2232. FOREIGN MILITARY MARKETS: VOLUME 3: SOUTH AMERICA / AUSTRALIA. *DMS, Inc. Monthly. 1981. $575.00.*

Monthly service detailing military markets in South America and Australia. Examines organization of government agencies; exports and imports; dollar values; foreign trade partners; government purchase procedures; present and future orders; U.S. manufacturers with licenses to sell to other countries; U.S. government funding of foreign military expenditures; and regulations governing sales in foreign markets.

2233. FRACTIONAL HORSEPOWER MOTORS IN THE U.S. *Frost and Sullivan, Inc. August 1980. $975.00.*

Analysis and sales forecast for 20 components in 17 key end user markets. Market share data by product for major competitors, including components for resale and captive. Impact of new electric drives. Questionnaire survey results.

2234. HOLTER MONITORING. *Theta Technology Corporation. no.118. November 1981. $295.00.*

Concentrating on the U.S. market, analysis and forecasts cover real time analysis monitoring, scanning services, company technological programs, promotional factors, and world sales.

2235. HOME OF THE FUTURE: A STRATEGIC ANALYSIS FOR CONSUMER ELECTRONIC FIRMS. *Venture Development Corporation. 1981. 250 pg. $2250.00.*

Report examines long-term future of major segments of home electronics. Covers effects of specific developments in home video, security and control, information services, and communication. Presents market forecasts for TV, VCR, cable TV, home security equipment, video disc systems, and home information hardware and service. Consumer survey includes price sensitivity for new and improved products, interest in home energy monitoring and control, and attitudes towards anticipated changes in home environment.

2236. HYBRID CIRCUIT AND THICK FILM MATERIAL MARKET IN THE U.S. *Frost and Sullivan, Inc. July 1982. 169 pg. $1100.00.*

Analysis of markets for hybrid circuit and thick film materials includes production costs and technology, market size, industry structure, end users, manufacturers, and demand forecasts to 1990.

2237. HYBRID INDUSTRY FORECAST. *Gnostic Concepts, Inc. 1981. $15000.00.*

A sector-by-sector analysis of production and end uses for hybrid circuits and factors driving demand and technology. The report includes circuit market analysis, circuit applications, interconnection analysis, component and materials analysis, manufacturing trends, technology trends, competitive environment, and economic environment.

2238. IMPACT OF GATE ARRAY TECHNOLOGY ON IC COMPONENTS. *Strategic Inc. July 1982. 140 pg. $1500.00.*

Analysis of market impact from gate array technology for IC's. Examines new products, manufacturers, utilizations, segmentation of end use markets, and market penetration. Forecasts to 1990.

2239. THE INDUSTRIAL DISPLAY MARKET IN THE U.S. *Frost and Sullivan, Inc. July 1980. 223 pg. $900.00.*

Analysis and sales forecast to 1985 for LED character displays and lamps, gas discharge displays, LCDs, vacuum fluorescence, and non-TV CRTs used in the test and measurement, analytical instrumentation, and process control / process analyzer segments of the industrial electronic equipment market. Examines display characteristics, features, and technologies; reviews the size and leading OEMs for the principal end-user markets; profiles major industrial display suppliers - including their product lines, sales and strategies.

2240. INDUSTRIAL ROBOTS: A DELPHI STUDY OF MARKETS AND TECHNOLOGY. *Society of Manufacturing Engineers. 1982. 230 pg. $65.00.*

Analysis and forecast to 1998 for industrial robot markets and new technology. Includes performance and capacity, market size, and end uses.

2241. INDUSTRIAL ROBOTS IN THE 1980'S. *Rudolph, William, editor. International Resource Development, Inc. 1979. 208 pg. $895.00.*

Market study of the industrial robot industry as of 1979, with forecasts through 1989. Describes current technology and applications. Profiles twenty leading manufacturers and briefly discusses fifty others. Analyzes market volume by segment, 1969-1989; and value of shipments, 1969-1979.

2242. INTEGRATED ELECTRONIC COMPONENTS. *Creative Strategies International. 1981. $1200.00.*

Analysis of the structure of the integrated electronic components industry, market segmentation and manufacturers' positions, prices, demand, capital requirements, and production growth trends. Also includes analysis of end use markets; shipments, 1980-1985, by product groups; competition factors; distribution; and 1980 market shares.

2243. INTERNATIONAL TECHNOLOGICAL COMPETITIVENESS: TELEVISION RECEIVERS AND SEMICONDUCTORS, PHASE 1. *Charles River Associates, Inc. U.S. National Science Foundation. 1979. 141 pg.*

Reviews technological developments in the semiconductor industry since 1960 and the sales position of U.S., Japanese, and European semiconductor firms. Strategies between the television receiver industry and the semiconductor industry are compared and the link between technology, strategies, and international competitive performance is analyzed. Finally, the report examines the role of government policy in influencing innovation, diffusion, and competitive performance in the industry.

2244. LIGHTING DEVICES. *Predicasts, Inc. November 1981. $900.00.*

Projects U.S. product shipments to the five major markets to 1995. Forecasts presented for over 50 products of the three major types of lighting. Analyzes technological changes and industry structure.

2245. LIGHTING FIXTURE INDUSTRY. *Specialists in Business Information, Inc. March 1981. 174 pg. $325.00.*

Market analysis and segmentation of the lighting fixture industry. Includes sales, shipment quantities and value, prices, end uses, industry structure, competition, foreign trade, manufacturers, and financial analysis.

2246. LIGHTING FIXTURES. *Business Trend Analysts, Inc. June 1981. $350.00.*

Analysis of markets for lighting fixtures with detailed market segmentation. Covers new products and technology, pricing, distribution, end uses, and manufacturers. Forecasts to 1990.

2247. LOW COST SENSORS: PRODUCTS, MARKETS AND OPPORTUNITIES. *Mackintosh Publications Ltd. June 1980. 14000.00 (United Kingdom).*

On the premise that declining microprocessor costs require the development of low-cost terminal sensors, this multi-client study explores technologies, applications, user needs and markets through 1985. Includes consultation with potential users in several industrial sectors, and with R & D personnel. Provides discussion of captive versus available markets. Geographic scope is U.S., Western Europe and Japan.

2248. THE MARKET FOR CRYSTALS, FILTERS AND LCD'S. *Marketdata Enterprises. April 1980. 166 pg. $225.00.*

Analysis of markets for crystals, filters and LCD's. Gives manufacturers' sales, 1961-1979, 1980 estimated sales, and forecasts for 1985 and 1990. Lists number of manufacturers and product lines, manufacturers' shipments, product mix and addresses of manufacturers. Covers end use consumption, 1972, 1977; historical data on production, 1958-1972; government purchases, 1977-1982; and operating performance.

2249. THE MARKET FOR ELECTRON TUBE PARTS, SOCKETS AND ENCLOSURES, DELAY LINES, PHONOGRAPH CARTRIDGES AND PICK-UPS. *Marketdata Enterprises. April 1980. 151 pg. $225.00.*

Analysis of markets for electronic tube parts, sockets and enclosures, delay lines, phonographic cartridges and pick-ups. Gives manufacturers' sales, 1961-1977, 1980 estimated sales, and forecasts, 1985, 1990. Includes manufacturers' shipments, product mix, numbers of manufacturers and product lines, and addresses. Covers end use consumption, 1972, 1977; historical data on production, 1958-1972; government purchases, 1977-1982 and operating performance.

2250. THE MARKET FOR ELECTRONIC RELAYS. *Marketdata Enterprises. April 1980. 162 pg. $225.00.*

Analysis of markets for electronic relays. Gives manufacturers' sales, 1961-1977, 1980 estimated sales, and forecasts to 1985 and 1990. Includes manufacturers' shipments; product mix; and listings of the number of manufacturers, addresses and product lines. Gives end uses consumption 1972, 1977; historical data on production, 1958-1972; government purchases, 1977-1982; market structure, 1977; and operating performance.

2251. THE MARKET FOR ELECTRONIC TRANSDUCERS. *Marketdata Enterprises. April 1980. 149 pg. $225.00.*

Analysis of markets for electronic transducers. Gives manufacturers' sales, 1961-1979, 1980 estimated sales, and forecasts, 1985, 1990. Includes manufacturers' shipments, product mix, numbers of manufacturers, product lines and addresses. Shows historical data on production, 1958-1972; consumption, 1972, 1977; government purchases, 1978-1982; 1977 market structure; and operating performance.

2252. THE MARKET FOR MAGNETIC AUDIO, VIDEO AND COMPUTER TAPE. *Marketdata Enterprises. April 1980. 150 pg. $225.00.*

Analysis of markets for magnetic audio, video and computer tape.

2253. THE MARKET FOR MICROWAVE COMPONENTS, DEVICES AND SUBASSEMBLIES. *Marketdata Enterprises. April 1980. 157 pg. $225.00.*

Analysis of markets for microwave components. Gives manufacturers' sales, 1961-1979, 1980 estimated sales, and forecasts for 1985 and 1990. Includes shipments, product mix, number of manufacturers and product lines, and addresses of manufacturers. Shows end use consumption, 1972, 1977; government purchases, 1977-1982; 1977 market structure; and operating performance.

2254. THE MARKET FOR PC BOARDS, MAGNETIC CORES, HARD AND SOFT FERRITES. *Marketdata Enterprises. April 1980. 162 pg. $225.00.*

Analysis of markets for PC boards, magnetic cores, and hard and soft ferrites. Gives manufacturers' sales, 1961-1979, 1980 estimated sales, and forecasts for 1985 and 1990. Includes shipments, product mix, number of manufacturers and product lines, and addresses of manufacturers. Shows end use consumption, 1972, 1977; historical data on production, 1958-1972; government purchases, 1977-1982; 1977 market structure; and operating performance.

2255. MARKET OPPORTUNITIES FOR ELECTRICALLY CONDUCTIVE POLYMERIC SYSTEMS: 1981-1987. *Schotland Business Research, Inc. September 1981. $4000.00.*

Examines competition among technologies for rendering conductivity to plastics and forecasts of markets for conductive polymers, 1981-1987. Examines utilization and manufacturers, suppliers and end use consumers.

2256. MARKET STUDY: TRENDS IN THE SELECTION OF CONDENSER - TUBE MATERIALS, 1961-1979. *Gaffoglio, C.J. Copper Development Association. no.702/0. April 1980. 17 pg.*

1961-1979 data on orders for utility condensers with breakdowns by tube material.

2257. MEDICAL LIGHTING EQUIPMENT MARKETS. *Theta Technology Corporation. August 1980. 170 pg. $750.00.*

Report on medical lighting markets covers four product groups: surgery and operating room lighting equipment, medical examining lights, fiber optic lighting systems, and heat treatment (IR/UV) lamp equipment. Data includes market size, forecasts to 1984, applications, trends, and companies involved. Report contains profiles of 21 leading companies, five-year market estimates by major product group and by end-user, and company rankings by product segment.

2258. MICROPROCESSOR DEVELOPMENT SYSTEMS MARKET IN THE U.S. *Frost and Sullivan, Inc. April 1982. 247 pg. $1250.00.*

Analysis of markets for microprocessor development systems. Examines the new technology of microprocessors and development systems, performance trends, shipments, and end uses. For vendors gives sales forecasts to 1986, product lines and marketing strategies.

2259. MICROPROCESSORS. *Predicasts, Inc. February 1982. $375.00.*

Analysis of the microprocessor industry structure worldwide, including historical data and forecasts. Covers technology, prices, markets by end uses and countries, and major manufacturers.

2260. MICROWAVE COMPONENTS. *Predicasts, Inc. August 1979. $325.00.*

Reviews the U.S. microwave components industry structure, technologies and applications in communications, test and measurement, military uses and civilian radar. Data included on microwave tubes, IC's, transistors, diodes, and passive and other components by type.

2261. MILITARY AND AEROSPACE MARKET FOR BATTERIES, FUEL CELLS AND PHOTOVOLTAICS IN THE U.S. *Frost and Sullivan, Inc. September 1981. 179 pg. $1100.00.*

Analysis of the military and aerospace market defines and categorizes the marketplace. Establishes the size of the present market. Profiles selected suppliers and gives competitive market share rankings. Provides market to 1986 of power sources consumption by product type and by end-user.

2262. MILITARY AND AEROSPACE POWER SUPPLY MARKET IN THE U.S. *Frost and Sullivan, Inc. June 1982. 271 pg. $1200.00.*

Analysis of aerospace and military markets for power supplies. Market segmentation includes end uses in defense contracts, research and development, and exports. Includes market value, sales forecasts to 1986, and rankings of manufacturers.

2263. THE MILITARY MICROWAVE COMPONENTS MARKET IN THE U.S. *Frost and Sullivan, Inc. June 1980. 233 pg. $950.00.*

1980 analysis of the military microwave components market which structures and traces the growth of the major end-user system groups for microwave components and isolates the total microwave portion of each; establishes the present market for 29 active and passive component products and forecasts sales of each to 1985; and discusses the captive / merchant aspects of the market. Examines evolutionary and revolutionary technical trends expected through 1985 and beyond, and possible displacement products moving into the microwave region. Profiles groups of microwave components manufacturers and system houses - their products, sales, and some of their marketing strategies. Compiled from original Frost and Sullivan-conducted surveys of both component and system houses.

2264. MILITARY OPTOELECTRONIC COMPONENT MARKET. *Frost and Sullivan, Inc. March 1982. $1200.00.*

Market report gives forecasts to 1986 for optoelectronic components; optoelectronic IC's; optoelectronic discretes; optocouplers and optoisolators; light emitting diodes; phototransisters; photodiodes; photovoltaic cells; photoconductive cells; and laser diodes. Profiles and ranks leading suppliers. Identifies appropriate military programs.

2265. MILITARY SEMICONDUCTOR MARKET IN THE U.S. *Frost and Sullivan, Inc. July 1981. 214 pg. $1000.00.*

Analysis of military markets for semiconductors, 1980. Examines market structure; manufacturers, suppliers, and the distribution system; utilization of semiconductors; major consumers; government budgets for military programs; products and sales of producers; and forecasts to 1985.

2266. NEAR-TERM HYBRID PASSENGER VEHICLE DEVELOPMENT PROGRAM, PHASE 1: APPENDICES A AND B, MISSION ANALYSIS AND PERFORMANCE SPECIFICATIONS STUDIES REPORT, VOLUME 1. *Minicars Inc., Goleta, Calif.; National Aeronautics and Space Administration. U.S. Department of Commerce. National Technical Information Service. October 1979. 468 pg.*

The three most promising vehicle use patterns (missions) for the near term electric hybrid vehicle were found to be all-purpose city driving, commuting, and family and civic business. The mission selection process was based principally on an analysis of the travel patterns found in the Nationwide Transportation Survey and on the Los Angeles and Washington, D.C. origin-destination studies data. Travel patterns in turn were converted to fuel requirements for 1985 conventional and hybrid cars. By this means, the potential fuel savings for each mission were estimated, and preliminary design requirements for hybrid vehicles were derived and revised. Sponsored in part by DOE.

2267. NEAR-TERM HYBRID VEHICLE PROGRAM, PHASE 1: APPENDIX A, MISSION ANALYSIS AND PERFORMANCE SPECIFICATION STUDIES REPORT. *General Electric Co., Schenectady, N.Y.; National Aeronautics and Space Administration. U.S. Department of Commerce. National Technical Information Service. October 1979. 136 pg.*

Results of a study leading to the preliminary design of a five passenger hybrid vehicle utilizing two energy sources (electricity and gasoline / diesel fuel) to minimize petroleum usage on a fleet basis are presented. The study methodology is described. Vehicle characterizations, the mission description, characterization, and impact on potential sales, and the rationale for the selection of the referenceinternal combustion engine vehicle are presented. Conclusions and recommendations of the mission analysis and performance specification report are included. Prepared for JPL.

2268. NEAR-TERM HYBRID VEHICLE PROGRAM, PHASE 1: APPENDIX D: SENSITIVITY ANALYSIS REPORT. *General Electric Co., Schenectady, N.Y.; National Aeronautics and Space Administration. U.S. Department of Commerce. National Technical Information Service. October 1979. 55 pg.*

Parametric analyses, using a hybrid vehicle synthesis and economics program (HYVELD) are described investigating the sensitivity of hybrid vehicle cost, fuel usage, utility, and marketability to changes in travel statistics, energy costs, vehicle lifetime and maintenance, owner use patterns, internal combustion engine (ICE) reference vehicle fuel economy, and driveline component costs and type. The lowest initial cost of the hybrid vehicle would be $1200 to $1500 higher than that of the conventional vehicle. For nominal energy costs ($1.00 / gal. for gasoline and 4.2 cents / kWh for electricity), the ownership cost of the hybrid vehicle is projected to be 0.5 to 1.0 cents / mi. less than the conventional ICE vehicle. To attain this ownership cost differential, the lifetime of the hybrid vehicle must be extended to 12 years and its maintenance cost reduced by 25 percent compared with the conventional vehicle. The ownership cost advantage of the hybrid vehicle increases rapidly as the price of fuel increases from $1 to $2 / gal. Prepared for JPL.

2269. NEW CONSUMER PRODUCT ELECTRONICS. *Business Communications Company. January 1980. 80 (est.) pg. $650.00.*

Adaptation of semi-conductor for technology creates important new applications directly relating to changing consumer lifestyles: calculators, personal communications equipment, watches, games, computers. Contents include: technology and the consumer; markets by product type, imports / exports; manufacturers by product / distribution; applications, old and new; who uses what and why; demographics, markets; state-of-the-art, changes; and future projections, criteria.

2270. NEW DIRECTIONS IN PHOTOVOLTAICS. *Business Communications Company. December 1980. $800.00.*

Market analysis of the photovoltaics industry. Discusses government and industry supported pilot projects and research, major U.S. and foreign companies' new ventures and acquisitions, and more.

2271. NEW DIRECTIONS IN ROBOTS FOR MANUFACTURING. *Business Communications Company. September 1979. 100 (est.) pg. $750.00.*

Study analyzes the robot industry. Applications in casting, forging, molding, stamping, material handling, forming and assem-

bly. Small industry listings with excellent growth potential. Contents include: manufacturing, economic justification, effects of foreign companies, types of robots / markets, robots by company / by type, and forecasts by type.

2272. THE NON DESTRUCTIVE TESTING (NDT) EQUIPMENT AND SUPPLIES MARKET IN THE U.S. AND CANADA. *Frost and Sullivan, Inc. January 1980. 260 pg. $1000.00.*

1980 industry analysis and five-year forecast of the market for NDT instruments and accessories in these disciplines: ultrasonics, eddy current, X-ray radiography, gamma ray radiography, penetrants, magnetic particles. Company sales and market share data are provided by product within these categories. In addition to the North American continent, sales estimates are provided for the international markets. The competitive situation is evaluated and in-depth company profiles are provided. Application trends are discussed and competitive advantages and disadvantages of each discipline are weighed. Development trends are investigated.

2273. NON-ENTERTAINMENT AUTOMOTIVE ELECTRONICS MARKET IN THE U.S. *Frost and Sullivan, Inc. May 1981. $1050.00.*

Forecasts and analyzes power management equipment; engine and transmission controls; and safety, comfort and convenience, and display equipment. Reviews status of the automotive industry, forecasts auto production trends, and examines product and production activity of major auto manufacturers. Identifies electronic systems and components in use, profiles major suppliers and competitively ranks the leaders. Assesses the impact of technology and foreign competition.

2274. OPTOELECTRONIC DEVICES AND EQUIPMENT. *Business Communications Company. October 1981. 150 pg. $950.00.*

Forecasts of markets for new optoelectronic products such as LED's, photodiodes, laser devices and optical characters, 1981-1985, 1990.

2275. OPTOELECTRONIC DEVICES MARKET IN THE U.S. *Frost and Sullivan, Inc. November 1980. $975.00.*

Analysis and forecast to 1986 of market by product types: displays (electroluminiscent, fluorescent, gas discharge, LED and laser diodes, LCD, plasma); lamps; couplers and isolators; photosensors (photoconductive, photovoltaic, phototransistors, photo diodes, solar cells). Identifies requirements of end user industries - industrial, commercial consumer, military, and distributor role. Covers technology impact, opportunities, threats, profiles and strategies of suppliers.

2276. THE OUTLOOK FOR THE CONNECTOR INDUSTRY AND THE ROLE OF AUGAT, INC. IN THE INTERCONNECTOR MARKET. *MSRA, Inc. 1981. 76 (est.) pg. $650.00.*

Market survey and analysis covering suppliers' opinions and ratings of manufacturer of connectors. Covers purchases, inventories, prices, competition, market shares, production growth, and new products.

2277. PASSIVE ELECTRONIC COMPONENTS MARKET (U.S.). *Frost and Sullivan, Inc. August 1980. $975.00.*

Analysis and market forecasts to 1985 for electronic capacitors, resistors, inductive components, and connectors used in military and aerospace, telecommunications, consumer, computer, and commercial end-user markets. Includes growth opportunities; supplies profiles, ranking, and strategies; market trends; and distribution channels.

2278. POWER SEMICONDUCTOR MARKET IN THE U.S. *Frost and Sullivan, Inc. September 1981. 210 pg. $1100.00.*

Market analysis with forecasts to 1986 for power semiconductors. Covers demand by product groups, total U.S. domestic production, new products and technologies, U.S. and Japanese industry structure, foreign trade, and manufacturers and suppliers, including product lines and sales.

2279. THE PRESSURE TRANSDUCER AND TRANSMITTER INDUSTRY: STRATEGIC ANALYSIS 1979-1984. *Klapfish, Maurice. Venture Development Corporation. January 1980. 189 pg. $1950.00.*

This report provides information on markets, shipments, technologies, trends, projections and mararket strategies of electronic pressure transducers and transmitters that provide an electronic output signal proportional to the unknown input pressure. A directory of manufacturers with addresses, telephone numbers and a brief description is provided.

2280. PRINTED WIRING AND FLEXIBLE CIRCUIT INDUSTRY FORECAST. *Gnostic Concepts, Inc. 1980. $6000.00.*

Printed wiring and flexible circuit market data for 1978-1984 includes markets, demand, prices, trends, costs, labor costs, competition, and production. Financial data is provided for each type of rigid printing wiring and flexible circuit. New developments discussed include assembly techniques, automatic insertion, chip carrier packaging, and direct chip attachment.

2281. PROFILE OF THE WORLD SEMICONDUCTOR INDUSTRY: VOLUME 2 - USA AND THE AMERICAS; JAPAN AND THE FAR EAST. *Benn Publications Ltd. 1982. 250 pg. $450.00.*

Analysis of semiconductor markets and the industry in the U.S. and the Americas, Japan, and the Far East includes forecasts to 1986, production, manufacturers and market shares, expansion of facilities and foreign investments, and new technology. Profiles of plants include addresses, equipment, product lines, number of employees, size of facilities, and 1981 sales.

2282. PROSPECTS OF THE ELECTRONIC COMPONENTS DISTRIBUTION BUSINESS IN THE UNITED STATES. *Meserve, Everett T. Arthur D. Little, Inc. March 19, 1979. 8 pg. $500.00.*

Part of market research service. Gives 1978 sales of 25 electronic component distributors and operating and financial ratios of 6 firms, 1973-77. Sources are Electronic News, Standard & Poor's Compustat Services and publisher's estimates.

2283. RELAY INDUSTRY FORECAST. *Gnostic Concepts, Inc. 1981. 750 pg. $7500.00.*

Report covers 1975-1985 relay markets, end-use analysis, technology outlook, competitive environment, economic outlook, and equipment production forecast. Relay categories include general purpose relays, telephone type relays, high performance relays, reed relays, solid state relays, and time delay relays. Volume Two consists of 5 private data bases divided into category lists, equipment data bases, relay market forecast, relay application requirements, and adjustments.

2284. RESISTOR INDUSTRY FORECAST. *Gnostic Concepts, Inc. 1981. $7500.00.*

Report discusses production of equipment which uses resistors for 1975-1985, use of resistors in this equipment, resistor market 1975-1985, and technological trends and competitive environment. Market analysis includes production, noncaptive production, shipments, domestic shipments, purchases, OEM purchases, noncaptive OEM purchases, OEM consumption, trade balance, inventory change, nonproduction use, demand, and value. Categories of resistors are fixed resistors, networks,

hybrids, potentiometers, elements and dials, trimmers, and non-linears.

2285. ROBOTICS. *Predicasts, Inc. February 1982. $900.00.*

Analyzes robotics industry and markets. Includes forecasts to 1985, effects of industrial productivity, end use industries, and new products and technologies.

2286. ROBOTICS TODAY: "82 ANNUAL EDITION. *Society of Manufacturing Engineers. Annual. 1982. 368 pg. $42.00.*

Annual analysis of the robotics industry and market. Covers new technology, end uses, and technical specifications for over 50 robot product lines. Includes addresses of manufacturers and suppliers and abstracts of relevant journal articles.

2287. SEMICONDUCTOR FABRICATION EQUIPMENT. *Frost and Sullivan, Inc. January 1982. 76 pg. $600.00.*

Market report analyzes the semiconductor fabrication equipment market. Examines types of manufacturers, technology, and trends. Categorizes equipment types, and analyzes operating characteristics, advantages and disadvantages, costs, market shares of major suppliers, and market penetration of newer equipment types. Forecasts worldwide sales of 26 fabrication equipment types to 1985. Profiles 14 major fabrication equipment suppliers. Includes products, sales, earnings, market strategies, and rankings.

2288. SEMICONDUCTOR INDUSTRY ECONOMETRIC SERVICE: DISCRETE DEVICES. *Gnostic Concepts, Inc. 1981. $15000.00.*

Semiconductor data includes economic outlook, end-use equipment forecast, discrete device market by end-use equipment, discrete device market forecast, competitive environment, technology analysis, and manufacturer profiles. Discrete devices include rectifiers, optoelectronics, thyristors, diodes, small signal transistors and power transistors. Special services include private data base, annual client strategy sessions, monthly economic impact reviews, shipments forecasts, and inquiry privileges.

2289. SEMICONDUCTOR PRODUCTS MARKET IN THE U.S. *Frost and Sullivan, Inc. July 1981. 217 pg. $1000.00.*

Analysis of semiconductor markets and industry structure. Examines market penetration and shares, competition with Japanese and European producers, foreign markets, manufacturers' product lines and sales, rankings, end uses, and shipments. Forecasts to 1986.

2290. SEMICONDUCTOR USER INFORMATION SERVICE. *Dataquest. Annual. 1982.*

Market and industry analysis service for semiconductor users includes costs, new technology, factors in purchases and deliveries, supply and demand, vendors' distribution channels, manufacturers' plant capacity, prices, production, technical specifications, and financial analysis of sources.

2291. SMALL MOTOR INDUSTRY FORECAST. *Gnostic Concepts, Inc. 1981. $9800.00.*

Identifies, defines, analyzes, and forecasts the technological and market trends for fractional and sub-fractional horsepower motors. Major small motor categories are single phase, non-synchronous; single phase, synchronous; polyphase, non-synchronous; servomotor; DC, wound field; DC, permanent magnets; and stepper motor. Report covers production of motor-using equipment, 1974-1980; analysis of the use of motors in this equipment; and competitive environment. Special services include strategy session and private data base.

2292. SOLAR HARDWARE SUPPLIES: A REVIEW. *Business Communications Company. January 1980. 115 pg. $650.00.*

Report on market potential for solar heating equipment (primarily "active" systems) and related materials, as of January 1980. Discusses solar collectors, controls, storage systems, fluids, insulation, auxiliary heating systems, piping and plumbing supplies. Considers distribution channels, leading manufacturers and recent and expected innovations.

2293. THE SOLID STATE MEMORY MARKET IN THE U.S. *Frost and Sullivan, Inc. 1979. 269 pg. $875.00.*

Market analysis and forecasts (1979-1985) of three major groups and nine specific types of solid state memory devices and growth rates to 1985. Examines the technology, reviews end markets, and profiles the major suppliers, products lines and strategies. Detailed reviews of 4K RAM. Additional data includes sales and prices (1979).

2294. THE SUPERCONDUCTIVITY INDUSTRY: OPPORTU-NITIES. *BCC, and Advance Technologies Enterprises, Inc. Business Communications Company. June 1980. 120 (est.) pg. $750.00.*

Description and analysis of 1980-2000 markets for technologies employing superconductivity (the nullification of electrical resistance at low temperatures). Applications discussed are in the fields of electrical generation and transmission, electric motors, magnetohydrodynamics, magnetic separation (in industrial processes, mining, and water purification), medical detection and surgery, and wheelless trains.

2295. SUPPLY / DEMAND OUTLOOK FOR THE U.S. SEMI-CONDUCTOR INDUSTRY. *Frost and Sullivan, Inc. January 1982. 34 pg. $300.00.*

Analyzes the supply - demand outlook for the U.S. semiconductor industry through 1983. Forecasts by device type and average selling price using two sets of economic assumptions. Analyzes and forecasts capacity and trends. Presents tabulated information on capital spending and production values for the U.S. and Japanese semiconductor industries as a whole, and for 10 Japanese suppliers for five years through 1981.

2296. SWITCH INDUSTRY FORECAST. *Gnostic Concepts, Inc. 1981. $7500.00.*

Switch data includes market forecast to 1982, end-use analysis, technology trends, competitive environment, economic outlook, and equipment production forecast. Switch categories include pushbutton, circuit breaker, toggle, rotary, position-sensing, slide, and rocker. Specialized service provides private database.

2297. SWITCHING POWER SUPPLY FORECAST. *Gnostic Concepts, Inc. 1981. $12000.00.*

Forecasts the total U.S. production of switching power supplies 1978-1986, in terms of total value, quantity and average price. Chapters include market analysis, applications analysis, competitive analysis, technological outlook, component demand forecast, economic outlook, and equipment production forecast. Private database offered as specialized service.

2298. SWITCHING POWER SUPPLY MARKET. *Frost and Sullivan, Inc. January 1982. $1150.00.*

Provides analysis and forecasts to 1986 for product categories by wattage range, frequency range, enclosed open-frame, and modular, custom, and standard. Gives manufacturer profiles and market shares.

2299. TIMING CONTROLS FORECAST. *Gnostic Concepts, Inc. 1981. 321 pg. $7500.00.*

Report focuses on evolving applications, price trends, and technological developments in timing controls. Chapters include market analysis, end-use trends, competitive environment, and equipment production forecasts to 1985.

2300. TOYS AND ELECTRONIC GAMES MARKET. *Frost and Sullivan, Inc. August 1981. $1200.00.*

Major questionnaire survey provides retailer and supplier profiles. Includes factors affecting demand and brand share analysis. Covers non-video electronic games, programmable cartridge video games, dedicated video games, hand-held minivideo games, cartridges, electronic learning aids, home pinball, and electronic chess.

2301. THE U.S. CONSUMER ELECTRONICS INDUSTRY: 1981 EDITION. *Venture Development Corporation. 2nd ed. 1981. 200 (est.) pg. $950.00.*

Forecasts U.S. manufacturers' unit shipments and value of shipments through 1983, for several categories of consumer electronics products. Covers audio and video equipment and software, calculators and home computers, telephones, fire / burglar alarms, automotive electronics, and electronic components of major home appliances.

2302. U.S. CONTRACT LIGHTING FIXTURES MARKETS - 1979-1984. *Stanley Smith and Company, Inc. 1980. 41 pg. $2000.00.*

Contract lighting fixtures report. Gives sales volume of products, prices, utilization by building type, market areas, factors influencing purchases, and forecastsof markets, 1979-1984.

2303. THE U.S. ELECTRONIC CAPACITORS INDUSTRY. *Marketdata Enterprises. July 1981. $295.00.*

Analysis of markets for electronic capacitors, 1961-1979, with 1981 estimated sales and forecasts for 1985 and 1990. Includes total sales; market shares; product mix; shipments, 1972, 1977; unit sales, 1961-1979; end use consumption, 1961-1980; foreign trade and competition. Covers exports, 1966-1980; imports, 1969-1980; industry structure, 1972, 1977; government purchases, 1977-1980; and bibliographic information.

2304. U.S. ELECTRONIC COILS, CHOKES, TRANSFORM-ERS AND REACTORS INDUSTRY. *Marketdata Enterprises. August 1981. 185 pg. $295.00.*

Analysis of markets for coils, transformers, chokes and reactors, 1961-1979, with 1981 estimated sales, and forecasts for 1985 and 1990. Includes total sales, market shares, product mix, and unit sales, 1976-1979; number of manufacturers, 1961-1979; end use consumption, 1961-1980; foreign trade and competition; industry structure, 1972, 1977; government purchases, 1977-1980; imports, 1969-1980; exports, 1966-1980; and bibliographic information.

2305. THE U.S. ELECTRONIC CONNECTORS INDUSTRY. *Marketdata Enterprises. July 1981. 203 pg. $295.00.*

Analysis of markets for electronic connectors, 1961-1979, with 1981 estimated sales, and forecasts 1985, 1990. Includes total sales; market shares; product mix; end use consumption, 1961-1980; foreign trade and competition. Covers exports, 1971-1980; imports, 1969-1980; industry structure, 1972-1977; government purchases, 1977-1980; and bibliographic information.

2306. THE U.S. ELECTRONIC RESISTORS INDUSTRY. *Marketdata Enterprises. August 1981. 205 pg. $295.00.*

Market analysis for resistors, 1961-1979, with 1981 estimated sales and forecasts, 1985-1990. Includes 1961-1979 unit sales and volume of sales; market shares, product mix, end use consumption, 1961-1980; foreign trade and competition. Covers industry structure, 1972-1977; government purchases, 1977-1980; and bibliographic information.

2307. THE U.S. LIGHTING FIXTURE INDUSTRY: AN ANALYSIS OF CURRENT PERFORMANCE AND FUTURE PROSPECTS. Business Trend Analysts, Inc. June 1981. 160 pg. $350.00.

Analysis of U.S. lighting equipment markets and industry structure, primarily during 1972-1980, with forecasts through 1990. By residential, commercial / institutional, and industrial sector, examines unit and dollar sales by product group. Discusses exports / imports by trading partner, market determinants, production costs (materials, labor, advertising), wholesale price and unit value trends. Covers industry-wide profitabilty and capital spending trends, and finances and activities of major manufacturers. Includes a manufacturers' directory.

2308. THE U.S. MILITARY MARKET FOR SEMICONDUCTOR COMPONENTS AND SUBSYSTEMS. Frost and Sullivan, Inc. January 1981. $1000.00.

Analysis and forecast to 1985 of U.S. military semiconductor requirements for IC's, discretes, optoelectronic devices, microwave, memories, microprocessors, etc. Identifies requirements for end-user systems markets (avionics, missiles, space, groundbased electronics, naval, ordinance, other). Discusses high usage programs and new technology (CCD, magnetic bubble memories, intermetallic componds, VHSIC/VLSI). Profiles major users (systems houses) and semiconductor suppliers.

2309. THE U.S. MILITARY POWER SUPPLY MARKET. Frost and Sullivan, Inc. May 1979. $850.00.

Five-year forecast and analysis of the military market for power supply equipment. Breakdowns by product type, by captive vs. non-captive, by custom vs. standard, and by end use (aircraft, missiles, ships, vehicles, electronic systems). Discusses product characteristics, major competitors and their business strategies.

2310. U.S. WELDING EQUIPMENT AND SUPPLIES. Predicasts, Inc. January 1982. $375.00.

Analysis of markets for welding equipment and supplies covers manufacturers, regulations, technology, and utilizations.

2311. VOICE INPUT / OUTPUT: MARKETS, TECHNOLOGIES, AND APPLICATIONS. Strategic Inc. July 1981. $950.00.

Forecasts 1985 sales of speech recognition and speech synthesis equipment in the information processing and consumer business. Defines composition of the two market segments.

2312. WORLD SEMICONDUCTOR INDUSTRY IN TRANSITION: 1978-1983. Rudenberg, H. Gunther. Arthur D. Little, Inc. February 1980. 77 pg. $2000.00.

1980 impact study by A.D. Little provides an in-depth look at the semiconductor industry, U.S. and abroad. Focuses on the industry structure, market size and growth, detailing technological trends and forecasts to 1983 of 6 major categories of semiconductor products: microprocessors and microcomputers, memory devices, integrated circuits, discrete, microwave and optoelectronic devices. Includes rankings of top U.S. manufacturers, company mergers during the 70's, information on services and suppliers to the industry and capital spending data. Provides summary discussions of major U.S. producers for the merchant market, U.S. captive producers and several foreign based producers. Statistics from industry associations and ADL estimates.

2313. WORLD TRENDS IN BATTERY POWER. Seitz, C. Ward. SRI International. Marketing Services Group. March 1980.

Discusses different types of batteries - primary, secondary and storage in terms of applications and markets. Discusses characteristics and variations of the different types of batteries. Information presented in table and chart format. Lists companies which have developed advanced storage batteries. Shows trends in world production for 1975, 1980, and forecasts for 1985 and 1990.

2314. 1979 FLEET MARKET FOR BATTERIES (MEDIUM AND HEAVY DUTY TRUCKS). National Aftermarket Audit Company. January 1980. 30 pg. $2900.00.

Survey of markets for truck batteries among 1500 fleets. Includes units in operation, brand names, performance data and market area analysis.

2315. 1979 HOUSEHOLD BATTERY MARKET (AUTOMOTIVE TYPE). National Aftermarket Audit Company. January 1980. 100 pg. $3400.00.

Survey of household markets for automotive type batteries. Tabulates units in operation, 1979 consumer purchases, prices, retail distribution sources, brand names, demographic statistics and market areas. Also covers battery chargers.

See 62, 67, 71, 72, 78, 82, 83, 84, 92, 98, 116, 124, 126, 127, 128, 129, 131, 132, 143, 144, 156, 163, 174, 878, 882, 892, 907, 912, 917, 941, 955, 979, 987, 1047, 1063, 1064, 1065, 1166, 1180, 1186, 1215, 1218, 3161, 3162, 3506, 3543, 3550, 3553, 3567, 3571, 3576, 4023, 4053, 4077.

Investment Banking Reports

2316. BEAR STEARNS AND CO.: INDUSTRY REPORT: CONNECTOR INDUSTRY: FOLLOW UP. Schlackman, Milton. Bear Stearns and Company. January 5, 1982. 11 pg.

Analysis of the connector industry and markets, including orders, prices and leading corporations. Shows corporations' stock prices; earnings per share, 1980-1982; price earnings ratio, 1980-1982; dividend and yield; sales; and orders, shipments and backlogs, 1979-1982.

2317. BEAR STEARNS AND CO.: RESEARCH DEPARTMENT HIGHLIGHTS: CONNECTOR INDUSTRY: FOLLOW-UP. Schlockman, Milton. Bear Stearns and Company. January 8, 1982. pg.2-3

Market analysis update for electronic connectors. Covers 1981-1982 orders, prices, projected growth rates, corporations, and competition. Shows stock prices; earnings per share, 1980-1982; price earnings ratio, 1980-1982; dividend and yield.

2318. BEAR STEARNS AND CO.: RESEARCH DEPARTMENT HIGHLIGHTS: SEMICONDUCTOR DISTRIBUTORS: AN UPDATE. Rosenthal, Harry K. Bear Stearns and Company. December 11, 1981. pg.30-31

Semiconductor supplier corporations report. Gives 1981 stock prices; earnings per share, 1980-1982; dividends; sales, 1980-1981; industry structure; product mix; and market shares.

2319. DONALDSON, LUFKIN AND JENRETTE: ACTION RECOMMENDATION: THE SEMICONDUCTOR INDUSTRY: CLIMBING A WALL OF WORRY. Vitolo, Aristide J. Donaldson, Lufkin and Jenrette Inc. April 14, 1981. 4 pg.

U.S. semiconductor market analysis. Comments on orders, prices, capacity and new products. For selected corporations, shows earnings per share, 1981-1983; price earnings ratio,

1982-1983; return on equity, 1981-1983; stock prices, 1980-1981; valuation, 1980; shares, 1980; market value, 1980; debt and stockholders' equity, 1980; net sales, 1978-1983; cost of sales, 1978-1983; research and development, 1978-1983; income, 1978-1983; and profit margins and taxes, 1978-1983. Also includes brief financial analysis comments and assessment of management.

2320. DONALDSON, LUFKIN AND JENRETTE: ACTION RECOMMENDATION: THE SEMICONDUCTOR INDUSTRY: THE EARLY GLIMMER OF RECOVERY. *Vitolo, Aristide J. Donaldson, Lufkin and Jenrette Inc. March 9, 1982. 5 pg.*

Provides outlook for the semiconductor industry. Shows stock price range, 1981-1982; earnings per share, 1981-1983; price earnings ratio, 1982-1983; return on equity, 1981-1983; dividend rate and yield; 1981 book value; and 1981 revenues per share, all for selected corporations. Also shows orders, 1976-1981; percent of orders filled by suppliers, 1976-1981; ratio of bookings to billings, 1976-1981; and forecasts of market value, 1980-1982.

2321. DONALDSON, LUFKIN AND JENRETTE: INDUSTRY VIEWPOINT: CONNECTOR INDUSTRY: FAST AND PROFITABLE GROWTH. *Donaldson, Lufkin and Jenrette Inc. May 24, 1982. 30 pg.*

Financial analysis of the connector industry and the leading corporations. Shows sales, profits and assets; costs, profits and average assets as percent of sales; profitability; growth rates; cash flow; capitalization; analytical ratios; and common stock data, all for 1967-1981, including breakdowns for corporations. Includes stock prices, e rnings per share, price earnings ratio, dividend and yield, sales profit margins, taxes, net income, and returns on equity, 1981-1983. Also includes market value of connector product groups in the U.S. and foreign markets, 1970-1971 and 1980-1981.

2322. DONALDSON, LUFKIN AND JENRETTE: RESEARCH BULLETIN: THE SEMICONDUCTOR INDUSTRY: TIME TO THROW IN THE TOWEL OR AVERAGE DOWN? *Vitolo, Austide J. Donaldson, Lufkin and Jenrette Inc. October 1, 1981. 11 pg.*

Semiconductor industry and market analysis. Graphs show new orders, 1976-1981; revenues and unit sales, 1980-1981; prices, 1980-1981; dollar values by product groups, 1980-1982; stock valuation, 1978-1981; and price earnings ratio, 1978-1981.

2323. DREXEL BURNHAM LAMBERT INC.: SEMICONDUCTOR MANUFACTURING EQUIPMENT: KEY TO SEMICONDUCTOR PRODUCTIVITY. *Schneider, William J. Drexel Burnham Lambert Inc. April 1982. 108 pg.*

Analysis of the semiconductor manufacturing equipment industry. Includes corporations' 1981 market sales, percent of sales in semiconductor machinery, 1981 market size and 3-5 year growth rates, and market shares by product groups; investment ratings; stock prices, 1981-1982; earnings per share, 1980-1983; price earnings ratio, 1981-1983; dividends and yield; and shares outstanding. Examines Japanese competition, the technologies, prices, and 1981-1982 shipments. Also includes financial analysis of individual corporations, with operating performance, 1974-1985; management and marketing strategies; balance sheets, including assets, liabilities and financial ratios, 1981-1982; working capital and funding requirements, 1981-1985; and other financial statements, 1974-1985.

2324. ELECTRONICS: TECHNOLOGY STOCK MONITOR. *Gumport, Michael A. Cyrus J. Lawrence Inc. Bi-monthly. May 7, 1980. 8 pg.*

Lists most attractive and least attractive technology stocks, based on historical performance and current price earnings ratios. Lists 33 stocks, with prices as of April 29, 1980: dividends, yield, price earnings ratio, five-year growth rate for sales and earnings, and five-year average retained to net worth.

2325. KIDDER, PEABODY AND CO.: TECHNOLOGY INDUSTRY PERSPECTIVES: ELECTRONICS. *Easterbrook, William D. Kidder, Peabody and Company Inc. Annual. 1982. $250.00.*

Analysis of the electronic industry and the effect of new technology. Examines sales, new-products, and marketing strategies of manufacturers of microwave components, electronic test equipment, electronic warfare systems and connectors.

2326. MORGAN STANLEY AND CO.: INVESTMENT PERSPECTIVES: ELECTRONICS: STRENGTH HERE AND THERE: RESEARCH COMMENT. *Richards, Gregory. Morgan Stanley and Company, Inc. June 14, 1982. pg.20-21*

Report on performance of selected electronics corporations. Includes 1981-1982 orders, shipments, capacity, employees, revenues and profits, and demand projections.

2327. MORGAN STANLEY AND CO.: INVESTMENT PERSPECTIVES: ELECTRONICS: TWO EARNINGS ESTIMATES CUT: RESEARCH COMMENT. *Richards, Gregory P. Morgan Stanley and Company, Inc. March 22, 1982. pg.15-16*

Reduction in 1982 earnings per share estimates for two electronics corporations. Reports layoffs, reduced new orders and business cycles.

2328. MORGAN STANLEY AND CO.: INVESTMENT PERSPECTIVES: RESEARCH COMMENT: AEROSPACE / ELECTRONICS: TAXES AND RESEARCH - A SHOT IN THE ARM. *Demisch, Wolfgang H. Morgan Stanley and Company, Inc. August 17, 1981. pg.4-5*

Analysis of impact of new legislation governing taxes and depreciation on research and development expenditures of aerospace and electronics firms in military and civil markets.

2329. MORGAN STANLEY AND CO.: INVESTMENT PERSPECTIVES: SEMICONDUCTORS: ORDERS STRONG IN MARCH: RESEARCH COMMENT. *Richards, Gregory P. Morgan Stanley and Company, Inc. April 12, 1982. pg.25-26*

Report on new orders for semiconductors in March, 1982. Market analysis includes foreign markets and results for individual corporations.

2330. MORGAN STANLEY AND CO.: RESEARCH BROADCAST: SEMICONDUCTORS: AT CYCLE BOTTOM. *Demisch, Wolfgang H. Morgan Stanley and Company, Inc. April 6, 1981. pg.24-25*

Analysis of the U.S. computer industry's position in the business cycle. Comments on mergers, layoffs, and other economic indicators.

2331. MORGAN STANLEY AND CO.: RESEARCH BROADCAST: SEMICONDUCTORS: STILL UNDER PRESSURE. *Demisch, Wolfgang, H. Morgan Stanley and Company, Inc. June 1, 1981.*

Report of impact on earnings of U.S. semiconductor companies from recent layoffs. Market analysis covers prospects for specific products and competition from the Japanese.

2332. MORGAN STANLEY AND CO.: RESEARCH MEETING COMMENT: SEMICONDUCTORS: IMPACT OF THE RECESSION. *Demisch, Wolfgang H. Morgan Stanley and Company, Inc. December 22, 1980.*

1980-1981 semiconductors market analysis. Comments cover supply and demand, inventory levels, capacity layoffs, and stock price indexes, 1974, 1980-1981.

2333. MORGAN STANLEY AND CO.: RESEARCH MEETING COMMENT: SEMICONDUCTORS: IN THE NARROW PASSAGE. *Demisch, Wolfgang H. Morgan Stanley and Company, Inc. January 12, 1981.*

Financial analysis of the U.S. semiconductors industry. Covers prices and profit margins; business cycles, 1973-1974, 1979-1980; and performance of stocks.

2334. MORGAN STANLEY AND CO.: RESEARCH MEETING COMMENT: SEMICONDUCTORS: ORDER WEAKNESS RESUMING. *Demisch, Wolfgang H. Morgan Stanley and Company, Inc. December 15, 1980.*

Semiconductor industry report. Covers orders, 1979-1981; market analysis 1981, including foreign markets; 1981 prices; supply, 1980-1982; and demand 1980-1981.

2335. MORGAN STANLEY AND CO.: RESEARCH MEETING COMMENT: SEMICONDUCTORS: THE SLOWDOWN BEGINS TO HURT. *Demisch, Wolfgang H. Morgan Stanley and Company, Inc. March 2, 1981.*

Semiconductor market analysis with commentary ranging over earnings, prices, backlogs in inventories, demand, profits, and foreign competition. Shows earnings per share estimate revisions, 1981-1982, and current stock prices.

2336. PAINE WEBBER MITCHELL HUTCHINS INC.: ELECTRICAL EQUIPMENT INDUSTRY PROSPECTS. *Paine Webber Mitchell Hutchins, Inc. March 10, 1980. 48 pg. $250.00.*

The status report on electrical equipment industry prospects includes recommendations; analysis of earnings and dividends prospects; and five-year comparative industrial financial performance with financial performance tables on nine leading firms within the industry.

2337. PAINE WEBBER MITCHELL HUTCHINS INC.: STATUS REPORT: ELECTRICAL EQUIPMENT QUARTERLY REVIEW, JUNE 1981. *Cornell, Robert T. Paine Webber Mitchell Hutchins, Inc. Quarterly. June 15, 1981. 27 pg.*

Quarterly review of the U.S. electrical equipment industry, covering recommendations for investments, economic indicators, 1978-1982, operating performance, activity in foreign markets, labor contracts, and stock valuations. Tables show 1981 stock price; earnings per share, 1980-1983; price earnings ratio, 1981-1983; sales and annual growth rates, 1979-1985; income and profit margins, 1979-1985; and taxes, 1979-1985.

2338. PAINE WEBBER MITCHELL HUTCHINS INC.: STATUS REPORT: ELECTRICAL EQUIPMENT QUARTERLY REVIEW: JUNE 1982. *Cornell, Robert T. Paine Webber Mitchell Hutchins, Inc. Quarterly. June 30, 1982. 27 pg.*

Quarterly financial analysis of electrical equipment corporations. Includes stock prices and valuation; marketing strategies, labor costs, and effect of economic indicators; performance indexes including return on equity; investment prospects for specific corporations; earnings per share, 1980-1983; price earnings ratio, 1982-1983; sales and growth rates, by lines of business, 1980-1986; profit margins, 1980-1086; and pretax income, 1980-1986.

2339. PAINE WEBBER MITCHELL HUTCHINS INC.: STATUS REPORT: ELECTRICAL EQUIPMENT QUARTERLY REVIEW: MARCH 1982. *Cornell, Robert T. Paine Webber Mitchell Hutchins, Inc. Quarterly. March 31, 1982. 27 pg.*

Quarterly review of electrical equipment corporations. Includes stock valuations; market analysis; financial analysis of individual corporations; sales, 1980-1985; profit margins, 1979-1985; income, 1979-1985; recent stock prices; earnings per share, 1980-1983; price earnings ratio, 1982-1983; and graphs of historical data on stock performance, 1960-1981.

2340. PAINE WEBBER MITCHELL HUTCHINS INC.: STATUS REPORT: PERSONAL TECHNOLOGY: IMPRESSIONS FROM 1981 SUMMER CONSUMER ELECTRONICS SHOW. *Isgur, Barbara S. Paine Webber Mitchell Hutchins, Inc. June 10, 1981. 3 pg.*

Personal technology industry analysis. Comments on corporations and new products.

2341. PAINE WEBBER MITCHELL HUTCHINS INC.: STATUS REPORT: ROBOTICS 1982: INTENSE COMPETITION PERVADES A WEAK MARKET. *Lustgarten, Eli S. Paine Webber Mitchell Hutchins, Inc. March 24, 1982. 4 pg.*

Analysis of the market for industrial robots and the operating performance of participating corporations. Covers installations and labor supply, new orders, competition, new technology, and foreign investments and mergers. For corporations, gives earnings per share, 1980-1983; market shares; and profit margins.

2342. PAINE WEBBER MITCHELL HUTCHINS INC.: STATUS REPORT: U.S. AND JAPANESE COMPETITION IN THE SEMICONDUCTOR INDUSTRY. *Goldwin, Seth C. Paine Webber Mitchell Hutchins, Inc. December 20, 1982. 13 pg.*

Analyzes competition between U.S. and Japanese semiconductor corporations on the basis of financial and industry structure. Examines debt and equity positions, labor costs, government funding for research and development, and interest rates. Shows revenues and average prices by product groups, 1981; short and long term debt, total debt, shareholders' equity, total capitalization, and debt-to-equity and retained earnings; production costs, 1980-1990; sales and profits, 1980-1981; and product lines in which there is competition.

2343. PAINE WEBBER MITCHELL HUTCHINS INC.: STATUS REPORT: VIDEO GAMES: A NEW GROWTH INDUSTRY. *Isgur, Lee S. Paine Webber Mitchell Hutchins, Inc. September 23, 1981. 5 pg.*

Analysis of the emerging industry of new video game products, 1976-1981. Examines earnings, market areas and game locations, growth rates, markets, and investment opportunities among manufacturers.

2344. SMITH BARNEY HARRIS UPHAM AND CO.: ELECTRICAL EQUIPMENT INDUSTRY COMMENTARY. *Leavitt, Russell L. Smith Barney, Harris Upham and Company. October 8, 1981. 19 pg. $200.00.*

Analysis of the electrical equipment industry and stock performance of major corporations. Shows stock price index comparisons, 1976-1981; earnings per share, 1979-1983; price earnings ratio, 1981-1983; current dividend and yield; research and development and capital expenditures as percent of total sales, 1975-1980; earnings growth rates, 1971-1980; 1980 operating performance; sales and profits by product lines, 1979-1982; and end use markets.

2345. SMITH BARNEY HARRIS UPHAM AND CO.: ELECTRICAL EQUIPMENT INDUSTRY MONITOR. *Leavitt, Russell L. Smith Barney, Harris Upham and Company. July 28, 1981. 66 pg. $1200.00.*

Quarterly review of markets and statistics for electric power moderation and electrical equipment. Shows 1981 stock price; 1978-1982 earnings per share; 1981 dividend and yield; returns; and growth rates for major companies. Gives stock price index performance, 1976-1981; indexes of power generation, 1976-1981; demand and consumption by end uses, 1973-1981; electricity sales, 1979-1981; revenues, 1973-1981; orders, 1974-1980, 1995; and expansion plans, 1981-2001. Includes capacity 1981-2001; peak load, 1973-2001; reserves, 1980-2001; power plant shipments, 1975-1986; capital expenditures, 1970-1980; and manufacturers' shipments of electrical products, 1967-1981. Covers inventories, 1979-1980; power company equipment installations, 1979-1980; wholesale distribution, 1978-1981; and cost and price indexes, 1967-1981.

2346. SMITH BARNEY HARRIS UPHAM AND CO.: SEMICONDUCTORS: SIA FLASH REPORT - FLASHES POSITIVE YEAR OVER YEAR COMPARISON ALTHOUGH SEQUENTIAL TREND REMAINS FLAT. *Barlage, James L. Smith Barney, Harris Upham and Company. July 17, 1981. 3 pg. $75.00.*

Analysis of trends in the U.S. semiconductor industry. Shows industry association data for monthly new order bookings, billings for shipments, and ratios of the two, all for 1981.

2347. SMITH BARNEY HARRIS UPHAM AND CO.: SEMICONDUCTORS: THE SUPPLY / DEMAND OUTLOOK FOR THE U.S. SEMICONDUCTOR INDUSTRY. *Barlage, James L. Smith Barney, Harris Upham and Company. November 6, 1981. 34 pg. $200.00.*

Analysis of the U.S. semiconductor industry with comparison to Japan. Includes plant capacity and manufacturers' shipments, 1978-1983; forecasts of economic indicators, 1980-1983; shipments by product groups, 1980-1982; worldwide sales of U.S. manufacturers, 1980-1981; capital expenditures of corporations, production and the ratios, 1977-1981; historical data, 1967-1971; and utilization of capacity, 1975-1982.

See 578, 1247, 1250.

Industry Statistical Reports

2348. BATTERY MANUFACTURERS, 1979: ANNUAL CENSUS OF MANUFACTURES: FABRICANTS D'ACCUMULATEURS, 1979: RECENSEMENT ANNUEL DES MANUFACTURES. *Canada. Statistics Canada. English/French. Annual. Cat. #43-208. June 1981. 18 pg. 5.40 (Canada).*

For manufacturers of dry cells, storage batteries and storage cells, and parts and supplies for batteries, tables give exports and imports, 1975-1980, as well as principal data on businesses, employees and ownership, wages and salaries, costs of fuel and of raw materials, value of shipments, and value added, 1971-1979, as well as inventories, 1979, all by province. Provides statistics on materials and supplies used and on manufacturers' shipment, by kind, for 1978-1979. Includes a list of companies with addresses.

2349. CERTAIN ELECTRIC MOTORS FROM JAPAN: DETERMINATION OF A REASONABLE INDICATION OF MATERIAL INJURY IN INVESTIGATION NO. 731-TA-7 (PRELIMINARY) UNDER SECTION 733(A) OF THE TARIFF ACT OF 1930. *Cates, Bruce. U.S. International Trade Commission. USITC Publication 1037. February 1980. 47 pg. Free.*

Report of Commission investigation into potential injury to domestic industry of imports of polyphase AC electric motors

from Japan. Text covers description and uses, tariff treatment, U.S. and Japanese industry, U.S. importers, channels of distribution, nature and extent of LTFV (less than fair value) sales. Consideration of injury analyzes U.S. production, consumption, shipments, exports, imports, employment, and U.S. and foreign producers' inventories. The financial position of U.S. producers is shown via examination of overall operations, profit-and-loss experience, research and development, and capital expenditures. Market penetration, prices to distributors, prices to OEM's, and lost sales are covered in consideration of casual relationship between LTFV sales and alleged injury. Tabular data, compiled primarily from industry response to Commission questionnaire, covers the 1976-1979 period.

2350. CURRENT INDUSTRIAL REPORTS: ELECTRIC LAMPS. *U.S. Department of Commerce. Bureau of the Census. Customer Service Branch. Quarterly. MQ-36B. November 1980. 20 (est.) pg. $4.30/yr.*

Report presents data on the quantity of production and end-of-quarter stocks, and quantity and value of shipments (total, domestic, and for export) of electric lamps, each by product, for the current and immediately preceding quarter. A summary table shows the quantity and value of shipments and end-of-quarter stocks of electric lamps for each of the 20 past quarters. Another table shows shipments, imports, and apparent consumption of electric lamps, each by product, for the current and immediately preceding quarter. An annual summary is issued.

2351. CURRENT INDUSTRIAL REPORTS: ELECTRIC LIGHTING FIXTURES, 1979. *U.S. Department of Commerce. Bureau of the Census. Customer Service Branch. Annual. MA-36L. September 1980. 10 pg. $.25.*

Report presents data on the quantity and value of shipments of electric lighting fixtures, and number of companies, by product, for current and preceding year. Statistics include manufacturers' shipments, exports, imports, and apparent consumption of electric lighting fixtures for current and preceding year.

2352. CURRENT INDUSTRIAL REPORTS: MOTORS AND GENERATORS, 1979. *U.S. Department of Commerce. Bureau of the Census. Customer Service Branch. Annual. MA-36 H. September 1980. 13 pg. $.30.*

Report presents statistics on quantity and value of shipments of motors and generators, by product, including number of companies, for current and preceding year. Data given for quantity and value of total shipments (including interplant transfers) and commercial shipments, and quantity of motors and generators produced and incorporated into other products at the same establishment, for current and preceding year. Statistics are presented comparing manufacturers' shipments, exports, imports, and apparent consumption for the current year.

2353. CURRENT INDUSTRIAL REPORTS: SWITCHGEAR, SWITCHBOARD APPARATUS, RELAYS, AND INDUSTRIAL CONTROLS: 1979. *U.S. Department of Commerce. Bureau of the Census. Customer Service Branch. Annual. MA-36A. October 1980. 12 pg. $.25.*

Report on switchgear and switchboard apparatus, relays, and industrial controls presents data on value of shipments, by product class for 1970 to 1979, and number of companies and value of shipments, by product, for current and preceding year. Includes data comparing manufacturers' shipments, exports, imports, and apparent consumption for current and preceding year.

2354. CURRENT INDUSTRIAL REPORTS: WIRING DEVICES AND SUPPLIES, 1979. *U.S. Department of Commerce. Bureau of the Census. Customer Service Branch. Annual. MA-36K. November 1980. 8 pg. $.25.*

Report presents statistics on quantity and value of shipments of wiring devices and supplies, including number of companies, by product, for current and preceding year. Data includes manufacturers' shipments, exports, imports, and apparent consumption for selected wiring devices for current and preceding year.

2355. EIS SHIPMENTS REPORT: CARBON AND GRAPHITE PRODUCTS. *Economic Information Systems, Inc. Quarterly. 1982. $75.00/yr.*

Report on carbon and graphite products industries. Arranged by state and county, report includes every plant in the industry with annual shipments over $500,000 and/or 20 or more employees. Plant listings detail address, telephone, estimated annual shipments, and percent of market. Similar statistics for each county and state.

2356. EIS SHIPMENTS REPORT: CATHODE RAY TV PICTURE TUBES. *Economic Information Systems, Inc. Quarterly. 1982. $75.00/yr.*

Report on cathode ray TV picture tube industries. Arranged by state and county, report includes every plant in the industry with annual shipments over $500,000 and/or 20 or more employees. Plant listings detail address, telephone, estimated annual shipments, and percent of market. Similar statistics for each county and state.

2357. EIS SHIPMENTS REPORT: COMMERCIAL LIGHTING FIXTURES. *Economic Information Systems, Inc. Quarterly. 1982. $110.00/yr.*

Report on commercial lighting fixtures industries. Arranged by state and county, report includes every plant in the industry with annual shipments over $500,000 and/or 20 or more employees. Plant listings detail address, telephone, estimated annual shipments, and percent of market. Similar statistics for each county and state.

2358. EIS SHIPMENTS REPORT: CURRENT - CARRYING WIRING DEVICES. *Economic Information Systems, Inc. Quarterly. 1982. $295.00/yr.*

Report on current-carrying wiring devices industries. Arranged by state and county, report includes every plant in the industry with annual shipments over $500,000 and/or 20 or more employees. Plant listings detail address, telephone, estimated annual shipments, and percent of market. Similar statistics for each county and state.

2359. EIS SHIPMENTS REPORT: ELECTRIC LAMPS. *Economic Information Systems, Inc. Quarterly. 1982. $80.00.*

Report on the electric lamp industry. Arranged by state and county, report includes every plant in the industry with annual shipments of over $500,000 and/or 20 or more employees. Plant listings detail address, telephone, estimated annual shipments, and percent of market. Similar statistics for each county and state.

2360. EIS SHIPMENTS REPORT: ELECTRICAL APPARATUS, NOT ELSEWHERE CLASSIFIED. *Economic Information Systems, Inc. Quarterly. 1982. $125.00.*

Report on miscellaneous electrical apparatus industries. Arranged by state and county, report includes every plant in the industry with annual shipments of over $500,000 and/or 20 or more employees. Plant listings detail address, telephone, estimated annual shipments, and percent of market. Similar statistics for each county and state.

2361. EIS SHIPMENTS REPORT: ELECTRICAL EQUIPMENT AND SUPPLIES, NOT ELSEWHERE CLASSIFIED. *Economic Information Systems, Inc. Quarterly. 1982. $160.00.*

Report on miscellaneous electrical equipment and supplies industry. Arranged by state and county, report includes every plant in the industry with annual shipments of over $500,000 and/or 20 or more employees. Plant listings detail address, telephone, estimated annual shipments, and percent of market. Similar statistics for each county and state.

2362. EIS SHIPMENTS REPORT: ELECTRON TUBES, RECEIVING TYPE. *Economic Information Systems, Inc. Quarterly. 1982. $75.00.*

Report on the (receiving type) electronic tube industry. Arranged by state and county, report includes every plant in the industry with annual shipments of over $500,000 and/or 20 or more employees. Plant listings detail address, telephone, estimated annual shipments, and percent of market. Similar statistics for each county and state.

2363. EIS SHIPMENTS REPORT: ELECTRON TUBES, TRANSMITTING. *Economic Information Systems, Inc. Quarterly. 1982. $75.00.*

Report on the transmitting electron tube industry. Arranged by state and county, report includes every plant in the industry with annual shipments of over $500,000 and/or 20 or more employees. Plant listings detail address, telephone, estimated annual shipments, and percent of market. Similar statistics for each county and state.

2364. EIS SHIPMENTS REPORT: ELECTRONIC CAPACITORS. *Economic Information Systems, Inc. Quarterly. 1982. $75.00.*

Report on the electronic capacitor industry. Arranged by state and county, report includes every plant in the industry with annual shipments of over $500,000 and/or 20 or more employees. Plant listings detail address, telephone, estimated annual shipments, and percent of market. Similar statistics for each county and state.

2365. EIS SHIPMENTS REPORT: ELECTRONIC COILS AND TRANSFORMERS. *Economic Information Systems, Inc. Quarterly. 1982. $150.00.*

Report on the electronic coil and transformer industry. Arranged by state and county, report includes every plant in the industry with annual shipments of over $500,000 and/or 20 or more employees. Plant listings detail address, telephone, estimated annual shipments, and percent of market. Similar statistics for each county and state.

2366. EIS SHIPMENTS REPORT: ELECTRONIC COMPONENTS, NOT ELSEWHERE CLASSIFIED. *Economic Information Systems, Inc. Quarterly. 1982. $350.00.*

Report on miscellaneous electronic components industries. Arranged by state and county, report includes every plant in the industry with annual shipments of over $500,000 and/or 20 or more employees. Plant listings detail address, telephone, estimated annual shipments, and percent of market. Similar statistics for each county and state.

2367. EIS SHIPMENTS REPORT: ELECTRONIC CONNECTORS. *Economic Information Systems, Inc. Quarterly. 1982. $90.00.*

Report on the electronic connector industry. Arranged by state and county, report includes every plant in the industry with annual shipments of over $500,000 and/or 20 or more employees. Plant listings detail shipments, and percent of market. Similar statistics for each county and state.

2368. EIS SHIPMENTS REPORT: ELECTRONIC RESISTORS. *Economic Information Systems, Inc. Quarterly. 1982. $75.00.*

Report on the electronic resistor industry. Arranged by state and county, report includes every plant in the industry with annual shipments of over $500,000 and/or 20 or more employees. Plant listings detail address, telephone, estimated annual shipments, and percent of market. Similar statistics for each county and state.

2369. EIS SHIPMENTS REPORT: ENGINE ELECTRICAL EQUIPMENT. *Economic Information Systems, Inc. Quarterly. 1982. $165.00/yr.*

Report on engine electrical equipment industries. Arranged by state and county, report includes every plant in the industry with annual shipments over $500,000 and/or 20 or more employees. Plant listings detail address, telephone, estimated annual shipments, and percent of market. Similar statistics for each county and state.

2370. EIS SHIPMENTS REPORT: INDUSTRIAL CONTROLS. *Economic Information Systems, Inc. Quarterly. 1982. $350.00/yr.*

Report on industrial control industries. Arranged by state and county, report includes every plant in the industry with annual shipments over $500,000 and/or 20 or more employees. Plant listings detail address, telephone, estimated annual shipments, and percent of market. Similar statistics for each county and state.

2371. EIS SHIPMENTS REPORT: LIGHTING EQUIPMENT, NOT ELSEWHERE CLASSIFIED. *Economic Information Systems, Inc. Quarterly. 1982. $145.00/yr.*

Report on miscellaneous lighting equipment industries. Arranged by state and county, report includes every plant in the industry with annual shipments over $500,000 and/or 20 or more employees. Plant listings detail address, telephone, estimated annual shipments, and percent of market. Similar statistics for each county and state.

2372. EIS SHIPMENTS REPORT: NONCURRENT - CARRYING WIRING DEVICES. *Economic Information Systems, Inc. Quarterly. 1982. $110.00/yr.*

Report on noncurrent-carrying wiring devices industries. Arranged by state and county, report includes every plant in the industry with annual shipments over $500,000 and/or 20 or more employees. Plant listings detail address, telephone, estimated annual shipments, and percent of market. Similar statistics for each county and state.

2373. EIS SHIPMENTS REPORT: PRIMARY BATTERIES, DRY AND WET. *Economic Information Systems, Inc. Quarterly. 1982. $75.00/yr.*

Report on dry and wet primary batteries industries. Arranged by state and county, report includes every plant in the industry with annual shipments over $500,000 and/or 20 or more employees. Plant listings detail address, telephone, estimated annual shipments, and percent of market. Similar statistics for each county and state.

2374. EIS SHIPMENTS REPORT: RESIDENTIAL LIGHTING FIXTURES. *Economic Information Systems, Inc. Quarterly. 1982. $195.00/yr.*

Report on residential lighting fixtures industries. Arranged by state and county, report includes every plant in the industry with annual shipments over $500,000 and/or 20 or more employees. Plant listings detail address, telephone, estimated annual shipments, and percent of market. Similar statistics for each county and state.

2375. EIS SHIPMENTS REPORT: SEMICONDUCTOR AND RELATED DEVICES. *Economic Information Systems, Inc. Quarterly. 1982. $225.00/yr.*

Report on semiconductor and related devices industries. Arranged by state and county, report includes every plant in the industry with annual shipments over $500,000 and/or 20 or more employees. Plant listings detail address, telephone, estimated annual shipments, and percent of market. Similar statistics for each county and state.

2376. EIS SHIPMENTS REPORT: STORAGE BATTERIES. *Economic Information Systems, Inc. Quarterly. 1982. $120.00/yr.*

Report on storage battery industries. Arranged by state and county, report includes every plant in the industry with annual shipments over $500,000 and/or 20 or more employees. Plant listings detail address, telephone, estimated annual shipments, and percent of market. Similar statistics for each county and state.

2377. EIS SHIPMENTS REPORT: VEHICULAR LIGHTING EQUIPMENT. *Economic Information Systems, Inc. Quarterly. 1982. $75.00/yr.*

Report on vehicular lighting equipment industries. Arranged by state and county, report includes every plant in the industry with annual shipments over $500,000 and/or 20 or more employees. Plant listings detail address, telephone, estimated annual shipments, and percent of market. Similar statistics for each county and state.

2378. EIS SHIPMENTS REPORT: WELDING APPARATUS, ELECTRIC. *Economic Information Systems, Inc. Quarterly. 1982. $110.00/yr.*

Report on electric welding apparatus industries. Arranged by state and county, report includes every plant in the industry with annual shipments over $500,000 and/or 20 or more employees. Plant listings detail address, telephone, estimated annual shipments, and percent of market. Similar statistics for each county and state.

2379. ELECTRIC LAMP AND SHADE MANUFACTURERS, 1980: ANNUAL CENSUS OF MANUFACTURERS: INDUSTRIE DES LAMPES ELECTRIQUES ET DES ABAT-JOUR, 1980: RECENSEMENT ANNUEL DES MANUFACTURES. *Canada. Statistics Canada. English/French. Annual. Cat. #35-214. June 1982. 6 pg. 5.40 (Canada).*

For manufacturers of electric table and floor lamps, as well as lamp shades of all types and of all materials, tables give statistics on establishments, employees, wages and salaries, fuel costs, costs of materials and supplies, value of shipments, value added, and ownership,1972-1980, and on inventories, 1980. Detailed data on materials and supplies used and on shipments of products is provided, by kind, for 1979-1980. Includes a list of company addresses.

2380. ELECTRIC LAMPS (LIGHT SOURCES): APRIL 1982: LAMPES ELECTRIQUES (SOURCES DE LUMIERE): AVRIL 1982. *Canada. Statistics Canada. English/French. Monthly. Cat. #43-009. June 1982. 4 (est.) pg. 18.00/yr. (Canada).*

Provides manufacturers' sales of electric lamps (light sources), in units and in dollars, for current month and year-to-date, by region. December issue carries a list of reporting firms.

2381. ELECTRICAL / ELECTRONIC CENSUS. *Lake Publishing Corporation. 1980. 19 pg.*

Provides 1979 data from census of subscribers to "Insulation / Circuits" covering products used, processing operations performed, buying influence, plant size, and other parameters alone and in combination with each other. Arranged by SIC code with extensive breakdowns for primary metal industries;

fabricated metal products, electrical and electronic equipment, transportation equipment, and instruments and related products.

2382. ELECTRONIC CHEMICALS USA. *Strategic Analysis, Inc. 1981. 400 (est.) pg. $10500.00.*

Data on electronic chemicals will include business opportunities (including acquisitions and spin-off targets), business overview, semiconductors, printed circuit boards, passive components, assembly materials and other suppliers, and future outlook. Discussions of the various industry elements will span market size; market maturity, future growth, competitiveness, profitability, technological developments, capital intensity, availability of acquisition candidates, and effects of government regulations.

2383. ELECTRONIC MARKET DATA BOOK, 1980. *Electronic Industries Association. Annual. 1980. 138 pg. $50.00.*

Annual statistics for the U.S. electronics industry, 1980. Covers consumer products, including televisions, radios, and CB radios; industrial and specialty communications equipment; CATV; broadcasting; computers; lasers; military electronics; solid state components; and tubes. Shows factory shipments, 1970-1979; production, 1973-1979; sales, 1970-1979; units in operation, 1954-1979; broadcasting facilities, 1961-1980; license applications, 1969-1979; CATV market penetration, 1979; military markets and other end uses, 1979; economic indicators, 1980-1981; government budgets and funding for research and development, 1979-1981; consumption, 1978-1979; and employment and average earnings, 1975-1979. Dates vary for various products, and not all products show the same data.

2384. ELECTRONIC MARKET TRENDS: EIA STATISTICS: MARCH 1981. *Marketing Services Department. Electronic Industries Association. Monthly. 36. March 1981. pg.26-30 $150.00/yr.*

Monthly statistical summary of sales in electronics products industries. Tables show U.S. factory sales of consumer electronic products for December 1979, December 1980, and year-to-date 1979 and 1980. Balance-of-trade figures for electronic products are arranged by product group and dollar value, for December 1980 and year-to-date through December 1980. Data is derived from U.S. government sources, and is adapted from the Associations monthly report "Electronics Foreign Trade."

2385. ELECTRONICS FOREIGN TRADE: JANUARY 1981 AND YEAR-TO-DATE. *Electronic Industries Association. Monthly. March 18, 1981. 25 pg. $150.00/yr.*

Monthly statistical analysis of value of foreign trade in electronics, prepared by the Electronics Industries Association from U.S. Commerce Department and Custom Bureau figures. Divided into the following categories: communications products, consumer electronics, electron tubes, electronic parts, industrial products, solid state products, and other electronic products. Tables show summary totals and detailed figures by TSUSA number. Arrangement is by unit and dollars, for current month and year-to-date.

2386. FACTORY SALES OF ELECTRIC STORAGE BATTERIES: APRIL 1982: VENTES A L'USINE DE BATTERIES D'ACCUMULATEURS ELECTRIQUES: AVRIL 1982. *Canada. Statistics Canada. English/French. Monthly. Cat. #43-005. June 1982. 3 (est.) pg. 18.00/yr. (Canada).*

Gives factory sales, in units and in dollars, of electric storage batteries, for current month and year-to-date. December issue carries a list of reporting firms.

2387. IEEE U.S. MEMBERSHIP SALARY AND FRINGE BENEFIT SURVEY 1981. *Abbott, Langer, and Associates. June 1981. $60.00.*

Based on data from 6700 Institute of Electrical and Electronic Engineers members, volume reports electrical and electronic engineering salaries by employer type, employer size, length of time with present firm, job level, job function, geographic region, level of education, length of experience, and supervisory repsonsibility, with numerous cross-tabulations. Includes information on pensions, insurance coverage, vacation time, tuition aid, and overtime.

2388. IMPORT TRENDS IN TSUS ITEMS 806.30 AND 807.00. *Katlin, Charles. U.S. International Trade Commission. USITC Publication 1029. January 1980. 83 pg. Free.*

A Commission research study on imports under items 806.30 and 807.00 of the Tariff Schedules of the United States (TSUS) which provide for duty-free treatment of the value of U.S. materials or parts sent abroad for processing or assembly. Statistics showing total imports under the two TSUS items for 1966-1978, imports by principal commodity groups, and country sources of imports for 1975-1978. Imports under 807.00 for 1978 were predominantly metal products such as motor vehicles, semi-conductors and parts, and television receivers, apparatus and parts, principally from West Germany, Mexico, Japan and Canada; imports under 806.30 for 1978 included steel and other metal mill products, semiconductors and parts, electronic and electrical articles, and vehicle parts, supplied principally by Canada, Malaysia, Mexico, and West Germany. Data compiled primarily from official figures of the U.S. Department of Commerce.

2389. INDUSTRY WAGE SURVEY: SEMICONDUCTORS: SEPTEMBER 1977. *U.S. Department of Labor. Bureau of Labor Statistics. Irregular. Bulletin 2021. April 1979. 23 pg.*

Report provides data on hourly wages and employee benefits, by occupation, of production workers, computer operators, and engineering technicians in the semiconductor manufacturing industry. Data is tabulated separately for the U.S., the Northeast, and the South.

2390. INTERNATIONAL COMPETITION IN ADVANCED INDUSTRIAL SECTORS: TRADE AND DEVELOPMENT IN THE SEMI-CONDUCTOR INDUSTRY. *Joint Economic Committee, 97th U.S. Congress, Second Session. U.S. Government Printing Office. su. doc. no. Y4.Ec7:In2/12. February 18, 1982. 183 pg. $6.50.*

Compares development of semiconductor industry in Japan and U.S., 1961-1979. Provides following information: foreign corporate investments in U.S. semi-conductor companies by investor and percent ownership, 1979; composition of domestic Japanese consumption; and dollar value of Japanese domestic computer sales by company, 1978, 1979. Ranks U.S. and Japanese 16K RAM producers, 1979. Gives dollar value of annual world IC production, 1978-1981; and world IC market share. Singles out U.S. and Japan. Includes major Japanese government support to information industry, 1971-1991; and annual number of electrical engineering graduates in Japan and U.S., 1969-1979. Marketing and trade data interspersed throughout text.

2391. MANUFACTURERS OF LIGHTING FIXTURES, 1980: ANNUAL CENSUS OF MANUFACTURES: FABRICANTS D'APPAREILS D'ECLAIRAGE, 1980: RECENSEMENT ANNUEL DES MANUFACTURES. *Canada. Statistics Canada. English/French. Annual. Cat. #43-211. April 1982. 6 pg. 5.40 (Canada).*

For manufacturers of incandescent and fluorescent residential, industrial, commercial and area and street lighting fixtures, statistics are given on exports and imports, 1976-1981, on establishments, employees and ownership, wages and salaries, costs of fuel and materials, value of shipments, and value added, 1972-1980, and on inventories, 1980. Tables provide materials and supplies used and manufacturers' shipments, by kind, 1979-1980. Includes a list of company addresses.

2392. MANUFACTURERS OF MISCELLANEOUS ELECTRICAL PRODUCTS, 1979: ANNUAL CENSUS OF MANUFACTURES: FABRICANTS DE PRODUITS ELECTRIQUES DIVERS, 1979: RECENSEMENT ANNUEL DES MANUFACTURES. *Canada. Statistics Canada. English/French. Annual. Cat. #43-210. July 1981. 20 pg. 5.40 (Canada).*

For manufacturers of miscellaneous electrical products, such as light bulbs and tubes, wiring devices, panel boards and low-voltage switch boards, carbon or graphite electrodes, conduit and fittings, tables give principal statistics on establishments, employees and ownership, wages and salaries, costs of fuel and materials, value of shipments, and value added, 1971-1979, as well as inventories, 1979, all by province. For 1978-1979, provides data on raw materials used and on shipments of good, by kind. A list of companies with addresses is included.

2393. PROBLEMS OF SMALL HIGH TECHNOLOGY FIRMS. *U.S. National Science Foundation. SuDoc no.1.2:P24/11. December 1981. 34 pg. Free.*

Gives results of survey of 13,000 high technology firms active in research and development, with fewer than 500 employees. Gives data on 11 problem areas by size and type of business, 1979.

2394. REPORT ON THE CANADIAN ELECTRICAL AND ELECTRONICS INDUSTRIES FOR THIRD QUARTER OF "81. *Canada. Department of Industry, Trade and Commerce. Quarterly. 1982.*

Examines rise in the trade deficit of Canada's office machine sector as well as 1981 deficits in electrical and electronics products. Also examines exports in the communications industry.

2395. SIA QUARTERLY STATISTICAL REVIEW. *Semiconductor Industry Association. Quarterly. 1980. $750.00.*

Analyzes semiconductor industry sales performance. Intended for financial and economic analysts.

2396. SIGNS AND DISPLAYS INDUSTRY. *Canada. Statistics Canada. French/English. Annual. 1981. 12 pg. 4.50 (Canada).*

Includes statistics on number of sign and display establishments; number and type of employees; salaries and wages; cost of fuel and electricity; value of shipped goods; and value added. Comparative statistics are given for Canada for earlier years, and by province for the preceding year. Data issued since 1960.

2397. SUMMARY OF TRADE AND TARIFF INFORMATION: IGNITION EQUIPMENT AND LIGHTING EQUIPMENT FOR MOTOR VEHICLES; BATTERIES. *U.S. Department of Commerce. International Trade Administration. pub. 841. April 1981. 34 pg.*

Presents 1981 U.S. tariff schedules for motor vehicle lighting and ignition equipment, and for all types of electrical batteries. Discusses industry structure and concentration, and analyzes domestic and international markets' segmentation, determinants and outlook. Includes 1975-1979 statistics on production, apparent consumption, and imports / exports.

2398. TANTALUM ELECTROLYTIC FIXED CAPACITORS FROM JAPAN: DETERMINATION OF NO INJURY IN INVESTIGATION NO. AA1921-159 UNDER THE ANTIDUMPING ACT, 1921, AS AMENDED. *U.S. International Trade Commission. USITC Publication 1092. August 1980. 16 pg. Free.*

A review of a 1976 investigation by the Commission of imports of tantalum electrolytic fixed capacitors from Japan, in the light of corrected statistics for the period January 1975 through June 1976. Table reflects new figures, compiled from official statistics of the U.S. Department of Commerce.

2399. WEIGHING MACHINERY AND SCALES FROM JAPAN: DETERMINATION OF NO MATERIAL INJURY OR THREAT THEREOF IN INVESTIGATION NO. 701-TA-7 (FINAL) UNDER SECTION 705(B) OF THE TARIFF ACT OF 1930, TOGETHER WITH THE INFORMATION OBTAINED IN THE INVESTIGATION. *Slingerland, David. U.S. International Trade Commission. USITC Publication 1063. May 1980. 88 pg. Free.*

Report of Commission investigation into potential injury to domestic industry of imports of electronic digital deli or service and counting scales from Japan. Text describes product and uses, U.S. tariff treatment, domestic producers and importers, U.S. production, shipments, exports, inventories, employment, financial experience, market penetration of imports from Japan, prices, and lost sales. Tables provide supporting details, for 1977-1979, from information supplied in response to Commission questionnaires and data from the U.S. Commerce Department's Census of Manufactures for various years.

2400. 1977 CENSUS OF MANUFACTURES: CARBON AND GRAPHITE PRODUCTS: PRELIMINARY REPORT: INDUSTRY SERIES. *U.S. Department of Commerce. Bureau of the Census. Customer Service Branch. Irregular. MC77-1-36A-6(P). 1979. 8 pg. $.35.*

This is a preliminary report from the Bureau of the Census 1977 Census of Manufactures. Tables with industry statistics from the U.S. for 1963-1977 and with statistics for selected geographic areas for 1972 and 1977 give data relating to employees, production costs, assets and expenditures, inventories, and ratios. Other tables supply 1972 and 1977 statistics on product classes giving quantity and value of shipments by all producers and statistics on materials consumed by kind.

2401. 1977 CENSUS OF MANUFACTURES: COMMERCIAL LIGHTING FIXTURES: PRELIMINARY REPORT: INDUSTRY SERIES. *U.S. Department of Commerce. Bureau of the Census. Customer Service Branch. Irregular. MC77-1-36C-5(P). 1979. 8 pg. $.35.*

This is a preliminary report from the Bureau of the Census 1977 Census of Manufactures. Tables with industry statistics from the U.S. for 1963-1977 and with statistics for selected geographic areas for 1972 and 1977 give data relating to employees, production costs, assets and expenditures, inventories, and ratios. Other tables supply 1972 and 1977 statistics on product classes giving quantity and value of shipments by all producers and statistics on materials consumed by kind.

2402. 1977 CENSUS OF MANUFACTURES: CURRENT-CARRYING WIRING DEVICES: PRELIMINARY REPORT: INDUSTRY SERIES. *U.S. Department of Commerce. Bureau of the Census. Customer Service Branch. Irregular. MC77-1-36C-2(P). 1979. 10 pg. $.35.*

This is a preliminary report from the Bureau of the Census 1977 Census of Manufactures. Tables with industry statistics from the U.S. for 1963-1977 and with statistics for selected geographic areas for 1972 and 1977 give data relating to employees, production costs, assets and expenditures, inventories, and ratios. Other tables supply 1972 and 1977 statistics on

product classes giving quantity and value of shipments by all producers and statistics on materials consumed by kind.

2403. 1977 CENSUS OF MANUFACTURES: ELECTRIC LAMPS: PRELIMINARY REPORT: INDUSTRY SERIES. *U.S. Department of Commerce. Bureau of the Census. Customer Service Branch. Irregular. MC77-1-36C-1(P). 1979. 9 pg. $.35.*

This is a preliminary report from the Bureau of the Census 1977 Census of Manufactures. Tables with industry statistics from the U.S. for 1963-1977 and with statistics for selected geographic areas for 1972 and 1977 give data relating to employees, production costs, assets and expenditures, inventories, and ratios. Other tables supply 1972 and 1977 statistics on product classes giving quantity and value of shipments by all producers and statistics on materials consumed by kind.

2404. 1977 CENSUS OF MANUFACTURES: ELECTRICAL EQUIPMENT AND SUPPLIES, N.E.C.: PRELIMINARY REPORT: INDUSTRY SERIES. *U.S. Department of Commerce. Bureau of the Census. Customer Service Branch. Irregular. MC77-1-36F-5(P). 1979. 8 pg. $.35.*

This is a preliminary report from the Bureau of the Census 1977 Census of Manufactures. Tables with industry statistics from the U.S. for 1963-1977 and with statistics for selected geographic areas for 1972 and 1977 give data relating to employees, production costs, assets and expenditures, inventories, and ratios. Other tables supply 1972 and 1977 statistics on product classes giving quantity and value of shipments by all producers and statistics on materials consumed by kind.

2405. 1977 CENSUS OF MANUFACTURES: ELECTRON TUBES, ALL TYPES: PRELIMINARY REPORT: INDUSTRY SERIES. *U.S. Department of Commerce. Bureau of the Census. Customer Service Branch. Irregular. MC77-1-36E-1(P). 1979. 9 pg. $.35.*

This is a preliminary report from the Bureau of the Census 1977 Census of Manufactures. Tables with industry statistics from the U.S. for 1963-1977 and with statistics for selected geographic areas for 1972 and 1977 give data relating to employees, production costs, assets and expenditures, inventories, and ratios. Other tables supply 1972 and 1977 statistics on product classes giving quantity and value of shipments by all producers and statistics on materials consumed by kind.

2406. 1977 CENSUS OF MANUFACTURES: ELECTRONIC CAPACITORS: PRELIMINARY REPORT: INDUSTRY SERIES. *U.S. Department of Commerce. Bureau of the Census. Customer Service Branch. Irregular. MC77-1-36E-5(P). 1979. 8 pg. $.35.*

This is a preliminary report from the Bureau of the Census 1977 Census of Manufactures. Tables with industry statistics from the U.S. for 1963-1977 and with statistics for selected geographic areas for 1972 and 1977 give data relating to employees, production costs, assets and expenditures, inventories, and ratios. Other tables supply 1972 and 1977 statistics on product classes giving quantity and value of shipments by all producers and statistics on materials consumed by kind.

2407. 1977 CENSUS OF MANUFACTURES: ELECTRONIC COILS AND TRANSFORMERS: PRELIMINARY REPORT: INDUSTRY SERIES. *U.S. Department of Commerce. Bureau of the Census. Customer Service Branch. Irregular. MC77-1-36E-7(P). 1979. 8 pg. $.35.*

This is a preliminary report from the Bureau of the Census 1977 Census of Manufactures. Tables with industry statistics from the U.S. for 1963-1977 and with statistics for selected geographic areas for 1972 and 1977 give data relating to employees, production costs, assets and expenditures, inventories, and ratios. Other tables supply 1972 and 1977 statistics on

product classes giving quantity and value of shipments by all producers and statistics on materials consumed by kind.

2408. 1977 CENSUS OF MANUFACTURES: ELECTRONIC COMPONENTS, N.E.C.: PRELIMINARY REPORT: INDUSTRY SERIES. *U.S. Department of Commerce. Bureau of the Census. Customer Service Branch. Irregular. MC77-1-36E-9(P). 1979. 11 pg. $.35.*

This is a preliminary report from the Bureau of the Census 1977 Census of Manufactures. Tables with industry statistics from the U.S. for 1963-1977 and with statistics for selected geographic areas for 1972 and 1977 give data relating to employees, production costs, assets and expenditures, inventories, and ratios. Other tables supply 1972 and 1977 statistics on product classes giving quantity and value of shipments by all producers and statistics on materials consumed by kind.

2409. 1977 CENSUS OF MANUFACTURES: ELECTRONIC CONNECTORS: PRELIMINARY REPORT: INDUSTRY SERIES. *U.S. Department of Commerce. Bureau of the Census. Customer Service Branch. Irregular. MC77-1-36E-6(P). 1979. 8 pg. $.35.*

This is a preliminary report from the Bureau of the Census 1977 Census of Manufactures. Tables with industry statistics from the U.S. for 1963-1977 and with statistics for selected geographic areas for 1972 and 1977 give data relating to employees, production costs, assets and expenditures, inventories, and ratios. Other tables supply 1972 and 1977 statistics on product classes giving quantity and value of shipments by all producers and statistics on materials consumed by kind.

2410. 1977 CENSUS OF MANUFACTURES: ELECTRONIC RESISTORS: PRELIMINARY REPORT: INDUSTRY SERIES. *U.S. Department of Commerce. Bureau of the Census. Customer Service Branch. Irregular. MC77-1-36E-4(P). 1979. 8 pg. $.35.*

This is a preliminary report from the Bureau of the Census 1977 Census of Manufactures. Tables with industry statistics from the U.S. for 1963-1977 and with statistics for selected geographic areas for 1972 and 1977 give data relating to employees, production costs, assets and expenditures, inventories, and ratios. Other tables supply 1972 and 1977 statistics on product classes giving quantity and value of shipments by all producers and statistics on materials consumed by kind.

2411. 1977 CENSUS OF MANUFACTURES: ENGINE ELECTRICAL EQUIPMENT: PRELIMINARY REPORT: INDUSTRY SERIES. *U.S. Department of Commerce. Bureau of the Census. Customer Service Branch. Irregular. MC77-1-36F-4(P). 1979. 9 pg. $.35.*

This is a preliminary report from the Bureau of the Census 1977 Census of Manufactures. Tables with industry statistics from the U.S. for 1963-1977 and with statistics for selected geographic areas for 1972 and 1977 give data relating to employees, production costs, assets and expenditures, inventories, and ratios. Other tables supply 1972 and 1977 statistics on product classes giving quantity and value of shipments by all producers and statistics on materials consumed by kind.

2412. 1977 CENSUS OF MANUFACTURES: INDUSTRIAL APPARATUS, N.E.C.: PRELIMINARY REPORT: INDUSTRY SERIES. *U.S. Department of Commerce. Bureau of the Census. Customer Service Branch. Irregular. MC77-1-36A-7(P). 1979. 8 pg. $.35.*

This is a preliminary report from the Bureau of the Census 1977 Census of Manufactures. Tables with industry statistics from the U.S. for 1963-1977 and with statistics for selected geographic areas for 1972 and 1977 give data relating to employees, production costs, assets and expenditures, inventories, and ratios. Other tables supply 1972 and 1977 statistics on

product classes giving quantity and value of shipments by all producers and statistics on materials consumed by kind.

2413. 1977 CENSUS OF MANUFACTURES: INDUSTRIAL CONTROLS: PRELIMINARY REPORT: INDUSTRY SERIES. *U.S. Department of Commerce. Bureau of the Census. Customer Service Branch. Irregular. MC77-1-36A-4(P). 1979. 8 pg. $.35.*

This is a preliminary report from the Bureau of the Census 1977 Census of Manufactures. Tables with industry statistics from the U.S. for 1963-1977 for selected geographic areas for 1972 and 1977 give data relating to employees, production costs, assets and expenditures, inventories, and ratios. Other tables supply 1972 and 1977 statistics on product classes giving quantity and value of shipments by all producers and statistics on materials consumed by kind.

2414. 1977 CENSUS OF MANUFACTURES: LIGHTING EQUIPMENT, N.E.C.: PRELIMINARY REPORT: INDUSTRY SERIES. *U.S. Department of Commerce. Bureau of the Census. Customer Service Branch. Irregular. MC77-1-36C-7(P). 1979. 9 pg. $.35.*

This is a preliminary report from the Bureau of the Census 1977 Census of Manufactures. Tables with industry statistics from the U.S. for 1963-1977 and with statistics for selected geographic areas for 1972 and 1977 give data relating to employees, production costs, assets and expenditures, inventories, and ratios. Other tables supply 1972 and 1977 statistics on product classes giving quantity and value of shipments by all producers and statistics on materials consumed by kind.

2415. 1977 CENSUS OF MANUFACTURES: MOTORS AND GENERATORS: PRELIMINARY REPORT: INDUSTRY SERIES. *U.S. Department of Commerce. Bureau of the Census. Customer Service Branch. Irregular. MC77-1-36A-1(P). 1979. 12 pg. $.35.*

This is a preliminary report from the Bureau of the Census 1977 Census of Manufactures. Tables with industry statistics from the U.S. for 1963-1977 and with statistics for selected geographic areas for 1972 and 1977 give data relating to employees, production costs, assets and expenditures, inventories, and ratios. Other tables supply 1972 and 1977 statistics on product classes giving quantity and value of shipments by all producers and statistics on materials consumed by kind.

2416. 1977 CENSUS OF MANUFACTURES: NONCURRENT-CARRYING WIRING DEVICES: PRELIMINARY REPORT: INDUSTRY SERIES. *U.S. Department of Commerce. Bureau of the Census. Customer Service Branch. Irregular. MC77-1-36C-3(P). 1979. 9 pg. $.35.*

This is a preliminary report from the Bureau of the Census 1977 Census of Manufactures. Tables with industry statistics from the U.S. for 1963-1977 and with statistics for selected geographic areas for 1972 and 1977 give data relating to employees, production costs, assets and expenditures, inventories, and ratios. Other tables supply 1972 and 1977 statistics on product classes giving quantity and value of shipments by all producers and statistics on materials consumed by kind.

2417. 1977 CENSUS OF MANUFACTURES: PRIMARY BATTERIES, DRY AND WET: PRELIMINARY REPORT: INDUSTRY SERIES. *U.S. Department of Commerce. Bureau of the Census. Customer Service Branch. Irregular. MC77-1-36F-2(P). 1979. 8 pg. $.35.*

This is a preliminary report from the Bureau of the Census 1977 Census of Manufactures. Tables with industry statistics from the U.S. for 1963-1977 and with statistics for selected geographic areas for 1972 and 1977 give data relating to employees, production costs, assets and expenditures, inventories, and ratios. Other tables supply 1972 and 1977 statistics on

product classes giving quantity and value of shipments by all producers and statistics on materials consumed by kind.

2418. 1977 CENSUS OF MANUFACTURES: RESIDENTIAL LIGHTING FIXTURES: PRELIMINARY REPORT: INDUSTRY SERIES. *U.S. Department of Commerce. Bureau of the Census. Customer Service Branch. Irregular. MC77-1-36C-4(P). 1979. 8 pg. $.35.*

This is a preliminary report from the Bureau of the Census 1977 Census of Manufactures. Tables with industry statistics from the U.S. for 1963-1977 and with statistics for selected geographic areas for 1972 and 1977 give data relating to employees, production costs, assets and expenditures, inventories, and ratios. Other tables supply 1972 and 1977 statistics on product classes giving quantity and value of shipments by all producers and statistics on materials consumed by kind.

2419. 1977 CENSUS OF MANUFACTURES: SIGNS AND ADVERTISING DISPLAYS: PRELIMINARY REPORT: INDUSTRY SERIES. *U.S. Department of Commerce. Bureau of the Census. Customer Service Branch. Irregular. MC77-1-39D-2(P). May 1979. 7 pg. $.35.*

This is a preliminary report from the Bureau of the Census 1977 Census of Manufactures. Tables with industry statistics from the U.S. for 1963-1977 and with statistics for selected geographic areas for 1972 and 1977 give data relating to employees, production costs, assets and expenditures, inventories, and ratios. Other tables supply 1972 and 1977 statistics on product classes giving quantity and value of shipments by all producers and statistics on materials consumed by kind.

2420. 1977 CENSUS OF MANUFACTURES: STORAGE BATTERIES: PRELIMINARY REPORT: INDUSTRY SERIES. *U.S. Department of Commerce. Bureau of the Census. Customer Service Branch. Irregular. MC77-1-36F-1(P). 1979. 9 pg. $.35.*

This is a preliminary report from the Bureau of the Census 1977 Census of Manufactures. Tables with industry statistics from the U.S. for 1963-1977 and with statistics for selected geographic areas for 1972 and 1977 give data relating to employees, production costs, assets and expenditures, inventories, and ratios. Other tables supply 1972 and 1977 statistics on product classes giving quantity and value of shipments by all producers and statistics on materials consumed by kind.

2421. 1977 CENSUS OF MANUFACTURES: VEHICULAR LIGHTING EQUIPMENT: PRELIMINARY REPORT: INDUSTRY SERIES. *U.S. Department of Commerce. Bureau of the Census. Customer Service Branch. Irregular. MC77-1-36C-6(P). 1979. 8 pg. $.35.*

This is a preliminary report from the Bureau of the Census 1977 Census of Manufactures. Tables with industry statistics from the U.S. for 1963-1977 and with statistics for selected geographic areas for 1972 and 1977 give data relating to employees, production costs, assets and expenditures, inventories, and ratios. Other tables supply 1972 and 1977 statistics on product classes giving quantity and value of shipments by all producers and statistics on materials consumed by kind.

2422. 1977 CENSUS OF MANUFACTURES: WELDING APPARATUS, ELECTRIC: PRELIMINARY REPORT: INDUSTRY SERIES. *U.S. Department of Commerce. Bureau of the Census. Customer Service Branch. Irregular. MC77-1-36A-5(P). 1979. 8 pg. $.35.*

This is a preliminary report from the Bureau of the Census 1977 Census of Manufactures. Tables with industry statistics from the U.S. for 1963-1977 give data relating to employees, production costs, assets and expenditures, inventories, and ratios. Other tables supply 1972 and 1977 statistics on

product classes giving quantity and value of shipments by all producers and statistics on materials consumed by kind.

2423. 1979 / 80 CANADIAN ELECTRONICS MARKET REPORT. *Maclean Hunter Ltd. 1980. 45 pg. 40.00 (Canada).*

Report details the Canadian electronics industry's major market segments. In addition to statistical treatment of specific electronic products, report also deals with relevant economic factors relating to capital expenditures, population growth, GNP, income and expenditures. Other basic statistics include radio and TV stations, cable TV, navigation aids, airports and aircraft, nuclear power generation and computer installations, as well as imports and exports.

2424. 1979 SIA SEMICONDUCTOR TRADE STATISTICS PROGRAM. *Semiconductor Industry Association. December 1979. $150.00.*

Semiconductor sales and orders reported according to geographical location, in U.S., Japan, Europe, or other areas of the world. Actual sale prices used to value all shipments and orders. Includes 34 product lines covering integrated circuits and discrete devices. Also includes 13 geographical areas and distributor summaries.

2425. 1981 ELECTRONIC MARKET DATA BOOK. *Electronic Industries Association. Annual. 1981. 117 pg. $50.00.*

Contains information based on production and sales figures provided by several hundred companies, supplemented by selected government and private sources. Data is provided on shipments, sales, production, and foreign trade, by product, for particular groups of years.

2426. THE 1982 ELECTRONIC MARKET DATA BOOK. *Electronic Industries Association. Annual. 1982. 150 (est.) pg. $55.00.*

Gives data based on production and sales figures provided to Electronic Industries Association by several companies and supplemented by selected government and private sources. Provides trends for following markets: consumer electronics, communications and industrial products, and electronic components.

See 204, 218, 1308.

Financial and Economic Studies

2427. COMPETITIVE FACTORS INFLUENCING WORLD TRADE IN INTEGRATED CIRCUITS: REPORT TO THE SUBCOMMITTEE ON INTERNATIONAL TRADE OF THE COMMITTEE ON FINANCE AND THE SUBCOMMITTEE ON INTERNATIONAL FINANCE OF THE COMMITTEE ON BANKING, HOUSING AND URBAN AFFAIRS OF THE UNITED STATES SENATE IN INVESTIGATION NO. 332-102 UNDER SECTION 332 OF THE TARIFF ACT OF 1930, AS AMENDED. *Hogge, Nelson J. United States International Trade Commission. U.S. International Trade Commission. USITC Publication 1013. November 1979. 140 pg. Free.*

A detailed study of the competitive position of the United States in the computer and integrated circuits industries, with particular emphasis on Japan and Western Europe (primarily West Germany, France, and the United Kingdom). Data is derived from questionnaire responses and from government source, covering 1974-1978. Text profiles U.S. and foreign producers, surveying investments, production, research and development, government support, employment and productivity, and trade flow. Appendix contains 72 tables of statistics on subjects in body of report.

2428. THE COMPLETE BOOK OF ELECTRIC VEHICLES. *Shacket, Sheldon R. Domus Books. 2nd ed. 1981. 224 pg. $9.95.*

Lists electric vehicles, equipment manufacturers, and associations including business addresses and product lines. Provides extensive historical and technological data. Includes legislation, energy supply projections, and bibliographic information.

2429. ELECTROCHEMICAL POWER SOURCES. *Barak, M., editor. Independent Battery Manufacturers Association, Inc. 1980. $68.00.*

Examines all types of primary and secondary batteries. Comparisons between different battery types for assorted applications.

2430. ELECTRONIC COMPONENTS, MEXICO: COUNTRY MARKET SURVEY CMS 79-011. *Industry and Trade Administration U.S. Department of Commerce. U.S. Government Printing Office. March 1979. 7 pg. $.50.*

One of a series of industry studies prepared for U.S. exporters. Covers the market for semiconductors, passive components, and electron tubes in Mexico through analyis of market for 1973-1976 and projections through 1980. Tables list production imports, exports, and market size by product and year, in U.S. dollars. Data is from U.S. government research study. Text surveys the competitive environment, end-user analysis, marketing practices, and regulations. Includes lists of best sales opportunities for U.S. exporters, and sources of additional information.

2431. ENERGY CONSIDERATIONS IN FIXED AND VEHICULAR LIGHTING: TRANSPORTATION RESEARCH CIRCULAR. *Stark, Richard E., editor. U.S. Transportation Research Board. National Academy of Sciences. no.228. May 1981. 9 pg.*

Summarizes current literature on the energy consumption and efficiency, tangible and intangible benefits, and costs of fixed roadway lighting and vehicular lighting. Emphasizes the comparative characteristics of these two lighting modes, with recommendations for implementing optimally energy-efficient systems.

2432. HIGH TECHNOLOGY GROWTH STOCKS. *Select Information Exchange. Monthly. $95.00/yr.*

Newsletter identifies emerging, fast growing companies with annual sales below 100 million dollars and growth in sales and earnings of 30 percent or more per year for the past three to five years. Each issue profiles a specific high technology market with recommendations and analyses of companies in that category. Previous recommendations are followed until closed out. Newsletter alerts subscribers to new issues with special investment merit.

2433. THE IMPACT OF MICROCHIP TECHNOLOGY ON EMPLOYMENT. *Metra Consulting Group Ltd. 1979. 300 pg. 35.00 (United Kingdom).*

Effects of microelectronics on economic indicators, employment and unemployment, and industry structure in end uses such as banking, insurance, data processing, education, manufacturing, office equipment, printing and publishing, industrial processing, telecommunications and materials handling.

2434. THE IMPLICATIONS OF MICROELECTRONICS FOR CANADIAN WORKERS. *Mather, Boris. Canadian Centre for Policy Alternatives. 1981. 10 pg.*

Analysis of the microelectronics industry in Canada, including employment and unemployment, women in the labor supply, shift work, legislation, and regulation.

2435. INDUSTRY RETIREMENT PRACTICES SERIES: VOLUME 7: ELECTRICAL EQUIPMENT AND ELECTRONICS. *Koetting, Ruth A., editor. Erisa Benefit Funds Inc. April 1981. 85 pg.*

Profiles deferred compensation plans for salaried personnel of 12 corporations that manufacture electrical and electronic equipment. Chart compares normal retirement benefits of the electrical equipment and electronics industry.

2436. INNOVATION, COMPETITION, AND GOVERNMENT POLICY IN THE SEMICONDUCTOR INDUSTRY. *Wilson, Robert W. Lexington Books. Division of D.C. Heath and Company. 1980. 240 pg. $21.95.*

Responding to the concern that decline in the rate of technological innovation contributes to inflation, slowed economic growth and less competitive performance by U.S. firms in international trade, the authors examine the semiconductor industry's research, pricing and marketing strategies, as well as the effects of government policies.

2437. THE LEADING COMPANIES IN THE U.S. ELECTRONICS INDUSTRY. *Wasserman, Jerry. Arthur D. Little, Inc. 1980.*

Analyzes financial and operating performance of leading U.S. electronics companies.

2438. MARKET AND ECONOMICS OF THE U.S. BATTERY INDUSTRY. *Business Information Services. 1982. $325.00.*

Presents industry overview and executive recommendations and overall domestic growth analysis from 1947 with forecasts to 1987. Also includes shipment and wholesale price index data; battery sales trends (1958-1987) by product line; end use market analysis (1950-1982); battery plant economic structure (1954-1980); industry labor situation (1958-1981); competitive environment; world trade (1958-1982); and financial information for approximately 30 firms.

2439. MEXICAN AUTOMOTIVE, STEEL, AND ELECTRONICS INDUSTRIES: TRADE EXPANSION OPPORTUNITIES. *Developing World Industry and Technology, Inc. February 1981. 74 pg. $35.00.*

Analysis of opportunities for expansion of Mexican foreign trade and production of automotives, steel and electronics. Includes government regulation of foreign investments, manufacturing facilities, exports, and economic indicators including transportation statistics, labor supply and capacity utilization.

2440. MICROELECTRONICS: BACKGROUND PAPERS FOR THE TASK FORCE TO THE GOVERNMENT OF ONTARIO. *Ontario Task Force on Microelectronics. Ontario. Ministry of Industry and Tourism. October 1981. 121 pg.*

Analysis of the performance and structure of Canada's electronic industry includes employment and unemployment, domestic consumption, shipments of products, value of products, net imports, competition and regulation. Examines production, competition, office automation, controls and instruments, hard and software, trade, industry performance for 1980.

2441. MICROS: A PERVASIVE FORCE: A STUDY OF THE IMPACT OF MICROELECTRONICS ON BUSINESS AND SOCIETY, 1946-1990. *Orme, Michael. Halstead Press. Division of John Wiley and Sons, Inc. 1979. 208 pg. $29.95.*

Sector-by-sector analysis of microelectronics past, present and future impact covers Japanese technology; strategy behind setting up of INMOS; IBM's market strategy; impact on information networks of the 1980's; social and political environment; the microprocessor; the computer; semiconductors; silicon; and satellites.

2442. A REPORT ON THE U.S. SEMICONDUCTOR INDUSTRY. *Industry and Trade Administration U.S. Department of Commerce. U.S. Government Printing Office. September 1979. 132 pg. $4.50.*

Report on domestic and international developments in the emergence and growth of the semiconductor industry from its beginnings in the late 1940's through 1977. Tabular data throughout text shows: manufacturing costs, yield ratios, materials used in production, employment and occupational information, domestic and world shipments and values, industry characteristics, growth rates, inventory-to-sales ratios, markets, prices and price indexes capital expenditures, structure of financial capital, international trading patterns, and list of innovations. Data is derived from Census Bureau and Bureau of Labor Statistics, and from industry sources.

2443. SEMICONDUCTOR INDUSTRY REVIEW. *Berdell, James R. Wall Street Transcript. June 1980. $20.00.*

U.S. semiconductor industry review and forecast (1980-1982). Industry bookings and durable goods expenditures (1980-1981) and capital spending and defense budget (1980-1982) are projected. Some topics include: microprocessors, auto industry needs, Japanese competition, pricing, multiples and overcapacity. Some 20 major businesses are reviewed.

2444. U.S. AND JAPAN SEMICONDUCTOR INDUSTRIES: A FINANCIAL COMPARISON. *Semiconductor Industry Association. 1981. $395.00.*

Results of research on the cost of capital for U.S. and Japanese semiconductor manufacturers.

2445. U.S. SEMICONDUCTOR INDUSTRY. *U.S. Dept. of Commerce, Bureau of Industrial Economics. U.S. Government Printing Office. SuDoc #003-009-00327-0. 1979. 131 pg. $5.00.*

Presents an overview of domestic and international developments influencing the U.S. semiconductor industry. Reviews production, research, product development, and competitive marketing strategies.

See 1330.

Forecasts

2446. AUTO AFTERMARKET - ELECTRICAL: PREDICASTS' INDUSTRY STUDY NO. E57. *Predicasts, Inc. February 1980. $725.00.*

Investigates markets for electrical automotive replacement parts. Historic (1963-1978) and projected (1985, 1990) data is presented for nine major groups of aftermarket parts and accessories. Replacement needs of the auto population are analyzed, as well as market opportunities expected to develop in the 1980's. A detailed industry structure analysis outlines trends in distribution and market shares, including profiles of major manufacturers and distributors.

2447. AUTOMOTIVE ELECTRONICS. *International Resource Development, Inc. October 29, 1980. 189 pg.*

1980 report analyzes the inclusion of electronic components and systems in the automobile industry to 1985. Components include semiconductors, microprocessors, sensors, transducers and actuators. Leading component suppliers and automobile manufacturers are analyzed in relation to shipments and applications of components and their costs.

2448. AUTOMOTIVE ENGINE AND TRANSMISSION ELEC-TRONICS FORECAST: A MULTICLIENT STUDY PROPOS-AL. *Gnostic Concepts, Inc. 1981. $15000.00.*

Analysis of the market and technological outlook for electronic systems and components in automotive power train controls. This study focuses on evolving applications, technological developments and price trends in controls of the engine, transmission and electrical system for U.S. production of automobiles, trucks, and buses from 1977 through 1986. Aftermarket requirements will also be forecasted. Objectives of this study are to identify, review, and assess current development efforts and production demand. Additionally, the study will define, analyze, and forecast the market trends for electronic power train control systems and components over the 1977 to 1986 period and will analyze the following major issues: industry / supplier trends, technological directions and competitive alignment.

2449. BATTERIES AND ELECTRIC VEHICLES: PREDI-CASTS' INDUSTRY STUDY NO. E59. *Predicasts, Inc. July 1980. $725.00.*

Provides a comprehensive investigation of primary and secondary battery markets and analyzes U.S. sales by type (lead-acid, nickel-cadmium, zinc-carbon, alkaline, etc.). Emphasizes changing technology (e.g. ultra long-life primary batteries) and new markets (e.g. rechargable batteries used in small consumer products) with projections to 1985 and 1990.

2450. THE CONSUMER ELECTRONICS INDUSTRY: 1981-1983. *FIND / SVP. 1981. 200 pg. $950.00.*

This study forecasts factory shipments to U.S. markets through 1983. The report identifies factors affecting industry categories and specific product groups. Attention is paid to opportunities in the automotive and the appliance industry.

2451. ELECTRONIC COMPONENTS MARKETS AND DIS-TRIBUTION: PREDICASTS' INDUSTRY STUDY NO. E58. *Predicasts, Inc. April 1980. $725.00.*

This study analyzes and projects U.S. markets to 1985 and 1990 for semiconductors, tubes, connectors, relays, switches, printed circuits, resistors and capacitors. Market shares, distribution channels and marketing strategies are reviewed for each. End-use markets are projected for each of the major product lines. Foreign competition and markets are also analyzed. Key companies are profiled, and the role of distributors is analyzed in detail.

2452. ELECTRONIC INDUSTRIES ASSOCIATION: TEN YEAR FORECAST OF GOVERNMENT MARKETS. *Gray, William E., editor. Electronic Industries Association. Annual. June 1980. 315 pg. $55.00.*

Annual trade association conference report and forecast, focusing on government markets for weapons, munitions, and particularly military electronics. Forecasts various dimensions of government military spending, including specific programs, and discusses effects of inflation, changing political goals, and different war scenarios. Presents general characteristics of, and Pentagon spending outlooks for several existing and proposed weapons systems, nuclear and conventional.

2453. ELECTRONIC MAIL IN THE 1980'S. *International Resource Development, Inc. December 1979. 233 pg. $895.00.*

This report describes the expected emergence in the 1980's of several different types of public and private electronic mail systems and services. Impact of new networks from Xerox, SBS, Hughes and AT&T. Relationship between facsimile and terminal-based electronic mail. Political considerations governing entry into electronic mail by consumables. Ten-year market projections, with assessments of the strategies of more than fifty suppliers or potential suppliers.

2454. ELECTRONICS IN THE AUTOMOTIVE INSTRUMENT PANEL: OPPORTUNITY AND RISK IN THE 1980'S. *Chase Econometrics. 1982.*

Examines the introduction of electronic instrument panels by U.S. and Japanese auto makers. Covers electronic gauges and warnings, trip computers, electronic service reminders, navigation systems, TV screen games, electronic radio, air conditioning, voice warnings, voice-activated controls, visual displays and touch controls.

2455. ELECTRONICS 1981 WORLD MARKETS: FORECAST DATA BOOK. *McGraw-Hill Book Company. Annual. 23rd ed. 1981. $125.00.*

Forecasts current and future demand for 200 component and equipment products in the U.S., Europe and Japan. Includes market estimates and growth rate tables (1979-1984); and product-by-product markets for 11 Western European countries.

2456. HOME AND COIN OPERATED ELECTRONIC GAMES MARKET. *Frost and Sullivan, Inc. April 1979. 237 pg. $800.00.*

Analyzes and forecasts the market to 1988 for home and commercial electronic games. Provides breakdowns in number of units, average selling price, sales value and growth rate for the following market segments: dedicated TV games, programmable TV games, cartidges, non-video electronic games, electronic learning aids, electronic chess and related games, and home pinball. Discusses technology trends, new product introductions, pricing, retailer and operator needs, company profiles, production rates and demographic considerations.

2457. THE IMPACT OF CMOS ON COMPETING IC TECH-NOLOGIES. *Strategic Inc. 1982. $1500.00.*

Reviews reasons for shift from NMOS to CMOS technology. Forecasts 1985 CMOS and NMOS price relationship.

2458. THE INDUSTRIAL ROBOT MARKET. *Frost and Sullivan, Inc. January 1979. 250 pg. $800.00.*

Sales forecasts for the industrial robot market through 1985 provided by end-user industry and by 5 basic robot types: PTP (point to point- servo controlled); NSPTP (point to point- non-servo controlled); CPT (continuous path- tape controlled); CPSS (continuous path- solid state controlled); and ADVDES (advanced design capacities). Also gives forecasts by type of power used to drive robot: hydraulic, electric and pneumatic. Includes responses to questionnaires sent to users, potential users in 24 industries, and suppliers.

2459. MOTORS AND GENERATORS: PREDICASTS' INDUS-TRY STUDY NO. E56. *Predicasts, Inc. December 1979. $725.00.*

Presents a detailed analysis of the electric motor and generator industry. Historical data (1963 to 1978) and forecasts (1985, 1990) are presented for shipments, imports, exports, and domestic sales of five major product categories. Projected markets are also shown for each category by product and by over 25 end uses. The structure of the industry is discussed, including company profiles and estimates of market shares.

2460. NEW BATTERY TECHNOLOGIES: PROGRESS AND PROBLEMS. *George, James H. B. Arthur D. Little, Inc. June 10, 1980. 13 pg.*

Examines developments in the effort to develop battery systems which promise to be technically superior to lead - acid. Lists companies involved in nickel zinc battery development.

2461. OUTLOOK FOR ELECTRONIC COMPONENTS AND SYSTEMS IN AUTOMOBILES IN THE 1980'S. *Mathews, Henry W. Arthur D. Little, Inc. August 3, 1979. 11 pg.*

Discusses the use, by automobile manufacturers, of outside sources for components and modules built to their specifications. States that it is a difficult market in which to compete because automotive OEM's (original equipment manufacturers), while providing business to outside sources, are anxious to produce in-house. Provides a list of selected supplies of automotive electronics by type of product - engine controls, monitor and display devices, and safety systems. Gives an overview of the market for each of these types of electronic components.

2462. PORTABLE ENERGY SOURCES: BATTERIES, FUEL AND SOLAR CELLS. *Business Communications Company. March 1980. 81 pg. $675.00.*

Market forecast through the 1980's for portable energy sources (excluding mechanical generators). Considers market structure, industrial secondary markets, consumer markets, and projected growth rates. Discusses such applications as electric vehicles and utility load levelling.

2463. STRATEGIC IMPACT OF INTELLIGENT ELECTRONICS IN U.S. AND WESTERN EUROPE; 1977-1987: VOLUME 1: EXECUTIVE SUMMARY. *Arthur D. Little, Inc. 1980. 35 pg. $35000.00.*

Volume 1 of 11-volume multi-client study on intelligent electronics in U.S. and Western Europe. (Pricing noted is for the complete set. For individual prices, contact Arthur D. Little.) Market areas included are automotive, business communication and consumer. Status, trends, consumption, value of shipments, product lines and general conclusions. Specific data on market areas available in later volumes.

2464. STRATEGIC IMPACT OF INTELLIGENT ELECTRONICS IN U.S. AND WESTERN EUROPE: 1977-1987: VOLUME 2: SOCIETAL ENVIRONMENT. *Ernst, Martin L. Arthur D. Little, Inc. March 1979. 195 pg. $35000.00.*

Volume 2 of 11-volume multi-client study discusses social and economic trends relevant to development and marketing of intelligent electronics equipment in U.S. and Western Europe. Demographic data, statistics through 1987 on families and households, incomes, trading, employment and unemployment, gross national products and projections by geographic area provided. Data on Western Europe also divided into the following countries; France, West Germany, and United Kingdom. Price listed is for the complete set. Contact publisher for individual prices. Data derived from Arthur D. Little, Inc. estimates, U.K. Treasury Model of the British Economy and the Economist Intelligence Unit and National Government statistics.

2465. STRATEGIC IMPACT OF INTELLIGENT ELECTRONICS IN U.S. AND WESTERN EUROPE: 1977-1987: VOLUME 3: TECHNOLOGY. *Curtis, David A. Arthur D. Little, Inc. March 1979. 129 pg. $35000.00.*

Third volume of 11-volume multi-client study discusses current status of the technology and makes technological forecasts for 1978-1987 in the multi-processor industry. Price projections, production and product development and design included. Price listed is for complete set. Contact publisher for individual prices. Data derived from Arthur D. Little estimates.

2466. STRATEGIC IMPACT OF INTELLIGENT ELECTRONICS IN U.S. AND WESTERN EUROPE: 1977-1987: VOLUME 4: AUTOMOTIVE. *Bishop, John. Arthur D. Little, Inc. March 1979. 89 pg. $35000.00.*

Volume 4 of 11-volume multi-client study discusses microprocessors in the automobile industry in U.S. and Europe. Includes microprocessor applications in vehicles, in service equipment and for replacement parts and products. Auto production; market areas, analyses, segmentation and forecast; sales; market shares; and prices provided. Analysis provided for U.S., Western Europe as a whole, and West Germany, France and United Kingdom individually. Price listed is for complete set. Contact publisher for individual prices. Data derived from Arthur D. Little, Inc., Environmental Protection Agency, & National Highway Traffic Safety Administration.

2467. STRATEGIC IMPACT OF INTELLIGENT ELECTRONICS IN U.S. AND WESTERN EUROPE: 1977-1987: VOLUME 5: BUSINESS COMMUNICATIONS. *Withington, Frederic G. Arthur D. Little, Inc. April 1979. 188 pg. $35000.00.*

Volume 5 of 11-volume multi-client study analyzes strategic impact of intelligent electronics on the business communications market. Includes typewriters, word processors, text processing terminals, copying and facsimile machines and voice recognition terminals. Countries discussed are U.S., United Kingdom, West Germany and France. Target years 1977, 1982, 1987 used to determine market environment and segments, product lines, shipments, competitor rankings, revenues and prices. Price listed is for complete set. Contact publisher for individual prices. Data derived from Arthur D. Little, Inc. estimates.

2468. STRATEGIC IMPACT OF INTELLIGENT ELECTRONICS IN U.S. AND WESTERN EUROPE: 1977-1987: VOLUME 7A: INDUSTRIAL / PROCESS CONTROLS. *Pastan, Harvey L. Arthur D. Little, Inc. April 1979. 82 pg. $35000.00.*

Volume 7A of 11-volume study discusses industrial electronics, including process controls. Instruments and applications provided. Market shares, production, imports / exports, sales and consumption of industrial electronic products provided. Supplier rankings, profiles, sales and earnings data given. Price listed is for complete set. Contact publisher for individual prices. Data derived from Arthur D. Little, Inc. estimates.

2469. STRATEGIC IMPACT OF INTELLIGENT ELECTRONICS IN U.S. AND WESTERN EUROPE: 1977-1987: VOLUME 7B: INDUSTRIAL / AUTOMATED MANUFACTURING. *Pastan, Harvey L. Arthur D. Little, Inc. April 1979. 171 pg. $35000.00.*

Volume 7B of 11-volume set discusses the strategic impact of intelligent electronics on automated manufacturing from 1977 to 1987 in USA and Western Europe (France, W. Germany and United Kingdom). Manufacturing functions, environment, market segments, products and costs; suppliers earnings; sales; rankings; and profiles given. Price listed is for complete set. Contact publisher for individual prices. Data derived is from Arthur D. Little, Inc. estimates.

2470. STRATEGIC IMPACT OF INTELLIGENT ELECTRONICS IN U.S. AND WESTERN EUROPE: 1977-1987: VOLUME 7C: INDUSTRIAL / ANALYTICAL INSTRUMENTS, AUTOMATED TEST EQUIPMENT, BUILDING AUTOMATION SYSTEMS. *Pastan, Harvey L. Arthur D. Little, Inc. April 1979. 164 (est.) pg. $35000.00.*

Volume 7C of 11-volume study divided into 3 sections on industrial electronic equipment industry: analytical instruments, automated test equipment, and building automation systems. Equipment environments, products, users, market forecasts, production, consumption, exports and growth rate provided. Supplier addresses, rankings and profiles included. Price listed is for complete set. Contact publisher for individual prices. Data derived is from Arthur D. Little, Inc. estimates.

2471. STRATEGIC IMPACT OF INTELLIGENT ELECTRONICS IN U.S. AND WESTERN EUROPE: 1977-1987: VOLUME 9: STRATEGIC CONSIDERATIONS. *Wasserman, Jerry. Arthur D. Little, Inc. April 1979. 94 pg. $35000.00.*

Volume 9 of 11-volume study descusses the implications of technology's role in strategic planning for the sectors of automobiles, business communication, industry and consumers desiring intelligent electronic equipment. Forecasts, 1977-1987. Market and product cycles 1977-1987, including feasibility, product function and functionality. Price listed is for complete set. Contact publisher for individual prices. Data derived is from Arthur D. Little, Inc. estimates.

2472. THE TRAFFIC CONTROL ROADWAY LIGHTING AND PARKING EQUIPMENT MARKETS. *Frost and Sullivan, Inc. March 1979. 252 pg. $800.00.*

Analyzes the traffic control, highway lighting, and parking equipment market, establishing quantities of major devices currently in operation and the annual unit volume purchase by local and state governments over the next decade. Quantitative and financial forecasts to 1988 are offered by product and end-user, subdivided by population center. Documents major manufacturers and their product lines with discussions of new equipment in the traffic control field. Based on interviews with industry personnel users and government statistics.

2473. THE U.S. SEMICONDUCTOR PRODUCTION AND TEST EQUIPMENT MARKETS. *Frost and Sullivan, Inc. August 1979. 227 pg. $900.00.*

Analyzes U.S. requirements through 1989 for 6 classes (wafer production, wafer masking, assembly equipment, pattern and mask generation, wafer processing, and test equipment) and 37 specific types of equipments. Forecasts of specific user dollar requirements for 1980, pre-emptive vs. merchant factor and price ranges for various types of equipment included. Contains profiles of top manufacturers with estimates on their production and test equipment requirements for 1980 as well as profiles for 14 production and test equipment suppliers.

2474. UPDATED PROJECTIONS OF AIR QUALITY IMPACTS FROM INTRODUCTION OF ELECTRIC AND HYBRID VEHICLES. *Charles River Associates. 1980. 172 pg. $30.00.*

Analysis of air pollution from electric vehicles and raw material markets. Examines reserves and short-term availability of nickel, cobalt, and lithium. Forecasts demographic statistics and industrial production growth.

2475. WORLD SEMICONDUCTOR FORECAST. *Semiconductor Industry Association. Annual. September 1979. 14 pg. $15.00.*

Annually published three-year forecast of U.S. semiconductor shipments and growth rates. Value of shipments given by market, by product group, and by product line. Markets are: total U.S., domestic O.E.M., domestic distributor, Western Europe, Japan, other international, and world total. Product groups are: total solid state; integrated circuits (digital bipolar, digital MO's, linear, and total); and discrete devices (diodes, small signal and power transistors, rectifiers, thyristors, optoelectronics, all other discrete, and total discrete).

See 1338, 1350, 1351, 1353, 1358, 1363, 1373, 3620.

Directories and Yearbooks

2476. AMERICAN ELECTRONICS ASSOCIATION DIRECTORY. *American Electronics Association. Annual. 1979. 220 pg. $50.00.*

Annual directory of manufacturers of electrical and electronic equipment, with emphasis on instruments, measuring devices, digital components, and software. Listings are alphabetic and by state or city and include addresses, chief executives, parent companies, subsidiaries, ownership structure, securities market, founding date, employment, product lines, and method of distribution.

2477. AMERICAN ELECTRONICS ASSOCIATION - MEMBERSHIP DIRECTORY. *American Electronics Association. Annual. 1981. 225 pg. $50.00.*

Alphabetical listings of 1000 electronics and high tech firms detail address, telephone, executives, products, services, number of employees, and type of firm. Includes geographic cross-referencing.

2478. AMERICAN ELECTRONICS DATA ANNUAL. *Pergamon Press, Inc. Annual. 1980.*

Annual review of U.S. microelectronic products and manufacturers.

2479. AUTOMOTIVE, BURGLARY PROTECTION, MECHANICAL EQUIPMENT DIRECTORY. *Underwriters Laboratories Inc. Annual. September 1981. 185 pg. $2.10.*

Addresses of manufacturers and product lines found in compliance with Underwriters Laboratories safety standards. Covers automotive equipment, burglary protection devices and miscellaneous mechanical equipment.

2480. BEST'S SAFETY DIRECTORY. *A.M. Best Company. Annual. 1981. 1500 pg. $30.00.*

Classified arrangement of 2000 manufacturers and suppliers of security, safety, pollution control, and industrial hygiene equipment.

2481. BEST'S SAFETY DIRECTORY: VOLUME 1: ADMINISTRATION, CONSULTANTS, APPAREL, NOISE, INDUSTRIAL HYGIENE. *A.M. Best Company. Annual. 1979. 778 pg.*

Two-volume 1980 directory provides summaries of important OSHA general industry, construction and maritime standards, as well as guidelines for implementation, and acts as a buyer's guide for related products and services. Information gathered into seven major areas: administration, apparel, noise, industrial hygiene, operational safety,pre-operational plant safety, and security. Various indexes.

2482. BEST'S SAFETY DIRECTORY: VOLUME 2: MACHINE GUARDING, TOOL HANDLING, MATERIALS HANDLING, PLANT MAINTENENCE, ELECTRICAL SAFETY, FIRE, SECURITY. *A.M. Best Company. Annual. 1979. 1532 pg.*

Two-volume 1980 directory provides summaries of important OSHA general industry, construction and maritime standards as well as guidelines for implementation, and acts as a buyers's guide for related products and services. Information gathered into seven major areas: administration, apparel, noise, industrial hygiene, operational safety, pre-operational plant safety and security. Various indexes.

2483. CANADA IN THE WORLD OF ELECTRONICS: 1980 / 1981 ELECTRONICS EXPORT DIRECTORY. *Maclean Hunter Ltd. Annual. May 1980. 96 pg. 17.00 (Canada).*

Listings of Canadian electronics manufacturers involved in export sales. Details company names, addresses, telephone, telex, twx, and product lines of exportable electronics products.

2484. THE CONSUMER ELECTRONICS DIRECTORY. *Marketing Development. Annual. 1979. 200 pg. $75.00.*

This directory lists information such as address, telephone number, products manufactured, key officers, number of employees, and sales on 1498 manufacturers of consumer electronics products that are sold to the U.S. market.

2485. D.A.T.A. BOOK ELECTRONIC INFORMATION SERIES: APPLICATION NOTES REFERENCE. *D.A.T.A., Inc. Semi-annual. v.26, no.33/ed.21. December 1981. 60 pg. $35.00.*

Tabulation of semiconductor products by application, with technial specifications, numbers, and uses. Indexed to addresses of manufacturers.

2486. D.A.T.A. BOOK ELECTRONIC INFORMATION SERIES: CONSUMER IC'S. *D.A.T.A., Inc. Semi-annual. v.26, no.31/ed.2. December 1981. 412 pg. $60.00.*

Tabulation of consumer IC products with technical specifications, numbers, and uses. Indexed to addresses of manufacturers.

2487. D.A.T.A. BOOK ELECTRONIC INFORMATION SERIES: DIGITAL IC'S. *D.A.T.A., Inc. Semi-annual. v.26, no. 32/ed.11. December 1981. 620 (est.) pg. $65.00.*

Tabulation of digital IC products with technical specifications, numbers, and uses. Indexed to addresses of manufacturers.

2488. D.A.T.A. BOOK ELECTRONIC INFORMATION SERIES: DIODES. *D.A.T.A., Inc. Annual. v.26, no.14/ed.48. June 1981. 662 pg. $60.00.*

Tabulation of diode products with technical specifications, numbers, and uses. Indexed to addresses of manufacturers.

2489. D.A.T.A. BOOK ELECTRONIC INFORMATION SERIES: DIODES: DISCONTINUED DEVICES. *D.A.T.A., Inc. Annual. v.25, no.15/ed.7. June 1981. 282 pg. $35.00.*

Tabulation of discontinued diode products with technical specifications, numbers, and uses. Indexed to addresses of manufacturers.

2490. D.A.T.A. BOOK ELECTRONIC INFORMATION SERIES: DISCONTINUED MICROWAVE. *D.A.T.A., Inc. Annual. v.26, no.27/ed.2. September 1981. 178 pg. $35.00.*

Tabulation of discontinued microwave products with technical specifications, numbers, and uses. Indexed to addresses of manufacturers.

2491. D.A.T.A. BOOK ELECTRONIC INFORMATION SERIES: DISCONTINUED OPTOELECTRONICS. *D.A.T.A., Inc. Annual. v.25, no.10/ed.4. May 1981. 69 pg. $35.00.*

Tabulation of discontinued optoelectronic products with technical specifications, numbers, and uses. Indexed to addresses of manufacturers.

2492. D.A.T.A. BOOK ELECTRONIC INFORMATION SERIES: DISCONTINUED TYPE LOCATOR. *D.A.T.A., Inc. Annual. v.25, no.29/ed.1. October 1981. 235 pg. $45.00.*

Tabulation of discontinued semiconductor products with technical specifications, numbers, and uses. Indexed to addresses of manufacturers.

2493. D.A.T.A. BOOK ELECTRONIC INFORMATION SERIES: INTEGRATED CIRCUIT: DISCONTINUED DEVICES. *D.A.T.A., Inc. Annual. v.25, no.20/ed.11. July 1981. 81 pg. $35.00.*

Tabulation of discontinued IC products with technical specifications, numbers, and uses. Indexed to addresses of manufacturers.

2494. D.A.T.A. BOOK ELECTRONIC INFORMATION SERIES: INTERFACE INTEGRATED CIRCUITS. *D.A.T.A., Inc. Semi-annual. v.27, no.5. March 1982. 625 pg. $65.00.*

Tabulation of interface integrated circuit products with technical specifications, serial numbers, and uses. Indexed to manufacturers' addresses.

2495. D.A.T.A. BOOK ELECTRONIC INFORMATION SERIES: MEMORY IC'S. *D.A.T.A., Inc. Semi-annual. v.27, no. 7/ed.22. April 1982. 360 (est.) pg. $65.00.*

Tabulation of digital IC products with technical specifications, numbers, and uses. Indexed to addresses of manufacturers.

2496. D.A.T.A. BOOK ELECTRONIC INFORMATION SERIES: MICROCOMPUTER SYSTEMS. *D.A.T.A., Inc. Semi-annual. v.25, no.37/ed.2. December 1981. 220 (est.) pg. $60.00.*

Tabulation of microcomputer sub-system products with technical specifications, numbers, and uses. Indexed to addresses of manufacturers.

2497. D.A.T.A. BOOK ELECTRONIC INFORMATION SERIES: MICROPROCESSOR IC'S. *D.A.T.A., Inc. Semi-annual. v.26, no.35/ed.2. December 1981. 300 pg. $60.00.*

Tabulation of microprocessor IC products with technical specifications, numbers, and uses. Indexed to addresses of manufacturers.

2498. D.A.T.A. BOOK ELECTRONIC INFORMATION SERIES: MICROWAVE. *D.A.T.A., Inc. Semi-annual. v.27, no.2, ed.48. March 1982. 345 pg. $60.00.*

Tabulation of microwave products with technical specifications, numbers, and uses. Indexed to addresses of manufacturers.

2499. D.A.T.A. BOOK ELECTRONIC INFORMATION SERIES: OPTOELECTRONICS. *D.A.T.A., Inc. Semi-annual. v.27, no.9/ed.15. April 1982. $60.00.*

Tabulation of optoelectronic products with technical specifications, numbers, and uses. Indexed to addresses of manufacturers.

2500. D.A.T.A. BOOK ELECTRONIC INFORMATION SERIES: POWER SEMICONDUCTORS. *D.A.T.A., Inc. Semi-annual. v.27, no.4/ed.16. March 1982. 740 (est.) pg. $65.00.*

Tabulation of power semiconductor products with technical specifications, numbers, and uses. Indexed to addresses of manufacturers.

2501. D.A.T.A. BOOK ELECTRONIC INFORMATION SERIES: THYRISTOR DISCONTINUED DEVICES. *D.A.T.A., Inc. Annual. v.26, no.23. September 1981. 225 pg. $35.00.*

Tabulation of discontinued thyristor products with technical specifications, numbers, and uses. Indexed to addresses of manufacturers.

2502. D.A.T.A. BOOK ELECTRONIC INFORMATION SERIES: THYRISTORS. *D.A.T.A., Inc. Semi-annual. v.26, no. 22/ed.17. September 1981. 415 (est.) pg. $60.00.*

Tabulation of thyristor products with technical specifications, numbers, and uses. Indexed to addresses of manufacturers.

2503. D.A.T.A. BOOK ELECTRONIC INFORMATION SERIES: TRANSISTOR: DISCONTINUED DEVICES. *D.A.T.A., Inc. Annual. v.26, no.21/ed.17. August 1981. 180 pg. $35.00.*

Tabulation of discontinued transistor products with technical specifications, numbers, and uses. Indexed to addresses of manufacturers.

2504. D.A.T.A. BOOK ELECTRONIC INFORMATION SERIES: TRANSISTORS. *D.A.T.A., Inc. Annual. 1982. $60.00.*

Tabulation of transistor products with technical specifications, numbers, and uses. Indexed to addresses of manufacturers.

2505. DESIGN NEWS ELECTRICAL / ELECTRONIC DIRECTORY. *Herman Publishing Company. Annual. 1980. $12.00.*

Directory of electrical and electronic component manufacturers. Lists suppliers by product type by area, and by trade name. Provides addresses and phone numbers of manufacturers and their local sales representatives.

2506. THE DIRECTORY OF DEFENSE ELECTRONIC PRODUCTS AND SERVICES: U.S. SUPPLIERS. *Fairchild Publications. Irregular. 1980. 228 pg. $40.00.*

Directory of electronic products, systems and services designed for defense and security. Includes data communications, defense equipment, field and optical communications and electronic components.

2507. EE: ELECTRICAL / ELECTRONIC PRODUCT NEWS. *Sutton Publishing Company, Inc. Monthly. June 1981. 56 pg.*

Monthly catalog of electronic components and analytic instruments of interest to electronics design engineers. Entries, supplied by manufacturers, include photo, description, specifications, and manufacturer's address. Contains a reader's service check-off card.

2508. ELECTRICAL BLUE BOOK OF NEW ENGLAND. *George D. Hall Company. Annual. April 1981. $22.50.*

Contains a cross-reference of electrical equipment and supplies, manufcturers, distributors, and representatives. Gives locations and phone numbers.

2509. ELECTRICAL CONSTRUCTION MATERIALS DIRECTORY. *Underwriters Laboratories Inc. Annual. May 1982. 486 pg. $5.25.*

Addresses of manufacturers of electrical construction materials and product lines found in compliance with Underwriters Laboratories safety standards.

2510. ELECTRONIC ENGINEERS MASTER CATALOG: VOLUME 1. *Hearst Business Communications, UTP Division. Annual. 23rd. 1980. pg.1-2648 $50.00.*

Volume 1 of 2-volume directory of product data of manufacturers in the electronic engineering industry. Divided into 44 product sections, lists manufacturers, sales offices and trademarks. Addresses, phone numbers of headquarters, and subsidiaries of manufacturers given.

2511. ELECTRONIC ENGINEERS MASTER CATALOG: VOLUME 2. *Hearst Business Communications, UTP Division. Annual. 23rd. 1980. pg.2649-4858 $50.00.*

Volume 2 of 2-volume directory on product data of manufacturers in U.S. electronics industry. Includes product directory, manufacturers directory, trademark directory and distributors

advertisors' index. Addresses, phone numbers, headquarters and subsidiaries of manufacturers given.

2512. ELECTRONIC ENGINEERS MASTER CATALOG: 1979-1980: VOLUME 1. *Hearst Business Communications, UTP Division. Annual. 1979. pg.1-2448*

Volume 1 of 2-volume directory on electronics for 1979-1980. Products, manufacturers and sales offices, trademarks, and advertisers are all listed alphabetically. Includes addresses of maufacturers, sales offices, and subsidiaries and catalog pages with product specifications, ratings and features.

2513. ELECTRONIC ENGINEERS MASTER CATALOG: 1979-1980: VOLUME 2. *Hearst Business Communications, UTP Division. Annual. 22nd. 1979. pg.2449-4530*

Volume 2 of 2-volume directory for the electronics industry continues the manufacturers' catalog containing product ratings, features and specifications, and alphabetical index to the catalog and distributors' index.

2514. ELECTRONIC INDUSTRIES ASSOCIATION 1978 TRADE DIRECTORY AND MEMBERSHIP LIST. *Electronic Industries Association. Irregular. March 1978. 176 pg.*

Membership data for the Electronic Industries Association updated to March 14, 1978. Provides committee data, councils, panels, group and division organization, departments and an alphabetical list of member businesses, with addresses, telephone numbers, executives, products and trade names.

2515. ELECTRONIC INDUSTRY TELEPHONE DIRECTORY: 1979-80. *Harris Publishing Company. Annual. 1979. 456 pg. $20.00.*

Arrangement is by advertisers' index, American, Canadian and foreign electronic companies, area city codes, area code map and classified product section.

2516. ELECTRONIC MARKETING DIRECTORY. *National Credit Office. Annual. June 1980. 1000 (est.) pg.*

1980 directory of electronic equipment provides statistics on manufacturers, products, markets, and sales. Provides a geographical and alphabetical listing of electronic equipment for data processing, transmission, wiring, semi conductors, components, scientific equipment, motors and controls, communications equipment, medical instruments and photographic instruments.

2517. ELECTRONIC NEWS FINANCIAL FACT BOOK AND DIRECTORY, 1979. *Fairchild Publications. Annual. 18th. 1979. 610 pg.*

This directory lists leaders in the electronics industry, with sales figures. Contains an index of corporations.

2518. ELECTRONIC NEWS FINANCIAL FACT BOOK AND DIRECTORY: 1981. *Fairchild Publications. Annual. 20th ed. September 1981. 543 pg. $90.00.*

Comprehensive directory of U.S. electrical / electronics corporations and their finances, with most statistics presented for either 1980 or 1981 plus the four preceding years. Covers operations by division, employment, plant size, revenues by operation, sales and earnings, income statements, stock exchange listings, number of shares, total and composite equity, earnings on common stock, dividends and splits, balance sheets, and operating ratios.

2519. ELECTRONICS AND COMMUNICATIONS DIRECTORY AND BUYERS GUIDE FOR CANADA. *Southam Business Publications, Ltd. Division of SouthamCommunications Ltd. Annual. 1979. 150 pg.*

Alphabetical and classified lists of manufacturers, suppliers, and products in the electronics field.

2520. ELECTRONICS BUYER'S GUIDE: 1980. *McGraw-Hill Book Company. Annual. May 1980. 1516 pg.*

A 1980 directory and buyer's guide to the electronics industry. Contains 4000 products and cross references to specific products, alphabetically listed. A separate listing of over 5000 manufacturers, sales offices and distributors, with addresses and phone numbers. Also included is an alphabetical list of product and brand trade names, and a directory of advertisers catalogs indexed by 18 product categories.

2521. ELECTRONICS INDUSTRY DIRECTORY OF MANUFACTURERS' REPRESENTATIVES: 1980-1981. *Electronics Representatives Association. Annual. 1980. 224 pg. $5.00.*

Annual directory of over 2500 electronics firms and branches, listed by region and cross-indexed alphabetically. Listings include addresses, size of firm, types of products handled, territory covered, and special facilities.

2522. ELECTRONICS PRODUCT ANNUAL DIRECTORY. *Frost and Sullivan, Inc. Annual. 1980. $350.00.*

Annual listing of current (and new) electronics products, accompanied by brief descriptions of each product. Listings from over 500 companies.

2523. ELECTRONICS REPRESENTATIVES DIRECTORY 1982. *Harris Publishing Company. Annual. 1982. 190 pg. $12.00.*

Arranged alphabetically and geographically, directory lists 3700 electronics firms and their representatives.

2524. GUIDE TO CANADIAN MANUFACTURERS: CHEMICALS, PETROLEUM, RUBBER, ELECTRICAL AND MISCELLANEOUS: VOLUME 3. *Dun and Bradstreet Canada Ltd. Market Services Division. Annual. May 1980. 255 pg.*

Canadian manufacturers' directory covering chemicals, petroleum, rubber, electrical and electronic equipment divided into 3 sections. Section 1 lists D-U-N-S numbers; manufacturer's names geo-alphabetically, with addresses and phone numbers; U.S. SIC codes; raw materials purchases; products; major capital equipment; square footage size of offices and plants; number of employees; executives. Section 2 lists manufacturers geo-alphabetically, bylines of business under SIC. Section 3 lists manufacturers, with addresses and SIC numbers.

2525. GUIDE TO CANADIAN MANUFACTURERS: PRIMARY AND FABRICATED METAL AND TRANSPORTATION EQUIPMENT: VOLUME 4. *Dun and Bradstreet Canada Ltd. Market Services Division. Annual. May 1980. 257 pg.*

Canadian manufacturers' directory covering iron and steel, metals, machinery and machine tools, electric and electronic equipment and transportation equipment divided into 3 sections. Section 1 lists D-U-N-S numbers; manufacturer's name geo-alphabetically, with addresses and phone numbers; U.S. SIC codes; raw materials purchases; products; major capital equipment; square footage size of offices and plants; number of employees; executives. Section 2 lists manufacturers geo-alphabetically, by lines of business under SIC. Section 3 lists manufacturers, with addresses and SIC numbers.

2526. IC (INTEGRATED CIRCUIT) MASTER. *Herman Publishing Company. Annual. January 1980. 1250 pg. $75.00.*

Directory of about 100 U.S. manufacturers of integrated circuits. Includes an alphabetic listing of manufacturers, a list of suppliers by product type, and a part number index and guide.

2527. INDUSTRIAL PRODUCTS MASTER CATALOG. *Herman Publishing Company. Annual. October 1980.*

Catalog and directory of electronic components and equipment available through distributors.

2528. INTERFERENCE TECHNOLOGY ENGINEERS' MASTER: DIRECTORY AND DESIGN GUIDE. *R and B Enterprises. Annual. 1982. 224 pg. $25.00.*

Design handbook and directory for electronic interference technology. Lists government agencies and employees, and business addresses of services and manufacturers of products.

2529. LIST OF CERTIFIED ELECTRICAL EQUIPMENT, 1980. *Canadian Standards Association. Annual. 27th ed. 1980. 1482 pg.*

Lists certified electrical equipment / products for use in ordinary and hazardous locations, as well as performance of certified electrical equipment, and electrical cables and wires. Also lists manufacturers and their city locations.

2530. MANUFACTURERS AND DISTRIBUTORS: WIND, WATER AND OTHER ELECTRIC. *Synerjy. Irregular. 1981. 6 pg. $3.00.*

List of U.S., Canadian, European, and other manufacturers / distributors of wind generators, fuel cells, solar photovoltaic collectors, high-capacity storage batteries, and electric vehicles. Entries, classified by technology and then by region, include addresses and briefly specify product lines, power generation capacities, maximum speed and cruising range (in the case of vehicles) and occasionally prices.

2531. MANUFACTURERS OF ELECTRICAL INDUSTRIAL EQUIPMENT, 1979: ANNUAL CENSUS OF MANUFACTURERS. *Canada. Statistics Canada. English/French. Annual. 43-207. August 1981. 21 pg.*

General statistical review, consolidated annually from 1971 to 1977 and for 1978 and 1979 by province, of establishments, manpower, energy and materials costs, value of shipments, value added, and wages and salaries. Includes opening and closing inventories, 1979, by province, and detailed breakdowns, by type, of materials and supplies and of shipments of manufactured goods, for 1978-1979. Includes a list of large companies.

2532. MASTER TYPE LOCATOR. *D.A.T.A., Inc. Annual. v.26, no.8/ed.3. April 1981. 537 pg. $40.00.*

Tabulation of semiconductor products by numbers indexed to addresses of manufacturers.

2533. THE PERSONAL ELECTRONICS BUYER'S GUIDE. *Sippl, C.J. Prentice-Hall, Inc. Irregular. 1979. 338 pg.*

Includes buyer's guide information on personal computers, calculators, microprocessors, kits, electronic games, solar energy, security appliances, voice recognition systems, music coding and synthesis, computer-assisted instruction, computer graphics, viewdata, other consumer products and portable data handlers. Provides addresses of businesses, and product information.

2534. PRODUCT DESIGN AND DEVELOPMENT. *Chilton Book Company. Monthly. 60 (est.) pg.*

Directory of material, component parts, products, and services for use by design engineers. Includes technical specifications, manufacturer's name and address, photo, and reader service card for further information.

2535. RECOGNIZED COMPONENT DIRECTORY. *Underwriters Laboratories Inc. Annual. March 1982. 2043 pg. $20.10.*

A listing of UL inspected factory-installed components of electrical equipment and machinery. These components have been designated as incomplete in construction features and not suitable as field-installed components. Information, classified by product group, includes manufacturer's address, brief product listing, and company catalog number. Product coverage includes fixtures, fittings, circuit breakers, switches, wires, and others. A supplement is issued every September.

2536. SLIG BUYERS GUIDE. *Independent Battery Manufacturers Association, Inc. Biennial. 7th ed. 1981. $7.00.*

Biennial directory of manufacturers, rebuilders, component products, equipment parts, suppliers, and services in the battery industry.

2537. SWEET'S 1980 CATALOG FILE: PRODUCTS FOR ENGINEERING; MECHANICAL, ELECTRICAL, CIVIL AND RELATED PRODUCTS: VOLUME 8. *McGraw-Hill Information Systems Company, Sweets Division. Annual. 1980. 850 (est.) pg.*

Volume 8 of a ten-volume 1980 directory consisting of a compilation of company catalogs offering products for engineering. Arranged by product classification, this volume covers: electrical products, such as; power transmission, conductors, raceways, conduits and trenches. Each catalog includes descriptions of equipment and products, prices, applications, illustrations, specifications and phone numbers and addresses. Indexes by company and catalog code, products, firms, divisional firms and trade names.

2538. SWEET'S 1980 CATALOG FILE: PRODUCTS FOR ENGINEERING; MECHANICAL, ELECTRICAL, CIVIL AND RELATED PRODUCTS: VOLUME 9. *McGraw-Hill Information Systems Company, Sweets Division. Annual. 1980. 600 (est.) pg.*

Volume 9 of a ten-volume 1980 directory consisting of a compilation of company catalogs offering products for engineering. Arranged by product classification, this volume covers: wiring devices, power generation, motors and motor controls, electrical and electronic controls and electrical switches. Each catalog includes descriptions of equipment and products, prices, applications, illustrations, specifications and phone numbers and addresses. Indexes by company and catalog code, products, firms, divisional firms and trade names.

2539. SWEET'S 1980 CATALOG FILE: PRODUCTS FOR ENGINEERING; MECHANICAL, ELECTRICAL, CIVIL AND RELATED PRODUCTS: VOLUME 10. *McGraw-Hill Information Systems Company, Sweets Division. Annual. 1980. 700 (est.) pg.*

Volume 10 of ten-volume 1980 directory consisting of a compilation of company catalogs offering products for engineering. Arranged by product classification, this volume covers: various types of lighting, heating and cooling devices, communications, public address equipment and systems, clocks and program equipment, electronic communications wire and cable. Each catalog includes descriptions of equipment andproducts, prices, applications, illustrations, specifications and phone numbers and addresses. Indexes by company and catalog code, products, firms, divisional firms and trade names.

2540. WHO'S WHO IN ELECTRONICS: 1980: DATA IN DEPTH. *Harris Publishing Company. Annual. 1980. 830 (est.) pg. $59.95.*

A 1980 directory of the electronics industry, consisting of: alphabetical and geographical listings of manufacturers with addresses, phone numbers and company statistics and executives; a purchasing index section arranged by product classifi-

cation; and an alphabetical listing of distributors with addresses and phone numbers. A geographical listing of company representatives is also included.

2541. WHO'S WHO IN ELECTRONICS 1982. *Harris Publishing Company. Annual. February 1982. $62.95.*

Includes 13,000 manufacturers, representatives, and distributors listed alphabetically by company name and geographically by city within state. Purchasing index shows manufacturers by 1600 product classifications. Includes names of personnel, annual sales, products, lines handled, and branch outlets.

2542. WORLD ELECTRONIC DEVELOPMENTS. *Prestwick International, Inc. Monthly. 0000. 10 pg. $155.00/yr.*

Weekly newsletter includes two-page analysis of new directions by sector (automation, control, and robotics; communications and information processing; computer technology and computer-aided design; integrated circuits and solid state technology; circuits and electronics; applications). Each issue announces approximately 50 new electronic products, processes, or techniques, with firm name and address and contact individual for the new development.

2543. THE 1978-1979 ELECTRONIC NEW PRODUCT DIRECTORY. *Concelmo, S., editor. Marketing Development. Annual. 1979. 286 pg.*

A directory listing new electronic products introduced between July, 1977 and August, 1978 by over 650 manufacturers as reported by stockholders reports and news releases. Listings of products by SIC codes, descriptions, prices, and manufacturers' addresses.

2544. 1980 AMERICAN ELECTRONICS ASSOCIATION DIRECTORY. *American Electronics Association. Annual. 1980. 280 pg. $55.00.*

Directory is an alphabetical listing of electronics and information technology companies belonging to the American Electronics Association. Listing includes addresses, telephone numbers and executive names. Geographical and product indexes are included.

2545. 1980 ASSEMBLY ENGINEERING MASTER CATALOG. *Hitchcock Publishing Company. Annual. 1979. 200 pg. $25.00.*

Directory covering adhesives, assembly equipment, electrical and electronic equipment, hardware, and tools and machinery for assembly engineering. Indexes provided for products (listed alphabetically) with supplier or manufacturers' addresses, trade names and product descriptions. Also includes an advertisers index.

2546. 1980 ELECTRONIC REPRESENTATIVES DIRECTORY. *Harris Publishing Company. Annual. 1980. 184 pg. $15.00.*

1980 annual directory reprinted from Who's Who in Electronics provides alphabetical and geographical listings of independent sales representatives in the electronics industry. Annotations include personnel, plant size, annual sales and products. Covers U.S., Canada and some foreign companies.

2547. 1980 GUIDE TO QUALITY CONSTRUCTION PRODUCTS. *Construction Products Manufacturers Council, Inc. Annual. 1980. 122 pg. Free.*

Directory of products, manufacturers, members and local representatives of the Producers' Council, Inc. Manufacturers represented include producers of structural steel, brick, concrete, roofing, interior furnishings, glass, power lighting, electronics and communications equipment. Entries provide name, address, contact personnel, telephone, major products and local sources of supply.

2548. 1980 - 81 SIA ANNUAL YEARBOOK AND DIRECTORY. *Semiconductor Industry Association. Annual. 1981. $40.00.*

Shows members, company executives, product lines, manufacturing locations, industry trade and performance, productivity and end uses of semiconductor devices.

2549. 1981 SOLID STATE PROCESSING AND PRODUCTION BUYERS GUIDE AND DIRECTORY. *Marshall, Samuel L. SYMCON. Annual. 9th ed. 1981.*

Annual, classified directory of about 1800 firms providing materials, equipment, and services contributing to the manufacture of solid state devices and circuits. Companies are listed within 60 major product / service groups and 2000 specific categories. Indexed by company name and brand name.

2550. 1982 DESIGN NEWS ELECTRICAL / ELECTRONIC DIRECTORY. *Control Data Corporation: Cybernet Services. Annual. 1981. $2.00.*

Lists a wide variety of electrical / electronic equipment and components available as of late 1981, specifying suppliers and their addresses.

See 253, 257, 1399, 1410, 1412, 1443, 1457, 1466, 1479, 3246, 3641.

Special Issues and Journal Articles

Advertising Age

2551. THE BIG BANK MARKETING OF HOME VIDEOGAMES. *Mansfield, Matthew F. Crain Communications, Inc. v.53, no.36. August 30, 1982. pg.M1-M3, M30*

Discusses state of the home video game industry and estimates market size. Provides 1982 estimated advertising budget for each manufacturer. Lists advertising agencies of each manufacturer. Gives 1982 estimated consoles already installed in U.S. homes; 1982 estimated consoles shipped; and 1982 estimated game cartridge shipments.

Appliance

2552. APPLIANCE: TRADE SHOW IN PRINT 1981. *Dana Chase Publications, Inc. Annual. June 1981. pg.59-74*

Showcases a variety of new hardware, electronic components, paints, and process machinery for manufacturers of home appliances. Entries include photos, descriptions with specifications, and manufacturers' names. Contains readers' service check-off card for product literature.

2553. FILLING A NEED WITH MECHANICAL SWITCHES. *Simpson, David E. Dana Chase Publications, Inc. Irregular. v.39, no.2. February 1982. pg.19-23*

Reviews available products and utilization of mechanical switches for electrical and electronic applications. New products are discussed with some technical specifications, and utilization is discussed by application group. Market trends are reviewed, along with prospects for the future mechanical switch market. Includes a listing of 30 mechanical switch manufacturers, with a description of product lines.

2554. INNOVATIONS IN MICROELECTRONICS. *Dana Chase Publications, Inc. June 1980. pg.27-31*

Surveys use of microprocessors in household appliances. Shows estimated value, 1978, 1984.

2555. THIRTIETH ANNUAL APPLIANCE INDUSTRY FORECASTS. *Steven, James. Dana Chase Publications, Inc. Annual. v.39, no.1. January 1982. pg.41-53*

Forecasts 1981-1987 unit shipments of individual categories of home appliances, heating and air conditioning equipment, electric housewares, and consumer electronics. Includes 1980 actual figures and 1981 estimates.

2556. TWENTY-NINTH ANNUAL REPORT: A TEN YEAR REVIEW 1972-1981. *Dana Chase Publications, Inc. Annual. v.39, no.4. April 1982. pg.45-48*

Provides detailed statistics for production, manufacturers' shipments and distributor sales of appliances between 1972 and 1981. Data broken down into major categories, including major appliances, comfort conditioning appliances, personal care appliances, electric housewares, consumer electronics, home care appliances, home security appliances, outdoor appliances, vending machines, and plumbing appliances / fixtures. Each of these major categories includes statistics for specific products.

2557. 1982 APPLIANCE INDUSTRY PURCHASING DIRECTORY. *Dana Chase Publications, Inc. Annual. v.39, no.1. January 1982. pg.D-1 - D-100*

Classified directory of U.S. suppliers of a side range of materials, supplies, machinery, components, subassemblies, and services to appliance manufacturers.

Appliance Manufacturer

2558. FORECASTS: MIDYEAR RISE IN INDUSTRIAL ORDERS IS RIGHT ON SCHEDULE. *Cahners Publishing Company. Semi-annual. v.30, no.10. October 1982. pg.27, 29*

Gives quarterly industrial orders of room and unitary air conditioners, gas and electric water heaters, 1981-1982; and 1983 projections. For month, year-to-date, and previous year's month gives unit sales of 20 major appliances, 17 environmental comfort products, and 5 consumer electronic products.

2559. VISION SYSTEMS BOOST ROBOT PRODUCTIVITY. *Klein, Art, ed. Cahners Publishing Company. v.30, no.8. August 1982. pg.50-54*

Examines the U.S. robot vision systems market in 1982. Discusses technologies and production procedures. Provides information on 14 robot vision systems including product name, manufacturer, and a description of the system.

The Battery Man

2560. THE BATTERY MAN: BATTERY SHIPMENTS. *Independent Battery Manufacturers Association, Inc. Tri-annually. July 1982. pg.18-20*

Tables give tri-monthly shipments of U.S. OEM batteries and replacement automotive batteries and monthly change and growth rate. Data from Battery Council International.

2561. BATTERY STATISTIC UPDATES, REPLACEMENT AND OEM SHIPMENT FIGURES THROUGH APRIL 1982. *Independent Battery Manufacturers Association, Inc. v.24, no.7. July 1982. pg.18, 19, 20*

Gives data on shipments of replacement automatic batteries and original equipment manufacturer (OEM) shipment for the first four months of 1982, with comparison data from 1981, and percentage increase or decrease.

2562. THE BATTERY MAN: PRODUCT NEWS. *Independent Battery Manufacturers Association, Inc. Monthly. v.24, no.3. March 1982. pg.28*

Lists new products related to the battery industry. Includes description of product, name of manufacturer, and contact for further information. Typical products listed include slide terminal bonder, oil sentry indicator and reserve capacity discharger.

2563. ELECTRIC VEHICLES. *Independent Battery Manufacturers Association, Inc. Monthly. v.24, no.3. March 1982. pg.25*

Document lists new developments and new products in the electric vehicle industry.

2564. INDUSTRIAL NEWS. *Independent Battery Manufacturers Association, Inc. Monthly. v.24, no.3. March 1982. pg.18, 22*

Lists industrial products and equipment of interest to the battery industry and other related areas. Includes description of product, name of manufacturer and contact for further information. Typical products listed include wire stripper cutters, finishing machines and twin-post automotive lifts.

Boating

2565. ELECTRONICS BUYERS GUIDE. *Ziff - Davis Publishing Company. Annual. April 1981. 218 pg. $2.00.*

Contains in-depth information about the most sophisticated electronics available to boating. All information was supplied and written by the manufacturers.

Broadcast Communications

2566. CRT GOOD, FLAT BETTER? *Globecom Publishing, Ltd. v.5, no.3. March 1982. pg.8*

Report on the market for flat-panel screens is summarized. Data on the size of the market for flat-panel displays is presented 1980, 1981 and manufacturers expecting to enter this market are identified.

Business Week

2567. ELECTRONIC GAME MAKERS BRACE FOR A SHAKE - OUT. *McGraw-Hill Book Company. October 27, 1980. pg.66T, 660, 66P*

Reviews the electronic games and toys industry. Shows demand, new products, and 1975-1980 wholesale sales.

2568. SEMICONDUCTOR EQUIPMENT TAKES ON ITS OWN GLOW. *McGraw-Hill Book Company. Irregular. November 17, 1980. pg.56E-56L*

Statistical article on the U.S. semiconductor industry. Shows production 1974-1984, by end uses, and rankings of companies acquired in mergers in 1979, by volume of sales.

2569. VIDEODISC MARKETS MAKE AN AMAZING ABOUT FACE. *McGraw-Hill Book Company. no.2727. September 20, 1982. pg.119-122*

Discusses videodiscs and videodisc players market size and prices to 1990. Discusses market, competition, and growth.

Canadian Aviation

2570. 1980 AVIONICS BUYER'S GUIDE. *Maclean Hunter Ltd. Annual. July 1980.*

Annual report on what is available in avionics in Canada, with listings of distributors and dealers.

Canadian Electronics Engineering

2571. ANNUAL BUYERS GUIDE AND DIRECTORY "80-'81. *Maclean Hunter Ltd. Annual. July 1980. 150 pg.*

Divided into 4 sections, guide contains an alphabetic listing of products and services (electronic and allied components) and a listing of their Canadian and foreign sources represented in Canada. Section 2 lists Canadian manufacturers and representatives (addresses and phone numbers); Section 3, U.S., U.K. and other manufacturers and their Canadian representatives; Section 4, the names and addresses of Canadian distributors. Includes a short products profile.

2572. ANNUAL REVIEW AND FORECAST: STATISTICAL. *Maclean Hunter Ltd. Annual. January 1981. pg.16-21 17.00 (Canada).*

Annual statistical review of the Canadian electronics industry. Lists electronic products by major categories (telecommunications equipment, semiconductors, components, computers, office machinery, instruments, consumer products). Covers exports (1976-1979, 1980 estimates), imports (1976-1979, 1980 estimates), production (1976-1978, 1980 estimates) and apparent consumption (1976-1979, 1980 estimates). Includes shipments by major category and type of product (1976-1978); Canadian imports of electronic products (1977-1980); and by Canadian exports of electronic equipment (1976-1978 with 1980 estimates).

2573. ANNUAL REVIEW AND FORECAST, STATISTICAL REVIEW, EXECUTIVE OUTLOOK. *Maclean Hunter Ltd. Annual. January 1982. pg.22-38*

Figures given for the performance of the Canadian electronics industry for 1981 include imports, exports, production, consumption of products 1977-1981, electronics consumption as a percentage of gross national product; shipments of electronic products by major category and type of product 1977-1979. Includes random samplings on forecasts and prospects for 1982 from senior industry and association executives.

2574. CANADIAN DOMESTIC ELECTRONIC INDUSTRY, 1980. *Maclean Hunter Ltd. Quarterly. September 1980. pg.13*

Statistical analysis of Canada's domestic electronics industry for fourth quarter 1979 and the first quarter 1980. Examines domestic markets, imports, number of shipments, exports and trade deficit.

2575. CANADIAN DOMESTIC ELECTRONICS INDUSTRY, 1981. *Maclean Hunter Ltd. Quarterly. March 1981. pg.14*

Statistical chart on the domestic electronics industry in Canada for second and third quarters of 1980. Examines domestic market, imports, shipments, exports, the trade deficit, and the number of employees in the electronics sector.

2576. CANADIAN ELECTRONICS ENGINEERING 81-82 ANNUAL BUYERS' GUIDE AND DIRECTORY. *Maclean Hunter Ltd. Annual. July 1981. 148 pg.*

Arranged as a product sources guide. Contains three basic sections: a listing of products and services; a manufacturers and distributors index; and foreign manufacturers having Canadian representation. Also contains a list of Canadian distributors. Includes business addresses and phone numbers.

2577. CANADIAN ELECTRONICS INDUSTRY. *Maclean Hunter Ltd. December 1981. pg.17*

Canada's domestic market in electronics is noted in comparison with the first and second quarters of 1981. Examines increase in shipments and exports.

2578. THE CANADIAN ELECTRONICS MARKET. *Maclean-Hunter Research Bureau. Maclean Hunter Ltd. Annual. August 1979. 44 pg. $40.00.*

Covers Canadian production, imports, exports and U.S. exports to Canada of electronic products. Surveys markets by industry: home entertainment products; telephone industry; cable TV industry; radio and television broadcasting; air and marine navigation radio aids; airports, aircraft and pilots; computer installations; and capital expenditures of new machinery and equipment by manufacturing and communication utilities.

2579. CEE ANNUAL BUYERS' GUIDE AND DIRECTORY. *Maclean Hunter Ltd. Annual. July 1979. 164 pg.*

Reference for the Canadian electronics buyer. Arranged as a products source guide, contains three basic sections listing products and services, manufacturers and distributors, and foreign manufacturers having Canadian representation. Also lists Canadian distributors and consultants. Gives names, addresses and telephone numbers.

2580. CEE REVIEW AND FORECAST, 1979-80. *Maclean Hunter Ltd. Annual. January 1980. pg.20-34*

Statistical review and forecasts, 1979-1980, for the electronics industry in Canada. General statistics include consumption, 1974-1979; imports, 1975-1979; production; exports; GNP; electronics consumption; shipments by major products categories and type, 1975-1977; imports of electronic products, 1976-1979; and Canadian exports of electronic equipment, 1975-1979. Includes executives' outlook and forecasts for the future.

2581. DOMESTIC MARKET - SHARE FALLS, ITC RESULTS CONFIRM. *Maclean Hunter Ltd. June 1980. pg.12*

A look at Canada's domestic electronics industry for 1979, based on a report by the Trade and Commerce Dept. Examines imports, production, employment and unemployment, and selling prices divided by sectors: radio and TV, communications and components, office machinery, and instrumentation.

2582. ELECTRONICS THIS MONTH. *Canadian Dept. of Industry Trade & Commerce. Maclean Hunter Ltd. Monthly. November 1980. pg.8*

Monthly graphs depicting Canadian electronics industry activity. Shows imports / exports, shipments and trade deficit for previous 2 quarters.

Canadian Fisherman and Ocean Science

2583. MARINE ELECTRONICS DIRECTORY. *Sentinel Publishing Company. Irregular. February 1979. pg.27-32*

Directory lists new fishing and oceanography electronics equipment. Listing is alphabetical and notes manufacturers, their names and addresses. Gives a brief resume of the products specified. Manufacturers are mainly U.S. with several Canadian noted.

Canadian Research

2584. MICROELECTRONICS RESEARCH: CANADIAN INITIATIVES AIM FOR A FOOTHOLD IN THIS SUPER INDUSTRY. *von Buchstab, Victor. Maclean Hunter Ltd. February 1982. pg.25-27, 30-34*

Analysis of Canada's research and development activities in the field of microelectronics. Examines Ontario's spending in the area, areas of new development, and projected and completed projects.

Canadian Welder and Fabricator

2585. WELDING INSTITUTE OF CANADA: MEMBERSHIP DIRECTORY 1982-1983. *Sanford Evans Publishing Ltd. Annual. v.73, no.6. June 1982. pg.54-86*

Annual membership directory of Welding Institute of Canada. Listing arranged geographically by chapters and alphabetically by chapter city name. Entries include name of company, or individual, and address. Also includes names, addresses and contact data for corporate members, honorary members and members-at-large.

CEE

2586. CEE: NEW PRODUCTS. *Sutton Publishing Company, Inc. Monthly. v.34, no.3. March 1982. pg.42-64*

Lists new products with photograph, brief description and manufacturer's name and address.

2587. CEE: NEW PRODUCTS. *Sutton Publishing Company, Inc. Monthly. December 1981. pg.36-47*

Monthly survey of electrical and electronic products with construction applications: wiring equipment, transformers, lighting equipment, heating and ventilation equipment, hand tools and instruments, and micro computers for energy management; plus related supplies and some construction equipment.

2588. CEE: PRODUCT PARADE. *Sutton Publishing Company, Inc. Monthly. December 1981. pg.28B-28F*

Monthly survey of new construction equipment and supplies, and of electrical and electronic equipment with construction applications (including lighting equipment, appliances, fans, and particularly intercom systems). Includes some specifications, and manufacturers' addresses.

2589. CEE SHOWCASE OF ELECTRICAL PRODUCTS: AT THE 1981 NATIONAL PLANT ENGINEERING AND MAINTENANCE SHOW. *Sutton Publishing Company, Inc. Annual. May 1981. pg.18-29*

Photo-illustrated survey of a wide range of new electrical and electronic equipment and components, relating to industrial plant construction. Products range from heavy generators and transformers, to HVAC equipment, wiring devices, motorized installations like doors and cranes, to power tools, digital measuring and control instruments, and electronic and non-electronic components for the foregoing. Provides manufacturers' addresses and a readers' service check-off card for company literature.

2590. ELECTRICAL CONSTRUCTION PRODUCT REFERENCE ISSUE. *Sutton Publishing Company, Inc. Annual. August 30, 1982. 318 pg.*

Annual buyers guide includes a manufacturers' index cross-referencing 1500 sources of electrical construction products to pages where their products and services are described. Contains listing of 600 free catalogs and technical data publications; product-locating index cross-referencing 2000 products to pages listing all sources of supply, company / product listings covering over 1500 sources of supply alphabetized within 35 product categories, and detailing address, telephone, contact person, products and special services, and local sales representation; and 1500 trademarks and tradenames referenced to their suppliers.

2591. 1982 NATIONAL ELECTRICAL EXPOSITION (NECA SHOW): SHOWCASE OF PRODUCTS. *Sutton Publishing Company, Inc. Annual. v.34, no.11. October 1982. pg.21-28*

Reviews products and tools for the electrical construction industry exhibited at the 1982 National Electrical Exposition. Includes brief product descriptions and provides manufacturers' addresses.

Ceramic Industry

2592. ANNUAL FORECAST FOR GLASS WHITEWARE, PORCELAIN ENAMEL AND ELECTRONIC / INDUSTRIAL AND NEWER CERAMICS. *Cahners Publishing Company. Annual. v.118, no.6. June 1982. 56 pg.*

Forecast for ceramic industry gives 1980-1983 indicators, economic housing starts, and automobile production. Provides 1981-1983 flat glass and glass container shipments. Gives 1981-1983 clay floor and wall tile shipments and U.S. imports, exports, total consumption, and import penetration for ceramic floor and wall tile. Provides 1981-1983 forecast for semiconductor and capacitor shipments.

2593. CAPACITOR VOLUME UP, VALUE DOWN. *Cahners Publishing Company. Annual. v.118, no.6. June 1982. pg.39*

Statistics on U.S. ceramic capacitors includes sales, imports and exports (1980, 1981); and shipments (1981, 1981, 1982). Briefly discusses closures of domestic manufacturing facilities.

2594. RESEARCH MAINTAINS TECHNICAL CERAMICS' GROWTH. *Cahners Publishing Company. Annual. v.118, no.6. June 1982. pg.36-37*

Examines the U.S. market for electronic ceramics to 1983 and gives statistics for domestic shipments of semiconductors. Discusses high-temperature materials and other technological developments.

Chemical Week

2595. LITHIUM BATTERIES: U.S. FIRMS JOIN IN. *McGraw-Hill Book Company. v.130, no.19. May 12, 1982. pg.53-54*

Reviews forecasted growth of lithium battery market. Cites three companies manufacturing the batteries.

Chilton's Instruments and Control Systems

2596. CHILTON'S INSTRUMENTS AND CONTROL SYSTEMS: 1983 BUYER'S GUIDE. *Chilton Book Company. Annual. v.55, no.10. October 1982. 162 pg. $14.95.*

1983 edition of annual buyers' guide for the U.S. instrument and control system industry features the following major sections: specification guides, product summary, product index, and directory of manufacturers including addresses and telephone numbers.

Circuits Manufacturing

2597. CIRCUITS MANUFACTURING: PC ''82 WRAP-UP. *Hastie, William M. Morgan Grampian Publishing Company. Annual. v.22, no.7. July 1982. pg.22-34*

The 1982 Printed Circuits Exposition is reviewed, with discussion of new technology, and a review of new products including equipment, material, and tools. Products are described with technical specifications and manufacturers are identified.

2598. CIRCUITS MANUFACTURING 1982 VENDOR DIRECTORY ISSUE. *Morgan Grampian Publishing Company. Annual. October 1981.*

The directory covers printed circuits, hybrid circuits and semiconductor circuits and the companies who provide assembly, cleaning, design, fabrication, testing, packaging and other components and services for the industry.

2599. CIRCUITS MANUFACTURING: 1983 BUYER'S GUIDE. *Morgan Grampian Publishing Company. Annual. v.22, no.2. October 1982. pg.26-172*

Lists U.S. vendors of 48 major electronic circuit products and services in seven categories: Assembly, cleaning, design, fabrication, packaging, services, and test / quality assurance. Also includes an alphabetical listing of vendors with their addresses and telephone numbers.

2600. FAB EQUIPMENT FORECAST. *Hugle, William B. Morgan Grampian Publishing Company. v.22, no.5. May 1982. pg.20. 24, 64-65*

Discusses trends in semiconductor fabrication and forecasts the expected market value, 1980 to 1984, of 16 types of manufacturing equipment.

2601. PC ''82 PREVIEW: INTERNATIONAL PRINTED CIRCUITS CONFERENCE: PRODUCT PREVIEW. *Benwill Publishing Corporation. Irregular. pg.34-38*

Product description of equipment presented at the International Printed Circuits Conference (New York, May 1982). Gives manufacturer, brief specifications, and photographs.

Coal Age

2602. COAL AGE: EQUIPMENT GUIDE - SURFACE MINING CABLE: SHD CABLE DELIVERS HIGH VOLTAGES. *Brezovec, David. McGraw-Hill Book Company. July 1981. pg.138-148*

Discusses types of SHD cable currently available for powering excavation and conveying equipment in strip-mining applications. Emphasis is on the relatively new higher-voltage (15 kv and 25 kv) cables. Chart lists specifications of cables on the market, by manufacturer: voltage and amperage capacity, size and number of conductor and ground wire strands, insulation thickness, overall diameter, and unit weight.

Computerworld

2603. TECHNOLOGY SPOTLIGHT: PERFECT CHIP NOT HERE YET. *Henkel, Tom. CW Communications Inc. v.16, no. 16. April 19, 1982. pg.1, 10-11*

Discusses semiconductor technologies and forecasts the use of future technologies. Includes a chart detailing the advantages and disadvantages of eight emerging technologies.

2604. U.S. SEMI VENDORS SHOULD MANUFACTURE IN JAPAN: REPORT. *CW Communications Inc. v.16, no.15. April 12, 1982. pg.69-70*

A report on the Japanese semiconductor industry, published by BA Asia, Ltd., is summarized. Presents data on Japanese production of integrated circuits (1980-1981), value of imports (1980-1981) and imports as a percent of consumption (1982-1984) with forecasts for the next few years. Activities of U.S. manufacture in Japan are outlined, and data is presented on percent of market held by Japanese producers, and Japanese expenditures for research and development (1980-1981).

Control Engineering

2605. TODAY'S CONTROL ENGINEER: MORE MONEY, MORE EDUCATION, MORE RESPONSIBILITY. *Pluhar, Kenneth. Technical Publishing Company. v.29, no.12. November 1982. pg.81-82*

Summarizes findings of a 1982 survey of U.S. control engineers. Reports data in six figures: job function, annual income, education level, expenditures for equipment and systems, expenditures specified / approved, and job responsibilities.

Datamation

2606. SEMICONDUCTOR INDUSTRY: AN OVERVIEW. *French, Michael B. Technical Publishing Company. April 1980. pg.164-170*

Analyzes the structure, financing, and market outlook of the U.S. semiconductor industry. Shows IC sales by 19 U.S. firms owned by foreign investors and one U.S. firm; distribution and suppliers; market segmentation; and unit costs of components.

Design Engineering

2607. CANADIAN MARKET FOR OEM COMPONENTS AND MATERIALS. *Maclean Hunter Ltd. Annual. July 1980. 49 pg. $25.00.*

A market report containing Canadian production, imports and exports of industrial products associated with the original equipment market.

Design News

2608. ELECTRICAL COMPONENTS REPORT. *Cahners Publishing Company. Irregular. May 25, 1981.*

Special issue reviews electrical components products.

2609. ELECTRONIC EXECUTIVES' SALARIES UP 9.6 PERCENT. *Cahners Publishing Company. v.38, no.18. September 27, 1982. pg.140*

Surveys U.S. electronic executives' 1982 salaries and compensation. Compares 1982 with 1981 percent increase for 9 categories of executives. Source is the American Electronics Association.

2610. 1980 ELECTRICAL / ELECTRONIC DIRECTORY. *Cahners Publishing Company. Annual. November 1979. 220 (est.) pg.*

Manufacturers of electrical and electronic devices and components are listed alphabetically, and by product group. Includes a trade-name index, an indexed and cross-referenced advertising section, and a check-off card providing access to product information and manufacturers' catalogs.

EDN Magazine

2611. DESIGNER'S GUIDE TO THICK - FILM HYBRID CIRCUITS (PART 2): MANUFACTURERS. *L. Jardine / Gnostic Concepts Inc. (Menlo Park, CA). Cahners Publishing Company. Irregular. v.26, no.21. October 28, 1981. pg.129-156*

Lists suppliers of electronic components (resistor networks, chip resistors, discrete semiconductors, chip capacitors, etc.), substrate materials and pastes used in the manufacture of thick-film hybrid circuits. Listings are by product classification, and include address and telephone number.

2612. EDN BOARD - LEVEL - MICROP - SYSTEM DIRECTORY: INCREASED CHIP CAPABILITIES EXPAND BOARD - MICROC POSSIBILITIES. *Teja, Edward R. Cahners Publishing Company. Annual. v.27, no.7. March 31, 1982. pg.70-140*

Updated seventh annual directory provides a sampling, rather than a complete catalogue, of available board microP's and microC's. Each half-page entry includes 3-part section listing memory, 3-part section listing software support, and comments. Entries are categorized by bus type, then CPU size and, finally, by supplier. Addresses for suppliers included. Additional table provides cross-reference, listing boards according to the CPU it uses.

2613. EDN: ELECTRONIC TECHNOLOGY - THE NEXT 25 YEARS: SILVER ANNIVERSARY ISSUE. *Cahners Publishing Company. Irregular. October 14, 1981.*

Special issue contains about 30 articles on new digital electronics components, systems, production methods, and applications anticipated through the year 2006.

2614. EDN: LEADTIME INDEX, ACTIVE COMPONENTS. *Cahners Publishing Company. Bi-weekly. v.27, no.21. 1982.*

For electronic component products within 12 categories indicates minimum and maximum amount of time necessary to allocate U.S. manufacturing capacity to build and ship a medium-sized order of a moderately popular item, 1982.

2615. EDN: LEADTIME INDEX: ACTIVE COMPONENTS. *Cahners Publishing Company. Monthly. v.26, no.21. October 28, 1981. pg.70*

Reports minimum and maximum leadtimes required to produce several categories of electronic components, as reported by a composite group of major manufacturers. Covers semiconductors, displays, electron tubes, integrated circuits, computer memory circuits and systems, optelectronic devices, panel meters, power supplies, and other components. Also reports apparent upward or downward trends in required leadtimes.

2616. EDN SPECIAL ISSUE: HIGHLIGHTING TRENDS AND SIGNIFICANT OFFERINGS IN SIX KEY PRODUCT AREAS: HARDWARE AND INTERCONNECT DEVICES, ICS AND SEMICONDUCTORS, COMPUTERS AND PERIPHERALS, INSTRUMENTS, POWER SOURCES, COMPONENTS. *Cahners Publishing Company. Irregular. July 22, 1981. 406 pg. $2.00.*

Survey of new electronic components and systems. Contains detailed descriptions including specifications, compatibility data, options and characteristics of other models in series, prices, manufacturer's / supplier's address, and photographs. Covers wiring boards and board-prototyping systems, integrated circuits and semiconductors, computers and peripherals (especially 'smart terminals'), in-circuit emulators, programmable power sources, and crystal oscillators.

2617. EDN SPECIAL REPORT: CMOS IC'S. *Cahners Publishing Company. Irregular. June 24, 1981. pg.89-100*

Surveys emerging complementary metal-oxide semiconductor technologies and their applications in integrated circuits. Discusses advantages in power consumption and retention, noise immunity, operating temperature range, and possible speed and complexity of operations. Discusses activities and products of major U.S. and Japanese manufacturers, and provides a one-page listing of U.S. manufacturers' addresses / phone numbers.

2618. ELECTRO / 82 PRODUCTS. *Cahners Publishing Company. Annual. v.27, no.10. May 12, 1982. pg.103-138*

Reviews new electronics and test products. Gives technical specifications and manufacturers' addresses.

2619. ELECTROMECHANICAL AND REED RELAYS. *McDermott, Jim, editor. Cahners Publishing Company. Irregular. v.27, no.1. January 6, 1982. pg.90-112*

Surveys new developments in traditional moving - contact relays and competition with new solid - state designs in printed circuit board applications. Lists technical characteristics and prices of representative devices. Also provides addresses and telephone numbers of manufacturers, indicating broad categories of relays produced.

2620. FLAT CABLES AND CONNECTORS. *Cahners Publishing Company. Irregular. August 4, 1982. pg.107-119*

Provides manufacturer listing of flat cables formed from multiple round or flat conductors. Gives name, address, and telephone number. Further information through reader card service.

2621. MANUFACTURERS OF FIXED CAPACITORS. *Cahners Publishing Company. Biennial. v.27, no.20. October 13, 1982. pg.118-122*

Provides names, addresses and telephone numbers of 120 fixed capacitor manufacturers.

Electric Light and Power

2622. ELECTRIC LIGHT AND POWER: NEW PRODUCTS. *Technical Publishing Company. Monthly. v.60, no.4. April 1982. pg.77*

Lists new products related to electric utility industry. Listing includes a brief description of product as well as name of manufacturer. Typical products listed include turnkeys, computer systems, cast coup

Electrical Apparatus

2623. ELECTRICAL APPARATUS: PRODUCT BRIEFS. *Barks Publications Inc. Monthly. v.35, no.5. May 1982. pg.49-51*

Reviews controls and control devices, electric motors and generators, and emergency power systems. Includes manufacturer's name and address, and product description, some photographs.

2624. ELECTROMECHANICAL BENCH REFERENCE: AFTERMARKET BUYING GUIDE, 1981. *Barks Publications Inc. Annual. 1981. 96 pg.*

Annual buyer's guide includes a product index (236 categories of shop equipment, insulation, parts, electronic components and replacement electrial apparatus); manufacturer's directory (1,000 firms' addresses and product lines); distributors' directory (geographic listing of 754 wholesale / jobbers in the U.S. and Canada); a remanufacturers' directory with addresses and

product groups; a directory of training and educational sources; and a listing of motor rebuilders that are certified by underwriters laboratories.

Electrical Business

2625. ELECTR-EX LITERATURE SHOWCASE. *Kerrwil Publications Ltd. Annual. September 1982. pg.21-27*

Describes catalogs and brochures available from exhibitors at Electr-Ex Ontario, 1982. Gives some illustrations. Includes readers' service card.

2626. ELECTR-EX ONTARIO. *Kerrwil Publications Ltd. Annual. September 1982. pg.12*

Alphabetically lists exhibitors at Electr-Ex Ontario, Canada 1982. Gives booth numbers and floor plan.

2627. LIST OF TCEM FAIR EXHIBITORS. *Kerrwil Publications Ltd. Annual. September 1982. pg.19*

Alphabetically lists exhibitors at Technical Conference on Electrical Maintenance, 1982. Gives booth numbers and floor plan.

Electrical Construction and Maintenance

2628. 1979 ELECTRICAL PRODUCTS YEARBOOK. *McGraw-Hill Book Company. Annual. 1979. 258 pg.*

1979 yearbook contains electrical product descriptions, photos, literature listings, and highlights of 50 items. Divided into 16 major categories. Each item lists name of manufacturer.

Electrical Consultant

2629. ELECTRICAL CONSULTANT: PRODUCT REVIEW. *Cleworth Publishing Company, Inc. Bi-monthly. v.62, no.3. May 1982. pg.58-62*

A bi-monthly listing of electrical and electronic equipment and components used throughout industry. Gives manufacturer's address, description, and photograph. Further information available through reader card service.

Electrical Contractor

2630. ELECTRICAL CONTRACTOR: ECONOMIC FORE-CAST. *National Electrical Contractors Association. Annual. v.47, no.1. January 1982. pg.21-32*

Electrical Contractor's 1982 forecast gives 1980 actual and 1981-1982 estimated sales. Shows 1972-1977 growth of contractors including sales, number of contractors, and number of employees by region.

2631. ELECTRICAL CONTRACTOR'S NEW PRODUCTS. *National Electrical Contractors Association. Monthly. v.46, no. 11. November 1981. pg.72-87*

Monthly list of new electrical products available. Provides a photograph and brief description for each product. Gives manufacturer's name.

2632. ELECTRICAL CONTRACTOR'S 1981 PRODUCT INFORMATION FILE. *Lead Industries Association, Inc. Annual. v.46, no.12. December 1981. pg.19-75*

1981 annual directory contains an alphabetical list of major U.S. electrical manufacturers and suppliers, with addresses and telephone numbers. A cross-indexed listing of manufacturers is arranged under 31 major product classes and numerous subdivisions. An additional section provides a directory of product literature including brochures, catalogs and product information offered free by the firms listed.

Electrical Contractor and Maintenance Supervisor

2633. ELECTRIC PRODUCTS PARADE DIRECTORY OF EXHIBITORS. *Maclean Hunter Ltd. Annual. January 1981. pg.18-23, 29-30 21.00 (Canada).*

Contains directory of exhibitors at the Electrical Products Parade held in Toronto Feb. 19-21. Notes the latest products and services for the electrical industry. No addresses or phone numbers available. Lists approximately 90 companies.

2634. ELECTRICAL BUYERS' GUIDE ''82. *Maclean Hunter Ltd. Annual. v.32, no.7. July 1982. pg.15-52*

Alphabetical listing of electrical manufacturers to the Canadian market. Includes head office addresses, phone and telex numbers, branches and/or Canadian agents. Products listed under 20 main product classifications.

2635. ELECTRICAL CONTRACTOR AND MAINTENANCE SUPERVISOR BUYERS' GUIDE. *Maclean Hunter Ltd. Annual. July 1981. pg.15-46*

More than 200 manufacturers, their products and their distributors are listed in this guide to electrical equipment in Canada. Companies and products listed in alphabetical order. Branch offices, Canadian agents also listed if applicable.

2636. ELECTRICAL CONTRACTOR AND MAINTENANCE SUPERVISOR: TCEM SHOWGUIDE. *Maclean Hunter Ltd. Annual. v.32, no.8. September 1982. pg.9-21*

Guide to twenty-seventh Technical Conference on Electrical Maintenance to be held in Toronto October 14-16, 1982. Includes new products to be shown, programs, listing of exhibitors, and employees at show.

2637. ELECTRICAL DEFICIT GROWS DESPITE GROWTH IN DEMAND. *Maclean Hunter Ltd. February 1982. pg.14*

Short analysis of the federal Department of Industry, Trade and Commerce's report on the electrical products industry. Includes growth in imports and exports, and existing deficit in electrical and electronic products in 1981.

2638. ELECTRICAL PRODUCTS BRAND PREFERENCE SURVEY. *Maclean Hunter Ltd. Annual. April 1979.*

Results of a survey conducted among Canadian electrical contractors, electrical maintenance personnel and consultants, to determine brand preference for 113 items of electrical products.

2639. FORECAST 82: ELECTRICAL INDUSTRY FACTS AND FIGURES. *Maclean Hunter Ltd. Annual. December 1981. pg.20-24, 32-33*

Statistics and forecasts on the electrical industry in Canada for 1982. Includes an electrical contracting industry review and outlook (value of all construction in Canada; total value of electrical construction; and total revenues and price indexes for electrical equipment and fixtures, 1971-1981); operating revenues for electrical contractors, 1971-1981; types of construction; number of businesses, 1972-1979, their operating revenues, wages and salaries, and number of employees; a provincial breakdown of businesses for 1979 (value of output, type of work, classification); financial ratios by size, 1979 (net profits, equity, capital employed, assets, liabilities); value of factory shipments, 1971-1981; number of employees, wages and salaries, 1972-1979, in the electrical equipment and sup-

plies industries; and value of imports, exports, and shipments for 1980.

2640. NEW PROGRAMMABLE CONTROLLER PRODUCTS. *Maclean Hunter Ltd. Annual. v.32, no.6. June 1982. pg.26-28*

Reviews new programmable controller products for the electrical construction and maintenance industry. Includes product description and manufacturer.

2641. 1981 ELECTRICAL INDUSTRY FORECAST. *Maclean Hunter Ltd. Annual. October 1980. pg.25-32, 41 18.00 (Canada).*

Statistical forecast for 1981 surveys the electrical industry in Canada. Notes industry review and outlook 1971-1981 with value of all construction, electrical construction in Canada, electrical contractors total revenues (current and 1971) and a price index (residential and non-residential) for electrical equipment and fixtures. Contains statistics on the electrical contracting industry (number of businesses, total operating revenue, wages and salaries and total employees); a province-by-province breakout of electrical contracting establishments for 1978 and their output values; and wholesale merchants - electrical supplies (number, volume of trading, purchases, inventories, wages and salaries). Includes electrical equipment and supplies industries (employees, wages and salaries 1971-1978); value of factory shipments; and the value of imports and exports shipments for 1979 and the first quarter of 1980.

Electrical World

2642. ELECTRICAL WORLD: EQUIPMENT ON DISPLAY. *McGraw-Hill Book Company. Irregular. v.195, no.9. September 1981. pg.167-177*

Listing of products displayed at the IEEE Overhead and Underground T & D Conference and Exposition in Minneapolis in Sept., 1981. Provides a photograph, description and manufacturer's name for electrical products, computers and tools and machinery for the electrical industry.

2643. ELECTRICAL WORLD'S NEW EQUIPMENT. *McGraw-Hill Book Company. Monthly. v.195, no.9. September 1981. pg.200-210*

Monthly listing of new electrical equipment and products. Contains a photograph, description and name of manufacturer for each new product.

Electronic Business

2644. BUSINESS BAROMETER: SEMICONDUCTOR SHIPMENTS INCH UP BUT DEMAND MAY NOT BE REAL; PLANT UTILIZATION; ECONOMIC INDICATORS. *Cahners Publishing Company. v.8, no.7. June 1982. pg.9-10, 12*

Gives IC shipments and semiconductor shipments with distributors' share for quarters of 1979-1983. Provides plant utilization for electrical machinery, instrumentation, and aerospace equipment manufacturers for quarters of 1979-1983; and 1981 growth rate and 1976-1981 profits for automatic test equipment. Gives U.S. Index of Industrial Production and Cahners Early Warning Indicator (housing starts, Dow Jones Industrials, and M-2), 1979-1983.

2645. A 'CALL TO ARMS' TO KEEP U.S. HIGH TECH LEADERSHIP. *Cahners Publishing Company. v.8, no.11. October 1982. pg.44, 49-50*

Summarizes recommendations of the California Commission of Industrial Innovation (CCII). Focuses on international competition in the high technology industries. Includes recommendations on trade barriers, research and development, computer education, and worker participation in U.S. industry. Tables provide data on the number of jobs in 12 industry sectors in California, 1970, 1980, and 1990. Indicates percent growth, 1980-1990. Source is the Center of Continuing Study of the California Economy, California Employment Development Department.

2646. CAPACITOR MAKERS SURVIVE ON R & D AND MARKETING INGENUITY. *Mead, Tim. Cahners Publishing Company. v.7, no.12. November 1981. pg.56-62*

Examines the U.S. capacitor market and discusses its world sales and growth markets. Graphs and tables show the top 14 capacitor manufacturers and their shipments during 1979 and 1980; U.S. market shares (1981) with forecasts for 1985, by type of capacitor; growth rates (1980

2647. CHOOSE A VENTURE CAPITALIST AS YOU WOULD A PARTNER. *Rind, Kenneth W. Cahners Publishing Company. v.8, no.9. August 1982. pg.64, 66*

Discusses the availability of growth capital to the U.S. electronics industry. Table provides annual data on venture capital funds (in millions of dollars), number of public offerings, and public offerings funds raised (in millions of dollars), 1975-1981. Source is Venture Capital, Capital Publishing Corp.

2648. COMPETITION MOUNTS IN THE U.S. CONNECTOR ARENA. *Domenicali, D. Cahners Publishing Company. Irregular. April 1981. pg.64-66*

1980 rankings of the top 10 connector manufacturers, according to 1980 domestic sales. Tables provide worldwide electronic industry consumption of connectors (1980) and U.S. connector production (1979-1981).

2649. DIRECTORY OF ELECTRONICS INDUSTRIES ORGANIZATIONS. *Cahners Publishing Company. Annual. November 1980. pg.106-110*

A directory of U.S. electronics industry associations alphabetically lists names, addresses, purpose, number of members, costs, elected officials and membership contracts.

2650. DIRECTORY OF ELECTRONICS INDUSTRY NEWSLETTERS. *Cahners Publishing Company. Irregular. March 1981. pg.124-132*

Compilation of more than 45 newsletters on the U.S. electronics industry. Each entry provides title, publisher, subject coverage, editor, frequency and price.

2651. DIRECTORY OF PLANT SITES FOR THE ELECTRONICS INDUSTRY. *Cahners Publishing Company. Annual. v.8, no.6. May 15, 1982. pg.21-111*

Directory of plant sites for electronics industry presents electronics map of the U.S. showing industry segments and concentration of plants. Includes map of semiconductor companies in Europe. U.S. site locator includes names of contacts, addresses, major appeal for electronics companies, educational facilities, financial and tax inducements, transportation, attractions for employees, and new plants.

2652. A DISAPPOINTING YEAR FOR TWO ATE LEADERS. *Cahners Publishing Company. v.8, no.7. June 1982. pg.110*

Gives the 1981 quarterly revenues for two ATE leaders and the Electronic Business ATE Index. Provides research and development costs, capital expenditures, earnings and sales data for two manufacturers, 1980, 1981.

2653. DISTRIBUTION: THE TOP 20 INDUSTRIAL ELECTRONIC DISTRIBUTORS. *Cahners Publishing Company. Irregular. July 1980. pg.40-52*

Ranks top 20 U.S. industrial electronic distributors according to 1979 sales and provides their net income, sales per employee, sales growth rate and size of field sales force for 1979. Additional table shows sales of components by industrial electronics distributors to OEM's (1979). Discusses IED / OEM relations, the basic economics of distribution, technical reps and in-plant distribution stores.

2654. ELECTRONIC BUSINESS BAROMETER. *Cahners Publishing Company. Monthly. July 1981. pg.9-15*

Monthly tables show trends in the U.S. electronics industry. Bar graphs display producer price indexes; industrial production; 90-day Treasury Bill rate; imports and exports of electronic goods and services; and economic indicators including shipments and new orders. Data covers 5 years past and 2 years forecast.

2655. ELECTRONIC BUSINESS MONTHLY INDICATORS. *Cahners Publishing Company. Monthly. January 1981. pg.7-10*

Monthly graphs provide data on the U.S. electronics market, including producer price indexes (1974-1980) with forecasts (1981-1982). Economic indicators include industrial production index and electrical machinery shipments, new orders and inventory-to-sales ratio (1974-1980) with forecasts to 1982.

2656. ELECTRONIC BUSINESS SECOND 100. *Cahners Publishing Company. Annual. v.7, no.12. November 1981. pg.43-52*

Annual rankings of the second-hundred largest companies in the U.S. electronics industry. Businesses are ranked according to the most recent fiscal year sales (1980). Also includes 1980 profits, profit as a percent of sales, cost of goods as a percent of sales, cost of goods as a percent of sales, total sales per employee, profit per employee, return on equity, current ratio and foreign sales as a percent of total sales.

2657. ELECTRONIC BUSINESS: THE ELECTRONIC BUSINESS 100 UPDATE. *Cahners Publishing Company. Annual. August 1980. pg.60-66*

The top 100 electronic businesses ranking for 1979 lists electronics revenues, total sales, and total net income.

2658. ELECTRONIC BUSINESS TOP 10: POWER-SUPPLY INDEPENDENTS BRACE FOR HIGHER COSTS. *Domenicali, D. Cahners Publishing Company. Irregular. February 1981. pg.56-61*

Reviews the top 10 U.S. power-supply independents. Charts and graphs display 1980 manufacturers' sales of top 10 businesses; and market shares (1980-1985) of switchers and ac-dc power supplies, switchers and dc-dc supplies, and switchers share of military power supplies.

2659. ELECTRONIC BUSINESS: TOP-PAID EXECUTIVES: A LISTING BY COMPANY. *Stallmann, Linda. Cahners Publishing Company. Irregular. v.8, no.10. September 1982. pg.152, 154, 156, 160-161, 163*

Lists the 100 top-paid U.S. electronics industry executives according to company. Supplements listing arranged by executive's name which appeared in 'Electronic Business,' August 1982.

2660. ELECTRONIC BUSINESS 100. *Cahners Publishing Company. Annual. v.8, no.9. August 1982. pg.47-62*

Annual ranking of the top 100 companies in the U.S. electronics industry provides the following data: 1981 rank, 1980 rank, company name, electronics sales, total sales, net income, return on equity, return on investment, cost of goods as percent of sales, R & D as percent of sales, sales per employee, net income per employee, 5-year compounded growth rate in sales, 5-year compounded growth rate in net income, capital outlays to net cash flow after dividends, debt as percent of total capital; and fiscal year-end. Also reports the following: comparisons by industry segments; composite performance of the 100, 1980-1981; growth of the 100, 1977-1981; and sales growth 5-years compounded, net income growth 5-years compounded, R & D as percent of sales, return on equity, sales per employee, and net income per employee for the top 10 and the bottom 5 companies.

2661. ELECTRONIC BUSINESS 100. *Cahners Publishing Company. Semi-annual. August 1981. pg.107-117*

1981 semi-annual analysis of the top 100 U.S.-based electronic businesses. Ranks businesses according to 1980 and 1981 sales. Details 1981 total revenues, profits, return on equity, return on investment, and cost of goods as percent of sales. Gives 5-year growth rate, total revenues and profits per employee, debt as percent of total capital and capital outlays to net cash flow after dividends.

2662. ELECTRONIC BUSINESS 1981 FORECAST FOR COMPONENTS AND EQUIPMENT PURCHASES. *Cahners Publishing Company. Annual. January 1981. pg.44-71*

Annual 1981 forecast for the U.S. and European electronics industries. Executives discuss distribution, new prospects in equipment markets, component forecast to 1982 and military markets. Tables show U.S. market sectors (1977-1981); producer price indexes (1979-1980); U.S. OEM purchases (1979-1981); and growth rates forecasts (1980-1986) for electronic equipment production. Includes data on total European electronic production and equipment production by country (1981, 1991).

2663. ELECTRONIC INDUSTRY 1981 EVENTS CALENDAR. *Cahners Publishing Company. Annual. January 1981. pg.73-80*

A month-by-month listing of conferences, meetings, exhibits, symposia and seminars in the electronics industry. Location, dates and contact telephone numbers given.

2664. EXECUTIVE COMPENSATION: IT'S MORE THAN BIG BUCKS. *Bond, George. Cahners Publishing Company. v.8, no.9. August 1982. pg.102-104*

Ranks the 100 top-paid executives in the U.S. electronics industry, 1982. Includes executive's name, position, company, and salary.

2665. LEADTIMES ARE DOWN FOR MIL-SPEC CONNECTORS. *Rozak, Laura. Cahners Publishing Company. v.8, no. 11. October 1982. pg.170, 172*

Surveys representative U.S. manufacturers' current (1982) leadtimes for mil-spec connectors. Graphs connector leadtimes (in weeks), 1979-1983.

2666. PC BOARDS: A TURBULENT $3.3B BUYERS' MARKET. *Furst, Al. Cahners Publishing Company. September 1981. pg.62-74*

Examines the printed circuit board market in the U.S. Discusses leading companies in the industry. Graphs and diagrams display the top 10 captive and independent U.S. printed circuit board manufacturers according to 1979 sales; and the growing market share (1981, 1985) of multilayers according to 1981 and 1985 sales.

2667. PLASMA: CHALLENGES AND OPPORTUNITIES. *Cahners Publishing Company. v.8, no.4. April 1982. pg.118, 120*

Provides statistical data on present and projected sales of plasma-processing equipment (1980-1990). Data given for projected percent of growth (1981-1990), increased use of

plasma-etch equipment, sales of IC plasma-processing equipment, opportunities for plasma-processing market expansion, and percent of present and projected applications accounting for plasma-processing sales in the U.S.

2668. POWER SEMICONDUCTOR SALES EXPECTED TO REACH $1.5 BILLION. *Cahners Publishing Company. v.8, no.9. August 1982. pg.110*

Discusses 1981-1986 sales of power semiconductors in the U.S. Identifies major markets. Provides data in millions of 1981 dollars on sales of bipolar power transistors, MOSFET's, power rectifiers, thyristors, and zener regulators, 1981-1983 and 1986. Also indicates total power semiconductor sales, 1981-1983 and 1986. Source is Frost and Sullivan Inc.

2669. POWER SUPPLY MARKET SWITCHES ON THE SWITCHERS. *Cahners Publishing Company. v.8, no.7. June 1982. pg.72-74*

Examines the 1982-1985 power supply market. Gives captive and merchant markets for U.S. switching power supplies, 1981, 1982, 1985; and top 16 makers of switching power supplies. Provides dollar sales and market percent of linears and switchers, 1981, 1985.

2670. PRINTED CIRCUITS: EQUIPMENT MANUFACTURERS TARGET PC-BOARD MANUFACTURERS. *Socolovsky, Alberto. Cahners Publishing Company. v.8, no.10. September 1982. pg.46-47, 52-53, 58, 62*

Discusses the growth of the U.S. printed-circuits industry, 1980-1982. Presents data on the 1980 and 1982 U.S. markets in millions of dollars for the following types of capital equipment for PC's: inspection and test, mechanical, chemical, patterning, and CAO. Examines the effect of technological change on fabrication equipment markets and on suppliers.

2671. PRINTED CIRCUITS: MULTILAYER SALES BRIGHTEN A GLOOMY PC-BOARD PICTURE. *Thames, Cindy. Cahners Publishing Company. v.8, no.10. September 1982. pg.76-77, 84, 86*

Examines sales of and markets for multilayer printed circuit boards, 1981-1986. Presents data on total U.S. production of printed-circuit boards, 1981 and 1986, and indicates market shares of six types of PC-boards including multilayer. Chart illustrates 1981 sales of the 10 leading independent makers of printed-circuit boards. Assesses applications for multi-layer, high density boards. Sources include Gnostic Concepts Inc. and Maine Electronics.

2672. RECESSION STRATEGIES EARN PAY-OFF FOR AMP. *Cahners Publishing Company. v.8, no.3. March 1982. 4 pg.*

Three charts provide statistical information on sales for connector manufacturers. Table one ranks leading connector and interconnection suppliers by estimated domestic sales for 1980 and 1981 ($ million). Table two charts U.S. connector and socket consumption and growth by original equipment manufacturers for 1981 and 1982 ($ million). Table three provides data on connector applications in the United States, providing figures for percent of sales attributable to specific markets, including electrical / industrial; consumer; business / retail / education; industrial; instrumentation; communications; computer; and government / military.

2673. SEMI - MATERIALS VENDORS SPLIT ON ECONOMIC OUTLOOK. *Arnold, William F. Cahners Publishing Company. Irregular. v.8, no.4. April 1982. pg.53-58*

Provides statistical data relative to future market trends in the semiconductor materials industry. Two tables include data on international semiconductor materials suppliers, including an alphabetized listing of companies and primary materials and world wide semiconductor materials forecasts, including figures for 1982 consumption ($ million) of listed package materials and listed water-fab materials (merchant and captive).

2674. THE SEMICONDUCTOR MARKET IN THE U.S. *Cahners Publishing Company. v.8, no.3. March 1982. pg.80*

Analyzes the U.S. semiconductor market. Statistical data is provided on forecasts of U.S. semiconductor shipments, including a breakdown of amount of integrated circuits and discrete components comprising those shipments; the Japanese semiconductor industry; analysis of the U.S. semiconductor market by device type and specification, including projected sales through 1986.

2675. TOP 10 CAPACITOR MANUFACTURERS. *Swartz, R. B. Gnostic Concepts. Cahners Publishing Company. Irregular. September 1980. pg.70, 75-78*

1979 rankings of top 10 capacitor manufacturers according to 1979 shipments to U.S. markets. Tables show 1979 capacitor market leaders by type of product and 1978-1983 growth rates, by market.

2676. THE TOP 10 CHIP MAKERS ENTER DECADE SUCCESSFUL AND TROUBLED. *Arnold, W.F. Cahners Publishing Company. Irregular. March 1980. pg.38-44*

Examines the U.S. market for semiconductors. Ranks top 10 U.S. and top 15 worldwide semiconductor manufacturers, according to 1979 sales, and provides 1978 sales with percent change. Ranks top 10 U.S. merchant manufacturers of worldwide IC shipments for 1979, with 1980 estimates. Lists top 10 merchant supplier semiconductor businesses according to 1979 sales, net income, total company revenues and cost of sales.

2677. TOP 10 CONNECTOR MAKERS. *Cahners Publishing Company. Irregular. April 1980. pg.50, 55-56, 60, 64*

Analysis of connector suppliers provides data on the top 15 companies; sales and growth rates for the interconnector industry; connector supplier growth rates; and financial data on leading connector suppliers: sales; net income; performance figures for the whole company, including per employee calculations of total revenue; and net income.

2678. THE TOP 10 IN SEMICONDUCTOR PRODUCTION EQUIPMENT. *Technical Ventures. Cahners Publishing Company. Irregular. March 1981. pg.42-60*

Reviews the top 10 in semiconductor production equipment. Provides 1980 rankings of top 10 VLSI capital equipment manufacturers with revenues (1979, 1980). Graphs display worldwide semi-conductor capital expenditures (1965-1985) and typical distribution of new-line investments in mid-1980.

2679. THE TOP 10 IN SEMICONDUCTORS: END TO DOUBLE-DIGIT GROWTH. *Arnold, W. F. Cahners Publishing Company. Irregular. March 1981. pg.56-60*

1979-1980 rankings of the top 10 semiconductor manufacturers, according to 1980 sales. Graph breaks down worldwide production (1980-1983) by total semiconductors, by discretes, and by IC's.

2680. THE TOP 10 MOS MEMORY SUPPLIERS: NEW PRICES AND DEVICES UPSET MOS MEMORY MARKET. *Domenicali, D. Cahners Publishing Company. December 1980. pg.84-95*

Ranks top 20 U.S. MOS memory suppliers according to 1979 estimated worldwide sales. Examines the market for MOS IC memory suppliers and RAM suppliers. Graphs display short-term (1978-1982) sales estimates and long-term (1970-1990) sales estimates for the worldwide market for MOS memories.

Shows 1980 MOS, RAM prices and worldwide MOS memory market shares by product (1979, 1980).

2681. THE TOP 20 RESISTOR MAKERS: RECESSION CATCHES UP WITH RESISTOR MAKERS. *Arnold, W.F. Gnostic Concepts, Inc. Cahners Publishing Company. November 1980. pg.54-58*

Examines U.S. market for resistors, and ranks top 10 resistor manufacturers according to 1979 estimated shipments to U.S. markets. Graphs show OEM consumption of resistors and resistor end uses (1979) with forecasts for 1984.

2682. TRANSISTOR - TRANSISTOR - LOGIC (TTL) POPULARITY UNDIMINISHED BY PROGRESS. *Arnold, William F. Cahners Publishing Company. v.8, no.9. August 1982. pg.88, 90*

Presents data on worldwide transistor-transistor-logic (TTL) shipments, in millions of dollars, 1981-1985. Compares Schottky shipments to Standard shipments. Source is the Semiconductor Industry Association.

2683. U.S. SEMICONDUCTOR SHIPMENTS EXPECTED TO DOUBLE BY 1986. *Cahners Publishing Company. v.8, no.3. May 1982. pg.80*

Summary of Frost and Sullivan's prediction of U.S. semiconductor shipments forecasts shipments to 1986. Provides table of top U.S. suppliers of integrated circuits (1980).

2684. WHO EARNS HOW MUCH? PAY OF KEY EXECUTIVES. *Cahners Publishing Company. v.8, no.3. March 1982. pg.38-48*

Two tables supply data on 100 top-paid executives in electronics. The first table lists executives beginning with the highest salaried; the second table lists executives alphabetically by company. Each list includes company, executive, title and salary.

Electronic Design

2685. ELECTRONIC DESIGN'S GOLD BOOK: THE WORLD'S ONLY COMPLETE ELECTRONICS DIRECTORY 1980 / 81: VOLUME 1: PRODUCTS, TRADE NAMES, MANUFACTURERS. *Hayden Book Company, Inc. Irregular. 1980. 1128 pg. $40.00.*

Directory lists over 5700 headings, over 10,000 manufacturers, over 50,000 local suppliers, 10,000 trade names, and 1600 distributors. Arranged by name and geographical location.

2686. GENERAL-PURPOSE MICROPROCESSORS: PERFORMANCE AND FEATURES. *Hayden Book Company, Inc. Annual. v.30, no.21. October 14, 1982. pg.118-139, 144-158*

Examines the state of U.S. microprocessor technology in 1982. Reports on new products and new technologies. Product selection guide lists

2687. 1980 TECHNOLOGY FORECAST. *Hayden Book Company, Inc. Annual. January 4, 1980. pg.65-107*

1980 forecast of new products in semiconductors, instruments, analog circuits, communications, passive components, computers, software and peripherals.

Electronic Engineering Times

2688. ELECTRO 82: SIGNIFICANT PRODUCT INTRODUCTIONS AT BOSTON SHOW. *CMP Publications, Inc. Annual. no.234. May 24, 1982. pg.26-38*

Illustrates and gives technical data for some of the products on display at Boston's Electro 82.

2689. ELECTRONIC ENGINEERING TIMES: COMPONENT PRICE / DELIVERY INDEX. *CMP Publications, Inc. Bi-weekly. 1982.*

Biweekly tabulation of price and delivery of major electronic products based on responses from manufacturers and distributors. Products are listed within the following categories: capacitors, connectors, discrete semis, integrated circuits, modules, printed circuits, relays, resistors, and switches. Number of weeks load time, and price stability indicator is given for each product.

2690. ELECTRONIC ENGINEERING TIMES NEW PRODUCTS SECTION. *CMP Publications, Inc. Bi-weekly. January 5, 1981. 12 (est.) pg.*

Biweekly section lists new products in the electronic engineering industry: components / modules, discrete semiconductors, materials / hardware / production, computers / peripherals and tests / measurement. Most entries include photograph, manufacturer, price ranges and brief description.

2691. LOGIC ANALYZERS AND SCOPES. *CMP Publications, Inc. Irregular. no.233. May 10, 1982. pg.61-63*

Product review discusses logic analyzers and scopes, with a review of new technology. New products are described with information on prices, technical specifications, and manufacturers and model numbers. A list of manufacturers of 'smart' logic analyzers and oscilloscopes is included, with addresses and telephone numbers.

2692. NCC 1982: PRODUCT HIGHLIGHTS. *CMP Publications, Inc. Annual. Issue 235. June 7, 1982. pg.16-22*

Describes new electrical and electronic products to be exhibited at NCC 1982. Includes product specifications and manufacturer's name and address.

2693. TECHNOLOGY FORECAST 1981. *CMP Publications, Inc. Annual. January 5, 1981. pg.1, 16-34.*

1981 forecasts for the U.S. electronic engineering industry. Interviews with industry leaders project technology innovations in semiconductors, computers, instruments and passive components.

2694. TOP 50 ELECTRONIC STOCKS OUTPERFORM MARKET IN RECENT WALL STREET REBOUND. *CMP Publications, Inc. no.245. October 25, 1982. pg.1, 6, 18*

Alphabetically lists 50 largest U.S. publicly traded electronic companies based on 1981 sales. Gives August 12 and October 15, 1982, stock market closing prices; and growth rate between those dates.

2695. TOP 50 FIRMS RAISE R & D SPENDING 18%. *CMP Publications, Inc. no.239. August 2, 1982. pg.5 ff.*

Reports on growth in R & D spending among 50 largest electronics firms in 1981. Contains figures on research and development expenditures (1981); percent increase (1980-1981); and R & D intensity. Table includes data on 10 speciic companies, showing R & D expenditures in dollars and as percent of sales. Includes discussion of 1981 legislation allowing tax credits for certain R & D costs.

Electronic News

2696. BONDING SYSTEMS: LONG JOURNEY HOME. *Snyderman, Nat. Fairchild Publications. v.28, no.1396. June 7, 1982. pg.80-81, 83*

Discusses the current trend of IC companies to move assembly operations back to the U.S. and the advancements being made in factory automation. Five tables show wire bonder market forecast, bonding equipment market forecast and sales, 1980 sales of 6 specific suppliers, and typical bonder purchasing and manufacturing practices.

2697. DOING BUSINESS IN 1981. *Fairchild Publications. Annual. January 5, 1981. 28 pg.*

1981 annual business forecast for the electronic industry, focusing on semiconductors and computers. Narrative discussion with interspersed statistics, written by industry leaders.

2698. ELECTRONIC NEWS FINANCIAL TABLES. *Fairchild Publications. Weekly. January 12, 1981. 4 pg. $22.00/yr.*

Weekly financial tables providing data on businesses on the New York and American Stock Exchanges and Over-the-Counter stocks. Gives week-end high and low stock prices, sales for NYSE and ASE, and bid, asked and previous bid for OTC.

2699. LOOKING AT THE LEADERS 1980: RANKING OF INDUSTRY BY SALES. *Brousell, David R. Fairchild Publications. Annual. July 14, 1980. pg.4-59*

1980 rankings on 50 U.S. and 13 foreign corporations by electronics sales. Corporations are arranged alphabetically with detailed financial data including electronic sales, gross sales, classification of products, markets, headquarters, major facilities, officers, directors, sales and earnings records, and forecasts. Foreign markets arranged alphabetically by country.

2700. SPOTLIGHT ON CAPACITORS. *Fairchild Publications. Annual. v.28, no.1407. August 23, 1982. 44 pg.*

Summarizes data on the 1982 U.S. capacitor market, as follows: dollar sales and unit sales by product group, 1971 and 1981; annual sales in dollars and units, 1977-1981, by product category; and manufacturers' 1981 market shares for four product categories. Discusses selected product groups and reviews technical innovations.

Electronic Packaging and Production

2701. ELECTRONIC PACKAGING AND PRODUCTION 1982 / 1983 VENDOR SELECTION GUIDE. *Cahners Publishing Company. Annual. v.22, no.7. July 1982. 346 pg.*

Vendor selection guide for 1982 and 1983 covers products for the U.S. electronic production and packaging industry. Product directory contains 1100 product group listings and names of vendors. Vendor directory provides names, addresses, telephone numbers, officers, product lines, and services. Discusses production techniques, new technologies, and testing products.

2702. MICROELECTRONICS BUYER'S GUIDE. *Milton S. Kiver Publications, Inc. Annual. November 1978. pg.67-113*

1978 buyer's guide listing addresses of U.S. manufacturers and suppliers of products, processing equipment, services, and materials for the semiconductor and hybrid industry.

2703. NEPCON NORTHWEST CONFERENCE / EXHIBITION: 1982. *Cahners Publishing Company. Annual. v.22, no. 10. October 1982. pg.60*

Provides an alphabetical listing of manufacturers exhibiting PC / microelectronics components, products, systems, and equipment at the 1982 Nepcon Northwest Conference / Exhibition (November 9-11, 1982, San Jose).

2704. PRINTED CIRCUITS BUYER'S GUIDE. *Milton S. Kiver Publications, Inc. Annual. December 1978. pg.81-122*

1978 directory of suppliers and manufacturers of processing equipment, services, materials, and new products for the U.S. circuit board manufacturing industry.

2705. STUDY FORECASTS PC MARKET GROWTH. *Markstein, Howard W. Cahners Publishing Company. v.22, no.10. October 1982. pg.12*

Summarizes findings of a 1982 study on printed wiring and backplane industry in the U.S. Forecasts printed wiring markets, 1981 and 1986. Indicates 1981 and 1986 use in application; and 1981-1986 average annual growth for nine end applications. Source is Gnostic Concepts, Inc.

2706. VENDOR SELECTION ISSUE 1979-1980. *Milton S. Kiver Publications, Inc. Annual. 1979. 326 pg.*

1979-80 directory of addresses and telephone numbers of U.S. vendors and suppliers of products and services used in the production and packaging of electronics. Includes production tools, production- machines, packaging or production materials and chemicals, production accessories, hybred circuit production equipment, hybred circuit processing materials and accessories, packaging or production services, test equipment, connectors and sockets, wire and cable, cabinets and enclosures, packaging hardware, circuit components and fasteners.

2707. WINDS OF CHANGE IN PRINTED WIRING MANUFACTURE. *Loeb, William E. Cahners Publishing Company. v.22, no.6. June 1982. pg.79-88*

Focuses on automation, productivity, and competition within the U.S. electronic industry. Includes 6 figures as follows: historical and forecasted growth of U.S. electronic equipment production, 1970-1990; computer power per dollar, 1970-1990; gross fixed private investment for equipment, 1970-1990; unit wire cost and number of wires used, 1970-1990; relative costs of computer electronic equipment, 1970-1990; and U.S. printed wiring production by noncaptive suppliers, 1980 and 1990.

Electronic Products Magazine

2708. CHIP CAPACITORS SERVICE LIST. *Hearst Business Communications, UTP Division. Irregular. February 4, 1982. pg.42*

Lists 31 manufacturers of tantalum, ceramic, MOS, and 8 mm tape chip capacitors. Gives city and telephone number. Further information available through reader card service.

2709. ELECTRONIC PRODUCTS / INTEGRATED CIRCUITS UPDATE. *Hearst Business Communications, UTP Division. Irregular. March 3, 1982. pg.81-85*

Lists new integrated circuits on the U.S. market, including digital, interface, linear, memory, and microprocessor types. Gives name and address of manufacturer; function specifications; and prices. Further information available through reader card service.

2710. ENCAPSULATED POWER SUPPLIES: SOURCE LIST. *Hearst Business Communications, UTP Division. Irregular. March 3, 1982. pg.69-70, 72*

Lists manufacturers of encapsulated power supplies that accept AC input and output logic level DC voltage. Gives address and telephone of manufacturer; input frequency range; output voltages; output power range; supply, construction, and mounting; maximum number of outputs; and range of case sizes.

2711. MINIATURE TRANSFORMERS. *Hearst Business Communications, UTP Division. Irregular. v.24, no.10. January 11, 1982. pg.65-68*

A directory of manufacturers of electronic miniature transformers including audio, high-voltage, I-F, isolation, matching, power, ac / dc converters, pulse, R-F, Scott-T, telephone converters, trigger transformers. Gives addresses of manufacturers and compliance with spec standards. Further information available through reader card service.

2712. PRODUCT SELECTION GUIDES FROM THE TWENTY-FIFTH EDITION OF ELECTRONIC ENGINEERS MASTER CATALOG. *Hearst Business Communications, UTP Division. Irregular. v.25, no.6. September 1982. 48 pg.*

Directory features electronic components available in the U.S. in 1982. Includes 12 product guides listing manufacturers' names and product designations. Additional directory lists manufacturers whose products meet military specifications. Includes guide to manufacturers' addresses.

2713. PUSHBUTTON SWITCHES: SOURCE LIST. *Hearst Business Communications, UTP Division. Irregular. November 30, 1981. pg.29-34*

A directory of pushbutton switch manufacturers in the United States. Information on lighted, unlighted, switch contacts, action and special feature switches is limited to company name and address.

2714. SOURCE LIST: CABLE-TO-CABLE CONNECTORS. *Hearst Business Communications, UTP Division. Irregular. v.25, no.1. June 8, 1982. pg.79-80*

Provides a breakdown of the product lines of over 50 U.S. manufacturers of cable-to-cable connectors. Includes manufacturer's name and address, and information on product shape; material; packaging; and interconnections, with all data provided by the manufacturer. Contians reader service number for further inquiries.

2715. SOURCE LIST: FLAT CABLES. *Hearst Business Communications, UTP Division. Irregular. December 1981. pg.55, 59*

A list of flat cable manufacturers in the U.S. Data includes number of conductors, gauge size, type of insulation, conductor spacing, wire identification and operating temperature. Gives address and telephone number of firm. Further information available through a reader card service.

Electronics

2716. ANNUAL TECHNOLOGY UPDATE ISSUE. *McGraw-Hill Book Company. Annual. October 23, 1980. pg.112-231*

Annual review of new products and technology in the electronics industry. Includes sections on semiconductors, computer memories, microsystems and software, components, test and measurement equipment, computers and peripherals, communications consumer products, and packaging and production.

2717. CADENCE SLOW FOR MILITARY SALES. *McGraw-Hill Book Company. v.55, no.17. August 25, 1982. pg.75-78*

Examines semiconductor sales to the military, 1981-1982. Gives 1980 sales of 2 firms.

2718. CAREER OUTLOOK: JOB-HOPPING LOSES APPEAL. *Costlow, Terry. McGraw-Hill Book Company. v.55, no.14. July 14, 1982. pg.226*

Trends in the employee turnover rate in the electronics industry are reviewed, with data on the turnover rate by U.S. region, 1979-1981. R

2719. COMPANIES STILL SHORT OF EE'S. *Connolly, Ray. McGraw-Hill Book Company. v.55, no.10. May 19, 1982. pg.105-110*

Analysis of the market for electrical engineers examines supply of employees by end use industries including university faculties.

2720. ELECTRONICS: CAREER OUTLOOK. *McGraw-Hill Book Company. v.55, no.8. April 21, 1982. pg.229*

Charts the employer preferences of bachelor-level and graduate-level engineering students. The source is Graduating Engineer, 1982.

2721. ELECTRONICS: NEW PRODUCTS. *McGraw-Hill Book Company. Bi-weekly. v.55, no.9. May 5, 1982. pg.193-258*

Reviews new electronics products featuring product description, cost, manufacturer's name, address, and telephone number.

2722. ELECTRONICS WORLD MARKET SURVEY AND FORECAST. *McGraw-Hill Book Company. Annual. January 3, 1980. pg.125-150*

Market forecasts (1978-1983) for the U.S., Japanese and Western Europe electronics industries. Arranged by product type, with statistics on industrial consumption and shipments.

2723. FIFTY YEARS OF ACHIEVEMENT: A HISTORY; LOOKING AHEAD TO THE YEAR 2000: ELECTRONICS SPECIAL COMMEMORATIVE ISSUE. *McGraw-Hill Book Company. Irregular. April 17, 1980. 646 (est.) pg.*

Fifty-year retrospective of the electronics industry discusses the evolution of specific technologies. Twenty-year forecast examines new technologies, potential growth industries, and the future role of engineers.

2724. MICROWAVES: ALL'S QUIET ON THE FRONT FOR MILITARY USERS. *McGraw-Hill Book Company. Irregular. September 11, 1980. pg.46*

Projections in military market for microwave components and systems through 1985. Microwave system sales to military, by type of system, 1980-1985, given. Data from Frost & Sullivan report.

2725. SEMICONDUCTORS FACE WORLDWIDE CHANGE: RECESSION APPEARS TO BE ENDING, BUT SOME ANXIETY PERSISTS. *Bierman, Howard. McGraw-Hill Book Company. v.55, no.10. May 19, 1982. pg.129-140*

Analyzes the semiconductor industry and market, the effects of captive production, and U.S. competition with Japan. Shows annual growth rates of worldwide shipments, 1979-1983; market size, 1980-1981 versus 1981-1982; utilizations, 1970, 1980, 1990; end use forecasts, 1981, 1980; rankings of U.S. captive suppliers, showing number of employees, 1981, and sales, 1981-1982; Japanese market penetration; and imports and exports, 1977-1981.

2726. THE WORLD MARKETS FORECAST. *McGraw-Hill Book Company. Electronics. Annual. January 13, 1981. pg.121-144*

A comparison study of equipment totals in dollars of the computer, electronic and communications equipment markets in the United States, Western Europe and Japan for 1979-1981. For the U.S., market forecasts of equipment are forecasted through 1984. Consumption and percent of market penetration of equipment groups in the U.S. are given for 1980-1985.

Electronics and Communications

2727. ELECTRONICS AND COMMUNICATIONS: EPIC ''81. *Southam Business Publications, Ltd. Division of SouthamCommunications Ltd. Annual. 1981. 152 pg.*

Electronic procurement index lists products and services available; names, addresses, and telephone numbers of suppliers; U.S. and overseas companies and their Canadian representatives; and key executives and associations in the industry.

2728. ELECTRONICS AND COMMUNICATIONS: FEATURE PRODUCTS. *Southam Business Publications, Ltd. Division of SouthamCommunications Ltd. Irregular. v.30, no.2. April 1982. pg.12, 40-43*

Provides descriptions of new electronic products, technical specifications and manufacturers.

2729. EPIC 80: ELECTRONIC PROCUREMENT INDEX FOR CANADA. *Southam Business Publications Ltd. Division of Southam Communications Ltd. Annual. 1979. 134 pg.*

Suppliers section contains, alphabetically, names, addresses and telephone numbers of both head office and branches of manufacturers and representatives who supply product groups and services to the Canadian electronics industry. Products section uses alphabetic coding system to identify different available types of a particular product (contains glossary of electronic equipment). Principals section comprises a list, by company name, of U.S. and overseas companies who supply products and services to the electronics industry through Canadian representatives.

2730. EPIC 82: ELECTRONIC PROCUREMENT INDEX FOR CANADA. *Southam Business Publications, Ltd. Division of SouthamCommunications Ltd. Annual. v.30, no.3. June 1982. pg.19-85*

Lists products and services available to the electronics industry in Canada, with suppliers' names and addresses and names of Canadian representatives of U.S. and overseas companies. Also includes industry and product news, and some feature articles. Includes reader service card for further information.

2731. FOCUS ON COMPONENTS. *Southam Business Publications, Ltd. Division of SouthamCommunications Ltd. Irregular. v.30, no.2. April 1982. pg.16-25*

Lists new electronic component products, manufacturers, and technical specifications.

2732. PROLIFERATION OF ELECTRONIC PRODUCTS: A CHALLENGE TO THE SEMICONDUCTOR MARKETER. *Springer, Glen. Southam Business Publications, Ltd. Division of SouthamCommunications Ltd. February 1982. pg.9*

Discusses semi-conductor marketing and research and development. Statistics on imports and exports, 1977-1980; and apparent domestic consumption, 1981.

Engineering Digest

2733. LIGHTING TRENDS OVER THE NEXT 20 YEARS. *Canadian Engineering Publications Ltd. October 1980. pg.25-30 18.00 (Canada).*

Research study conducted by Westinghouse Electric Corp. in U.S. to identify major trends in lighting technology through the year 2000. Based on survey of over 100 lighting experts from U.S. Predicts cost of electricity (operating ratios) will triple by 2000 but the cost of lighting average home, office, department store or factory will actually be a lower percentage of electricity budgets equivalent for use in Canada.

Financial World

2734. THE WILTED BEAUTIES OF HIGH TECH. *Kessler, Jeffrey. Macro Communications, Inc. v.151, no.11. June 1, 1982. pg.16-21*

Article on investment in high tech industries shows performance of high tech stocks. Gives 1981 actual and 1982-1983 forecasted earnings per share, recent stock prices, current and projected price earnings ratios, dividends and yields for 21 high tech companies.

Fortune

2735. ATARI AND THE VIDEO - GAME EXPLOSION. *Bernstein, Peter W. Time, Inc. July 27, 1981. pg.40-46*

Article focuses on the development since 1972 of industry leader Atari, but also discusses other manufacturers of coin-operated video games and their best-selling products. Includes estimates of 1981 sales for six firms, plus estimates of 1973 and 1981 revenues and profits for Atari.

2736. JAPAN'S OMINOUS CHIP VICTORY. *Bylinsky, George. Time, Inc. v.104, no.12. December 14, 1981. pg.52-57*

Reviews new technology in the semiconductor industry, comparing progress of the U.S. and Japan in developing semiconductor memories. Presents market shares worldwide by country and by corporations within each country. Discusses decreasing prices and increasing memory capacity, with an overview of new products being developed by leading U.S. manufacturers. Reviews capital investment required for research and development along with current world sales and projections for 1985 sales.

Hardware Age

2737. ELECTRICAL: A LINE WITH POWER. *Chilton Book Company. v.219, no.6. June 1982. pg.43-45, 48*

Analyzes sales of electrical hardware. Gives value of shipments of electrical supplies and lighting fixtures, 1967-1995, residential lighting fixtures and electrical supplies share of market by 3 store types, 1971-1990.

High Technology

2738. HIGH TECHNOLOGY STOCK INDEX. *Technology Publishing Company. Monthly. v.2, no.4. July 1982. pg.92*

Reviews monthly performance of high technology stocks in relation to both Dow Jones Industrials and Standard and Poor's 500. Lists 35 groups within high technology industries and gives percent change in stock prices from March to April.

Discusses best and worst performing groups. Mentions representative stocks and price earnings ratio of stocks in the index.

Hot Rod

2739. IGNITION SAFARI. *Davis, Marlon. Petersen Publishing Company. Irregular. May 1982. pg.93*

Reviews automotive ignition systems available for purchase giving details on performance characterisitcs at different levels of competition and other technical data. Lists products and manufacturer's address.

IEEE Spectrum

2740. IEEE SPECTRUM: NEW PRODUCT APPLICATIONS. *Institute of Electrical and Electronics Engineers, Inc. Monthly. v.19, no.4. April 1982. pg.80*

Listing of new products includes product name, model numbers, technical specifications, price, brief description and manufacturer's name and address.

2741. IEEE SPECTRUM: TECHNOLOGY "82. *Christiansen, Donald, editor. Institute of Electrical and Electronics Engineers, Inc. Annual. v.19, no.1. January 1982. pg.30-75*

Reviews U.S. developments in electrical and electronic technologies during 1981, discussing individual manufacturers' activities and innovations. Product groups covered are computers, communications electronics, solid state components / devices, instrumentation, industrial electronics, power / energy systems, and consumer electronics.

Industry Week

2742. A CALL TO ARMS FOR U.S. CHIPMAKERS. *Patterson, William. Penton / IPC. Subsidiary of Pittway Corporation. v.214, no.3. August 9, 1982. pg.51-57*

Reviews U.S. and Japanese chipmakers showing 1976-1980 average profit margins for four Japanese and four U.S companies. Lists leading U.S. chipmakers 1981 rank and production figures. Graphs show 1970-1982 U.S. - based merchant sales and U.S. - based captive supply value; 1980-1984 production and consumption of U.S. firms, Japanese firms, Western Europe, and the rest of the world.

2743. UPTICK OR UPTURN FOR SEMICONDUCTORS? THAT IS THE QUESTION. *Penton / IPC. Subsidiary of Pittway Corporation. v.213, no.7. June 28, 1982. pg.79-80*

Discusses May Semicon West Industry Trade Show, U.S. chipmakers vs. Japanese competitors, and forecasts of 1982 shipments.

Insulation / Circuits

2744. CIRCUIT EXPO "82 EXHIBITS. *Lake Publishing Corporation. Annual. v.28, no.11. October 1982. pg.50-55*

Previews circuit board manufacturing products exhibited at the 1982 Circuit Expo. Lists U.S. manufacturers and provides brief product descriptions.

2745. COIL WINDING / ELECTRICAL MANUFACTURING EXPO 1982: PREVIEW. *Lake Publishing Corporation. Annual. v.28, no.10. September 1982. pg.53-62*

Gives information on CW / EME 1982. Describes workshop and technical sessions. Names leaders and lists manufacturers, cities and products.

2746. ELECTRO / 82: EXHIBITORS. *Lake Publishing Corporation. Irregular. May 1982. pg.69-90*

An alphabetical listing of electronic equipment exhibited at the Electronic Show and Convention (Boston, May 25-17, 1982). Gives manufacturer, booth number, description, and photograph.

2747. INSULATION / CIRCUITS DESK MANUAL 1980. *Lake Publishing Corporation. Annual. 1980. 344 pg.*

1980 handbook and directory of suppliers, equipment, instruments, and accessories used in the manufacturing of electrical and electronic products in the U.S. and Canada. Lists product lines, addresses of vendors and manufacturers, and advertisers.

2748. INSULATION / CIRCUITS DESK MANUAL: 1982-1983. *Lake Publishing Corporation. Annual. v.28, no.7. June 1982. 408 pg. $25.00.*

Contains alphabetical lists of technical societies and trade associations, terms, concepts, definitions, materials, test instruments, components, equipment, tools, accessories, supplies, and suppliers. Charts show properties, characteristics, and uses. Includes subject index, advertisers' index, and free product information.

2749. INSULATION / CIRCUITS: NEW INSTRUMENTS AND EQUIPMENT. *Lake Publishing Corporation. Monthly. v.28, no.4. May 1982. pg.115-129*

Lists new instruments and equipment related to the insulation / circuits industry. Includes brief description of product as well as name and address of manufacturer. Typical products listed include lamination assembly machines, screen printer set-up, digital earth resistance tester and portable sander / lapper.

2750. INSULATION / CIRCUITS: NEW MATERIALS AND COMPONENTS. *Lake Publishing Corporation. Monthly. v.28, no.4. April 1982. pg.100-114*

Lists new materials and components related to the insulation / circuits industry. Includes brief description as well as name and address of manufacturer. Typical products listed include high-temp silicone potting material, clear epoxy casting resin, cable mounting bases and fiber optics alignment systems.

2751. SEMICON / EAST "82 PREVIEW. *Lake Publishing Corporation. Annual. v.28, no.9. August 1982. pg.53-54*

Previews products and equipment for the semiconductor industry to be displayed at the 1982 Semicon East Show. Provides a product description and manufacturer's name and address.

2752. WESCON / 82 PREVIEW. *Lake Publishing Corporation. Annual. v.28, no.9. August 1982. pg.46-50*

Previews electronic components and products to be displayed at the 1982 Wescon Show. Provides product description and manufacturer's name and address.

2753. THE 1982 INTERNATIONAL MICROELECTRONICS SYMPOSIUM (ISHM "82): EXHIBITORS. *Lake Publishing Corporation. Annual. v.28, no.11. October 1982. pg.113-117*

Lists U.S. exhibitors at the 1982 International Microelectronics Symposium. Indicates product lines.

Interiors

2754. 1982 A-Z GUIDE TO INTEGRATED FURNITURE LIGHTING. *Billboard Ltd. Annual. v.13, no.3. October 1982. pg.38-44*

Directory of integrated furniture lighting lists companies along with type of lighting, light source, mounting and closure.

Iron and Steelmaker

2755. ELECTRIC ARC FURNACES IN STEEL PLANTS, UNITED STATES 1982. *Iron and Steel Society of AIME. v.9, no.5. May 1982. pg.46-67*

Chart details electric arc furnaces in U.S. steel plants (1982) alphabetically by steel company. The following data is included for each plant location: plant capacity; start-up date; furnace manufacturer; regulator manufacturer power usage; and technical specifications on shell diameter, maximum transformer capacity, electrode diameter, and electrode usage.

LD and A

2756. LDA DIRECTORY: LIGHTING EQUIPMENT AND ACCESSORIES. *Illuminating Engineering Society. Annual. July 1981. pg.16-61*

Annual directory of lighting equipment and accessories lists new products, with manufacturers and addresses. Includes cross-indexes.

Machine Design

2757. ELECTRICAL AND ELECTRONICS REFERENCE. *Penton / IPC. Subsidiary of Pittway Corporation. Irregular. May 14, 1981.*

Special reference issue on electrical and electronics components.

2758. MACHINE DESIGN: 1982 ELECTRICAL AND ELECTRONICS REFERENCE ISSUE. *Penton / IPC. Subsidiary of Pittway Corporation. Annual. v.54, no.11. May 13, 1982. 323 pg.*

Updates products and technology in electrical and electronics aspects of machine design. Includes descriptions, technical specifications, and charts in nine major areas: motors; motor controls and protectors; machine and process control systems; electromechanical and solid-state switches; transducers; indicators and displays; power supplies; interconnections; and test, measurement, and development instrumentation. Contains news briefs; advertising index; and reader service cards for further product information.

Mart

2759. CONSUMER ELECTRONICS: "82 SALES SEEN AT RECORD LEVELS DESPITE GROWING PAINS. *Morgan-Grampian, Inc. v.28, no.9. May 1982. pg.64*

Forecast of 1982 electronics market estimates market growth, VCR sales, car stereo sales, color TV sales, monochrome TV shipments, projection TV sales, and video disc sales.

2760. VIDEO-GAME SALES SKYROCKETING AS MORE SUPPLIERS ENTER FIELD. *Ciccolella, Cathy. Morgan-Grampian, Inc. v.28, no.10. July 1982. pg.26-28*

Article on video games estimates 1982 market size and 1982-

1983 market penetration. Discusses prices, advertising expenditure, and distribution.

Merchandising

2761. EIGHTH ANNUAL ELECTRONICS STATISTICAL AND MARKETING REPORT. Gralla Publications. Annual. v.7, no.8. August 1982. pg.17-30

Examines sales of consumer electronic products, 1982. Gives percent of sales by type of product or retail outlet, 1981-1982, in the following categories: video, electronics furniture, personal electronics, audio, and software.

2762. MERCHANDISING: FIFTY-NINTH ANNUAL STATISTICAL AND MARKETING REPORT. Gralla Publications. Annual. March 1981. 94 pg.

Provides a five-year analysis of domestic shipments and retail sales, and imports / exports, of several houseware, consumer electronics, and major home appliance products.

2763. MERCHANDISING: PRODUCT PICTURE: PERSONAL ELECTRONICS. Gralla Publications. Monthly. March 1982. pg.68

A monthly listing of new electronic products for personal use including digital watches, calculators, videogames, small computers, etc. Information includes address of manufacturer, price, description, and photograph.

2764. TENTH ANNUAL CONSUMER SURVEY. Rath, Lee. Gralla Publications. Annual. v.7, no.5. May 1982. pg.17-39

Reports annual consumer survey for appliance, electronics, and housewares retailers. Gives, for 20 different products, regional responses, market shares, purchases planned, and reasons for purchasing. Survey conducted at select shopping locations in January and February 1982. Provides interviewee characteristics.

Microwaves

2765. 1979-1980 MICROWAVES PRODUCT DATA DIRECTORY. Hayden Book Company, Inc. Annual. 1979. 544 pg. $18.00.

1979 microwaves product data directory contains over 245 product categories, with names, addresses, telephone numbers of manufacturers. Provides alphabetical listing of manufacturers under each product category. Provides product data such as specs, outline drawings, schematics and pricing information.

2766. 1980-1981 PRODUCT DATA DIRECTORY. Hayden Book Company, Inc. Annual. August 27, 1980. 624 pg.

Directory of products for the microwave industry including performance specifications and addresses and telephone numbers of manufacturers.

Modern Machine Shop

2767. SME CONFERENCE / EXPOSITION: LIST OF EXHIBITORS; ON DISPLAY AT THE SME TOOL EXPO. Gardner Publications Inc. Irregular. v.54, no.12. May 1982. pg.182-262

Alphabetically lists company names and booth numbers for exhibitors at SME's Golden Anniversary Exposition (May 17-20, 1982). Also reviews selected product displays. Data includes product name, booth number, manufacturer's name and address, and product description. Some photographs.

Motor Age

2768. MOTOR AGE TOOLS AND EQUIPMENT ILLUSTRATED BUYERS' GUIDE. Chilton Book Company. Annual. March 1981. pg.39-95

1981 directory of tools and equipment (T & E) used in the automotive repair industry. Includes hand tools; power tools; engine rebuild tools; starting / charging system T & E; tune-up T & E; suspension and brake T & E; body shop T & E; transmission / rear axle T & E; and general service shop T & E. Provides product, model, price where available, features and manufacturer.

National Safety News

2769. SAFETY PRODUCT INDEX: 1981 EDITION. National Safety Council. Annual. March 1981. pg.129-221

Contains product index and brand name index to manufacturers, suppliers and distributors of safety products and devices for plant design, maintenance, and operation; industrial hygiene, first aid and medical treatment; hazard controls; and personal protection.

Nelson Survey of Industry Research

2770. ANALYST VIEWPOINT: OUTLOOK FOR THE ELECTRONICS INDUSTRY. Levine, Elliot. First Manhattan Co. Nelson Publications. Irregular. September 1980. pg.47-61

Electronics industry review by a leading Wall Street analyst. Covers earnings estimates, new products, production growth, industrial plants and equipment, industry capacity, and Japanese competition. Table shows the number of businesses in the industry and 1980 price earnings ratio, profit margin, dividend yield, return on equity, and stock price performance. Lists executives of brokerage firms investment research services and the corporations that they follow.

2771. VIEWPOINT: ELECTRICAL EQUIPMENT INDUSTRY REVIEW. McCoy, Robert W. Kiddir Peabody & Co. Nelson Publications. Irregular. June 1981. pg.43-50

Analysis of markets for electrical products, corporations and economic indicators affecting investments. Table shows number of corporations, price earnings ratio, profit margins, dividends, yields, return on equity, and stock price performance. Also lists corporations, executives, and executives of Wall St. investment research services.

Oil and Gas Journal

2772. OIL AND GAS JOURNAL QUARTERLY COSTIMATING. Farrar, G.L. Penn Well Publishing Company. Quarterly. January 5, 1981. pg.78-79

Quarterly report examines costs for several selected equipment items used in refinery construction and operations, including construction and oil field machinery, motors and generators, switchgear and transformers. Tables show indexes for machinery (1976-1979) and itemized refining cost indexes.

Ontario Business

2773. TOTAL CANADIAN ELECTRONIC MARKET 1979. Ontario Chamber of Commerce. March 1982. pg.30

Lists communications and components sales, 1979; radio and TV (consumer products); office machinery and computers; and instruments and controls. Also includes estimates of types of office automation equipment available, with forecasts for 1985.

Play Meter

2774. TOP VIDEOS AT AMOA. *Bucki, Mike. Skybird Publishing Company Inc. v.8, no.1. January 1982. pg.22-24*

A review of the new products in the video game marketplace. Play Meter reviewers select ten games forecast to be the top games in 1982, from those reviewed at the 1981 AMOA show. Games are ranked by each of five reviewers. Descriptions are included with some discussion of competition in the field and performance of the 'top games' forecast in last year's review.

Playthings

2775. SALES FUTURE LIES WITH ADULTS: FROST AND SULLIVAN SURVEY INDICATES 50 PERCENT OF ELECTRONIC USERS ARE ADULTS. *Geyer-McAllister Publications, Inc. v.80, no.5. May 5, 1982. pg.40*

Summary of Frost and Sullivan survey on electronic toys and games presents buyer / user mix of electronic games. Provides 1980-1981 European sales and graphic illustration of electronic games market.

2776. VIDEO LEADS CHARGE IN GAMES BATTLE: MANUFACTURERS' SHIPMENTS SWELL TO $1.2 BILLION IN 1981. *Leccese, Donna. Geyer-McAllister Publications, Inc. v.80, no.4. April 1982. pg.46-50, 81*

Forecasts 1985 video game market size. Tables include distribution of unit sales by outlet, 1981; distribution of unit sales by age of recipient, 1981; average price, 1981; unit growth, 1980-1981; dollar groth, 1980-1981; and age and sex of purchaser by type of product, 1981.

Power Engineering

2777. POWER ENGINEERING: NEW EQUIPMENT. *Technical Publishing Company. Monthly. v.86, no.4. April 1982. pg.114-124*

Lists new products and equipment related to power engineering. Listing includes brief description of product and/or equipment as well as name and address of manuacturer. Equipment listed according to the following categories: coal, ash handling; coatings; construction, maintenance; cooling towers; electrical; instruments, controls; materials handling; pollution, noise control; pumps, compressors; safety, security; seals, gaskets; values, traps, piping; and water, waste treatment.

Purchasing Magazine

2778. ELECTRONICS: SOME COMPONENT LEADS WILL STRETCH. *Cahners Publishing Company. v.92, no.11. June 10, 1982. pg.19, 23*

Discusses inventories of electronic components, 1982. Gives 1981, 1982 price indexes for resistors (base 1967), linear integrated circuits (base 1974) and capacitors (base 1967).

Purchasing World

2779. SEMICONDUCTORS: A BAD SITUATION WORSENS. *Technical Publishing Company. June 1980. pg.54-58*

Presents purchasing executives' views on shortages of semiconductors in the U.S. Shows 1966-1977 domestic shipments

and 1970-1979 foreign investment equity in domestic manufacturers.

RNM Images

2780. DIGITAL R / F INDUSTRY UPDATE "82: PART 2. *W.G. Holdsworth and Associates. Irregular. v.12, no.2. April 1982. pg.6-10*

Second article in a series summarizing the digital subtraction angiography industry. Listings are organized by manufacturer and include a brief product description, and technical specifications. Listings are indexed numerically to a reader response card.

2781. DIGITAL RADIOGRAPHY 1982: A CLINICAL OVERVIEW. *Stakun, D.J. W.G. Holdsworth and Associates. v.12, no.2. April 1982. pg.18-24*

Presents an overview of the 1982 conference on digital radiography. Topics include analog vs. digital, cardiovascular evaluation, hybrid subtraction, parametric imaging, retrofitting, organ quantitation and ventriculography.

2782. XERORADIOGRAPHY - AN ECONOMICAL ALTERNATIVE TO DIGITAL TECHNOLOGY. *W.G. Holdsworth and Associates. v.12, no.2. April 1982. pg.27-32*

Compares the technical and economic advantages and disadvantages of xeroradiography and computerized digital radiography.

Robotics Today

2783. ROBOTS VI TRIGGERS SPURT IN NEW TECHNOLOGY. *Stauffer, Robert N. Robot Institute of America. Irregular. v.4, no.2. April 1982. pg.45-49*

Product review describes new robotics products demonstrated at the Robots VI Conference. Data is included on manufacturers, end uses, and new technology employed, with technical specifications and selected photographs.

Rubber World

2784. MICROWAVE HEATING ADVOCATED FOR PRODUCTIVITY IMPROVEMENT. *Kastein, Ben. Bill Communications, Inc. v.186, no.1. April 1982. pg.50*

Summarizes papers presented at the Northeast Ohio Rubber Group winter meeting on the advantages of using microwave equipment in rubber processing. Examines improved productivity, reduced energy costs, and technology. A separate paper reviews energy demands with forecasts of total energy demand by source to the year 2000. Energy demand from nuclear, coal, gas, oil, and solar / hydro power is reviewed by percentage, 1980-2000.

Semiconductor International

2785. MAGNETRON SPUTTERING SYSTEMS. *Burggraaf, Pieter S. Cahners Publishing Company. Irregular. v.5, no.10. October 1982. pg.37-52*

Discusses the development of the magnetron cathode and its contribution to semiconductor metallization and other thin-film deposition processes. Table lists vendors and top-of-the-line sputtering systems for production wafer processing. Document provides brief description of products. Photographs also supplied for selected products.

2786. 1979-80 BUYER'S GUIDE. *Milton S. Kiver Publications, Inc. Annual. 1979. pg.81-158*

Buyer's guide for semiconductor manufacturers. Includes manufacturers' addresses and telephone numbers and section on new products.

Solid State Technology

2787. SEMICON/WEST "82: TECHNICAL PROGRAM. *Technical Publishing Company. Annual. v.25, no.4. April 1982. pg.7, 9-13, ff.*

Manufacturers and suppliers exhibiting at the 1982 SEMICON / WEST convention are listed alphabetically with booth numbers. Technical program is also included.

2788. SOLID STATE TECHNOLOGY: NEW PRODUCTS. *Technical Publishing Company. Monthly. November 1981.*

Monthly journal feature provides information on new electronic equipment including photograph, model, brief specifications, and manufacturer. Further information available through reader card service.

2789. SOLID STATE TECHNOLOGY: PATENT REVIEW. *Technical Publishing Company. Monthly. v.25, no.2. February 1982. pg.87-89*

Monthly journal listing of patents for electronic components such as MOS devices and semiconductors. Includes patent number, title, authors and company, and date filed.

2790. SOLID STATE TECHNOLOGY: PATENT REVIEW. *Technical Publishing Company. Monthly. November 1981.*

Monthly journal feature lists patents in the field of solid state electronics. Data includes number, title, author, U.S. claim number, brief description, company, and date.

Transmission and Distribution

2791. TRANSMISSION AND DISTRIBUTION: NEW LITERATURE. *Cleworth Publishing Company, Inc. Monthly. v.34, no.1. January 1982. pg.85-87 $3.00.*

Monthly listing of new electric power transmission / distribution equipment manufacturers' literature. Specifies company address, and types / capacities of products described.

2792. TRANSMISSION AND DISTRIBUTION: NEW PRODUCTS. *Cleworth Publishing Company, Inc. Monthly. v.34, no.1. January 1982. $3.00.*

Monthly survey of new electric power transmission and distribution equipment, and associated excavation / erection machinery, hardware, and instruments. Includes manufacturers' addresses, specifications and photographs.

Utility Purchasing and Stores

2793. ELECTRIC UTILITY BUYERS' REVIEW OF CONSTRUCTION AND MAINTENANCE EQUIPMENT AND TOOLS. *Pritchard Publishing Company. Irregular. May 1980. pg.11-20 $1.00.*

Electric utility buyers' review of construction and maintenance equipment and tools lists latest developments under major headings. Provides names and addresses of companies and sources of supply.

Video Systems

2794. USING VIDEODISC TECHNOLOGY. *Nugent, Ron. Intertec Publishing Corporation. v.8, no.3. May 1982. pg.16-21*

Videodisc technology is reviewed; recent developments and applications are described. Product groups are identified with an explanation of product types and features. Tables present major suppliers and manufacturers for each product group, videodiscs available from major manufacturers with technical specifications, some prices, and notes on advantages and disadvantages; and summary of the forecast for total annual expenditures (1981, 1985, 1990) for videodiscs by type of market (consumer, education, business and industrial training, and information storage and retrieval.)

Welding Design and Fabrication

2795. FORTY-ONE THOUSAND ARC WELDING ROBOTS BY 1995. *Predicasts, Inc. Penton / IPC. Subsidiary of Pittway Corporation. v.55, no.10. October 1982. pg.18*

Gives numbers of arc welding and spot welding robots on line in U.S., 1977, 1980, 1985, 1990, 1995. Includes annual growth rate and total sales, 1985. Gives name and location of robotic engineering society.

See 315, 337, 341, 355, 361, 365, 375, 394, 676, 703, 1483, 1484, 1508, 1537, 1545, 1554, 1593, 1595, 1600, 1636, 1643, 1647, 1667, 1671, 1673, 1679, 1687, 1690, 1691, 1693, 1694, 1695, 1697, 1701, 1706, 1709, 1712, 1755, 1810, 1813, 1815, 2748, 3677, 3678, 3679, 3680, 3681, 3683, 3684, 3694, 3697, 3716, 3728, 3735, 3763.

Numeric Databases

2796. COMPUTERIZED SYSTEM OF THE ADVISORY GROUP ON ELECTRONIC DEVICES (AGED). *Advisory Group on Electronic Devices. (Available from Direct from Producer).*

Provides data on research programs in the area of electron devices supported by the U.S. government. Data extracted from government contractor quarterly and final reports, work unit summaries, foreign contractor reports and other sources. Data base includes 8000 citations to research programs on electron devices of the laser, display microwave, and power types. Publications include monthly government sponsored list, biennial project briefs, and biennial comprehensive list of microwave tubes. Available only to qualified contractors and consultants.

Monographs and Handbooks

2797. AUTOMOTIVE ELECTRICAL EQUIPMENT. *Crouse, William H. Independent Battery Manufacturers Association, Inc. 1979. $20.00.*

Examines storage batteries and maintenance; cranking-motor, generator, and ignition system fundamentals; trouble-shooting; checking and adjustment; and service and testing instruments.

2798. IMPACT OF THE MICROELECTRONICS INDUSTRY ON THE STRUCTURE OF THE CANADIAN ECONOMY. *McLean, J. Michael. Institute for Research on Public Policy. 1979. 50 pg.*

Electronics industry overview includes absence of facilities for large-scale manufacturing of components; and high competitiveness in some subsectors such as telecommunications.

2799. INDUSTRIAL ROBOTS - A SUMMARY AND FORE-CAST FOR MANUFACTURING MANAGERS. *Tech Tran Corporation. 1982. 167 pg. $50.00.*

Report designed to aid production managers, engineers and others in robot selection and end uses. Provides a summary of all aspects of industrial robotics inlcuding the technology of robots, capabilities and applications, economic factors affecting use, eqiupment justification, selection and installations, and a forecast of future development and applications.

2800. INTERCONNECT: WHY AND HOW. *Kuehn, Dick. Telecom Library Inc. 2nd Rev. February 1982. 75 pg. $15.00.*

Contents include voice terminal equipment; types of customer-owned telephone systems; types of interconnect suppliers; system feasibility and equipment features; projected system growth; system features; system traffic; types of switching equipment; maintenance of cost control; cost analysis; interconnect contract checklist; and other topics.

2801. MARKETING POLICY: ELECTRONIC COMPONENTS. *Venture Development Corporation. 1981. 112 pg. $275.00.*

A framework for formulating marketing policy for electronic component companies. Includes management structures, the role of market analysis, distribution channels, and executive selection and compensation guidelines.

2802. MICROELECTRONICS: REPORT OF THE TASK FORCE TO THE GOVERNMENT OF ONTARIO. *Ontario Task Force on Microelectronics. Ontario. Ministry of Industry and Tourism. October 1981. 10 pg.*

Lists series of recommendations relating to the industrial sector. Includes need for policies and programs that will enable Ontario to exploit fully the industrial benefits produced by the microelectronics industry. Considers investments and labor supply.

2803. MICROPROCESSORS AND MICROCOMPUTERS. *Capece, Raymond P. McGraw-Hill Book Company. 1980. 482 pg. $13.95.*

State-of-the-art survey of the latest microelectronic hardware and software. Analyzes recent advances of manufacturers.

2804. REPORTERO INDUSTRIAL. *Keller Publishing Corporation. Spanish. Monthly. November 1980. 36 (est.) pg.*

Monthly survey of new industrial equipment products manufactured by U.S., Western and Eastern European, and other producers. Includes machine tools, hand tools, materials handling equipment, electric generating and transmission equipment. Examines pumps and allied devices, measuring and control instruments, some drilling and earth-moving equipment, and assorted accessories and supplies. General descriptions, photographs and manufacturers' addresses are provided by the manufacturers.

2805. REPORTERO INDUSTRIAL MEXICANO. *Keller Publishing Corporation. Spanish. Monthly. December 1980. 60 (est.) pg.*

Monthly survey of new industrial equipment products manufactured by U.S., Mexican, Western and Eastern European, and other producers. Includes machine tools, hand toóls, materials handling equipment, electric generating and transmission equipment. Examines pumps and allied devices, measuring and control instruments, some drilling and earth-moving equipment, and assorted accessories and supplies. General descriptions, photographs, and manufacturers' addresses are provided by the manufacturers.

2806. SPECIFIER'S GUIDE TO MECHANICAL AND ELECTRICAL PRODUCTS 1981-82. *Cahners Publishing Company. Annual. 7th ed. September 30, 1981. 238 pg. $3.00.*

Annual guide gives a review of developments in each major end use and descriptions of new products in heating, ventilating and air conditioning; energy management systems; plumbing; liquid, gas and air handling systems. Covers fire protection; electrical distribution; motors, starters and controls; power generation; lighting; and security. Contains a manufacturers' index.

2807. SUMMARY OF TRADE AND TARIFF INFORMATION: ELECTRICAL CAPACITORS AND RESISTORS. *Graves, Harold M. U.S. Department of Commerce. International Trade Administration. Irregular. Control no.6-5-3. May 1981. 24 pg.*

Presents current (1981) U.S. tariff schedules fore electrical capacitors and resistors; discusses domestic and foreign market outlook. Reviews value of 1976-1979 domestic shipments and imports and exports (by trading partner); estimates number of U.S. manufacturers.

2808. U.S. INDUSTRIAL COMPETITIVENESS: A COMPARISON OF STEEL, ELECTRONICS, AND AUTOMOBILES. *U.S. Government Printing Office. Y3.T 22/2:2 C73/5. 1981. 206 pg. $7.00.*

Various factors influencing industrial competitiveness are identified. The current status and future prospects of the three industries are evaluated and compared.

2809. U.S. MICROELECTRONICS INDUSTRY. *Hazewindus, Nico. Pergamon Press, Inc. November 1982. 232 pg. $25.00.*

Discusses structure of the microelectronics industry and potential effects of technical change. Contents include examples of integrated circuits and their impact; three application areas in microelectronics; integrated circuits (technology; products; industry forecasts); labor supply problems; microelectronics and employment; and federal government policies.

2810. WORLD INDUSTRIAL REPORTER. *Keller Publishing Corporation. Monthly. November 1980. 24 (est.) pg.*

Monthly survey of new industrial equipment products, manufactured by U.S., Western European, Eastern European, and other producers. Includes machine tools, materials handling equipment, pumps and allied devices, electric generating and transmission equipment. Examines measuring and control instruments, some drilling and earth-moving equipment, and assorted accessories and supplies. General descriptions, photographs, and manufacturers' addresses are provided by the manufacturers.

2811. WORLD INDUSTRIAL REPORTER: ARABIC EDITION. *Keller Publishing Corporation. Arabic/English. Bi-monthly. December 1980.*

Bimonthly, bilingual Arabic / English survey of new industrial equipment products. Includes machine tools, materials handling equipment, pumps and allied devices, electric generating and transmission equipment, measuring and control instruments. Lists some drilling and earth-moving equipment, and assorted accessories and supplies. Entries are provided by U.S., Eastern and Western European, and other manufacturers and include general product description, photographs, and manufacturer's address.

See 399, 402, 1839, 3362.

Conference Reports

2812. AN AMERICAN RESPONSE TO FOREIGN INDUSTRIAL CHALLENGE IN THE HIGH TECHNOLOGY INDUSTRIES. *Semiconductor Industry Association. June 1980. 173 pg. $95.00.*

Conference papers from executives of high technology industries and government agencies. Covers foreign trade, taxes and new technology.

2813. PROCEEDINGS OF THE DOD ELECTRONICS MARKET: FORECAST FOR THE "80'S: VOLUME 1. *Electronic Industries Association. 1980. 230 pg. $55.00.*

Forecast and market analysis conference proceedings for the U.S. electronics industry. Includes new products and government expenditures.

See 1856, 4274.

Newsletters

2814. ELECTRONICS OF AMERICA. *Electronics of America. Weekly.*

Weekly newsletter contains abstracts, and data on recent electronic industry developments.

2815. ICECAP REPORT. *Eklund, Mel H. Integrated Circuit Engineering Corporation. Monthly. 1981. $395.00.*

Monthly newsletter covering the semiconductor industry in the U.S. Examines technical conferences, technology trends, process equipment, new IC structures / circuits, business events, IC applications, vendor profiles and international developments. Includes 1-page flyers on industry trends issued periodically.

2816. MICROELECTRONICS DIGEST. *Girard Associates Inc. Monthly.*

Monthly newsletter provides processed information on tools, techniques, processes, active and passive elements and metallizing in electronics. Covers technical and economic aspects of the microelectronic industry.

2817. NEW AMERICAN ELECTRONICS LITERATURE AND TECHNICAL DATA. *Electronics of America. Monthly.*

Monthly newsletter provides processed information on recent developments and products in the electronics industry.

Services

2818. "AN" EQUIPMENT. *DMS, Inc. Monthly. 1981. $585.00.*

Monthly service analyzing military markets and over 5,000 military electronic systems. Details government contracts, installations, end use applications, and manufacturers.

2819. ELECTRONIC SYSTEMS. *DMS, Inc. Monthly. 1982. $650.00.*

Detailed reports on 100 major U.S. electronic programs for military and civil applications include data on the electronic portions of weapons systems, plus budgets, contractors, systems makeup, sponsoring agency, and forecasts.

See 1875, 1878, 3371.

Indexes and Abstracts

2820. CURRENT PAPERS IN ELECTRICAL AND ELECTRONICS ENGINEERING (CPE). *Institute of Electrical and Electronics Engineers, Inc. Quarterly. 1981.*

Quarterly list of recent documents in the field includes detailed annotations.

2821. VENTURECASTS: THE 1982 ELECTRONIC COMPONENTS DATA BOOK. *Venture Development Corporation. 1982. $595.00.*

Provides abstracts of market analysis and forecasts from 1981 electronics and business periodicals, and associations' and government statistics covering electronic components in the Americas. Includes sales, shipments, revenues, installed base, annual growth rates, consumption, expenditures, market size, production, and imports and exports by product groups.

Central and South America

Market Research Reports

2822. EQUIPMENT AND MATERIALS FOR ELECTRONIC COMPONENTS PRODUCTION, BRAZIL: A COUNTRY MARKET SURVEY. *U.S. Government Printing Office. Irregular. CMS 79-113. February 1979. 7 pg. $.50.*

Report analyzes the current and projected markets for the sale of equipment and materials for electronic components production in Brazil. Data is provided on the total market (production, imports, exports, market size); imports by type, total and amount from U.S.; end user profile of domestic manufacturers; best sales opportunities for U.S. exporters. Describes the production equipment and materials market, the electronic components industry, business practices and tariffs.

See 414, 415.

Financial and Economic Studies

See 3783.

Western Europe

Market Research Reports

2823. ACTIVE COMPONENTS IN EUROPE TO 1986. *Larsen Sweeney Associates Ltd. 1981. 238 pg. 595.00 (United Kingdom).*

Market research report on active components in the following countries: Austria, Belgium, Luxembourg, Denmark, Eire, Finland, France, West Germany, Italy, Netherlands, Norway, Portugal, Spain, Sweden, Switzerland and the U.K. Covers market analysis to 1986, market forecast to 1986, distribution, production, suppliers, market segmentation, employment and unemployment, profits, sales, and end uses. Active components include: valves and tubes, semiconductors, and integrated circuits.

2824. THE AUTOMOTIVE ELECTRONICS MARKET IN EUROPE: VOLUME 1. *Frost and Sullivan, Inc. April 1980. 214 (est.) pg. $1250.00.*

Volume 1 of a two-volume market analysis of electronic components for the auto industry in Western Europe. Forecasts demand to 1985; analyzes manufacturers and industry structure; evaluates competition from U.S. and Japanese producers; investigates end uses and market areas; and covers products and production; 1980-1985.

2825. THE AUTOMOTIVE ELECTRONICS MARKET IN EUROPE: VOLUME 2. *Frost and Sullivan, Inc. April 1980. 214 (est.) pg. $1250.00.*

Volume 2 of a two-volume market analysis for automotive electronics in Western Europe. Covers new products, 1980-1985, industry structure; and manufacturers and subsidiaries.

2826. THE BUSINESS OF PRINTED CIRCUITS: TECHNICAL AND COMMERCIAL FACTORS DETERMINING THE PCB BUSINESS IN EUROPE 1980-1985. *BPA Management Services. May 1981. 104 pg. 425.00 (United Kingdom).*

Structure of the printed circuit board industry in Western Europe and the market, 1980-1985. Gives technical aspects of products including performance, production costs, end uses, manufacturers, and competition from the Far East and the U.S.

2827. CAPACITOR MARKET IN WESTERN EUROPE. *Frost and Sullivan, Inc. September 1981. 254 pg. $1400.00.*

Market analysis for capacitors in Western Europe. Includes analysis of end use industries, product groups, new product trends, industry structure, manufacturers and product lines, competition, and demand forecasts, 1986-1990.

2828. CAPACITORS IN EUROPE TO 1986. *Larsen Sweeney Associates Ltd. 1981. 248 pg. 595.00 (United Kingdom).*

Market research report on capacitors to 1986 in the following countries: Austria, Belgium, Luxembourg, Denmark, Eire, Finland, France, West Germany, Italy, Netherlands, Norway, Portugal, Spain, Sweden, Switzerland and the U.K. Gives market forecast to 1986, distribution, production, suppliers, market segmentation, employment and unemployment, profits, sales, and end users. Types of capacitors include: aluminum electrolytic, paper, mica, plastic, tantalum electrolytic, ceramic, glass and chip.

2829. CAPACITORS IN WESTERN EUROPE: VOLUME 2: CAPACITOR MARKET TRENDS. *Mackintosh Consultants, Inc. January 1980. $8500.00.*

Completed multi-client study analyzes capacitor markets in Western Europe (especially West Germany, the U.K., France and Italy) for the period 1978-1983. Quantity and value of sales are forecast, with breakdowns by capacitor type and end-use industry.

2830. CATHODE RAY TUBES. *Larsen Sweeney Associates Ltd. 1981. 238 pg. 795.00 (United Kingdom).*

Market research report on cathode ray tubes in the following countries: Austria, Belgium, Luxembourg, Denmark, Eire, Finland, France, West Germany, Italy, Netherlands, Norway, Portugal, Spain, Sweden, Switzerland and the U.K. Covers market analysis, forecasts to 1986, production, suppliers, market segmentation, employment and unemployment, profits, sales and end users of all types of cathode ray tubes.

2831. CHIPS IN THE 1980'S: THE APPLICATION OF MICROELECTRONIC TECHNOLOGY IN PRODUCTS FOR CONSUMER AND BUSINESS MARKETS. *Economist Intelligence Unit Ltd. June 1979. 58 pg.*

Discusses development of the microprocessor and its applications in office equipment, and consumer electronics fields. Projections and past statistics (1977-1985) are given in all markets, and names of suppliers are given. Examines future trends.

2832. COILS IN EUROPE TO 1986. *Larsen Sweeney Associates Ltd. 1981. 239 pg. 695.00 (United Kingdom).*

Market report to 1986 on coils including aerials, crossovers, deflection plate, field, flyback, focusing, line loading, magnet, oscillator, pick-up, PPI deflection, radio frequency, saturable reactor, scanning, solenoid, subminiature, toroidal and transformer in the following countries: Austria, Belgium, Luxembourg, Denmark, Eire, Finland, France, West Germany, Italy, Netherlands, Norway, Portugal, Spain, Sweden, Switzerland and the U.K. Covers forecasts to 1986, market analyses, production, market segmentation, sales, distribution, employment and unemployment, profits, end users, and suppliers.

2833. COLOUR TV TUBES. *Larsen Sweeney Associates Ltd. 1981. 216 pg. 595.00 (United Kingdom).*

Market research report on color TV tubes in the following countries: Austria, Belgium, Luxembourg, Denmark, Eire, Finland, France, West Germany, Italy, Netherlands, Norway, Portugal, Spain, Sweden, Switzerland and the U.K. Covers forecasts to 1986, distribution, production, suppliers, market segmentation, employment and unemployment, profits, sales, and end users. Miniature tubes of nine sizes are discussed and there is a separate report for monochrome TV tubes.

2834. COMPONENTS IN EUROPE TO 1986. *Larsen Sweeney Associates Ltd. 1981. 242 pg. 595.00 (United Kingdom).*

Market research report on components such as valves and tubes, semiconductors, integrated circuits, passive components, audio components, controls, sensors, actuators and consumer electronics components. Covers forecasts for markets to 1986, distribution, production, suppliers, market segmentation, employment and unemployment, profits, sales and end uses. Covers the following countries: Austria, Belgium, Luxembourg, Denmark, Eire, Finland, France, West Germany, Italy, Netherlands, Norway, Portugal, Spain, Sweden, Switzerland and U.K.

2835. CONNECTORS IN EUROPE TO 1986. *Larsen Sweeney Associates Ltd. 1981. 231 pg. 695.00 (United Kingdom).*

Market research on connectors in the following countries: Austria, Belgium, Luxembourg, Denmark, Eire, Finland, France, West Germany, Italy, Netherlands, Norway, Portugal, Spain, Sweden, Switzerland, and the U.K. Gives forecasts to 1986, distribution, production, suppliers, market segmentation, employment and unemployment, profits, sales and end uses. 28 connector product groups are analyzed.

2836. CONSUMER ELECTRONICS IN THE U.K. - NOW AND THE 80'S. *Acumen-System Three. January 1980. 375.00 (United Kingdom).*

Analysis of consumer electronics in the United Kingdom. Includes sections on ownership levels for different products, market size and development, distribution patterns, product group analysis, brands, views of the retail trade, trade strategy, and major technical trends.

2837. DATA ACQUISITION COMPONENTS MARKET IN W. EUROPE. *Frost and Sullivan, Inc. December 1981. 228 pg. $1500.00.*

Analysis of the data acquisition components market in Western Europe categorizes and defines the components under study; and differentiates between parts, modules, hardware and software, and stand-alone data loggers and components used in standard data acquisition systems. Provides insight into areas of opportunity created by technological change; analyzes West European current and expected demand for DAC by forecasting consumption and growth rates within five national markets and for eight component types, 19 end-user segments and 11 applications to 1985; and assesses current and expected supply by providing a West European production forecast by national markets and by product type to 1985. Examines the DAC import / export picture. Discusses current and expected applications and economic bases upon which the DAC market will grow. Presents a competitive analysis of suppliers. Covers analog to digital converters, multiplexes, sample / hold circuits, operational amplifiers, isolation amplifiers and interface circuits.

2838. THE DATA ACQUISITIONS SYSTEMS MARKET IN W. EUROPE. *Frost and Sullivan, Inc. May 1981. 215 pg. $1250.00.*

Containsanalysis and production / consumption forecast to 1990 for signal modules (multiplexers, converters, conditioners); support modules (alarms, data highway, interfaces, output, power suppliers, and recording devices); and data organization (scanners, loggers, rack panel, controllers, processors, and software) in W. Germany, U.K., France, Benelux, and other European countries. Analyzes end-users and applications and state of the art. Profiles major suppliers and considers imports and exports.

2839. DISCRETE SEMICONDUCTORS IN EUROPE TO 1986. *Larsen Sweeney Associates Ltd. 1981. 229 pg. 595.00 (United Kingdom).*

Market research report on discrete semiconductors to 1986 in Austria, Belgium, Luxembourg, Denmark, Eire, Finland, France, West Germany, Netherlands, Norway, Portugal, Spain, Sweden, Switzerland and the U.K. Covers forecasts to 1986 and distribution, production, suppliers, market segmentation, employment and unemployment, profits, sales, and end users. Markets covered include diodes, indicating devices, rectifiers, throysistors, and transistors.

2840. DMS (DEFENSE MARKETING SERVICES) MARKET INTELLIGENCE REPORT: NATO WEAPONS. *DMS, Inc. Monthly. 1980.*

Market analysis of more than 80 NATO procurement and R&D programs that affect the drive for NATO standardization. Updated monthly. Index lists reports in 6 sections: military aircraft, warships, missiles, vehicles ordinance and electronic systems. Each report includes designation, country, executive, contractor, status, mission, type, characteristics, performance, timetable, and funding. Additional information on background material, current activity and DMS's forecasts provided. Lists of major contractors included.

2841. EEC MARKETS FOR FRACTIONAL HP MOTORS. *Frost and Sullivan, Inc. March 1982. $1500.00.*

Market report includes forecasts of FHP motors by size and by type (AC, DC, stepper), by country. Profiles top 20 suppliers with market share data. Includes industry structure and assessment of Far Eastern competition.

2842. ELECTRICAL SECTOR RESEARCH. *Laing and Cruickshank. Monthly. 1981.*

Monthly review gives trends, forecasts and market analysis for the U.K. electrical industry.

2843. ELECTRONIC COMPONENT DISTRIBUTION: A STRATEGIC MARKET ANALYSIS: VOLUME 1: FRANCE. *Mackintosh Publications Ltd. 1981. $750.00.*

Market survey of electronic component distributors in the main European markets examines local conditions and assesses growth of various market segments. Surveys availability of products from stock and meeting delivery dates. Examines current performance of distributors and projects trends to 1983.

2844. ELECTRONIC COMPONENT DISTRIBUTION: A STRATEGIC MARKET ANALYSIS: VOLUME 2: GERMANY. *Mackintosh Publications Ltd. 1981. $750.00.*

Market survey of electronic component distributors in the main European markets examines local conditions and assesses growth of various market segments. Surveys availability of products from stock and meeting delivery dates. Examines current performance of distributors and projects trends to 1983.

2845. ELECTRONIC COMPONENT DISTRIBUTION: A STRATEGIC MARKET ANALYSIS: VOLUME 3: UNITED KINGDOM. *Mackintosh Publications Ltd. 1981. $750.00.*

Market survey of electronic component distributors in the main European markets examines local conditions and assesses growth of various market segments. Surveys availability of products from stock and meeting delivery dates. Examines current performance of distributors and projects trends to 1983.

2846. ELECTRONIC COMPONENT DISTRIBUTORS. *Keynote Publications Ltd. 1st ed. September 1979. 10 pg.*

Market report on electronic component distributors in the U.K. Gives industry structure, market size and trends, sales through distributors in the U.K., imports, recent developments, and future prospects. Financial data for 1978 given for selected companies.

2847. ELECTRONIC COMPONENTS. *Keynote Publications Ltd. 1st ed. February 1979. 8 pg.*

Market report on electronic components in the U.K. gives industry structure, market background, recent developments, future prospects, and financial data, 1976-1978, for selected British companies. Shows employees, 1972-1976, exports 1976-1978, and imports, 1976-1978.

2848. THE ELECTRONIC COMPONENTS MARKET IN BELGIUM. *Systec. U.S. Department of Commerce. International Trade Administration. NTIS #ITA 81-11-501. July 1981. 54 pg. $15.00.*

Survey of the market for electronic components in Belgium describes the market size and sources of supply, end users, and market access. Gives data on the current and projected total market by major product category, by country of origin. Lists principal domestic, U.S., and third country component manufacturers. Gives sales by major component types and analyzes the market for each. Profiles end user industries.

2849. THE ELECTRONIC COMPONENTS MARKET IN THE NETHERLANDS. *Systec Consultants Ltd. U.S. Department of Commerce. International Trade Administration. NTIS #ITA 81-11-507. July 1981. 83 pg. $15.00.*

Provides data on current and projected (1985) market size by major subcategories, current and projected (1985) imports by major subcategories by country of origin, sales by products (1979) and by end users (1979-1980 and projected 1985), and size of end user industries. Lists U.S., domestic, and third country components manufacturers trading in the Netherlands. The market for individual products is analyzed and profiles are provided of major end user industry sectors. Business practices, tariffs, and standards are described. The trade lists include: potential buyers of U.S. electronic components; distributors and agents; trade associations; and journals and magazines.

2850. THE ELECTRONIC COMPONENTS MARKET IN THE UNITED KINGDOM. *Systec Consultants Ltd. U.S. Department of Commerce. International Trade Administration. NTIS #ITA 81-08-508. March 1981. 87 pg. $15.00.*

Survey of the market for electronics components in the United Kingdom provides information on total market size (production, imports, exports) by suppliers and product category divisions. Provides data on sales for individual products and by end user industry. End user industry analysis provides profiles on each sector, containing outlook for purchases and an assessment of supplier countries and manufacturers. Business practices, tariffs, and standards are covered. Trade lists cover potential buyers of U.S. electronic components; distributors and agents; trade associations; journals and magazines.

2851. ELECTRONIC COMPONENTS, NORWAY: A COUNTRY MARKET SURVEY. *U.S. Government Printing Office. Irregular. CMS 79-012. March 1979. 11 pg. $.50.*

Report analyzes the current and projected markets for the sale of electronic components in Norway. Data is provided on the total market (production, imports, exports, market size); manufacturers' sales of selected products to domestic and export markets; current number of businesses and estimated purchases by major end-user sectors; and best sales opportunities for U.S. exporters. Describes the domestic market and competitive environment, and provides an end-user analysis and information on marketing practices and trade and technical regulations.

2852. ELECTRONIC GAMES MARKET IN EUROPE. *Frost and Sullivan, Inc. September 1981. 359 pg. $1425.00.*

Analyzes and forecasts the electronic games market in Europe through 1989. Assesses state-of-the-art and analyzes cost - price projections for electronic games by type of product. Appraises production techniques. Estimates market size, and forecasts future shipments by each major application usage. Evaluates market position, product lines and marketing strategies.

2853. ELECTRONIC INFORMATION DISPLAYS: TECHNOLOGY AND MARKET OPPORTUNITIES FOR FLAT PANEL DISPLAYS IN WESTERN EUROPE. *Mackintosh Publications Ltd. June 1981. 11500.00 (United Kingdom).*

This multi-client study analyzes European markets for light emitting diodes, liquid crystal displays, and other electronic information displays, through 1986. Forecasts are presented in the context of emerging new applications (automobile instrument panels, banking and hand-held terminals, electronic toys, etc.), and technological developments in the U.S., Japan and Europe. Focus is on U.K., France and West Germany. Includes discussion of competitive position of major manufacturers, inter-technology competition, and price trends.

2854. ELECTRONIC INSTRUMENTS. *Keynote Publications Ltd. 1st ed. February 1979. 15 pg.*

Market report on electronic instruments in the U.K. gives industry structure, market background, recent developments, future prospects, financial data, and financial appraisal for selected U.K. companies and agents.

2855. ELECTRONICS. *Hoare and Govett Ltd. October 1979. 75 pg. 40.00 (United Kingdom).*

Reviews the U.K. electronics industry including analyses of telecommunications equipment; electronic computers; elec-

tronic components; electronic capital goods; and consumer electronics. The prospects for six major corporations are assessed.

2856. ELECTRONICS COMPONENTS SECTOR WORKING PARTY: PROGRESS REPORT 1980. *National Economic Development Office. Annual. 1980. 12 pg.*

1980 report on electronic components in the U.K., giving forecasts to 1984 on sales, imports and exports, and an analysis of the market in 1980.

2857. EMERGENCY LIGHTING SYSTEMS IN EUROPE TO 1986. *Larsen Sweeney Associates Ltd. Quarterly. 1981. 213 pg. 595.00 (United Kingdom).*

Market report on emergency lighting equipment in Europe. Covers market analysis, forecasts to 1986, sales, production trends, profits, assets, end users, employment and unemployment, products, capital expenditures, investments and distribution. Updated quarterly.

2858. EQUIPMENT AND MATERIALS FOR ELECTRONIC COMPONENTS PRODUCTION, DENMARK: A COUNTRY MARKET SURVEY. *U.S. Government Printing Office. Irregular. CMS 79-107. February 1979. 9 pg. $.50.*

Analyzes the current and projected markets for the sale of equipment and materials for electronic components production in Denmark. Data is provided on the total market (production, imports, exports, market size); profile of domestic manufacturers; best sales opportunities for U.S. exporters. Describes the production equipment and materials market, the electronic components industry, business practices and tariffs.

2859. EQUIPMENT AND MATERIALS FOR ELECTRONIC COMPONENTS PRODUCTION, GERMANY: A COUNTRY MARKET SURVEY. *U.S. Government Printing Office. Irregular. CMS 79-104. February 1979. 9 pg. $.50.*

Report analyzes the current and projected markets for the sale of equipment and materials for electronic components production in Germany. Data is provided on the total market (production, imports, exports, market size); imports by type, total and amount from U.S.; end-user profile of domestic manufacturers; best sales opportunities for U.S. exporters. Describes the production equipment and materials market, the electronic components industry, business practices and tariffs.

2860. EQUIPMENT AND MATERIALS FOR ELECTRONIC COMPONENTS PRODUCTION, ITALY: A COUNTRY MARKET SURVEY. *U.S. Government Printing Office. Irregular. CMS 79-108. February 1979. 10 pg. $.50.*

Report analyzes the current and projected markets for the sale of equipment and materials for electronic components production in Italy. Data is provided on the total market (production, imports, exports, market size); imports by type, total and amount from U.S.; end-user profile of domestic manufacturers; best sales opportunites for U.S. exporters. Describes the production equipment and materials market, the electronic components industry, business practices and tariffs.

2861. EQUIPMENT AND MATERIALS FOR ELECTRONIC COMPONENTS PRODUCTION, NORWAY: A COUNTRY MARKET SURVEY. *U.S. Government Printing Office. Irregular. CMS 79-101. February 1979. 8 pg. $.50.*

Analyzes the current and projected markets for the sale of equipment and materials for electronic components production in Norway. Data is provided on the total market (production, imports, exports, market size); imports by type, total and amount from U.S.; end-user profile of domestic manufacturers; and best sales opportunities for U.S. exporters. Describes the production equipment and materials market, the electronic components industry, and business practices and tariffs.

2862. EQUIPMENT AND MATERIALS FOR ELECTRONIC COMPONENTS PRODUCTION, SPAIN: A COUNTRY MARKET SURVEY. *U.S. Government Printing Office. Irregular. CMS 79-103. February 1979. 17 pg. $.50.*

Report analyzes the current and projected markets for the sale of equipment and materials for electronic components production in Spain. Data is provided on the total market (production, imports, exports, market size); imports by type, total and amount from U.S.; end-user profile of domestic manufacturers; best sales opportunities for U.S. exporters. Describes the production equipment and materials market, the electronic components industry, business practices and tariffs.

2863. EQUIPMENT AND MATERIALS FOR ELECTRONIC COMPONENTS PRODUCTION, SWEDEN: A COUNTRY MARKET SURVEY. *U.S. Government Printing Office. Irregular. CMS 79-112. February 1979. 10 pg. $.50.*

Report analyzes the current and projected markets for the sale of equipment and materials for electronic components production in Sweden. Data is provided on the total market (production, imports, exports, market size); imports by type, total and amount from U.S.; end-user profile of domestic manufacturers; best sales opportunities for U.S. exporters. Describes the production equipment and materials market, the electronic components industry, business practices and tariffs.

2864. EQUIPMENT AND MATERIALS FOR ELECTRONIC COMPONENTS PRODUCTION, UNITED KINGDOM: A COUNTRY MARKET SURVEY. *U.S. Government Printing Office. Irregular. CMS 79-102. February 1979. 7 pg. $.50.*

Analyzes the current and projected markets for the sale of equipment and materials for electronic components production in the United Kingdom. Data is provided on the total market (production, imports, exports, market size); imports by type, total and amount from U.S.; end-user profile of domestic manufacturers; and best sales opportunities for U.S. exporters. Describes the production equipment and materials market, the electronic components industry, and business practices and tariffs.

2865. ERC MARKET MONITOR STUDIES: LIGHTING EQUIPMENT. *E.R.C. Statistics International Ltd. Annual. November 1981. 1800.00 (United Kingdom).*

Market analysis for lighting equipment in Western Europe. Shows consumption, sales, production, retail distribution and prices, domestic and foreign trade, and brand names, 1975-1981. Also examines consumer behavior and attitudes, advertising, industry structure, manufacturers, imports and exports, and new products.

2866. EUROPEAN BATTERIES AND ELECTRIC VEHICLES. *Predicasts, Inc. July 1979. $325.00.*

Reviews European industry structure, new technologies and applications for batteries and electric vehicles. Historical and projected markets included for motor vehicle batteries, primary and other secondary batteries.

2867. THE EUROPEAN ELECTRONICS INDUSTRY 1982-1984. *Venture Development Corporation. 1982.*

Analysis and forecasts for the European electronics industry, 1982-1984. Covers production, consumption, imports, exports, and market structure by country. Profiles major manufacturers.

2868. EUROPEAN INTEGRATED CIRCUITS MARKETS. *Admerca AG. 1981. 500 pg. $16500.00.*

Market analysis for integrated circuits in Europe. Includes for-

eign trade, production, demand, competition, industrial plants and equipment, distribution, prices, end uses, and suppliers.

2869. FILTERS, NETWORKS, AND DELAY LINES IN EUROPE TO 1986. *Larsen Sweeney Associates Ltd. 1981. 795 pg.*

Market research report on filters, networks and delay lines to 1986 in Austria, Belgium, Luxembourg, Denmark, Eire, Finland, France, West Germany, Italy, Netherlands, Norway, Portugal, Spain, Sweden, Switzerland and the U.K. Covers forecasts to 1986 and distribution, production, suppliers, market segmentation, employment and unemployment, profits, sales, and end users. 28 product types are analyzed.

2870. FLEXIBLE CIRCUITRY IN WESTERN EUROPE. *BPA Management Services. March 1981. 117 pg. 800.00 (United Kingdom).*

Market analysis for flexible circuits in Western Europe with forecasts to 1985. Gives value of markets by product type, country, and end use sectors. Covers production costs and technologies.

2871. THE FUTURE OF ELECTRICAL GOODS DISTRIBUTION IN FRANCE UP TO 1990. *Metra Consulting Group Ltd. 1981. 70000.00 (France).*

The study aims to define and describe the primary factors in the development of France's electrical goods distribution system; to rank these factors in order of importance and to evaluate their interaction; and to determine the most likely course of developments in distribution through 1990. Scope includes products which can be distributed by electrical goods wholesalers: domestic appliances, electronic consumer goods, electric heating, electrical equipment, lighting, cable and wire, industrial electronics, accessories, other products.

2872. HEATING, COOLING AND LIGHTING PRODUCT MARKETS FOR DOMESTIC, COMMERCIAL AND INDUSTRIAL BUILDINGS IN EUROPE. *Frost and Sullivan, Inc. April 1982. 346 pg. $1450.00.*

Analysis of European markets for heating, cooling and lighting equipment. Includes installed base of commercial buildings and housing units, 1980-1981; market value forecasts by product groups and end uses, 1982-1983, 1985, 1987; installations of equipment, 1972-1980; new building construction, 1964-1978; market and economic indicators by country; industry structure, manufacturers, and suppliers; and distribution channels. Lists addresses of businesses.

2873. HIGH DENSITY INTERCONNECTION - THE FUTURE OF THE PRINTED CIRCUIT. *BPA Management Services. 1980. 259 pg. 2200.00 (United Kingdom).*

Report on high density interconnection technologies in Western Europe and the future of the printed circuit. Covers products and processes, performance, production costs and competition among the options. Analyzes demand including market structure, military markets and other end uses.

2874. HYBRID CIRCUITS MARKETS (EUROPE). *Frost and Sullivan, Inc. February 1982. $1500.00.*

Market report provides analysis and sales forecasts through 1986 for: hybrid substrates, components, , connection materials; thick film substrates; and custom thick and thin film circuits. Discusses technological advancements. Includes company profiles.

2875. HYBRID IC'S IN EUROPE TO 1986. *Larsen Sweeney Associates Ltd. 1981. 235 pg. 695.00 (United Kingdom).*

Market research report on hybrid IC's to 1986 in Austria, Belgium, Luxembourg, Denmark, Eire, Finland, France, West Germany, Italy, Norway, Netherlands, Portugal, Spain, Sweden,

Switzerland and the U.K. Gives forecasts to 1986 and covers distribution, production, suppliers, market segmentation, employment and unemployment, profits, sales, and end users. Seven major product areas are covered and eleven application sectors.

2876. IC PACKAGING AND EQUIPMENT INTERCONNECT PRACTICE. *BPA Management Services. July 1982. 247 pg. (2 vols.) $3900.00.*

Report on effects of new technology for mounting electronic chips examines the industrial plants and equipment, production costs, and manufacturers. Forecasts world consumption of chips, 1985, 1990.

2877. IC'S AND MICROELECTRONICS IN EUROPE TO 1986. *Larsen Sweeney Associates Ltd. 1981. 231 pg. 595.00 (United Kingdom).*

Market report to 1986 on IC's and microelectronics covering product groups: hybrid IC's, linear IC's, logic circuits, memory circuits, and microprocessors in the following countries: Austria, Belgium, Luxembourg, Denmark, France, Finland, West Germany, Netherlands, Italy, Norway, Portugal, Spain, Sweden, Switzerland and the U.K. Gives forecasts to 1986, sales, production, market segmentation, distribution, employment and unemployment, profits, end uses, and suppliers.

2878. INDUSTRIAL X-RAY. *Larsen Sweeney Associates Ltd. 1981. 245 pg. 795.00 (United Kingdom).*

Market report to 1986 in industrial x-ray covering low, medium, high and super voltage equipment in the following countries: Austria, Belgium, Luxembourg, Denmark, Eire, Finland, France, West Germany, Italy, Netherlands, Norway, Portugal, Spain, Sweden, Switzerland and the U.K. Deals with market analyses, forecasts to 1986, production, market segmentation, sales, distribution, employment and unemployment, profits, end users and suppliers.

2879. INTEGRATED CIRCUIT PACKAGING AND INTER-CONNECTION. *BPA Management Services. September 1981. 2500.00 (United Kingdom).*

Market analysis report on new products for integrated circuit packaging and interconnection, 1981-1990. Covers competition with existing technolgoies, market shares, performance, producton costs, demand forecasts 1981-1986, and end uses. Includes military markets, raw material consumption, and the structure of the interconnection industry in Western Europe.

2880. INTEGRATED INDUSTRIAL FIRE AND ENERGY SYS-TEMS MARKETS IN EUROPE. *Frost and Sullivan, Inc. March 1982. 305 pg. $1400.00.*

Contains 1980-1985 annual estimates and 1991 forecasts for markets in W. Germany, France, U.K., Italy, Benelux, Sweden and rest of Europe. Subdivided by system-fire protection, security; fire security / energy; and integrated total communications in three types of industrial end user. Provides product trends including demand analysis and supplier company profiles.

2881. KEY NOTE REPORTS: ELECTRONIC COMPONENTS. *Keynote Publications Ltd. 1981. 20.00 (United Kingdom).*

A report on electronic components in the U.K. covering the structure of the industry, distribution networks, major manufacturers, markets, and finances. Analysis includes products by sales volume and value, market shares, foreign trade, new products, mergers, legislation, financial analysis, operating ratios, and forecasts of trends.

2882. KEY NOTE REPORTS: PRINTED CIRCUITS. *Keynote Publications Ltd. 1981. 20.00 (United Kingdom).*

A report on printed circuits in the U.K. covering the structure of the industry, distribution networks, major manufacturers, markets, and finances. Analysis includes products by sales volume and value, market shares, foreign trade, new products, mergers, legislation, financial analysis, operating ratios, and forecasts of trends.

2883. LAMINATES FOR PRINTED CIRCUITRY: ARE THEY SATISFACTORY FOR FUTURE INTERCONNECTION TECHNOLOGIES? *Donnelly, Ted, editor. BPA Management Services. August 1981. 2500.00 (United Kingdom).*

Market analysis for laminated boards for printed circuits in Western Europe, with forecasts to 1985. Covers technologies and new products, markets by country and end use sectors, raw materials consumption, and foreign market comparisons.

2884. LINEAR IC'S IN EUROPE TO 1986. *Larsen Sweeney Associates Ltd. 1981. 232 pg. 695.00 (United Kingdom).*

Market report to 1986 on linear IC's covering 16 types in Austria, Belgium, Luxembourg, Denmark, Eire, Finland, France, West Germany, Italy, Netherlands, Norway, Portugal, Spain, Sweden, Switzerland and the U.K. Gives market analysis, forecasts to 1986, production, market segmentation, sales, distribution, employment and unemployment, profits, end users and suppliers.

2885. LOGIC CIRCUITS IN EUROPE TO 1986. *Larsen Sweeney Associates Ltd. 1981. 248 pg. 795.00 (United Kingdom).*

Market report to 1986 on logic circuits in Austria, Belgium, Luxembourg, Denmark, Eire, Finland, France, West Germany, Italy, Netherlands, Norway, Portugal, Spain, Sweden, Switzerland and the U.K. Ten types of logic circuits are covered. Gives market analysis and forecasts to 1986, production, market segmentation, sales, distribution, employment and unemployment, profits, end users and suppliers.

2886. MACHINE TOOL CONTROL EQUIPMENT IN EUROPE TO 1986. *Larsen Sweeney Associates Ltd. 1981. 229 pg. 695.00 (United Kingdom).*

Market report to 1986 on machine tool control equipment in Austria, Belgium, Luxembourg, Denmark, Eire, Finland, France, West Germany, Italy, Netherlands, Norway, Portugal, Spain, Sweden, Switzerland and the U.K. Twenty-two types of machine control equipment are covered. Gives market analysis, forecasts to 1986, production, market segmentation, sales, distribution, employment and unemployment, profits, end users, and suppliers.

2887. THE MARKET FOR ELECTRICAL PRODUCTS IN GREECE. *Goulden, O.A. O.A. Goulden and Partners. August 1981. 90 pg. 150.00 (United Kingdom).*

Analysis of markets for electrical products in Greece, including motors and alternators, cables, transformers, switches and controls, and accessories. Covers the markets; power generation; industry structure; breakdown by product groups; manufacturers; distribution; and sales, foreign trade, and production statistics.

2888. THE MARKET FOR ELECTRICAL PRODUCTS IN PORTUGAL. *Goulden, O.A. O.A. Goulden and Partners. January 1981. 82 pg. 150.00 (United Kingdom).*

Analysis of the power generation industry in Portugal and markets, by product groups, including generation and transmission equipment, cables, motors and alternators, and controls. Profiles manufacturers; compares statistics for sales, consumption, production, exports, and imports; summarizes foreign trade situation; and examines sales and distribution of electricity.

2889. THE MARKET FOR ELECTRICAL PRODUCTS IN SPAIN. *Goulden, O.A. O.A. Goulden and Partners. May 1981. 100 pg. 150.00 (United Kingdom).*

Analysis of the structure of the power generation industry in Spain and the markets for products including alternators and motors, cables, transformers, switchgear, controls, and accessories. Includes profiles of manufacturers and statistics showing consumption, production, foreign trade, production growth, and costs.

2890. MEMORY CIRCUITS IN EUROPE TO 1986. *Larsen Sweeney Associates Ltd. 1981. 234 pg. 695.00 (United Kingdom).*

Market report to 1986 on 14 types of memory circuits in Austria, Belgium, Luxembourg, Denmark, Eire, Finland, France, West Germany, Italy, Netherlands, Norway, Portugal, Spain, Sweden, Switzerland and the U.K. Gives forecasts to 1986, production, market segmentation, sales, distribution, employment and unemployment, profits, end users and suppliers.

2891. MICROELECTRONICS: THE VIEWS OF SENIOR BRITISH MANAGEMENT 1979. *Market and Opinion Research International. December 1979. 20.00 (United Kingdom).*

This report is based on 754 interviews with British managers in industry. Covers managers' knowledge of micro-electronics technology and sources of information; expertise in their companies; seminars and workshops attended; and implementation of the new technology.

2892. MICROPROCESSORS IN MANUFACTURED PRODUCTS. *Northcott, J. Policy Studies Institute. November 1980. 48 pg. 3.25 (United Kingdom).*

Study assesses the form, extent and effects of microprocessor applications and appraises the government's support measures in the U.K. Based on case studies of 90 manufacturing companies in 5 industries: electrical appliances, heating and ventilating, cars, toys and games, and physical test instruments.

2893. MICROPROCESSORS IN WESTERN EUROPE. *Creative Strategies International. March 1982. $1450.00.*

Provides analysis of Western European markets for microprocessors and the industry structure. Forecasts unit sales and dollar sales by end uses and market areas, 1975-1985. Examines mergers and foreign ownership in Europe, product mix, research and development of new technologies and products, price trends, and competition for market shares among foreign and European manufacturers.

2894. MICROWAVE DIODES IN EUROPE TO 1986. *Larsen Sweeney Associates Ltd. 1981. 230 pg. 595.00 (United Kingdom).*

Market report to 1986 on microwave diodes, mixers, oscillators, switching and video in Austria, Belgium, Luxembourg, Denmark, Eire, Finland, France, West Germany, Italy, Netherlands, Norway, Portugal, Spain, Sweden, Switzerland and the U.K. Gives forecasts to 1986, production, market segmentation, sales, distribution, employment and unemployment, profits, end users and suppliers.

2895. MICROWAVE VALVES IN EUROPE TO 1986. *Larsen Sweeney Associates Ltd. 1981. 247 pg. 695.00 (United Kingdom).*

Market research report with forecast to 1986 on microwave valves in Austria, Belgium, Luxembourg, Denmark, Eire, Finland, France, West Germany, Italy, Netherlands, Norway, Por-

tugal, Spain, Sweden, Switzerland and the U.K. Covers eleven product groups individually. Gives market analyses, production, market segmentation, sales, distribution, employment and unemployment, profits, end uses, and suppliers.

2896. THE MILITARY AND AVIATION DISPLAY MARKET IN WESTERN EUROPE. *Frost and Sullivan, Inc. October 1980. 311 pg. $1250.00.*

Analysis and forecast (1980-1995) of display requirements in 17 countries and 12 end-use applications to 1990. Coverage by technology (i.e., C/CRT, LED, LCD, PDP, lasers, electroluminescent, electrolytic, electrochromatic). Examines display technology and expected evolvement to 1995; pinpoints high dollar programs and requirements; and reviews products and capabilites of all European industry suppliers.

2897. MILITARY AND COMMERCIAL MICROWAVE PRODUCTS IN WESTERN EUROPE. *Frost and Sullivan, Inc. November 1980. $1000.00.*

Analysis and forecast to 1990 of microwave tubes, solid-state discrete devices, passive components, integrated circuits, and solid-state systems for applications in radar, nav-aids, EW, communications and consumers. Includes industry and technological trends, company profiles, and data on Western European economics and tariff structures.

2898. MILITARY ELECTROMECHANICAL COMPONENTS MARKET IN WEST EUROPE. *Frost and Sullivan, Inc. December 1981. 314 pg. $1500.00.*

Examines the military electromechanical components market in West Europe. Defines and establishes the present day market; overviews national and NATO policies; surveys state-of-the-art technology and trends in the market, and discusses key military end-equipment programs now in development and assesses the significance of their impact on the electromechanical component market. Profiles and competitively ranks both major electromechanical component suppliers and end-users; and provides a forecast of European military demand.

2899. MILITARY SEMICONDUCTOR MARKET IN EUROPE. *Frost and Sullivan, Inc. February 1981. 248 pg. $1350.00.*

Analysis and forecast to 1986 of military semiconductor sales in 13 western European nations by device type, including discretes, mos, digital IC's, bipolar digital, linear, hybrids, other IC's, and optoelectronic devices. Includes future technological and marketing trends, projected defense electronic systems expenditures, profiles, and competitive standings of device suppliers and OEM equipment manufacturers (users).

2900. MONOCHROME TV TUBES IN EUROPE TO 1986. *Larsen Sweeney Associates Ltd. 1981. 216 pg. 595.00 (United Kingdom).*

Market research report with forecasts to 1986 on production, market segmentation, sales, distribution, employment and unemployment, profits, end users and suppliers. Covers nine types of monochrome TV tubes in Austria, Belgium, Luxembourg, Denmark, Eire, Finland, France, West Germany, Italy, Netherlands, Norway, Portugal, Spain, Sweden, Switzerland and the U.K.

2901. THE NON-ENTERTAINMENT AUTOMOTIVE ELECTRONICS MARKET IN EUROPE. *Frost and Sullivan, Inc. June 1980. 428 pg. $1250.00.*

Market analysis and forecast to 1985 of national end-user markets in West Germany, France, the United Kingdom, Italy, Spain, and other markets. Includes five-year automobile production forecasts by country, technological market trends (power management, engine management, safety related systems, display and convenience systems) and principal automotive electronic manufacturers and their strategies.

2902. OPTOELECTRONIC COMPONENTS MARKET IN EUROPE. *Frost and Sullivan, Inc. July 1981. 346 pg. $1350.00.*

Market analysis for optoelectronic components, including fiber optic cables and connectors, in Western Europe. Examines economic indicators by country, 1978-1976; imports, exports, and tariffs; structure of demand by end uses; new product trends; and production forecasts to 1986 for major manufacturers, product lines, and sales.

2903. OPTOELECTRONICS IN EUROPE. *Admerca AG. 1981. 100 pg. $14000.00.*

Continuous service covering LED's and other new optoelectronic products in Europe. Includes demand production, competition, and market analysis.

2904. PANEL METERS IN EUROPE TO 1986. *Larsen Sweeney Associates Ltd. 1981. 216 pg. 595.00 (United Kingdom).*

Market research report with forecasts to 1986 on the four main categories of panel meters. The following areas of industry application are covered: communications, telecommunications, consumer, EDP, industrial, instrumentation, military / navigation, custom / specials and others. Covers Austria, Belgium, Luxembourg, Denmark, Eire, Finland, France, West Germany, Italy, Netherlands, Norway, Portugal, Spain, Sweden, Switzerland and the U.K. Gives market analyses, production statistics, market segmentation, sales, distribution, employment and unemployment, profits, end users and suppliers.

2905. PASSIVE COMPONENTS IN EUROPE TO 1986. *Larsen Sweeney Associates Ltd. 1981. 256 pg. 795.00 (United Kingdom).*

Market research report with forecasts to 1986 on 15 categories in the passive component sector in Austria, Belgium, Luxenbourg, Denmark, Eire, Finland, France, West Germany, Italy, Netherlands, Norway, Portugal, Spain, Sweden, Switzerland and the U.K. Gives market analyses, production statistics, market segmentation, sales, distribution, employment and unemployment, profits, end users and suppliers.

2906. PCB CONNECTORS: THE IMPACT OF HIGH INTERCONNECTION DENSITY. *BPA Management Services. September 1981. 2500.00 (United Kingdom).*

Report on PCB connectors in Western Europe and the impacts of high interconnection density. Covers new products 1980-1985; market analysis including end uses, imports, exports and distribution, 1980-1985; raw material consumption and labor costs; and investments in existing technologies.

2907. PHOTOELECTRIC CONTROLS IN EUROPE TO 1986. *Larsen Sweeney Associates Ltd. 1981. 221 pg. 695.00 (United Kingdom).*

Market research report with forecasts to 1986 on 28 photoelectric controls and controllers end user markets covered individually in Austria, Belgium, Luxembourg, Denmark, Eire, Finland, France, West Germany, Italy, Netherlands, Norway, Portugal, Spain, Sweden, Switzerland and the U.K. Gives market analyses, production statistics, market segmentation, sales, distribution, employment and unemployment, profits, end users and suppliers.

2908. POWER FACTOR CONTROLLERS - THEIR MARKETS AND PERFORMANCE. *Rattle, J.D. ERA Technology Ltd. August 1981. 990.00 (United Kingdom).*

Analysis of the size and structure of markets in the U.K. for power factor controls for electric induction motors. Examines

the installed base, user attitudes, trends in utilization by various industries, and price, energy cost, and performance factors.

2909. THE POWER SEMICONDUCTOR MARKET IN EUROPE. *Frost and Sullivan, Inc. June 1980. 248 pg. $275.00.*

In-depth 248-page report analyzes the power semiconductor market in Western Europe. Isolates and establishes the size of the present day power segment of the semiconductor industry, and forecasts consumption through 1985 by national markets, by device types, by OEM categories within country, and by selective, high-growth applications within those OEM categories. The report examines detailed structure of each major OEM segment including power semiconductor product requirements, applications, trends, and end-equipment sales forecasts. Report assesses the advances in power discrete technology and profiles the leading suppliers in the power semiconductor business.

2910. PRINTED CIRCUIT BOARD MARKET IN WEST EUROPE. *Frost and Sullivan, Inc. December 1982. 255 pg. $1500.00.*

Analysis of markets for printed circuit boards in Western Europe. Examines end uses, including military markets; OEM consumption; product groups; market areas; and impact of new technology and new products. Forecasts to 1987. Profiles manufacturers.

2911. PRINTED CIRCUIT BOARDS: EUROPE AND THE MIDDLE EAST. *BPA Management Services. 1979. 73 pg. 450.00 (United Kingdom).*

Report provides analysis of printed circuit board markets for Europe, Middle East and Eastern Europe, for 1979 and forecasts to 1984. Covers prices, imports and exports to 1984.

2912. PRINTED CIRCUIT BOARDS IN EUROPE TO 1986. *Larsen Sweeney Associates Ltd. 1981. 217 pg. 595.00 (United Kingdom).*

Market report to 1986 on printed circuit boards in Western Europe. Covers forecasts to 1986 on production, distribution, sales, end users, employment and unemployment. Eight types of printed circuit board are dealt with individually and eleven application sectors are covered: communications, telecommunications, consumer, EDP, industrial, instrumentation, automotive, military / defense, governmental, custom / specials, and unspecified / unallocated.

2913. PRINTED CIRCUITS IN WESTERN EUROPE. *BPA Management Services. March 1981. 74 pg. 500.00 (United Kingdom).*

Market analysis for printed circuits in Western Europe, 1980-1985. Gives market value by end uses, countries, and end use industries.

2914. PROFILE OF EUROPEAN CAPACITOR MANUFACTURERS: AND EUROPEAN MARKET FORECAST TO 1983. *Mackintosh Consultants, Inc. Irregular. 1980. 96 pg. $450.00.*

Analyzes size and company shares of the 1978-1979 U.S., Japanese, and Western European fixed capacitor markets. Includes breakdowns by capacitor type / technology, and forecasts through 1983. Profiles of 18 major manufacturers present (1979) addresses, telephone / telex, plant size, employment, capacity, product lines, W. European sales, parent company, and key executives' names.

2915. PROFILE OF THE EUROPEAN SEMICONDUCTOR INDUSTRY. *Benn Publications Ltd. 1981. $450.00.*

Forecasts markets to 1984 for discrete devices and integrated circuits, with U.S. and Japanese comparisons. Profiles of each manufacturer detail market share and a plant by plant survey. Other sections analyze single board computer market, distribution, custom design houses, and strategic factors affecting growth to 1991.

2916. PROFILE OF THE WORLD SEMICONDUCTOR INDUSTRY: VOLUME 1 - EUROPE. Benn Publications Ltd. 1982. 150 pg. $375.00.

Analysis of semiconductor markets and the industry in Europe includes forecasts to 1986, production, manufacturers and market shares, expansions of facilities and foreign investments, and new technology. Profiles of plants include addresses, equipment, product lines, number of employees, size of facilities, and 1981 sales.

2917. RECTIFIERS IN EUROPE TO 1986. Larsen Sweeney Associates Ltd. 1981. 219 pg. 795.00 (United Kingdom).

Market report to 1986 on the following types of rectifiers: germanium, high temperatures, high voltage, ignition, mercury arc, mercury vapor, metal, selenium, silicon, thyratran, thyrister, vacuum tube, and xenon-filled in Western Europe. Covers forecasts to 1986 for production, sales, distribution, end users, employment and unemployment.

2918. REPORT ON ELECTRONIC COMPONENTS STUDY - FRANCE: SUPPLEMENTARY DATA. Berry, D.J. Sira Institute Ltd. U.S. Department of Commerce. International Trade Administration. NTIS #ITA 81-09-501. June 30, 1981. 42 pg.

Supplement to two-volume survey on the market for electronic components in France provides additional company data (staff, capital, products) on French manufacturers listed in the original survey. Also provides separate reports on major manufacturers.

2919. REPORT ON ELECTRONIC COMPONENTS STUDY - FRANCE: VOLUME 1. Berry, D.J. Sira Institute Ltd. U.S. Department of Commerce. International Trade Administration. NTIS #ITA 81-09-501. April 24, 1981. 96 pg. $15.00.

Survey on the market for electronic components in France lists domestic manufacturers and provides data on the companies. Gives figures on the market size for the major product categories and imports by country as a percentage of the total in each cagegory. Provides data on sales and production by end user industry sectors. Lists best sales prospects, by product for U.S. manufacturers in 1981-1985. Contains a competitive assessment of U.S., domestic, and third country suppliers.

2920. REPORT ON ELECTRONIC COMPONENTS STUDY - FRANCE: VOLUME 2. Berry, D.J. Sira Institute Ltd. U.S. Department of Commerce. International Trade Administration. NTIS # ITA 81-09-501. April 24, 1982. 80 pg.

Survey on the market for electronic components in France describes business practices, tariffs, and standards. Contains trade lists which cover agents and distributors; trade associations; potential buyers; publications; and trade shows. Appendices contain import statistics in BTN 6-digit classification, by country of origin, and a summary for imports and exports in 4-digit BTN codes, 1978-1980.

2921. THE RESISTOR MARKET IN EUROPE. Frost and Sullivan, Inc. September 1981. 256 pg. $1400.00.

Market report contains forecasts through 1985 and for 1991 for potentiometers (wirebound and non) rheostats, wirewound (fixed) carbon, metal film, thermistors, networks (thick and thin film) and others for use in key end user markets in Benelux, France, U.K., F.R.G. and the rest of Europe. Includes state of the art and new product developments. Major manufacturers profiled.

2922. RESISTORS IN EUROPE TO 1986. Larsen Sweeney Associates Ltd. Quarterly. 1981. 267 pg. 595.00 (United Kingdom).

Market report to 1986 on 17 types of resistors in Western Europe. Covers forecasts to 1986 for sales, production trends, deliveries, distribution costs, marketing costs, capital investments, advertising costs, employment and unemployment. Report updated quarterly.

2923. SEMICONDUCTOR PRODUCTION AND TEST EQUIPMENT MARKET IN WESTERN EUROPE. Frost and Sullivan, Inc. July 1981. 472 pg. $1500.00.

Analysis of European markets for plants and equipment for the production of semiconductors, with forecasts to 1986 and 1990. Examines economic indicators and the semiconductor industry, major end use consumers and manufacturers, distribution, credit, regulations on trade and production trends. Lists addresses of suppliers.

2924. THE SEMI-CUSTOM REVOLUTION AND YOUR BUSINESS. BPA Management Services. 1982. 170 pg. $1700.00.

Report on effects of semi-custom IC's examines production costs, development costs, new technology, and performance. Includes analysis and forecast of world markets, 1980-1985; and worldwide manufacturers and suppliers.

2925. SPECIAL PURPOSE CIRCUITS / BIPOLAR AND MO'S IN EUROPE TO 1986. Larsen Sweeney Associates Ltd. Quarterly. 1981. 235 pg. 695.00 (United Kingdom).

Market report to 1986 on special purpose circuits / bipolar and MO's in Western Europe. The following application sectors are covered: communications, telecommunications, consumer, EDP, industrial, instrumentation, automotive, military / defense; governmental. 25 end user sectors are analyzed individually for each country. Gives forecasts to 1986 for sales, production trends, deliveries, distribution costs, marketing costs, capital investments, advertising costs, employment and unemployment. Report updated quarterly.

2926. STEPPER MOTOR MARKETS IN EUROPE. Frost and Sullivan, Inc. July 1981. 336 pg. $1500.00.

Market analysis and forecast for Stepper Motors in Europe, 1980-1990. Product groups analyzed by price and performance. Also examines end use markets, manufacturers, industry structure and foreign trade. Shows purchases and market value by groups, 1980-1990. Surveys suppliers.

2927. A STUDY OF THE SIZE OF THE EEC MARKET FOR PRIMARY AND SECONDARY BATTERIES. Techonomics Ltd. October 1980. $420.00.

Market analysis for primary and secondary batteries in Western Europe. Gives government statistics on foreign trade 1979; values of domestic production, 1977-1979, (varies by country); import market shares; and U.S. manufacturers' shipments, 1972.

2928. SURVEY OF THE MARKET IN GERMANY FOR ELECTRONIC COMPONENTS: FOREIGN MARKET SURVEY REPORT. Kindel (K.W.) (Germany, F.R.). U.S. Department of Commerce. National Technical Information Service. DIB-80-03-500. 1979. 51 pg. $10.00.

The market research was undertaken to study the present and potential U.S. share of the market in West Germany for electronic components; to examine growth trends in West German end-user industries over the next few years; to identify specific product categories that offer the most promising export potential for U.S. companies; and to provide basic data which will assist U.S. suppliers in determining current and potential sales and marketing opportunities. The trade promotional and marketing techniques were also reviewed.

2929. SWEDEN: ELECTRONIC COMPONENTS MARKET SURVEY. *Konsulterna. U.S. Department of Commerce. International Trade Administration. NTIS # ITA81-12-503. September 1981. 72 pg. $15.00.*

Analyzes the total market by major category and imports by major category (both 1977-1981 and projected 1984). An analysis is presented of each product category with an assessment of U.S., domestic, and third country suppliers. Each end user sector is profiled and data is given on sales by products and by end user. Details business practices, tariffs and standards, and lists principal potential agents and prospective customers.

2930. SWITCHES IN EUROPE TO 1986. *Larsen Sweeney Associates Ltd. Quarterly. 1981. 263 pg. 895.00 (United Kingdom).*

Market report to 1986 on 37 types of switches in Western Europe. Covers forecasts to 1986 for sales, production trends, deliveries, distribution costs, marketing costs, capital investments, advertising costs, end users, employment and unemployment. Report updated quarterly.

2931. SWITCHING EQUIPMENT IN EUROPE TO 1986. *Larsen Sweeney Associates Ltd. Quarterly. 1981. 221 pg. 695.00 (United Kingdom).*

Market report to 1986 on 6 types of switching equipment in Western Europe. Gives forecasts to 1986 for sales, production trends, deliveries, distribution costs, marketing costs, capital investments, advertising costs, end users and employment and unemployment. 24 end user sectors are analyzed individually for all 16 countries. Report updated quarterly.

2932. TERMINALS, TERMINAL BLOCKS, HOLDERS AND SOCKETS IN EUROPE TO 1986. *Larsen Sweeney Associates Ltd. 1981. 219 pg. 695.00 (United Kingdom).*

Market report to 1986 on 10 product groups of terminals, terminal blocks s holders and sockets in Western Europe. Covers forecasts to 1986 for sales, production trends, deliveries, distribution costs, marketing costs, capital investments, advertising costs, end users and employment and unemployment. Report updated quarterly.

2933. THICK FILM HYBRIDS AND NETWORKS IN TELE-COMMUNICATIONS. *BPA Management Services. January 1981. 136 pg. 2200.00 (United Kingdom).*

Analysis of markets for thick film hybrids and networks in the Western European telecommunications industry, with forecasts to 1984, 1988 and beyond. Covers technology, production costs, competition with IC's, demand and end uses including a telecommunications industry analysis, and manufacturers.

2934. THYRISTORS, SILICON CONTROL RECTIFIERS AND FOUR LAYER DIODES IN EUROPE TO 1986. *Larsen Sweeney Associates Ltd. Quarterly. 1981. 217 pg. 595.00 (United Kingdom).*

Market report to 1986 on seven types of thyristors, silicon control rectifiers / four layer diodes in Western Europe. Gives forecasts to 1986 for sales, production trends, deliveries, distribution costs, marketing costs, capital investments, advertising costs, end users, employment and unemployment. Report updated quarterly.

2935. TRANSISTORS IN EUROPE TO 1986. *Larsen Sweeney Associates Ltd. Quarterly. 1981. 227 pg. 795.00 (United Kingdom).*

Market report to 1986 on 19 main types of transistors available in Western Europe. Covers forecasts to 1986 on sales, produc-

tion trends, deliveries, distribution costs, marketing costs, capital investments, advertising costs, end users, employment and unemployment. Report updated quarterly.

2936. THE UNITED KINGDOM MARKET FOR BLANK AND PRE-RECORDED AUDIO CASSETTES, CARTRIDGES AND TAPES. *Economist Intelligence Unit Ltd. 1979. 233 pg. $1200.00.*

Analysis of U.K. markets for blank and pre-recorded tapes, cassettes and cartridges. Includes 1978 market size; domestic production, imports, exports, and apparent consumption, market structure and shares by size and brand; consumer purchase patterns, equipment ownership and behavior; marketing strategies; prices and profit margins; distribution; product groups and new products; manufacturers; sales; forecasts to 1983; and extensive deomgraphic statistics section.

2937. UNITED KINGDOM PRINTED CIRCUIT BOARD INDUSTRY. *Systec Consultants Ltd. July 1980. 60 pg. 250.00 (United Kingdom).*

Study of the U.K. printed circuit board market analyzes 1980 production, sales, market size, investment trends, imports and exports.

2938. VALVES AND TUBES IN EUROPE TO 1986. *Larsen Sweeney Associates Ltd. Quarterly. 1981. 238 pg. 795.00 (United Kingdom).*

Market report to 1986 on 28 types of thermionic valves and tubes on the market in Western Europe. The report is updated quarterly and covers market analysis, forecasts to 1986, sales, production trends, profits, assets, end users, employment and unemployment, products, capital expenditures, investments, and distribution.

2939. WELDING CONTROLS IN EUROPE TO 1986. *Larsen Sweeney Associates Ltd. Quarterly. 1981. 221 pg. 595.00 (United Kingdom).*

Market report to 1986 on 7 types of welding controls and equipment available in Western Europe. Report is updated quarterly and covers: market analysis, forecasts to 1986, sales, production trends, profits, assets, end users, employment and unemployment, products, capital expenditures, investments and distribution.

2940. X-RAY EQUIPMENT IN EUROPE TO 1986. *Larsen Sweeney Associates Ltd. Quarterly. 1981. 240 pg. 695.00 (United Kingdom).*

Market report to 1986 on the following x-ray equipment: diagnostic; industrial; laboratory; military and security; portable; and therapeutic in Western Europe. 27 end users markets are covered individually, giving market analysis, forecasts to 1986, sales, production trends, profits, assets, end users, employment and unemployment, products, capital expenditures, investments, and distribution. Updated quarterly.

See 419, 442, 445, 448, 458, 1912, 1913, 1914, 1930, 1931, 1949, 1969, 1970, 1985, 1986, 3390, 3801, 3811, 3812, 3813, 3816, 3817, 3818, 3819, 3822, 3838, 3842, 3852, 3853, 4302.

Investment Banking Reports

2941. ELECTRICALS SECTOR TRENDS SURVEY. *L. Messel and Company. April 7, 1981. 22 pg.*

Stockbrokers report summarizes performances of companies in the British electrical and electronics sector. Gives sales, profits, yields, exports, imports, and market trends for electrical equipment, wires and cables, telecommunications, capital electronics, components, and industrial engines. Provides commentaries and forecasts for each.

2942. WARBURG PARIBAS BECKER - A.G. BECKER: INDUSTRY COMMENT: THE VIEW FROM EUROPE - SEMICONDUCTOR TRENDS: A SYNOPSIS OF CONVERSATION WITH AMD. Orlansky, Aharon. Warburg Paribas Becker - A.G. Becker, Inc. June 7, 1982. 2 pg.

Brief analysis of market trends for semiconductors in the U.S. and Western Europe. Includes original equipment manufacturers' new ordersrom vendors versus suppliers, prices, and product mix.

Industry Statistical Reports

2943. ANNUAL STATISTICAL SURVEY OF THE ELECTRONICS INDUSTRY. National Economic Development Office. Annual. 1980.

Tables provide data on production, sales, trade outside Great Britain, employment, earnings, investment, and research and development.

2944. ANNUAL STATISTICAL SURVEY OF THE ELECTRONICS INDUSTRY. National Economic Development Office. Annual. 1979.

Annual statistical survey of the British electronics industry contains tables showing data on: production, sales, overseas trade, employment, earnings, investments, and scientific research and development.

2945. BRITAIN'S TOP 500 ELECTRONIC COMPANIES. Jordan and Sons (Surveys) Ltd. Annual. 1st ed. 1979. 50 pg.

U.K. survey provides an up-to-date summary of companies in the electronics industry. Tables include: 20 largest quoted companies; 20 largest private companies; 20 largest foreign-owned companies; companies with largest pre-tax profits; highest profitability; exports; highest proportion of sales; and highest wage payers. Financial tables show sales, exports, profits, U.K. employees, wages, current assets, liabilities, bank overdraft and short term loans, and net fixed assets. Covers component manufacturers, component distributors, instruments and communications, computers and data processing, printed circuit board manufacturers, and consumer electrical and electronic products.

2946. BUSINESS MONITOR: RADIO AND ELECTRONIC COMPONENTS. United Kingdom. Central Statistics Office. Quarterly. 1981. 16 pg.

Quarterly U.K. government statistics covering 1976-1980 sales and market value for radio and electronic components produced by U.K. manufacturers.

2947. BUSINESS MONITOR: REPORT ON THE CENSUS OF PRODUCTION - ELECTRIC LAMPS, ELECTRIC LIGHT FITTINGS, WIRING ACCESSORIES, ETC. Her Majesty's Stationery Office. Annual. 1981. 10 pg.

U.K. government census of production statistics covering electric lamps, electric light fittings, wiring accessories, etc. Gives production statistics, 1975-1979; capital expenditures, 1975-1979; analysis of corporations by size, 1979; gross value added; and operating ratios, 1978-1979.

2948. BUSINESS MONITOR: REPORT ON THE CENSUS OF PRODUCTION: ELECTRICAL EQUIPMENT FOR MOTOR VEHICLES, CYCLES AND AIRCRAFT. Her Majesty's Stationery Office. Annual. 1981. 10 pg.

Government of the U.K. report on the Census of Production covering electrical equipment for motor vehicles, cycles and aircraft. Gives production statistics, 1975-1979; capital expen-

ditures, 1975-1979; stocks and work in progress; corporations by size, 1979; employment and unemployment; and operating ratios.

2949. BUSINESS MONITOR: REPORT ON THE CENSUS OF PRODUCTION - ELECTRICAL MACHINERY. Her Majesty's Stationery Office. Annual. 1980. 8 pg.

Government of the U.K. report on the census of production for electrical machinery. Gives production statistics, 1975-1979; capital expenditures, 1975-1979; inventories, 1975-1979; employment and unemployment; operating ratios, 1978-1979; and number of employees.

2950. ELECTRICAL COMPONENT MANUFACTURERS. ICC Business Ratios. Annual. 1981. 95.00 (United Kingdom).

One in a series of 150 annual industry reports comparing performance between up to 100 individual leading companies, subsectors, and sectors over a three-year period. Fourteen ratio tables by company detail profit margin profitability, asset utilization, return on capital, liquidity, stock turnover, credit period, exports, profit per employee, average remuneration, sales per employee, capital employed per employee, and other data. Six annual growth rate tables cover sales, total assets, capital employed, average wages, total wages, and exports. Other company data includes names of directors and secretary and the registered office address. Gives analyses of current problems and likely developments within the sector and industry as a whole.

2951. ELECTRICAL INSTALLATION EQUIPMENT MANUFACTURERS. ICC Business Ratios. Annual. 36. 1981. 80.00 (United Kingdom).

One in a series of 150 annual industry reports comparing performance between up to 100 individual leading companies, subsectors, and sectors over a three-year period. Fourteen ratio tables by company detail profit margin, profitability, asset utilization, return on capital, liquidity, stock turnover, credit period, exports, profit per employee, average remuneration, sales per employee, capital employed per employee, and other data. Six annual growth rate tables cover sales, total assets, capital employed, average wages, total wages, and exports. Other company data includes names of directors and secretary and theregistered office address. Gives analyses of current problems and likely developments within the sector and industry as a whole.

2952. ELECTRICAL MOTORS AND CONTROL EQUIPMENT. ICC Business Ratios. Annual. 1981. 80.00 (United Kingdom).

One in a series of 150 annual industry reports comparing performance between up to 100 individual leading companies, subsectors, and sectors over a three-year period. Fourteen ratio tables by company detail profit margin, profitability, asset utilization, return on capital, liquidity, stock turnover, credit period, exports, profit per employee, average remuneration, sales per employee, capital employed per employee, and other data. Six annual growth rate tables cover sales, total assets, capital employed, average wages, total wages, and exports. Other company data includes names of directors and secretary and the registered office address. Gives analyses of current problems and likely developments within the sector and industry as a whole.

2953. ELECTRONIC INSTRUMENT MANUFACTURERS. ICC Business Ratios. Annual. 1981. 80.00 (United Kingdom).

One in a series of 150 annual industry reports comparing performance between up to 100 individual leading companies, subsectors, and sectors over a three-year period. Fourteen ratio tables by company detail profit margin, profitability, asset utilization, return on capital, liquidity, stock turnover, credit period, exports, profit per employee, average remuneration, sales per employee, capital employed per employee, and other data. Six annual growth rate tables cover sales, total assets, capital employed, average wages, total wages, and exports. Other company data includes names of directors and secretary and the registered office address. Gives analyses of current problems and likely developments within the sector and industry as a whole.

2954. GESCHAFTSBERICHT 1979: BROWN BOVERI. [Annual Report 1979: Brown Boveri.]. *Osterreichische Brown Boveri-Werke AG. German. Annual. April 1980. 49 pg.*

1979 annual report of the Austrian joint stock company Brown Boveri, active in the fields of power production and distribution, certain electrical applications, techniques and industrial utilization of heat. Most figures and tables cover 1970-1979. Emphasizes the company's 1979 research activities.

2955. GESCHAFTSBERICHT 1980: BROWN BOVERI. [Annual Report 1980: Brown Boveri.]. *Osterreichische Brown Boveri-Werke AG. German. Annual. April 1981. 45 pg.*

1980 annual report of the Austrian joint-stock company Brown Boveri, active in the fields of power production and distribution, certain electrical applications, drive mechanisms, and industrial use of heat. Most figures and tables cover 1971-1980. Emphasizes the company's 1980 R & D activities.

2956. INDUSTRIES ELECTRIQUES ET ELECTRONIQUES STATISTIQUES 1979. [Statistiques for the Electrical and Electronic Industries, 1979.]. *Insee O-E-P. French. June 1980. 31 pg.*

Tables indicate the percent of production for each branch of the electrical and electronics industry, exports, and imports. Shows results since 1970 for the electrical and electronics industries as a whole (data processing equipment not included).

2957. LIGHTING EQUIPMENT. *ICC Business Ratios. Annual. 1981. 95.00 (United Kingdom).*

One in a series of 150 annual industry reports comparing performance between up to 100 individual leading companies, subsectors, and sectors over a three-year period. Fourteen ratio tables by company detail profit margin, profitability, asset utilization, return on capital, liquidity, stock turnover; credit period, exports, profit per employee, average remuneration, sales per employee, capital employed per employee, and other data. Six annual growth rate tables cover sales, total assets, capital employed, average wages, total wages, and exports. Other company data includes names of directors and secretary and the registered office address. Gives anlayses of current problems and likely developments within the sector and industry as a whole.

2958. MACKINTOSH YEARBOOK OF WEST EUROPEAN ELECTRONICS DATA 1980. *Mackintosh Publications Ltd. Annual. 7th. 1980. 135 pg.*

Comprehensive directory gives production and market data (1978-1980), with some 1981-1983 projections for the electronics industry in West European countries. Covers data processing, control and instrumentation, communications, telecommunications consumer electronics and components. Appendices give translations of product headings into French, German, and Italian; list of sources; basic census statistics about each country; gross domestic product by country; and employment in the electronics industry.

2959. MACKINTOSH YEARBOOK OF WEST EUROPEAN ELECTRONICS DATA, 1981. *Mackintosh Publications Ltd. Annual. 8th ed. 1981. 160 pg. 130.00 (United Kingdom).*

Annual publication presents planning information on West European electronics industry, by country. Statistics and eco-

nomic data, 1978-1984, are given for office equipment, control and instrumentation, medical equipment, industrial equipment, communication and military equipment, telecommunications equipment, and equipment in the following fields: video, audio, electronic watches and clocks, and components.

2960. MACKINTOSH YEARBOOK OF WEST EUROPEAN ELECTRONICS DATA, 1983. *Benn Publications Ltd. Annual. 10th ed. 1982. $375.00.*

Country by country statistics cover production, markets, imports, and exports with breakdowns by equipment type (EDP, office equipment, control and instrument, medical and industrial, communications and military, telecommunications, and consumer video, audio, other, and components, including active, passive and audio). Appendices include line-by-line definitions of product headings; economic data for Europe, Japan, and U.S.; four-language glossary; and electronics employment sources. Gives narrative evaluation of trends in the electronics industry, 1973-1993.

Financial and Economic Studies

2961. ANWENDUNGEN, VERBREITUNG UND AUSWIRKUNGEN DER MIKROELEKTRONIK IN OSTERREICH. [Economic Consequences of the Applications of Microelectronics in Austria.]. *Federal Ministry of Research and Sciences, Vienna. Osterreichisches Institut fur Wirtschaftsforschung (Wifo). German. April 1981. 182 pg.*

The research, ordered by the Federal Ministry of Sciences and Research, surveys in-depth all the relevant aspects of the applications of microelectronics in Austria: domestic production and market surveys, microelectronics in industry, in office, macroeconomic aspects and forecast for the 1980's. Consumer electronics applications such as video recorders and chess computers and the consequences of microelectronics applications on hours of labor terminate the study.

2962. DEFENCE ELECTRONICS. *Tysoe, John. Laurie Milbank and Company. November 1980. 86 pg.*

Report covers major electronics corporations in Britain involved in the Defence Programme. Shows the emphasis of U.K. defense spending in the 1980's and its relevance to these corporations.

2963. ELECTRICAL BULLETIN. *Scrimgeour, Kemp-Gee and Company. Irregular. February 1981. 65 pg.*

Private circulation report covers consumer electronics, television rental, household electrical goods, electronic and electrical capital goods and components. Lists U.K. companies active in these areas and gives statistics on earnings, yields, and sales, 1976-1980; raw material prices and profits forecasts, 1980-1983; and earnings per share and earnings forecasts, 1980-1984.

2964. ELECTRICAL CHARTS. *Hodgkinson, K. L. Messel and Company. November 1980. 40 pg.*

Chart book shows the price and relative performance of the major electrical and electronic companies on the U.K. Stock Exchange. In addition to plotting the share price against the FT Actuaries 500 index since 1970, it shows the performance against the relative Messel Sector Index. The companies covered are grouped as follows: TV contractors; TV rentals; industrial electricals; consumer electricals; and office equipment.

2965. ELECTRICAL INSTALLATION EQUIPMENT MANUFACTURERS. *Inter Company Comparisons Ltd. January 1980. 49.00 (United Kingdom).*

Report looks at the financial performance of 59 U.K. electrical installation equipment manufacturers. Examines profitability, efficiency, and growth, to March 1979.

2966. ELECTRICAL MACHINERY, TELECOMMUNICA-TIONS EQUIPMENT AND APPARATUS - GREECE. *ICAP Hellos SA. 1979. 127 pg. $275.00.*

Report on the electrical machinery and telecommunications equipment industry in Greece. Examines employment, number of businesses, value added compared to all manufacturing, production, installed base of plants and equipment, and raw materials. Gives financial and other, analysis of manufacturers, foreign trade, total market value analysis of size and growth of markets, distribution, and investment opportunities.

2967. ELECTRICAL REVIEW. *L. Messel and Company. Bimonthly. June 30, 1981. 5 pg.*

The Electrical Review is published at the end of each stock exchange account period and gives market comment and company results, including sales and profits, for the electronics and electrical sector companies. Includes individual company news, industry news and the electrical sector weekly report, showing share prices, profits, yields, and price earnings ratios.

2968. ELECTRICAL SECTOR ANALYSES. *Hickey, P. Laurie Milbank and Company. Monthly. 1981. 3 pg.*

Monthly research report on individual corporations in the electrical and electronics sector on the U.K. stock market. Gives share prices, dividends, yields, corporation analysis, price earnings ratios, profits and sales.

2969. ELECTRICALS SECTOR TRENDS SURVEY. *Hodgkinson, K. L. Messel and Company. Weekly. April 7, 1981. 22 pg.*

Sector trends survey giving shares in the electrical sector on the U.K. stock exchange. Analyzes trends, forecasts for 1981, sales (1975-1980) and deliveries, (1975-1981). Analyses of products are given for the following sectors: electrical machinery, wires and cables, industrial engines, telecommunications, capital electronics and components.

2970. ELECTRONIC EQUIPMENT MANUFACTURERS: AN INDUSTRY SECTOR ANALYSIS. *Inter Company Comparisons Ltd. 6th ed. 1981. 50 pg.*

Report giving comparisons by means of business ratios of growth of sales, assets, capital employed, employment and unemployment, profits and stocks (1976-1979) of 99 leading U.K. electronic equipment manufacturers. Includes statistics for 1976-1979 for return on capital, liquidity, exports, profit per employee, and average wages paid.

2971. FINANCIAL SURVEYS: ELECTRIC COMPONENT MANUFACTURERS AND DISTRIBUTORS. *Inter Company Comparisons Ltd. 1981. 59.80 (United Kingdom).*

Annual financial analysis, with two year comparative figures, for electronic component manufacturers and distributors in the U.K. For all applicable businesses, tabulates turnover of sales; total assets, including fixed assets, current assets, intangible assets, and investments; current liabilities; profits before taxes; executive compensation; capital employed; and subsidiaries.

2972. FINANCIAL SURVEYS: ELECTRONIC MANUFAC-TURERS AND DISTRIBUTORS. *Inter Company Comparisons Ltd. 1981. 59.80 (United Kingdom).*

Annual financial analysis, with two year comparative figures, for electronic manufacturers and distributors in the U.K. For all applicable businesses, tabulates turnover of sales; total assets, including fixed assets, current assets, intangible assets, and investments; current liabilities; profits before taxes; executive compensation; capital employed; and subsidiaries.

2973. FINANCIAL SURVEYS: LIGHTING DEVICES AND SYSTEMS. *I.C.C. Information Group Ltd. Annual. 1981.*

One in a series of annual financial surveys for quoted and unquoted companies in particular fields. Listings for past two years cover turnover, total assets, current liabilities, profit before tax and group relief (excluding extraordinary items), and payment to directors. Listings detail whether firm is part of a group account or a subsidiary of another company. Companies with name changes are noted.

2974. ICC BUSINESS RATIO REPORT: ELECTRICAL INSTALLATION EQUIPMENT MANUFACTURERS. *ICC Business Ratios. Annual. 4th ed. 1982. 97.00 (United Kingdom).*

Financial analysis of 60 U.K. electrical installation equipment corporations includes business addresses, executives, lines of business, and ownership. Company data for 1978-1981 includes fixed and intangible assets, stocks, total current assets, short-term debt, liabilities, net assets, capital employed, sales, exports, profits, interest paid, wages and salaries, and number of employees. Also includes 19 ratios of profits, liquidity, capital utilization, gearing, productivity, exports and employees; growth rates; and performance forecasts.

2975. INDUSTRIAL ADJUSTMENT AND POLICY II: TECHNI-CAL CHANGE AND SURVIVAL: EUROPE'S SEMICONDUC-TOR INDUSTRY. *Dosi, Giovanni. B.V. Uitgevers-Maatschappy Bonaventura. 1981. 103 pg. 4.00 (United Kingdom).*

Covers semiconductor technology and international competition; world industry growth; nature and direction of technological change; factors in U.S. technological lead; Europe's adjustment problem; arguments for public intervention; public policies; and future of European semiconductors (constraints on future developments; possible scenarios for the future). Tables show France, 1952-1979 semiconductor production, exports and imports (U.K., U.S., Italy, Japan); 1964-1979 structural changes and government interventions in the electronics industry; 1974-1980 world semiconductor markets; 1973, 1976, 1978 world produciton of semiconductors by region and by ownership; 1973 and 1978 value of semiconductor shipments per employee for various nations; and 1974-1978 production of integrated circuits by Western European nations.

2976. MACKINTOSH EUROPEAN ELECTRONICS COMPA-NIES FILE: 1981 / 82. *Mackintosh Consultants, Inc. Monthly. 1981. 300 (est.) pg. $325.00.*

Identifies and profiles the 100 leading European electronics manufacturers / suppliers by sales, plus the top 10 in each of several product sectors, in monthly releases and three annual supplements. Covers significant developments involving other corporations (new products, expansions, takeover bids, etc.). Profiles include balance-sheet and income data, number of employees, plant sizes, capital and R & D expenditures. There is also a parent-subsidiary index including about 4000 subsidiaries.

2977. OVERVIEW OF THE ELECTRONIC INDUSTRY IN EUROPE. *United Nations. Industrial Development Organization Industrial Information Section. Industrial and Technological Information Bank. 1981. 6 pg. Free.*

Covers the background of development (including experiences of U.S., U.S.S.R., and Japan); worldwide growth of electronics; the Western European industry (development potential and prospects to 1991); the integrated circuits industry; and market aspects and investment. Includes statistical tables.

2978. SOCIAL AND EMPLOYMENT IMPLICATIONS OF MICROELECTRONICS. *U.K. Central Policy Review Staff. May 1979. 24 pg.*

Paper by the Central Policy Review Staff of the U.K. Government reviewing the government's activities in the field of microelectronics. Examines the employment and social consequences.

Forecasts

2979. SEMICONDUCTOR MARKET IN EUROPE. *Frost and Sullivan, Inc. July 1979. $975.00.*

Analyzes and forecasts (1987) market for discrete semiconductor devices and integrated circuits and diodes, transistors and bipolars. Forecasts given by end product segments (data processing, instrumentation and control, communications, defense, and consumer electronics) and by European country. Reviews manufacturers, technology trends and import tariff structure.

2980. STRATEGIC IMPACT OF INTELLIGENT ELECTRONICS IN U.S. AND WESTERN EUROPE: 1977-1987: VOLUME 8: WESTERN EUROPE. *Roelter, Martyn. Arthur D. Little, Inc. April 1979. 63 pg. $35000.00.*

Volume 8 of 11-volume study discusses intelligent electronic equipment for automobiles, business communications, consumer applications, and industrial applications for Western Europe and USA, with Japan cited throughout. Production, consumption, exports / imports, and value of shipments of electronic equipment products given. Market shares, rankings, sales of suppliers provided. Forecasts 1977 through 1987. Price listed is for complete set. Contact publisher for individual prices. Data derived is from Arthur D. Little, Inc. estimates, U.S. Dept. of Commerce, and Market for Process Control Instrumentation.

2981. WEST GERMAN ELECTRONIC INDUSTRY FORECAST 1981. *Gnostic Concepts, Inc. 1981. 600 pg. $10000.00.*

Reviews the industry in 1981 and forecasts 1982, identifying the segments expected to show the highest growth rates.

Directories and Yearbooks

2982. ANNUAIRE GENERAL DE L'ELECTRONIQUE ET DES INDUSTRIES CONNEXES. [General Yearbook of Electronics and Related Industries.]. *Compagnie Francaise d'Edition. French. Annual. 1980.*

Annual listing of electronics firms, agencies, and equipment.

2983. ANNUARIO DELL'INDUSTRIA COMMERCIO RADIO TELEVISIONE ELETTRONICA ELETTRODOMESTICA. [Yearbook of Radio, Television, Electronics, and Household Appliances Industries and Trade.]. *Angelotti Editore. Italian. Annual. 1980.*

Lists manufacturers, companies, wholesalers, retailers, products, and trade names in Italy's radio, TV, electronics, and appliances industries.

2984. EI DISTRIBUTOR SURVEY. *Lesterstar Ltd. Annual. 1980. 152 pg. 3.50 (United Kingdom).*

Directory of electronic component and distributors. Includes address; telephone; products carried; and key executives (directors, sales executives, manufacturing principals).

2985. ELECTRICAL AND ELECTRONICS TRADES DIRECTORY 1981. *Peter Peregrinus Ltd. Annual. 99th ed. 1981. 749 pg. 27.50 (United Kingdom).*

This is a comprehensive directory organized in seven sections: manufacturers, suppliers, and servicing companies; manufacturers' representatives; wholesalers and distributors; institutions and societies; electricity undertakings; brand names;

ABC guide to products and materials and an index section of companies and institutions.

2986. ELECTRICAL AND ELECTRONICS TRADES DIRECTORY 1982. *Peter Peregrinus Ltd. Annual. 100th ed. February 1982. $72.50.33.00 (United Kingdom).*

Directory for electrical and electronic products in the U.K. Gives business addresses for manufacturers, suppliers, services and subsidiaries. Includes product lines, power generation undertakings, associations, and brand names.

2987. ELECTRO: ANNUAIRE DE L'ELECTRICITE ET DE L'ELECTRONIQUE. [Electro: Yearbook of the Electric and Electronic Industries.]. *Societe Nouvelles d'Editions Publicitaires. French. Annual. 1980.*

Lists manufacturers, wholesalers, retailers, contractors, and trade names.

2988. EUROPEAN ELECTRONIC COMPONENT DISTRIBUTOR DIRECTORY: 1980 / 81. *Mackintosh Consultants, Inc. Annual. 2nd ed. 1980. 200 (est.) $75.00.*

Lists about 1200 distributors of electronic components in France, West Germany, Italy and the U.K. (which together represent about 75% of the European market). Alphabetic listings for each nation include addresses, telephone / telex, subsidiaries (including foreign), parent / associate company, manufacturers represented, year incorporated, and product lines. Distributors are cross-referenced by detailed product group.

2989. EUROPEAN ELECTRONICS COMPONENT DISTRIBUTOR DIRECTORY, 1981/82. *Mackintosh Publications Ltd. Annual. 3rd ed. 1981. 30.00 (United Kingdom).*

Directory of European electronic component distributors. Financial analysis includes prices, sales, and operating performance. Also lists brand names. Includes industry structure, manufacturers and distribution channels, franchises, mergers and bankruptcies.

2990. FIRMENPORTRATS DER SAP - MITGLIEDER. [Portraits of SAP - Companies.]. *Verlag Schweizer Automotik Pool,Allgemeine Treuhaud AG. German. Irregular. August 1982. 170 (est.) pg.*

Directory of the members of the Swiss Automation Pool (SAP), listing names, addresses, information of general interest and the programs of manufacture, distribution and representation.

2991. GUIDE TECHNIQUE DE L'ELECTRONIQUE PROFESSIONELLE. [Technical Guide to Professional Electronics.]. *Guides Techniques Professionelles. French. Annual. 1981.*

Lists French manufacturers of electronic equipment and foreign manufacturers registered in France.

2992. HANDBOOK 1979-1980 OF BEAMA (BRITISH ELECTRICAL AND ALLIED MANUFACTURERS' ASSOCIATION LTD.): A GUIDE TO THE ORGANIZATIONS OF THE ASSOCIATION. *Meigh, E. British Electrical and Allied Manufacturers Association Ltd. Irregular. December 1979. 60 pg. 2.50 (United Kingdom).*

Handbook provides a guide to BEAMA, the British Electrical and Allied Manufacturers' Association Ltd. organization. Includes a directory of membership, details of centrally-organized committees, and the membership of the federated associations. BEAMA is the leading trade association of the electrical, electronic and allied industries in Britain.

2993. IEA DIRECTORY OF INSTRUMENTS ELECTRONICS AUTOMATION. *Morgan Grampian Book Publishing Company Ltd. Annual. 15th ed. 1981. 333 pg. 15.00 (United Kingdom).*

Directory of information on suppliers, business activities, organizations, products, agents and brand names in the interrelated fields of instrumentation, electronics and automation. Includes a comprehensive list of companies in the field with addresses and telephone numbers. Lists brand names used by companies whose products are included in the buyer's guide.

2994. INSTITUTE OF ELECTRICAL ENGINEERS LIST OF MEMBERS. *Institute of Electrical and Electronics Engineers, Inc. Annual. 1981. 592 pg. $100.00.*

Directory contains 65,500 entries for all classes of members of Britain's Institute of Electrical Engineers.

2995. LIGHTING INDUSTRY FEDERATION LTD. BUYER'S GUIDE. *Lighting Industry Federation. Irregular. 1979. Free.*

Membership directory includes chart indicating the product types of each of the 75 member firms in the Lighting Industry Federation.

2996. MACKINTOSH EUROPEAN ELECTRONICS COMPA-NIES FILE 1982/83. *Benn Publications Ltd. Annual. 4th ed. April 1982. $295.00.*

Loose-leaf guide to corporate structure of European electronics industry analyzes the 100 largest electronics groups whose sales exceed $100 billion, profiling sales by product, area, leading financial indicators, corporate structure, and European subsidiaries, with European market leaders' summaries. Annual subscription includes a monthly news bulletin and other supplements.

2997. MACKINTOSH YEARBOOK OF WEST EUROPEAN ELECTRONICS DATA: 1982. *Mackintosh Publications Ltd. Annual. 9th ed. 1981. $350.*

On a country by country basis, the yearbook gives the total European production import / export market. Includes production by country, output of electronics equipment and components, and consumption of electronics equipment. Provides a forecast.

2998. PROFILE OF EUROPEAN SEMICONDUCTOR MANU-FACTURERS. *Mackintosh Publications Ltd. Irregular. 1979. 98 pg.*

Lists major suppliers in the semiconductor market, new ventures in the industry, and major suppliers of electronic components. Indexes by country, by company and by manufacturing locations. Contains directory of manufacturers, telephone, address, capabilities, process technology, principal products, employees, plant area and 1978 sales. Information obtained from individual companies by the publisher.

2999. SWITZERLAND - YOUR PARTNER: VOLUME 6: ENERGY GENERATION AND DISTRIBUTION. *Swiss Confederation. Office for the Development of Trade. Irregular. 1980. 144 pg. 6.00 (Switzerland).*

Presents an overall analysis of the energy generation and distribution industries in Switzerland. Includes indexes to products and services, suppliers, and a list of professional associations and groupings. Most of the book consists of company-by-company descriptions presenting the range of products exported and consultants available.

3000. WHERE TO BUY ELECTRICAL PLANT SUPPLIES AND SERVICES. *Loader, H.F. Where to Buy, Ltd. Annual. 62nd ed. 1981. 121 pg. 5.00 (United Kingdom).*

Annual directory giving an index to the principal sources of supply of power plants, household appliances, electronic and radio apparatus, and most types of electrical and electronic equipment. Gives names, addresses and telephone numbers of companies.

3001. WHERE TO BUY EVERYTHING ELECTRICAL. *Fuel and Metallurgical Journals Ltd. Annual. 1980. 176 pg. .40 (United Kingdom).*

Annual buyers guide to British suppliers of power plant, household appliances, and electronic and radio equipment.

3002. ZVEI ELECTRO BUYERS' GUIDE. *Zentralverband der Elektrotechnischen Industrie. Verlag W. Sachon. English/German/French/Spanish. Annual. 1981. 1300 (est.) pg. 35.00 (West Germany).*

ZVEI is the German Electrical and Electronic Manufacturers Association. This annual directory lists member firms by a wide range of detailed product classifications: generating, transmission and storage equipment, industrial equipment, machinery and appliances, power tools, domestic appliances, illumination, communications equipment, measuring and control instruments, medical electronics, and insulating and other materials. Specifies addresses, trade marks, and specific product lines. Content in German, English, French and Spanish; cover and introductory material available in any of the above languages.

3003. ZVEI ELECTRO BUYERS' GUIDE 1979. *Verlag W. Sachon. Irregular. 1979. 650 (est.) pg.*

ZVEI is the German Electrical and Electronic Manufacturers Association. Directory covers products and manufacturers of equipment for power engineering, heat treating, domestic appliances, lighting, entertainment, electronics, communications, data processing, measurement and process automation.

See 3877, 3883, 3887.

Special Issues and Journal Articles

British Business

3004. ENGINEERING INDUSTRIES: PRODUCTION UP TO JUNE. *Her Majesty's Stationery Office. v.9, no.4. October 1, 1982. pg.178*

Provides statistics for the level of output of the combined U.K. engineering industries for the second quarter of 1982. Table lists indices of production for engineering industries (seasonally adjusted volume indices), 1976-1982. (Quarterly figures included).

3005. INDICES OF PRODUCTION FOR ENGINEERING INDUSTRIES. *Her Majesty's Stationery Office. v.7, no.19. April 30, 1982. pg.817*

Statistics include 1976-1982 indices of production for engineering industries.

Business and Finance

3006. IRELAND - NO SILICON VALLEY. *Belenes Publications Ltd. v.18, no.2. April 22, 1982. pg.2*

Summarizes results of Report no. 64 of the National Economic and Social Council, better known as the Telesis Report. Two charts provide data on world major producers of integrated circuits with plants in Ireland and on selected electronics companies in Ireland, including geographical location of research and development, marketing and fabrication for products produced in Ireland for each company listed.

Business Week

3007. EUROPEAN CHIP MAKERS' LAST CHANCE?
McGraw-Hill Book Company. November 17, 1980. pg.152A-152B

Analysis of the market for semiconductors in Western Europe. Shows 1978-1979 sales, and estimated sales for 1980, 1982, and 1985.

Chemistry and Industry

3008. THE ROLE OF MICROPROCESSORS IN FOOD PRO-CESSING AND ANALYSIS. *Coleman, Ronald F. Society of Chemical Industry. no.17. September 4, 1982. pg.652-658*

Reports on the effect of microelectronics on U.K. agriculture and horticulture. Discusses mechanization and electrification, the emergence of farm electronics, the current status of farm electronics, future developments, and conclusions. Tables provide data, as follows: utilization of U.K. land, 1979; quantity and output of U.K. farm produce, 1980 provisional figures; inputs and net income, U.K. farms; distribution of farm sizes in the U.K., 1979; land labor force in agriculture as a percentage of total labor force (Western Europe), 1950-1980.

Circuits Manufacturing

3009. CIRCUIT NEWS: THE BALANCE OF INTERCONNEC-TION. *Morgan Grampian Publishing Company. v.22, no.8. August 1982. pg.15-16*

Summarizes the results of two reports from BPA (Technology and Management) Ltd., 'The Business of Thick Film' and 'Automatic Discrete Wiring.' Two figures provide statistical data for value comparison: component-loaded interconnection technologies in Western Europe, 1960-1990; and for total world multilayer market versus 6-plus and 'Multiwire' penetration, 1970-1980. (Source: BPA Ltd.)

3010. U.S. ELECTRONICS TOPS IN EUROPE. *Morgan Grampian Publishing Company. v.22, no.2. October 1982. pg.18-20*

Discusses the impact of U.S.-owned firms on the European electronics market. Table lists the top 10 electronics firms in Europe, 1980-1981. Includes country, firm, electronics sales, and profit before taxes / sales ratio. Source is Mackintosh Publications, Ltd.

Control and Instrumentation

3011. MONITOR: LOAD CELLS. *Morgan Grampian Book Publishing Company Ltd. Irregular. v.14, no.6. June 1982. pg.31-37*

Updates various types of load cells currently available in Great Britain, including manufacturer's or supplier's name, address, product specifications, and reader inquiry card for further information.

Diagramm

3012. DIE MODERNSTE FERTIGUNGSSTALTE EUROPAS FUR INTEGRIERTE SCHALTUNGEN IN VILLACH EROFF-NET. [Europe's Most Modern Production Plant for Integrated Circuits Inaugurated in Villach.]. *Herbert Muck Ges. m.b.H. German. February 1981. 22 pg.*

Article describes the production of MOS-Circuits in VLSI-Technique (150,000 parts/chip). By means of graphs, shows the evolution of the chips markets from 1975 to 1980, with forecasts to 1985.

The Dock and Harbour Authority

3013. LEADING LIGHTS. *Foxlow Publications Ltd. Irregular. February 1981. pg.311-313*

Lists lights and light systems used as navigational aids in the shipping industry such as buoy and other marine lanterns. Information includes photograph, brief technical specifications, and address of manufacturer.

Electrical Review

3014. ELECTRICAL BUSINESS FAILURES SOAR BY 87 PER CENT. *IPC Electrical-Electronic Press Ltd. Division of IPC Industrial Press. v.211, no.4. August 1982. pg.12*

Discusses electrical business liquidations. Gives 1981-1982 bankruptcies by quarter.

3015. ELECTRICAL REVIEW: ANNUAL BUYERS GUIDE. *IPC Electrical-Electronic Press Ltd. Division of IPC Industrial Press. Annual. 1981. 98 pg.*

A directory to manufacturers and wholesalers listed under 75 electrical product groups, such as wires, cables, switchgear, instruments, batteries, etc. Data includes address of company, telex, telephone, contact person, association membership, and product specifications. Provides a list of trade associations.

3016. ELECTRICAL REVIEW: NEW ELECTRICAL PROD-UCTS. *IPC Business Press Ltd. Monthly. 1981.*

A monthly feature listing electrical products and equipment. Includes photograph, description, and name and address of manufacturer. Further information available through reader card service.

3017. ELECTRICAL REVIEW: NEW PRODUCTS. *IPC Electrical-Electronic Press Ltd. Division of IPC Industrial Press. Monthly. v.211, no.2. July 1982. pg.51-53*

Directory of new electrical products summarizes technical details in selection of new electrical equipment. Provides names and addresses of U.K. manufacturers.

3018. ELECTRICAL REVIEW: SILICON JUNCTION. *IPC Electrical-Electronic Press Ltd. Division of IPC Industrial Press. Monthly. v.210, no.25. June 25, 1982. pg.30-31*

Lists, describes, and illustrates silicon products. Provides names of manufacturers.

3019. MANUFACTURERS, SUPPLIERS, SERVICES. *IPC Electrical-Electronic Press Ltd. Division of IPC Industrial Press. Monthly. v.211, no.2. July 1982. pg.61-62*

Monthly directory includes names, addresses, telephone numbers, and contacts for U.K. manufacturers, suppliers, and services by products.

Electronic Business

3020. ACADEMIA - INDUSTRY LINKS LURE COMPANIES TO ULSTER. *Black, Derek. Cahners Publishing Company. v.8, no.11. October 1982. pg.146, 148, 150*

Evaluates Northern Ireland as a site for electronics and microelectronics manufacture. Surveys multinational high-technology companies with locations in Northern Ireland. Discusses government incentives and educational resources.

3021. CAPITALISTS FROM BRITAIN VENTURE INTO U.S. MARKET. *Jones, Keith. Cahners Publishing Company. v.8, no. 11. October 1982. pg.110, 112*

Surveys U.K. financial institutions currently (1982) providing venture capital to U.S. electronics companies. Discusses the entry of these U.S. businesses into European markets.

3022. EUROPEAN ELECTRONICS MARKED BY MANY CHANGES. *Cahners Publishing Company. v.8, no.8. July 1982. pg.79-89 ff.*

Examines the electronics industry of Western Europe giving figures for European and U.S. electronic sales. Ranks the top 100 manufacturers in 1981 giving sales, electronics sales, profits, investments, sales per employee, R and D, capital expenditures, return on gross capital employed, and country of origin.

3023. THE EUROPEAN TOP 50 ELECTRONICS COMPANIES. *Mackintosh International. Cahners Publishing Company. July 1980. pg.56-59*

Rankings of the top 50 European electronics companies, based on either 1978 or 1979 revenues. Gives electronic sales and total sales for fiscal year ending either 1978 or 1979, with country of origin. Details sales, capital expenditures and profits of top 10 European businesses, and compares them to the top 10 U.S. businesses.

3024. INCENTIVES, NOT BLARNEY, DRAW COMPANIES TO IRELAND. *Furst, Al. Cahners Publishing Company. v.8, no. 10. September 1982. pg.130, 132, 134*

Reports on the Republic of Ireland as the fastest-growing site in Europe for electronics expansion by foreign companies. Ranks the top 10 electronics companies in Ireland by number of employees (1982). Indicates product lines. Map illustrates the locations of 94 electronic companies in Ireland. Source is the Industrial Development Authority - Ireland.

3025. WEST EUROPEAN CAPACITOR MARKET TO HIT $1.6 BILLION IN 1983. *Machintosh Consultants, Inc. Cahners Publishing Company. Irregular. February 1981. pg.112-114*

Reviews and forecasts (1981-1983) the West European market for capacitors. Graph displays market shares (1978-1983) for electrolytic, plastic, ceramic and paper and other capacitors.

3026. WEST GERMANY OFFERS STRONG MARKET FOR U.S. COMPONENT SUPPLIERS. *Cahners Publishing Company. October 1980. pg.120-124*

Reviews the West German electronic components market and their imports of U.S. supplies. Summarizes the competitive environment, end user needs, marketing practices and demand. One graph displays West German production, imports, exports and market size (1978-1982) for semiconductors, passive components and electron tubes.

Elektronik Report

3027. ELEKTRONIK INTERNATIONAL 1982. [Electronic Report 1982.]. *Verlag Technik Report Ges. m.b.H. German. Irregular. September 1982. 98 pg.*

Summarizes new developments in the electronics industry, including computers worldwide, with a special stress on West Europe.

3028. IE 79 - MESSEVORSCHAU. [Industrial Electronics Fair - Preview.]. *Verlag Technik Report Ges. m.b.H. Annual. October 1979. pg.16-97*

Special issue devoted to the fifth International FaFair of Industrial Electronics and Electrotechnique, Vienna, October 1979. Includes a detailed survey of all the electronic building blocks

available on the national market, both domestically produced and imported goods.

3029. PRODUKTE. [Products.]. *Verlag Technik Report Ges. m.b.H. German. Annual. October 1980. pg.23-88*

Special issue devoted to the Munich Fair Electonica "80, November 1980. Includes a detailed survey of all the electronic building blocks available on the market, both domestic and imported; market trends; and related figures.

Engineer

3030. REPORT URGES SELECTIVE APPROACH. *Lee, Allan. Morgan Grampian Book Publishing Company Ltd. v.254, no. 6579. April 29, 1982. pg.13*

Summarizes a report on the electronics industry in Great Britain published by the National Economic Development Council in April, 1982. Discusses marketing strategies and possible government actions to help the industry. Identifies 5 key areas to concentrate on. Includes a table of figures comparing production, imports, exports, percent of foreign imports, and employment in the industry in 1970, 1975, 1980.

The Industrial Robot

3031. SWEDISH INDUSTRIES' EXPERIENCE WITH ROBOTS. *Carlson, Jan. IFS (Publications) Ltd. v.9, no.2. June 1982. pg.88-91*

Analysis of utilization of industrial robots and other new technology in Swedish industry shows installations, 1970, 1973, 1977, 1979, 1984, 1990; average annual growth rates, 1970-1990; productivity comparisons between numerical control (NC) machines and robots; percent of value added comparisons; and percent of engineering production by product group.

Informes y Estudios de CEOE

3032. BIENES DE EQUIPO: SECTOR CLAVE PARA EL DESARROLLO. [Machinery and Equipment: Key for the Development Sector.]. *Confederacion Espanola de Organizaciones Empresariales. Spanish. no.1. December 1980. pg.19-22 Free (Spain).*

Statistical and textual analysis of machinery and equipment production in Spain. Tables by the organization SERCOBE show statistics for 1970 through 1979 on production, imports, exports, financial indexes, and advances and declines in the machinery and equipment production sector for the year 1970 through 1979 and forecast for 1980.

Insulation / Circuits

3033. UPDATE ON ELECTRONIC INDUSTRY IN THE UNITED KINGDOM. *Judd, Mike. Lake Publishing Corporation. v.28, no.11. October 1982. pg.56-58*

Assesses the electronic industry in the United Kingdom in 1982. Analyzes the market for six leading electronics industries' components. Reports 1981 production for six industrial sectors. Indicates amount of sales for component product groups in 1981.

3034. AN UPDATE ON SPAIN'S ELECTRONIC INDUSTRY. *Riguera, Jose. Lake Publishing Corporation. May 1982. pg.15-16*

Discusses production volumes in Spain of communications equipment, consumer electronic goods, components, resisters, printed circuits, semiconductors, connector producers, and others. Examines the degree to which Spain must rely on foreign suppliers for advanced types of equipment and components.

3035. UPDATE ON THE ELECTRONIC INDUSTRY IN BELGIUM. *Albertus, Dick. Lake Publishing Corporation. v.28, no.9. August 1982. pg.13-14*

Assesses the electronics industry in Belgium. Discusses the role of foreign corporations. Examines four product categories and reports value of production, sales, and growth rates from 1978 through 1985. Table presents data on end-user electronic component purchases by component type for 1978, 1980, and 1985.

Inter Electronique

3036. COMPOSANTS ELECTRONIQUES PLUS DE ONZE MILLIARDS. [Electronic Components: More than Eleven Thousand Million.]. *Yugoslavia. Centre for Technical and Scientific Documentation. French. March 30, 1981. pg.16*

Reviews the French electronics industry in 1980, in terms of growth, turnover, demand, and the export / import balance. Tables show production, exports and imports, 1977-1980.

International Business Week

3037. LAST CHANCE TACTICS OF EUROPE'S CHIP MAKERS. *McGraw-Hill Book Company. no.2745-76. June 28, 1982. pg.49-54*

Examines the European semiconductor industry and the influence of American and Japanese makers. Gives 1981 dollar sales of six subsidiaries of U.S. IC makers and of top European IC makers. Provides total annual dollar sales of ICs in Europe, 1980-1984.

MINTEL Market Intelligence Reports

3038. ELECTRONIC GAMES. *Mitel Publications Ltd. May 1981. pg.61-69*

Analysis of the electronic games market. Gives market value, manufacturers, profile of incidence in the home by area, age of consumers, profile of recent purchases, list of advertisers, distribution, and future outlook.

Ordenacion de la Economia Espanola

3039. RECONVERSION INDUSTRIAL DEL SECTOR FABRICANTE DE EQUIPO ELECTRICO PARA LA INDUSTRIA DE AUTOMACION. [Industrial Reconversion of the Electrical Equipment Manufacturing Sector for the Automation Industry.]. *Banco Hispano Americano. Servicio De Estudios. Spanish. no.32. January 1982. pg.25-32 Free (Spain).*

Decree as published in the Official Spanish State Bulletin, November 30, 1981, establishes objectives, benefits and resources for industrial reconversion in the electrical machinery industry. The reconversion plan is effective through 1988 and contains details of Spanish government aid by subsidies, credits, and guarantees to participating businesses.

Retail Business

3040. DRY CELL BATTERIES: SPECIAL REPORT NO. 3. *Economist Intelligence Unit Ltd. July 1981. pg.28-32*

Examines the 4 principal types of dry cell batteries in use in the United Kingdom (zinc carbon, alkaline manganese, mercu-

ric oxide and silver oxide). Discusses market size, usage of batteries, manufacturers and brands, prices, margins and pricing policy. Tables detail manufacturers' sales, imports / exports and advertising expenditures (1976-1980); market shares (1977, 1980); and prices for Vidor batteries.

ServEx: El Semanario del Comercio Exterior

3041. EL COMERCIO EXTERIOR DE ESPANA. [Spain's Foreign Trade.]. *Banco de Bilbao. Service de Estudios. Spanish. No.1362. July 23, 1981. pg.1-3 Free.*

Study provides 3 tables of Spain's foreign trade. Value of shipments of imports and exports are compared by month and total to date in 3 major product groups, for first 4 months of 1980 and 1981. Shows geographic distribution of Spanish foreign trade in 6 world sectors with value of shipments and comparison by world sectors. Value of shipments of imports and imports of 21 product sections are listed by Spanish customs product groups. All statistics cover first 4 months of 1980 and 1981. Text discusses advances and declines of Spain's foreign trade.

3042. LISTA - APPENDICE DEL ARANCEL DE ADUANAS: BIENES DE EQUIPO. [Supplement Customs Duties List: Machinery and Equipment.]. *Banco de Bilbao. Service de Estudios. no.1379. March 4, 1982. pg.7-10*

Spanish royal decree published in the Official Spanish State Bulletin on February 11, 1982, amplifies, modifies and extends customs duties on 60 machinery and equipment product groups. Electrical and electronically-controlled machinery is included. Alpha-numeric Spanish customs classifications, descriptions, percent of import duties and term in effect are shown in table form.

3043. MINISTERIO DE ECONOMIA Y COMERCIO: ARANCEL DE ADUANAS. [Ministry of Economy and Commerce: Customs Duties.]. *Banco de Bilbao. Service de Estudios. Spanish. No.1343. February 19, 1981. pg.11-15 Free.*

Tables show detailed classifications of electric and electronic equipment and respective Spanish customs duties as published in the official Spanish state bulletin (Boletin Oficial del Estado) Dec. 18, 1980. Items are classified with official Spanish customs alpha-numeric codes.

Strategies

3044. LE MARCHE DES PILES: FACE AUX AMERICANS, LES FRANCAIS N'ONT PAS DIT LEUR DERNIER MOT. [The Battery Market.]. *Vernes, A. Strategies. French. November 2, 1981. pg.56-64*

Reviews the French battery market. Shows 1980 turnover and exports, imports, consumption and investments in advertising for 1976-1980. Gives trademarks and advertising agencies.

L'Usine Nouvelle

3045. EQUIPEMENTS ELECTRONIQUES: LES PIONNIERS FRANCAIS. [Electronic Equipment: French Pioneers.]. *Le Boucher, Eric. Usine Nouvelle. French. no.28/1982. July 8, 1982. pg.42-47*

Examines French electronic equipment manufacturers' activities, and their attempts at market penetration. Graphs illustrate projected world market trends for microlithography equipment (used in the manufacture of integrated circuits) through 1985, and major producing nations' overall market shares through 1990.

3046. FILS ET CABLES - LES FRANCAIS VISENT LE RANG MONDIAL. [Wires and Cables - The French Aim for World Class.]. *Dalem, Didier. Edi Edition Documentation Industrielle. no.11/1982. March 11, 1982. pg.71-72*

Briefly discusses the structure of the French electrical wire and cable industry and its market. Quotes recent sales revenues, and employment figures for major manufacturers and for the industry as a whole and evaluates industry-wide production growth rates since 1971 and outlook through 1986. Identifies the 10 top-ranked producers on the world market, and the top 8 on the European market.

3047. PILES ELECTRIQUES: GIPELEC ET WONDER, DEUX ENTREPRISES SOUS TENSION. [Electric Batteries.]. *Dalem, Didier. Usine Nouvelle. French. no.39/1982. September 1982. pg.97-99*

Survey of the French market for batteries covers total unit sales (1979-1981), total shipments value (1976-1981), 1981 market shares of major domestic and foreign (i.e., U.S. and West German) suppliers, and 1976-1981 imports and exports. Discusses activities and market strategies of the 2 leading French manufacturers.

3048. TECHNOLOGIES 82 (SPECIAL): UNE SELECTION DE 500 PRODUITS INDUSTRIELS NOUVEAUX. [Technologies "82 (Special): A Selection of 500 New Industrial Devices.]. *Usine Nouvelle. French. Irregular. September 1982. 326 pg.*

Describes some 500 examples of state-of-the-art industrial equipment, in 17 product or end-use categories. Includes photographs and scattered specifications. A reader's service card provides access to manufacturers' literature, although manufacturers are not identified in print.

Wereldmarkt

3049. NEDERLANDSE METALEKTRO AKTIEF OP DE DUITSE MARKT. [Dutch Electric Motors Active on the German Market.]. *N.R.C. B.V. Afd. Tydschriften. Netherlandish. no.8. February 25, 1982. pg.16-17*

Examines activities of the Dutch electrotechnical industry in and exports to West Germany. Shows sales, 1980; exports, 1980; employees, 1980; and imports, and exports, 1979-1981.

Wireless World

3050. WIRELESS WORLD: NEW PRODUCTS. *IPC Electrical-Electronic Press Ltd. Division of IPC Industrial Press. Monthly. v.88, no.1559. August 1982. pg.82*

Monthly directory feature describes and reviews new electronic products. Provides names and addresses of manufacturers.

3051. WIRELESS WORLD: NEW PRODUCTS. *IPC Electrical-Electronic Press Ltd. Division of IPC Industrial Press. Monthly. November 1980.*

Lists electronic products associated with the radio, TV, and electrical industry including transistors, computer equipment, converters, scopes, measurement devices, etc. Data includes photograph, brief technical specifications, and addresses of manufacturers. Further information available through a reader card service.

See 488, 489, 490, 496, 501, 2042, 2047, 2050, 2063, 3425, 3902.

Numeric Databases

3052. COMPELEC. *G. Cam-Serveur A. Annual.*

Presents annual economic indicators for electronic components from 1974 forward.

3053. EDF - DOC. [EDF - DOC.]. *CISI On-Line Information Service. French. Monthly.*

Contains information relevant to the French electrical industry including source components, economics, management, applications, distribution, and social aspects. From 1972 forward.

Monographs and Handbooks

3054. CHIP TECHNOLOGY AND THE LABOUR MARKET. *Metra Consulting Group Ltd. 1980. 400 pg. 35.00 (United Kingdom).*

Analysis of the jobs and skills that will be affected by microelectronics and the possible gains in productivity and types of new work that may be created. Identifies sectors of office work and manufacturing industries where over 500,000 jobs may disappear by 1990.

3055. SPON'S MECHANICAL AND ELECTRICAL SERVICES PRICE BOOK. *Davis Belfield and Everest, Chartered Quantity Surveyors (London, Eng.). E. and F.N. Spon Ltd. 11th ed. 1979. 438 pg. 11.75 (United Kingdom).*

Provides extremely detailed listings of average materials / components / costs, wage rates, and professional fees involved in electrical and heating / ventilation / air conditioning building installations in Britain. Includes an estimating guide (generally providing per-meter or per-square-meter costs), and a contracting and estimating guide specifically for large industrial projects.

3056. ZAHN UND COMPANY, 200 JAHRE. [Zahn and Company, 200 Years.]. *Zahn and Company. German. March 1980. 30 pg.*

Commemorative of the 200th anniversary of the foundation of Zahn and Company, the Austrian manufacturer of chandeliers and electroliers, both metallic and crystal. Shows historical development of the company, different famous chandeliers manufactured, and a reference list of churches, castles and villas where such chandeliers hang.

Conference Reports

3057. WORLD ELECTRONICS - STRATEGIES FOR SUCCESS. *Mackintosh Publications Ltd. 1st ed. 1981. 65.00 (United Kingdom).*

Proceedings of the Mackintosh / Financial Times Conference at Monte Carlo, May 1980. Collection of personal views by leading executives in the electronics industry on three main themes: industry strategies; the challenge to U.S. domination from France, Japan, Korea, and the Arab World; and the 'Information Society.'

Newsletters

See 2069.

Services

3058. ELECTRICALS MONTHLY REVIEW. *Cullen, A.S. De Zoate and Bevan. Monthly. August 22, 1979. 21 pg.*

Monthly review for private circulation gives financial analyses and corporate reports for main U.K. electrical and electronic corporations. Shows share prices, profits and sales.

3059. MICROPROCESSORS AND THE SMALL BUSINESS. *Hulse, S. U.K. Department of Industry. Irregular. 1st ed. April 1981. 23 pg.*

This is a service by the Department of Industry, intended for use by small companies. Gives an introduction to microelectronics for office or industrial use.

See 506.

Indexes and Abstracts

3060. ELECTRONICS AND COMMUNICATIONS ABSTRACTS. *Multi-Science Publishing Company Ltd. Monthly. 1980.*

Monthly abstract service covers electronics and communication science and their applications.

3061. VENTURECASTS: THE 1982 EUROPEAN DATA BOOK. *Venture Development Corporation. 1982. $595.00.*

Provides abstracts of market analysis and forecasts from 1981 electronics and business periodicals, and associations' and government statistics covering electronics, computers and communications equipment in Europe. Includes sales, shipments, revenues, installed base, annual growth rates, consumption, expenditures, market size, production, and imports and exports by product groups.

Eastern Europe

Indexes and Abstracts

3062. BILTEN DOKUMENTACIJE: ELEKTROTEHNIKA. [Bulletin of Documentation: Electrical Engineering.]. *Yugoslavia. Centre for Technical and Scientific Documentation. Monthly. 1981.*

Monthly guide to electrical engineering articles and books.

3063. REFERATIVNYI ZHURNAL: ELEKTRONIKA I EE PRIMENENIE: ABSTRACTS JOURNAL: ELECTRONICS AND ITS UTILIZATION. *Vsesoiuznyi Institut Nouchnoi Informatsii. Monthly. 1981.*

Monthly electronics abstracts journal, in Russian with contents in English.

See 4334.

Asia

Market Research Reports

3064. THE ASIAN ELECTRONICS INDUSTRY 1982-1984. *Venture Development Corporation. 1982.*

Analysis of the electronics industry in Asia, with forecasts, 1982-1984. Covers computers, components, instrumentation and consumer goods. Gives data for production, consumption, imports, exports, markets, and manufacturers.

3065. THE CONSTRUCTION INDUSTRY IN THE FAR EAST: PRODUCT REPORT: LIGHTING AND LIGHTING FIXTURES. *Business Management and Marketing Consultants Ltd. Industrial Market Research Ltd. July 1981. $2000.00.*

Analysis of the construction industry in South East Asia and the market for lighting and lighting fixtures. Examines imports, end uses, suppliers, distribution, prices, user attitudes, legislation and regulations, and market penetration strategies. Includes a directory of manufacturers.

3066. CURRENT ELECTRONIC COMPONENTS INDUSTRY IN JAPAN. *Mackintosh Publications Ltd. 1981. 155 pg. 110.00 (United Kingdom).*

Examines the present structure and performance of the industry, detailing sales by product, 1975-1980; sales / profits by manufacturer (200 companies); end-user markets; exports / imports, captive production / sub-contractors / new entrants: technical trends and future demand prospects, for semiconductor and passive components in Japan. Provides an analysis of the electronic components market.

3067. CURRENT INDUSTRIAL ROBOTS INDUSTRY IN JAPAN. *Yano Economic Research Institute Co., Inc. Information Researchers, Inc. November 1981. 180 pg. $426.00.*

Analysis of the industrial robots industry in Japan. Includes industry structure; distribution channels; demand trends; new products; and production, and market shares of manufacturers.

3068. ELECTRONIC COMPONENTS, HONG KONG: A COUNTRY MARKET SURVEY. *U.S. Government Printing Office. Irregular. CMS 79-016. March 1979. 7 pg. $.50.*

Report analyzes the current and projected markets for the sale of electronic components in Hong Kong. Data is provided on the total market (production, imports, exports, market size); manufacturers' sales of selected products to domestic and export markets; current number of businesses and estimated purchases by major end-user sectors; best sales opportunities for U.S. exporters. Describes the domestic market and competitive environment, and provides an end-user analysis and information on trade and technical regulations.

3069. ELECTRONIC COMPONENTS, SINGAPORE: A COUNTRY MARKET SURVEY. *U.S. Government Printing Office. Irregular. CMS 79-015. March 1979. 7 pg. $.50.*

Report analyzes the current and projected markets for the sale of electronic components in Singapore. Data is provided on the total market (production, imports, exports, market size); current number of businesses and estimated purchases by major end-user sectors; manufacturers' sales of selected products to domestic and export markets; best sales opportunities for U.S. exporters. Describes the domestic market and competitive environment, and provides an end-user analysis and information on marketing practices.

3070. EQUIPMENT AND MATERIALS FOR ELECTRONIC COMPONENTS PRODUCTION, HONG KONG: A COUNTRY MARKET SURVEY. *U.S. Government Printing Office. Irregular. CMS 79-114. February 1979. 7 pg. $.50.*

Report analyzes the current and projected markets for the sale of equipment and materials for electronic components production in Hong Kong. Data is provided on the total market (production, imports, market size); imports by type, total and amount from U.S.; end-user profile of domestic manufacturers; best sales opportunities for U.S. exporters. Describes the production equipment and materials market, the electronic components industry, business practices and tariffs.

3071. EQUIPMENT AND MATERIALS FOR ELECTRONIC COMPONENTS PRODUCTION, JAPAN: A COUNTRY MARKET SURVEY. *U.S. Government Printing Office. Irregular. CMS 79-110. February 1979. 8 pg. $.50.*

Report analyzes the current and projected markets for the sale of equipment and materials for electronic components production in Japan. Data is provided on the total market (production, imports, exports, market size); imports by type, total and amount from U.S.; end-user profile of domestic manufacturers; best sales opportunities for U.S. exporters. Describes the production equipment and materials market, the electronic components industry, business practices and tariffs.

3072. EQUIPMENT AND MATERIALS FOR ELECTRONIC COMPONENTS PRODUCTION, TAIWAN: A COUNTRY MARKET SURVEY. *U.S. Government Printing Office. Irregular. CMS 79-106. February 1979. 7 pg. $.50.*

Analyzes the current and projected markets for the sale of equipment and materials for electronic components production in Taiwan. Data is provided on the total market (production, imports, exports, market size); imports by type, total and amount from U.S.; end-user profile of domestic manufacturers; and best sales opportunities for U.S. exporters. Describes the production equipment and materials market, the electronic components industry, and business practice and tariffs.

3073. THE JAPANESE MARKET FOR AUTOMOTIVE MAINTENANCE EQUIPMENT AND ACCESSORIES: FOREIGN MARKET SURVEY REPORT. *Kearney (A.T.) International, Inc. Tokyo (Japan) Industry and Trade Administration. U.S. Department of Commerce. National Technical Information Service. DIB-80-02-503. 1979. 54 pg. $10.00.*

The market research was undertaken to study the present and potential U.S. share of the market in Japan for automotive maintenance equipment and accessories to examine growth trends in Japanese end-user industries over the next few years; to identify specific product categories that offer the most promising export potential for U.S. companies; and to provide basic data which will assist U.S. suppliers in determining current and potential sales and marketing opportunities. The trade promotional and marketing techniques which are likely to succeed in Japan were also reviewed.

3074. JAPANESE PROSPECTS IN THE WORLD IC MARKET. *Nomura Securities Co., Ltd. Frost and Sullivan, Inc. April 1980. 76 pg. $600.00.*

Market analysis for Japanese involvement in the world semiconductor industry. Profiles manufacturers, including product lines and market shares. Covers industry structure, production, technology, and raw material consumption. End uses analyzed.

3075. THE JAPANESE SEMICONDUCTOR INDUSTRY 1981-82. *Silin, Robert. Consulting Group of Bank of America's BA Asia Ltd. Japanese Semiconductor Study. 1982. 300 pg. $690.00.*

Discusses production and consumption in Japan by product subgroups with projections of future trends. Includes import / export data with short-term projections for imported products and export volume. Data on production, market shares and general activities of Japanese products are detailed. Additional information includes: sales breakdowns, new products, major foreign semiconductor manufacturers selling in Japan, and manufacturers and importers of equipment.

3076. MARKET RESEARCH STUDY ON MOTOR VEHICLE MAINTENANCE EQUIPMENT AND REPLACEMENT PARTS, TAIWAN: FOREIGN MARKET SURVEY REPORT. *China Credit Information Service ltd., Taipei (Taiwan) Industry and Trade Administration. U.S. Department of Commerce. National Technical Information Service. DIB-80-02-515. 1979. 96 pg. $10.00.*

The market research was undertaken to study the present and potential U.S. share of the market in Taiwan for motor vehicle maintenance equipment and replacement parts; to examine growth trends in Taiwan end-user industries over the next few years; to identify specific product categories that offer the most promising export potential for U.S. companies; and to provide basic data which will assist U.S. suppliers in determining current and potential sales and marketing opportunities. The trade promotional and marketing techniques which are likely to succeed in Taiwan were also reviewed.

3077. NEW CERAMIC MARKET IN JAPAN. *Yano Economic Research Institute. Benn Publications Ltd. 1982. 230 pg. $695.00.*

Analyzes Japanese ceramic market, detailing major manufacturers and their research and development activities. Covers market size and shares by type of ceramics (i.e., sensors, IC packages, piezoelectric, condensor, thermistor, varistor) as well as by type of user markets; patent and technical trends and applications by major manufacturers. Profiles of major manufacturers include historical sales, number of employees, sales outlets, and materials and products.

3078. PRINTED CIRCUITS IN JAPAN. *BPA Management Services. February 1981. 55 pg. 250.00 (United Kingdom).*

Analysis of the printed circuit industry in Japan with forecasts to 1985. Covers production costs, operating performance, productivity, government aid, investments, market analysis including prices, research and development, and other economic indicators, Shows exports, imports, demand, production and 1976-1980, 1985 sales, all by product.

3079. PRODUCTION MACHINERY, TOOLS AND SPECIAL MATERIALS FOR ELECTRONIC COMPONENTS (KOREA): FOREIGN MARKET SURVEY REPORT. *ASI Market Research, Inc. (Korea). U.S. Department of Commerce. National Technical Information Service. DIB-80-07-503. 1980. 31 pg. $10.00.*

The market research was undertaken to study the present and potential U.S. share of the market in Korea for production machinery, tools and special materials for electronic components; to examine growth trends in Korean end-user industries over the next few years; to identify specific product categories that offer the most promising export potential for U.S. companies; and to provide basic data which will assist U.S. suppliers in determining current and potential sales and marketing opportunities. The trade promotional and marketing techniques which are likely to succeed in Korea were also reviewed.

3080. PROFILE OF ELECTRONIC COMPONENTS MANU-FACTURERS IN JAPAN. *Yano Economic Research Institute Co., Ltd. Information Researchers, Inc. March 1981. 180 pg. $260.00.*

Survey of electronic components manufacturers in Japan. Gives performance data, sales, net profits, exports, ratio of final demands, distribution channels, and end users.

3081. SOLID STATE MEMORY MARKET IN THE FAR EAST. *Frost and Sullivan, Inc. December 1980. 584 pg. $1600.00.*

Solid state memory market analysis and forecasts to 1990 for 8 Far Eastern countries including Japan. Includes exports to the U.S., major consumers and vendors, new products, economic indicators, market segmentation, end uses, foreign trade and distribution.

3082. SURVEY OF JAPAN MARKET FOR MICROWAVE COMPONENTS, SYSTEMS AND TEST EQUIPMENT: FOREIGN MARKET SURVEY REPORT. *Pacific Projects Ltd. Tokyo (Japan). U.S. Department of Commerce. National Technical Information Service. DIB-80-02-504. 1979. 121 pg. $10.00.*

The market research was undertaken to study the present and potential U.S. share of the market in Japan for microwave components, systems and equipment; to examine growth trends in Japanese end-user industries over the next few years; to identify specific product categories that offer the most promising export potential for U.S. companies; and to provide basic data which will assist U.S. suppliers in determining current and potential sales and marketing opportunities. The trade promotional and marketing techniques which are likely to succeed in Japan were also reviewed.

See 529, 2072, 2077, 2080, 2081, 2084, 2085, 3925, 3935.

Financial and Economic Studies

3083. EXPORTING TO JAPAN: ENERGY CONVERSION / CONSERVATION EQUIPMENT. *Metra Consulting Group Ltd. 1981. 10.00 (United Kingdom).*

One in a series of 13 reports motivated by the Commission of the European Community's decision to support European exports to enter the Japanese market. Series examines specific sectors offering growth and export potential. Reports are available only to companies with bases within the EEC.

Forecasts

3084. SEMICONDUCTOR / MICROELECTRONIC INDUSTRY IN JAPAN. *Electronic Trend Publications. 1982. $495.00.*

Report covers the semiconductor and integrated circuit industry and gives market value, production and growth rates. Lists investment in facilities and market shares, since 1976.

Directories and Yearbooks

3085. ASIAN ELECTRICAL AND ELECTRONICS TRADE DIRECTORY. *Croner Publications, Inc. Annual. 1980. 512 pg. $78.00.*

Covering eight nations, directory details address, telephone, cable and/or telegraphic address, executives' and agents' names, annual sales, issued capital, and other data for electrical and electronic goods manufacturers, suppliers, wholesalers, agents, and importers.

3086. ASIAN ELECTRICAL AND ELECTRONICS TRADES DIRECTORY 1979/80. *Maruzen Company Ltd. Irregular. 1979. $148.00.*

Coverage of 3800 firms in Indonesia, Hong Kong, Korea, Malaysia, Philippines, Singapore, Taiwan, and Thailand. Spans executives' names, annual sales, products, services, address, telephone, telex, number of employees, capital (U.S. $), and banking source.

3087. ASIAN SOURCES: ELECTRONICS COMPONENTS. *Pacific Subscription Service. Monthly. 1982. 300 pg. $185.00/yr.*

One in a series of six illustrated monthly guides to export products availabel from Asian nations. Prices and other trade information for integrated circuits displays, resistors, PCB's, capacitators, tape mechanisms, testing machinery, measuring instruments, power supplies, and parts and accessories.

3088. INDIAN ELECTRICAL YEARBOOK. *M. Largo-Alfonso. Annual. 1980. 560 pg. $5.50.*

Directory details electrical goods manufacturers and their representatives, suppliers, exporters, and contractors, with a classified index.

3089. JAPAN EBC ELECTRONICS BUYERS GUIDE. *DEMPA Publications, Inc. Annual. 1980.*

Annual volume includes a directory of about 1,800 electronic products; manufacturers' directory of about 1,200 companies; directory of 520 trading firms; Japaneses firms' branches and agents. Provides trade names; foreign firms acting as representatives; and organizations involved in the electronic field.

3090. JAPAN FACT BOOK ''80. *DEMPA Publications, Inc. Annual. 1980. $25.00.*

1980 annual directory lists the facts on over 60 of Japan's electronics leaders. Provides summaries of businesses, business outlook, and latest products and technological achievements. Includes statistics on trade patterns, home appliances, electronic materials, parts and components, microprocessors, facsimiles and computers.

3091. TAIWAN CONSUMER AND INDUSTRIAL PRODUCTS: ELECTRONICS. *U.S. International Marketing Company. Monthly. 1981. 120 (est.) pg. $50.00/yr.*

One in a series of nine illustrated buyers' guides, directory includes prices and manufacturers of electronics.

3092. TAIWAN CONSUMER AND INDUSTRIAL PRODUCTS: ELECTRONICS COMPONENTS. *U.S. International Marketing Company. Quarterly. 1981. 120 (est.) pg. $50.00/yr.*

One in a series of nine illustrated buyers' guides, directory includes prices and manufacturers of electronics components.

3093. TAIWAN ELECTRONICS COMPONENTS. *U.S. International Marketing Company. Quarterly. 1982. 120 pg. $50.00/yr.*

One in a series of nine illustrated buyer's guides to Taiwanese consumer and industrial products.

See 2087, 3940.

Special Issues and Journal Articles

Asian Finance

3094. PROFILE JAPAN: THE TOKYO BOOM: A KILLING AT KABUTOCHO. *Asian Finance Publications Ltd. August 15, 1981. pg.60-61*

Examines foreign investment in Japan's high technology industries. Tables show foreign stock ownership in 40 selected businesses as of March 31, 1981, and the increase in the number of foreign shares from 1980; stock price indices (1975-1980); foreigner's acquisition of stocks, debentures and beneficiary certificates (1981); and foreign investment (1974-1980) by type of security.

Aviation Week and Space Technology

3095. JAPANESE PRODUCTION RUNS LIMIT ROBOTIC INVESTMENT. *McGraw-Hill Book Company. v.112, no.5. August 2, 1982. pg.91-92*

Discusses investments in industrial robots by Japanese companies. Figures illustrate 1980 robot procurements, in millions of dollars, by 16 Japanese industry segments. Additional figure illustrates sales and production trends of industrial robots in Japan, 1968-1980.

Business Japan

3096. BRIGHT OUTLOOK FOR JAPAN'S BATTERY MANU-FACTURERS. *Yamashita, Sadao. Nihon Kogyo Shimbun Publications. v.27, no.1. January 1982. pg.117-122*

Examines the Japanese battery industry. Provides data on the following: production of battery-operated devices in Japan, 1975-1980; production by battery type, 1976 and 1980; production by battery type and size, 1979-1981; and battery exports by destination, 1976-1980. Additional tables give descriptions of primary batteries generally used. Source is Japan Dry Battery Industries Association.

Circuits Manufacturing

3097. CIRCUITS NEWS: ROBOTS HITTING THE BEST SELLER LIST. *Morgan Grampian Publishing Company. v.22, no.8. August 1982. pg.18*

Summarizes report of several marketing firms concerning the robot and automated machine tool industry. Table provides figures for projected U.S. robot shipments, 1982-1992. (Source: International Resource Development).

3098. JAPAN'S SEMICONDUCTOR INDUSTRY. *Morgan Grampian Publishing Company. Irregular. March 1981. pg.58-77*

Ranks production, imports, exports, and shipments of Japanese companies, by sales and profits. Shows sales by U.S. companies to Japan, by value and percent of market. Examines consumption by type of device and outlook for trade opportunities.

Computer and Electronics Marketing

3099. JAPAN PC BOARD INDUSTRY REPORT. *Kubota, Satoshi. Morgan-Grampian, Inc. October 1981. pg.6-8*

Gives total Japanese production, value in dollars and yen, and per cent of total for various types of industrial and consumer printed circuit boards. Forecasts 1981 production, and ana-

lyzes value and production based on company size; volume of exports since 1974; 1979-1980 exports by country of destination; average sales price of PC boards; and plant / production equipment investments, 1978-1980.

Eastern Economist

3100. DEMAND FOR CAPACITORS. *Eastern Economist Ltd. Irregular. v.79, no.2. July 11, 1982. pg.83*

Includes figures showing 1982 capacity, 1982 exports, and 1978-1979 production. Examines shortages. Provides names, addresses, and products manufactured for manufacturers of ceramic and electrolytic capacitors.

3101. TOWARD BETTER ELECTRONIC COMPONENTS. *Singh, Hardev. Eastern Economist Ltd. May 16, 1980. pg.1003-1004*

Reviews Indian production and trade of electronic components, such as radios and televisions. Annual data includes production (1960, 1965, 1970-1979); demand (1971-1978); exports (1971-1978); imports (1971-1978); growth and potential growth (1971-1983). Also gives production data (1975-1979) of electronic components by U.S., Japan, W. Europe, and India. Discusses the potential of semiconductor production.

Economic Review

3102. FUTURE PROSPECTS OF ELECTRONICS INDUSTRY IN THE REPUBLIC OF CHINA. *International Commercial Bank of China. October 1980. pg.1-4*

Reviews the electronics industry in Taiwan and assesses its future prospects. Statistics given for number of companies (including foreign firms) and average capital (1978); number of employees (1978); production (1977, 1978); value added (1977); production by major product (1977); employment structure (1977); average monthly salaries (1974-1978); principal sources of R & D (1978); and R & D expenditures (1978).

Economic World

3103. QUALITY AND VOLUME PRODUCTION: KEY TO EXPANDING JAPANESE SHARE IN IC MARKET. *Misra, Prashanta. Economic Salon Ltd. March 1981. pg.26-30*

Examines Japan's share of the U.S. semiconductor market. Discusses the different types of RAMs; major competitors; top businesses in the market; Japanese R & D expenditures from the government; and industry cooperation. Forecasts (1980-1983) the outlook for the world semiconductor market, including solid state, integrated circuits, and discrete devices.

The Economist

3104. TOMORROW'S LEADERS: A SURVEY OF JAPANESE TECHNOLOGY. *Economist Newspaper Ltd. v.283, no.7242. June 19, 1982. 21 pg.*

Surveys the Japanese electronic industry. Gives Japan's percent share of consumer electronics world production, and dollar exports of 4 products and export restraint index, 1963-1985. Provides estimated 1982 shipments of 64K computers by brand to U.S. and other markets; and volume and yen value of shipments of office computers, 1967-1980. Shows percentage of inventions for U.S., Great Britain, France and West Germany, 1750-1950; and percent of foreign patents filed in U.S. by Japan, Great Britain, France and West Germany, 1883-1978.

3105. THE MICROCHIP BATTLE: THE 64K QUESTION. *Economist Newspaper Ltd. v.283, no.7237. May 15, 1982. pg.94-95*

Includes chart providing statistical and graphical data for the world market for 64k rams. Chart includes figures for total value in million dollars, units in million dollars and average selling price in dol

Electronic Business

3106. ELECTRONICS PRODUCTION ESCALATES IN KOREA. *Cahners Publishing Company. November 1980. pg.86-87*

Examines the electronics market in Korea. Graphs and diagrams display Korean production, exports and imports, and growth of foreign investment in Korea (1971-1979). Gives Korean production by major markets and exports by major products (1979).

3107. THE JAPANESE INVASION: CHIPS NOW, COMPUTERS NEXT? *Cahners Publishing Company. July 1981. pg.84-88*

Presents highlights from a report by Quantum Science Corp. on Japan's domination of the world semiconductor market by 1985. Graphs show Japanese shipments (1976-1979) to the U.S.; and semiconductor production (1979-1983) by U.S., Japan and Western Europe. Ranks 7 Japanese computer businesses by R & D expenditures.

3108. PRINTED CIRCUITS IN JAPAN. *Cahners Publishing Company. v.8, no.4. April 1982. pg.120*

Provides statistical data for present and projected Japanese printed-circuit-board market. Projections also provided for Japanese industrial and consumer electronics market.

3109. SEMICONDUCTOR MARKET IN JAPAN TAKES ON GREATER SOPHISTICATION. *Cahners Publishing Company. Irregular. September 1980. pg.115, 118-124*

Examines the semiconductor market in Japan. Tables show Japanese consumption (1977-1979) of semiconductors; rankings of semiconductor manufacturers according to 1979 total sales, with total market shares and discrete market shares (1979); and rankings of U.S. semiconductor businesses in Japan according to 1978 sales.

3110. STREAM OF HONG KONG EXPORTS CONTINUES TO WORLD MARKETS. *Cahners Publishing Company. October 1980. pg.126*

Tables rank Hong Kong electronics exports (1979) by 15 world markets with value of end products and parts and components. Gives electronic exports (1978-1979) by type of product.

Indonesian Commercial Newsletter

3111. DEVELOPMENTS IN THE ELECTRIC LIGHT INDUSTRY; A COMMODITY SURVEY. *P.T. Data Consult Inc. no.176. June 22, 1981. pg.4-14*

Covers production and production capacity, components, imports of bulbs, projected consumption, and marketing. Appendix contains a list of electric lightbulb and fluorescent lamp companies.

Insulation / Circuits

3112. THE ROAD TO SINGAPORE LEADS TO ELECTRONICS. *Goh, T.C. Electover, Ltd. (Singapore). Lake Publishing Corporation. July 1982. pg.15-16*

Reviews the electronic industry of Singapore giving 1981 data for employment; value added per worker; output volume and value; trade and trade directions; semiconductor and other component production. Discusses government policies and future outlook.

Journal of the Electronics Industry

3113. JEI STATISTICAL INDICATORS. *DEMPA Publications, Inc. Monthly. January 1981. pg.78-79*

Monthly tables provide value and volume for products in Japan's electronics industry, including consumer and industrial electronics and general parts. Data is shown for 4 months previous to date of issue. One graph displays production for the past 2 years.

3114. JEI VITAL EXPORT STATISTICS. *DEMPA Publications, Inc. Monthly. January 1981. supplement*

Monthly supplement insert provides export statistics for Japan's consumer and industrial electronics industry. Gives volume and value for products exported from Japan. Data shown for 3 months previous to date of issue.

Modern Asia

3115. JAPANESE MOVE FAST TO CATCH UP IN THE IC BUSINESSES. *Fujii, John. Johnston International Publishing Corporation. v.15, no.4. May 1981. pg.109-112*

Examines Japan's semiconductor industry. Discusses major manufacturers, new technology, foreign trade, and growth. Two tables show Japan's semiconductor exports and imports worldwide and with the U.S. (1976-1980).

Semiconductor International

3116. JAPANESE CONTINUE EXPANDING SEMICONDUCTOR FACILITIES. *Cahners Publishing Company. v.5, no.6. June 1982. pg.24*

Discusses capital spending of 5 corporations including allocations. Compares April 1982 to previous year.

3117. PHILIPPINE UPDATE. *Iscoff, Ron. Cahners Publishing Company. v.5, no.8. August 1982. pg.97-114*

The growth in exports of electronic products from the Philippines is discussed. Tables detail U.S.-owned semiconductor companies with Philippine-based assembly and test facilities; and semiconductor assemblers ranked by capacity, with U.S. marketing subsidiaries as representatives. Data on exports (1981) and capacity of selected facilities is also presented. Issues in management and labor are discussed, with comments from industry executives.

Solar Age

3118. MY FACT-FINDING TOUR OF JAPAN: PHOTOVOLTAICS. *Maycock, Paul. Solar Vision, Inc. v.7, no.9. September 1982. pg.14*

Reports results of fact-finding tour to major Japanese PV producers, various research labs and many of the project managers working with the government's Sunshine Project. Data provided for Japanese share of world market in PV industry; Japanese module production, 1984; and Japanese and American competition in the amorphous silicon industry.

See 2091, 2097, 2100, 2101.

Monographs and Handbooks

3119. SEMICONDUCTOR INDUSTRY AND R & D IN INDIA. *United Nations. Industrial Development Organization Industrial Information Section. Industrial and Technological Information Bank. 1981. 16 pg. Free.*

With emphasis on semiconductor components, report covers facilities, capacities, and production planning in major public enterprises and private enterprises; semiconductor research and development; and research centers and laboratories.

Newsletters

3120. JET (JAPAN ELECTRONICS TODAY) NEWS. *Mackintosh Publications Ltd. Semi-monthly. $495.00/yr.*

Newsletter summarizes recent information on the Japanese electronics industry. Includes new products and technologies, applications, company results, foreign trade statistics, and announcements of exhibitions, conferences and publications.

See 3945.

Indexes and Abstracts

3121. AMFETEX JOURNAL OF JAPANESE ELECTRONICS AND POWER ENGINEERING. *Amfetex, Inc. Monthly. 1981.*

Monthly periodical provides English - language abstracts of electronics and power engineering articles in Japanese journals.

3122. VENTURECASTS: THE 1982 ASIAN DATA BOOK. *Venture Development Corporation. 1982. $595.00.*

Provides abstracts of market analysis and forecasts from 1981 electronics and business periodicals, and associations' and government statistics covering electronics, computers, and communications equipment, in Asia. Includes sales, shipments, revenues, installed base, annual growth rates, consumption, expenditures, market size, production, and imports and exports by product groups.

Oceania

Market Research Reports

3123. ELECTRONIC COMPONENTS, AUSTRALIA: A COUNTRY MARKET SURVEY. *U.S. Government Printing Office. Irregular. CMS 79-014. March 1979. 10 pg. $.50.*

Report analyzes the current and projected markets for the sale of electronic components in Australia. Data is provided on the total market (production, imports, exports, market size); manufacturers' sales of selected products to domestic and export markets; current number of businesses and estimated purchases by major end-user sectors; best sales opportunities for U.S. exporters. Describes the domestic market and competitive environment, and provides an end-user analysis and information on marketing practices) and trade and technical regulations.

3124. EQUIPMENT AND MATERIALS FOR ELECTRONIC COMPONENTS PRODUCTION, AUSTRALIA: A COUNTRY MARKET SURVEY. *U.S. Government Printing Office. Irregular. CMS 79-109. February 1979. 7 pg. $.50.*

Analyzes the current and projected markets for the sale of equipment and materials for electronic components production in Australia. Data is provided on the total market (production, imports, exports, market size); imports by type, total and amount from U.S.; end-user profile of domestic manufacturers; and best sales opportunities for U.S. exporters. Describes the production equipment and materials market, the electronic components industry, and business practices and tariffs.

Industry Statistical Reports

3125. PRICE INDEX OF ELECTRICAL INSTALLATION MATERIALS. *Australian Bureau of Statistics. Monthly. 1981. 2 pg.*

Price indexes for conductors; conduits and accessories; switchboard and switchgear materials; and all groups of electrical installation materials.

Financial and Economic Studies

3126. THE AUSTRALIAN ELECTRONICS INDUSTRY: A REPORT BY THE ELECTRONICS INDUSTRY ADVISORY COUNCIL. *Australian Government Publishing Service. 1980. 106 pg. 3.80 (Australia).*

Report by the Electronics Industry Advisory Council to the Australian government analyzes the present and future prospects for the Australian electronics industry. Topics covered include a history of the industry; main sectors; research and development; government purchasing; other government policies; employment, education, and training; and the industry's future.

Directories and Yearbooks

3127. AUSTRALIAN ELECTRONICS DIRECTORY. *Technical Indexes Pty. Ltd. Annual. 1980. 750 pg. $60.00.*

Australian suppliers directory covers domestic and imported electronic investments, components, and systems.

3128. DIRECTORY OF ELECTRONICS AND INSTRUMENTATION, 1980. *Beckett, Robin, ed. Associated Group Media Ltd. Annual. March 1980. 140 pg. $7.00.9.00 (New Zealand).*

New Zealand directory includes indexes of product categories, standard specifications, overseas principals and laboratories in the field of electronics and instrumentation. Includes members of National Electronics Development Assn. (NEDA) and the Automatic Control and Instrumentation Society (ACIS).

3129. THE ELECTRICAL DIRECTORY. *Technical Publications Ltd. Annual. 1980. 168 pg. 3.50 (New Zealand).*

Manufacturers and suppliers directory includes a trade name index and listings of foreign manufacturers with New Zealand representation; consulting engineers, electrical contractors, government agencies, and electrical supply authorities.

3130. MINGAY'S PRICE SERVICE: ELECTRICAL EDITION. *Thomson Publications. Semi-annual. 1981. 240 pg. 40.00 (Australia).*

Provides prices and brand names of electrical products.

Special Issues and Journal Articles

Insulation / Circuits

3131. ELECTRONICS DOWN UNDER: AN UPDATE ON THE ELECTRONICS INDUSTRY IN AUSTRALIA. *Mong, Francis. Lake Publishing Corporation. v.28, no.10. September 1982. pg.23-24*

Discusses Australian electronics industry giving 1981 market size figures for telephone and communications equipment market, radio communications market, consumer and industrial markets. Table shows 1981 domestic sales figures for consumer and industrial products. Provides production, export, import, and demand figures for components. Discusses semiconductors and future of industry.

Middle East

Market Research Reports

3132. BATTERIES RETAIL AUDIT REPORT. *Middle East Marketing Research Bureau Ltd. Bi-monthly. 1981. 45 (est.) pg.*

Bimonthly market analysis and audit of retail sales of batteries in Saudi Arabia.

3133. ELECTRONIC COMPONENTS, ISRAEL: A COUNTRY MARKET SURVEY. *U.S. Government Printing Office. Irregular. CMS 79-013. March 1979. 7 pg. $.50.*

Report analyzes the current and projected markets for the sale of electronic components in Israel. Data is provided on the total market (production, imports, exports, market size); manufacturers' sales of selected products to domestic and export markets; current number of businesses and estimated purchases by major end-user sectors; and best sales opportunities for U.S. exporters. Describes the domestic market and competitive environment, and provides an end-user analysis and information on marketing practices and trade and technical regulations.

See 537.

Africa

Financial and Economic Studies

3134. DEVELOPMENT OF ELECTRONICS INDUSTRY IN KENYA: FINAL REPORT. *United Nations. Industrial Development Organization Industrial Information Section. Industrial and Technological Information Bank. 1981. 80 pg. Free.*

Report covers industry history in Kenya and present industry structure; economic and political constraints; domestic market for electronic products; short-term strategy (manufacture of certain components to increase domestic content); proposed actions (government policy, creation of a development center); training requirements, technical assisstance, and sources of know-how; recommendations; tables.

Office Equipment

International

Market Research Reports

3135. FACSIMILE AND ELECTRONIC MAIL. *Creative Strategies International. June 1980. $1500.00.*

Examines markets for convenience and operational facsimile. Includes a description and prospects for the various forms of "electronic mail." Forecasts by product type for facimile systems in units and dollars through 1984. Competitive market shares also provided. Evaluates "electronic mail" features that will contribute to facsimile growth and those that will compete. Provides detailed analysis of worldwide competitors.

3136. POINT-OF-SALE SYSTEMS. *Predicasts, Inc. June 1980. $325.00.*

International market analysis for point-of-sale (POS) equipment for supermarkets, department stores, and other retail establishments. Covers manufacturers and new products. Forecasts installations, by end users.

3137. TYPEWRITERS / AUTOMATIC AND WORD PROCESSORS. *Larsen Sweeney Associates Ltd. Annual. 1981. 995.00 (United Kingdom).*

Market analysis for automatic typewriters and word processors, with forecasts to 1986. Covers 8 product groups. Includes analysis of marketing and the industry by country, and surveys of end users and producers.

3138. WORD PROCESSING EQUIPMENT. *Predicasts, Inc. December 1980. $325.00.*

International market analysis for word processing equipment. Reviews manufacturers activities, product performance, and economic indicators.

See 7, 773, 783.

Financial and Economic Studies

3139. WORD PROCESSING INDUSTRY: A STRATEGIC ANALYSIS. *Venture Development Corporation. 1981. 250 pg. $2250.00.*

Market, industry, and technology analysis and forecasts within and outside the U.S. cover installed base at end of 1980; shipments through 1986 by type of system and end user segment; manufacturers' market shares; current and potential users including type of system owned, purchase plans, and appeal of new and proposed product features. Strate

Directories and Yearbooks

3140. DIRECTORY OF SELF-ADHESIVE AND HEATSEAL LABELLING MACHINES AND IMPRINTERS. *Zusammengestellt von Reclige par: C.G. Wedgwood & Co. (London). FINAT. English/French/German. Annual. 2nd ed. 1979. 223 pg.*

Contents include a directory of manufacturers; list of products and manufacturers and product directories (with such data as manufacturer, model, type, function, label size, printup, output, machine dimensions, facilities, electrical supply) for applications, applicators for self-adhesive labels, overprinting machines for labels, label dispensers, price-weight labelling scales, computer-output address labelling machines and hand-held labelling guns.

Special Issues and Journal Articles

Buro Report

3141. BUROFRUHLING IN WIEN: IFABO "82. [Office Spring in Vienna: IFABO "82.]. *Verlag Technik Report Ges. m.b.H. German. Irregular. May 1982. pg.63-83*

Reviews IFABO "82 (The Intern. Fair on Office Equipment held in Vienna, May 12-15, 1982). Contains a comprehensive review of new products of 294 exhibitors from 12 countries.

3142. IFABO "80. [IFABO "80.]. *Verlag Technik Report Ges. m.b.H. German. Annual. April 1980. pg.29-75*

Special issue is devoted to IFABO "80 (International Fair of Office Organization in Vienna May 7-10, 1980). Exhibited products were minicomputers, display terminals, telecommunications, register and mail handling equipment, general office equipment and office furniture from 221 manufacturers all over the world. Includes descriptions of products and names and addresses of manufacturers.

3143. IFABO "81: 250 DIREKTAUSSTELLER. [IFABO "81: 250 Direct Participants.]. *Verlag Technik Report Ges. m.b.H. German. Annual. May 1981. pg.42-78*

Special issue is devoted to IFABO "81 (International Fair of Office Organization) in Vienna (May 13-16, 1981). Exhibited products were minicomputers, display terminals, telecommunications, register and mail handling equipment, general office equipment and office furniture from over 250 manufacturers all over the world. Includes description of products and names and addresses of manufacturers.

See 42.

Monographs and Handbooks

See 838.

Newsletters

See 864.

Services

See 866.

Indexes and Abstracts

See 875.

North America

Market Research Reports

3144. THE ADVANCED WORK STATION: BEYOND WORD PROCESSING. *Yankee Group. December 1979. $9500.00.*

An analysis of developments in the integrated office automation field. Focuses on the word processing industry (installed

base 1978, shipments 1978-1979, and vendor market shares 1978-1979) and looks at future products and systems. Tables include shipments, 1978-1979; value of shipments, 1978-1979; revenues, 1977-1978; costs, 1979-1985; retail distribution, 1977-1979; maintenance costs, 1978-1984; and usage, 1978. Volume 3 of Yankee Group's Communications Information System Planning Service and available only to subscribers.

3145. AUTOMATED OFFICE OVERVIEW. *International Data Group. February 1980. 59 pg.*

Market data on office automation includes product lines (1979), value of shipments (1978-1984), shipments (1979), and vendors (1979). Areas covered are word processing, electronic mail, facsimiles, teleconferencing, and printers. Discusses technological trends.

3146. CHANGING DISTRIBUTION CHANNELS FOR OFFICE EQUIPMENT. *Creative Strategies International. April 1981. $1200.00.*

Analysis of the retail market for office equipment. Examines distribution strategies including services. From a survey of dealers.

3147. CHANNELS OF DISTRIBUTION FOR OFFICE AUTOMATION EQUIPMENT: CONTINUOUS INFORMATION SERVICES. *Interactive Data Corporation. August 1980. 90 (est.) pg.*

1980 report on distribution channels of office automation equipment includes dictating, copier, word processors, facsimile and micrographics equipment. Vendor distributions, product lines, market areas and average advertising expenditures for 1980 given. Publication is part of I.D.C.'s Continuous Information Service.

3148. COMMUNICATING WORD PROCESSORS. *International Resource Development, Inc. April 1979. 202 pg. $1285.00.*

Analysis of the market for 'electronic typewriters' with word processing and communications capabilities. Impact of new networks, including ACS, SBS, XTEN. Review of expected changes in usage of business forms, fine papers, envelopes, etc. Present and future markets with 10-year forecasts of market sizes and discussion of present and potential suppliers of communicating word processors.

3149. COPIER EQUIPMENT AND SUPPLIES MARKET IN THE U.S. *Frost and Sullivan, Inc. July 1981. 217 pg. $1100.00.*

Market analysis for copier equipment and supplies, with forecasts to 1985. Examines several product lines, market structure, raw material consumption, units in operation, revenues from sales, new products, industry structure and manufacturers. Includes survey of user attitudes and purchases.

3150. COPIERS AND DUPLICATORS. *Creative Strategies International. June 1980. $1500.00.*

Examines U.S. and world market for copier and duplicator equipment. Industry analyzed by product segments: plain paper copiers (low-, medium-, and high-speed) coated paper copiers, and duplicators. Includes forecasts in units and dollars through 1984 by equipment segment and industry classification (SIC). Competitive market shares also provided. Diverse avenues of distribution used by different suppliers and analyzed and compared in cost and effectiveness. Major technical issues such as single part dry toner and laser light sources evaluated.

3151. COPIERS AND DUPLICATORS. *Predicasts, Inc. May 1979. $775.00.*

Market research report on copying and duplicating machines, and associated supplies (paper, chemicals, etc.), with discussion of accruing revenues and operation costs. Examines competition for specific markets among five major copying, and three major duplicating processes; leading manufacturers and their product lines are evaluated. Data is for 1963-1978, with forecasts for 1983 and 1990 included.

3152. CPT AND THE WORD PROCESSING INDUSTRY 1980 - 1985. *MSRA, Inc. 1981. 50 (est.) pg. $400.00.*

Focusing on the CPT Corporation, this report examines the market environment for and performance of major word processor vendors during the late 1970's, with forecasts through 1985. Discusses shipment volume and value, market shares, installed equipment breakdown, revenues, pretax income, earnings per share, and other common stock performance data.

3153. DATEK PRINT HEAD REPORT. *MSRA, Inc. November 9, 1981. 92 pg. $325.00.*

Report on dot-matrix NLQ (near letter quality) printers. Reviews new technology; profiles manufacturers' product lines, including performance and technical specifications; and lists patents.

3154. DICTATION SYSTEMS IN THE AUTOMATED OFFICE. *Creative Strategies International. July 1979. $1250.00.*

Examinesthe U.S. market for portable, desktop, and central dictation systems. Detailed analysis of major competitors, their strategies, and market shares. Forecasts for dictation systems by product segment through 1983. Includes discussion of the role of dictation in the office of the future.

3155. DISPLAY WORD PROCESSORS. *Creative Strategies International. February 1981. $1200.00.*

Analyzes U.S. market for display word processors. Forecasts in units and dollars through 1985 provided by product class (standalone / shared resource, shared logic, computer-based terminal systems, and personal computers sold primarily as word processors). Explores user characteristics, competitive factors, and the impact of future technologies. Analyzes the increasing importance of avenues of marketing.

3156. ELECTRONIC MAIL EQUIPMENT. *Predicasts, Inc. October 1981. $825.00.*

Market analysis for electronic mail equipment. Examines demand by end uses and performance categories; competition with other modes; installed base and annual installations forecast, 1985, 1990, 1995; regulations; costs; industry structure; and imports.

3157. THE ELECTRONIC TYPEWRITER AND WORD PROCESSING MARKET; CONTINUOUS INFORMATION SERVICES. *Interactive Data Corporation. January 1979. 24 pg.*

1979 report on electronic typewriters and word processing markets. Covered are price and performance comparisons, shipments, installations and dollar values of products, through 1983. Publication is part of I.D.C.'s Continuous Information Service.

3158. THE ELECTRONIC TYPEWRITER: AUTOMATED BUSINESS COMMUNICATIONS. *Interactive Data Corporation. August 1980. 46 pg.*

1980 research report on electronic typewriter equipment, vendors and forecasts. Equipment market analysis includes prices, performance, maintenance costs of manufacturers products, profiles of leading vendors, and forecasts of market shares, shipments, installations and dollar value of typewriters

(1979-1984). Publication is part of I.D.C.'s Automated Businesses Communications series.

3159. FACSIMILE: ITS ROLE IN FUTURE OFFICE SYSTEMS. *MSRA, Inc. February 1981. 132 pg. $850.00.*

Analysis of facsimile markets includes technology, regulations, market size, growth rates, manufacturers, products competition, and profiles of corporations.

3160. FACSIMILE MARKETS. *International Resource Development, Inc. 1981. 185 pg.*

Forecasts value of U.S. market for document facsimile transmission equipment, through 1990. Examines market growth and competition from communicating word processors, intelligent copier / printers, and other new technologies.

3161. THE FUTURE OF ELECTRONICALLY CONTROLLED PRINTING SYSTEMS FOR BUSINESS COMMUNICATIONS. *Institute for Graphic Communications, Inc. August 1980. $4250.00.*

Market analysis and technology assessment of electronically controlled printing systems includes manufacturing cost, architectural features, image quality, reliability, cost, printing system performance, impact printing electrophotography, electrostatic printing, ink jet printing; thermal printing, magnetic printing, and electrosensititve printing. Details applications, relationship between applications and technologies, and vendor strategies.

3162. FUTURE OFFICE SYSTEMS, EQUIPMENT AND ASSOCIATED SOFTWARE MARKETS. *Frost and Sullivan, Inc. March 1980. 493 pg. $1000.00.*

Market analyses and forecasts through 1990 of the automated office equipment industry. Emphasis is on existing technology, both hardware and software, which heralds the future office including the human interface, output only devices, processors, storage media and communications facilities. Discusses the 1980 to 1984 period for which forecasts in dollars and units are given for word processing equipment; text preparation and editing equipment; intelligent typewriters; and various type computers used in the office. A separate breakout for automated business office equipment expenditures through 1984 is provided. The specific equipment forecasts are preceded by a forecast on the total office automation market. In the 1985-1989 period, an assessment is made of the expected new structure of the office resulting from the introduction of truly intergrated systems.

3163. THE HARD COPY GRAPHICS INDUSTRY: A STRATEGIC ANALYSIS. *Venture Development Corporation. 1982.*

Analysis of markets for hard copy graphics equipment and the industry structure. Forecasts average annual growth rates, shipments and value of shipments to 1986. Examines market segmentation, new technologies, competition with Japanese manufacturers, installed base, end users, and market shares of devices.

3164. THE HIGH QUALITY COMPUTER PRINTER INDUSTRY: A STRATEGIC ANALYSIS. *Venture Development Corporation. 1981. 200 pg. $2250.00.*

Analysis of markets for high quality computer printers, with forecasts to 1985. Covers the industry structure, including competiton, products, manufacturers and market shares; and the market, including demographic statistics, end uses, purchase decision criteria, market segmentation and foreign markets. Examines distribution channels; marketing strategies; the 1980 installed base and 1980-1985 shipments and dollar value.

3165. HUMAN RESOURCES AND WORD PROCESSING. *International Word Processing Association. Biennial. December 1979. 58 pg. $10.00.*

Biennial market survey examining utilization of word processing equipment in U.S. and Canadian business, government and schools, 1978. Covers effects on management, employment and labor turnover, 1974-1980.

3166. THE INFORMATION PROCESSING SUPPLIES MARKET. *Frost and Sullivan, Inc. March 1981. 219 pg. $1000.00.*

Market analysis and forecasts to 1985 for six office information processing product groups. Includes installed base, distribution methods, manufacturers and market shares.

3167. IN-PLANT PRINTING MARKET. *Frost and Sullivan, Inc. January 1982. $1150.00.*

Provides market forecasts through 1986 for number of in-plant operation units, phototypesetting supplies, photographic films and papers, plates by type, chemicals, electrostatic copy paper, offset papers, toners and ink, composition equipment, cameras, duplicators, presses, and bindery purchases. Discusses impact on the printing industry and on internal functions.

3168. INTELLIGENT AND OFFICE ELECTRIC TYPEWRITERS. *Creative Strategies International. January 1981. $1200.00.*

Analysis of markets for intelligent and office electric typewriters, with forecasts, in unit sales and dollar value, to 1984. Examines markets by product groups and manufacturers; impact of new products on market structure; direct sales, wholesale and retail distribution cost comparisons; and competition.

3169. MARKET TRENDS IN REPROGRAPHICS AND FACSIMILE. *MSRA, Inc. 1982. 225 pg. $300.00.*

Conference papers analyzing reprographic and facsimile markets and technologies. Includes market size and growth rates, shipments, market shares, corporations' marketing strategies and technical specifications for new products.

3170. MICROGRAPHICS IN THE OFFICE: A MARKET STUDY. *International Data Group. August 1979. 85 pg.*

Analysis of office micrographics includes value of shipments (1978), usage (1978-1980), market shares of word processors installed by type of media used (1978, 1983), average purchase price of facsimile equipment (1979, 1985), number of employees (1978), and industry distribution. Identifies source document cameras and word processing media and provides case studies of micrographic use and interfacing technologies. Discusses equipment configurations and user analysis.

3171. NBI AND THE WORD PROCESSING INDUSTRY. *MSRA, Inc. 1981. 30 (est.) pg. $400.00.*

Focusing on NBI, Inc. and other leading word processor vendors (Wang, CPT, Lanier), discusses industry shipments and revenues, 1976-1980, with projections through 1985. Covers installed equipment product mix, 1980 and 1985; U.S. market shares; and leading vendors' recent and projected earnings per share growth rates and other common stock data, as well as R and D expenditures.

3172. NEW DIRECTIONS FOR WORD PROCESSING: 1978-1983. *Creative Strategies International. June 1979. $7000.00.*

A strategic analysis of the word processing market through 1983. Analyzes major W.P. competitors and market shares. Includes survey results identifying present and future end-user preferences.

3173. NEW DIRECTIONS IN OFFICE IMAGING. *MSRA, Inc. 1981. 375 pg. $550.00.*

Opinions of industry executives, market and industry analysis and new product trends in office imaging. Covers competition for market shares among vendors and utilization requirements.

3174. OFFICE AUTOMATION. *Frost and Sullivan, Inc. June 1981. 14 pg. $175.00.*

Gives profiles of three selected companies. 1980 and 1985 market estimates are given for four examples of office input and output sectors (word processing, facsimile, microfilm and electric typewriters). Discusses earnings per share through 1982, new products offered, marketing strengths, and financial status. Forecasts market development.

3175. OFFICE AUTOMATION: DIRECTIONS FOR THE 1980'S. *MSRA, Inc. 1981. 382 pg. $550.00.*

Opinions of industry executives from a seminar. Covers new products and trends, utilization requirements, shipments and reviews to 1985, installed base, market shares of vendors, distribution channels and general industry analysis.

3176. OFFICE AUTOMATION IN THE BANKING INDUSTRY. *Frost and Sullivan, Inc. August 1980. 332 pg. $1000.00.*

Assessment of office automation systems in financial institutions in the U.S. Forecast through 1990 of office automation system components by work stations; central processor units; manager CRT's; printers; high speed printers; telecommunications interfaces; photocomposition equipment; applications software. Bank questionnaire analysis. Vendor profiles and market data share.

3177. OFFICE AUTOMATION MARKETS. *International Resource Development, Inc. October 1981. 170 pg. $985.00.*

Analysis of markets for office automation equipment and forecasts of manufacturers' shipments to 1990. Projects utilization of new products for information storage and distribution and improvements in executive labor productivity. Includes profiles of corporations and trends in industry structure.

3178. OFFICE AUTOMATION OPPORTUNITIES. *SRI International. Marketing Services Group. 1981.*

Projection of market volume, growth rates, new product opportunities, and ten-year trends in automated office systems industry based on input from 4000 business locations. Concerns such office technologies as word processing, electronic mail, office data processing, facsimile, intelligent copiers, in-house reprographics, and photo type setting.

3179. OFFICE AUTOMATION SERVICE. *Creative Strategies International. Monthly. 1981. $24000.00.*

Service provides continuous information about office automation including trends; technology; market strategies; competition; company profiles; end uses; and applications. Separated into 5 volumes which can be purchased separately: Office of the Future; the Marketplace; Competitive Analysis and Company Profiles; Europe; and Japan.

3180. OFFICE AUTOMATION / WORD PROCESSING INDUSTRY. *Frost and Sullivan, Inc. June 1982. 43 pg. $475.00.*

Analysis by a Wall Street firm of the word processing and office automation markets. Examines changing market structure; installed base, 1978-1981; market penetration percentage and units in operation forecasts by end uses, 1978-1990, and by product groups, 1978-1986; dollar and unit sales, 1978-1981, 1982-1986; manufacturers' market shares, distribution and marketing strategies; prices and performance; and technology trends.

3181. OFFICE CONSUMABLES AND THE AUTOMATED OFFICE. *International Resource Development, Inc. 1982. 168 pg. $1285.00.*

Analysis of markets for consumable office products such as paper supplies, toner, magnetic media and typewriter ribbons. Forecasts revenues by product groups and average annual growth rates, 1982-1991. Examines industry structure and leading manufacturers' marketing strategies.

3182. OFFICE DICTATION EQUIPMENT 1982-1986: A STRATEGIC ANALYSIS. *Venture Development Corporation. 1982. 170 pg. $375.00.*

Analysis of dictation equipment markets, technology, and competition. Examines consumer purchases and utilizations; forecasts of shipments to 1986; historical data; industry structure; distribution; market shares, 1981-1980; and marketing strategies. Profiles manufacturers, suppliers, dealers and consumers.

3183. THE OFFICE EQUIPMENT AND SUPPLIES MARKET IN CANADA. *Maclean Hunter Ltd. May 1981. 50 pg. $60.00.*

Canadian office equipment and supplies market analysis. Includes domestic production plus imports, minus exports to yield domestic consumption, 1960-1980; breakdown by products; shipments, 1977-1980; foreign trade partners; number of manufacturers and value of shipments, 1966-1981; computer installations, 1965-1979; number of businesses in selected sectors, size, and percent of total shipments; construction value, 1959-1980; employment in selected sectors, 1957-1978; and forecasts to 1985 including personal and family income and personal consumption expenditures, 1968-1985, and population and gross national product, 1985.

3184. OFFICE EQUIPMENT LEASING. *International Resource Development, Inc. 1981. 220 pg. $1285.00.*

Analysis of changes in the market for leased office equipment, 1981-1991. Assesses impact of new tax legislation and utilization of new products on manufacturers' and vendors' strategies. Forecasts market shares and market penetration of product groups, 1981-1991.

3185. THE OFFICE OF THE FUTURE. *Predicasts, Inc. April 1982. $900.00.*

Analysis of the markets for computer, word processing, teleconferencing, duplicating and other products and services for the office of the future. Forecasts to 1985, 1990, 1995. Includes manufacturers and technology.

3186. THE OFFICE PRODUCTS ANALYST (OPA). *Martin Simpson & Co., Inc. Martin Simpson and Company, Inc. Monthly. 1980. $115.00.*

A monthly independent newsletter devoted to cost / performance analysis of office products - specializing in word processing, reprographics (copiers) and related new technologies. The OPA provides comparisons of competing equipment, including current user experience for product reliability, service and costs, in-depth research and market statistics.

3187. OFFICE PRODUCTS FOR THE HOME AND FOR COTTAGE INDUSTRIES. *Business Communications Company. September 1981. 125 (est.) pg. $850.00.*

Analysis of the market for office supplies among households and cottage industries. Covers typewriters and office equipment, stationery supplies, and furniture. Examines market structure, distribution channels, and major manufacturers.

3188. OFFICE SYSTEMS INDUSTRY - THE EMERGING OPPORTUNITIES IN THE OFFICE OF THE FUTURE. *Frost and Sullivan, Inc. June 1981. 18 pg. $375.00.*

Office systems market analysis with forecasts to 1985. Covers the industry structure including the industry size in 1980 and 1985, manufacturers and other suppliers, market shares, revenues, mergers, and market shares. Analyzes market segmentation, reporting the number of consumers among small and larger offices, the percent of shipments to each segment, and the percent of employees.

3189. OFFICE SYSTEMS: THE OFFICE OF THE FUTURE MARKET. *FIND / SVP. April 1981. 17 pg. $375.00.*

Study reviews the components of the office systems market, discusses requirements and risks, and examines the strategies of suppliers of each type of product. Distribution, breadth of product line, and software capabilities of major companies are studied.

3190. OFFICE SYSTEMS: TRENDS AND OUTLOOK. *MSRA, Inc. 1982. 317 pg. $300.00.*

Conference papers analyzing markets and technology trends for office systems. Includes marketing strategies, new technology, vendors and product lines, revenues, installed base, shipments, market shares, market surveys, financial analysis, forecasts to 1985, advantages of retail distribution through dealers and market structure.

3191. OPPORTUNITIES IN COMMUNICATIONS SERVICES FOR DIGITAL INFORMATION: A STUDY OF USER NETWORKS AND NEEDS. *INPUT. January 1981. $15000.00.*

Market analysis report on corporate voice, data, image and text processing / distribution systems and networks, with forecasts to 1985. Drawn in part from a survey of vendors and users. Analyzes present and future installations and end uses; projects units in operation; analyzes competition among major manufacturers; and discusses regulations and effects on management structures.

3192. THE OUTLOOK FOR XEROX AND THE REPROGRAPHIC INDUSTRY. *MSRA, Inc. May 1980. 68 pg. $495.00.*

Analysis of the reprographics industry and the major vendors in the U.S. and Japan examines competition; market shares, 1974-1984; foreign markets; and effects of inflation on earnings of manufacturers.

3193. PRINTERS, INTELLIGENT COPIERS AND HARD COPY DEVICES FOR THE AUTOMATED OFFICE. *Yankee Group. June 1980. 267 pg. $9500.00.*

Provides overview of printing technologies including formed-character impact printing, dot matrix display, and non-impact printing. Offers a list of important office imaging suppliers and products by company. Also includes current (1979) survey of usage, complaints and needs. Additional chapters include evolving alternatives and new product opportunities. Volume 9 of Yankee Group's Communication Information Systems Planning Service and available only to subscribers.

3194. REPORT TO MANAGEMENT ON OFFICE EQUIPMENT SUPPLIERS. *Alltech Publishing Company. June 1980. $30.00.*

1980 report covering 54 suppliers of word processors, printers, data entry and other automated office equipment products. Directory of names, addresses and telephone numbers of companies surveyed included.

3195. ROBOTS AND AUTOMATED MAIL HANDLING SYSTEMS. *Predicasts, Inc. 1980. $300.00.*

Analysis of markets for robots and automated mail handling systems, with historical data and forecasts. Covers trends in end uses, manufacturers' research and development, production, and marketing strategies.

3196. SCALES AND WEIGHING EQUIPMENT MARKET. *Frost and Sullivan, Inc. September 1982. 213 pg. $1250.00.*

Analysis of scales and weighing equipment markets. Includes forecasts of manufacturers' shipments, imports, exports, and consumption for 15 product groups, 1972-1982; industry structure; market size and growth rates; manufacturers' market shares, product groups, marketing strategies, ranking, facilities, and mergers; sales, 1982-1987; and a market survey of dealers.

3197. THE SEYBOLD REPORT ON WORD PROCESSING. *Seybold Publications Inc. Monthly. 0000. 16 (est.)pg. $96.00/yr.*

Newsletter analyzes system capabilities via reports on individual systems and, in 'Buyers' Guide' issues, cross-comparisons of system futures. Issues include system updates and general industry news.

3198. THE STEELCASE NATIONAL STUDY OF OFFICE ENVIRONMENTS: DO THEY WORK? *Louis Harris & Associates. Steelcase, Inc. 1979. 127 pg.*

National survey conducted by Louis Harris polled 1047 office workers, 209 business executives, and 225 office design professionals on how office workers perceive their offices and how their offices contribute or detract from job satisfaction and productivity.

3199. THE STEELCASE NATIONAL STUDY OF OFFICE ENVIRONMENTS, NO. 2: COMFORT AND PRODUCTIVITY IN THE OFFICE OF THE 80'S. *Louis Harris and Associates. April 1980. Free.*

The thrust of the office equipment study is the relationship between perceived comfort and job performance. Data comes from personal interviews in January 1980 with 1004 white collar workers in offices of 25 or more employees, and with 203 executives who have responsibilities for office planning in major corporations.

3200. THE TECHNICAL OFFICE: ANALYSIS AND RESEARCH. *Yankee Group. 1980. 100 pg. $625.00.*

Basic study (1980) with quarterly reports discussing the automated office includinng terminals, printers and typewriters. Forecasts, market analyses 1979-1980, case studies, prices 1979-1980, competition analysis 1979-1980, and assessment of technologies provided. Includes product compatibility and product announcements by companies.

3201. TONER AND TONING: TECHNOLOGY KEYS TO NON-IMPACT PRINTING. *DATEK. MSRA, Inc. September 1981. 172 pg. $840.00.*

Analysis of toning technologies and the printers for non-impact printing includes market analysis and market survey data on user attitudes, purchases, costs, and performance.

3202. U.S. DICTATION MARKETS, 1982-1986. *Venture Development Corporation. 1982.*

Report identifies factors in the sales of nine dictation equipment product segments; market shares; and user and non-user attitudes towards dictation equipment. Examines distribution strategy, and includes the results of a national survey of retailers.

3203. THE U.S. MARKET FOR WORD PROCESSING EQUIPMENT IN THE 80'S. *Frost and Sullivan, Inc. March 1981. 238 pg. $1100.00.*

U.S. word processing equipment market analysis and forecasts to 1986. Examines competition, manufacturers, new products and utilizations, and industry structure.

3204. THE U.S. OFFICE EQUIPMENT INDUSTRY 1982-1985: A STRATEGIC ANALYSIS. *Venture Development Corporation. 1982.*

Analysis and forecast for electronic office equipment, 1982-1986. Includes shipments, 1982-1986; growth rates; rankings of product lines; vendors; and new technology.

3205. USER RATINGS OF WP SYSTEMS. *Datapro Research, Inc. (McGraw-Hill). 1981. $19.00.*

Contains an 'honor roll' of 19 stand alone word processors and multi-terminal systems based on a survey of 3745 respondents out of 13,300 WP users.

3206. WORD PROCESSING MARKETPLACE: 1979. *Interactive Data Corporation. October 1979. 126 pg.*

1979 research report on the word processing marketplace covers typewriters, non-display stand-alones, displays, printers and processors, Shipments, dollar value and growth rates and forecasted, 1978-1984. Businesses, facilities, prices, and manufacturers of computer products given. Vendors' products and market shares of peripherals to 1983 given. Vendors and manufacturers are listed alphabetically with revenues and product lines for 1978. Publication is part of I.D.C's Continuous Information Services.

3207. WORD PROCESSING MARKETS. *Business Communications Company. July 1979. 79 pg. $750.00.*

Document examines information processing users by type, suppliers and installed base by major vendors. Report includes forecasts, patterns of sales and installations emerging for the 1980's. Contents include: the automated office, issues; systems, manufacturers, costs; markets, forecasts; and installations.

3208. WORD PROCESSING TYPEWRITERS. *Strategic Inc. January 1982. $1200.00.*

Analysis of word processing typewriters, the market, technology and product demand. Includes market size and shares, 1981, 1987; utilizations; competition; vendors and product lines; performance comparisons, factors in purchases, and forecasts to 1987. Shows installed base and revenues, 1981-1987.

3209. WORLDWIDE STUDY OF ELECTRONIC FUNDS TRANSFER: SYSTEMS AND EQUIPMENT REQUIREMENTS (EFTS III). *Battelle Memorial Institute. April 1979. $18000.00.*

Study, intended primarily forEFT product / service suppliers, examines markets for those products and services in three regions: North America, Japan, and Western Europe. Demand models are developed for each nation covered. Technological trends are also discussed.

3210. XEROX PRICING: THE CUTTING EDGE OF COMPETITIVE STRATEGY. *Martin Simpson & Co., Inc. Martin Simpson and Company, Inc. October 1979. 137 pg. $495.00.*

1979 "Technology Note" provides a review of the word processor and office copier industry's pricing structure, including details of more than 80 different machines and pricing plans. The report includes comparative pricing of a variety of machines from Xerox, IBM, Kodak and Savin.

See 67, 92, 110, 939, 944, 959, 997, 998, 1010, 1027, 1054, 1067, 1074, 1103, 1123, 1126, 1143, 1147, 1149, 1166, 1176, 1180, 1194, 1199, 1204, 1216, 3562.

Investment Banking Reports

3211. PAINE WEBBER MITCHELL HUTCHINS INC.: STATUS REPORT: MAINFRAME / OFFICE EQUIPMENT MONTHLY, MAY 1981. *Garrett, Sandy. Paine Webber Mitchell Hutchins, Inc. Monthly. May 7, 1981. 17 pg.*

Monthly financial analysis report for corporations in the U.S. mainframe computer and office equipment industry. Gives earnings per share, 1977-1982; revenues, 1977-1981; 1981 dividends and stock prices; price earnings ratio, 1981; relative valuation, 1980-1986; and graphs showing relative price earnings ratio, earnings per share, and return on equity, 1960-1980.

See 1238, 1241, 1242, 1290, 1296.

Industry Statistical Reports

3212. CANADIAN SECRETARY SURVEY. *Maclean Hunter Ltd. June 1980.*

Provides a profile of 'Canadian Secretary' readers and information on their purchasing influence across a broad range of office equipment, supplies and services.

3213. COUNTY BUYING PATTERNS FOR OFFICE PRODUCTS, INCLUDING PERCENTAGES BY STANDARD METROPOLITAN STATISTICAL AREAS. *National Office Products Association. Irregular. 10th ed. 1980. 23 pg. $28.00.*

Listing of percentage of buying patterns for all office products sold in the U.S., by county, up to four decimal places. Includes summary by state, and data for SMSA's. Source is U.S. Census Bureau's County Business Patterns.

3214. CURRENT INDUSTRIAL REPORTS: COMPUTERS AND OFFICE AND ACCOUNTING MACHINES, 1979. *U.S. Department of Commerce. Bureau of the Census. Customer Service Branch. Annual. MA-35R. November 1980. 14 pg. $.30.*

Report presents data on the value of shipments of office, computing, and accounting machines compared to 1977 Census of Manufactures; value of shipments by class of product from 1970 to 1979; and number of companies, quantity and value of shipments for each product for current and preceding year. Includes data on quantity and value of manufacturers' shipments, exports, imports, and apparent consumption, by product, for current and preceding year.

3215. EIS SHIPMENTS REPORT: CALCULATING AND ACCOUNTING MACHINES. *Economic Information Systems, Inc. Quarterly. 1982. $75.00/yr.*

Report on calculating and accounting machine industries. Arranged by state and county, report includes every plant in the industry with annual shipments over $500,000 and/or 20 or more employees. Plant listings detail address, telephone, estimated annual shipments, and percent of market. Similar statistics for each county and state.

3216. EIS SHIPMENTS REPORT: OFFICE MACHINES, NEC. *Economic Information Systems, Inc. Quarterly. 1982. $135.00/yr.*

Report on NEC office machine industries. Arranged by state and county, report includes every plant in the industry with annual shipments over $500,000 and/or 20 or more employees. Plant listings detail address, telephone, estimated annual shipments, and percent of market. Similar statistics for each county and state.

3217. EIS SHIPMENTS REPORT: TYPEWRITERS. *Economic Information Systems, Inc. Quarterly. 1982. $75.00/yr.*

Report on typewriter industries. Arranged by state and county, report includes every plant in the industry with annual shipments over $500,000 and/or 20 or more employees. Plant listings detail address, telephone, estimated annual shipments, and percent of market. Similar statistics for each county and state.

3218. INDUSTRY NEWS: CBEMA: EXPORTS OF COMPUTERS / BUSINESS EQUIPMENT SHOW DECREASE IN FIRST QUARTER 1982. *Computer and Business Equipment Manufacturers Association. June 11, 1982. 3 pg.*

Report on imports and exports of business equipment and computers. Includes trade balance with Japan and Western Europe, 1976-1982. Gives CBEMA membership data including member corporations.

3219. NOPA DEALER OPERATING RESULTS 1980. *Management Services Dept., NOPA. National Office Products Association. Annual. 51st ed. 1981. 56 pg. $28.00.*

Annual survey results from dealer members of the National Office Products Association. Includes highlights and vital statistics, profit and loss statements, and balance sheets for 1979/1980, with some data for 1978 and earlier years. Arrangement is by volume of business in eight categories from $100,000 - $250,000 range to over $10 million range; by regional breakdown in four geographic areas; in type-of-product breakdown; and 14 financial ratios. Contains chart of accounts and worksheets.

3220. OFFICE AND STORE MACHINERY MANUFACTURERS. *Canada. Statistics Canada. French/English. Annual. 1981. 9 pg. 4.50 (Canada).*

Includes statistics on number of office machinery establishments; number and type of employees; salaries and wages; cost of fuel and electricity; value of shipped goods; and value added. Comparative statistics are given for Canada for earlier years, and by province for the preceding year. Data issued since 1960.

3221. OFFICE AND STORE MACHINERY MANUFACTURERS, 1980: ANNUAL CENSUS OF MANUFACTURES: FABRICANTS DE MACHINES POUR LE BUREAU ET LE COMMERCE, 1980: RECENSEMENT ANNUEL DES MANUFACTURES. *Canada. Statistics Canada. English/French. Annual. Cat. #42-216. February 1982. 6 pg. 5.40 (Canada).*

For manufacturers of typewriters, cash registers, vending machines, mechanical computing machines, scales and balances, as well as electronic computers, data processors and control devices, statistics are provided on capital and repair expenditures, 1978-1981, as well as on employees and ownership, wages and salaries, costs of fuel and of raw materials, value of shipments, and value added, 1972-1980. Gives tables on inventories, 1980, and on materials and supplies used and shipments, by kind, 1979-1980. Includes a list of companies with addresses.

3222. OFFICE PRODUCTS DEALER: PROFILE OF A DEALER "82. *Hitchcock Publishing Company. Annual. 1982.*

Contains information on office products dealers including products and services sold, numbers of outside salespeople employed, how dealer businesses are organized, facilities used and their cost, average account sizes, and plans for future development. Covers catalogs used, advertising expenditures, profit margins by product catagory, average annual sales volume, sales compensation plans, commercial versus walk-in sales volume, number of employees, inventory turnover and product ordering lead times.

3223. PORTABLE ELECTRIC TYPEWRITERS FROM JAPAN: DETERMINATION OF MATERIAL INJURY IN INVESTIGATION NO. 731-TA-12 (FINAL) UNDER SECTION 735(B) OF THE TARIFF ACT OF 1930. *Cutchin, John. U.S. International Trade Commission. USITC Publication 1062. May 1980. 56 pg. Free.*

Report of Commission investigation into potential injury to domestic industry of imports of portable electric typewriters from Japan. Text describes product and its uses, U.S. tariff treatment, nature and extent of sales at LTFV (less than fair value), and survey of the industry. Covers U.S. and foreign producers, U.S. importers, production, capacity, capacity utilization, shipments and exports, inventories, employment, financial performance, investment and capital expenditures, U.S. consumption and market penetration of imports, prices and lost sales. Tables supply supporting details for 1976 through 1979, compiled primarily from industry response to Commission questionnaires.

3224. SALARY INFORMATION FOR WORD PROCESSING PERSONNEL: U.S., CANADA, 1980. *International Word Processing Association. Annual. 6th ed. 1980. 80 pg. $25.00.*

Annual survey of wages and salaries for word processing personnel in U.S. and Canada, 1980. For 15 positions, shows number of businesses, number of employees, and average weekly earnings.

3225. 1977 CENSUS OF MANUFACTURES: CALCULATING AND ACCOUNTING MACHINES: PRELIMINARY REPORT: INDUSTRY SERIES. *U.S. Department of Commerce. Bureau of the Census. Customer Service Branch. Irregular. MC77-1-35F-3(P). August 1979. 8 pg. $.35.*

This is a preliminary report from the Bureau of the Census 1977 Census of Manufactures. Tables with industry statistics from the U.S. for 1963-1977 and with statistics for selected geographic areas for 1972 and 1977 give data relating to employees, production costs, assets and expenditures, inventories, and ratios. Other tables supply 1972 and 1977 statistics on product classes giving quantity and value of shipments by all producers and statistics on materials consumed by kind.

3226. 1977 CENSUS OF MANUFACTURES: CARBON PAPER AND INKED RIBBON: PRELIMINARY REPORT: INDUSTRY SERIES. *U.S. Department of Commerce. Bureau of the Census. Customer Service Branch. Irregular. MC77-1-39C-4(P). June 1979. 7 pg. $.35.*

This is a preliminary report from the Bureau of the Census 1977 Census of Manufactures. Tables with industry statistics from the U.S. for 1963-1977 and with statistics for selected geographic areas for 1972 and 1977 give data relating to employees, production costs, assets and expenditures, inventories, and ratios. Other tables supply 1972 and 1977 statistics on product classes giving quantity and value of shipments by all producers and statistics on materials consumed by kind.

3227. 1977 CENSUS OF MANUFACTURES: MARKING DEVICES: PRELIMINARY REPORT: INDUSTRY SERIES. *U.S. Department of Commerce. Bureau of the Census. Customer Service Branch. Irregular. MC77-1-39C-3(P). March 1979. 7 pg. $.35.*

This is a preliminary report from the Bureau of the Census 1977 Census of Manufactures. Tables with industry statistics from the U.S. for 1963-1977 and with statistics for selected geographic areas for 1972 and 1977 give data relating to employees, production costs, assets and expenditures, inventories, and ratios. Other tables supply 1972 and 1977 statistics on

product classes giving quantity and value of shipments by all producers and statistics on materials consumed by kind.

3228. 1977 CENSUS OF MANUFACTURES: OFFICE MACHINES, N.E.C. AND TYPEWRITERS: PRELIMINARY REPORT: INDUSTRY SERIES. *U.S. Department of Commerce. Bureau of the Census. Customer Service Branch. Irregular. MC77-1-35F-5(P). August 1979. 9 pg. $.35.*

This is a preliminary report from the Bureau of the Census 1977 Census of Manufactures. Tables with industry statistics from the U.S. for 1963-1977 and with statistics for selected geographic areas for 1972 and 1977 give data relating to employees, production costs, assets and expenditures, inventories, and ratios. Other tables supply 1972 and 1977 statistics on product classes giving quantity and value of shipments by all producers and statistics on materials consumed by kind.

3229. 1980 OFFICE PRODUCTS MARKET REPORT. *National Office Products Association. Irregular. 5th ed. 1981. 21 pg. $28.00.*

National Office Products Association analysis of the market based on the 1977 Census of Manufactures, with actual data for 1972 and 1977 and projections through 1981. Arrangement is by product classifications: office supplies, wood office furniture, metal office furniture, public building and related furniture, partitions and fixtures, office machines, photocopying equipment, microfilming equipment, mailing and parcel post scales, and safes and chests. Tables are arranged by SIC number and show value and percent of total in category for each segment, plus annual experience growth factor and compound growth factor utilized.

See 1302, 2394, 2399.

Financial and Economic Studies

3230. MICROGRAPHICS IN THE OFFICE: MARKET PROSPECTS. *Frost and Sullivan, Inc. September 1981. 284 pg. $1100.00.*

Assesses factors influencing market growth and direction. Probes basic factors at work to understand why the market will develop a certain way and what suppliers can do to meet user requirements. Examines industry evolution. Provides sales forecasts by equipment and supplies product lines by major application. Source data includes 8/6 respondents to a Frost and Sullivan user survey.

3231. OFFICE EQUIPMENT INDUSTRY REVIEW. *Bellace, Joseph J. Wall Street Transcript. July 1979. $20.00.*

Review and forecasts (1979-1984) for the office equipment industry including data on earnings per share growth rates (1979-1984) for computer mainframes, copying / duplicating, computer graphics, computer services, distributed data processing, minicomputers and telecommunications equipment. Discusses worldwide market growth, Japanese competition, software development and IBM aggressiveness. Nine major companies reviewed.

3232. SCIENCE AND ELECTRONICS: TRADE BALANCE REPORTS: FAVORABLE BALANCE OF TRADE IN BUSINESS MACHINES INCREASES IN FIRST HALF 1979. *U.S. Department of Commerce. Bureau of Industrial Economics. Office of Public Affairs and Publications. Semi-annual. ITA 79-147. September 21, 1979. 2 pg.*

One in a series of semiannual publications detailing U.S. exports and imports (total and by region or nation). Data for first halves of 1978 and 1979. Companion set of releases gives data for entire year and two previous years.

3233. SCIENCE AND ELECTRONICS: TRADE BALANCE REPORTS: FAVORABLE BALANCE OF TRADE INCREASES IN BUSINESS MACHINES DURING 1979. *U.S. Department of Commerce. Bureau of Industrial Economics. Office of Public Affairs and Publications. Semi-annual. BIE 80-5. March 5, 1980. 2 pg.*

One in a series of semiannual publications detailing annual U.S. exports and imports (total and by region or nation), 1977-1979. Companion set of releases gives data for half of year of coverage and previous year.

3234. STANDARD AND POOR'S INDUSTRY SURVEYS: OFFICE EQUIPMENT, SYSTEMS AND SERVICES: CURRENT ANALYSIS. *Standard and Poor's Corporation. Semi-annual. September 13, 1979. 6 pg.*

Financial report on the office equipment industry in the U.S. Includes common stock and stock price indexes of the industry for 1976-1979; market and economic indicators for 1977-1979; leading U.S. manufacturers' shipments, 1978-1979; and profiles. Gives vendors' and manufacturers' financial positions.

3235. WARNING: HEALTH HAZARDS FOR OFFICE WORKERS. *Working Women Education Fund. 1981. $4.00.*

Discusses evidence linking clerical work using cathode ray tube (CRT) computer displays with a variety of ailments including eyestrain, migraines, nausea, back pain, muscular distress, and occupational stress. Investigates the potential of job redesign to alleviate these problems.

See 1329.

Forecasts

3236. FOURTH MARTIN SIMPSON TECHNOLOGY SEMINAR: COPIER AND WORD PROCESSING TRENDS IN THE 1980'S. *Martin Simpson & Co., Inc. Martin Simpson and Company, Inc. May 1979. 460 pg. $600.00.*

The edited transcripts contain the thoughts of leading industry representatives from Xerox Data Systems, Savin, Hunt Chemical, Saxon, Eastman Kodak, Digital Equipment, Wang Labs, Qyx, CPT and the "Office Product Analyst," as well as leading-edge end-users from AT&T, ITT, Gibbs & Hill and Price Waterhouse. The prospects for copiers, computer printers, intelligent copiers, word processing systems and intelligent typewriters are discussed in detail. The copier and word processing sections may be ordered individually.

3237. THE FUTURE OF THE OFFICE PRODUCTS INDUSTRY: COMPLETE REPORT. *Stanford Research Institute International, Menlo Park, CA. National Office Products Association. Annual. November 1979. 276 pg. $1000.00.*

Complete analysis of the total office products industry, using 1977 actual figures as a basis for extrapolating short-term (1980 and 1983) and long-term (1986 and 1989) patterns. Study analyzes technological trends and their impact on office machine designs and subsequently on office procedures, environment and services; estimates of the office machine population; office product usage patterns; current and future distribution practices; office product areas significantly impacted by technology; and risks and opportunities for the industry. In-text tables show parcel and dollar value for each type of business. Appendix shows estimates of sales by units 1974-1989 by type of product and type of purchaser. Other appendices show composition of market segments analyzed, composition of the office supply categories analyzed (with SIC), research methodology and sources, list of participants, and glossary.

3238. THE FUTURE OF THE OFFICE PRODUCTS INDUSTRY: MEMBERSHIP SUMMARY. *Stanford Research Institute International, Menlo Park, CA. National Office Products Association. Annual. 1979. 30 pg. $20.00.*

Summary of the detailed study of the office products market through 1989, based on 1977 analysis. Charts indicated end-user purchases for 1977, with projections for 1983 and 1989 arranged by type of product. Text summarizes results of the complete study, arranged for office supplies, office furniture, and office machines. Covers impact of technology, changes in product mix, and changes in distribution. Includes contents pages of the complete report.

3239. OFFICE AUTOMATION: KEY TO THE INFORMATION SOCIETY. *Russel, Robert. Institute for Research on Public Policy. 1982. 22 pg. 3.00 (Canada).*

Analysis of office automation emphasizes electronic filing cabinets. Covers videotex terminals and videodisc players.

3240. THE WORD PROCESSING INDUSTRY: A STRATEGIC ANALYSIS. *FIND / SVP. 1981. 250 pg. $2250.00.*

Study of the market, technology, competition, and trends for word processors. Presents data on the installed base at year-end 1980 and industry shipments through 1986. Forecasts shipments by type of system - standalone, shared logic, shared resource, and multifunction, domestic and foreign markets, and end user segments assesses manufacturers' market shares. Includes analysis of current and potential users with emphasis on the type of systems owned, future purchase plans and the appeal of new and proposed product features.

3241. XEROX NEW PRODUCTS: PRICING COMPARISONS AND PROSPECTS. *Martin Simpson & Co., Inc. Martin Simpson and Company, Inc. November 26, 1979. 14 pg. $95.00.*

1979 "Technology Note" provides a review of four new products by Xerox - the 3300, 5600, 8200 and 9500 - and how they compare with competitive products in terms of pricing and performance.

See 233, 1352, 1354, 1360, 1363, 1364, 1365.

Directories and Yearbooks

3242. ALL ABOUT 150 WORD PROCESSORS. *Datapro Research, Inc. (McGraw-Hill). Annual. 1981. 110 pg. $19.00.*

Survey summarizes the characteristics of over 150 models of standalone and multi-terminal text editing systems from 56 vendors. Comparison charts describe each model's basic configuration, pricing, storage medium and capacity, editing features and printer characteristics.

3243. ALL ABOUT 245 COPIERS. *Datapro Research, Inc. (McGraw-Hill). Annual. 1981. 127 pg. $19.00.*

Provides comparison charts detailing the pricing and characteristics of 245 office copier models. Includes plain paper, coated paper, thermal and dual spectrum paper units offered by 40 vendors in the office copier market.

3244. ALL ABOUT 70 DATA COLLECTION DEVICES. *Datapro Research, Inc. (McGraw-Hill). Annual. 1981. 25 pg. $15.00.*

Report surveys characteristics of 70 portable and stand-alone devices plus data collection systems available from 29 vendors. Includes consumer ratings. Equipment covered ranges from central recording systems to hand-held tape recorders.

3245. ALL ABOUT 90 WORD PROCESSING SOFTWARE PACKAGES. *Datapro Research, Inc. (McGraw-Hill). Annual. 1981. 52 pg. $15.00.*

Report provides details on the name of vendor, computer model and processing requirements, operating system requirements, source language or compiler, source listing availability, documentation, training and maintenance support. Provides pricing and the number of users of 90 word processing software packages. Alphabetical listing with addresses and phone numbers of the 73 vendors who supply the packages.

3246. AUERBACH'S ELECTRONIC OFFICE: MANAGEMENT AND TECHNOLOGY: VOLUME 2. *Auerbach Information, Inc. Quarterly. 1980. 1000 (est.) pg.*

Volume 2 of 2-volume set on the electronic office includes a directory of companies and analyses and product reports for the following: word processors, dictation equipment, office copiers, integrated office systems, COM and CIM, microform reader / printer and retrieval systems, communications facilities, network control systems and facsimile equipment. Comparisons among competitors, user reactions, specifications, performance, price data, maintenance rates are given for each product.

3247. A BUYER'S GUIDE TO ELECTRONIC CASH REGISTERS. *Datapro Research, Inc. (McGraw-Hill). Annual. 1981. $15.00.*

Buyer's guide includes comparison charts covering assorted electronic cash registers' characteristics and prices.

3248. CENTRAL DICTATING SYSTEMS REPORT: BUYERS LABORATORY INC. *Buyers Laboratory, Inc. Irregular. 1979. 60 pg.*

A 1979 report and directory on central dictating systems. Each entry includes information on components, prices, special features and special uses. Illustrations provided throughout.

3249. DATABOOK 70: COPIERS AND DUPLICATORS. *Maclean Hunter Ltd. Monthly. 1980. $430.00/yr.*

Product reports on Canadian copying and duplicating equipment. Published in 2 volumes and updated monthly.

3250. DATAPRO REPORTS ON COPIERS AND DUPLICATORS. *Datapro Research, Inc. (McGraw-Hill). Annual. 1980.*

2-volume work with monthly supplements covers the spectrum of reproduction equipment, including copiers, duplicators, and offset printers. Contains product profiles, user evaluations, and comparison charts on copiers, duplicators, copy / duplicating systems and suppliers, and auxiliary equipment and systems.

3251. DIRECTORY OF AUTOMATED OFFICE CONSULTANTS. *Datapro Research, Inc. (McGraw-Hill). Irregular. 1981. 25 pg. $15.00.*

Profiles 103 automatic office consulting firms in the U.S. and Canada. Lists each firm's principals, addresses, telephone, date established, number of consultants on staff, geographical area served, services offered, and special areas of expertise.

3252. THE DIRECTORY OF INFORMATION / WORD PROCESSING EQUIPMENT AND SERVICES: A GUIDE TO OFFICE AUTOMATION. *Information Clearing House. Irregular. July 1, 1982. 240 pg. $29.95.*

Directory of office data processing and communications equipment manufacturers, software vendors, services, and consultants. Lists business addresses, contact executives, and product lines or services. Includes market analysis.

3253. ELECTRONIC MAIL EXECUTIVES DIRECTORY: SECTION 1: A-J. *International Resource Development, Inc. Semi-annual. 1980. 550 (est.) pg. $95.00/2 vols.*

Electronic Mail Executives Directory identifies executives from within top U.S. companies who are responsible for office automation, telecommunications, word processing, data communications and inplant printing supplies and procurement. Includes alphabetical list of companies (A-J) and geographic index. Each entry contains name, address, telephone number of company, key personnel, sales volume, employees, principal business, short description of type of equipment used, and name, address, telephone number of key executives.

3254. ELECTRONIC MAIL EXECUTIVES DIRECTORY: SECTION 2: K-Z. *International Resource Development, Inc. Semi-annual. 1980. 550 (est.) pg. $95.00/2 vols.*

Electronic Mail Executives Directory identifies executives from within top U.S. companies who are responsible for office automation, telecommunications, word processing, data communications and inplant printing supplies and procurement. Includes alphabetical list of companies (K-Z) and a geographical index. Each entry contains name, address, telephone number of company and of key executives; key personnel, sales volume, employees, principal business, and short description of type of equipment used.

3255. GEYER'S WHO MAKES IT: 1980. *Geyer-McAllister Publications, Inc. Annual. 50th. 1980. 242 pg.*

Annual directory of paper products, office equipment and printing machinery. Listings for products and for corresponding manufacturers given alphabetically. Alphabetical list of trade names and associations and advertisers' index also included.

3256. OFFICE SUPPLIES MARKETPLACE. *Maclean Hunter Ltd. Quarterly. September 1981. 1.50 (Canada).*

Listing of new products on the market. Includes description and names of manufacturers.

3257. SELECTED OFFICE COPIERS: HIGH VOLUME. *Datapro Research, Inc. (McGraw-Hill). Annual. 1981. 67 pg. $15.00.*

Analyses of the offerings of three major copier vendors. Designed for users requiring 30,000 to 60,000 copies per month. Units included are IBM Series III 10 and Model 20; IBM Series III Model 30 and Model 40; Eastman Kodak Ektaprint 100 Series; Eastman Kodak Ektaprint 150 Series; and Xerox 8200.

3258. SELECTED OFFICE COPIERS: LOW VOLUME. *Datapro Research, Inc. (McGraw-Hill). Annual. 1981. 27 pg. $15.00.*

Analyses of five major copiers intended for users requiring up to 15,000 copies per month. Included are: Canon NP-200; Minolta EP 310; Royal RBC 115; Savin 770/780; and Xerox 2600.

3259. SPECIAL REPORT: NOPA CONVENTION REPORT. *National Office Products Association. Irregular. September 1979. pg.114-142 $5.00.*

Survey of 152 new products scheduled for presentation at the 75th National Office Products Association Convention, 1979. Listings include brief descriptions, photographs and manufacturers or vendors.

3260. SURVEY OF ELECTRONIC MAIL SYSTEMS. *Hanagan and Associates. Annual. 1981. $95.00.*

Overviews 20 systems' history, availability, reliability, current user base, system support, user commands available, price structure, comparative advantages and disadvantages, and vendor characteristics. Written from the viewpoint of the user, considers such aspects as ease of operation and diagnostics and documentation available.

3261. WORD PROCESSING: ELECTRONIC TYPEWRITERS. *Datapro Research, Inc. (McGraw-Hill). Annual. 1981. $15.00.*

Examines three vendors' electronic typewriter lines in terms of product characteristics and prices.

3262. WORD PROCESSING SYSTEMS USER RATINGS. *Datapro Research, Inc. (McGraw-Hill). Annual. 1981. $15.00.*

Report provides an analysis of how over 1700 word processing users rated over 70 makes and models of equipment from 28 vendors of standalone and multi-terminal word processors. Contains a narrative of results excerpted from the survey and summary tables containing an analysis of each system's performance as rated by the users as well as overall vendor summaries.

See 265, 612, 1392, 1393, 1399, 1403, 1425, 1426, 1430, 1434, 1442, 1455, 1476, 1478, 2476.

Special Issues and Journal Articles

ABA Banking Journal

3263. BANKING BUYERS GUIDE 1981-82. *Simmons-Boardman Publishing Company. Annual. May 1981. pg.83-183*

1981 annual buyers guide classifies and lists products and services available to the banking industry. Gives manufacturers' and suppliers' business address and telephone numbers. Includes a trade name listing with company name.

Bank Systems and Equipment

3264. ATM'S FIVE-YEAR MARATHON TRIGGERS NETWORK EXPANSION. *Moore, James B. Gralla Publications. October 1981. pg.95-97*

Analyzes development and growth rates of automatic teller machine (ATM) networks. Discusses the rate at which ATM networks have been introduced, along with expansion rates; motives for expansion; terminal installation rates; service expectation; and criteria for evaluation. Tables detail the impact of network maturation on the average number of terminals per institution; average percentage of regular users; and average daily transaction volumes.

3265. BANK SYSTEMS AND EQUIPMENT: 1982 DIRECTORY AND BUYER'S GUIDE. *Gralla Publications. Annual. v.19, no.1. January 1982. pg.82-302*

Annual buyer's guide covers new products and product development. Lists addresses for manufacturers, distributors and employees associated with computers, office equipment, and commercial banking. Specifically discusses electrical equipment for banking. List products alphabetically and by manufacturer.

3266. CHECK PROCESSING EQUIPMENT AND SUPPLIES: GROWTH TRIGGERS SEARCH FOR COST - EFFICIENT TECHNOLOGY. *Gralla Publications. September 1981. pg.65-70*

Product review covering 47 systems for check processing. Includes check sorters, encoders, encoding services, filing systems, endorsing machines, MICR readers, MICR reject processing systems, de-encoders, printing and proving units; and software. Reviews specifications for each product and highlights unique characteristics.

3267. CONVENTION PREVIEW: EXAMINATION OF DELIVERY SYSTEM HIGHLIGHTS NAMSB CONFERENCE. *Gralla Publications. v.19, no.2. February 1982. pg.74-75*

Lists names of exhibitors at the 1982 annual National Association of Mutual Savings Banks Conference. Product groups, product lines, and services provided are included for some companies.

3268. MICROGRAPHIC SYSTEMS, EQUIPMENT AND SUPPLIES: PRESENT TECHNOLOGIES LAY GROUNDWORK FOR 1990'S: SPECIAL PRODUCT REVIEW. *Gralla Publications. Annual. v.18, no.12. December 1981. pg.73-81*

Surveys microfilm and microfiche equipment currently available, including splicers, copiers, and automated retrieval systems. Descriptions include manufacturer's name and some specifications.

Canadian Datasystems

3269. WHAT'S NEW IN WORD / TEXT PROCESSING. *Fistell, Linde. Maclean Hunter Ltd. Irregular. April 1980. pg.32-62*

A review of the word processing industry includes a short list of suppliers and manufacturers of WP products and equipment in Canada and the U.S. Contains a comprehensive listing of businesses with newly-introduced systems. These include name of business, model, display, printer capacities, configuration, media, editing features, communications compatible, EDP capability, peripherals and purchase prices.

3270. WORD PROCESSING: NEW PRODUCTS SURVEY. *Fistell, Lendie S.A. Maclean Hunter Ltd. Annual. September 1981. pg.51-71*

Tables provide a condensed look at some of the latest word processing equipment now on the Canadian market. Listing includes names of businesses, model, display, printer, configuration, media, editing features, communications compatible, EDP capacity, peripherals and purchase prices in Canadian dollars.

Canadian Grocer

3271. AVERAGE NUMBER OF CHECKOUT COUNTERS. *Maclean Hunter Ltd. September 1981. pg.13*

Lists the number of mechanical or electronic checkout counters in grocery stores by chain or independent.

Canadian Office

3272. CANADIAN OFFICE REDBOOK ANNUAL DIRECTORY. *Whitsed Publishing Ltd. Annual. v.11, no.6. June 1980. 92 pg.*

Lists Canadian supply sources in office equipment including 80 product categories and manufacturers names, addresses and phone numbers. Gives branches and dealers, including products handled.

3273. CANADIAN OFFICE REDBOOK: ANNUAL DIRECTORY OF PRODUCTS AND SUPPLIERS FOR THE OFFICE ENVIRONMENT. *Whitsed Publishing Ltd. Annual. April 1981.*

Lists Canadian office equipment manufacturers and their products.

3274. CANADIAN OFFICE: REDBOOK 1982. *Whitsed Publishing Ltd. Annual. v.13, no.5. May 1982. pg.12-52*

Provides listing of sources of supply for thousands of products used in the office. Includes names of manufacturers, their addresses and telephone numbers, names of key sales employees, as well as an alphabetical listing of products available and associations.

3275. CANADIAN OFFICE: 1981 DEALER DIRECTORY: FIFTH ANNUAL DIRECTORY OF CANADIAN DEALERS AND DISTRIBUTORS OF OFFICE EQUIPMENT AND FURNITURE. *Whitsed Publishing Ltd. Annual. v.12, no.11. November 1981. pg.19-76*

Annual directory of Canadian office equipment and furniture dealers / distributors, by province. Specifies, for most dealerships, address, telephone number, branch location(s) (if any), manager or representative, and manufacturers represented (brand names carried).

3276. DEALER DIRECTORY ISSUE: CANADIAN DEALERS AND DISTRIBUTORS OF OFFICE EQUIPMENT AND FURNITURE. *Whitsed Publishing Ltd. Annual. September 1979. pg. 31-60*

Lists independent office dealers and distributors of machines, supplies and furnishings in alphabetical order by province. Listings include addresses, phone numbers, chief executives and products. Provides a manufacturers' listing in the products section for each dealer.

Canadian Office Products and Stationery

3277. CANADIAN OFFICE PRODUCTS AND STATIONERY DEALERS GUIDE 1982. *Southam Business Publications, Ltd. Division of SouthamCommunications Ltd. Annual. v.15, no.6. December 1981. pg.25-66*

Annual directory of suppliers, manufacturers and distributors of Canadian office products. Arranged alphabetically by company name, each entry includes address and telephone number. For U.S. manufactureres, the name of principal Canadian distributo is given. Also features alphabetical listing of trade names.

Canadian Secretary

3278. COPIER MARKETPLACE. *Maclean Hunter Ltd. Quarterly. May 1981. pg.16-18*

Lists new products put out by Canadian and U.S. copier manufacturers. Includes descriptions and names and addresses of manufacturers.

Chain Store Age, Supermarkets

3279. WINN-DIXIE TOPS SCANNING LIST. *Lebhar-Friedman, Inc. Irregular. November 1980. pg.11*

Brief statistical article showing trends in installations of checkout scanners for U.S. supermarkets. Covers 2 manufacturers.

CIPS Review

3280. MICROELECTRONICS AND EMPLOYMENT IN CANADIAN PUBLIC ADMINISTRATION. *Wilkins, Russell. Canadian Information Processing Society. 1982. pg.22-23*

Studies the effects of microelectronics and computers on municipal employment and unemployment in three major Ontario cities (Oshawa, Toronto, and Ottawa). Examines productivity as a result of computer applications and cost reductions for certain operations.

Communication Technology Impact

3281. OFFICE AUTOMATION SALES TO SOAR. *Credicasts, Inc. Elsevier Science Publishers. v.4, no.7. October 1982. pg.13-14*

Summarizes an office automation study. Gives annual dollar value of sales of 5 types of office automation equipment, net imports, and volume of shipments, 1967-1995.

Computer Business News

3282. THE WORLD OF LOW - SPEED CHARACTER PRINTERS - PRINTER VENDORS. *CW Communications Inc. Irregular. v.5, no.6. February 8, 1982. pg.18-20*

Lists addresses of low-speed character printer manufacturers.

ComputerData

3283. INTEGRATED OFFICE SYSTEMS - WHO SELLS THEM AND WHAT YOU GET. *Whitsed Publishing Ltd. v.7, no.2. February 1982. pg.38-39*

Lists manufacturers of integrated, automated office systems, along with their products, services and Canadian locations.

Fortune

3284. WHAT'S DETAINING THE OFFICE OF THE FUTURE. *Uttal, Bro. Time, Inc. v.105, no.9. May 3, 1982. pg.176-196*

Analysis and forecast of office automation market provides statistical data on U.S. market for office automation equipment and includes 1981 and 1986 figures for shipments of word processors, electronic typewriters, professional work stations, intelligent copiers and digital PBXs. Indicates companies capturing the word processor business (percent).

Graphic Arts Monthly and the Printing Industry

3285. DUPLICATORS OFFER FAST, HIGH QUALITY COPY REPRODUCTION FOR SMALL PRINTER. *Technical Publishing Company. Irregular. v.54, no.7. July 1982. pg.43-48*

Describes the product lines of 14 duplicator manufacturers, 1982. Gives numbers, technical data, and prices of newer models.

Infosystems

3286. INFOSYSTEMS: SECOND ANNUAL WP SALARY SURVEY: WP SALARY INCREASES RANGE FROM 10 TO 23 PERCENT. *Miller, Frederick W. Hitchcock Publishing Company. Annual. June 1981. pg.52-53*

Based on a national survey sample, reports representative weekly salaries (high, low, median, mean, mode, 25th and 75th percentiles) for seven job titles which encompass use of word processing equipment. Average weekly salaries are examined by size of total word processing budget. Percentage of respondents using word processing equipment is analyzed by industry group; product mix (memory typewriter versus in-house mainframe computer, minicomputer, time-sharing, etc), and in-house versus contracted WP services, are tabulated for the aggregate sample.

IRM

3287. COST - EFFECTIVENESS IN MICROFORM STORAGE. *Poli, Bruce. Information and Records Management Inc. Irregular. v.16, no.2. February e, 1982. pg.24-26*

Product review features devices for storage of microfilm and microfiche. Describes types of units, discussing the suitability of each to certain microform collections and storage needs. Lists manufacturers of microform storage products and addresses. Includes selected products in photographs with information about manufacturer and product features.

3288. DOCUMENT DESTRUCTION OVERVIEW AND DIRECTORY. *Information and Records Management Inc. Irregular. February 1981. 40 pg.*

Directory of corporations manufacturing paper shredders and related devices.

3289. FEEDING THE AUTOMATED OFFICE. *Mueller, Robert R. Information and Records Management Inc. v.16, no.2. February 1982. pg.20-23, 36*

Examines the market for supplies and accessories for word processing, equipment, typewriters, and computers. Includes charts and tables detailing market shares of various types of office supplies, vendors; dealers, equipment manufacturers, supplies manufacturers, and specialists; and estimates of usage for products used for word processers, computers, copiers, typewriters, and dictation equipment. Details estimated use in number of units and by monthly cost. Gives factors to be considered in choosing a vendor. Features comments from manufacturers and dealers.

3290. INFORMATION AND RECORDS MANAGEMENT: NEW PRODUCTS AND EQUIPMENT. *PTN Publishing Corporation. Monthly. November 1981. pg.12-14*

Monthly listing of new office products introduced in the U.S. market. Provides manufacturer's name and a description of the product.

3291. INFORMATION AND RECORDS MANAGEMENT: PRODUCTS. *Information and Records Management Inc. Monthly. January 1982. pg.44-47*

Monthly listing of new office equipment including word processors, disk systems, and business computers includes photograph, description, and name of company. Further information available through reader card service.

3292. MICROFILM READERS AND READER PRINTERS: STATE OF THE ART. *Mainiero, Ellen. PTN Publishing Corporation. Annual. v.16, no.4. April 1982. pg.30-34, 70*

Product review, featuring microfilm readers and reader-printers, discusses new products, technological advances, and industry trends. Technical specifications and descriptions are included for product types, with additional information on selected models and manufacturers' new products.

3293. 1981 FILING STORAGE AND INFORMATION RETRIEVAL GUIDEBOOK. *PTN Publishing Corporation. Annual. January 1981. pg.14-63*

Listings in the information and records management system buyers' guide are broken down by category: filing, storage and retrieval; records center operations; microfilm housing; forms, DP storage, handling; filing and retrieval aids; and office systems. Listings include lengthy product descriptions and company name.

Lodging Hospitality

3294. LODGING PRODUCTS DATA GUIDE. *Penton / IPC. Subsidiary of Pittway Corporation. March 1981. pg.37-98*

Classified product listing of furniture, offifice equipment, laundry and maintenance supplies, electrical fixtures, food and beverage equipment and bathroom fixtures available to the lodging industry. Provides manufacturers and a description of their products.

Magazine of Bank Administration

3295. DIRECTORY OF CASH DISPENSING DEVICES AND AUTOMATED TELLERS. *Bank Administration Institute. Annual. May 1980.*

Special issue includes a directory of cash dispensing devices and automated tellers.

3296. DIRECTORY OF COIN AND CURRENCY PROCESSING EQUIPMENT AND SUPPLIES. *Palmer, Bruce, editor. Bank Administration Institute. Irregular. September 1981. pg.54-58*

Alphabetical listing of 32 manufacturers / suppliers of coin and currency processing equipment and supplies.

Management World

3297. FIRST ANNUAL GUIDE TO COPIERS. *Polaski, Iris. Administrative Management Society. Annual. September 1981. pg.12-25*

First annual survey of the U.S. copier market. Three articles discuss the types and brands of popular copiers; various copier capabilities; considerations in selection; and how to analyze office copier needs. Also presents highlights from an AMS member survey on types and brands of copiers commonly used in offices. Gives rankings of the top 10 U.S. copier manufacturers, with their share of the market. One table details average monthly copier volume. Includes listing of 35 copier manufacturers, their types of equipment, and business addresses.

3298. THE INFORMATION REVOLUTION AND THE OFFICE OF 1990. *Administrative Management Society. January 1982.*

Forecasts to 1990 of changes in business resulting from the information revolution. Examines impact of changes in employee composition and expectations, installations of new technology and management innovations. Projects expenditures for office equipment to 1990.

3299. MANAGEMENT WORLD'S FIRST ANNUAL GUIDE TO DICTATION. *Steinmetz, Nonie. Administrative Management Society. Annual. v.10, no.11. November 1981. pg.8-17*

1981 survey of office dictating equipment. Discusses new products, major businesses, and presents a survey on who's using what. The survey presents results of interviews with 500 panelists on the types of systems used. Lists 9 popular equipment manufacturers and their share of the market; and types of systems and media used and the percentage of panelists using them. Additional information includes a directory of 18 manufacturers, with addresses and telephone numbers and special equipment features.

3300. SECOND ANNUAL GUIDE TO COPIERS. *Polaski, Iris. Administrative Management Society. Annual. v.11, no.9. September 1982. pg.9-17*

Annual analysis of the office copier market. Includes trends and guides for purchases, market segmentation and prices, new products, manufacturers' addresses, market survey of utilization of copiers and manufacturers, and bibliographic information.

3301. SECOND ANNUAL WORD PROCESSING GUIDE. *Fitzwater, Teresa. Administrative Management Society. Annual. v.11, no.4. April 1982. pg.7-18*

Guide to the U.S. word processing market includes explanation of terms; description of typical systems; discussion of factors involved in successful WP operations and operator job satisfaction; survey of most used equipment types, manufacturers, average prices, and integration with other equipment; and a compilation of word processing bibliographic references, manufacturers, and equipment. Supplies reader reply card for further information. Data provided by publisher survey.

3302. WHAT'S ON THE MARKET: WP EQUIPMENT REVIEW. *Unley, Shirley M. Administrative Management Society. Annual. April 1981. pg.16-22, 48*

First annual guide to available word processing equipment. Discusses various types, price ranges, and options. Lists 34 manufacturers, showing types of equipment produced and business addresses and telephone numbers.

3303. WHO'S USING WHAT: WP SURVEY REPORT. *Thomas, Edward. Administrative Management Society. Annual. April 1981. pg.27-29*

Highlights the results of an AMS member survey on who is using word processing equipment. Presents methodology (380 usable questionnaires); background of respondents; types of equipment being used or considered; methods of purchasing equipment; and tasks handled by equipment.

Mart

3304. WHAT'S NEW IN DIGITALS. *Morgan-Grampian, Inc. Monthly. March 1982. pg.22*

Monthly listing of digital calculators for office and personal use gives manufacturer's name, description, photograph, and price when available. Further information available through reader card service.

Modern Office Procedures

3305. THE BRIDGE BETWEEN THE USER AND THE MICROGRAPHICS SYSTEM. *Penton / IPC. Subsidiary of Pittway Corporation. Annual. v.27, no.4. April 1982. pg.72-95*

Directory (1982) presents technical specifications for reader / reader - printer products of 44 U.S. manufacturers. Includes models and prices.

3306. MODERN OFFICE PROCEDURES 1982 PHOTOTYPESETTING EQUIPMENT SPECIFICATIONS. *Penton / IPC. Subsidiary of Pittway Corporation. Annual. v.27, no.10. October 1982. pg.88-89*

Listing includes name of company, model, methods of input, screen, capabilities, and output.

3307. READER SURVEY: THE SWING TO ELECTRONIC FURNITURE. *Penton / IPC. Subsidiary of Pittway Corporation. v.27, no.10. October 1982. pg. 94-100*

Reports results of a recent survey sent to readers of Modern Office Procedures. Data provided for sources most helpful in gathering information on electronic systems furniture; buying activity in the electronics furnishings marketplace, percentage of respondents who boucht and/or will buy selected products; and top twenty companies (conventional, open plan and electronic) in office furnishings.

3308. SHOW PREVIEW: MICROGRAPHICS. *Penton / IPC. Subsidiary of Pittway Corporation. Annual. v.27, no.4. April 1982. pg.128-138*

Previews micrographic products to be exhibited at the 1982 National Micrographics Association Conference and Exhibition.

3309. 1980 BUYERS PLANBOOK. *Whaley, John H. Penton / IPC. Subsidiary of Pittway Corporation. Annual. January 1980. pg.67-133*

Series of articles devoted to the following product areas for offices: office design (furniture); data processing (mini - micro - large computers, input / output systems, communication terminals, data communication equipment); storage and retrieval (data binders, files, record systems); micrographics (cameras, readers, duplicators); communications (audio-visual equipment, facsimile, intercoms, wire / message services); mail processing (addresses, letter openers, postage scales); word processing (shared-terminals, edit devices phototypesetting); reprographics (copies, duplicators); paper handling (binding machines, cutters, laminators); office products, supplies and services (paper, carbon sets, calculators). Each product listing includes corporations (and addresses) which sell the product.

The Nielsen Researcher

3310. MERCHANDISING IN THE 80'S: HOW RETAILERS WILL USE SCAN DATA. *Scott, Steven E. A.C. Nielsen Company. no.1. 1982. pg.1-15*

Presents an overview of the development, spread, and utility of Universal Product Code scanning. Data covers number 1977-1982; importance, 1979-1981; and projected number of scanning stores, 1980-1985. Includes a promotion analysis of personal care items and the effect of scanning data on potential sales.

The Office

3311. BUYERS' GUIDE TO COMPUTER - OUTPUT - MICROFILM RECORDERS. *Office Publications, Inc. Irregular. April 1982. pg.190*

Lists computer - output - microfilm recorders. Includes manufacturer's name, models available, prices, and machine characteristics. Further information available through reader card service.

3312. BUYERS' GUIDE TO COPYING EQUIPMENT. *Office Publications, Inc. Annual. v.95, no.7. July 1982. pg.140-148*

Provides manufacturers, model, purchase or rental price, and user information for copying equipment, 1982.

3313. BUYERS' GUIDE TO COPYING EQUIPMENT. *Office Publications, Inc. Irregular. v.94, no.1. July 1981. pg.121-122, 126*

1981 directory lists manufacturers and model numbers of more than 100 different types of duplicating machines. Also provides prices and feature descriptions.

3314. BUYERS' GUIDE TO DICTATING MACHINES. *Office Publications, Inc. Irregular. v.94, no.5. November 1981. pg.184, 186*

Directory lists 16 U.S. manufacturers of dictating equipment. Includes model numbers, prices and special features for each.

3315. BUYER'S GUIDE TO DICTATING MACHINES. *Office Publications, Inc. Annual. November 1980. pg.154, 159*

The buyer's guide to dictating machines is broken down by company and then by specific product. Details include model number; price; recording medium; recording time (minutes); recorder, transcriber or both; configuration; playback; speaking device; transcribing device; method of connection; unlimited review; audible scanning; and telephone attachment.

3316. BUYERS' GUIDE TO MICROGRAPHIC READERS AND READER - PRINTERS. *Office Publications, Inc. Irregular. April 1982. pg.178, ff.*

Lists micrographic readers and reader - printers. Includes manufacturer's name, models available, price, and machine characteristics. Further information available through reader card service.

3317. BUYERS' GUIDE TO WORD PROCESSORS. *Office Publications, Inc. Irregular. v.94, no.4. October 1981. pg.158-166*

Directory lists about 55 U.S. manufacturers of word processing equipment. Includes prices, model numbers, and brief descriptions of each.

3318. BUYER'S GUIDE TO WORD PROCESSORS. *Office Publications, Inc. Annual. November 1980. pg.140-152*

The buyer's guide to word processors is broken down by company and then by specific product. Details include model number, price, information stored on storage capacity, words per minute, auxiliary input, auxiliary output, stored information locator, visual display, line counter, automatic underlining, changeable type, half spacing, keyboard error corrector and justification.

3319. DATA PROCESSING PRODUCTS: NEW EQUIPMENT IN AUTOMATION. *Office Publications, Inc. Monthly. April 1982. pg.193-202*

A regular listing supplying information on new DP equipment includes name and address of manufacturer, description, and photograph. Further information available through reader card service.

3320. THE OFFICE: NEW PRODUCTS. *Office Publications, Inc. Monthly. April 1982. pg.204-236*

Monthly listing of new office equipment including printers, transceivers, telephone systems, etc. Information includes name and address of manufacturer, description, and photograph. Further information available through reader card service.

3321. THE OFFICE: NEW PRODUCTS. *Office Publications, Inc. Monthly. v.94, no.2. August 1981. pg.150-172*

Monthly listing of new office products available from U.S. manufacturers. Includes addresses of manufacturers and a description of each product and prices, where available.

3322. PAPER SHREDDERS PROVIDE AN ADDED MEASURE OF CONFIDENCE. *Office Publications, Inc. Irregular. v.94, no.2. August 1981. pg.121-132*

Directory lists and describes some 20 paper shredders used to destroy confidential documents available from U.S. manufacturers.

3323. POLL REVEALS BUYING HABITS, PREFERENCE OF COPIER USERS. *Office Publications, Inc. v.95, no.7. July 1982. pg.44, 46*

Highlights of survey of copier users give acquisition method and selection criteria. Provides advantages and problems of copiers by volume usage.

Office Equipment and Methods

3324. BUYERS' GUIDE "82: OFFICE EQUIPMENT AND METHODS. *Maclean Hunter Ltd. Annual. v.27, no.10. December 1981. 27 pg.*

Contains information for Canadian users of office equipment / products and supplies. Lists names and addresses of Canadian and U.S. suppliers and their products and services. Includes an associations directory noting contacts, addresses and telephone numbers.

3325. OFFICE EQUIPMENT AND METHODS: 1981 BUYERS' GUIDE. *Maclean Hunter Ltd. Annual. December 1980. 60 pg.*

Annual report on the office equipment and furniture markets, giving forecasts for the industry for 1981. Contains listing of associations, office equipment suppliers and manufacturers, products and product lines available in Canadian markets.

3326. OFFICE EQUIPMENT INDUSTRY FORECAST: INDUSTRY LOOKS FOR CLOUT. *Wallace, Mary Beth. Maclean Hunter Ltd. December 1981. 4 pg.*

Forecasts by trade association presidents on the growth and need for research and development in Canadian high technology in a number of product categories.

3327. 1981 WP REPORT. *Maclean Hunter Ltd. Annual. September 1981. pg.35-46*

Survey of the word processing market includes a look at new products (electronic typewriters, multi-purpose microcomputers, office automation systems), education and job enrichment, vendor training, and office skills.

Office Products Dealer

3328. COMPUTER SYSTEMS DEALER: SYSTEMS PRODUCTS. *Hitchcock Publishing Company. Monthly. 1982.*

Monthly directory of products available for the office, with special emphasis on furniture and equipment useful in a computerized environment. Each entry includes a brief description with technical specifications, brand name, manufacturer, and photograph.

3329. HOW ARE THE COPIER SPECIALISTS DOING? *Majorowicz, Sandra. Hitchcock Publishing Company. v.109, no.7. July 1981. pg.40-42*

Presents a review of the market for copiers, with comments and industry forecasts from three copier dealers. Discusses the growth rate of the market along with increasing copier prices and marketing trends. Identifies market segments and reviews marketing strategies.

3330. HOW ARE THE WORD PROCESSING SPECIALISTS DOING? *Hitchcock Publishing Company. v.109, no.7. July 1981. pg.34-36*

Reviews the market for word processing equipment, profiling three major vendors. Discusses competition, the size of the market in total sales volume; and market penetration. Forecasts market growth.

3331. NATIONAL OFFICE MACHINE DEALERS ASSOCIATION SHOW ISSUE. *Hitchcock Publishing Company. Annual. July 1981.*

Special issue covering the National Office Machine Dealers Association annual show.

3332. NATIONAL OFFICE PRODUCTS ASSOCIATION SHOW ISSUE. *Hitchcock Publishing Company. Annual. October 1981.*

Special issue covering the National Office Products Association show.

3333. OFFICE PRODUCTS DEALER: PRODUCTS. *Hitchcock Publishing Company. Monthly. 1982.*

Monthly review of products for the office including furniture, stationery, copies equipment and supplies and small electronic machines. Each entry includes brief description, brand name, technical specifications, photograph, and name of manufacturer or supplier.

3334. OFFICE PRODUCTS DEALER: 1981 PRODUCT REVIEW. *Hitchcock Publishing Company. Annual. v.109, no. 12. December 1981. pg.32-37*

Lists and describes the most popular office products in 1981, according to the readers of the magazine. Includes description, photograph, and manufacturers' names for office machines, furnishings and computer hardware.

3335. OFFICE PRODUCTS DEALER: "82 AND BEYOND. *Plant, Janet. Hitchcock Publishing Company. Annual. v.109, no.10. October 1981. pg.30-35*

Industry leaders from the dealer, wholesaler and manufacturer segments of the office products industry identify the product, and buying and marketing trends which will prevail in 1982 and throughout the decade.

3336. OPD SURVEY: PROFILE OF A DEALER 1979. *Hitchcock Publishing Company. Irregular. 1979.*

1979 survey profiles the U.S. office products dealer. Part 1 details who office products dealers are and how they run their businesses. Part 2 details the products dealers sell and gives the percentage of dealers within a specified grouping who sell a particular product. Uses the 3 major product categories of supplies, machines, and furniture. Previous "OPD Profile" was published in 1976. Charts describe dealers in terms of profit margins, compensation plans, commissions paid, commercial and retail sales, accounts receivable, advertising costs and media, locations, account size, stock turns, shows attended.

3337. WHAT DEALERS PAY THEMSELVES. *Hitchcock Publishing Company. Irregular. February 1981. pg.36-39*

Survey of salaries of office equipment dealers. Tables display total cash compensation of dealer management; median compensation of top executives by equity held and lines carried; equity held by top executives; top executive fringe benefits and prerequisites; and components and median amounts of top executive compensation.

3338. 1979 DIRECTORY AND DEALER BUYING GUIDE. *Hitchcock Publishing Company. Annual. December 1978. 192 pg.*

A 1979 directory and dealer buying guide to office supplies and equipment. Arrangement is by type of product followed by suppliers and manufacturers with separate address and phone listings. Wholesalers and associations also given.

Office World News

3339. COPIER INDUSTRY SEES DRAMATIC CHANGES IN MARKET, PRICING AND DISTRIBUTION. *Hearst Business Communications, UTP Division. February 1, 1982. pg.1, 10*

Reviews the market for paper copiers and examines trends in manufacture and replacement. Statistics given for plain paper copier copy volume by market segments; PPC placements (1975, 1980, 1985); growth rate (1975-1985); PPC placements

by area (1979, 1980, 1981); and annual worldwide Japanese shipments (1977-1981). Discusses competitive pricing and distribution patterns.

3340. DEALERS START SMALL, GROW INTO IP SALES. *McCooey, Eileen. Hearst Business Communications, UTP Division. v.10, no.19. October 15, 1982. pg.1, 25*

Summarizes findings of a 1982 spot survey of U.S. business equipment dealers regarding their sales of complex, office automation systems. Charts illustrate breakdown (in percent) of dealers selling electronic typewriters, word processing systems, and small business computers. Presents comments by selected dealers.

3341. DICTATION / TRANSCRIPTION EQUIPMENT. *Hearst Business Communications, UTP Division. Irregular. v.10, no.2. January 15, 1982. pg.13-14, 16*

Directory lists dictation and transcription equipment, including technical specifications and special features. Listing is arranged alphabetically by manufacturer, with product listings for each. Prices are included.

3342. ELECTRIC / ELECTRONIC TYPEWRITERS 1982. *Hearst Business Communications, UTP Division. Annual. v.10, no.19. October 15, 1982. pg.11-23*

Chart provides technical specifications for electric / electronic typewriters available from 22 U.S. vendors in 1982. Includes the following information: vendors, models, nine categories of specifications, and list prices.

3343. A PREVIEW OF THE NOMDA EXHIBITS. *Hearst Business Communications, UTP Division. Irregular. July 1, 1982. pg.14, 16, 18, 22*

Lists exhibited office products at NOMDA "82 (July 14-17, Kansas City) including calculators, copiers, printers, typewriters, etc. Gives manufacturer, description, price, booth number, and photograph.

3344. SPOTLIGHT ON NATIONAL OFFICE PRODUCTS ASSOCIATION (NOPA) EXHIBITS. *Hearst Business Communications, UTP Division. Annual. v.10, no.106. October 1, 1982. pg.23-41*

Alphabetically lists companies exhibiting products at the 1982 NOPA show. Includes a product description, telephone number, name of contact person (if available), and booth number.

Plan and Print

3345. INTERNATIONAL REPROGRAPHIC ASSOCIATION MEMBERSHIP ROSTER. *International Reprographic Association, Inc. Irregular. December 1981. pg.27-30*

By state and city / town, lists member companies active in blueprint and graphics reproduction services, and in the manufacture and furnishing of relevant equipment and supplies.

Purchasing Magazine

3346. MAJOR SHAKEOUT ROCKING THE COPIER BUSINESS. *Cahners Publishing Company. v.93, no.6. September 23, 1982. pg.80-81*

Examines U.S. plain paper copier market, 1981-1982. Gives number of units shipped by 7 manufacturers, 1981 and number of units installed by same manufacturers, 1982.

The Southern Banker

3347. THE SOUTHERN BANKER: EQUIPMENT AND SUP-PLY NEWS. *McFadden Business Publications. Monthly. v.157, no.3. March 1982. pg.22-26*

Reviews new products and services in the commercial banking field. Includes descriptions of products and services, and names and addresses of manufacturers and vendors.

Today's Office

3348. DICTATION / TRANSCRIPTION EQUIPMENT: BUY-ERS CHECKLIST. *Montross, Teri. Hearst Corporation. Irregular. January 1982. pg.69-80*

Charts survey currently-available dictation / translation equipment, indicating functions, probability, cassette size(s) employed, maximum recording time, features and options, and prices.

3349. ELECTRONIC MAIL: A NEW MEDIUM FOR THE MES-SAGE. *Pomerantz, David. Hearst Corporation. v.17, no.3. August 1982. pg.41-47*

Analysis of markets and technology for electronic mail includes competition among media; vendors; installations, utilization, and sales in 1980, with forecasts to 1995 and 1980-1995 growth rates; and projected operating budgets for communications to 1990.

3350. ELECTRONIC PRINTING SYSTEMS. *Dooley, Brian J. Hearst Corporation. v.17, no.3. August 1982. pg.37-39, 82*

Market and technology analysis for electronic printing systems. Examines performance and lists manufacturers.

3351. READERS AND READER / PRINTERS CHART. *Hearst Corporation. Annual. v.16, no.12. May 1982. pg.61-67*

Lists readers and reader / printers for office automation. Includes vendor, models, microforms accepted, and features and options. Gives prices for most systems.

3352. SMALL BUSINESS AND OA: THE ROMANCE OF THE EIGHTIES. *Hearst Corporation. v.17, no.2. July 1982. pg.30-36*

Examines use of various size computers, word processors, and other electronic equipment by firms with less than 500 or 100 employees, 1982. Gives current and planned use of 14 office automation types; and present and planned data processing usage (in-house or service firm) in small enterprises, 1980-1983.

3353. TODAY'S OFFICE, 1982: COPIERS. *Hearst Corporation. Annual. v.17, no.2. July 1982. pg.34-60*

Discusses reprographic equipment state-of-the-art, 1982. Lists manufacturers, models, technical specifications, and retail prices.

Toronto Business Magazine

3354. TORONTO SECRETARYS' EFFICIENCY GUIDE. *Zanny Ltd. Annual. February 1982. 30 pg.*

Listing of services, supplies products and equipment purchased by secretaries. Categories broken down into five geographical areas - City, Scarborough, North York, Etobicoke, Mississauger.

Typeworld

3355. WORD PROCESSING INTERFACE: PART 2. *Blum Publications. v.6, no.6. April 6, 1982. pg.16-17, 23*

Reviews point-to-point technology in word processing industry. Forecasts cumulative U.S. sales of pagination systems and equipment, 1982-1986. Gives end uses for pagination systems and equipment analyzed.

United States Banker

3356. TOOLS FOR THE BANKER. *Cleworth Publishing Company, Inc. Irregular. v.92, no.11. November 1981. pg.68-69*

Review of products for use in the banking industry. Each entry includes product description and manufacturer, with photographs for selected items.

What to Buy for Business

3357. TABLETOP PRINTING AND DUPLICATING MACHINES. *Oppenheim Derrick, Inc. Irregular. February 1982. pg.3-16*

Review of table top copiers, including offset presses, stencil and spirit duplicators and photo-copiers. Reviews the technology, costs and features of utilization. Shows manufacturers, addresses, brand names and prices.

Word Processing Systems

3358. WORD PROCESSING AND INFORMATION SYS-TEMS: REVENUES AND EARNINGS REPORT. *Geyer-McAllister Publications, Inc. v.9, no.9. September 1982. pg.10*

Lists revenues and earnings for 10 U.S. word processor manufacturers. Provides data for first 6 months of 1981 and 1982. Includes revenues, earnings, and percent change.

3359. WORD PROCESSING SYSTEMS ANNUAL SOURCE GUIDE AND REFERENCE ISSUE. *Geyer-McAllister Publications, Inc. Annual. December 1979. pg.1-104*

1979 annual guide to word processing systems, arranged by manufacturers with product and service sources, by product and service categories, and by equipment trade names. A special advertising section with information on companies and their products included. Reference section gives information resources and schools which teach word processing.

See 1489, 1536, 1557, 1573, 1597, 1641, 1661, 1676, 1677, 1683, 1708, 1723, 1724, 1765, 1778, 1781, 1783, 1784, 1785, 1789, 1793, 1798, 1819, 1823, 1827, 1828, 2571, 2609, 2660, 2773, 3650, 3756.

Monographs and Handbooks

3360. THE MANAGER'S GUIDE TO PHOTOCOPIER SELEC-TION. *3M Canada Inc. 1981. 12 pg.*

Offers advice for determining copying volume and features that are needed, assessing hidden copying costs, increasing efficiency and comparing copiers.

3361. THE OFFICE OF THE FUTURE. *Uhlig, R.P. Elsevier-North Holland Publishing Company. 1979. 380 pg. $35.00.*

Discusses potential uses for computers and communications equipment in offices, the technology which would make these changes possible, and the impact on individuals, groups, and organizations.

3362. TEST REPORTS ON ELECTRONIC TYPEWRITERS. *Buyers Laboratory, Inc. 1981.*

Report provides test results of six competing models of electronic typewriters. IBM, Exxon, Olympia, Olivetti, and Brother machines are rated for features, ease of use, downtime, and durability.

3363. THE WORD PROCESSING HANDBOOK. *International Self-Council Press Ltd. 1981. 201 pg.*

Step-by-step guide to office automation includes types of equipment and products available, case studies, glossary of terms and suggested readings.

3364. WORD PROCESSING: SELECTED SHARED LOGIC SYSTEMS. *Datapro Research, Inc. (McGraw-Hill). 1981. $15.00.*

Analyzes AM Jacquard J-100, Four-Phase Foreword, and DEC Word Processing Systems. Selected from Datapro Reports on Word Processing.

See 1836, 1843, 1854, 4265, 4266, 4267, 4268.

Conference Reports

See 4273.

Newsletters

3365. OFFICEMATION PRODUCT REPORTS. *Management Information Corporation. Monthly. $295.00/yr.*

Monthly newsletter evaluates office automation systems (i.e., word processors, electronic mail, intelligent typewriters, computerized records management). Descriptions cover hardware, software, functions, operation, prices, and advantages and disadvantages.

Services

3366. DATABOOK 70: OFFICE SYSTEMS. *Maclean Hunter Ltd. Monthly. 1980. $605.00/yr.*

Information on the full spectrum of office products, systems, and techniques available in Canada. Published in 3 volumes and updated monthly.

3367. DATABOOK 70: RETAIL AUTOMATION. *Maclean Hunter Ltd. Bi-monthly. 1980. 400 (est.) pg. $395.00.*

Complete data on current POS / retail automation equipment and systems. Contains detailed reports and comparison tables. Published in 1-volume and updated bi-monthly.

3368. DATAPRO REPORTS ON WORD PROCESSING. *Datapro Research, Inc. (McGraw-Hill). Monthly. $515.00/yr.*

Covering word processing hardware and software and dictation equipment and supplies, annual subscription service includes two loose-leaf volumes, monthly report supplements, monthly newsletter, and a telephone / telex inquiry service.

3369. DATAPRO REPORTS ON WORD PROCESSING: INTERNATIONAL EDITION. *Datapro Research, Inc. (McGraw-Hill). Monthly.*

International coverage spans word processing hardware and software and dictation equipment and supplies. Annual subscription service includes loose-leaf volume, monthly report supplements, monthly newsletter, and a telephone / telex inquiry service.

3370. OFFICE PRODUCTS SERVICE. *Buyers Laboratory Inc. Frost and Sullivan, Inc. Bi-monthly. 1981. $425.00.*

A yearly subscription service providing testing of office products and purchasing guidelines. Includes performance ratings of products, user attitudes, prices, and new product reports.

3371. TYPELINE. *Buyers Laboratory, Inc. Bi-monthly. $150.00/yr.*

Bimonthly service covers manual, type-bar electric, single element, and electronic typewriters. Initial volume provides reports and rating charts for 34 models from 12 manufacturers. Subsequent bimonthly test reports on new models detail price, weight, maximum paper width, writing line, features, general appraisal (estimated downtime, ability to make many good carbon copies, test typists' reactions, users' reactions), test observations (numerous categories), electrical system, and guarantee.

See 1877, 1878, 1880, 1882.

Dictionaries

3372. GLOSSARY OF COPIER AND REPROGRAPHICS TERMS. *Datapro Research, Inc. (McGraw-Hill). 1981. 27 pg. $15.00.*

Definitions of terms common to in-house reprographics, photocomposition, plate-making, and other processes. Selected from 'Datapro Reports on Copiers and Duplicators.'

3373. GLOSSARY OF 492 WORD PROCESSING TERMS. *Datapro Research, Inc. (McGraw-Hill). 1982. 12 pg. $19.00.*

Glossary defines 492 terms concerning features and characteristics of word processing systems as well as word processing procedures and operations.

3374. OFFICE AUTOMATION: A GLOSSARY AND GUIDE. *Shaw, Carmine. Knowledge Industry Publications. 1982. 500 pg. $59.50.*

Glossary of 7500 terms covers data processing, word processing, telecommunications, reprographics, micrographics and other record management systems, video, viewdata, videotext.

See 1900.

Western Europe

Market Research Reports

3375. AUTOMATIC TYPEWRITERS AND WORD PROCESSORS IN EUROPE TO 1986. *Larsen Sweeney Associates Ltd. Quarterly. 1981. 235 pg. 995.00 (United Kingdom).*

Market report to 1986 on electric, electronic, and automatic typewriters, word processors and computer - based word processor systems in Western Europe. Updated quarterly, the report covers the following aspects in 25 end user sectors: market analysis, forecasts to 1986, sales, production trends, profits, assets, end users, employment and unemployment, products, capital expenditures, investments, and distribution.

3376. CALCULATORS IN EUROPE TO 1986. *Larsen Sweeney Associates Ltd. 1981. 242 pg. 995.00 (United Kingdom).*

Market research report on calculators to 1986 in the following countries: Austria, Belgium, Luxembourg, Denmark, Eire, Finland, France, West Germany, Italy, Netherlands, Norway, Portugal, Spain, Sweden, Switzerland, and the U.K. Covers market forecasts to 1986, distribution, production, suppliers, market segmentation, employment and unemployment, profits, sales, and end uses. Calculators covered are: battery operated, battery and mains operated, mains operated, internal memory, removable magnetic memory, and programmable. There are 24 end user markets covered.

3377. CASH REGISTERS IN EUROPE TO 1986. *Larsen Sweeney Associates Ltd. 1981. 233 pg. 895.00 (United Kingdom).*

Market research report on electronic cash registers to 1986 in the following countries: Austria, Belgium, Luxembourg, Denmark, Eire, Finland, France, West Germany, Italy, Netherlands, Norway, Portugal, Spain, Sweden, Switzerland and the U.K. Covers market forecasts to 1986, distribution, production, suppliers, market segmentation, employment and unemployment, profits, sales and end users. Types of cash registers include online, independent, coumputerized, and laser scanning point of sale terminals. Twenty end user markets are discussed.

3378. THE COM MARKET 1980 IN THE U.K. AND THE CONTINENT OF EUROPE. *G.G. Baker and Associates. July 1980. 55 pg.*

Research based on questionnaires and interviews on the COM market throughout Europe. Covers market value of recorders, recorder film, duplicating film, forecasts to 1990 and consumer usage.

3379. DICTATION EQUIPMENT IN EUROPE TO 1986. *Larsen Sweeney Associates Ltd. 1981. 231 pg. 695.00 (United Kingdom).*

Market research report on dictation equipment of all types in Austria, Belgium, Luxembourg, Denmark, Finland, Eire, France, West Germany, Italy, Netherlands, Norway, Portugal, Spain, Sweden, Switzerland and the U.K. Covers forecasts to 1986, distribution, production, suppliers, market segmentation, employment and unemployment, profits, sales and end users. 25 end user sectors are analyzed individually.

3380. EFTS AND SECURITY PRINTING: (EFTS V). *Hirsch, P. Battelle - London. 1982. 30000.00 (West Germany).*

Study will examine the effect of electronic fund transfer systems (EFTS) on the security paper and printing industry. It will focus on changes in the demand for cash, checks, and traveler's checks in Western European countries.

3381. ELECTRONIC TYPEWRITERS. *Frost and Sullivan, Inc. February 1982. $1500.00.*

Gives analysis and forecasts to 1990 for electro-mechanical type basket and single element typewriters; electronic typewriters, including enhanced features and internal storage types; and supplies and equipment. Covers distribution channels and market for supplies and sub-assemblies.

3382. ERC MARKET MONITOR EUROPE: OFFICE EQUIPMENT. *E.R.C. Statistics International Ltd. December 1981. 2400.00 (United Kingdom).*

Western European market analysis for office equipment. Includes sales, brand names, prices, advertising, consumer behavior, retail distribution, industry structure, production manufacturers, imports and exports, new products, and foreign

3383. THE EUROPEAN COMPUTER, WORD PROCESSING AND COPIER MARKETS: OPPORTUNITIES AND CHALLENGES FOR THE MAJOR MULTINATIONAL COMPANIES. *Martin Simpson and Company, Inc. European Technology Comments #35. December 1980. 66 pg. $475.00.*

Analysis of the computer, word processor and copier industry in Western Europe. Evaluates strategies, performance and outlook for major vendors. Includes revenues, installed base, competition analysis and economic indicators.

3384. EUROPEAN ENVIRONMENT FOR OFFICE SYSTEMS PRODUCTS. *Frost and Sullivan, Inc. April 1982. 257 pg. $1250.00.*

Analysis of European markets for office products and systems. Forecasts unit and dollar sales, 1981-1990. Examines technology, manufacturers, revenues, market penetration, cost trends, end uses and product groups, and employees and labor productivity.

3385. EUROPEAN MARKET MONITOR: OFFICE EQUIPMENT. *E.R.C. Statistics International Ltd. E.303. July 1980. 1800.00 (United Kingdom).*

Western European market analysis and forecast for industrial consumption of office equipment, 1974-1980.

3386. KEY NOTE REPORTS: OFFICE MACHINERY. *Keynote Publications Ltd. 1981. 20.00 (United Kingdom).*

A report on office machinery in the U.K. covering the structure of the industry, distribution networks, major manufacturers, markets, and finances. Analysis includes products by sales volume and value, market shares, foreign trade, new products, mergers, legislation, financial analysis, operating ratios, and forecasts of trends.

3387. THE MARKET FOR BUSINESS EQUIPMENT AND SYSTEMS IN BELGIUM. *Systec Consultants Ltd. U.S. Department of Commerce. International Trade Administration. NTIS #ITA 81-10-504. May 1981. 71 pg. $15.00.*

Survey of the market for business equipment and systems in Belgium describes market size and sources of supply, end users, and market access. Provides data on current and projected market size by product category. Gives import data by major subcategories, by country of origin and market share percent. Lists major U.S. and third country suppliers. Profiles data on estimated expenditures for new instatllations and replacements. Marketing, regulations, and technical requirements are described. Trade lists cover principal agents and distributors; prospective customers; trade and professional bodies; publications; and trade events.

3388. THE NETHERLANDS BUSINESS EQUIPMENT MARKET: A FOREIGN MARKET SURVEY REPORT. *Konsulterna. U.S. Department of Commerce. International Trade Administration. NTIS #ITA 82-05-508. May 1981. 72 pg. $15.00.*

Survey of the market in the Netherlands for business equipment covers word processing systems, photo composers, small business computer systems, micrographic equipment, facsimile equipment, and copying equipment. Analyzes market organization and structure and provides a competitive assessment of domestic, U.S., and third country suppliers. Gives profiles of each end user sector, including current equipment and projected expenditures. A section of the report provides information on business practices, trade restrictions, and technical requirements. Trade lists include major imports and distributors, major importers and distributors; major prospective customers; trade publications.

3389. THE OFFICE AUTOMATION SERVICE: VOLUME IV: EUROPE. *Creative Strategies International. Monthly. 1982.*

Monthly service providing analysis of the European office automation market and industry. Includes installed base, shipments, forecasts and market segmentation by countries. Examines competition, manufacturers, suppliers, and retail distribution, including financial analysis.

3390. OFFICE EQUIPMENT MARKET 1981-1987. *IDC Europa Ltd. 1982. 150 (est.) pg. 1250.00 (United Kingdom).*

Report on office equipment market in Western Europe 1981-1987, examines principal office products including word processors, typewriters, PABX, copiers and facsimile markets. Analyzes market size and projection by country with full analysis by product category and manufacturer market share. Examines penetration of communications facilities within the office, such as electronic mail, teletext and videotex / viewdata. Also provides supplier profiles and strategies.

3391. OFFICE SUPPLIES MARKET IN EUROPE. *Frost and Sullivan, Inc. October 1979. $950.00.*

Market analysis and forecast to 1989 for office equipment, paper products and supplies in Western Europe. Shows production, imports, exports, and net apparent consumption; profiles of leading manufacturers with market shares; distribution channels and end-use markets.

3392. THE RANK ORGANIZATION LIMITED AND THE OTHER OFFICE EQUIPMENT COMPANIES. *Scott, Goff, Hancock and Co., Research Dept. MSRA, Inc. 1981. 105 pg. $600.00.*

Analysis of the European reprographics industry and market structure. Includes investment recommendations and national economic indicators, 1980 installed base of copiers, revenues by product lines, financial analysis, and marketing strategies of manufacturers.

3393. SWEDEN BUSINESS EQUIPMENT MARKET. *Konsulterna. U.S. Department of Commerce. International Trade Administration. NTIS # ITA81-08-512. February 1981. 105 pg. $15.00.*

Survey of the market for business equipment in Sweden provides data on market size and imports by major product classification. Gives a competitive assessment of domestic, U.S., and third country suppliers in each product category. Profiles end user sectors, including projected needs for specific products. Business practices, trade restrictions, and technical requirements are described. Trade lists cover principal potential agents and distributors; major prospective customers; trade, industrial, professional associations; publications; and major trade exhibitions.

3394. U.K. COMPUTER INSTALLATION CENSUS: SERIES 1980 / 1981: VOLUME 1 : WORD PROCESSING SYSTEMS. *Pedder Associates Ltd. Annual. 8th ed. 1981. 130 (est.) pg. 800.00 (United Kingdom).*

Analyzes U.K. installed base of word processor systems by number and value for each model type under six product group headings: stand-alone hard copy systems without display, stand-alone thin-window display systems, stand-alone CRT display based systems, multi-work station systems, electric typewriters, and word processing software packages in use on general purpose computer systems. Provides market definitions, overview of market, summary of total market showing suppliers' market shares by number and value of installations, trends and developments for each market sub-classification, analysis by market sub-classification, and figures showing average value of installation and number of installations. 1979 / 1980 report covered 111 model types from 39 suppliers.

3395. U.K. COMPUTER INSTALLATION CENSUS: SERIES 1980 / 1981: VOLUME 4: CENSUS SUMMARY, WORD PROCESSOR AND GENERAL PURPOSE SYSTEMS. *Pedder Associates Ltd. Annual. 8th ed. 1981. 130 (est.) pg. $350.00 (ukd).*

Provides summaries of market shares for the three general product groups covered in volumes 1-3: word processing systems, small general purpose computer systems valued at less than 15,000 pounds, and large general purpose computer systems valued at more than 15,000 pounds. Gives overall summaries and commentary but does not give model by model data.

3396. UNITED KINGDOM MARKET FOR BUSINESS EQUIPMENT AND SYSTEMS. *John Martin Publishing Ltd. September 1979. 95.00 (United Kingdom).*

Analysis of markets for business equipment and systems in the U.K., 1978, with forecasts to 1983. Examines market structure and growth, end uses, production of domestic manufacturers, and trade associations.

3397. WORD PROCESSING. *BIS Marketing Research. 1981. 9000.00 (United Kingdom).*

Word processing market analysis for the U.K., West Germany, and France, 1980-1985. Includes the 1980 installed base, market penetration opportunities, market segmentation and growth rates, end uses, demand, user attitudes toward suppliers, ratings including price and performance factors, and distribution channels.

3398. WORD PROCESSING: FRANCE. *BIS Marketing Research. 1981. 3500.00 (United Kingdom).*

Word processing markets analysis for France, 1980-1985. Includes the 1980 installed base, market penetration opportunities, market segmentation and growth rates, end use demand, user attitudes toward suppliers, ratings including price and performance factors, and distribution channels.

3399. WORD PROCESSING MARKET IN EUROPE. *Frost and Sullivan, Inc. January 1981. 377 pg. $1,450.00.*

Covers market penetration, with forecasts to 1990; market segmentation (replacement purchases, purchases by new customers); growth patterns by product group (thin window, partial page, full page, shared resource); and factors customers consider before purchasing.

3400. WORD PROCESSING MARKETPLACE - WESTERN EUROPE, 1978-1983. *IDC Europa Ltd. 2nd ed. February 1982. 130 pg. 1250.00 (United Kingdom).*

Market research report providing forecasts and analysis of Western European word processing markets, 1978-1983. Gives consumers, vendors, market structure, installed base, value of shipments, unit sales, forecasts, sales, market areas, demand, user attitudes, supply, and market size. Considers shared logic and stand-alone. Devotes major section to vendor strategies.

3401. WORD PROCESSING: UNITED KINGDOM. *BIS Marketing Research. 1981. 3500.00 (United Kingdom).*

Word processing markets analysis for the U.K., 1980-1985. Includes the 1980 installed base, market penetration opportunities, market segmentation and growth rates, end uses, demand, user attitudes toward suppliers, ratings including price and performance factors, and distribution channels.

3402. WORD PROCESSING: WEST GERMANY. *BIS Marketing Research. 1981. 3500.00 (United Kingdom).*

Word processing equipment market analysis for West Germany. Includes the 1980 installed base, market penetration opportunities, market segmentation and growth rates, end use demand, user attitudes toward suppliers, ratings including price and performance factors, and distribution factors.

See 428, 1926, 1935, 1953, 1954, 1977, 4295.

Industry Statistical Reports

3403. BRANCHEN SERVICE: HERSTELLER VON BURO-MASCHINEN, ADV-GERATEN UND EINRICHTUNGEN: INDUSTRIE: HERST V. BUROMASCH., ADV-GERATEN U-EINRICHTUNGEN. [Branch Service: Producers of Business Machines, Computers and Office Furniture.]. *Institut fur Wirtschaftsforschung. German. Monthly. December 1980. 4 pg.*

Offers monthly statistics on demand, production, price expectations, imports and exports, and orders for the German office furniture, business machines; and computer industries.

3404. BUSINESS MONITOR: OFFICE MACHINERY: QUARTERLY STATISTICS. *Government Statistical Service. Her Majesty's Stationery Office. Quarterly. 1981. 9 pg. 5.10 (United Kingdom).*

Quarterly manufacturer's sales, export, import and wholesale price statistics in pounds sterling are given for the United Kingdom business machinery industry. Figures have been acquired from U.K. manufacturers employing more than 25 people and from H.M. Customs and Excise. Included are statistics concerning non-electronic data processing and handling equipment, duplicators, offset - litho weighing less than 2,000 lbs., cash registers, ticket issuing machines, addressing machines, document handling machines, coin handling and check writing machines, typewriters, and calculating machines, in addition to parts and accessories.

3405. BUSINESS MONITOR: OFFICE MACHINERY: REPORT ON THE CENSUS OF PRODUCTION, 1976. *Government Statistical Service. Her Majesty's Stationery Office. Annual. 1979. 7 pg. .60 (United Kingdom).*

Gives production, production costs, wages and salaries, capital expenditures, stocks, size of businesses, employment, and employees for United Kingdom office machinery manufacturers. Provides seven tables under the following headings: output and costs, 1973-1976; capital expenditure, 1973-1976; stocks and work in progress, 1976; analysis of establishments by size, 1976; regional distribution of employment, net capital expenditure, net output and gross value added at factor cost, 1976; percentage analysis of twelve-month periods covered by returns received from United Kingdom establishments employing 20 or more people, 1976; and percentage analysis of employees, by full and part-time employment and sex, 1976. Report makes use of figures gathered from U.K. manufacturers employing 20 or more people.

3406. BUSINESS MONITOR: OFFICE MACHINERY: REPORT ON THE CENSUS OF PRODUCTION, 1978. *Government Statistical Service. Her Majesty's Stationery Office. Annual. 1980. 8 pg. 1.50 (United Kingdom).*

Gives production, wages and salaries, production costs, capital expenditures, stocks, size of businesses, employment, employees and operating ratios for United Kingdom office machinery manufacturers. Provides eight tables under the following headings: output and costs,1974-1978; capital expenditure, 1974-1978; stocks and work in progress, 1974-1978; analysis of establishments by size, 1978; regional distribution of employment, net capital expenditure, net output and gross value added at factor cost, 1978; percentage analysis of twelve-month periods covered by returns received from United Kingdom establishments, 1978; percentage of employees by full and part-time employment and sex, 1977; and operating ratios, 1977-1978. Report covers all U.K. business machinery

manufacturers employing 50 or more people, half of the manufacturers employing 20 to 49 people, and a small sample of those employing 11 to 19 people.

3407. BUSINESS MONITOR: OFFICE MACHINERY: REPORT ON THE CENSUS OF PRODUCTION, 1979. *Government Statistical Service. Her Majesty's Stationery Office. Annual. 1981. 8 pg. 2.50 (United Kingdom).*

Gives production, wages and salaries, production costs, capital expenditure, stocks, size of businesses, employment, employees and operating ratios for United Kingdom office machinery manufacturers. Provides eight tables: output and costs, 1975-1979; capital expenditure, 1975-1979; stocks and work in progress, 1975-1979; analysis of businesses by size, 1979; regional distribution of employment, net capital expenditure, net output and gross value added at factor cost, 1979; percentage analysis of twelve-month periods covered by returns received from United Kingdom establishments, 1979; percentage analysis of employees, by full and part-time employment and sex, 1977; and operating ratios, 1978-1979. Figures have been gathered from half of the business machinery manufacturers employing between 20 and 49 people and from all manufacturers employing more than 50 people.

3408. BUSINESS MONITOR: PHOTOGRAPHIC AND DOCUMENT COPYING EQUIPMENT: QUARTERLY STATISTICS. *Her Majesty's Stationery Office. Quarterly. 1982. 6.65/yr. (United Kingdom).*

Gives quarterly statistics of the United Kingdom photographic and document copying industry. Includes sales by U.K. manufacturers, exports, imports, and employment. Sales, imports and exports figures are provided for over 40 types of equipment.

3409. ELECTRONIC OFFICE EQUIPMENT: EUROPEAN MARKET TRENDS 1973-1983: VOLUME 1. *Mackintosh Publications Ltd. 1980. 225.00 (United Kingdom).*

Market analysis, 1973-1983, for electronic office equipment in Europe. Examines market trends for photocopying machines and word processors.

3410. OFFICE MACHINERY SWP: PROGRESS REPORT 1980. *Office Machinery Sector Working Party. National Economic Development Office. Annual. 1980. 11 pg.*

Report provides the Office Machinery Sector Working Party's forecasts concerning manufacturers' sales, employment and unemployment, imports and exports of the United Kingdom office machinery industry. Tables show U.K. manufacturers' sales for office machinery (excluding computers and peripherals), and U.K. balance of trade for office equipment including computers and peripherals. The U.K. balance of trade for photocopiers is presented as background statistics for the industry forecasts.

See 2959, 2960.

Financial and Economic Studies

3411. FINANCIAL SURVEYS: STATIONERY AND OFFICE EQUIPMENT MANUFACTURERS AND DISTRIBUTORS. *Inter Company Comparisons Ltd. 1981. 59.80 (United Kingdom).*

Annual financial analysis, with two year comparative figures, for stationary and office equipment manufacturers and distributors in the U.K. For all applicable businesses, tabulates turnover of sales; total assets, including fixed assets, current assets, intangible assets, and investments; current liabilities;

profits before taxes; executive compensation; capital employed; and subsidiaries.

3412. OFFICE MACHINERY. *Keynote Publications Ltd. 1980. 10 pg.*

Economic report of the office machinery industry in the United Kingdom. Examines industry structure, market size and trends, recent developments, future prospects, financial appraisal, financial data, recent press articles, and further sources. Provides statistical information regarding number of employees, sales, imports, exports, market shares, profits, wages and salaries, return on capital, current assets, and current liabilities.

See 2005, 2964.

Forecasts

3413. END USER MARKETS FOR WORD PROCESSORS, 1981-1985. *IDC Europa Ltd. 1982. 1250.00 (United Kingdom).*

Market research report provides forecasts and analysis of the West European end user markets for word processors, 1981-1985. Gives size, share, projection by country, and analysis by product category of the word processor market; analysis of user consideration of cost / benefits, workforce acceptance and applications; end user profiles, including size and nature of business, and job title of decision maker; and examination of current products and how they meet the needs of the automated office.

See 2007, 2012.

Directories and Yearbooks

3414. ANNUAIRE GENERAL DES FOURNISSEURS EN INFORMATIQUE ET EN BUREAUTIQUE: O1 DIGEST. [Yearbook of Suppliers for Informatics and Offices.]. *Groupe Tests. French. Annual. December 1981. 58 pg.*

Directory of manufacturers, suppliers, consultants and services for all office and information needs, e.g. central processing units and peripherals, accessories and furniture, counseling and services for data handling and processing.

3415. BUSINESS EQUIPMENT GUIDE. *B.E.D. Business Books Ltd. Semi-annual. 1980. 404 pg. 9.50 (United Kingdom).*

Published semi-annually, directory covers manufacturers and concessionaires of office equipment and machinery, reprographic equipment, and computers and peripheral equipment.

3416. BUSINESS EQUIPMENT TRADE ASSOCIATION LIST OF MEMBERS. *Business Equipment Trade Association. Annual. 1980. 40 pg. 3.00 (United Kingdom).*

Listings of member firms detail address, telephone, and contact person. Contains trade names index.

Special Issues and Journal Articles

Business Equipment Digest

3417. BUSINESS EFFICIENCY EXHIBITION 1982. *B.E.D. Business Books Ltd. Irregular. v.22, no.6. June 1982. pg.67-79*

Gives descriptions of manufacturers' new products and regular product lines of office automation equipment to be exhibited at the Business Efficiency Exhibition, 1982.

3418. BUSINESS EQUIPMENT DIGEST: NEW IN THE OFFICE. *B.E.D. Business Books Ltd. Monthly. February 1982. 12 pg.*

Monthly listing of business office equipment including telephones, modems, computers, peripherals, calculators, and machines. Includes manufacturer's name, retail price, description and photograph. Further information available through reader card service.

3419. BUSINESS PRINT SHOW. *B.E.D. Business Books Ltd. Annual. v.22, no.10. October 1982. pg.101-105*

Illustrates and describes some in-house printing equipment displayed at Repro Workshop "82, the London Print Show. Lists exhibitors.

3420. INFO 82 - THE ELECTRONIC OFFICE SHOW: GUIDE TO EXHIBITORS. *B.E.D. Business Books Ltd. Irregular. February 1982. pg.22-48*

Reviews products and exhibitors at the Fifth European Information Technology and Management Exhibition and Conference (Feb. 9-12, 1982) in London. Equipment includes word processors, data processors, communications equipment, copiers, etc. Information includes company name, booth number, and products presented. Further information available through reader card service.

3421. NEW IN THE OFFICE. *B.E.D. Business Books Ltd. Monthly. v.21, no.11. November 1981.*

Monthly product review featuring office products, including copiers, computer terminals, word processors, typewriters and calculators. Each entry includes product description and manufacturer. Some items also include technical specifications, photographs, and retail prices.

3422. PREVIEW LONDON BUSINESS SHOW. *B.E.D. Business Books Ltd. Annual. v.22, no.9. September 1982. pg.25-54*

Features products exhibited at the 1982 London Business Show. Provides manufacturer's name and product description.

Marketing in Europe

3423. ELECTRONIC CALCULATORS IN BELGIUM: SPECIAL REPORT NO. 1. *Economist Intelligence Unit Ltd. June 1981. pg.38-51*

Reviews the market for hand-held and desk-top calculators in Belgium. Discusses consumption patterns, suppliers, product ranges, 9 major businesses involved in the market, retail distribution, prices and margins. Tables detail imports and exports (1974-1979) and imports and exports (1979) by countries of origin and destination; apparent consumption (1974-1979); major brands and distribution; sales (1981) by type of outlet; retail prices (1981) of selected brands; and advertising expenditures (1976-1980).

3424. ELECTRONIC CALCULATORS IN ITALY: SPECIAL REPORT NO. 3. *Economist Intelligence Unit Ltd. June 1980. pg.67-73*

Covers the Italian market for hand-held and desk-top calculators. Tables provide statistics on production, imports / exports, consumption and advertising expenditures (1976-1979). Details retail distribution (1979) and retail prices (May 1980) by manufacturer and brand. Summarizes major manufacturers and brands.

3425. ELECTRONIC CALCULATORS IN THE NETHERLANDS. *Economist Intelligence Unit Ltd. no.226. September 1981. pg.36-44*

Examines the market for pocket and desk-top electronic calculators in the Netherlands. Discusses production, foreign trade, consumption and consumption patterns, distribution, major manufacturers and brands, prices and advertising and promotion. Numerous tables detail imports / exports (1977-1980) by countries of origin and destination; apparent consumption (1977-1980); retail prices in 1981; and advertising expenditures (1980, 1981).

3426. TYPEWRITERS IN ITALY. *Economist Intelligence Unit Ltd. no.225. August 1981. pg.66-71*

Reviews production, foreign trade, consumption patterns and distribution of typewriters in Italy. Tables show production, consumption, imports and exports (1976-1980) of typewriters. Discusses major manufacturers brand names and market shares (1980) of 6 brands. Details retail prices (1981) of selected brands.

Retail Business

3427. LASER SCANNING AT THE CHECKOUT: SPECIAL REPORT NO.1. *Economist Intelligence Unit Ltd. October 1981. pg.16-22*

Examines the use of laser scanning devices in the United Kingdom as a means of increasing efficiency in supermarket checkouts. Discusses the numbering system; equipment manufacturers; existing operations in the U.K. as of July, 1981; kinds of equipment; gains and savings at checkout; and consumer attitudes.

L'Usine Nouvelle

3428. LES MARCHES DE LA BUREAUCRATIQUE: (II) TELE-COPIE - LE QUITTE OU DOUBLE FRANCAIS. [Office Markets: (II) Telecopiers - France's Double or Nothing.]. *Hennion, Blaudine. Usine Nouvelle. French. v.45/1982. November 4, 1982. pg.100-101*

Briefly examines the world market for telecopiers, with selected forecasts through 1983-1986 for major market areas. Discusses products, output levels, and market shares of individual manufacturers and producing nations, with the focus on French and Japanese producers. Some technical specifications are included.

See 2042, 2048, 2056, 2057, 3045.

Asia

Market Research Reports

3429. OPERATIONS MARKET RESEARCH: BUSINESS EQUIPMENT AND SYSTEMS IN SINGAPORE. *U.S. International Trade Commission. 1980. 53 pg. $10.00.*

Analysis of markets for business equipment and computers in Singapore, 1980. Covers demand, manufacturers supplying equipment, employment, trade associations, regulations, and imports.

See 2080, 2081, 2085.

Special Issues and Journal Articles

Business Japan

3430. AUTOMATION REVOLUTIONIZES OFFICE PROCE-DURES. *Furusawa, Akira. Nihon Kogyo Shimbun Publications. v.26, no.12. December 1981. pg.101-106*

Examines the growth of Japan's business machines industry and automation of office tasks. Discusses office computers, copiers, facsimile, word processors and personal computers. Tables show shipments (1976-1980) of office computers, with forecasts for 1981-1990; and facsimile production (1973-1980).

3431. ECR RECORDS HIGH GROWTH WITH AUTOMATION PROGESS. *Ikeda, Mitsuo. Nihon Kogyo Shimbun Publications. October 1981. pg.95, 99-104*

Reviews the growth of the electronic cash register sector and examines production and trade trends. Statistics, in tables and graphs, include production and exports (1975-1981); system ECR production (1978-1980); diffusion rate of source marketing devices in Europe (1980, 1981); production and export value (1975-1981); exports to the U.S. (1976-1980); exports to W. Germany (1976-1980); and number of stores using hand scanners in the U.S. and Canada (1975-1981).

See 2090.

Oceania

Market Research Reports

3432. OFFICE EQUIPMENT IN AUSTRALIA - PHOTOCO-PIERS. *BIS Marketing Research. Biennial. 1979. 4000.00 (Australia).*

Analysis of the Australian market for photocopiers examines end users, user attitudes toward brand names, present and future demand, and suppliers' reputations.

3433. THE OFFICE EQUIPMENT MARKET IN AUSTRALIA - TYPEWRITERS. *BIS Marketing Research. Biennial. 1979. 4000.00 (Australia).*

Analysis of the Australian market for typewriters examines end users, attitudes toward brand names, present and future demand, and suppliers' reputations.

3434. A SURVEY OF THE AUSTRALIAN BUSINESS EQUIP-MENT AND SYSTEMS MARKET. *P.A. Consulting Services Pty Ltd. U.S. Department of Commerce. International Trade Administration. NTIS #DIB 79-07-507. April 1979. 114 pg. $15.00.*

Provides data on the total market by major subcategories (1975, 1977, 1978, 1982) and on imports from the U.S. by major subcategories (1975, 1977, 1978, 1982). Specific product markets are analyzed with a competitive assessment given of U.S., third country, and domestic suppliers. Data is provided on estimated expenditures for equipment and systems by major user sectors, 1977 and projected 1982, and each sector is profiled in the text. The business equipment and systems configuration of one significant company in each sector is described. Market access data details marketing practices, trade restrictions, and tecnical requirements. Trade lists include: principal trade and professional associations; publications; major fairs, exhibitions and conferences.

3435. THE WORD PROCESSING MARKET IN AUSTRALIA. *BIS Marketing Research. Annual. 1981. 5500.00 (Australia).*

Analysis of Australian markets for word processing equipment. Examines current installations and incidence of combination with data processing, future growth rates and market determinants. Second volume contains market survey of user attitudes toward manufacturers and suppliers.

Directories and Yearbooks

3436. RYDGE'S OFFICE EQUIPMENT BUYERS' GUIDE. *Rydge Publications Pty. Ltd. Annual. 1981. 174 pg. 12.50 (Australia).*

Manufacturers and suppliers of office equipment, ancillary products, and services listed under product groups.

Scientific and Technical Instruments

International

Market Research Reports

3437. AIR POLLUTION EQUIPMENT. *Predicasts, Inc. May 1981. $375.00.*

Worldwide market analysis for air purification equipment. Includes manufacturers' activities, regulations and technologies.

3438. ANALYTIC INSTRUMENTS. *Predicasts, Inc. October 1980. $325.00.*

Worldwide analysis of markets for analytic instruments such as spectrographs, spectrophotometers, electrochemical instruments, and electron microscopes. Covers technology of products, manufacturers, and end uses.

3439. AUTOMATIC TEST EQUIPMENT (ATE) WORLD MARKET FORECAST. *Gnostic Concepts, Inc. 1981. 600 pg. $12000.00.*

The world market for semiconductor and board automatic test equipment is forecast from 1982 to 1987. Market shares are projected throughout the forecast period by geographic area, and by product group. Data on market size for selected product groups is also included. The dynamics of an expanding market are reviewed.

3440. BIOMEDICAL ENGINEERING. *Predicasts, Inc. January 1981. $325.00.*

Worldwide market analysis for biomedical equipment. Covers products, manufacturers, and research and development.

3441. FLOWMETERS. *Larsen Sweeney Associates Ltd. Annual. 1981. 795.00 (United Kingdom).*

Market analysis for flowmeters, with forecasts to 1986. Covers

20 product groups. Includes analysis of marketing and industry by country and surveys of end users and suppliers.

3442. LASER REPORT. *Advanced Technology Publications. Monthly. $36.00/yr.*

Gives the market outlook for lasers and opto-electronics.

3443. NAVIGATION EQUIPMENT. *Predicasts, Inc. August 1981. $375.00.*

Analysis of manufacturers of navigational equipment worldwide. Includes technology of new products and market analysis.

3444. POLLUTION TEST AND MEASUREMENT EQUIPMENT. *Predicasts, Inc. June 1981. $375.00.*

Worldwide market analysis for air, water and noise pollution testing and measurement equipment. Examines new products and utilizations and activities of manufacturers.

3445. PRODUCTION INSTRUMENTATION. *Predicasts, Inc. January 1982. $375.00.*

Analysis of markets for production instrumentation worldwide. Includes historical data and forecasts, manufacturers' activities, technology trends and end uses.

3446. THE SAUDI ARABIAN BUSINESS EQUIPMENT AND INFORMATION SYSTEMS MARKET. *U.S. International Trade Commission. 1980. 44 pg. $10.00.*

Market analysis for office equipment and information systems in Saudi Arabia. Includes users attitudes.

3447. SOIL TESTING EQUIPMENT. *Larsen Sweeney Associates Ltd. Annual. 1981. 695.00 (United Kingdom).*

Analysis of farm markets for soil testing equipment, with forecasts to 1986. Includes marketing and industry analysis by country and surveys of suppliers and farm users' attitudes.

3448. TIMEKEEPING IN THE 1980'S: AN INDUSTRY IN TURMOIL. *H.K. Lake & Associates. FIND / SVP. February 1980. 244 pg. $1500.00.*

Analyzes the technical and consumer aspects of the timepiece industry, both watches and clocks, through the 1980's. Projections are provided for the years 1978 through 1985, plus 1990 for the United States, Europe, Far East, Communist Bloc and other areas. Display types, types of movements and prices are considered, as are component and assembly technologies. New features, functions and consumer preferences are detailed along with an analysis of buyer attitudes. Marketing concepts and distribution patterns anticipated during the 1980's are reviewed.

3449. WATCHES. *Predicasts, Inc. October 1979. $325.00.*

Worldwide analysis of markets and manufacturers of electric, electronic, and mechanical watches and clocks. Reviews technology and products, by country and region.

3450. WORLD AND U.S. MARKETS FOR CONVENTIONAL AND AUTOMATIC TEST AND MEASUREMENT EQUIPMENT. *Frost and Sullivan, Inc. January 1981. 228 pg. $1000.00.*

Market analysis through 1985 by segment: automatic test equipment (ATE) and conventional test and measurement (non-ATE) and subgroups within these segments. Coverage spans market trends and end user surveys.

3451. WORLD WATER POLLUTION CONTROL EQUIPMENT. *Predicasts, Inc. October 1980. $775.00.*

Expenditures for water pollution control equipment by 55 nations (and 8 world regions) are surveyed during the period

1970-1979, and forecast for 1985, 1990 and 1995. Topics discussed include water quality standards, end-use consumption of water, control equipment, market end-use segmentation, comparative process efficacy and cost-effectiveness, production index and yearly shipments,and industry structure and market shares.

See 11, 759, 769, 783.

Industry Statistical Reports

3452. BULLETIN OF STATISTICS ON WORLD TRADE IN ENGINEERING PRODUCTS, 1978. *U.N. Economic Commission for Europe. United Nations Publications. Sales Section. Annual. July 1980. 387 pg. $22.00.*

Compilation of 1978 trade tables, covering most of the major nations of the world, for a galaxy of high-technology products: industrial machinery, machine tools, transportation equipment, electrical generating and transmission equipment, electronic instruments and components, digital electronics, and some consumer and commercial machinery. Presents value of trade in U.S. dollars, by exporters, by importer, by detailed product group (i.e., by 5-digit SITC code within section 7, and for groups 69, 861 and 864). Includes totals for the world, and for nine world regions and subregions.

See 2142.

Directories and Yearbooks

3453. CLOCKS AND WATCHES. *World Wide Trade Services. Irregular. 1979. $3.00.*

Directory lists 70 export suppliers of numerous kinds of clocks and watches.

3454. ISA DIRECTORY OF INSTRUMENTATION: INTERNATIONAL VOLUME: PRODUCTS, MANUFACTURERS, TRADE NAMES. *Instrument Society of America. Annual. 1979. 234 pg. $30.00.*

International directory contains the following sections: products used in the instrumentation and process control industry, arranged alphabetically by product categories; manufacturers listed in the product section with addresses, telephone numbers and TLX or TWX for the U.S. and Canada; and trade names provided by manufacturers.

3455. ISA DIRECTORY OF INSTRUMENTATION, 1981-1982. *Instrument Society of America. Irregular. 1981. 1584 pg. $74.50.*

Directory gives over 39,000 listings of product categories by vendor; names, addresses, sales offices, international offices, and manufacturers' representatives of manufacturers. Includes trade names and product specifications grouped by product category.

3456. THE OPTICAL INDUSTRY AND SYSTEMS PURCHASING DIRECTORY: VOLUME 1. *Laurin, Teddi C., editor. Optical Publishing Company, Inc. Annual. 26th ed. 1980. 1000 (est.) pg. $41.00.*

1980 international handbook and directory for the optics industry. Contains indexes of advertisers, manufacturers, vendors and services, with addresses. Also lists products and brand names. Covers much technical information and specifications.

3457. THE OPTICAL INDUSTRY AND SYSTEMS PURCHASING DIRECTORY: VOLUME 2. *Laurin, Teddi C., editor. Optical Publishing Company, Inc. Annual. 26th ed. 1980. 1000 (est.) pg. $41.00.*

1980 international handbook and directory for the optics industry. Contains indexes of advertisers, manufacturers, vendors and services, with addresses. Also lists products and brand names. Covers much technical information and specifications.

3458. QUALITY CONTROL BUYERS GUIDE. *W.R.C. Smith Publishing Company. Irregular. October 1981. $3.00.*

A listing of the top 128 quality control equipment makers, their products, and the tests which each device will perform.

See 25, 26, 2157.

Special Issues and Journal Articles

Air Transport World

3459. SPECIAL SURVEY REPORT: AIRLINES SPENT $265 MILLION ON AVIONICS IN 1980. *Henderson, Danna K. Penton / IPC. Subsidiary of Pittway Corporation. April 1981. pg.39-40*

Reviews world airlines' expenditures (1980) on avionics equipment and forecasts 1981 expenditures. Gives major types of equipment and a list of desired improvements.

American Laboratory

3460. 1980 LABORATORY BUYERS' GUIDE EDITION. *International Scientific Communications Inc. Annual. 1979. 280 pg.*

An international listing of manufacturers and suppliers of scientific laboratory equipment.

Electronic Business

3461. RECESSION TAKES ITS TOLL ON AUTOMATIC TEST EQUIPMENT (ATE) MANUFACTURERS. *Thames, Cindy. Cahners Publishing Company. v.8, no.11. October 1982. pg.60-66, 74*

Examines the world market for automatic test equipment (ATE), 1981-1982. Provides data on ATE's share of worldwide test instrumentation market, 1981; revenues of 10 leading manufacturers of automatic test equipment, 1981; financial data on the 10 ATE leaders, 1981-1982; ATE market segments for semiconductor testing, circuit board testing, and interconnect verification analysis, 1980-1981; semiconductor technology's impact on testing, 1970-1990; and applications of ATE, 1981. Sources include Dataquest Inc. and the Japan Electric Measurement Instruments Manufacturers' Association.

High-Speed Surface Craft

3462. MARINE EQUIPMENT. *Kalerghi Publications. Monthly. June 1981.*

Monthly feature lists marine equipment, including hardware, communications, and electrical products. Data includes photograph, brief specifications, and manufacturers.

Industrial World

3463. INDUSTRIAL WORLD'S NEW PRODUCT REVIEW. *Johnston International Publishing Corporation. Monthly. v.206, no.1. January 1981. pg.30-32*

Monthly listing of new products available from international manufacturers supplying industrial manufacturing industries. Provides a description and photograph for each.

Interavia

3464. INTERAVIA: NEW PRODUCTS AND TECHNOLOGY. *Interavia SA. Monthly. October 1981. pg.1037*

A monthly listing of aviation equipment, instrumentation, products, etc. Information includes address of manufacturer, technical specifications, and photograph.

Norwegian Shipping News

3465. NORWEGIAN SHIPPING NEWS: NEW PRODUCTS. *K/S Selvig Publishing A/S. Monthly. December 1981.*

Monthly feature gives information on new maritime products. Provides a photograph of the product and address, telephone, and telex number of manufacturer.

Textile Industries

3466. A HARD LOOK AT QC TESTING. *Seidel, Leon E. W.R.C. Smith Publishing Company. Irregular. October 1981. 7 pg.*

Reprint of a report on new technology for quality control testing of textiles. Lists addresses of 128 manufacturers.

See 833, 2175, 2178.

Monographs and Handbooks

3467. CURRENT TRENDS IN OPTICS. *Arecchi, F.T. John Wiley and Sons Canada Ltd. 1981. 200 pg. $34.95.*

A collection of papers which reviews a variety of subjects in the field of optics. Topics covered include: prospects of solar energy; overview of astronomical optics in China; interferometric methods in optical astronomy; options for next generation telescopes; unconventional imaging; interactive image processing for image restoration and enhancement; image reconstruction for stellar interferometry; optical figuring by elastic relaxation methods; phase conjugation; real time holography and degenerate four wave mixing in photoreactive BSO crystals; telecommunications through optical fibres; electrooptic photosensitive media for image recording and processing; interferometric methods in astronomy.

3468. FIBER OPTICS TECHNOLOGY TRANSFER 1981. *Information Gatekeepers, Inc. Annual. lst ed. January 1981.*

Abstracts of U.S. and international patents with brief assessments of potential markets and applications. Covers patented fiber optics technologies.

3469. INSTRUMENTOS Y CONTROLES INTERNACIONALES. [International Instrumentation and Controls]. *Keller Publishing Corporation. Spanish. Bi-monthly. November 1980. 34 (est.) pg.*

Bimonthly, Spanish-language survey of new measuring and control equipment. Advertisements, placed by U.S., Western and Eastern European manufacturers contain photographs,

general product descriptions, and in some cases specifications and manufacturers' or distributors' addresses. Includes a product information checkoff card.

3470. INTERNATIONAL INSTRUMENTATION AND CONTROLS. *Keller Publishing Corporation. Bi-monthly. November 1980. 40 (est.) pg.*

Bimonthly survey of new measuring and control instruments. Advertisements, containing photographs and general product descriptions, are placed by U.S., British, other Western European, and some Eastern European manufacturers. Includes a checkoff product information card.

3471. MONITORING / SAMPLING MANUAL. *McIlvaine Company. Annual. 1981. $275.00/yr.*

Manual for smokestack monitoring and sampling devices worldwide. Industry analysis includes utilization of all product groups, technical aspects of plants and equipment, finances and operations of manufacturers, patents and legislation. Also includes buyers guide listing suppliers by country.

3472. ROLE AND PLACE OF ENGINEERING INDUSTRIES IN NATIONAL AND WORLD ECONOMIES: UPDATING TO 1970-1975 OF THE ANALYTICAL PART. *Unipub. 1980. 92 pg. $9.00.*

Contents include: general developments; main trends in world output of engineering industries; the metal and engineering products industries' role in national economies; structure of the engineering industries; general aspects of world trade in engineering products; the pattern of international trade in engineering products and geographical distribution; and national developments and country statements.

Conference Reports

3473. METERING APPARATUS AND TARIFFS FOR ELECTRICITY SUPPLY. *Institute of Electrical and Electronics Engineers, Inc. October 26, 1982.*

Conference proceedings cover system economics and tariff structure; metering equipment design; metering data collection and data transmission; legal requirements and standards; maintenance and testing; and consumers' viewpoint.

Newsletters

3474. MONITORING / SAMPLING NEWSLETTER. *McIlvaine Company. Monthly. l40.00/yr.*

Monthly newsletter for the smokestack sampling and monitoring industry worldwide. Industry analysis includes interviews with executives, new orders and contracts, installations, new products and patents, market analysis and manufacturers' activities.

Indexes and Abstracts

3475. ION SELECTIVE ELECTRODES. *Sira Institute Ltd. 1981. 20.00 (United Kingdom).*

Bibliography of over 1000 selected references to aspects of ion-selective electrodes derived from journals worldwide, published from 1977-1981.

3476. KEY ABSTRACTS: ELECTRICAL MEASUREMENT AND INSTRUMENTATION. *Institute of Electrical and Electronics Engineers, Inc. 1981. $72.00/yr.*

Arranged by topic, abstracts include author, author's address, publication, nation where published, volume and number, pages, publication date, synopsis, and number of references.

3477. KEY ABSTRACTS: PHYSICAL MEASUREMENT AND INSTRUMENTATION. *Institute of Electrical and Electronics Engineers, Inc. 1981. $72.00/yr.*

Arranged by topic, abstracts include author, author's address, publication, nation where published, volume and number, pages, publication date, synopsis, and number of references.

3478. MONITORING / SAMPLING FACT FINDER. *McIlvaine Company. Monthly. 1981. $130.00/yr.*

Abstract service for the smoke stack monitoring and sampling equipment industry worldwide. Includes journal articles, government reports and interviews with industry executives. Industry analysis includes news of products, plants, equipment and manufacturers.

3479. PREDI-BRIEFS: MEASURING AND CONTROL DEVICES. *Predicasts, Inc. Monthly. 1981.*

One in a series of 29 industry-intensive abstract periodicals derived from Predicasts' PROMT. Abstracts 200 to 400 articles per month covering such topics as acquisitions, capacities, end uses, government regulation, market shares and statistics, and new technology.

3480. PREDI-BRIEFS: OPTICAL AND REPROGRAPHIC EQUIPMENT. *Predicasts, Inc. Monthly. 1981.*

One in a series of 29 industry-intensive abstract periodicals derived from Predicasts' PROMT. Abstracts 200 to 400 articles per month covering such topics as acquisitions, capacities, end uses, government regulation, market shares and statistics, and new technology.

3481. PREDI-BRIEFS: POLLUTION CONTROL. *Predicasts, Inc. Monthly. 1981.*

One in a series of 29 industry-intensive abstract periodicals derived from Predicasts' PROMT. Abstracts 200 to 400 articles per month covering such topics as acquisitions, capacities, end uses, government regulation, market shares and statistics, and new technology.

See 869, 2201.

North America

Market Research Reports

3482. AIR POLLUTION CONTROL IN CANADA. *McIlvaine Company. September 1980. 36 (est.) pg. $95.00.*

Analysis of markets for air pollution control equipment in Canada. Examines economic indicators and regulation; expenditures for new plants and equipment by end use industries; production growth in mines and minerals; shipments and equipment costs; costs of complete installations; manufacturers' sales; U.S. subsidiaries; and forecasts of purchases, by products, through 1985.

3483. THE AIR POLLUTION CONTROL MARKETPLACE. *Robert E. De La Rue Associates. December 1980. 105 pg.*

Market analysis and forecasts for air pollution control equipment. Discusses market penetration opportunities.

3484. THE AMERICAN MARKET FOR SCIENTIFIC AND INDUSTRIAL INSTRUMENTS. *Sira Institute Ltd. December 1980. 62 pg. 25.00 (United Kingdom).*

Analysis of the U.S. market for scientific and industrial instruments, aimed at manufacturers and suppliers from the U.K. Includes growth rate forecasts to 1984 and sources of information on exports, including trade associations.

3485. ANAEROBIC METHODS. *Luning Prak Associates, Inc. Irregular. 1981. 20 pg. $1500.00.*

Market survey of 1000 microbiologists. Gives utilization of anaerobic methods and products.

3486. ANALYTICAL INSTRUMENT MARKET ANALYSIS SERVICE. *Creative Strategies International. Quarterly. 1981. $495.00.*

Quarterly market analysis service for analytical instruments. Includes the Pittsburgh Conference Review, Annual Market Wrap Up and other reports. Topics include end use market segmentation, growth rates, economic indicators, market shares, new products, regulations, sales, foreign trade, and government expenditures. Forecasts to 1984.

3487. ANALYTICAL INSTRUMENTATION: GROWTH MARKETS. *Advanced Technologies, Inc. Business Communications Company. February 1980. 100 (est.) pg. $750.00.*

Markets for analytical instruments are evaluated through 1990, based on trends disclosed in a 1979 survey of about 2000 U.S. laboratories. Discusses product mix, applications and market penetration of various devices, and leading manufacturers and their market shares. Product groups covered include chromatographs, magnetic and mass spectrometers, microscopes, balances, and centrifuges. Markets discussed are U.S., Japan, Germany, U.K., and Italy.

3488. ANALYTICAL INSTRUMENTATION MARKET. *Frost and Sullivan, Inc. May 1981. 238 pg. $1050.00.*

Contains analysis establishing AI market size in terms of shipments and consumption through 1980. Assesses industry structure. Provides competitive analysis of top manufacturers: market shares, profiles, new and current products, and pricing. Gives forecast to 1989 by 9 product classifications. Provides user questionnaire survey results.

3489. AUTOMATED CLINICAL CHEMISTRY ANALYZERS AND CLINICAL CHEMISTRY REAGENTS. *Robert S. First, Inc. April 1980. $11000.00.*

Analysis of markets for automated clinical chemistry instruments and reagents.

3490. AUTOMATIC TEST EQUIPMENT USERS REPORT AND FORECAST. *Gnostic Concepts, Inc. 1981. $12000.00.*

Market analysis and forecasts, 1979-1987 for automatic equipment for testing integrated circuits and circuit boards. Covers uses, technology of new products, costs, manufacturers, and demand factors.

3491. BIOMEDICAL INSTRUMENTATION. *Theta Technology Corporation. October 1979. 175 pg. $600.00.*

Market analysis for biomedical instrumentation includes 1979 figures for sales, market shares, products, new products, company profits. Also gives market size (1979-1983) and forecasts (1979-1983). Instrumentation includes acute care facilities, cardiac monitors, defibrillators, blood pressure monitors, and pacemakers.

3492. BUILDING CONTROL SYSTEMS (BCS) MARKET - U.S. *Frost and Sullivan, Inc. September 1982. 210 pg. $1275.00.*

Analysis of markets for building control systems (BCS). Includes market segmentation by building types, market areas and existing versus new construction; examples of energy conservation; vendors and market shares; installations and shipments, 1974-1982; and sales forecasts, 1980-1983, 1987, 1991.

3493. CIT / NDT: AN INVESTIGATION INTO THE USES AND TECHNOLOGY OF COMPUTED INDUSTRIAL TOMOGRAPHY. *White, William. White (R&D) Consultants. Technology Marketing Group Ltd. March 1981. 200 (est.) pg. $4750.00.*

Analysis of computed industrial tomography for non - destructive testing: the industry and the market. Covers technical aspects of industrial plants and equipment, manufacturers and end uses, user attitudes, products, patents, and regulations.

3494. CLINICAL AUTOMATED MICROBIOLOGY. *Theta Technology Corporation. no.108. December 1981. $295.00.*

Market analysis and forecast covers types of tests and methods; lab needs and concerns about automating, making automated equipment, and products; and dollar volume in equipment and supplies sales.

3495. CLINICAL CHEMISTRY ANALYZERS. *Theta Technology Corporation. no.292. November 1982. $600.00.*

Market research report with forecasts through 1986 covers automated and semiautomated analyzers, batch analyzers, the true replacement market, and the group practice office labs market.

3496. CLINICAL CHEMISTRY INSTRUMENTS. *Theta Technology Corporation. No. 103. April 1981. $295.00.*

Analysis of markets for clinical chemistry instruments, industry structure, new products, and manufacturers. Includes mergers, market shares, and the competition.

3497. CLINICAL HPLC. *Theta Technology Corporation. no. 106. June 1981. $295.00.*

Market analysis and forecast covers competing technologies, HPLC influence on U.S. market, and projected influence of new systems.

3498. CLINICAL LABORATORY AUDIT - RADIOIM-MUNOASSAY. *Luning Prak Associates, Inc. Semi-annual. 1982. $4000.00.*

A semiannual clinical laboratory market survey service. Gives market shares and utilization of radioimmunoassay tests and products.

3499. CLINICAL LABORATORY AUDIT - URINE AND FECES. *Luning Prak Associates, Inc. Semi-annual. 1982. $6500.00.*

Clinical laboratory market survey. Gives urine and feces testing products utilization data and market shares.

3500. CLINICAL LABORATORY INSTRUMENTS MARKET IN THE U.S. *Frost and Sullivan, Inc. September 1981. 277 pg. $1200.00.*

Analysis of markets for clinical laboratory instruments, with forecasts to 1985. For ten product groups, gives historical data, manufacturers' product lines, sales, prices, and forecasts. Also includes units in operation, 1980 installations, and dollar values to 1985.

3501. CLINICAL MICROSCOPE MARKET. *Luning Prak Associates, Inc. Irregular. 1981. $4250.00.*

Survey of the clinical microscope market in medical laboratories. Includes brand shares, brand name considerations in consumer purchases, user attitudes and sales forecasts by end uses within the laboratory.

3502. COMBUSTION OPTIMIZATION ANALYTICAL INSTRUMENTATION MARKET IN THE U.S. *Frost and Sullivan, Inc. no.A1159. September 1982. 272 pg. $1275.00.*

Analysis of the market for combustion optimization analysis instruments. Includes growth rates and market forecast to 1986; market size; end uses; technical specifications; capacity of boiler installations; energy conservation, energy costs, and fuel consumption savings; competition among methods and new technology for analysis; manufacturer sales, product lines, prices, marketing strategies, market shares, and distribution; and industry concentration ratios.

3503. THE COMMERCIAL / INDUSTRIAL LASER MARKET IN THE U.S. *Frost and Sullivan, Inc. 1979. 298 pg. $900.00.*

Market analysis (1978) of lasers includes lasers for commercial and industrial application (graphics, medical, communications, construction, etc.), industry structure (1978), company profiles (1978), and products (1978). Company profiles include sales and income figures, 1969-1978.

3504. COMMUNICATION TEST EQUIPMENT MARKET. *Frost and Sullivan, Inc. February 1982. $1200.00.*

Contains analysis and forecasts 1979-1990 for data test sets; test equipment (portable, microwave radio, telephone equipment, PBX and key systems); fiber optics; and microwave. Includes supplier profiles and market shares.

3505. DIAGNOSTIC MARKET - HOSPITAL AND PRIVATE LABORATORIES. *IMS America Ltd. Quarterly. 1981. $19500.00.*

Survey of diagnostic equipment markets from responses of hospital and private clinical laboratories.

3506. DIGITAL ANALYSIS AND THE PROCESS CONTROL INDUSTRY. *Frost and Sullivan, Inc. March 1980. 92 pg. $700.00.*

Impact of digital technology on process controls. Industry outlook and competitive aspects. Components of process control. Role of the types of computers used. Marketing methods. Market size and characteristics. Company profiles on: AccuRay; Analog Devices; Fischer & Porter; General Signal; Honeywell; Measurex, and others.

3507. DYNAMIC SMALL COMPANIES IN THE BIOMEDICAL FIELD. *Robert S. First, Inc. January 1981. $15000.00.*

Survey of new businesses in the biomedical field.

3508. ELECTRIC UTILITY CUSTOMER - SIDE LOAD MANAGEMENT EQUIPMENT. *Frost and Sullivan, Inc. January 1981. $1050.00.*

Examines the use of load management in utility operations - reduction of capacity, capital, scarce fuel. Sales forecast for hardware: communication and load control systems (radio, powerline carrier, ripple, hybrid, telephone, local controllers, others) and energy storage (thermal, "cool," hybrid). Company profiles. Market shares.

3509. ELECTROLYTES. *Luning Prak Associates, Inc. October 1979. $6800.*

In this electrolyte market analysis, 706 clinical chemists and 604 chemistry supervisors and chief MT's describe current instrumentation, buying plans, brand exposure, brand leanings, feature evaluation, request configurations and demographics.

3510. ELECTROLYTES, AACC. *Luning Prak Associates, Inc. October 1979. $4000.*

In this electrolyte (AACC) market analysis, 706 clinical chemists indicated current instrumentation, buying plans, brand exposure, brand leanings, feature evaluation, request configuration and demographics.

3511. ELECTROLYTES, ASMT. *Luning Prak Associates, Inc. October 1979. $4000.*

In the electrolyte (ASMT) market analysis, 604 chemistry supervisors and chief MT's describe current instrumentation, buying plans, brand exposure, brand leanings, feature evaluation, request configurations and demographics.

3512. ELECTRONIC WATCH AND CLOCK FORECAST. *Gnostic Concepts, Inc. 1981. $9800.00.*

Analyzes market data, consumer demand trends, timepiece issues, competitive environment, international profile, component developments, and economic outlook of electronic watches and clocks. Competitive environment includes data on 30 companies in the field. Component developments center on use of modules, integrated circuits, displays, quartz crystals and batteries. Market data includes production, sales, prices and consumption.

3513. ELECTRO-OPTICAL INSTRUMENTATION MARKET FOR QUALITY AND PRODUCTION CONTROL IN THE U.S. *Frost and Sullivan, Inc. January 1982. 188 pg. $1100.00.*

Analyzes the quality and production control applications for electro-optical instrumentation in manufacturing machined parts, product assembly, continuous manufacturing systems, printed circuit and semiconductor manufacturing and other U.S. production systems. Defines and categorizes the principles of operation of the classes of electro-optical instrumentation. Establishes installed base of such equipment and identifies size of present day market. Discusses marketing and technological trends shaping future growth. Surveys potential end users and suppliers. Provides dollar forecast of U.S. demand for 11 electro-optical product groupings to 1986.

3514. EMERGING DIAGNOSTIC TECHNOLOGY. *Channing Weinberg. 1981. $10000.00.*

Opportunities for new businesses from new diagnostic technologies. Analysis of effects on existing markets from new products.

3515. ENERGY CONSERVATION EQUIPMENT FOR THE RESIDENTIAL / COMMERCIAL SECTORS. *Business Communications Company. May 1981. 208 pg. $850.00.*

Covers six major areas: insulation; heating; ventilating and air conditioning; solar energy; glazing and fenestration; energy management systems; and lighting. Market projections for each area include residential and commercial sectors.

3516. THE ENERGY EFFICIENT RESIDENTIAL HEATING, COOLING AND ASSOCIATED CONTROLS EQUIPMENT MARKET. *Frost and Sullivan, Inc. 1981. 352 pg. $1050.00.*

Evaluates U.S. markets through 1990 or 1995 for heat pumps, high efficiency heating and air-conditioning equipment, solar air and water heating systems, wood and coal stoves and furnaces, photovoltaics, and associated control instruments. Discusses new and retrofit markets on the basis of regional trends in housing stock age, new construction, residential energy usage; and costs of conventional fuels and alternative systems. Discusses major manufacturers and their R & D activities, industry structure, and market penetration / saturation.

3517. THE EXPANDING LASER INDUSTRY. *Business Communications Company. December 1980. 138 pg. $850.00.*

Technical / economic report outlines lasers by type and analyzes the industry and technological developments. Discusses the market by major application, such as the industry for measurement and controls, precision work, cutting, communications, optics, energy, military, medical, etc.

3518. FIBER OPTICS IN COMMERCIAL COMMUNICATIONS. *Gnostic Concepts, Inc. 1981. 290 pg. $10000.00.*

Analysis of markets for fiber optics components for U.S. commercial communications, with forecasts to 1990. Includes production and annual growth rates, 1981-1990; the technology; consumption by end uses; competition among manufacturers and market shares; price trends; and government regulation.

3519. FIBER OPTICS IN DATA PROCESSING. *Gnostic Concepts, Inc. 1981. 258 pg. $19750.00.*

Analysis of markets for fiber optics in data processing end uses. Includes technology reviews; forecasts to 1990; consumption, 1982-1990; price trends; and U.S. production.

3520. FIBER OPTICS IN GOVERNMENT COMMUNICATIONS. *Gnostic Concepts, Inc. 1981. 240 pg. $10000.00.*

Analysis of markets for fiber optics components in government communications including military markets, with forecasts to 1990. Includes technology, consumption by end uses, government expenditures, competition among manufacturers, production, and price trends.

3521. FIBER OPTICS IN GOVERNMENT DATA BUS. *Gnostic Concepts, Inc. 1981. 350 pg. $10000.00.*

Analysis of markets for fiber optics components for government data bus, with forecasts, 1981-1990. Includes consumption by end uses; military markets, government purchases by agency, manufacturers, production and prices.

3522. FIBER OPTICS: TRENDS, DYNAMICS. *Business Communications Company. G-044R. February 1982. $975.00.*

Analysis of new developments and new markets for fiber optics covers such applications as telephone, computer, and CATV industries. Report considers sales and current and future technology, and markets for optical fibers (lasers, LED's, photodiodes and transistors, and systems using such parts).

3523. FIBER-OPTIC CONNECTORS. *International Resource Development, Inc. February 1980. 103 pg. $895.00.*

This is an up-to-date and complete analysis of the U.S. fiber optic connectors market. Provides connector market overview, technological developments and market segmentation by industry. Discusses supplier industry structure, including product strategies of major suppliers, market shares, and company profiles.

3524. FIBEROPTIC SENSORS. *Kessler Marketing Intelligence. 1981.*

Analysis of markets for fiberoptic sensors. Forecasts market value, 1981, 1986, 1991. Examines military markets and industrial consumption.

3525. THE FLOWMETER INDUSTRY: A STRATEGIC ANALYSIS. *Venture Development Corporation. 1981. 258 pg. $2490.00.*

Analysis of markets for flowmeters, technology and manufacturers. Forecasts dollar sales, unit sales and prices 1981-1986. Includes 1981 shipments, market shares, end uses, industry structure and marketing strategies.

3526. GAS MONITORING INSTRUMENT MARKETS - 1982 UPDATE. *William T. Lorenz and Company. 1981. 90 pg. $395.00.*

This report provides forecasts as to the markets involved, the competitors in the field, and what testing / monitoring applications are foreseen to 1985.

3527. HAZARDOUS WASTE: TREATMENT CHEMICALS, TESTING LABS, ANALYTIC INSTRUMENTS AND ASSOCI-ATED CHEMICAL REAGENTS. *Frost and Sullivan, Inc. August 1980. $975.00.*

An in-depth look at this industry: assessing volume of hazardous waste generating compounds; its effect on industries; projections of analytical instruments and testing chemicals; testing labs; environmental and waste management firms; federal legislation and programs.

3528. HEMATOLOGY. *Luning Prak Associates, Inc. Semiannual. 1982. $2750.00.*

Market survey reporting utilization of hematology equipment.

3529. HEMATOLOGY INSTRUMENTS. *Theta Technology Corporation. No. 105. May 1981. $295.00.*

Analysis of markets for automated hematology instruments, the technology, and the industry structure. Profiles manufacturers and new products.

3530. INCOMING WORKLOAD IN THE CHEMISTRY LAB. *Luning Prak Associates, Inc. 1979. 700 pg. $12000.00.*

Market survey for lab supplies. Gives statistics on workloads from hospitals and clinical chemists.

3531. INDUSTRIAL ENERGY CONTROLS MARKET. *Frost and Sullivan, Inc. March 1982. $1200.00.*

Includes market analysis and 10-year forecasts for combustion energy controls (controllers, analyzers, other combustion instruments); fluid energy controls (flow transmitters, controllers and other fluid energy instruments); electrical energy controls (power demand controllers, other electrical controls); and energy management systems. Profiles leading suppliers.

3532. INDUSTRIAL MACHINE CONTROLS. *Frost and Sullivan, Inc. March 1979. $850.00.*

Forecast for measurement products, decision devices, actuators, data acquisition systems, packaged control systems and robots. Analysis by end user market. Survey results.

3533. INDUSTRIAL NON-DESTRUCTIVE TESTING INSTRU-MENTATION MARKET (U.S.). *Frost and Sullivan, Inc. February 1980. $1000.00.*

Market forecast for these approaches: ultrasonics and search units; eddy current; x-ray; gamma ray; acoustic emissions; magnetic particle and penetrants; immersion and other systems. Examines product lines, end uses, suppliers, market shares and market strategies and technical development for each. Company profiles.

3534. INDUSTRIAL ON-STREAM PROCESS ANALYZER MARKET. *Frost and Sullivan, Inc. March 1982. $1200.00.*

Market report contains analysis and forecast of on-stream analyzers, by major domestic process industry and exports; and by product classification. Determines features, functions and specification desires by end-users. Evaluates present suppliers and possible entrants. Examines technology trends and other market forces.

3535. INDUSTRIAL PROCESS TEMPERATURE MONITOR-ING AND CONTROL INSTRUMENTATION. *Frost and Sullivan, Inc. 1981. 260 pg. $1100.00.*

Evaluates markets for 23 categories of temperature / monitoring instruments through 1985, in terms of unit sales to ten end-user industries. Projects total dollar sales and shipments through 1990. Discusses major manufacturers and their product lines, industry structure and potential entry points, and sales strategies. Presents results of a survey of user needs and users' company rankings.

3536. INSTITUTIONAL ENERGY CONSERVATION HEAT-ING, COOLING, LIGHTING MATERIALS AND CONTROL MARKET (U.S.). *Frost and Sullivan, Inc. April 1979. $800.00.*

Report on the market for energy efficient boilers, heat pumps, solar systems, air conditioning systems, lighting systems, insulation, storm windows and doors, combustion control and monitoring equipment and energy management systems. Supplier outlook.

3537. THE INSTRUMENTATION AND CONTROL EQUIP-MENT MARKET IN THE POWER INDUSTRY. *Frost and Sullivan, Inc. May 1980. 287 pg. $975.00.*

Market report contains forecasts and analysis to 1995 for instrumentation and control products and services for new construction, retrofit, instrument racks, display instruments, field-mounted instruments, computers, secondary processing electronic hardware, major I and C systems, and analysis instruments. Includes end elements, bulk material and installation contract services. Includes power demand growth rates, future generation mixes, technological I and C trends, and a review of the industry exchanges between the utility, architect - engineer, system supplier, OEM, and I and C supplier.

3538. INSTRUMENTATION IN THE U.S. *Larsen Sweeney Associates Ltd. Annual. 1981. 395.00 (United Kingdom).*

Reports for 9 subsectors of the U.S. instrumentation industry. Each provides markets and manufacturers' installed base of plants and equipment, by product; market and industry analysis; market penetration opportunities; and a survey of end users.

3539. THE INSTRUMENTS AND CONTROL SYSTEMS MAR-KET FOR OFFSHORE OIL EXPLORATION AND PRODUC-TION (U.S.). *Frost and Sullivan, Inc. April 1980. 218 pg. $775.00.*

Analysis of the market for instruments and control systems for offshore oil drilling installations, with forecasts to 1990. Analyzes market structure in terms of four end uses in the development and exploration process and identifies key executives. Covers products, manufacturers, regulations, and new product trends.

3540. INTRAOCULAR LENS. *Theta Technology Corporation. no.120. July 1981. $295.00.*

Market analysis and forecasts cover U.S. and international markets for intraocular lenses. Data includes cost differences between injection molded lenses and lathe cut lenses, FDA and other government actions, corporate acquisitions, and cataract patient information.

3541. LABORATORY INSTRUMENTS: MEXICO: COUNTRY MARKET SURVEY. *U.S. Department of Commerce. Sales Branch. Irregular. CMS 81-606. February 1981. 7 pg. $.50.*

Analyzes the current and potential market for laboratory instruments in Mexico. Describes current and future research in government, educational, nonprofit, and industrial laboratories and provides data on laboratory expenditures and purchases of instruments for 1979 and 1983 (projected). Includes data on imports of selected laboratory instruments, by country of origin, for 1977-1979, and sales of selected instruments by U.S. and by all others, for 1979 and 1983 (projected).

3542. LASER EQUIPMENT. *Theta Technology Corporation. January 1979. 154 pg. $600.00.*

Analysis of the laser equipment market includes products (1978), manufacturers (1978), growth rates (1978-1983), sales (1978, 1983), and government regulations (1978). Also includes market structure (1978), impact of technology, marketing strategies, and product designs.

3543. LASER FOCUS 1981 BRAND RECOGNITION STUDY: REVISED. *Advanced Technology Publications. Annual. September 1981. 15 pg.*

Lists name-recognition percentages for manufacturers of laser, fiber optic communications, and associated equipment and components, based on 1250 responses to a Spring 1981 survey of Laser Focus magazine subscribers. Manufacturers are listed by detailed product group, in order of descending recognition percentages. Includes gas, solid-state, and dye lasers; positioning equipment; photomultiplier, pyroelectric and silicon detectors; optics and filters; capacitors and modulators, etc.

3544. LASERS IN INDUSTRY, ENERGY AND PHOTOCHEMISTRY. *International Resource Development, Inc. September 1979. 189 pg. $895.00.*

Future markets for lasers and laser systems in machining operations, including cutting, drilling, material removal, and in annealing and welding. Technology trends and supplier strategies. Laser-driven photochemical applications, including isotype separation and nuclear fusion discussed. Expected developments in excimer lasers, and discussions of CO_2 vs. solid-state lasers for industrial applications.

3545. THE LEVEL INSTRUMENTATION INDUSTRY: A STRATEGIC ANALYSIS. *Venture Development Corporation. 1982. 210 pg. $2750.00.*

Analysis and forecast for level instruments, 1982-1986. Divides product groups by technology. Includes industry structure, shipments and values, 1981-1986; manufacturers; market shares; 1981 imports and exports; prices; and marketing strategies.

3546. THE LOW COST TEST INSTRUMENTATION MARKET IN THE U.S. *Frost and Sullivan, Inc. October 1981. $1000.00.*

Contains a 5-year analysis and forecast of demand levels by end-users for ATE, analog and digital measuring instruments, oscilloscopes, engine test equipment, microwave test instruments, signal generators, fault injectors, and others. Defines spectrum of instruments, establishes price ranges, identifies both traditional and non-traditional users and their expectations. Discusses technology advances, new products, and distribution. Profiles major suppliers for each class of equipment.

3547. THE MACHINE TOOLS PROGRAMMABLE CONTROLS MARKET. *Frost and Sullivan, Inc. March 1981. 303 pg. $1000.00.*

Covers both machine tools and associated controls. Gives forecast to 1985 by type of machine tool - controlled by NC vs. PC. Provides market outlook for numerical controls and computer numerical controls; direct numerical controls; and programmable controllers. Includes market shares of leading suppliers, and questionnaire survey results.

3548. MANAGING THE NATION'S CHEMICAL WASTE STREAMS. *FIND / SVP. December 1980. 27 pg. $500.00.*

As a result of government regulations developed over the last ten years, the industry that provides the equipment and services to chemical waste generators for managing their effluent streams should expand its current annual sales rate. This study analyzes the two major sectors of this market: testing and treatment systems, and processing and disposal services. Each analysis reports on regulatory directives, describes the products involved, segments individual market subsectors, and presents finances of the leading publicly-owned companies.

3549. THE MARKET FOR DIGITAL AND ANALOG CONTROL VALVES (U.S.). *Frost and Sullivan, Inc. January 1981. $1000.00.*

Sales forecasts to 1986 for: pressure, flow, temperature and other regulators; pneumatic and electric actuated control valves in key end user and OEM markets. Competitive analysis, market share data and company profiles.

3550. THE MARKET FOR PRODUCTION AND TEST EQUIPMENT IN THE U.S. SEMICONDUCTOR INDUSTRY. *Frost and Sullivan, Inc. September 1979. $900.00.*

Forecast and analysis to 1989 for the equipment used to produce and test semiconductor devices such as discretes (transistors, diodes, rectifiers, light sensitive devices, light emitting devices); and intergrated circuits (monolithic & hybrid IC's). Over 60 types of production and test equipments covered, including aligners, bonders, cameras, clean room equipment, computer-aided design equipment, furnaces, molding systems, scribing machines, ATE, centrifuges, counters, analyzers, spectrometers, microscopes, plotters, etc. Report considers long-term technology and demand trends in end-equipment; and the leading suppliers.

3551. MARKET STUDIES OF THE UNITED STATES: A REPORT ON A STUDY OF THE MARKET IN THE MID-ATLANTIC STATES FOR CLINICAL LABORATORY AND DIAGNOSTIC PRODUCTS. *Wind Associates Inc. Canada. Department of External Affairs. 1982.*

Market analysis completed by Wind Assoc., Inc., 1981, in association with the Canadian Consulate General in Philadelphia contains market potential for various products and possible guidelines for long-term Canadian export development as well as marketing strategies.

3552. MEDICAL LAB EQUIPMENT. *Theta Technology Corporation. December 1979. 170 pg. $600.00.*

Analysis of markets for medical lab equipment, with forecasts to 1985 by product groups and markets.

3553. THE MEXICAN MARKET FOR INDUSTRIAL PROCESS CONTROLS: FOREIGN MARKET SURVEY REPORT. *Servicios de Mercadeo Industrial, Mexico City. U.S. Department of Commerce. National Technical Information Service. DIB-80-02-507. 1979. 88 pg. $10.00.*

The market research was undertaken to study the present and potential U.S. share of the market in Mexico for industrial process controls; to examine growth trends in Mexican end-user industries over the next few years; to identify specific product categories that offer the most promising export potential for U.S. companies; and to provide basic data which will assist U.S. suppliers in determining current and potential sales and marketing opportunities. The trade promotional and marketing techniques which are likely to succeed in Mexico were also reviewed.

3554. MICROBIOLOGICAL TESTING. *Theta Technology Corporation. December 1979. 170 pg. $600.00.*

Analysis of microbiological testing market includes growth rate (1979-1985), product groups (1979), market shares (1979), distribution (1979), patents (1979), government regulations (1979) and forecasts to 1985. System technologies include enterotubes, micro-differential identification systems, immunological testing, and automated testing systems.

3555. MICROBIOLOGY SYSTEM UPDATE. *Luning Prak Associates, Inc. Irregular. 1981. 40 pg. $8400.00.*

Market survey examining automation in microbiology laboratories. Includes utilization of instruments, user attitudes, units in operation, market shares, and brand name considerations in consumer purchases.

3556. NEW PRODUCT NEEDS IN THE CLINICAL MICROBIOLOGY MARKET. *Luning Prak Associates, Inc. October 1980. $9600.*

This analysis of new product needs in the clinical miciobiology market is based on interviews with over 1,000 microbiologists. Problems encountered in the clinical microbiology lab are rated according to seriousness and frequency. The study generates and screens needs, taking the product development team to the point of concept design (need satisfaction). The investigation covers over 450 "problems."

3557. NEW RESIDENTIAL, COMMERCIAL HVAC AND MONITOR SYSTEMS. *Business Communications Company. March 1981. 159 pg. $750.00.*

Report examines instruments to monitor and control energy as well as more efficient HVAC systems for residential and commercial applications. Outlines applications, equipment, technology, manufacturers, and projections.

3558. NON-ISOTOPIC IMMUNOASSAY. *Theta Technology Corporation. May 1979. 288 pg. $600.00.*

Market data for non-immunoassay testing includes company profiles (1978), growth rates (1978-1985), products (1978), suppliers (1978), market segmentation (1978), market shares (1978), distribution (1978), usage (1978), and forecasts to 1985. System technologies include enzyme immunoassay, immunofluorescence, immunodiffusion, immunoelectrophoresis, and immunoprecipitation. Forecasts classified by technology.

3559. NORTH AMERICAN FIBER OPTICS - 3. *Gnostic Concepts, Inc. 1981. 630 pg. $10000.00.*

Market analysis and forecasts to 1990 for fiber optics. Includes technology of products, consumption by end uses, consumption, market shares, production, and economic indicators including national accounts and government expenditures.

3560. OPTICAL INSTRUMENTATION FOR PRODUCTION AND QUALITY CONTROL MARKETS. *Frost and Sullivan, Inc. February 1982. $1100.00.*

Includes forecasts to 1985 for borescopes, fiber optics instruments, optical comparators, microscopes, optical laser systems, spectrophotometers, and other optical systems. Examines new technological advances. Gives supplier profiles and market shares. Contains broad-based user questionnaire survey.

3561. THE OPTICAL INSTRUMENTS AND LENS INDUSTRY: AN ECONOMIC, MARKETING AND FINANCIAL STUDY - INVESTIGATION. *Survey Force Ltd. 1979. 265 pg. 195.00 (United Kingdom).*

Analysis of the U.S. and world optical instruments and lenses industry and market, with forecasts to 1987. Covers domestic shipments, 1963-1979; sales, 1958-1976; consumption, 1967-1977; product line breakdown; number of facilities; end uses; economic indicators for demand; industry structure; capital investment, 1958-1976; inventories, costs, 1958-1972; labor, 1958-1978; foreign trade, world production, 1975-1980; mergers; manufacturers; addresses; and new products.

3562. PHOTORECEPTORS: CRITICAL TECHNOLOGY FOR A COMPLEX MARKET: VOLUME 1. *Diamond Research Corporation. November 1982. $1900.00.*

Analysis of photoreceptor markets for end uses in photocopiers and non-impact printers. Forecasts new technology trends and revenues. Includes performance, costs and technical specifications; market size and structure; foreign markets; manufacturers' sales and market shares; production costs and yields; raw materials; market penetration; distribution, retail sales, and profits; and the installed base of duplication devices.

3563. PROCESS CONTROL EQUIPMENT FOR THE PULP AND PAPER INDUSTRY. *Frost and Sullivan, Inc. September 1981. $1100.00.*

Market report presents sales forecasts to 1985 for: control room instruments, magnetic flow meters, bortex meters, boiler systems, analog and digital control room instruments, and valves. Includes CO_2 analyzers, PH analyzers, and hierarchical systems. Contains market analysis and supplier profiles. Examines industry by process.

3564. PROCESS CONTROL EQUIPMENT MARKET, 1977-1989. *Frost and Sullivan, Inc. January 1981. $1050.00.*

Provides product analysis of measuring instruments, control actuators, digital and analog controllers, control systems. Market analysis for process and MFGRG industries, electric and gas utilities, PCE exports. Supplier analysis. Sales projections by product, by market, by supplier.

3565. PROCESS CONTROL EQUIPMENT MARKETS: PREDICASTS INDUSTRY STUDY NO. E61. *Predicasts, Inc. December 1980. $775.00.*

Forecasts (1995) markets for process control equipment, control valves and actuators, and large and microprocessing computers used to measure, analyze and adjust process variables. Markets are analyzed by type of control and by end use. Industry structure and selected leading businesses are presented.

3566. PROFILED KEY MARKETS FOR U.K. PROCESS CONTROLS EXPORTERS: VOLUME 1: PXCA. *Tactical Marketing Ltd. 1981. 120 pg. 95.00 (United Kingdom).*

Market research report of markets for United Kingdom process control exports to United States, Taiwan, Malaysia, Egypt, Nigeria, and Sweden. Covers process instrumentation, equipment, valves, and computers. Also includes market size, forecasts to 1985, product definitions, economic effects, development plans, major industrial users, usage trends, domestic manufacturers, other competition, buyer attitudes, distributors, regulations and constraints, and contracts and opportunities.

3567. PULP AND PAPER PROCESS CONTROLS IN THE U.S. MARKET. *Frost and Sullivan, Inc. July 1981. 215 pg. $1050.00.*

Market analysis for process controls for the U.S. pulp and paper industry, with forecasts, 1980-1985. Examines utilization of controls in industrial plants and equipment; manufacturers and products; market shares; industry structural trends; government regulation; energy costs; new products; capital expenditures, including rankings of biggest spenders; and user attitudes.

3568. QUANTITATIVE NON-ISOTOPIC IMMUNOASSAY. *Theta Technology Corporation. no. 151. September 1981. $750.00.*

Market analyses and forecasts to 1985 cover market segments (hospital clinical labs, commercial labs, commercial clinical labs, and M.D. labs); diagnostic values of quantitative non-isotopic immunossay; EIA, FIA, and NIA; RIA testing saturation; and testing for patient care, diagnosis, and therapeutic monitoring.

3569. RADIOASSAY INSTRUMENTATION. *Theta Technology Corporation. no.012. March 1981. $495.00.*

Covers market saturation of gamma and beta counters and semi- and fully-automated systems.

3570. REFERENCE LABORATORY MARKET. *Luning Prak Associates, Inc. October 1980. $2500.*

Based on a survey of 706 clinical chemists, the reference laboratory market study discusses market shares by type, size and location of account. Image and service ratios are provided.

3571. RESIDENTIAL ENERGY CONSERVATION, HEATING, COOLING AND LIGHTING PRODUCTS. *Frost and Sullivan, Inc. June 1981. $1075.00.*

Gives market share information on boiler and furnace controls, heat pumps, and electrical resistance heating units. Includes pricing information (gas vs. oil vs. electric) and distribution channels. Includes company profiles. Contains installed vs. appliance microprocessor based control systems for lighting, heating, cooling, and security.

3572. SEMICONDUCTOR ATE, HANDLING AND ENVIRON-MENTAL STRESS EQUIPMENT. *Frost and Sullivan, Inc. January 1982. $1200.00.*

Gives analysis and forecast to 1990 for test systems (general purpose logic, memory, linear, discrete, process control); integrated circuits handlers (ambient and high and low temperature); discrete handlers; integrated circuit burn-in equipment; discrete burn-in equipment; and environmental chambers. Gives supplier profiles and rankings.

3573. SENSORS - OPPORTUNITIES IN THE 80'S. *Craft, Steve, editor. Mackintosh Consultants, Inc. 1981.*

Market analysis and forecast to 1986 for electronic sensors. Examines end use industry trends, demand, production costs, research and development, market shares and new products.

3574. STACK MONITOR MARKETS. *William T. Lorenz and Company. June 1980. 90 pg. $250.00.*

1980 analysis of the U.S. market for combustion control and compliance stack monitoring instrumentation, including forecasts of demand to 1985. Covers economic indicators in the energy field, effects of federal regulation, descriptions and assessments of technologies. Applications include energy, chemicals, stone-concrete-glass, food, petroleum, coal, metals, paper and electric utilities. Discusses competition and consulting and industrial markets. Includes a directory of manufacturers and producers.

3575. STAND ALONE LOGGERS. *Frost and Sullivan, Inc. January 1982. $1200.00.*

Provides analysis and forecast to 1990 for power suppliers, interface units, control units, and memory devices. Discusses technological trends. Provides supplier profiles and market shares.

3576. TEMPERATURE AND APPLIANCE CONTROLS. *Predicasts, Inc. August 1980.*

Discusses the U.S. temperature and appliance control market, including company activities, new products, and applications. Reviews energy controls, temperature and climate controls, boiler and combustion controls, appliance thermostats, timers, and other controls.

3577. TEMPERATURE SENSORS 2: A STRATEGIC ANALYSIS. *Venture Development Corporation. October 1981.*

Market study on temperature sensors includes an analysis of the current U.S. electronic temperature sensor market by technology, applications, type of users, and sensor configurations. Gives projections and trends through 1985, by technology and applications, along with strategic marketing recommendations. Describes industry participants.

3578. TEST AND MEASUREMENT INDUSTRY. *Frost and Sullivan, Inc. May 1980. 78 pg. $500.00.*

Two reports by a Wall Street firm analyze market trends, developing products and new techniques in test and measurement industy. Breaks down current sales by major product groups. Corporate analysis by product line, market share and profitability. Analysis of financial statements and 1981 projection of sales and capital expenditures. Companies: H-P, Tektronics, John Fluke, GenRad, Teradyne & Wavetek.

3579. TEST EQUIPMENT. *Creative Strategies International. March 1981. $1200.00.*

Analyzes U.S. markets for manufacturers of test equipment. Provides forecasts (1980-1985) for 5 product segments: commodity instruments, timers and counters, signal analyzers, signal sources and others (GPIB controllers, communications test equipment and microprocessor development systems). Examines industrial R&D, manufacturing production, testing, and field maintenance. Gives profiles of major manufacturers.

3580. ULTRASONIC EQUIPMENT. *Theta Technology Corporation. January 1979. 186 pg. $600.00.*

Market analysis (1978) of ultrasonic equipment includes profiles of leading manufacturers (1978), government regulations (1978), market shares (1978), user impact (1978), and medical liability (1978).

3581. U.S. FEDERAL GOVERNMENT - CONTRACT AWARDS FOR PHARMACEUTICALS, MEDICAL SUPPLIES AND LABORATORY DIAGNOSTICS. *IMS America Ltd. Monthly. 1982. $5100.00.*

Census of government contract purchases of pharmaceuticals, medical supplies and laboratory diagnostics.

3582. THE U.S. MARKET FOR PROCESS CONTROL INSTRUMENTATION IN THE FOOD AND BEVERAGE INDUSTRY. *Frost and Sullivan, Inc. October 1980. 330 pg. $1000.00.*

Gives 10-year forecast by product: measuring instruments, controllers and indicators, and control actuators, by type. Analyzes end user market. Evaluates specifications, features and functions of PCI. Examines marketing channels and strategies. Analyzes major suppliers of food and beverage process instrumentation. Discusses strategies for market entry. Provides end-user questionnaire survey.

3583. THE U.S. TEST AND MEASUREMENT EQUIPMENT MARKET. *Frost and Sullivan, Inc. January 1981. $1000.00.*

Analysis and forecast to 1989 for over 20 classes of equipment including generators, synthesizers, counters, meters, oscilloscopes, analyzers, recorders, and ATE. Identifies major trends, problems, new technology and equipment. Overviews end-users. Profiles major suppliers to the marketplace, including competitive position by market segment, product lines, new product introductions, and marketing strategies.

3584. WATCHES AND CLOCKS. *Specialists in Business Information, Inc. July 1979. 160 pg. $350.00.*

Investigation of the U.S. watch and clock industry includes an analysis of import penetration by product type. Examines effects of electronics and discusses digital watch market and clock sales.

3585. 1981 UPDATE: AIR POLLUTION CONTROL INDUSTRY OUTLOOK. *William T. Lorenz and Company. January 1981. 350 pg. $700.00.*

Industrial markets for air pollution control are identified and discussed. Major product areas are discussed and key competitors analyzed. Areas covered include carbon adsorption, various electrostatic precipitator technologies, thermal and catalytic incineration, and control of fine particulate and nitrogen oxide emissions. Detailed market information is provided for utility and industrial flue gas desulfurization equipment, and for stationary source and ambient air monitoring instruments. Competitors are identified and the market for instruments is projected through 1985. New monitoring techniques and the status of state monitoring programs are analyzed. Also discussed are legislative and regulatory trends, and market effects of increasing coal and synthetic fuel combustion.

3586. 1981 UPDATE - WATER POLLUTION CONTROL INDUSTRY OUTLOOK. *William T. Lorenz and Company. 1981. 400 (est.) pg. $675.00.*

Market and industry analysis and forecasts to 1985 for water pollution control equipment, chemicals, and instruments. Examines plants and equipment by end use industries, expenditures, government funding, specific products, manufacturers, new products, and legislation.

See 64, 67, 71, 72, 83, 131, 132, 133, 187, 909, 955, 1008, 2240, 2251, 2258, 2264, 2286, 4039.

Investment Banking Reports

3587. BEAR STEARNS AND CO.: INDUSTRY REPORT: THE INSTRUMENTATION INDUSTRY: CYCLICAL WEAKNESS LIKELY TO PERSIST INTO 1982. *Fleming, Gerald S. Bear Stearns and Company. December 14, 1981. 22 pg.*

Analysis of markets and performance of instrumentation corporations. Gives stock prices; earnings per share, 1980-1982; price earnings ratio, 1980-1982; growth rates, 1970-1985; production and orders, 1968-1981; economic indicators, 1981-1982; and financial statements for individual corporations, including capitalization and 1979-1982 income statements.

3588. BEAR STEARNS AND CO.: RESEARCH DEPARTMENT HIGHLIGHTS: THE INSTRUMENTATION INDUSTRY: CYCLICAL WEAKNESS LIKELY TO PERSIST INTO 1982. *Fleming, Gerald S. Bear Stearns and Company. December 11, 1981. pg.2-5*

Report on economic indicators for instrument corporations. Shows stock prices, 1979-1981; earnings per share, 1980-1982; and price earnings ratio, 1982. Also includes analysis of markets, business cycles and investment opportunities.

See 1250, 2324, 2325, 2326, 2341.

Industry Statistical Reports

3589. CURRENT INDUSTRIAL REPORTS: SELECTED INSTRUMENTS AND RELATED PRODUCTS, 1978. *U.S. Department of Commerce. Bureau of the Census. Customer Service Branch. Annual. MA-38B. March 1980. 19 pg. $.30.*

Report includes statistics on instruments such as regulating and control valves, engineering and scientific instruments, controls, fluid meters and counting devices, instruments to measure electricity, measuring and controlling devices, analytical and optical instruments. Data is given for quantity and value

of shipments of selected instruments and related products, including number of companies, by product, for current and preceding year. Data includes export shipments, imports for consumption, and apparent consumption, for the current year.

3590. EIS SHARE OF MARKET REPORT: ENGINEERING AND SCIENTIFIC INSTRUMENTS. *Economic Information Systems, Inc. Quarterly. 1982. $350.00.*

Analysis of scientific and engineering instruments includes marketplace concentration and ownership structure. Companies in industry are listed and then ranked by production. Other data includes market share (dollars and percent), with groupings of all plants a company operates in the industry, with the parent company's market share broken down on a per-plant basis.

3591. EIS SHARE OF MARKET REPORT: ENVIRONMENTAL CONTROLS. *Economic Information Systems, Inc. Quarterly. 1982. $135.00.*

Covers marketplace concentration and ownership structure. Companies in industry are listed and then ranked by production. Other data includes market share (dollars and percent), with groupings of all plants a company operates in the industry, with the parent company's market share broken down on a per-plant basis.

3592. EIS SHARE OF MARKET REPORT: FLUID METERS AND COUNTING DEVICES. *Economic Information Systems, Inc. Quarterly. 1982. $75.00.*

Report addresses marketplace concentration and ownership structure. Companies in industry are listed and then ranked by production. Other data includes market share (dollars and percent), with groupings of all plants a company operates in the industry, with the parent company's market share broken down on a per-plant basis.

3593. EIS SHARE OF MARKET REPORT: INSTRUMENTS TO MEASURE ELECTRICITY. *Economic Information Systems, Inc. Quarterly. 1982. $315.00.*

Analysis of markets for instruments to measure electricity includes marketplace concentration and ownership structure. Companies in industry are listed and then ranked by production. Other data includes market share (dollars and percent), with grouping of all plants a company operates in the industry, with the parent company's market share broken down on a per-plant basis.

3594. EIS SHARE OF MARKET REPORT: MEASURING AND CONTROLLING DEVICES, NOT ELSEWHERE CLASSIFIED. *Economic Information Systems, Inc. Quarterly. 1982. $260.00.*

Statistical analysis covers marketplace concentration and ownership structure. Companies in industry are listed and then ranked by production. Other data includes market share (dollars and percent), with groupings of all plants a company operates in the industry, with the parent company's market share broken down on a per-plant basis.

3595. EIS SHARE OF MARKET REPORT: OPTICAL INSTRUMENTS AND LENSES. *Economic Information Systems, Inc. Quarterly. 1982. $200.00.*

Coverage of optical instruments and lens market includes marketplace concentration and ownership structure. Companies in industry are listed and then ranked by production. Other data includes market share (dollars and percent), with groupings of all plants a company operates in the industry, with the parent company's market share broken down on a per-plant basis.

3596. EIS SHARE OF MARKET REPORT: PROCESS CONTROL INSTRUMENTS. *Economic Information Systems, Inc. Quarterly. 1982. $245.00.*

Study covers marketplace concentration and ownership structure. Companies in industry are listed and then ranked by production. Other data includes market share (dollars and percent), with groupings of all plants a company operates in the industry, with the parent company's market share broken down on a per-plant basis.

3597. EIS SHARE OF MARKET REPORT: WATCHES, CLOCKS AND WATCHCASES. *Economic Information Systems, Inc. Quarterly. 1982. $90.00.*

Analysis of watch, clock, and watchcase markets covers marketplace concentration and ownership structure. Companies in industry are listed and then ranked by production. Other data includes market share (dollars and percent), with groupings of all plants a company operates in the industry, with the parent company's market share broken down on a per-plant basis.

3598. EIS SHIPMENTS REPORT: ENGINEERING AND SCIENTIFIC INSTRUMENTS. *Economic Information Systems, Inc. Quarterly. 1982. $350.00.*

Report on the scientific and technical instruments industry. Arranged by state and county, report includes every plant in the industry with annual shipments of $500,000 and/or 20 or more employees. Plant listings detail address, telephone, estimated annual shipments, and percent of market. Similar statistics for each county and state.

3599. EIS SHIPMENTS REPORT: ENVIRONMENTAL CONTROLS. *Economic Information Systems, Inc. Quarterly. 1982. $135.00.*

Report on the environmental controls industry. Arranged by state and county, report includes every plant in the industry with annual shipments of $500,000 and/or 20 or more employees. Plant listigns detail address, telephone, estimated annual shipments, and percent of market. Similar statistics for each county and state.

3600. EIS SHIPMENTS REPORT: FLUID METERS AND COUNTING DEVICES. *Economic Information Systems, Inc. Quarterly. 1982. $75.00.*

Report on the fluid meter and counting devices industries. Arranged by state and county, report includes every plant in the industry with annual shipments of $500,000 and/or 20 or more employees. Plant listings detail address, telephone, estimated annual shipments, and percent of market. Similar statistics for each county and state.

3601. EIS SHIPMENTS REPORT: INSTRUMENTS TO MEASURE ELECTRICITY. *Economic Information Systems, Inc. Quarterly. 1982. $315.00.*

Report on the electricity measurement instruments industry. Arranged by state and county, report includes every plant in the industry with annual shipments of over $500,00 and/or 20 or more employees. Plant listings detail address, telephone, estimated annual shipments, and percent of market. Similar statistics for each county and state.

3602. EIS SHIPMENTS REPORT: MEASURING AND CONTROLLING DEVICES, NOT ELSEWHERE CLASSIFIED. *Economic Information Systems, Inc. Quarterly. 1982. $260.00.*

Report on miscellaneous measuring and controlling devices industry. One in a series of reports on particular four-digit SIC code industries. Arranged by state and county, report includes every plant in the industry with annual shipments of over $500,000 and/or 20 or more employees. Plant listings detail address, telephone, estimated annual shipments, and percent of market. Similar statistics for each county and state.

3603. EIS SHIPMENTS REPORT: OPTICAL INSTRUMENTS AND LENSES. *Economic Information Systems, Inc. Quarterly. 1982. $200.00.*

Report on the optical instrument and lens industry. Arranged by state and county, report includes every plant in the industry with annual shipments of over $500,000 and/or 20 or more employees. Plant listings detail address, telephone, estimated annual shipments, and percent of market. Similar statistics for each county and state.

3604. EIS SHIPMENTS REPORT: PROCESS CONTROL INSTRUMENTS. *Economic Information Systems, Inc. Quarterly. 1982. $245.00.*

Report on the process control instruments industry. Arranged by state and county, report includes every plant in the industry with annual shipments of over $500,000 and/or 20 or more employees. Plant listings detail address, telephone, estimated annual shipments, and percent of market. Similar statistics for each county and state.

3605. EIS SHIPMENTS REPORT: WATCHES, CLOCKS AND WATCHCASES. *Economic Information Systems, Inc. Quarterly. 1982. $90.00.*

Report on the watches, clocks and watchcases industries. Arranged by state and county, report includes every plant in the industry with annual shipments of over $500,000 and/or 20 or more employees. Plant listings detail address, telephone, estimated annual shipments, and percent of market. Similar statistics for each county and state.

3606. JEWELRY STORE SALES OF WATCHES AND CLOCKS. *Chilton Book Company. 1979. $10.00.*

Features maps which mark Standard Metropolitan Statistical Areas in each state. Includes number of jewelry stores in each state and SMSA total sales volume, number of stores carrying a specific product line, and their relationship to total sales.

3607. SCIENTIFIC AND PROFESSIONAL EQUIPMENT INDUSTRIES 1979: ANNUAL CENSUS OF MANUFACTURES. *Canada. Statistics Canada. English/French. Annual. September 1981. 40 pg. 7.20 (Canada).*

1979 survey of Canadian manufacturers of scientific, technical and medical instruments. Tables show level of overall activity, 1971-1979; percentage change, 1977-1979; production costs in relation to the size of enterprise, 1979; and quantity and value of materials used and goods produced, 1978-1979.

3608. 1977 CENSUS OF MANUFACTURES: ENGINEERING AND SCIENTIFIC INSTRUMENTS: PRELIMINARY REPORT: INDUSTRY SERIES. *U.S. Department of Commerce. Bureau of the Census. Customer Service Branch. Irregular. MC77-1-38A-1(P). September 1979. 9 pg. $.35.*

This is a preliminary report from the Bureau of the Census 1977 Census of Manufactures. Tables with industry statistics from the U.S. for 1963-1977 and with statistics for selected geographic areas for 1972 and 1977 give data relating to employees, production costs, assets and expenditures, inventories, and ratios. Other tables supply 1972 and 1977 statistics on product classes giving quantity and value of shipments by all producers and statistics on materials consumed by kind.

3609. 1977 CENSUS OF MANUFACTURES: ENVIRONMEN-TAL CONTROLS: PRELIMINARY REPORT: INDUSTRY SERIES. *U.S. Department of Commerce. Bureau of the Census. Customer Service Branch. Irregular. MC77-1-38A-2(P). May 1979. 8 pg. $.35.*

This is a preliminary report from the Bureau of the Census 1977 Census of Manufactures. Tables with industry statistics from the U.S. for 1963-1977 and with statistics for selected geographic areas for 1972 and 1977 give data relating to employees, production costs, assets and expenditures, inventories, and ratios. Other tables supply 1972 and 1977 statistics on product classes giving quantity and value of shipments by all producers and statistics on materials consumed by kind.

3610. 1977 CENSUS OF MANUFACTURES: FLUID METERS AND COUNTING DEVICES: PRELIMINARY REPORT: INDUSTRY SERIES. *U.S. Department of Commerce. Bureau of the Census. Customer Service Branch. Irregular. MC77-1-38A-4(P). August 1979. 9 pg. $.35.*

This is a preliminary report from the Bureau of the Census 1977 Census of Manufactures. Tables with industry statistics from the U.S. for 1963-1977 and with statistics for selected geographic areas for 1972 and 1977 give data relating to employees, production costs, assets and expenditures, inventories, and ratios. Other tables supply 1972 and 1977 statistics on product classes giving quantity and value of shipments by all producers and statistics on materials consumed by kind.

3611. 1977 CENSUS OF MANUFACTURES: INSTRUMENTS TO MEASURE ELECTRICITY: PRELIMINARY REPORT: INDUSTRY SERIES. *U.S. Department of Commerce. Bureau of the Census. Customer Service Branch. Irregular. MC77-1-38A-5(P). August 1979. 11 pg. $.35.*

This is a preliminary report from the Bureau of the Census 1977 Census of Manufactures. Tables with industry statistics from the U.S. for 1963-1977 and with statistics for selected geographic areas for 1972 and 1977 give data relating to employees, production costs, assets and expenditures, inventories, and ratios. Other tables supply 1972 and 1977 statistics on product classes giving quantity and value of shipments by all producers and statistics on materials consumed by kind.

3612. 1977 CENSUS OF MANUFACTURES: MEASURING AND CONTROLLING DEVICES, N.E.C.: PRELIMINARY REPORT: INDUSTRY SERIES. *U.S. Department of Commerce. Bureau of the Census. Customer Service Branch. Irregular. MC77-1-38A-6(P). September 1979. 9 pg. $.35.*

This is a preliminary report from the Bureau of the Census 1977 Census of Manufactures. Tables with industry statistics from the U.S. for 1963-1977 and with statistics for selected geographic areas for 1972 and 1977 give data relating to employees, production costs, assets and expenditures, inventories, and ratios. Other tables supply 1972 and 1977 statistics on product classes giving quantity and value of shipments by all producers and statistics on materials consumed by kind.

3613. 1977 CENSUS OF MANUFACTURES: OPTICAL INSTRUMENTS AND LENSES: PRELIMINARY REPORT: INDUSTRY SERIES. *U.S. Department of Commerce. Bureau of the Census. Customer Service Branch. Irregular. MC77-1-38A-7(P). August 1979. 8 pg. $.35.*

This is a preliminary report from the Bureau of the Census 1977 Census of Manufactures. Tables with industry statistics from the U.S. for 1963-1977 and with statistics for selected geographic areas for 1972 and 1977 give data relating to employees, production costs, assets and expenditures, inventories, and ratios. Other tables supply 1972 and 1977 statistics on product classes giving quantity and value of shipments by all producers and statistics on materials consumed by kind.

3614. 1977 CENSUS OF MANUFACTURES: PROCESS CONTROL INSTRUMENTS: PRELIMINARY REPORT: INDUSTRY SERIES. *U.S. Department of Commerce. Bureau of the Census. Customer Service Branch. Irregular. MC77-1-38A-3(P). May 1980. 10 pg. $.35.*

This is a preliminary report from the Bureau of the Census 1977 Census of Manufactures. Tables with industry statistics from the U.S. for 1963-1977 and with statistics for selected geographic areas for 1972 and 1977 give data relating to employees, production costs, assets and expenditures, inventories, and ratios. Other tables supply 1972 and 1977 statistics on product classes giving quantity and value of shipments by all producers and statistics on materials consumed by kind.

3615. 1977 CENSUS OF MANUFACTURES: WATCHES, CLOCKS, AND WATCHCASES: PRELIMINARY REPORT: INDUSTRY SERIES. *U.S. Department of Commerce. Bureau of the Census. Customer Service Branch. Irregular. MC77-1-38B-6(P). April 1979. 10 pg. $.35.*

This is a preliminary report from the Bureau of the Census 1977 Census of Manufactures. Tables with industry statistics from the U.S. for 1963-1977 and with statistics for selected geographic areas for 1972 and 1977 give data relating to employees, production costs, assets and expenditures, inventories, and ratios. Other tables supply 1972 and 1977 statistics on product classes giving quantity and value of shipments by all producers and statistics on materials consumed by kind.

See 2384, 2385.

Financial and Economic Studies

3616. ENERGY CONSERVATION - ENERGY CONSERVATION CONTROL; HEATING, COOLING AND LIGHTING PRODUCTS; AUTOMOTIVE DIAGNOSTIC TEST EQUIPMENT. *Cyrus J. Lawrence. Frost and Sullivan, Inc. September 1980. 54 pg. $350.00.*

Financial analysis of U.S. manufacturers of energy conservation control devices, heating and cooling machinery, waste recovery, systems and diagnostic instruments, in light of pressures for energy conservation. Covers expenditures for new plants and equipment, energy consumption, energy costs, savings and regulations. Provides financial statements for industry leaders, listing earnings, returns on equity, price earnings ratios, market shares, sales, income, revenues, profits, new orders, and end uses. Includes forecasts to 1985.

3617. SCIENCE AND ELECTRONICS: TRADE BALANCE REPORTS: FAVORABLE TRADE BALANCE IN ENGINEERING, ELECTRICAL, OPTICALS AND ANALYTICAL INSTRUMENTS INCREASES 64 PERCENT. *U.S. Department of Commerce. Bureau of Industrial Economics. Office of Public Affairs and Publications. Semi-annual. ITA 79-155. October 3, 1979. 2 pg.*

One in a series of semiannual publications detailing U.S. exports and imports (total and by region or nation). Data for first halves of 1978 and 1979. Companion set of releases gives data for entire year and two previous years.

3618. SCIENCE AND ELECTRONICS: TRADE BALANCE REPORTS: FAVORABLE TRADE BALANCE POSTED FOR MEASURING AND CONTROLLING INSTRUMENTS IN 1979. *U.S. Department of Commerce. Bureau of Industrial Economics. Office of Public Affairs and Publications. Semi-annual. BIE 80-10. April 3, 1980. 2 pg.*

One in a series of semiannual publications detailing annual U.S. exports and imports (total and by region or nation), 1977-1979. Companion set of releases gives data for first half of year of coverage and previous year.

Forecasts

3619. ANALYTICAL INSTRUMENTATION. *Theta Technology Corporation. July 1980. $750.00.*

A market research report on analytical instrumentation (spectroscopes and spectrometers, nuclear instruments, chromatography, electrophoresis systems and microscopes) and its use in medicine, industrial quality and process control, and environmental pollution monitoring. Provides market shares, analysis and segmentation; company profiles, forecasts and product evaluation.

3620. AUTOMATIC TEST EQUIPMENT. *Creative Strategies International. February 1980. $1200.00.*

1980 market analysis for U.S. produced non-captive automatic test equipment. Forecasts through 1984 are provided for major market segments: component testing, subsystem testing, and system testing. Each of these segments is further broken down by user category. Component testing is segmented by type of component: discrete, linear ICs, SSI and MSI, LSI memory, microprocessor LSI and VLSI. Major competitors in each market segment are identified along with the share of the market that each holds.

3621. CLINICAL IMMUNOLOGY PRODUCTS MARKET. *Frost and Sullivan, Inc. January 1981. 222 pg. $900.00.*

Forecast and analysis of the market for clinical immunology instruments, reagents, equipment and supplies. Analysis of the impact of new techniques on various medical specialties.

3622. PROCESS CONTROL EQUIPMENT. *Creative Strategies International. October 1980. $1200.00.*

Examines the U.S. market for process control equipment including the "controls" market segment. Forecasts are presented by segment through 1984. Study includes a discussion of packaged systems and sub-component technologies. Covers trends in custom building-block systems and end-user requirements. Analyzes competition and market shares with write-ups on major competitors and listings of all participating companies. Discusses the overall effect of modular-based systems on the future of process control industries including the effect of microprocessor and computer-controlled systems.

3623. THE RESOURCE RECOVERY MARKET: SOLID WASTE FUEL, MATERIALS RECOVERY WASTE AND MATERIALS HANDLING EQUIPMENT, AND PROCESSING AND ENVIRONMENT CONTROL EQUIPMENT. *Frost and Sullivan, Inc. May 1979. 252 pg. $775.00.*

Analyzes and forecasts the market in 11-21 year periods for the resource recovery industry in 8 categories (fuel; raw materials from refuse; waste handling equipment; baling and compaction equipment; mixing, size reduction and separation equipment; processing equipment; environmental control equipment; miscellaneous equipment and instrumentation; services). Also forecasts for the next 11-21 years potential markets for various raw materials. Supplies operating and capital cost data for cost comparisons of the various resource recovery approaches.

Directories and Yearbooks

3624. AIR DIFFUSION COUNCIL - DIRECTORY OF CERTIFIED PRODUCTS AND MEMBERSHIP ROSTER. *Air Diffusion Council. Irregular. 1980. Free.*

Lists manufacturers of air control and distribution devices, providing company name and address. Gives air diffusion and distribution products tested and certified by council, along with manufacturer name and model number.

3625. AMCA DIRECTORY OF LICENSED PRODUCTS. *Air Movement and Control Association. Annual. January 1980. 40 pg. Free.*

List of member manufacturers of air movement and control products. Gives manufacturers' names, addresses, products, model numbers, catalog numbers and sizes. Product index provided.

3626. AUTOMATED BLOOD CELL AND DIFFERENTIAL COUNTERS: INSTRUMENT DATA SERVICE. *Luning Prak Associates, Inc. Annual. December 1980. $2400.00.*

For each of the 15,000 entries in the automated blood cell and differential counters data bank, information includes location, brand, model number, year acquired, application, tests / month acquisition mode, reagent source and service provider.

3627. AUTOMATED CHEMICAL ANALYZERS: INSTRUMENT DATA SERVICE. *Luning Prak Associates, Inc. Annual. December 1980. $3600.00.*

For each of the 15,000 instrument copies in the automated chemical analyzer data bank, information includes location, brand, model number, year acquired, application, tests / month acquisition mode, reagent source and service provider.

3628. BLOOD GAS ANALYZERS: INSTRUMENT DATA SERVICE. *Luning Prak Associates, Inc. Irregular. December 1980. $2400.00.*

For each of the 15,000 entries in the blood gas analyzer data bank, information includes location, brand, model number, year acquired, application, tests / month acquisition mode, reagent source and service provider.

3629. CHEMISTRY (ALL CLASSES, INCLUDING COMPUTERS): INSTRUMENT DATA SERVICE. *Luning Prak Associates, Inc. Irregular. 3811. December 1980. $7000.00.*

For each of the 15,000 entries in the data bank on chemistry (including computers) instrumentation in the health care industry, information includes location, brand, model number, year acquired, application, tests / month acquisition mode, reagent source and service provider.

3630. CHILTON'S CONTROL EQUIPMENT MASTER. *Chilton Book Company. Annual. 1979. 1000 (est.) pg.*

Directory of control instruments and equipment for 1980. Includes directories of manufacturers and of products, listed alphabetically, manufacturers' catalog pages desciding products, listed by company; and manufacturers and sales office index, listed alphabetically, with addresses of headquarters and subsidiary offices.

3631. CLINILAB PRODUCTS COMPANY DIRECTORY - USA - 1980. *Robert S. First, Inc. Irregular. September 1980. 479 pg. $300.00.*

Listing of 350 U.S. manufacturers and suppliers of laboratory reagents and instruments. Gives sales volume and percent of sales going to exports.

3632. COAGULATION ANALYZERS / PROTHROMBIN TIMERS: INSTRUMENT DATA REPORT. *Luning Prak Associates, Inc. Annual. December 1980. $2400.00.*

For each of the 15,000 entries in the coagulation analyzers / prothrombin timers data bank, information includes location, brand, model number, year acquired, application, tests / month acquisition mode, reagent source and service provider.

3633. DENSITOMETERS: INSTRUMENT DATA SERVICE. *Luning Prak Associates, Inc. Irregular. December 1980. $1500.00.*

For each of the 15,000 entries in the densitometer data bank, information includes location, brand, model number, year acquired, application, test / month acquisition mode, reagent source and service provider.

3634. DIRECTORY OF INSTRUMENTATION. *Instrument Society of America. Irregular. 2nd ed. 1981. 1282 pg. $65.00.*

Two volumes with company listings for approximately 1900 manufacturers with their sales offices and representatives arranged by territory. Listings are arranged by about 260 product categories, with the companies that manufacture each. A separate section provides a complete listing of sales representatives arranged geographically, followed by the manufacturers each represents. A specifications section contains detailed information on more than 1000 instrumentation and control products.

3635. ELECTROLYTE ANALYZER / FLAME PHOTOMETERS: INSTRUMENT DATA SERVICE. *Luning Prak Associates, Inc. Annual. December 1980. $2400.00.*

For each of the 15,000 instrument copies in the electrolyte analyzer / flame photometers data bank, information includes location, brand, model number, year acquired, application, test / month acquisition mode, reagent source and service provider.

3636. EQUIPMENT FOR FLOOD AND FLASH FLOOD WARNING SYSTEMS. *U.S. Dept. of Commerce, National Weather Service. U.S. Department of Commerce. National Technical Information Service. Irregular. NTIS #PB81-224131. 1981. 416 pg.*

Describes equipment potentially useful in flood and flash flood warning systems. Entries are primarily components from which an equipment system can be assembled.

3637. THE FIBER OPTIC COMMUNICATIONS REPORT. *JMA Research and Consulting. Irregular. 1981. 100 pg. $97.00.*

Report on fiber optic transmission systems with a specifications guide to the systems. Also includes listings of vendors and suppliers of lightguide cable and components.

3638. FISHER SCIENTIFIC 81: APPARATUS, CHEMICALS, FURNITURE AND SUPPLIES FOR INDUSTRIAL, CLINICAL, COLLEGE AND GOVERNMENT LABORATORIES. *Fisher Scientific Company. Annual. 1980. 2000 pg.*

Two-part buyers' guide lists scientific and medical laboratory equipment. Gives pictures, descriptions and prices, and chemicals by name (lot analysis, size, catalog number, purity grade, safety data, lot numbers, and OSHA classification).

3639. GAMMA COUNTERS / AUTOMATED RIA ANALYZERS. *Luning Prak Associates, Inc. Annual. December 1980. $3600.00.*

For each of the 15,000 entries in the gamma counters / automated RIA analyzers data bank, information includes location, brand, model number, year acquired, application, tests / month acquisition mode, reagent source and service provider.

3640. GUIDE TO SCIENTIFIC INSTRUMENTS: 1979-80. *American Association for the Advancement of Science. Annual. 1979. 270 (est.) pg.*

A guide to laboratory instruments and equipment with names and addresses of their manufacturers. Arranged in alphabetical order by names of instruments within major categories.

3641. GUIDE TO SCIENTIFIC INSTRUMENTS: 1980-81. *Nelson, Philip H., editor. American Association for the Advancement of Science. Annual. 1980. 270 pg. $10.00.*

Annual buyer's guide to scientific and technical instruments. Detailed product group headings are listed alphabetically, and manufacturers and their addresses are listed under each heading. Many manufacturers supply photo-illustrated advertisements, and these are cross-referenced in the product-category listing. Manufacturers are separately indexed (with addresses and telephone numbers), and there are several product-information check-off cards.

3642. HEALTH DEVICES SOURCEBOOK. *ECRI Shared Service. Annual. 3rd ed. July 1981. $95.00.*

1981 annual directory lists 5000 products including patient care equipment, medical and surgical instruments, clinical laboratory supplies, disposables and test instruments available from U.S. and Canadian suppliers. Includes over 2000 manufacturers and selected distributors and importers.

3643. HEALTH DEVICES SOURCEBOOK, 4TH ED. *ECRI Shared Service. Annual. 4th ed. 1982. $110.00.*

Two-volume annual directory of 6000 medical devices and their manufacturers and suppliers covers patient care equipment, medical instruments, special furniture, disposables, and related test instruments. Includes list of manufacturers, distributors, and importers; trade name section; and references to evaluations published in 'Health Devices.'

3644. HIGH PRESSURE LIQUID CHROMATOGRAPHS: INSTRUMENT DATA SERVICE. *Luning Prak Associates, Inc. Irregular. December 1980. $1500.00.*

For each of the 15,000 entries in the high pressure liquid chromatograph data bank, information includes location, brand, model number, year acquired, application, tests / month acquisition mode, reagent source and service provider.

3645. ISA DIRECTORY OF INSTRUMENTATION: NORTH AMERICAN VOLUME: SALES OFFICES, SPECIFICATIONS, REPRESENTATIVES. *Instrument Society of America. Annual. 1979. 729 pg. $30.00.*

North American directory of instrumentation contains: sales offices-representatives (arranged by manufacturer followed by that company's sales offices and / or sales representatives); specifications grouped by product category; and sales representatives with addresses, telephone, TLX or TWX numbers and key contact.

3646. LABORATORY BUYERS' GUIDE 1981. *Southam Business Publications, Ltd. Division of SouthamCommunications Ltd. Annual. November 1980. 146 pg.*

Company directory alphabetically lists most major suppliers of laboratory equipment in Canada. Each listing gives, in order, company name, head-office address and telephone, branch offices (including dealer telephones), and names of foreign manufacturers represented by the Canadian suppliers. Principals section lists manufacturing companies and their Canadian sales agents. Most such manufacturers listed are U.S. and European. Made-in-Canada products are noted throughout the hardware and chemicals listings.

3647. PLATELET COUNTERS / PLATELET AGGREGOMETERS: INSTRUMENT DATA SERVICE. *Luning Prak Associates, Inc. Irregular. December 1980. $15.00.*

For each of the 15,000 entries in the platelet counters / platelet aggregometers data bank, information includes location, brand, model number, year acquired, application, tests / month acquisition mode, reagent source and service provider.

3648. THE TEST AND MEASUREMENT EQUIPMENT MANU-FACTURERS' DIRECTORY. *Anderson, Bud. Marketing Development. Irregular. 1980. 190 pg. $75.00.*

Lists manufacturers by type of test and measurement equipment. Also gives company profiles.

3649. 1982 LASER FOCUS BUYERS' GUIDE. *Penn Well Publishing Company. Annual. January 1982. 608 pg.*

Buyers' guide for laser and fiber optical products and services. Lists business address for manufacturers, suppliers and associations; lines of business; and affiliated and subsidiary corporations and sales representatives. Also includes charts of technological specifications and technological and bibliographic information.

See 257, 1479, 2479, 2480, 2481, 2482, 2499, 2507, 2528, 2543.

Special Issues and Journal Articles

AIA Journal

3650. AIA JOURNAL: PRODUCTS. *American Institute of Architects. Monthly. v.71, no.1. January 1982. pg.100*

Reviews new products for the architectural field. Includes product description and gives manufacturer's name and address.

Air Conditioning, Heating and Refrigeration News

3651. AIR CONDITIONING, HEATING AND REFRIGERA-TION NEWS: WHAT'S NEW. *Business News Publishing Company. Weekly. January 18, 1982. pg.12*

This weekly survey of heating, ventilation, air-conditioning and refrigeration equipment emphasizes measuring / control instruments, and components / accessories.

3652. AIR CONDITIONING, HEATING AND REFRIGERA-TION NEWS: WHAT'S NEW. *Business News Publishing Company. Weekly. October 26, 1981. pg.38-41 $1.50.*

Weekly survey of new heating, ventilation and air-conditioning equipment. Covers relevant components, production and hand tools, and measuring / control devices. Descriptions generally include some product dimensions and/or specifications, manufacturer's address, and often a photograph.

Air Transport World

3653. AIRLINES BEGIN TO MOVE INTO NEW TYPES OF AVIONICS. *Penton / IPC. Subsidiary of Pittway Corporation. April 1980. pg.44-65*

Surveys airlines' use of new avionics products. Shows expenditures for new equipment, 1978-1981; purchases, 1978-1981; and total investment in avionics. Gives manufacturer, aircraft, and equipment type for installations.

Analytical Chemistry

3654. ANALYTICAL CHEMISTRY: LABORATORY GUIDE ISSUE. *American Chemical Society. Annual. 1981. 250 pg. $4.00.*

Entries detail address, telephone numbers, for manufacturers of instruments, equipment, chemicals, and related items for chemical labs. Includes product, service, chemical and trade name indexes.

3655. ANALYTICAL CHEMISTRY 1979-80 LAB GUIDE. *American Chemical Society. Annual. August 1979. 258 pg.*

A directory of suppliers and manufacturers of laboratory supplies, services and equipment in the U.S., 1979-1980, includes products, brand names, and advertising indexes.

3656. ANALYTICAL CHEMISTRY 1980-81 LAB GUIDE. *American Chemical Society. Annual. August 1980. 292 pg.*

A directory of suppliers and manufacturers of laboratory supplies, services and equipment in the U.S., 1980-1981. Includes products, brand names and an advertising index.

3657. ANALYTICAL CHEMISTRY: 1982-83 LAB GUIDE. *American Chemical Society. Annual. v.54, no.10. August 1982. 292 pg.*

Divided into nine sections, document lists advertised products and services in alphabetical order; main dealer locations and main branch offices by region and state; instruments, equipment and supplies, including name, address and telephone numbers of manufacturers; chemicals and name of manufacturers; research and analytical services followed by an alphabetical list of suppliers; meetings and seminars, including date, place, subject of meeting, and sponsoring organization; new books reviewed by The American Chemical Society from July, 1981-July, 1982; commonly used trade names of products and services with names of companies claiming them or their exclusive rights; and manufacturers of products and suppliers of services.

AOPA Pilot

3658. 1982 AVIONICS DIRECTORY AND BUYERS' GUIDE. *Aircraft Owners and Pilots Association. Annual. v.25, no.6. June 1982. pg.53-95*

Gives manufacturer, specifications, and price for 1982 navigation, communication, and other avionics equipment.

Appliance Manufacturer

3659. WIDE CHOICE OF HUMIDISTATS EASES HUMIDITY CONTROL TASK. *Klein, Art. Cahners Publishing Company. v.30, no.4. April 1982. pg.48-50*

Product review discusses humidistats with a review of product types, and technical drawings to illustrate principles of operation. Data is included on shipments and unit sales of dehumidifiers and humidifiers (1976-1980). A listing details models available by manufacturer, with data on humidity range covered, sensing element, and additional technical specifications.

Canadian Aviation

3660. 1982 AVIATION DIRECTORY OF CANADA. *Maclean Hunter Ltd. Annual. April 1982. pg.76-138*

Lists major Canadian air carriers, their head office addresses, operating fleet, number of employees, chief executives; fixed wing air carriers, rotary wing air carriers, and foreign air carrier offices in Canada. Includes manufacturers, producers, and distributors of products for the industry; associations and publications, academic institutions; government departments and agencies; sources of supplies and services.

3661. 1982 AVIONICS BUYERS GUIDE. *Maclean Hunter Ltd. Annual. v.55, no.7. July 1982. pg.55-58*

Annual guide listing manufacturers of avionics equipment in 27 product categories. Listing is arranged by manufacturer with products by model number and includes technical description, weight, size and price. An additional listing features all manufacturers with addresses, telephone numbers, and product groups manufactured.

Canadian Chemical Processing

3662. BUYERS' GUIDE TO PROCESS EQUIPMENT, CONTROLS, INSTRUMENTATION 1980 / 81. *Southam Business Publications, Ltd. Division of SouthamCommunications Ltd. Annual. July 1979. pg.13-69*

Directory of distributors who supply Canadian chemical processors with equipment, controls and instrumentation, 1979. Lists addresses and product lines.

3663. CANADIAN CHEMICAL PROCESSING BUYERS' GUIDE TO PROCESS EQUIPMENT, CONTROLS, INSTRUMENTATION 1980 / 81. *Southam Business Publications, Ltd. Division of SouthamCommunications Ltd. Annual. July 1980. 74 pg.*

1980 buyers' guide for the Canadian chemical processing industry. A company directory alphabetically lists names, addresses and telephone numbers of process equipment suppliers. An equipment directory provides names of suppliers for specific process equipment, arranged by product category.

3664. CANADIAN CHEMICAL PROCESSING: NEW PRODUCTS. *Southam Business Publications, Ltd. Division of SouthamCommunications Ltd. Monthly. May 1982. pg.47-58*

Monthly listing of tools, equipment, and instruments used in Canadian chemical processing operations includes name of manufacturer, description, and photograph. Further information through reader card service.

3665. CANADIAN CHEMICAL PROCESSING: 1982 PROCESS EQUIPMENT GUIDE. *Southam Business Publications, Ltd. Division of SouthamCommunications Ltd. Annual. July 28, 1982. pg.27-79*

1982 directory lists Canadian companies that supply process equipment, process controls, and process instrumentation. Company section provides supplier's name, address, and branches. Product section provides company name under each product category.

Canadian Controls and Instruments

3666. CANADIAN CONTROLS AND INSTRUMENTS BUYERS' GUIDE 80/81. *Maclean Hunter Ltd. Annual. July 1980.*

Annual classified listing of instrumentation, control and automation products, grouped in categories ranging from sensors and components through instruments, recorders, controllers, computers, and valves. Suppliers addresses, branches and representatives are given.

3667. CANADIAN CONTROLS AND INSTRUMENTS BUYERS' GUIDE 1981 / 82. *Maclean Hunter Ltd. Annual. August 1981. 140 pg.*

Buyers' guide includes a product index, and a products listing complete with their suppliers, manufacturers and representatives (addresses, phone and telex numbers, branches and representatives) in process control, manufacturing, automation and measurement instrumentation industries.

3668. CANADIAN CONTROLS AND INSTRUMENTS BUYERS GUIDE 1982 / 83. *Maclean Hunter Ltd. Annual. August 1982. 18.00 (Canada).*

Annual classified listing of control and instrumentation products includes a directory of manufacturers, branches and representatives.

Canadian Research

3669. PREVIEW OF "81 SCIENTIFIC SUPPLIERS' EXPOSITION: EXHIBITORS LISTING. *Maclean Hunter Ltd. Annual. v.14, no.5. September 1981. pg.52-73*

Lists names of exhibitors at the Advances in Analytical Instrumentation and Techniques Conference held in Toronto Sept. 22-24. Scientific suppliers are listed alphabetically, together with employees at show and their products.

Chemical and Engineering News

3670. PROCESS CONTROL EQUIPMENT IGNORES RECESSION. *Stinson, Stephen C. American Chemical Society. v.59, no.49. December 7, 1981. pg.15-19*

Examines the companies that produce process control equipment and the industries that use it. Sales and growth rates to 1985 are forecast. Covers computers and programmable controllers; temperature, pressure, flow rate, and liquid level sensors, transmitters, and controllers; materials analyzers; and valves.

Chemical Engineering

3671. CHEMICAL ENGINEERING EQUIPMENT BUYERS' GUIDE: 1981. *McGraw-Hill Book Company. Annual. July 1981.*

Annual directory of products and services available to the chemical engineering industry. Lists businesses providing instruments and controls equipment, chemical processing equipment, and materials handling equipment.

3672. CHEMICAL ENGINEERING EQUIPMENT BUYERS' GUIDE 1981: THE COMPLETE GUIDE TO SOURCES OF SUPPLY FOR EQUIPMENT, MATERIALS OF CONSTRUCTION, AND SERVICES. *McGraw-Hill Book Company. Annual. July 28, 1980. 910 pg.*

A listing of U.S. suppliers of equipment and services to the chemical process industry. Includes trade names.

Chemical Industry Product News

3673. CHEMICAL INDUSTRY PRODUCT NEWS. *Putman Publishing Company. Bi-monthly. v.21, no.2. April 1982. pg.35*

Bimonthly digest of new products for the chemical processing industries. Includes manufacturers and bibliographic information.

Chilton's Instruments and Control Systems

3674. COMPANY RECOGNITION STUDY. *Chilton Book Company. Biennial. 1980. $35.00.*

Study conducted to determine recognition for approximately 160 different products in the instruments and control systems industry. Identifies which companies are recognized as the leading manufacturers.

3675. FIBER OPTICS FOR DATA COMMUNICATIONS. *Persun, Terry. Chilton Book Company. v.55, no.8. August 1982. pg.49-54*

Reviews recent developments in the U.S. fiber optics industry. Discusses technological developments for each component of a fiber optic system and indicates new products in each component category. Lists manufacturers and their product lines.

Civil Engineering ASCE

3676. CIVIL ENGINEERING: PRODUCT NEWS. *American Society of Civil Engineers. Monthly. October 1981.*

A monthly description of new products, machinery, and tools in the field of civil engineering. Data includes photograph, description, manufacturer's name, and technical specifications. Further information available through reader card service.

Control and Instrumentation

3677. MONITOR: LIQUID LEVEL CONTROLLERS. *Amivest Corporation. Monthly. v.14, no.8. August 1982. pg.23-25*

Reviews well established and new systems in the area of process control. Systems briefly described; name of manufacturer included.

Control Engineering

3678. CONTROL ENGINEERING'S GUIDE TO THE CHEMICAL PLANT EXPOSITION: CPE EXPO ''82. *Morris, Henry M. Technical Publishing Company. Annual. v.29, no.6. April 1982. pg.80-81*

Lists new products to be exhibited at the Chemical Plant Equipment Exposition, June 7-10, 1982, Anaheim, CA. Listing is alphabetical by product and includes company name, city and state as well as booth number.

3679. CONTROL PRODUCTS PREVIEW. *Pharmaceutical Manufacturers Association. Irregular. v.29, no.5. April 1982. pg.111-156*

Previews products for the control and instrumentation field. Includes product description and manufacturer's name and address.

3680. EXHIBITS TO FEATURE WIDE RANGE OF CONTROL PRODUCTS. *Morris, Henry M. Technical Publishing Company. Irregular. v.29, no.5. April 1982. pg.81-86*

Previews product exhibits at Control Expo/82. Includes manufacturer's name and address and booth number, and distinguishes new from standard products.

3681. RECORDERS AND LOGGERS; DISTRIBUTED CONTROL DATA BASES; SERVO MOTORS AND AMPLIFIERS. *Dun-Donnelley Publishing Corporation. Irregular. July 1981.*

Special issue on control instruments: recorders, loggers, distributed control data bases, servo motors and amplifiers.

Design Engineering

3682. CANADA'S FLUID POWER AND CONTROLS BUYER'S GUIDE 1980 / 1981. *Maclean Hunter Ltd. Annual. July 1980. pg.30-44, 73-86*

Guide to 123 major types of fluid power and controls components and devices. Includes air logic devices, air compressors, filter - lubricator - regulators, hydraulic manifolds, hydraulic constant displacement pumps and hydraulic valves. Listing is divided into an alphabetical products directory and manufacturers' names; manufacturers' addresses and distributors' names, and an alphabetical list of distributors, addresses.

3683. DESIGN ENGINEERING: NEW PRODUCTS. *Morgan-Grampian, Inc. Monthly. v.53, no.1. January 1982. pg.73-81*

Lists and briefly describes new electronic, electrical, fluid power, and mechanical products, and new engineering equipment. Includes manufacturers' addresses and coded reply cards.

3684. DESIGN ENGINEERING SHOW: AN ENGINEER'S SUPERMARKET: MARCH 29 - APRIL 1, 1982 (CHICAGO). *Morgan-Grampian, Inc. Irregular. March 1982. pg.47-66*

Lists exhibitors at the Design Engineering Show and Conference, Chicago, March 29-April 1, 1982 and provides a sample of mechanical components; electrical / electronic products; materials and processes; fluid power items; and design, manufacturing and test equipment.

Design News

3685. MARKET EXPANDS FOR A/D AND D/A CONVERTERS. *Cahners Publishing Company. v.38, no.18. September 27, 1982. pg.140*

Assesses the U.S. market for A/D and D/A converters. Discusses end uses, technology, and market segmentation. Indicates annual growth rates, 1982-1986. Charts manufacturers' shipments (in dollars), 1981-1986. Source is Venture Development Corp.

EDN Magazine

3686. MANUFACTURERS OF IEEE - 488 PRODUCTS. *Cahners Publishing Company. Irregular. v.26, no.21. October 28, 1981. pg.91-94*

Lists U.S. and Western European manufacturers of instruments, components and micro / minicomputers used for interfacing, testing and controlling electronic instruments attuned to the IEEE - 488 standard. Specifies U.S. address, head office address in the case of European manufacturers, and types of equipment manufactured.

EE - Electrical Equipment

3687. MOTORS AND RELATED PRODUCTS. *Sutton Publishing Company, Inc. Irregular. v.41, no.12. December 1981. pg.15-27*

Surveys motors, motor controls (including digital controllers) and relays, motor components and accessories, and relevant test instruments currently on the market. Indentifies manufacturers and provides manufacturers' descriptions and specifications.

Electronic Business

3688. ATE MARKET PREPARES FOR NEW TECHNOLOGIES, SLOWER GROWTH. *Cahners Publishing Company. October 1980. pg.57, 60, 64-66*

Examines the U.S. automatic test equipment (ATE) market. Ranks top 8 manufacturers of ATE by estimated 1979 worldwide sales. Provides financial data for 15 U.S. manufacturers including 1980 sales, net income, revenues, cost of sales and R & D as a percent of total revenues.

3689. FIBER - OPTIC DEMAND IS SET TO SOAR. *Tsantes, John. Gnostic Concepts, Inc. Cahners Publishing Company. Irregular. April 1981. pg.80-82*

Forecasts U.S. demand for fiber-optics through 1990. Tables show U.S. fiber optic component production and consumption (1980-1990).

3690. HOW TO SAVE JAPANESE INSTRUMENT MARKET FOR U.S SUPPLIERS. *Cahners Publishing Company. v.8, no.7. June 1982. pg.36, 40*

Examines U.S. balance of trade with Japan in the electronic measuring instrument industry. Gives dollar value of Japan's instrument imports and exports, 1979-1981; and U.S. dollar sales to Japan, 1980, 1981.

3691. SYSTEM DESIGNERS ADDRESSING THE HIGH COST OF ATE USE. *Thames, Cindy. Cahners Publishing Company. October 1981. pg.40-48*

Examines the U.S. market for automatic test equipment (ATE). Ranks the top 10 suppliers of ATE according to 1980 sales and forecasts the 1981 and 1987 top 35 leaders. Provides financial data including net income, revenues and sales for 7 businesses (for fiscal year ending May 31, 1981). Discusses the Japanese share of the market.

3692. THE TOP 10 MANUFACTURERS IN TEST AND MEASUREMENT. *Dataquest Inc. Cahners Publishing Company. 382. October 1980. pg.56-57*

Examines the top 10 U.S. manufacturers of test and measurement equipment and ranks them according to 1979 worldwide sales. Graphs display market shares for test instruments, automatic test equipment and MDS and sales by test application (1979) with forecasts for 1984. Provides growth rates (1979-1980) for selected products.

Electronic Design

3693. FOCUS ON OPTOISOLATORS: THEY ADVANCE IN ISOLATION AND CTR: SPECIAL REPORT. *Grossman, Morris. Hayden Book Company, Inc. Irregular. v.30, no.11. May 27, 1982. pg.203-212*

Discusses development and characteristics of optoisolators. Provides statistics for isolation levels and a comparison to incandescent and neon-lamp operated couplers. Contains diagrams of circuitry, a life-cycle growth curve, photographs of optocouplers and 6-pin DIPs, and an optoisolator manufacturers' directory, with input, output, and packaging information.

Electronic Engineering Times

3694. TECHNOLOGY STRATEGIES: SEMICONDUCTOR ATE. *Miklosz, John. CMP Publications, Inc. Annual. no.245. October 25, 1982. pg.39-57*

Describes specific LSI / VLSI test systems. Gives names and addresses of U.S. makers of larger LSI / VLSI and memory test systems, 1982.

Electronic Products Magazine

3695. PRESSURE TRANSDUCERS: SOURCE LIST. *Hearst Business Communications, UTP Division. Irregular. November 30, 1981. pg.76-77*

A directory of U.S. pressure transducer manufacturers, which is limited to manufacturers address by product type. Transducer types are strain gage, piezoelectric, capacitance, force balance, poteutiometric, differential transformer, vibrating wire and tube, and variable reluctance.

3696. TEMPERATURE TRANSDUCERS. *Hearst Business Communications, UTP Division. Irregular. March 26, 1982. pg.75-77*

Directory indicates manufacturers of thermocouples, thermistors, RD's, and semi-conductors. Gives address and telephone

numbers. Updates a previous list contained in the November 30, 1981 issue.

3697. TEMPERATURE TRANSDUCERS: SOURCE LIST. *Hearst Business Communications, UTP Division. Irregular. November 30, 1981. pg.85-86*

A directory of U.S. temperature transducer manufacturers, giving manufacturer's address by type of product. Temperature transducers include thermocouples, thermistors, RTDs, and semiconductors.

Electro-Optical Systems Design

3698. DIMENSIONAL MEASUREMENT WITH HELIUM - NEON LASERS. *Tebo, Albert T. Cahners Publishing Company. v.14, no.10. October 1982. pg.21-27*

Reviews helium - neon lasers and their applications including optical alignment, laser ranging, particle sizing and counting, optical micrometry, wafer flatness testing, laser Doppler velocimetry, ellipsometry, and interferometry. Lists 25 manufacturers of helium - neon lasers available in the U.S.

3699. ELECTRO-OPTICAL SYSTEMS DESIGN: NEW PRODUCTS. *Cahners Publishing Company. Monthly. v.14, no.4. April 1982. pg.57-67*

Reports on recent products in optical and laser technologies. Gives manufacturers, their cities and states, and some illustrations.

3700. ELECTRO-OPTICAL SYSTEMS DESIGN VENDOR SELECTION ISSUE, 1979-80. *Milton S. Kiver Publications, Inc. Annual. November 1979. 162 pg.*

1979 directory of over 600 products and 1200 suppliers of electro- optical systems. Includes radiation detectors, image sensors and detectors, lasers, incoherent radiation sources, image sources and information displays, active optical components and assemblies, electronic support components and equipment, testers and test systems, meters and indicators, support equipment, hardware, materials and chemicals, services, systems, and a list of companies, sales offices, representatives and distributors.

Energy User News

3701. ENERGY MANAGEMENT SYSTEMS. *Fairchild Publications. Irregular. February 16, 1981. pg.12-13.*

Technology report on energy management systems. Lists major suppliers, with addresses, of energy management systems and new products introduced by manufacturers in 1981.

3702. HEAT SCANNERS. *Fairchild Publications. Irregular. February 2, 1981. pg.12-13*

Technology report on infrared heat scanning equipment. Lists new products introduced in 1981 by 8 manufacturers. Includes a listing of major suppliers of equipment and services with addresses and phone numbers.

3703. MAJOR SUPPLIERS OF INDUSTRIAL PROCESS EMS. *Fairchild Publications. Annual. v.7, no.34. August 23, 1982. pg.14*

Alphabetically lists major industrial process energy management system suppliers. Includes addresses.

Foundry Management and Technology

3704. FOUNDRY MANAGEMENT AND TECHNOLOGY'S PRODUCT SPOTLIGHT: FOUNDRY INSTRUMENTATION. *Penton / IPC. Subsidiary of Pittway Corporation. Irregular. v.109, no.7. August 1981. pg.58-60*

Provides a brief description of scientific instruments available from U.S. manufacturers that are used in the foundry industry.

3705. FOUNDRY MANAGEMENT AND TECHNOLOGY'S 1981 WHERE-TO-BUY DIRECTORY. *Penton / IPC. Subsidiary of Pittway Corporation. Annual. v.109, no.9. September 1981. 172 (est.) pg.*

1981 annual directory of products and equipment available to the U.S. foundry industry. Directory is arranged in 6 parts: manufacturers' addresses and phone numbers; classified products list; trade names directory; advertisers' directory; agents, dealers and local sources; and a catalog file reference directory.

Gas Industries

3706. GAS INDUSTRIES: ANNUAL METERING AND INSTRUMENTATION ISSUE. *Gas Industries. Annual. August 1980.*

Annual review of developments for gas Btu measurements, metering, and instrumentation, with attention to new products for measurement, regulation, and control.

3707. GAS INDUSTRIES: NEW PRODUCT REVIEW AND RESEARCH ISSUE. *Gas Industries. Annual. December 1980.*

Annual special issue reporting on new products in the gas industry. Includes instrumentation, measurement, hydraulic tools, communication, construction equipment, plus a review of gas industry research projects.

In Tech Instrumentation Technology

3708. IN TECH: ISA CONFERENCE AND EXHIBIT 1980. *Instrument Society of America. Annual. September 1980. 334 pg.*

1980 annual conference report of the Instrument Society of America includes a section on products with manufacturers' names. Brief descriptions and illustrations provided. Section on exhibitors indicates company name, with addresses and specialty.

Industrial Research and Development

3709. LABCON "82 EXHIBITORS AND PLAN OF EXHIBIT AREA. *Technical Publishing Company. Irregular. v.24, no.8. August 1982. pg.107*

Lists manufacturers and suppliers exhibiting at the Labcon "82 conference, including manufacturers and suppliers of scientific and technical instruments.

3710. 1979 YELLOW PAGES OF INSTRUMENTATION EQUIPMENT AND SUPPLIES. *Dwyer, William J., editor. Technical Publishing Company. Annual. January 1979. 210 pg. $15.00.*

Lists manufacturers alphabetically, addresses and telephone numbers. Includes detailed product listing.

3711. 1980'S: TRENDS IN ANALYTICAL INSTRUMENTS. *Technical Publishing Company. Irregular. February 1980. pg.160-164*

Statistics, on the use of 33 types of analytical instruments and equipment, supplied by the readers of Industrial Research / Development. Present use and plans to acquire within one year are detailed in charts.

3712. THE 1981 TELEPHONE DIRECTORY OF INSTRU-MENTATION, EQUIPMENT AND SUPPLIERS. *Technical Publishing Company. Annual. January 15, 1981. 286 pg.*

Annual 1981 directory containing alphabetical listings of instrumentation, equipment, supplies, components, materials and services. Provides manufacturers' addresses and telephone numbers.

Industrial Water Engineering

3713. INDUSTRIAL WATER ENGINEERING: NEW PRODUCTS. *Wakeman / Walsworth Inc. Bi-monthly. November 1981. pg.30-31*

A bi-monthly listing of water engineering equipment and products such as filters, boilers, degasifiers, flowmeters, pumps, etc. Includes manufacturer, description, and photograph. Further information available through reader card service.

3714. INDUSTRIAL WATER ENGINEERING: SPECIAL CODING TOWER ISSUE: NEW PRODUCTS. *Wakeman / Walsworth Inc. May 1981. pg.31-33*

Surveys new coding tower components and processes, such as aqueous chemicals, tower metering and controlling devices, packings, pumps, and electronic ozone generators.

InTech

3715. GUIDE TO SELECTING COMBUSTION STACK GAS ANALYZERS. *Instrument Society of America. Irregular. v.29, no.4. April 1982. pg.25-51*

Guide reviews combustion stack gas analyzers. Gives technical specifications and manufacturers for each product group. Source is a survey by InTech, 1982.

3716. IN TECH: PRODUCTS. *Instrument Society of America. Monthly. v.29, no.4. April 1982. pg.63-85*

Reviews new products for the instrumentation and control field. Includes product description and manufacturer's name and address.

3717. INSTRUMENT SOCIETY OF AMERICA / 82: INDEX TO EXHIBITORS. *Instrument Society of America. Annual. v.29, no.9. September 1982. pg.295-323*

Directory lists exhibitors at the Instrument Society of America's 1982 conference. Includes manufacturer's name and address. Additional listings present manufacturers by product and service categories.

Laboratory Management

3718. LABORATORY MANAGEMENT: 1980 GOLD BOOK. *United Business Publications, Inc. Annual. December 1979. 70 (est.) pg.*

Annual directory and buyers guide for clinical laboratory supplies and services. Lists products and addresses of manufacturers.

Laser Focus with Fiberoptic Communications

3719. FIBEROPTIC PRODUCTS. *Advanced Technology Publications. Monthly. 1981. $33.00/yr.*

Monthly product review including devices and instruments for use in fiberoptic technology. Each entry includes description and specifications, photograph, price and manufacturer.

3720. FIBEROPTICS PRODUCTS. *Advanced Technology Publications. Monthly. v.18, no.8. August 1982. pg.98-103*

Directory of new fiber optic products provides descriptions of products, names of manufacturers, and pricing information.

3721. LASER PRODUCTS. *Advanced Technology Publications. Monthly. v.18, no.8. August 1982. pg.58-66*

Directory of new laser products provides descriptions of products, names of manufacturers, and pricing information.

3722. LASER PRODUCTS. *Advanced Technology Publications. Monthly. 1981. $33.00/yr.*

Monthly product review including devices and instruments for use in laser research. Each entry includes description and specifications, price, and manufacturer.

Lloyd's Ship Manager

3723. LLOYD'S SHIP MANAGER: NEW EQUIPMENT. *Lloyd's of London Press Ltd. Bi-monthly. 1981.*

Lists new maritime products and equipment, including radio navigation systems, communications equipment, etc. Data includes address of manufacturer, photograph and description, and prices. Further information available through reader card service.

3724. PRODUCT FOCUS: COMPASSES. *Lloyd's of London Press Ltd. Irregular. August 1981. pg.13*

A directory of maritime compass manufacturers. Includes gyrocompasses, magnetic, and digital repeaters. Data includes address of manufacturer / supplier; trade name and specifications; price in July 1981; and service representative.

3725. PRODUCT FOCUS: SPEED LOGS. *Lloyd's of London Press Ltd. Irregular. September 1981. pg.23-30*

A directory of maritime speed logs including single-axis doppler and electromagnetic systems. Data includes address of manufacturer / supplier; trade name, model number, and specifications; price in August 1981; and international service and sales representative.

Marine Engineering / Log

3726. MARINE EQUIPMENT NEWS. *Simmons-Boardman Publishing Company. Monthly. v.87, no.3. March 1982. pg.129-143*

Reviews new equipment in the marine engineering field. Includes manufacturer's name and address and product description with photograph.

Materials Evaluation

3727. MATERIALS EVALUATION: NEW NDT PATENTS. *American Society for Nondestructive Testing. Monthly. December 1981.*

Covers new patents for nondestructive testing equipment and products. Data includes patent number, inventor, and description.

Measurements and Control

3728. ELECTRO 1982 / HIGHLIGHTS. *Measurements and Data Corporation. v.16, no.2, issue 92. April 1982. pg.102-123*

Exhibitors at the Electro / 82 Convention. Lists manufacturers, product lines, and technical specifications. Includes bibliographic information.

3729. M & C BUYERS GUIDE, 1980: PART 1. *Measurements and Data Corporation. Annual. February 1980. pg.145-217*

Annual measurements and control equipment buyers guide includes manufacturing data on the following equipment: control systems and elements; digital plotters; integrated-circuit thermometers; moisture and humidity instruments; oscilloscopes; potentiometric pressure transducers; radiation thermometers and pyrometers; test meters, analog and panel meters; torque measurements; and turbine flowmeters. Addresses and phone numbers given.

3730. M & C BUYERS GUIDE, 1980: PART 2. *Measurements and Data Corporation. Annual. April 1980. pg.162-236*

A 1980 buyers guide to measurements and control equipment surveying the following areas: counters, frequency, tabulators; digital pressure transducers; electromagnetic flowmeters; galvanometer direct writing recorders; level measurement and control; lock-in amplifiers; pH meters, industrial; resistance thermometers, thermistors; strain-gage pressure transducers; and timing or programming equipments. Manufacturers data is given with addreses and phone numbers.

3731. M & C BUYERS GUIDE, 1980: PART 3. *Measurements and Data Corporation. Annual. June 1980.*

Annual measurements and control equipment buyers guide including: anemometers; bridges and potentiometers; capacitive pressure transducers; control valves; linear array recorders; magnetic measurements, materials, shielding; recorder charts, pens, accessories; spectrum analyzers; thermocouples; viscometers; and vortex flowmeters. Manufacturers' data including addresses and phone numbers given.

3732. M & C BUYERS GUIDE, 1980: PART 4. *Measurements and Data Corporation. Annual. September 1980.*

An annual measurements and control equipment buyers guide covering the following: acoustics; differential pressure flowmeters; encoders, digitizers; filled-system thermometers; filters; hardness testers; light beam recorders; mass flowmeters; reluctive and LVDT pressure transducers; rotameter and variable-area flowmeters; and standard R / L / C / EMF. Manufacturers addresses and phone numbers given.

3733. M & C BUYERS GUIDE, 1980: PART 5. *Measurements and Data Corporation. Annual. October 1980. pg.151-211*

A 1980 buyers guide to measurements and control equipment including: annunciators; bimetallic thermometers; flow detectors; mass or force, load cells; manometers; piezoelectric pressure transducers; positive displacement flowmeters; tape recorders, data acquisition; transient or surge protectors; ultrasonic flowmeters; and X-Y recorders. Addresses and phone numbers of manufacturers supplied.

3734. M & C BUYERS GUIDE, 1980: PART 6. *Measurements and Data Corporation. Annual. December 1980.*

An annual buyers guide to measurements and control equipment, covering the following: accelerometers and vibration; bourdon pressure gauges; deadweight testers and tensile testers; digital panel meters, voltmeters, multimeters; dimensional gauging and proximity sensors; open-channel flowmeters; potentiometric (stereo) recorders; signal generators; temperature-sensitive, paints, crayons, etc; and vacuum. Addresses and phone numbers of manufacturers given.

3735. MEASUREMENTS AND CONTROL BUYERS GUIDE 1982: PART 2. *Measurements and Data Corporation. Annual. v.16, no.2, issue 92. April 1982. pg.179-268*

Buyers guide for measurement and control devices. Reviews the technology. Lists manufacturers, addresses, executives, product lines, technical specifications, and prices.

3736. MEASUREMENTS AND CONTROL BUYERS GUIDE: PART 3. *Measurements and Data Corporation. Annual. v.16, no.3. June 1982. pg.196-276*

Surveys new U.S. measurement and control products including: anenometers; bridges / potentiometers; capacitive pressure transducers; control valves; linear array recorders; magnetic measurements materials, shielding; recorder charts, pens, accessories; spectrum analyzers; thermocouples; viscometers; and vortex flowmeters. Lists manufacturers' names, addresses, and phone numbers. Includes technical specifications with basic ranges, principles and starting prices. Also provides a 1982 market analysis.

3737. MEASUREMENTS AND CONTROL BUYERS' GUIDE - PART 4. *Measurements and Data Corporation. Monthly. Issue 94. September 1982. pg.193-268*

Directory presents measurement and control equipment available in the U.S. in 1982. Provides estimated 1982 sales for 11 product groups. Lists products, technical specifications, manufacturers' addresses, and contact personnel.

Medical Laboratory Observer

3738. MLO CLINICAL LABORATORY REFERENCE. *Medical Economics Company. Annual. 7th ed. June 1980.*

A reference directory for clinical laboratories. Lists products and manufacturers.

Metal Progress

3739. METAL PROGRESS TESTING AND INSPECTION BUYERS GUIDE AND DIRECTORY 1981. *American Society for Metals. Annual. 2nd ed. February 1981. 178 pg.*

For the U.S. and Canada, gives addresses, phone numbers, products and services in a company directory, equipment and supplies directory and commercial services directory. Includes glossary of terms relating to testing and inspection.

3740. METAL PROGRESS 1980 TESTING AND INSPECTION BUYERS GUIDE AND DIRECTORY. *American Society for Metals. Annual. 1st ed. 1980. 138 pg.*

Annual buyers guide and directory for metal testing and inspection equipment and supplies. Includes an alphabetical listing of companies that sell testing and inspection equipment and supplies; a directory of sources of specific types of equipment and supplies; an alphabetical listing by state of commercial laboratories offering services; and a directory of sources of specific types of laboratory services.

Microwaves

3741. 1980 GAAS FET PRODUCT GUIDE. *Hayden Book Company, Inc. Annual. February 1980. pg.62-80*

Product guide with technical information on GaAs FET's, per-

formance, and new products. Includes directory of manufacturers with addresses.

Mining Engineering

3742. ALPHABETICAL GUIDE TO THE EXHIBITORS. *Society of Mining Engineers of AIME. Annual. v.33, no.10. October 1981. pg.1489-1494*

Alphabetical listing of exhibitors at the 1981 SME-AIME meeting. Products and services are listed for each company, including machinery, instruments and engineering services of interest to the mining industry.

Money

3743. THE BEST: WHICH WATCH. *Trunzo, Candace E. Time, Inc. v.10, no.11. November 1982. pg.181-182*

Product review features high quality watches. Watches in the $500 to $9000 range are reviewed and ranked according to quality and investment value, based on opinions of jewelers, watch repairers, horologists and time measurement specialists. Popular product groups are discussed.

Motor Age

3744. STATE OF THE ART: HAND HELD TESTERS. *Chilton Book Company. Irregular. v.101, no.6. June 1982. pg.44-46*

Illustrates and describes small automotive testing machines. Includes manufacturers and some prices.

Multichannel News

3745. REPORT SAYS TEST EQUIPMENT MANUFACTURERS TO BENEFIT FROM AT AND T RULING. *Frost and Sullivan. Fairchild Publications. v.3, no.39. October 4, 1982. pg.36*

Describes Frost and Sullivan report on communications test equipment, 1982. Gives dollar value and market share, 1980, 1981; and annual growth rate, 1982-1986. Provides 3 fastest growing market segments.

National Mall Monitor

3746. EUN READER RATINGS OF ENERGY MANAGEMENT SYSTEMS. *National Mall Monitor, Inc. v.12, no.4. September 1982. pg.41*

Survey, with a total of 1036 responses, asked readers to rate energy management system manufacturers based on their personal contact and experience with these companies. Companies are listed under three headings: high recognition companies (more than 100 respondents); medium recognition (40 or more respondents); and low recognition (39 or few respondents). Companies ranked according to highest scores (arrived at by averaging the joint value of all responses).

Optical Engineering

3747. OPTICAL ENGINEERING: NEW PRODUCTS. *International Society of Optical Engineers. Bi-monthly. v.20, no.5. September 1981.*

Reviews new products in the field of optical engineering, including fiber optics, lasers, image sensors; video; etc. Data includes photograph, descriptions, and name and address of manufacturer. Gives technical specifications and price, when available.

Pharmaceutical Technology

3748. PHARMACEUTICAL TECHNOLOGY: NEW PRODUCTS. *Pharmaceutical Technology. Monthly. v.6, no.4. April 1982. pg.106-113*

Lists new products related to pharmaceutical technology. Listing includes brief description of product as well as name and address of manufacturer. Typical products listed include airless spray system, solid glass spheres, abrasive mats and incubators.

3749. PHARMACEUTICAL TECHNOLOGY: 1982 BUYERS' GUIDE. *Pharmaceutical Technology. Annual. v.6, no.7. July 1982. 204 pg. $4.00.*

List manufacturers and suppliers of chemical raw materials and pharmaceutical ingredients, manufacturing equipment and supplies, packaging materials, lab instrumentation. Includes alphabetical directory of manufacturers with addresses and telephone numbers.

Photographic Trade News

3750. DIGITAL IMAGING - REQUIEM FOR A SNAPSHOT? *Fernandez, Brian. First Manhattan Co. PTN Publishing Corporation. v.26, no.1. January 4, 1982. pg.26-27, 29*

Discusses the technology, costs, and expected applications and picture quality of digital imaging systems. Examines the flexibility of image manipulation and correction; and market potential for the instant - picture, scientific, and medical sectors.

Plastics Design and Processing

3751. PLASTICS DESIGN AND PROCESSING: NEW EQUIPMENT. *Lake Publishing Corporation. Bi-monthly. v.22, no.4. April 1982. pg.42-47*

A bi-monthly listing of equipment used in plastics processing operations. Gives manufacturer's address, description, and photograph. Further information available through reader card service.

Plastics Technology

3752. PROCESS ENGINEERING NEWS. *Bill Communications, Inc. Monthly. v.28, no.5. May 1982. pg.41-45, 148-165*

Regular monthly directory lists and describes new process engineering equipment. Gives names and addresses of manufacturers, brand names, and prices.

Pollution Engineering

3753. 1980 ENVIRONMENTAL YEARBOOK AND PRODUCT REFERENCE GUIDE. *Technical Publishing Company. Annual. December 1979. pg.83-226*

1980 environmental yearbook and product reference guide provides a listing of products used in pollution control in the U.S., and their manufacturers. Product guide, divided into 5 sections, lists instrument and control products, air pollution control products, wastewater treatment products, noise control products, and waste treatment products. Company names, addresses, and telephone numbers are listed alphabetically in manufacturers section. Provides manufacturers literature, fed-

eral plans and standards and key personnel in agencies, information sources, environmental books, and upcoming meetings.

Power Engineering

3754. PLANT ENGINEERING EQUIPMENT. *Britton, Phyllis A., ed. Technical Publishing Company. Irregular. v.36, no.10. May 13, 1982. pg.26-50*

Gives descriptions of new products for plant engineering, including manufacturers.

Product Design and Development

3755. PRESSURE TRANSDUCERS. *Chilton Book Company. Irregular. July 1981.*

Special issue on pressure transducer products.

Progressive Grocer

3756. EQUIPMENT GUIDE 1982. *Progressive Grocer. Annual. v.60, no.12. December 1981. pg.33-85*

Annual equipment guide featuring reviews of new equipment for the retail food store. Statistics include percent of purchases in each of 31 equipment categories; overview of the market for scanners, with market shares by manufacturers and sales, 1975-1981; overview of the market for energy management systems (EMS) with system prices and cost of operation; and results of a survey showing number of installations by type of system and applications. Includes an equipment catalog featuring new products, with descriptions and photographs; guide to equipment manufacturers, with company names; addresses and product lines, arranged by product group.

Quality

3757. EXPO TIME EXHIBITOR PRODUCTS SECTION. *Hitchcock Publishing Company. Annual. v.21, no.4. April 1982. pg.97-150*

Reviews products to be displayed at Quality Expo TIME 1982. Includes manufacturer's name, booth number, and product description.

3758. ISA "82 PREVIEW. *Hitchcock Publishing Company. Irregular. v.21, no.10. October 1982. pg.58-62*

Preview of the Instrument Society of America conference and exhibition includes new products and regular product lines to be exhibited by manufacturers, with technical specifications.

3759. QUALITY EXPO TIME PROGRAM: EXHIBITOR INFO. *Hitchcock Publishing Company. Annual. v.21, no.4. April 1982. pg.Q31-Q95*

Lists manufacturers exhibiting products in the quality control field at Quality Expo TIME 1982. Includes products to be displayed; manufacturer's name, address, and telephone number; booth number; and representatives.

3760. QUALITY: INSTRUMENTATION SCOUTING REPORT. *Hitchcock Publishing Company. Monthly. v.21, no.4. April 1982. pg.83-84*

Reviews new products in the quality control field. Gives technical specifications, product description, and manufacturer's name.

Sea Technology

3761. SEA TECHNOLOGY: PRODUCT DEVELOPMENT. *Compass Publications Inc. Monthly. v.23, no.8. August 1982. pg.86-87*

Monthly directory of new technical and scientific instruments describes new products and their end uses and manufacturers.

Seaway Review

3762. MARITIME PRODUCTS FOR THE GREAT LAKES. *Great Lakes Press. Quarterly. v.11, no.1. September 1981. pg.131-132*

Surveys new products of interest to commercial shipping companies, with product descriptions and manufacturers' addresses. Navigational instruments, coatings, and freight handling equipment are featured.

Semiconductor International

3763. AMERICAN VACUUM SOCIETY EXHIBITORS AND BOOTH NUMBERS. *Cahners Publishing Company. Annual. v.5, no.10. October 1982. pg.105-116*

Lists exhibitors and their booth numbers for the American Vacuum Society 29th National Symposium (Baltimore, Maryland, November 16-18, 1982). Many of the products are briefly described in a special product preview section. Price of products described also included.

Sound and Vibration

3764. BUYER'S GUIDE TO SYSTEMS FOR NOISE AND VIBRATION CONTROL. *Acoustical Publications, Inc. Irregular. August 1980.*

A listing of manufacturers and suppliers of systems for the control of noise, vibration, and mechanical shock including sound absortive systems, sound barriers, composite systems, silencers, and vibration isolation systems. Gives company's address and telephone number.

3765. DYNAMIC MEASUREMENT INSTRUMENTATION BUYER'S GUIDE. *Acoustical Publications, Inc. Irregular. March 1980. pg.14-17*

Lists manufacturers and suppliers of dynamic measurement instrumentation grouped as analyzers, calibrators, control systems, generators, meters, recorders and displays, signal conditioners, special purpose systems, and transducers. Gives addresses and telephone numbers of all companies.

3766. DYNAMIC TESTING EQUIPMENT BUYER'S GUIDE. *Acoustical Publications, Inc. Annual. November 1981. pg.28-29*

Lists companies providing structural analysis and dynamic testing hardware or software, including finite element analysis systems, physical testing machines, signal generators, simulators, structural analysis systems, test fixtures and accessories, transducers, vibration exciters, etc. Data includes company name and address, arranged by hardware or software type.

3767. HEARING CONSERVATION EQUIPMENT BUYER'S GUIDE. *Acoustical Publications, Inc. Irregular. January 1980. pg.21*

Lists names and addresses of firms producing audiometers, audiometric systems, hearing protection devices, noise expo-sure monitors, noise dosimeters, sound level meters, and sound level calibrators.

Water and Pollution Control

3768. WATER AND POLLUTION CONTROL: DIRECTORY AND ENVIRONMENTAL HANDBOOK: 1981. *Southam Business Publications, Ltd. Division of SouthamCommunications Ltd. Annual. November 1980. pg.24-110 $2.00.*

Annual directory of Canadian and U.S. suppliers of pollution control equipment and services to the Canadian market. Firms are listed alphabetically, with head office. Provides subsidiaries' and distributors' addresses and phone numbers. Contains a cross-listing by product group. Separate sections list government agencies and their personnel, engineering consultants, and environmental and trade associations.

Wire Journal

3769. WIRE JOURNAL MEASURING, TESTING AND INSPECTING EQUIPMENT DIRECTORY. *Wire Journal, Inc. Irregular. March 1981. pg.70-85*

Directory of testers, monitors, gauges, controls and similar equipment used in the manufacture of wire and cable.

See 272, 273, 315, 316, 318, 350, 356, 357, 370, 377, 1535, 1569, 1636, 1643, 1662, 1709, 1712, 1800, 1804, 1810, 2589, 2596, 2610, 2613, 2615, 2618, 2622, 2645, 2687, 2690, 2691, 2693, 2698, 2716, 2721, 2723, 2728, 2738, 2741, 2748, 2749, 2758, 2763, 2765, 2766, 2769, 2773, 2777, 2783, 3345.

Monographs and Handbooks

3770. FIBER OPTICS PATENT DIRECTORY: 1881-1979. *Patent Data Publications. 1980. 161 pg.*

Lists and cross-references all U.S. patents for fiber optics devices issued from 1881 to October 30th, 1979, by the 25 relevant Patent Office subclasses (class 350, subclasses 96.1 to 96.34). Within subclasses, patents are listed alphabetically by patent holder. Listings contain patent number and descriptive title.

3771. FIBER OPTICS USER SERVICE. *Gnostic Concepts, Inc. Annual. 1981. $15000.00.*

Annual service for users of fiber optics. Covers products, suppliers, technology forecasts, technical specifications, prices, and utilizations.

3772. THE PROBE GUIDE TO OPTICAL CABLE. *Probe Research, Inc. January 1981. 25 pg. $50.00.*

Guide to optical cable products and manufacturers. Gives performance and technical specifications, utilization for data, telephone and CATV, market analysis, guidelines for suppliers, determinants of purchases, and regulations.

See 2799, 2804, 2805, 2810, 2811.

Newsletters

3773. FIBER / LASER NEWS. *Phillips Publishing Inc. Biweekly. $147.00/yr.*

Biweekly newsletter covering fiber optics and lasers. Includes management and marketing strategies for new technologies.

Indexes and Abstracts

3774. PROCESS ENGINEERING UPDATE. *Predicasts, Inc. Weekly. 1982.*

Weekly abstracts of articles on process engineering in chemical and allied industries. Includes new technology in plants and equipment, capital expenditures, equipment costs, energy conservation and analysis of markets.

3775. VENTURECASTS: THE 1982 INSTRUMENTATION DATA BOOK. *Venture Development Corporation. 1982. $595.00.*

Provides abstracts of market analysis and forecasts from 1981 electronics and business periodicals, and associations' and government statistics covering instruments in the Americas. Includes sales, shipments, revenues, installed base, annual growth rates, consumption, expenditures, market size, production, and imports and exports by product groups.

Dictionaries

3776. ILLUSTRATED GLOSSARY OF PROCESS EQUIPMENT. *Paruit, Bernard H., ed. Gulf Publishing Company. English/French/Spanish. January 1982. 317 pg. $39.95.*

Tri-lingual glossary of terms for process plants, equipment and technology.

Central and South America

Market Research Reports

3777. CLINICAL LABORATORY MARKETS IN LATIN AMERICA: BRAZIL, ARGENTINA AND CHILE. *Frost and Sullivan, Inc. September 1981. 231 pg. $1400.00.*

Provides comprehensive background of the clinical laboratory field and detailed historic and forecast unit and dollar markets. Offers analysis of national and regional characteristics as they relate to marketing strategies, pricing, and competition. Estimates number of instruments in the field by model and manufacturer. Gives historic and forecast unit and dollar sales for instruments by market segment.

3778. INDUSTRIAL PROCESS CONTROLS, BRAZIL: A COUNTRY MARKET SURVEY. *U.S. Government Printing Office. Irregular. CMS 79-205. January 1979. 11 pg. $.50.*

Analyzes the growth of key industries in Brazil and the current and projected markets for the sale of industrial process controls in that country. Data is provided on the total market (production, imports, exports) by major category; demand prospects for specific controls by selected end user industries; imports by category by country of origin; domestic establishments manufacturing controls; composition of the process industries; output, capital expenditures, plant capacity utilization; market by user industries; products with good sales potential for U.S. manufacturers. Describes the competitive environment, major industrial users, and market access.

3779. INDUSTRIAL PROCESS CONTROLS, VENEZUELA: A COUNTRY MARKET SURVEY. *U.S. Government Printing Office. CMS 79-214. February 1979. 12 pg. $.50.*

Analyzes the current and projected markets for the sale of industrial proccess controls in Venezuela. Data is provided on the total market (production, imports, exports, market size); imports by country of origin, output, capital expenditures, plant capacity, utilization by user industries; size and composition of

process industries; market by user industries; size and demand for specific controls by selected industries; products with good sales potential for U.S. manufacturers. Describes the domestic market, competitive environment, major industrial users, market access data.

3780. LABORATORY INSTRUMENTS: VENEZUELA: COUNTRY MARKET SURVEY. *U.S. Department of Commerce. Sales Branch. Irregular. CMS 79-603. December 1979. 12 pg. $.50.*

Analyzes the current and potential market for laboratory instruments in Venezuela. Describes current and future research in government, educational, nonprofit, and industrial laboratories and provides data on laboratory expenditures and purchases of instruments for 1977 and 1982 (projected). Provides data on imports of selected instruments, by country of origin, for 1975-1977, and sales of selected instruments for U.S. and for all others for 1977 and 1982 (projected).

3781. LASER AND ELECTRO-OPTICAL EQUIPMENT (BRAZIL). *Lindsey, Richard P. U.S. Department of Commerce. International Trade Administration. NTIS #ITA 82-02-503. 1981. 6 pg.*

Summary of the current and future market situation for laser and electro-optical equipment lists specific types of equipment with good sales potential. Data provided on the total market size (domestic production, imports, exports) for 1979-1980 and forecast for 1985.

3782. PROFILED KEY MARKETS FOR U.K. PROCESS CONTROLS EXPORTS: VOLUME 2: PXCB. *Tactical Marketing Ltd. 1981. 120 pg. 95.00 (United Kingdom).*

Market research report of markets for United Kingdom process control exports to Brazil, France, Israel, Singapore, South Korea, and West Germany. Covers process instrumentation, equipment, valves, and computers. Also includes market size, forecasts to 1985, process definitions, economic effects, usage trends, domestic manufacturers, other competition, buyer attitudes, distributors, regulations and constraints, and contacts and ipportunities.

See 410, 2822.

Financial and Economic Studies

3783. LABORATORY INSTRUMENTS, BRAZIL: COUNTRY MARKET SURVEY CMS 79-601. *Industry and Trade Administration U.S. Department of Commerce. U.S. Government Printing Office. August 1979. 9 pg. $.50.*

One of a series of industry studies to assist U.S. exporters. Covers all types of laboratory instruments used in Brazil by government, educational and nonprofit institutions, independent commercial laboratories, medical organizations, and major industries. Tables show production, imports worldwide and from U.S., exports, market size, purchases by country and by type of equipment, for 1975 or 1976-1977, with forecasts for 1982. Data is derived primarily from Brazilian government sources. Text covers competitive environment, major purchasers, and trade and technical regulations.

Western Europe

Market Research Reports

3784. ANALOGUE METERS IN EUROPE TO 1986. *Larsen Sweeney Associates Ltd. 1981. 231 pg. 595.00 (United Kingdom).*

Market research report on analogue meters in the following countries: Austria, Belgium, Luxembourg, Denmark, Eire, Finland, France, West Germany, Italy, Netherlands, Norway, Portugal, Spain, Sweden, Switzerland and the U.K. Covers market analysis to 1986, market forecasts to 1986, distribution, production, suppliers, market segmentation, employment and unemployment, profits, sales and end uses. Analogue meters covered include: communications, telecommunications, consumer, EDP, industrial, instrumentation, governmental / defense and others unspecified.

3785. ANALYTICAL INSTRUMENTS IN EUROPE TO 1986. *Larsen Sweeney Associates Ltd. 1981. 229 pg. 695.00 (United Kingdom).*

Market research analysis on analytical instruments to 1986 in the following countries: Austria, Belgium, Luxembourg, Denmark, Eire, Finland, France, West Germany, Italy, Netherlands, Norway, Portugal, Spain, Sweden, Switzerland and the U.K. Gives market forecast to 1986 and covers distribution, production, suppliers, market segmentation, employment and unemployment, profits, sales and end uses. Analytical instruments sectors covered include: electromagnetics, gas, liquids, light, noise, particle, ultrasonic and vibration.

3786. AUTOMATIC TEST EQUIPMENT MARKETS IN WESTERN EUROPE. *Frost and Sullivan, Inc. February 1980. $1250.00.*

Analysis of markets for automatic test equipment in Western Europe. Forecasts utilization by end use industries. Profiles manufacturers.

3787. AUTOMOTIVE DIAGNOSTICS EQUIPMENT MARKETS IN EUROPE. *Frost and Sullivan, Inc. 1979. 278 pg. $950.00.*

Market analysis for automotive diagnostics equipment in Western Europe, 1977-1980, with forecasts for 1983 and 1987. For countries, shows imports, domestic production, exports, and net apparent consumption, 1976-1987. Includes end uses, units in operation, and transportation statistics. Analyzes market penetration opportunities showing manufacturers, trade associations, and tariffs.

3788. BUILDING CONTROL SYSTEMS MARKETS IN EUROPE. *Frost and Sullivan, Inc. September 1981. $1450.00.*

Gives analysis with 1979 base year forecasts 1981, 1985, and 1989 in U.K., France, West Germany, Belgium, Denmark, Sweden, and the Netherlands for building automation systems (multiple, large and small structures), and power management systems (controls and controllers) in five end user market segments. Provides outlook and prospects in a competitive environment.

3789. BUILDING SERVICES AUTOMATION. *Bull, R.S., editor. ERA Technology Ltd. June 1981. 990.00 (United Kingdom).*

Market analysis for automated building services in the U.K. Covers types and utilization patterns for plants and equipment, energy usage, and costs, user attitudes toward installations, and forecasts.

3790. CHART RECORDERS IN EUROPE TO 1986. *Larsen Sweeney Associates Ltd. Quarterly. 1981.*

Market report to 1986 on chart recorders in Western Europe. Report is updated quarterly and covers market analysis, forecasts to 1986, sales, production trends, profits, assets, end users, employment and unemployment, products, capital expenditures, investments and distribution.

3791. CLINICAL LABORATORY APPARATUS MARKETS IN WESTERN EUROPE. *Frost and Sullivan, Inc. April 1982. 150 pg. $1175.00.*

Analysis of markets for 22 clinical laboratory apparatus product groups in Western Europe, with forecasts, 1979-1985. Includes product technology; laboratory industry structure; healthcare professionals; expenditures by end uses; regulations; associations; distribution channels; manufacturers; marketing strategies; and prices.

3792. CLINICAL LABORATORY INSTRUMENTATION MARKETS IN WESTERN EUROPE. *Frost and Sullivan, Inc. April 1982. 150 pg. $1200.00.*

Analysis and forecast of European markets for clinical laboratory instrumentation, 1979-1985. Includes technology reviews for all product groups, distribution and marketing strategies, maintenance services, prices, and business addresses of companies. For each country, includes hospital facilities, domestic production and imports, manufacturers, suppliers, and end uses.

3793. CLINICAL LABORATORY TESTING IN EUROPE - VOLUME 1 - FRANCE. *Robert S. First, Inc. November 1979. 830 pg. 30000.00 (Switzerland).*

Analysis of clinical laboratory testing markets in France.

3794. CLINICAL LABORATORY TESTING IN EUROPE - VOLUME 2 - GERMANY. *Robert S. First, Inc. April 1980. 30000.00 (Switzerland).*

Analysis of clinical laboratory testing markets in Germany.

3795. CLINICAL LABORATORY TESTING IN EUROPE - VOLUME 3 - ITALY. *Robert S. First, Inc. February 1981. 30000.00 (Switzerland).*

Analysis of clinical laboratory testing markets in Italy.

3796. COUNTERS IN EUROPE TO 1986. *Larsen Sweeney Associates Ltd. 1981. 225 pg. 795.00 (United Kingdom).*

Market Analysis of counters in Austria, Belgium, Luxembourg, Denmark, Eire, Finland, France, West Germany, Italy, Netherlands, Norway, Portugal, Spain, Sweden, Switzerland and the U.K. Covers market forecasts to 1986 and data on distribution, production, suppliers, market segmentation, employment and unemployment, profits, sales and end uses. 26 counter product groups are covered and there is a breakdown of 18 electric counters.

3797. DIGITAL METERS IN EUROPE TO 1986. *Larsen Sweeney Associates Ltd. 1981. 221 pg. 495.00 (United Kingdom).*

Market research report on digital meters in Austria, Belgium, Luxembourg, Denmark, Finland, France, West Germany, Italy, Netherlands, Norway, Portugal, Spain, Sweden, Switzerland, and the U.K. Covers forecasts to 1986 for distribution, production, suppliers, market segmentation, employment and unemployment, profits, sales, and end uses. The following application sectors are analyzed: communications, telecommunications, consumer, EDP, industrial, instrumentation, automotive, military / defense, governmental and others.

3798. ELECTRONIC COMPONENT AND CIRCUIT BOARD ATE AND MDS MARKETS IN W. EUROPE. *Frost and Sullivan, Inc. April 1981. 258 pg. $1350.00.*

Contains analysis and forecast to 1990 for consumption of ATE equipment in Benelux, France; U.K.; W. Germany; and the rest of Europe. Includes equipment for testing passive / active components, bareboards, incircuits, and functionals. Includes forecasts by end users including computer, commercial, consumer, government, and components. Contains analysis of high growth markets, MDS, state of the art, and new product developments. Profiles major end user segments and principal suppliers.

3799. ELECTRONIC RECORDERS IN EUROPE TO 1986. *Larsen Sweeney Associates Ltd. Quarterly. 1981.*

Market report to 1986 on electronic recorders in Western Europe. Updated quarterly, the report covers: market analysis, forecasts to 1986, sales, production trends, profits, assets, end users, employment and unemployment, products, capital expenditures, investments and distribution.

3800. ELECTRONICS INDUSTRY PRODUCTION AND TEST EQUIPMENT SWEDEN: MARKET BRIEF. *Konsulterna AB. U.S. Department of Commerce. International Trade Administration. NTIS # ITA 82-04-504. February 1982. 17 pg. $15.00.*

Analyzes the Swedish market for electronics industry production and test equipment. Discusses the market environment, including a list of products representing the potential for U.S. exporters. Examines the competitive environment and marketing channels. Gives end user profiles of the 3 large Swedish industrial groups accounting for production of most of the semiconductors and integrated circuits. Gives data on the market by product categories and projections for 1985. Lists trade associations, government agencies, trade shows, and trade publicat

3801. ENERGY CONSERVATION CONTROLS MARKETS IN THE PROCESS INDUSTRIES IN EUROPE. *Frost and Sullivan, Inc. December 1981. 354 pg. $1450.00.*

Establishes and forecasts consumption of industrial energy conservation controls over the period 1980-1990 in the five major West European countries. Forecasts consumption by major end-user industries. Identifies and ranks (where possible) leading suppliers to Western Europe for each product category.

3802. ERC MARKET MONITOR EUROPE: CLOCKS. *E.R.C. Statistics International Ltd. February 1982. 1000.00 (United Kingdom).*

Western European market analysis for clocks. Includes sales, brand names, prices, advertising, consumer behavior, retail distribution, industry structure, production, manufacturers, imports and exports, new products, and foreign trade.

3803. ERC MARKET MONITOR EUROPE: WATCHES. *E.R.C. Statistics International Ltd. February 1982. 1000.00 (United Kingdom).*

Western European market analysis for watches. Includes sales, brand names, prices, advertising, consumer behavior, retail distribution, industry structure, production manufacturers, imports and exports, new products, and foreign trade.

3804. EXPORTING TO JAPAN: ENERGY SAVING AND CONVERTING EQUIPMENT. *Metra Consulting Group Ltd. 1980. 10.00 (United Kingdom).*

Survey of markets in Japan for exports of energy conservation and conversion equipment. Available only to Western European companies.

3805. FIBRE OPTICS FOR PROCESS CONTROL AND BUSINESS COMMUNICATIONS. *Baker, P.D.W. ERA Technology Ltd. June 1982. 1450.00 (United Kingdom).*

Analysis of markets in the U.K. for short-range fibre optic lines and equipment for industrial processes and business communications. Examines patterns of utilizaton, current installations, user attitudes, performance costs, and manufacturers.

3806. FLOW RECORDERS IN EUROPE TO 1986. *Larsen Sweeney Associates Ltd. Quarterly. 1981.*

Market report to 1986 on flow recorders in Western Europe. Updated quarterly, report covers: market analysis, forecasts to 1986, sales, production trends, profits, assets, end users, employment and unemployment, products, capital expenditures, investments, and distribution.

3807. IMAGING AND LIGHT SENSING TUBES IN EUROPE TO 1986. *Larsen Sweeney Associates Ltd. 1981. 239 pg. 795.00 (United Kingdom).*

Market research report on imaging and light sensing tubes to 1986 in Austria, Belgium, Luxembourg, Denmark, Eire, Finland, France, West Germany, Italy, Netherlands, Norway, Portugal, Spain, Sweden, Switzerland and the U.K. Gives forecasts to 1986 and covers distribution, production, suppliers, market segmentation, employment and unemployment, profits, sales and end uses. Products analyzed include image converters, image intensifiers, image orthicans, and T.V. camera pick ups.

3808. INDICATING DEVICES IN EUROPE TO 1986. *Larsen Sweeney Associates Ltd. 1981. 234 pg. 695.00 (United Kingdom).*

Market research report on indicating devices to 1986 in Austria, Belgium, Luxembourg, Denmark, Eire, Finland, France, West Germany, Italy, Netherlands, Norway, Portugal, Spain, Sweden, Switzerland and the U.K. Gives forecasts to 1986 and covers production, distribution, suppliers, market segmentation, employment and unemployment, profits, sales and end uses. The following types of indicting devices are analyzed: electrochromic, electrophoretic, electroluminescent, fibre-optic, incandescent, LED, LCD, plasma, lamps, cathode ray tubes, edge lit displays and vacuum fluorescence.

3809. INDUSTRIAL CONTROL IN EUROPE TO 1986. *Larsen Sweeney Associates Ltd. 1981. 247 pg. 795.00 (United Kingdom).*

Market research report to 1986 on industrial controls in Austria, Belgium, Luxembourg, Denmark, Eire, Finland, France, West Germany, Italy, Netherlands, Norway, Portugal, Spain, Sweden, Switzerland and the U.K. Gives forecasts to 1986 and distribution, production, suppliers, market segmentation, employment and unemployment, profits, sales, and end uses. Product groups are: data logging, monitoring, display, recording, controllers, equipment and peripherals. Eleven markets are covered including: chemicals and plastics, food / drink / tobacco, nucleonics, steel, motors and motor components.

3810. INDUSTRIAL PROCESS CONTROLS AND ROBOTICS EQUIPMENT MARKET IN THE UNITED KINGDOM. *Tactical Marketing Ltd. 1982. 100 pg. 95.00 (United Kingdom).*

Market research report provides analysis and forecasts of the industrial process controls and robotics equipment market in the United Kingdom. Covers instrumentation, valves, computers, peripherals, and robotics. Subjects discussed include market size, forecasts to 1985, product definitions, the challenge of robotics, economic effects, leadingmakers and products, domestic manufacturing base, foreign inputs, market shares, user segment profiles, trends and developments, access to the market factors, regulations, constraints and special opportunities.

3811. INDUSTRIAL PROCESS CONTROLS, BELGIUM: A COUNTRY MARKET SURVEY. *U.S. Government Printing Office. Irregular. CMS 79-218. September 1979. 5 pg. $.50.*

Analyzes the process industries in Belgium and the current and projected markets for the sale of industrial process controls. Data is provided on the market for controls (production, imports, exports) by major category, and total imports and imports from U.S. by major category. Describes the domestic industry, competitive environment, major industrial users, and market access.

3812. INDUSTRIAL PROCESS CONTROLS, FRANCE: A COUNTRY MARKET SURVEY. *U.S. Government Printing Office. Irregular. CMS 79-215. February 1979. 11 pg. $.50.*

Report analyzes the current and projected markets for the sale of industrial process controls in France. Data is provided on the total market (production, imports, exports) by major category; imports by type by country of origin; size and number of domestic manufacturers; output, capital expenditures, and plant capacity utilization by user industries; size and composition of process industries; market by user industries; demand by end-user industries; and products with good sales potential for U.S. manufacturers. Discusses the competitive environment, major industrial users, and market access data.

3813. INDUSTRIAL PROCESS CONTROLS, GERMANY: A COUNTRY MARKET SURVEY. *U.S. Government Printing Office. Irregular. CMS 79-208. February 1979. 12 pg. $.50.*

Report analyzes the current and projected markets for the sale of industrial process controls in West Germany. Data is provided on the total market (production, imports, exports) by major category; demand prospects by selected end-user industries; imports by country of origin; output, capital expenditures, plant capacity utilization by user industries; size and composition of process industries; market by user industries; and products with good sales potential for U.S. manufacturers. Describes the domestic market, competitive environment, major industrial users, and market access data.

3814. INDUSTRIAL PROCESS CONTROLS IN THE UNITED KINGDOM. *Systec Consultants Ltd. U.S. Department of Commerce. International Trade Administration. NTIS #DIB 80-06-502. February 1980. 122 pg. $15.00.*

Survey of the market for industrial process controls in the United Kingdom provides data on market size by major category (1976-1980, forecast 1983) and on imports both of specific controls by country of origin (1976-1978) and by major product group by country of origin (1978, 1983). Gives size and composition of the domestic manufacturing industry. A table provides the U.K. sales outlook for specific U.S. products and major competing source countries, 1978-1983. Profiles end user sectors and provides data on industry size, production, capital expenditures, capacity utilization, demand, and new construction of plants. Analyzes oil and gas extraction, electrical power generation, chemicals, and petroleum refining sectors in detail. Business practices, technical standards, tariffs are detailed. Trade lists include associations; in all user industries; trade events in all user industries; and journals in all user industries.

3815. INDUSTRIAL PROCESS CONTROLS, ITALY: COUNTRY MARKET SURVEY CMS 81-222. *International Trade Administration. U.S. Department of Commerce. U.S. Government Printing Office. February 1981. 11 pg. $.50.*

Survey of the market for U.S. manufacturers for industrial process controls sales to Italy. Text surveys competitive environment, major industrial users, and market access. Seven tables quantify the industrial process control market by type of product, size of establishment, analysis of the industrial process control industry and market for 1976-1978, with projections for

1983. Includes list of recommended opportunities for U.S. manufacturers.

3816. INDUSTRIAL PROCESS CONTROLS, NORWAY: A COUNTRY MARKET SURVEY. *U.S. Government Printing Office. Irregular. CMS 79-212. February 1979. 12 pg. $.50.*

Report analyzes the current and projected markets for the sale of industrial process controls in Norway. Data is provided on the total market (production, imports, exports); demand prospects for specific controls by selected end-user industries; imports by country of origin; output, capital expenditures, plant capacity utilization by user industries; market by user industries; and products with good sales potential for U.S. manufacturers. Describes the domestic market, competitive environment, major industrial users, and market access data.

3817. INDUSTRIAL PROCESS CONTROLS, SWEDEN: A COUNTRY MARKET SURVEY. *U.S. Government Printing Office. Irregular. CMS 79-213. February 1979. 13 pg. $.50.*

Report analyzes the current and projected markets for the sale of industrial process controls in Sweden. Data is provided on the total market (production, imports, exports); imports by country of origin; output, capital expenditures, plant capacity utilization by user industries; size and composition of process industries; market by user industries; demand prospects for specific controls by end user industries; and products with good sales potential for U.S. manufacturers. Describes the domestic market, competitive environment, major industrial users, and market access data.

3818. INDUSTRIAL PROCESS CONTROLS, SWITZERLAND: A COUNTRY MARKET SURVEY. *U.S. Government Printing Office. Irregular. CMS 81-225. July 1979. 11 pg. $.50.*

Analyzes the current and projected markets for the sale of industrial process controls in Switzerland. Data is provided on the total market (production, imports, exports, market size); imports by country of origin; products with good sales potential for U.S. manufacturers; size and number of domestic manufacturers of process controls; size and composition of process industries; output, capital expenditures, plant capacity, utilization by user industries; market by user industries; demand prospects for specific controls by selected end-user industries. Describes the domestic market, competitive environment, major industrial users, market access data.

3819. INDUSTRIAL PROCESS CONTROLS, UNITED KINGDOM: A COUNTRY MARKET SURVEY. *U.S. Government Printing Office. Irregular. CMS-81-223. April 1979. 11 pg. $.50.*

Analyzes the current and projected markets for the sale of industrial process controls in the United Kingdom. Data are provided on the total market (production, imports, exports, market size); imports by country of origin; products with good sales potential for U.S. manufacturers; size and number of domestic manufacturers; size and composition of process industries; output, capital expenditures, plant capacity, utilization by user industries; market by user industries; demand prospects for specific controls by selected industries. Describes the domestic market, competitive environment, major industrial users, and market access data.

3820. INDUSTRIAL SENSORS MARKETS IN EUROPE. *Frost and Sullivan, Inc. January 1981. 262 pg. $1400.*

Contains analysis and forecast of 22 types of temperature and humidity, pressure, position, force, flow, level and other sensors, 1979 to 1990. Includes estimates for 15 processes and manufacturing industries. Includes W. Germany, France, U.K., Benelux and the rest of Europe. Contains growth products, application trends, and 36 company profiles.

3821. INDUSTRIAL TEMPERATURE CONTROL MARKET IN EUROPE. *Frost and Sullivan, Inc. December 1979. 414 pg. $1250.00.*

Two-volume market report contains analysis and forecast through 1988 in Germany, France, Netherlands, U.K., and Italy for primary sensors - thermocouples, thermowells, RTD, thermistors, pyrometers, transmitters - TC and RTD to current filled system; controllers - discrete, GP, recorder, trip amplifiers, mechanical regulation; and programmables - PLC's and PC's, temperature controllers. Includes end user section breakdown for process industries. Gives supplier profiles and market shares by product and company. Contains comparison of U.S. and Japan with market opportunities and future dangers.

3822. INDUSTRIAL ULTRASONIC MACHINERY MARKETS IN EUROPE. *Frost and Sullivan, Inc. April 1982. 225 pg. $1550.00.*

Analysis of industrial ultrasonic machinery in Europe. Forecasts, 1981-1987, for end uses in plastics, metals, marine instruments, process industries, detection equipment and switches, covering 20 ultrasound product groups. Describes the technology, industry structure, manufacturers, national markets, and foreign trade.

3823. INSTRUMENTATION RECORDERS IN EUROPE TO 1986. *Larsen Sweeney Associates Ltd. Quarterly. 1981.*

Market report to 1986 on instrumentation recorders in Western Europe. Updated quarterly, the report covers: market analysis, forecasts to 1986, production trends, sales, profits, assets, end users, employment and unemployment, products, capital expenditures, investment, and distribution.

3824. KEY NOTE REPORTS: ELECTRONIC INSTRUMENTS. *Keynote Publications Ltd. 1981. 20.00 (United Kingdom).*

A report on electronic instruments in the U.K. covering the structure of the industry, distribution networks, major manufacturers, markets, and finances. Analysis includes products by sales volume and value, market shares, foreign trade, new products, mergers, legislation, financial analysis, operating ratios, and forecasts of trends.

3825. LAB AMPLIFIERS IN EUROPE TO 1986. *Larsen Sweeney Associates Ltd. 1981. 251 pg. 995.00 (United Kingdom).*

Market report to 1986 on lab amplifiers covering 45 types in Austria, Belgium, Luxembourg, Denmark, Eire, Finland, France, West Germany, Italy, Netherlands, Norway, Portugal, Spain, Sweden, Switzerland and the U.K. Gives market analyses, forecasts to 1986, production, market segmentation, sales, distribution, employment and unemployment, profits, end users and suppliers.

3826. LABORATORY INSTRUMENTATION IN BELGIUM: A FOREIGN MARKET SURVEY REPORT. *Sira Institute Ltd. U.S. Department of Commerce. International Trade Administration. NTIS# ITA 82-03-508. August 1981. 167 pg. $15.00.*

Survey on the market in Belgium for laboratory instruments describes the total market for laboratory instruments and then analyzes specific product markets with future projections. Provides data on imports of each instrument, giving major country suppliers, actual value, percent market share, percent for lab use, and amount for lab use. Gives a competitive assessment of U.S. firms and third country suppliers, with important firms and their products listed. A section of the report provides an end-user analysis for both research and development and by industry group. Information is provided on trade practices and regulations and technical requirements. Lists trade events; exhibitions, conferences, and seminars; trade, industrial, and professional associations; trade publications; distributors and

agents for laboratory instruments; and members of trade association UDIAS.

3827. LABORATORY INSTRUMENTATION IN THE NETHERLANDS. *Sira Institute Ltd. U.S. Department of Commerce. International Trade Administration. NTIS #ITA 81-1-506. July 1981. 181 pg. $15.00.*

Provides data on current (1978-1981) and projected (1985) market size, imports by type of instrument and country of origin (1978-1980), actual and projected imports from U.S., sales by geographic origin of instrument (1980, 1985), government and private research and development funds, estimated laboratory expenditures and equipment purchases by major end users. Analyzes specific product markets and provides a competitive assessment of U.S., domestic, and third country suppliers. End user industry sectors are profiled and data is given on employment, exports, value added and industrial production. Market access information describes trade practices, regulations, and technical requirements. The trade lists inclucle: trade events; trade, industrial, professional associations; exhibitions, conference, seminars; publications, list of NEMEC members; and list of member firms in the instrumentation association.

3828. LABORATORY INSTRUMENTATION IN THE UNITED KINGDOM: A MARKET RESEARCH STUDY. *Sira Institute Ltd. U.S. Department of Commerce. International Trade Administration. NTIS #ITA 81-10-502. May 1981. 153 pg. $15.00.*

Survey of the market for laboratory instruments in the United Kingdom provides figures on total market size and wholesale price index by major product category. Analyzes imports by specific piece of equipment, by country of origin, with figures on actual value, percent market share, percent for lab use, and amount for lab use. Analyzes specific product markets with statistics for sales in U.K., including U.S. market share. Provides domestic manufacturers. End user analysis provides data on government and privately-funded research and development, employment, and sales. End user sectors are individually profiled, including sales, imports, exports, employment, research and development expenditures, major companies, market for laboratory instruments, trade practices, regulations, and technical requirements are described. Trade lists cover trade fairs, exhibitions, conferences, and seminars; trade, industrial, and professional associations; and publications.

3829. LABORATORY INSTRUMENTS: FRANCE: COUNTRY MARKET SURVEY. *U.S. Department of Commerce. Sales Branch. Irregular. CMS 81-609. September 1981. 12 pg. $.50.*

Analyzes the laboratory instrument industry in France and current and future spending on research and development by government, educational, nonprofit, and industrial laboratories. Provides data on imports of selected instruments; sales of selected instruments by U.S., foreign, and domestic producers; and laboratory expenditures by major end user. Describes projected research and the market for specific instruments that will be needed in each industry.

3830. LABORATORY INSTRUMENTS: SPAIN: COUNTRY MARKET SURVEY. *U.S. Department of Commerce. Sales Branch. Irregular. CMS 81-605. February 1981. 9 pg. $.50.*

Analyzes the current and potential market for laboratory instruments in Spain. Describes current and future research in government, educational, non-profit, and industrial laboratories, and provides data on laboratory expenditures and purchases of instruments for 1979 and 1983. Provides data on imports of selected instruments, by country of origin, for 1977-1979; and sales of selected instruments, by U.S., other foreign, and domestic producers, for 1979 and 1983 (projected).

3831. LABORATORY PURCHASE AUDIT. *Technical and Medical Studies Ltd. Quarterly. 1982.*

Market survey for laboratory supplies purchased in the U.K. from audit of orders.

3832. THE MARKET FOR LABORATORY INSTRUMENTS IN FRANCE. *P.A. International Management Consultants Ltd. U.S. Department of Commerce. International Trade Administration. NTIS #ITA 81-08-513. February 1981. 108 pg. $15.00.*

Survey of the market for laboratory instruments in France provides data on the market size as a whole and on imports by type of instrument (BTN number) and country of origin. Provides current and projected data on sales for specific product markets, with percent from domestic, U.S., and third country suppliers. Gives a competitive assessment of those suppliers, with leading manufacturers. End user analysis profiles individual industry sectors, describing growth, products, R & D, demand for laboratory instruments, established suppliers, and computerization. Trade practices, regulations, technical requirements are described. Trade lists cover trade, industrial, professional associations; publications; and trade events.

3833. MARKET OF ENVIRONMENTAL PROTECTION EQUIPMENT IN GERMANY. *Infratest-Industria GmbH. February 1980. 50 pg. $23000.00 (weg).*

Analysis of 60 industrial market segments for pollution control equipment in West Germany. Examines government regulations, current investments in pollution control installations and forecasts of investments to 1985, and technology of plants and equipment.

3834. MATERIALS TEST EQUIPMENT IN EUROPE TO 1986. *Larsen Sweeney Associates Ltd. 1981. 241 pg. 695.00 (United Kingdom).*

Market report to 1986 on materials test equipment in Europe. 28 types are covered individually for Austria, Belgium, Luxembourg, Denmark, Eire, Finland, France, West Germany, Italy, Netherlands, Norway, Portugal, Spain, Sweden, Switzerland and the U.K. Gives market analysis to 1986, production, market segmentation, sales, distribution, employment and unemployemnt, profits, end users and suppliers.

3835. MEASURING AND TEST INSTRUMENTS IN EUROPE TO 1986. *Larsen Sweeney Associates Ltd. 1981. 243 pg. 595.00 (United Kingdom).*

Market report to 1986 on twenty types of measuring and test instruments in Austria, Belgium, Luxembourg, Denmark, Eire, Finland, France, West Germany, Italy, Netherlands, Norway, Portugal, Spain, Sweden, Switzerland and the U.K. Gives forecasts to 1986, production, market segmentation, sales, distribution, employment and unemployment, profits, end users and suppliers.

3836. MICROPROCESSOR BASED PROCESS CONTROL INSTRUMENTATION IN EUROPE. *Frost and Sullivan, Inc. September 1979. 201 pg. $900.00.*

This report provides an outlook on the process control industry by major product category and forecasts the penetration of microprocessor-based systems for 1979-1989. Market prospects are established by end-user industry in the principal European countries. Economic and end-user industry process and capacity trends are discussed. Applications for microprocessor-based systems are identified.

3837. MICROWAVE TEST AND MEASUREMENT EQUIPMENT IN EUROPE TO 1986. *Larsen Sweeney Associates Ltd. 1981. 255 pg. 695.00 (United Kingdom).*

Report on microwave test and measurement equipment to 1986 in Austria, Belgium, Luxembourg, Denmark, Eire, Finland, France, West Germany, Italy, Netherlands, Norway, Portugal, Spain, Sweden, Switzerland and the U.K. Gives forecasts to 1986, market analysis, production, market segmentation, sales, distribution, employment and unemployment, profits, end users and suppliers. This is a multi-client study and cannot be made available to some companies.

3838. THE MILITARY AND AVIATION DISPLAY MARKET IN EUROPE. *Frost and Sullivan, Inc. September 1980. 311 pg. $1250.00.*

Market analysis report including forecasts, 1980-1990, for the aviation and military markets for electro-optical display instruments. Surveys demand and upcoming government expenditures; industry structure, including profiles of manufacturers; new products and utilizations; and market size and stuctures, by end uses.

3839. MILITARY MARKET FOR SIGNAL PROCESSORS AND SPECTRUM ANALYSERS IN WEST EUROPE. *Frost and Sullivan, Inc. September 1982. 279 pg. $1500.00.*

Analysis of Western European military markets for signal processors and spectrum analyzers, 1982-1990, forecasts demand by end uses. Examines market size and growth rates; technology; manufacturers' product lines, capacity, sales, and market shares; and government expenditures, inventories of vehicles and weapons systems, and defense contracts.

3840. NON-DESTRUCTIVE TEST (NDT) EQUIPMENT MARKETS IN EUROPE. *Frost and Sullivan, Inc. September 1981. 350 pg. $1400.00.*

Market analysis for non-destructive test equipment in Western Europe, 1980, including ultrasonics, X-rays, and penetrants. Forecasts markets to 1983 and 1990 by country and product groups. Examines research and development, end uses and foreign trade. Includes rankings of manufacturers and market shares.

3841. NORWAY INDUSTRIAL PROCESS CONTROLS MARKET. *U.S. Department of Commerce. International Trade Administration. NTIS #ITA 81-10-501. May 1981. 95 pg. $15.00.*

Survey of the market for industrial process controls in Norway provides data on current and projected market size by product category. Gives data on imports for specific process controls by country of origin. Examines the size and composition of the domestic industry and gives a competitive assessment of U.S. and third country suppliers and the major companies. Analyzes the sales outlook for specific U.S. products and major competing source countries, 1979-1983. Data on end user industries, by sectors includes output, capital expenditures, capacity utilization, new installations and replacements, and demand prospects for specific controls. Describes business practices, regulations, and technical requirements. Trade lists cover trade, industrial, professional associations, and trade journals. Appendix provides an analysis of the Norwegian petrochemical industry.

3842. NUCLEONIC INSTRUMENTS IN EUROPE TO 1986. *Larsen Sweeney Associates Ltd. 1981. 227 pg. 695.00 (United Kingdom).*

Market research report with forecasts to 1986 on 13 product groups of nucleonic instruments in Austria, Belgium, Luxembourg, Denmark, Eire, Finland, France, West Germany, Italy, Netherlands, Norway, Portugal, Spain, Sweden, Switzerland and the U.K. Covers production, market segmentation, sales, distribution, employment and unemployment, profits, end uses, and suppliers.

3843. THE ON-STREAM ANALYZERS MARKET IN EUROPE. *Frost and Sullivan, Inc. July 1980. 344 pg. $1350.00.*

Market analysis and forecasts for the process control industry in Western Europe. Includes U.S. and Japanese manufacturers. Breaks down analysis by products, market areas, and end use industries. Projects investments and expenditures for new plants and equipment. Analyzes demand; economic indicators; and industry structure, including rankings and market shares of manufacturers.

3844. OPTICAL INSTRUMENTS, ELECTRONIC, IN EUROPE TO 1986. *Larsen Sweeney Associates Ltd. 1981. 249 pg. 795.00 (United Kingdom).*

Market research report with forecasts to 1986 on eleven types of electronic optical instruments in Austria, Belgium, Luxembourg, Denmark, Eire, Finland, France, West Germany, Italy, Netherlands, Norway, Portugal, Spain, Sweden, Switzerland and the U.K. Gives market analyses, production, market segmentation, sales, distribution, employment and unemployment, profits, end users and suppliers.

3845. OSCILLATORS IN EUROPE TO 1986. *Larsen Sweeney Associates Ltd. 1981. 331 pg. 695.00 (United Kingdom).*

Market research report with forecasts to 1986 on sixteen types of oscillators in Austria, Belgium, Luxembourg, Denmark, Eire, France, West Germany, Italy, Netherlands, Norway, Portugal, Spain, Sweden, Switzerland and the U.K. Gives market analyses, production, market segmentation, sales, distribution, employment and unemployment, profits, end users and suppliers.

3846. OSCILLOSCOPES IN EUROPE TO 1986. *Larsen Sweeney Associates Ltd. 1981. 243 pg. 695.00 (United Kingdom).*

Market research report with forecasts to 1986 on twelve types of oscilloscopes in Austria, Belgium, Luxembourg, Denmark, Eire, Finland, France, West Germany, Italy, Netherlands, Norway, Portugal, Spain, Sweden, Switzerland and the U.K. Gives production statistics, market analyses, market segmentation, sales, distribution, employment and unemployment, profits, end users and suppliers.

3847. POLLUTION MONITORING IN EUROPE TO 1986. *Larsen Sweeney Associates Ltd. 1981. 262 pg. 795.00 (United Kingdom).*

Market report to 1986 on pollution monitoring equipment including monitoring, analysis and sensory equipment, data logging and recording equipment, display, controllers, peripherals and other equipment in Western Europe. Covers forecasts to 1986 for production, distribution, sales, end users, employment and unemployment, in industry sectors such as chemicals and plastics; food, drink and tobacco; oil, gas and petroleum; steel and metals; transport equipment; and manufacturing industries.

3848. POTENTIOMETRIC RECORDERS IN EUROPE TO 1986. *Larsen Sweeney Associates Ltd. 1981. 226 pg. 595.00 (United Kingdom).*

Market report to 1986 on potentiometric recorders in Western Europe. Gives forecasts to 1986 for production, distribution, sales, end users, employment and unemployment. The following end user markets are covered: chemicals and plastics; food, drink and tobacco; gas industry; nucleonics; petroleum refining; steel; manufacturing industries; utilities and water treatment.

3849. PRESSURE FLOW AND LEVEL PCI MARKETS IN EUROPE. *Frost and Sullivan, Inc. April 1981. 301 pg. $1350.00.*

Market report includes pressure measuring equipment and transmitters (5 types), flowmeters and transmitters (7 types), and level measuring instruments (3 types) in W. Germany, France, U.K., Italy, and the Netherlands, 1980-1990, in 11 end use markets. Contains market shares of the top 20 companies and includes market opportunities.

3850. PROCESS CONTROL INSTRUMENTATION MARKETS IN EUROPE. *Frost and Sullivan, Inc. September 1982. 372 pg. $1500.00.*

Analysis of European markets for process control instruments. Forecasts consumption and growth rates, 1980-1987. Includes rankings of manufacturers by country, product lines, market shares, investments in end use industries, new technology, industry structure, foreign trade, and industrial production index.

3851. PROCESS CONTROL INSTRUMENTS IN EUROPE TO 1986. *Larsen Sweeney Associates Ltd. 1981. 243 pg. 795.00 (United Kingdom).*

Market report to 1986 on process control instruments in Western Europe. Covers forecasts to 1986 for production, sales, distribution, end users and employment and unemployment. Thirty-one industry sectors are covered individually.

3852. RECORDERS IN EUROPE TO 1986. *Larsen Sweeney Associates Ltd. 1981. 222 pg. 795.00 (United Kingdom).*

Market report to 1986 on recorders of the following types: chart; strip; computer data; electrostatic; event; magnetic tape; polar; sound level; transient; vibration and video recorders in Western Europe. Covers forecasts to 1986 for production, sales, distribution, end users, employment and unemployment.

3853. RELAYS IN EUROPE TO 1986. *Larsen Sweeney Associates Ltd. 1981. 243 pg. 795.00 (United Kingdom).*

Marketing report to 1986 on the following relay equipment: choppers; coaxial; counting; differential; electrostatic; encapsulated; flameproof; frequency sensitive; latch; mercury switch; mercury wetted; miniature; moving coil; photoelectric; plug-in; polarised; reed; solid state; subminiature; voltage-sensing; solid state and thermionic in Western Europe. Gives forecasts to 1986 for production. Distribution, sales, end users, employment and unemployment.

3854. SCIENTIFIC ANALYTICAL INSTRUMENTATION MARKETS IN EUROPE. *Frost and Sullivan, Inc. June 1980. 237 pg. $1350.00.*

In-depth 23-page report analyzes and forecasts European markets from 1979 to 1985 for scientific analytical instruments. Countries covered are West Germany, U.K., Belgium, Switzerland, France, Netherlands, and Italy. The range of products covered in the study include 24 individual instruments headed under electro-magnetic radiation, nuclear - magnetic resonance spectrometers, mass spectrometers, chromatographs, electro-chemical devices, thermal analyzers, nuclear analytical, electron microscopes, surface analyzers, special purpose, and data acquisition and processing. Products are broken down by 10 end-user sectors with forecasts to 1985. Market shares by company are provided for 11 important market segments.

3855. THE SCIENTIFIC INSTRUMENTS AND EQUIPMENT INDUSTRY IN EUROPE. *Market Studies International. Division of European InterCompany Comparisons Ltd. 1982. 150.00 (United Kingdom).*

Provieds analysis of scientific instrument and equipment markets in Western Europe, including market size and growth rates. Tablulates government statistics on production, sales, imports, exports, apparent consumption, industry structure and market shares, 1974-1980.

3856. SENSORS IN MANUFACTURING AND PROCESS TECHNOLOGY. *Battelle-Frankfurt. 1983. 18000.00 (West Germany).*

Analysis and forecast of markets for sensors, especially for use in microelectronic measurement and instrumentation. Examines new technology, economic indicators and new utilizations.

3857. SIGNAL GENERATORS IN EUROPE TO 1986. *Larsen Sweeney Associates Ltd. Quarterly. 1981. 224 pg. 795.00 (United Kingdom).*

Market report to 1986 on 13 types of signal generators in Western Europe. Gives forecasts to 1986 for production, sales, end users, distribution costs, marketing costs, capital investments, advertising costs, and employment and unemployment. Report updated quarterly.

3858. SPECTRUM ANALYSERS IN EUROPE TO 1986. *Larsen Sweeney Associates Ltd. Quarterly. 1981. 211 pg. 595.00 (United Kingdom).*

Market report to 1986 on 6 types of spectrum analyzers in Western Europe. Gives forecasts to 1986 for sales, production trends, deliveries, distribution costs, marketing costs, capital investments, advertising costs, end users, employment and unemployment. Report updated quarterly.

3859. STAND-ALONE DATA LOGGER MARKET IN WEST EUROPE. *Frost and Sullivan, Inc. April 1982. 267 pg. $1500.00.*

Analysis of European markets for stand-alone data loggers. Forecasts consumption, growth rates, domestic production, imports, and exports, by end uses and product groups, 1981-1986. Includes technology review, rankings of manufacturers, market shares, and marketing strategies. Also includes market survey results.

3860. SURVEY OF THE MARKET FOR INDUSTRIAL PROCESS CONTROLS IN GERMANY. *Kindel, Karl-Wilhelm. U.S. Department of Commerce. International Trade Administration. NTIS # ITA81-11-505. August 1981. 105 pg. $15.00.*

Provides data on current (1980) and projected market size by major subcategory, imports of specific controls (1978-1980) by country of origin and projected (1985) imports by major subcategory, and size and composition of the domestic manufacturing industry. Includes lists of major U.S., domestic, and third country suppliers and the sales outlook (1980-1985) for specific U.S. products and major competing source countries. Individual end user industries are profiled, with tables providing data on the size and composition of the process industries; current (1980) and projected (1985) output, capital expenditures, and current percent of plant capacity utilization; new installations and replacements (1980 and 1985) by industry; demand prospects for specific controls by industry. Market access data describes business practices, regulations and technical requirements. The trade lists include: 1) trade and industrial associations; 2) journals; 3) trade events.

3861. SURVEY OF THE MARKET FOR LABORATORY INSTRUMENTS IN GERMANY. *Kindel, Karl-Wilhelm. U.S. Department of Commerce. International Trade Administration. NTIS # ITA81-11-502. July 1981. 103 pg. $15.00.*

Survey of the market for laboratory instruments in West Germany describes the market size, sources of supply, end users, and market access. Gives data on market size (production, imports, exports) for the total market. Provides import data by type of instrument, by country of origin, and includes percent of market share, percent for lab use, and amount for lab use. Specific product markets are analyzed and figures given for sales of selected instruments by geographic origin. An assessment of the competitive position of U.S. and third country suppliers is given. Each major end user sector is analyzed in terms of products, sales, research and development, laboratories, leading German companies, and demand for instruments. Trade practices, regulations, and technical requirements are described. Trade lists cover principal trade associations; list of publications; and trade events.

3862. SWEDEN: INDUSTRIAL PROCESS CONTROLS MARKET. *Hashmi, Kalle. U.S. Department of Commerce. International Trade Administration. NTIS # ITA81-08-511. April 1981. 98 pg. $15.00.*

Survey of the market for industrial process controls in Sweden provides information on the market size (production, imports, exports), sources of supply, end users, and market access data. Lists U.S. sales potential of specific products. Profiles principal end user industries, including projected capital expansion and demand for specific industrial process controls. The need for contractors, consultants, and engineers is analyzed. Business practices, regulations, and technical requirements are covered, and a list of trade, industrial, and professional associations is included.

3863. TECHNOLOGY ASSESSMENT AND FORECAST OF THE INDUSTRIAL DISPLAY MARKET IN EUROPE. *Frost and Sullivan, Inc. January 1980. 326 pg. $1000.00.*

Analysis of Western European markets for industrial process display devices. Covers products, end use industries, and countries. Forecasts sales. Examines new products and technology, user attitudes, manufacturers, economic indicators by country, and distribution channels.

3864. THERMOMETERS IN EUROPE TO 1986. *Larsen Sweeney Associates Ltd. Quarterly. 1981. 230 pg. 795.00 (United Kingdom).*

Market report to 1986 on 17 types of thermometers available in Western Europe. Covers forecasts to 1986 for sales, production trends, deliveries, distribution costs, marketing costs, capital investments, advertising costs, end users, employment and unemployment. 16 industry sectors as end users are analyzed individually. Complete updates quarterly.

3865. THERMOSTATS IN EUROPE TO 1986. *Larsen Sweeney Associates Ltd. Quarterly. 1981. 235 pg. 695.00 (United Kingdom).*

Market report to 1986 on 13 types of thermostats available in Western Europe. Covers forecasts to 1986 for sales, production trends, deliveries, distribution costs, marketing costs, capital investments, advertising costs, end users, and employment and unemployment. 15 industry sectors are individually analyzed for each country. Complete updates quarterly.

3866. UNITED KINGDOM MARKET FOR PROCESS CONTROL INSTRUMENTATION. *John Martin Publishing Ltd. 1979. 125.00 (United Kingdom).*

Covers market quantification; market development; market characteristics; end user data; the U.K. manufacturing industry; market access data; and trade lists and publications.

3867. THE U.K. MARKET FOR WRIST WATCHES. *Economist Intelligence Unit Ltd. 1982. 635.00 (United Kingdom).*

Analysis of U.K. market for wrist watches. Includes demographic statistics, 1971-1991; gross domestic product and personal consumption expenditures, 1976-1980; retail sales, 1976-1980; retail price index, 1976-1981; foreign trade, 1976-1980; market size, 1972-1980; production, 1970-1980; imports, 1976-1980; exports, 1970-1980; balance of trade,

1976-1980; apparent consumption, 1972-1980; prices and product groups, 1975-1980; employees, 1975-1980; consumer purchases, 1977-1980; financial analysis of manufacturers and dealers, 1977-1980; brand names and market shares, 1977-1980; advertising expenditures, 1976-1980; forecasts to 1986; and trade associations.

3868. THE WATCH AND CLOCK REVOLUTION TO 1985. *Mackintosh Consultants, Inc. 1979. $950.00.*

Forecasts worldwide markets through 1985 for traditional and electronic watches and clocks. Focuses on the U.K., France, West Germany, the U.S. and Japan.

See 422, 425, 442, 458, 460, 471, 1949, 2854, 2858, 2859, 2860, 2861, 2862, 2863, 2864, 2902, 2923.

Industry Statistical Reports

3869. ANALYTICAL TABLES OF FOREIGN TRADE - NIMEXE - 1980; OPTICAL PRECISION INSTRUMENTS: VOLUME L: CHAPTERS 90-99. *Organisation for Economic Cooperation and Development. 1981. 1500.00 (Belgium).*

One in a series of 13 volumes of external trade statistics of the European Community and of the member states in the Nimexe nomenclature. Gives breakdowns into products by country order for each six-figure Nimexe heading in twelve volumes (vols.A-L) by commodity group, and into country by products order by Nimexe chapter (two-figure code) in a thirteenth volume (vol.Z).

3870. BUSINESS MONITOR: WATCHES AND CLOCKS. *Her Majesty's Stationery Office. Quarterly. January 1982. 6.65 (United Kingdom).*

Quarterly business monitor provides U.K. production figures for watches and clocks.

3871. ENGINEER'S HAND AND SMALL TOOLS MANUFAC-TURERS. *ICC Business Ratios. Annual. 1981. 80.00 (United Kingdom).*

One in a series of 150 annual industry reports comparing performance between up to 100 individual leading companies, subsectors, and sectors over a three-year period. Fourteen ratio tables by company detail profit margin, profitability, asset utilization, return on capital, liquidity, stock turnover, credit period, exports, profit per employee, average remuneration, sales per employee, capital employed per employee, and other data. Six annual growth rate tables cover sales, total assets, capital employed, average wages, total wages, and exports. Other company data includes names of directors and secretary and the registered office address. Gives analyses of current problems and likely developments within the sector and industry as a whole.

3872. OPTICAL INDUSTRY. *ICC Business Ratios. Annual. 1981. 95.00 (United Kingdom).*

One in a series of 150 annual industry reports comparing performance between up to 100 individual leading companies, subsectors, and sectors over a three-year period. Fourteen ratio tables by company detail profit margin, profitability, asset utilization, return on capital, liquidity, stock turnover, credit period, exports, profit per employee, average remuneration, sales per employee, capital employed per employee, and other data. Six annual growth rate tables cover sales, total assets, capital employed, average wages, total wages, and exports. Other company data includes names of directors and secretary and the registered office address. Gives analyses of current problems and likely developments within the sector and industry as a whole.

3873. POLLUTION CONTROL. *ICC Business Ratios. Annual. 1981. 95.00 (United Kingdom).*

One in a series of 150 annual industry reports comparing performance between up to 100 individual leading companies, subsectors, and sectors over a three-year period. Fourteen ratio tables by company detail profit margin, profitability, asset utilization, return on capital, liquidity, stock turnover, credit period, exports, profit per employee, average remuneration, sales per employee, capital employed per employee, and other data. Six annual growth rate tables cover sales, total assets, capital employed, average wages, total wages, and exports. Other company data includes names of directors and secretary and the registered office address. Gives analyses of current problems and likely developments within the sector and industry as a whole.

3874. SCIENTIFIC INSTRUMENT MANUFACTURERS. *ICC Business Ratios. Annual. 1981. 80.00 (United Kingdom).*

One in a series of 150 annual industry reports comparing performance between up to 100 individual leading companies, subsectors, and sectors over a three-year period. Fourteen ratio tables by company detail profit margin, profitability, asset utilization, return on capital, liquidity, stock turnover, credit period, exports, profit per employee, average remuneration, sales per employee, capital employed per employee, and other data. Six annual growth rate tables cover sales, total assets, capital employed, average wages, total wages, and exports. Other company data includes names of directors and secretary and the registered office address. Gives analyses of current problems and likely developments within the sector and industry as a whole.

See 2953, 2959, 2960.

Financial and Economic Studies

3875. DIE FEINMECHANISCHE UND OPTISCHE INDUS-TRIE AUS DER SICHT DER SIEBZIGER JAHRE. [The Precision Mechanical and Optical Industries from the Point of View of the 70's.]. *Duncker und Humblot. German. 1980. 286 pg.*

Examines development trends in the precision mechanical and optical industries (photography, slide and film projectors, optical instruments, measuring and regulating instruments, medical precision machinery) during the 1970's in Western Europe and especially in West Germany. Includes a survey of the trade in those articles. 140 tables document production (1960-1976) and trade (1970-1976).

3876. ICC BUSINESS RATIO REPORT: ELECTRONIC INSTRUMENT MANUFACTURERS. *ICC Business Ratios. Annual. 7th ed. 1982. 97.00 (United Kingdom).*

Financial analysis of 60 U.K. electronic instrument corporations includes business addresses, executives, lines of business, and ownership. Company data for 1978-1981 includes fixed and intangible assets, stocks, total current assets, short-term debt, liabilities, net assets, capital employed, sales, and number of employees. Also includes 19 ratios of profits, liquidity, capital utilization, gearing, productivity, exports and employees; growth rates; and performance forecasts.

Directories and Yearbooks

3877. AUTOMATIK UND INDUSTRIELLE ELEKTRONIK. [Automation and Industrial Electronics.]. *Max Binkert and Company. German. Annual. 1980.*

Annual listing of manufacturers and equipment, associated

companies, and products. Gives technical data and trade names.

3878. CHEMICAL ENGINEERING INDEX PRODUCT DATA BOOKS. *Technical Indexes Pty. Ltd. Semi-annual. 1982. $75.00.22.00 (United Kingdom).*

Data book (directory) lists addresses, telephone and telex numbers of major manufacturers and suppliers, UK agents for overseas companies, association and trade organizations. Details products, tradenames, new products.

3879. CLINILAB PRODUCTS COMPANY DIRECTORY - WEST GERMANY - 1981. *Robert S. First, Inc. Irregular. January 1981. 156 pg. $275.00.*

Listings of over 100 West German manufacturers and suppliers of laboratory chemicals and instruments. Includes sales volume and exports as percent of sales.

3880. THE DIRECTORY OF INSTRUMENTS, ELECTRONICS, AUTOMATION. *Morgan Grampian Book Publishing Company Ltd. Annual. 16th ed. 1982. 330 pg.*

Directory of instrumentation contains manufacturers' addresses, telephone numbers, U.K. agents of overseas organizations represented in the United Kingdom, trade names, products, association addresses, distributors, and index to advertisers.

3881. DIRECTORY OF POLLUTION CONTROL EQUIPMENT COMPANIES IN WESTERN EUROPE. *Inter Company Comparisons Ltd. Annual. 4th ed. 1982. 35.00 (United Kingdom).*

Sixteen nations' listings of 6000 manufacturers and suppliers include address, telephone, telex, names of chief executives and production and sales directors, numberr of employees, sales figures for past financial year, and product descriptions. Product index identifies manufacturers under 52 product and service categories. Listings of consultancies and associations.

3882. DIRECTORY OF POLLUTION CONTROL EQUIPMENT COMPANIES IN WESTERN EUROPE. *Inter Company Comparisons Ltd. Annual. 3rd ed. 1981. 30.00 (United Kingdom).*

Directory of 6,000 manufacturers of pollution control equipment in the U.K. and Western Europe. Lists addresses of manufacturers, executives, last year's sales, and products. Also includes analysis of the industry and markets.

3883. ELECTRONIC ENGINEERING INDEX PRODUCT DATA BOOK. *Technical Indexes Pty. Ltd. Tri-annually. 1982. $115.00.27.00 (United Kingdom).*

Data book (directory) listing addresses, telephone and telex numbers of major manufacturers and suppliers, UK agents for overseas companies, association and trade organizations. Details products, tradenames, new products.

3884. ENGINEER BUYERS' GUIDE. *Morgan Grampian Book Publishing Company Ltd. Annual. 1980. 650 pg. 7.50 (United Kingdom).*

Annual directory of 2000 manufacturers and suppliers includes addresses and telephone numbers; 1,000 non-British firms with their U.K. agents; 2,000 registered trade names. Buyers guide details 3,000 products and services, with a total of 75,000 listings.

3885. ENGINEERING COMPONENTS AND MATERIALS INDEX PRODUCT DATA BOOK. *Technical Indexes Pty. Ltd. Semi-annual. 1982. $75.00.22.00 (United Kingdom).*

Data book (directory) lists addresses, telephone and telex numbers of major manufacturers and suppliers, UK agents for overseas companies, association and trade organizations. Details products, trade names, new products.

3886. INSTRUMENT MANUFACTURERS: ALPHABETICAL LISTING / UNITED KINGDOM 1980. *David Rayner. Annual. 1979. 202 pg. 5.30 (United Kingdom).*

Coverage of 550 scientific and industrial equipment and systems manufacturers includes names of executives; addresses; telephone and telex numbers; number of employees; sales contacts; and affiliated firms.

3887. INSTRUMENTS, ELECTRONICS, AUTOMATION. *Morgan Grampian Book Publishing Company Ltd. Annual. 1979. 336 pg. 11.00 (United Kingdom).*

Annual listing includes 3,000 manufacturers and distributors (with addresses and telephone and telex numbers); 1,500 British agents for foreign firms; 3,000 registered trade names; 180 organizations; and 150 U.K. and non-U.K. conferences and exhibitions. Buyers guide has 5,000 product groups and 100,000 listings.

3888. LABORATORY EQUIPMENT DIRECTORY. *Morgan Grampian Book Publishing Company Ltd. Annual. 8th ed. 1981. 260 pg. 15.00 (United Kingdom).*

Directory of laboratory equipment in the United Kingdom lists manufacturers and distributors: addresses, telephone and telex numbers, a listing of overseas organizations with products in the laboratory field and their U.K. representation, trade names, products.

3889. LABORATORY EQUIPMENT DIRECTORY. *Morgan Grampian Book Publishing Company Ltd. Annual. 1980. 229 pg. 8.50 (United Kingdom).*

Annual directory covers approximately 1,700 manufacturers and distributors (including addresses and telephone numbers); 3,300 varieties of lab equipment and services (60,000 listings); 1,000 organizations and institutions; 1,000 non-U.K. equipment suppliers and their agents in Great Britain; and 2,100 brand names.

3890. LABORATORY EQUIPMENT INDEX PRODUCT DATA BOOK. *Technical Indexes Pty. Ltd. Annual. 1982. $40.00. 15.00 (United Kingdom).*

Directory lists addresses, telephone and telex numbers of major manufacturers and suppliers, UK agents for overseas companies, association and trade organizations. Details products, trade names and new products.

3891. MADE IN EUROPE: TECHNICAL EQUIPMENT CATALOG. *Pacific Subscription Service. Monthly. 1982. 100 pg. $150.00/yr.*

One in a series of three illustrated monthly guides to export products available from European nations. Includes sections on technological transfer, products and processes, and European trade fairs and exhibbitions.

3892. QUID HORLOGER. [Swiss Watchmaking Yearbook.]. *Croner Publications, Inc. German. Annual. 1980. $25.00.*

Alphabetical listings of Swiss watchmakers detail address, telephone, telex, executives, and products. Includes product listings, trademark directory and a 'Who's Who.'

3893. SWITZERLAND - YOUR PARTNER: VOLUME 8: ENVIRONMENT CONSERVATION. *Swiss Confederation. Office for the Development of Trade. Irregular. 1981. 128 pg. S. Fr. 6 (Switzerland).*

Presents an overall analysis of the environment conservation and pollution control industries in Switzerland. Includes indexes to products and services, suppliers, and a list of professional associations and groupings. Company-by-company descriptions present the range of products exported in the following categories: drinking water supply and treatment; treatment of wastewater, sludge and effluents; air purification; noise abatement; disposal of wates and refuse; fire protection; and civil defense.

See 2999, 3002.

Special Issues and Journal Articles

Bulletin Credit Suisse

3894. THE FUTURE OF SWITZERLAND'S WATCHMAKING INDUSTRY. *Baillod, Gil. Credit Suisse. Economics, Public Relations and Marketing Division. Summer. June 1981. pg.2-4*

Examines the Swiss watch industry giving figures for exports of watches and parts (1960, 1974, 1980). Assesses worldwide industry developments and discusses necessary changes required for continued growth.

3895. SWISS WATCH INDUSTRY: ON THE ROLLER-COASTER OF WORLD ECONOMIC EVENTS. *Retornaz, Rene. Credit Suisse. Economics, Public Relations and Marketing Division. Winter. December 1981. pg.21-22*

Reviews the 1981 Swiss watchmaking industry giving figures for production, workers employed, and exports (1979-1981). Examines stocks position, self-financing, and future prospects.

Control and Instrumentation

3896. ACHEMA 82: A PREVIEW. *Morgan Grampian Book Publishing Company Ltd. Triennial. v.14, no.5. May 1982. pg.43-46*

Describes and illustrates products on display at ACHEMA, 1982. Gives manufacturers of equipment for the petrochemical and allied industries.

3897. CONTROL AND INSTRUMENTATION: PRODUCTS. *Morgan Grampian Book Publishing Company Ltd. Monthly. December 1981.*

Regular feature lists and describes new electronic control instruments and products. Gives manufacturers' names. Further information available through a reader card service.

3898. TEMPERATURE CONTROLLERS: MONITOR. *Morgan Grampian Book Publishing Company Ltd. Irregular. March 1982. pg.33-35*

A survey of temperature controllers available on the market today includes a listing of 14 manufacturers and descriptions of their products. Further information through reader card service.

The Dock and Harbour Authority

3899. CARGO LOAD INDICATORS. *Foxlow Publications Ltd. Irregular. July 1981. pg.76-77*

A directory of technology available for weighing loads handled by cargo handling equipment including weigh bridges, crane systems, floor scales, etc. Information includes photograph, brief technical specifications, and address of manufacturer.

3900. METER INSTRUMENTATION. *Foxlow Publications Ltd. Irregular. June 1981. pg.42-45*

A guide to equipment available to monitor required safety levels in port and harbor facilities such as sound monitors, dust recorders, radiation meters, gas detectors, etc. Information includes photograph, brief technical specifications, address of manufacturers, and trade name.

Jewelers' Circular - Keystone

3901. "81 AN UP AND DOWN YEAR FOR SWISS PRODUCERS. *Thompson, Joe. Chilton Book Company. v.153, no.5. May 1982. pg.64-65*

Overview of the 1981 Swiss watch industry includes performance of brand name watches and new electronic watches, and the effect of world economic problems and competition from manufacturers in the Far East. Data is presented on production in units, 1980-1981; U.S. sales of Swisss watches, 1980-1981; sales of watches in selected price ranges, 1981; and Swiss watch exports to the United States in units and value, 1977-1981.

Journal of the Chartered Institute of Building Services

3902. BUILDING SERVICES: PRODUCTS. *Building Services Publications Ltd. Monthly. 1981.*

Lists products and equipment associated with building services, including pumps, burners, refrigerators, etc. Data includes manufacturers, brief technical specifications, and photograph. Further information available through a reader card service.

Laser Focus with Fiberoptic Communications

3903. USES OF HIGH-POWER LASERS IN FRENCH INDUSTRY. *Mangin, Charles Henri. Advanced Technology Publications. v.18, no.7. July 1982. pg.32-42*

Provides names of domestic and foreign suppliers of lasers. Forecasts growth rates and assesses 1981 market situation. Gives 1981 laser capacity. Provides table showing industrial sectors, laser applications, and number and power of lasers. Estimates market size (1981-1985).

Lloyd's Ship Manager

3904. PRODUCT FOCUS: LEVEL GAUGES. *Lloyd's of London Press Ltd. Irregular. November 1981. pg.27-30*

A directory of manufacturers producing level gauges for use in shipping machinery and equipment. Data includes manufacturer / supplier, model and trade name, and technical specifications.

Manufacturing Chemist

3905. MANUFACTURING CHEMIST'S LABORATORY EQUIPMENT. *Morgan Grampian Book Publishing Company Ltd. Monthly. v.53, no.10. October 1982. pg.83, 87*

Lists and briefly describes new laboratory equipment. Name and address of U.K. manufacturer supplied. Photographs included for selected products. Typical products listed include infrared noncontacted moisture analyzer, air compressors, and lab-size screw feeder.

Marketing in Europe

3906. WATCHES IN FRANCE. *Economist Intelligence Unit Ltd. no.232. March 1982. pg.50-61*

Report on watches in France includes statistics on production, 1970-1981; imports, 1980-1981; exports, 1980-1981; consumption, 1970-1981; sales, 1975-1981; market shares, 1981; retail sales, 1980; and consumer prices, 1982. Discusses market segmentation, manufacturers and brands, distribution, advertising, and prospects.

NDT International

3907. NDT INTERNATIONAL: NEW EQUIPMENT. *IPC Industrial Press Ltd. Bi-monthly. December 1981.*

Lists new equipment and products in the field of non-destructive testing, including modules, control units, analyzers, and various other inspection units. Includes photographs, descriptions, and addresses and names of manufacturers.

Reed's Special Ships

3908. REED'S SPECIAL SHIPS: NEW EQUIPMENT. *Thomas Reed Publications Ltd. Bi-monthly. December 1981.*

Lists, describes, and provides a photograph of new maritime products and equipment. Gives manufacturer's name.

Uhren Juwelen

3909. 50 JAHRE UHREN JUWELEN. [50 Years of Watches and Jewelry.]. *Osterreichischer Wirtschaftsverlag. German. June 1982. 66 pg. 30.00 (Austria).*

Celebrates the fiftieth anniversary of the (Austrian) Federal Association of Jewellers. Relates the historical development of the profession in Austria, pointing out the basic changes in professional and consumer attitudes.

L'Usine Nouvelle

3910. BIENS D'EQUIPEMENT: RECONQUERIR LA FRANCE? [Capital Goods: Reconquering the French Market?]. *Loreal, Annick. Usine Nouvelle. French. no.36/1982. September 2, 1982. pg.48-53*

Examines statistics on France's 1981 trade with the U.S., West Germany, the Netherlands, and Japan, identifying in each case leading deficit and surplus product categories, and the corresponding deficits or surplusses. Profiles 3 capital goods sectors: packaging machinery, measuring instruments, and steelmaking equipment. Discusses activities of leading manufacturers.

3911. COMMENT CHOISIR UN ANALYSEUR RAPIDE DE PROTEINES. [How to Choose a Rapid Protein Analyzer.]. *Gordon, Elisabeth. Usine Nouvelle. French. Irregular. October 1982. pg.156-166*

Provides specifications on several rapid protein analyzers available in France. Information includes supplier's address, method of analysis used, uses (laboratory and/or industrial), analysis time, range of error, sample capacity, and capital and operating costs. Text discusses the major processes employed.

Water Services

3912. WATER SERVICES: PRODUCTS AND PROCESSES. *Fuel and Metallurgical Journals Ltd. Monthly. 1981.*

Contains descriptions of products used in the water services industry, such as pumps, calibration devices, pipe jointing systems, measurement controls, etc. Data includes brief technical specifications, addresses of manufacturers, and photographs where available.

See 2063, 3048, 3051.

Monographs and Handbooks

3913. EUROPE BUYER'S GUIDE: TECHNICAL EQUIPMENT. *U.S. International Marketing Company. Monthly. 1981. 160 (est.) pg. $30.00/yr.*

Monthly guide to new technical equipment produced in 18 European nations. Specifies manufacturers / suppliers and prices, and provides color photographs. Subscription includes an annual directory supplement.

3914. WATER INDUSTRY CONTROL SYSTEMS: A REPORT ON A U.K. STRATEGY FOR MEETING THE REQUIREMENTS OF THE WATER INDUSTRY FOR INSTRUMENTATION, CONTROL, AND AUTOMATION EQUIPMENT SYSTEMS FROM U.K. SOURCES. *ERL. December 1979. 150 pg. 10.00 (United Kingdom).*

Examines the demand for water instrumentation, control, and automation systems in the U.K. The technical performance of U.K. manufacturers is discussed and suggestions for improved performance are presented. Data on sales volume in the water industry is presented, 1979-1984; export sales market volume, 1979; and percent of the investment, productivity, quality control, and profit margins are discussed in addition to the effect of imports on the U.K. market.

Conference Reports

3915. PAPERS PRESENTED AT THE SIXTH FLUID POWER SYMPOSIUM, CAMBRIDGE, ENGLAND, APRIL 1981. *British Hydromechanics Research Association. 1981. 430 pg. $69.00. 30.00 (United Kingdom).*

28 papers cover research and development; basic theory and experimentation; and applications of oil hydraulics and pneumatics in control systems, servo systems, filtration, contamination, noise, pumps and motors, transmission systems, valves instrumentation and testing, circuit / system design, and sales in fluid power.

Asia

Market Research Reports

3916. THE CONSTRUCTION INDUSTRY IN THE FAR EAST: PRODUCT REPORT: HEATING SYSTEM CONTROLS. *Business Management and Marketing Consultants Ltd. Industrial Market Research Ltd. July 1981. $2000.00.*

Analysis of the construction industry in South East Asia and the market for heating system controls. Examines imports, end uses, suppliers, distribution, prices, user attitudes, legislation and regulations, and market penetration strategies. Includes a directory of manufacturers.

3917. INDUSTRIAL PROCESS CONTROLS, HONG KONG: A COUNTRY MARKET SURVEY. *U.S. Government Printing Office. Irregular. CMS 80-219. September 1980. 13 pg. $.50.*

Report analyzes the current and projected markets for the sale of industrial process controls in Hong Kong. Data is provided on the total market (production, imports, exports) by major category; imports by country of origin; products with good sales potential for U.S. manufacturers; size and composition of process industries; output, capital expenditures, plant capacity utilization by user industries; market by user industries; and demand prospects by selected end-user industries. Describes the competitive environment, major industrial users, and market access data.

3918. INDUSTRIAL PROCESS CONTROLS, INDIA: A COUNTRY MARKET SURVEY. *U.S. Government Printing Office. Irregular. CMS 79-216. August 1979. 11 pg. $.50.*

Report analyzes the current and projected markets for the sale of industrial process controls in India. Data is provided on the total market (production, imports, exports); demand prospects for specific controls by selected end-user industries; imports by country of origin; output, capital expenditures, plant capacity utilization by user industries; market by user industries; and products with good sales potential for U.S. manufacturers. Describes the domestic market, competitive environment, major industrial users, and market access data.

3919. INDUSTRIAL PROCESS CONTROLS, JAPAN: COUNTRY MARKET SURVEY CMS 81-223. *International Trade Administration. U.S. Department of Commerce. U.S. Government Printing Office. March 1981. 11 pg. $.50.*

Survey of the market for U.S. manufacturers for sales of industrial process controls to Japan. Text surveys competitive environment, major industrial users, and market access. Seven tables quantify the industrial process control market by type of product, size of establishment, analysis of the industrial process control industry and market for 1976-1978, with projections for 1983. Includes list of recommended opportunities for U.S. manufacturers.

3920. INDUSTRIAL PROCESS CONTROLS, KOREA: A COUNTRY MARKET SURVEY. *U.S. Government Printing Office. Irregular. CMS 79-203. January 1979. 13 pg. $.50.*

Report analyzes the current and projected markets for the sale of industrial process controls in Korea. Data is provided on the total market (production, imports, exports); imports by country of origin; output, capital expenditures, plant capacity utilization by user industries; size and composition of process industries; market by user industries; demand prospects for specific controls by end user industries; and products with good sales potential for U.S. manufacturers. Describes the domestic market, competitive environment, major industrial users, and market access data.

3921. INDUSTRIAL PROCESS CONTROLS, MALAYSIA: A COUNTRY MARKET SURVEY. *U.S. Government Printing Office. Irregular. CMS 79-202. January 1979. 14 pg. $.50.*

Report analyzes the current and projected markets for the sale of industrial process controls in Malaysia. Data is provided on the total market (production, imports, exports); demand prospects for specific controls by selected end-user industries; imports by country of origin; output, capital expenditures, plant capacity utilization by user industries; market by user industries; and products with good sales potential for U.S. manufacturers. Describes the domestic market, competitive environment, major industrial users, and market access data.

3922. INDUSTRIAL PROCESS CONTROLS, PHILIPPINES: A COUNTRY MARKET SURVEY. *U.S. Government Printing Office. Irregular. CMS 79-217. August 1979. 11 pg. $.50.*

Analyzes the current and projected markets for the sale of industrial process controls in the Philippines. Data is provided on the market by general category; market by user industries; imports by country of origin; output, capital expenditures, plant capacity utilization by user industries; size and composition of process industries; and demand prospects for specific controls by selected end-user industries; products with good sales potential for U.S. manufacturers. Describes the domestic market, competitive environment, major industrial users, and market access data.

3923. INDUSTRIAL PROCESS CONTROLS, SINGAPORE: A COUNTRY MARKET SURVEY. *U.S. Government Printing Office. Irregular. CMS 79-2007. January 1979. 12 pg. $.50.*

Analyzes the current and projected markets for the sale of industrial process controls in Singapore. Data is provided on the total market (production, imports, exports); imports by country of origin; output, capital expenditures, plant capacity utilization by user industries; size and composition of process industries; market by user industries; demand prospects for specific controls by end-user industries; and products with good sales potential for U.S. manufacturers. Describes the domestic market, competitive environment, major industrial users, and market access data.

3924. INDUSTRIAL PROCESS CONTROLS, TAIWAN: A COUNTRY MARKET SURVEY. *U.S. Government Printing Office. Irregular. CMS79-206. January 1979. 11 pg. $.50.*

Analyzes the current and projected markets for the sale of industrial process controls in Taiwan. Data is provided on the total market (production, imports, exports, market size); imports by country of origin; size and number of domestic manufacturers of process controls; size and composition of process industries; the market by user industries; output, capital expenditures, plant utilization by user industries; demand prospects for specific controls by selected user industries; products with good sales potential for U.S. manufacturers. Describes the domestic market, competitive environment, major industrial users, and market access data.

3925. THE JAPANESE MARKET FOR LASERS AND ELECTRO-OPTICAL EQUIPMENT: FOREIGN MARKET SURVEY REPORT. *Kearney (A.T.) International, Inc. Tokyo (Japan). U.S. Department of Commerce. National Technical Information Service. DIB-80-02-522. 1979. 99 pg. $10.00.*

The market research was undertaken to study the present and potential U.S. share of the market in Japan for lasers and electro-optical equipment; to examine growth trends in Japanese end-user industries over the next few years; to identify specific product categories that offer the most promising export potential for U.S. companies; and to provide basic data which will assist U.S. suppliers in determining current and potential sales and marketing opportunities. The trade promotional and marketing techniques which are likely to succeed in Japan were also reviewed.

3926. KOREAN MARKET SURVEY: INDUSTRIAL INSTRUMENTATION. *ASI Market Research, Inc. (Korea). U.S. Department of Commerce. International Trade Administration. NTIS #DIB-80-02-511. June 1979. 34 pg. $15.00.*

Survey of the market for industrial instrumentation in Korea assesses the 1978 market for each major product category. U.S. products with increasing sales potential are listed and sales ratings for 1979-1983 are given to specific products in each major category. Each end user sector is analyzed with current and future needs described. Market access information covers import regulations and tariffs.

3927. LABORATORY INSTRUMENTS: HONG KONG: COUNTRY MARKET SURVEY. *U.S. Department of Commerce. Sales Branch. Irregular. CMS 80-604. September 1980. 10 pg. $.50.*

Analyzes the current and potential market for laboratory instruments in Hong Kong. Describes current and future research in government, educational, nonprofit, and industrial laboratories and provides data on capital investments and purchases of laboratory instruments for 1979 and 1983 (projected). Includes data on imports of selected laboratory instruments, by country of origin, for 1977-1979, and sales of selected instruments by U.S. and by all others, for 1979 and 1983 (projected).

3928. LABORATORY INSTRUMENTS: JAPAN: COUNTRY MARKET SURVEY. *U.S. Department of Commerce. Sales Branch. Irregular. CMS 79-602. December 1979. 11 pg. $.50.*

Analyzes the laboratory instrument industry in Japan and current and future spending on research and development by government, educational, nonprofit, and industrial laboratories. Provides data on imports of selected instruments; sales of selected instruments by U.S., foreign, and domestic manufacturers; and laboratory expenditures by major end user. Describes projected research and the market for specific instruments that will be needed in each industry.

3929. LABORATORY INSTRUMENTS: PHILIPPINES: COUNTRY MARKET SURVEY. *U.S. Department of Commerce. Sales Branch. Irregular. CMS 81-607. June 1981. 7 pg. $.50.*

Analyzes the current and potential market for laboratory instruments in the Philippines. Describes current and future research in government, educational, nonprofit, and industrial laboratories and provides data on laboratory expenditures and purchases of instruments for 1979 and 1983 (projected). Includes data on imports of selected laboratory instruments, by country of origin, for 1977-1979, and sales of selected instruments by U.S. and by all others, for 1979 and 1983 (projected).

3930. LABORATORY INSTRUMENTS: SINGAPORE: COUNTRY MARKET SURVEY. *U.S. Department of Commerce. Sales Branch. Irregular. CMS 81-608. August 1981. 10 pg. $.50.*

Analyzes the current and potential market for laboratory instruments in Singapore. Describes current and future research in government, educational, nonprofit, and industrial laboratories and provides data on laboratory expenditures and purchases of instruments for 1979 and 1983 (projected). Provides data on imports of selected laboratory instruments, by country of origin, for 1979, and sales of selected instruments by U.S. and by all others for 1979 and 1983 (projected).

3931. MARKET STUDY ON POLLUTION AND WASTE TREATMENT EQUIPMENT AND INSTRUMENTS (THAILAND). *SGV-NA Thalang and Co. Ltd. U.S. Department of Commerce. International Trade Administration. NTIS # DIB-80-02-520. October 1979. 18 pg. $15.00.*

Survey of the market for pollution and waste treatment equipment and instruments in Thailand provides a competitive market assessment and indicates potential for sales of specific U.S. products. Gives data on market size and imports by major product category, for 1978-1979 and forecast to 1980 and 1985. Provides end user industry sectors and describes needs. Business practices, standards and regulations, and tariff rates are detailed.

3932. MARKET SURVEY OF LABORATORY EQUIPMENT IN SINGAPORE. *Applied Research Corp., Singapore. U.S. International Trade Commission. 1980. 177 pg. $10.00.*

Analysis of markets for laboratory equipment in Singapore. Estimates potential market shares for exports from U.S. manufacturers; analyzes growth rates of end use industries; specifies promising products; and assesses sales potential.

3933. POLLUTION CONTROL AND WASTE TREATMENT EQUIPMENT - SINGAPORE: A FOREIGN MARKET SURVEY REPORT. *Lim, Albert. U.S. Department of Commerce. International Trade Administration. NTIS # ITA-82-03-515. December 1981. 62 pg. $15.00.*

Survey on the market in Singapore for pollution control and waste treatment equipment analyzes current apparent market (local production, imports, exports) and forecasts the market for 1986, by major product category. Gives data on imports by major product category, by country of origin, with market share

indicated. Provides a competitive assessment of U.S. and third country suppliers, by major product category and for consulting services. Analyzes end user industries, with profiles of each industry and its current and projected need for the equipment, in both the government and private sectors. A section of the survey describes marketing practices, sales representation, trade fairs, trade regulations, technical requirements and credit terms. Appendices list leading consulting engineering firms; leading suppliers; products with good sales potential for U.S. suppliers.

3934. POLLUTION CONTROL MARKET PHILIPPINES: FOREIGN MARKET SURVEY REPORT. *PA Management Consultants Ltd. (Hong Kong). U.S. Department of Commerce. National Technical Information Service. DIB-79-08-502. 1979. 89 pg. $10.00.*

The market research was undertaken to study the present and potential U.S. share of the market in the Philippines for pollution control equipment; to examine growth trends in the Philippine end-user industries over the next few years; to identify specific product categories that offer the most promising export potential for U.S. companies; and to provide basic data which will assist U.S. suppliers in determining current and potential sales and marketing opportunities. The trade promotional and marketing techniques which are likely to suceed in the Philippines were also reviewed.

3935. PROCESS CONTROL INSTRUMENTATION MARKET RESEARCH IN THE PHILIPPINES: FOREIGN MARKET SURVEY REPORT. *Mobius (Philippines). U.S. Department of Commerce. National Technical Information Service. DIB-79-09-504. 1979. 105 pg. $10.00.*

The market research was undertaken to study the present and potential U.S. share of the market in the Philippines for process control instruments; to examine growth trends in the Philippine end-user industries over the next few years; to identify specific product categories that offer the most promising export potential for U.S. companies; and to provide basic data which will assist U.S. suppliers in determining current and potential sales and marketing opportunities. The trade promotional and marketing techniques which are likely to succeed in the Philippines were also reviewed.

3936. PROCESS CONTROL INSTRUMENTATION MARKETS IN THE FAR EAST. *Frost and Sullivan, Inc. December 1981. 378 pg. $1400.00.*

Analysis of markets for process control instruments in countries of the Far East including Australia. Includes forecasts, 1982-1986; technology; growth rates and market size by product groups and end uses; domestic production and demand by country; import market penetration; distribution channels; manufacturers; and listings of business addresses.

3937. SENSOR MARKET AND MID-TERM PROSPECTS IN JAPAN AND TECHNOLOGY TRENDS IN FIBRE OPTICS. *Yano Economic Research Institute. Benn Publications Ltd. 1982. 200 (est.) pg. $880.00.*

Study covers concept and strategy of each manufacturer, market transition in terms of volume and value, market share, application area for sensors, future trends, development trends, marketing factors, and recent research and development. Report analyzes, by type, sensor and technological applications and includes technical analysis of fiber optics.

See 509, 515, 524, 525, 526, 527, 528, 2084, 3070, 3071, 3072, 3082.

Directories and Yearbooks

3938. ENGINEERING BUYERS GUIDE. *Indian Export Trade Journal. Annual. 1980. 300 pg. $15.00.*

Lists exporters, importers, and manufacturers of engineering-related products.

3939. GUIDE BOOK OF JAPANESE OPTICAL PRECISION INSTRUMENTS 1978/79. *Maruzen Company Ltd. Irregular. 1979. 186 pg. $21.60.*

Directory of optical precision instruments made in Japan.

3940. JEMIMA "81/82: JAPAN MEASURING INSTRUMENTS MANUFACTURERS ASSOCIATION CATALOG. *Japan Electric Measuring Instruments Manufacturers Association. Biennial. 1981. 248 pg.*

Catalog of products manufactured by Japanese companies belonging to the Japan Electric Measuring Instruments Manufacturers' Association. A statistical profile of the industry is followed by descriptions of products in the following categories: indicating meters, electrcity meters, test and measuring equipment, process measuring and control instruments, measuring instruments for environmental quality, nuclear instruments, and electrical measuring instruments for medical application. Also inlcuded is a directory of members of JEMIMA.

Special Issues and Journal Articles

Business Japan

3941. ELECTRIC MEASURING INSTRUMENT INDUSTRY MEETS CHANGING INDUSTRIAL NEEDS. *Nihon Kogyo Shimbun Publications. v.26, no.9. September 1981. pg.87-97*

Reviews the Japanese electric measuring instrument industry which includes indicating meters; electricity meters; process and test measures; nuclear measures, and others. Statistics include output (1976-1980) by instrument type; market share (1980) of output; exports and imports by instrument type; trade by trading partner; and ratio of imports to domestic demand (1976-1980).

3942. ELECTRIC MEASURING INSTRUMENTS ASSURED ESSENTIAL ROLE IN FUTURE-RELATED INDUSTRIES. *Kishino, Nobuo. Nihon Kogyo Shimbun Publications. no.9. September 1982. pg.109-110*

Lists measuring instruments manufactured in Japan. Shows 1977-1981 production, and 1977-1981 imports and exports.

Diamond's Industria

3943. JAPAN'S TIMEPIECE INDUSTRY: COPING WITH CHANGING REQUIREMENTS. *Diamond Lead Company, Ltd. v.12, no.6. June 1982. pg.16-24*

Examines the Japanese timepiece industry. Discusses current and historical economic conditions affecting the industry. Describes problems the industry faces from recession and from foreign competition. Data includes domestic production, 1960-1981; number of units and value of exports, 1981; and world production by type, 1981. Sources are the Japan Clock and Watch Association and the Citizen Watch Co.

Jewelers' Circular - Keystone

3944. HONG KONG SWAMPS U.S. WATCH MARKET; ELECTRONICS DOMINATE IMPORT TOTALS. *Chilton Book Company. v.153, no.11. November 1982. pg.96*

Discusses Hong Kong competition in U.S. watch markets. Provides figures for imports and watch prices.

See 2101, 3094.

Newsletters

3945. JEMIMA NEWSLETTER. *Japan Electric Measuring Instruments Manufacturers Association. Quarterly. 8 pg.*

Contains articles on the electric measuring instruments industry in Japan. Topics covered include marketing notes, new products, and reports on new technology. Statistics provided include exports, imports, and production.

Oceania

Market Research Reports

3946. MARKET RESEARCH STUDY FOR AUSTRALIA IN THE FIELD OF INDUSTRIAL PROCESS CONTROLS. *Price Waterhouse Associates Pty. U.S. Department of Commerce. International Trade Administration. NTIS #DIB-80-02-514. October 1979. 129 pg. $15.00.*

Survey of the market for industrial controls in Australia describes and lists domestic, U.S., third country suppliers and provides a competitive assessment of their position. Profiles end user industry sectors and needs for specific equipment. Provides data on the size of the total market (1976-1979, estimated 1980-1984) and imports of specific controls by country of origin (1976-1979). Gives sales outlook for specific U.S. products and major competing source countries (1978-1983). End user sector data includes industry size, 1976-1977, new installations and replacements (1978-1979, 1983-1984), production and capital expenditures (1976-1977, 1983-1984). Also includes a list of major projects which may affect demand for equipment (1979-1989). Business practices, regulations, technical requirements, and tariff classifications are detailed. Trade lists include associations, publications, and events.

See 3124.

Directories and Yearbooks

3947. AUSTRALIAN ENGINEERING DIRECTORY. *Technical Indexes Pty. Ltd. Annual. 1980. 750 pg. $60.00.*

Directory of Australian suppliers of domestic and imported engineering materials and products. Available on magnetic tape.

See 3128.

Middle East

Market Research Reports

3948. INDUSTRIAL PROCESS CONTROLS, ISRAEL: A COUNTRY MARKET SURVEY. *U.S. Government Printing Office. Irregular. CMS 79-204. January 1979. 11 pg. $.50.*

Report analyzes the current and projected markets for the sale of industrial process controls in Israel. Data is provided on the total market (production, imports, exports); imports by country of origin; output, capital expenditures, plant capacity utilization by user industries; size and composition of process industries; market by user industries; demand prospects for specific controls by end user industries; and products with good sales potential for U.S. manufacturers. Describes the domestic market, competitive environment, major industrial users, market and access data.

3949. INDUSTRIAL PROCESS CONTROLS, SAUDI ARABIA: A COUNTRY MARKET SURVEY. *U.S. Government Printing Office. Irregular. CMS 81-224. April 1981. 11 pg. $.50.*

Analyzes the current and projected markets for the sale of industrial process controls in Saudi Arabia. Data is provided on the market by general category; imports by country of origin; products with good sales potential for U.S. manufacturers; output, capital expenditures, plant capacity utilization by user industries; market by user industries; size and composition of process industries; and demand prospects for specific controls by selected end-user industries. Describes the domestic market, competitive environment, major industrial users, and market access data.

See 534.

Africa

Market Research Reports

3950. INDUSTRIAL PROCESS CONTROLS, EGYPT: A COUNTRY MARKET SURVEY. *U.S. Government Printing Office. Irregular. CMS 79-201. January 1979. 10 pg. $.50.*

Analyzes the current and projected markets for the sale of industrial process controls in Egypt. Data is provided on the market by types of controls; demand prospects by end user industries; imports by country of origin; size and composition of process industries; output, capital expenditures, percent of capacity utilized by industries using process controls; market by user industries; products with good sales potential for U.S. manufacturers. Describes the competitive environment, major industrial users, and market access data.

3951. INDUSTRIAL PROCESS CONTROLS, NIGERIA: A COUNTRY MARKET SURVEY. *U.S. Government Printing Office. Irregular. CMS 79-211. February 1979. 8 pg. $.50.*

Report analyzes the current and projected markets for the sale of industrial process controls in Nigeria. Data is provided on the total market (production, imports, exports); demand prospects for specific controls by selected end-user industries; imports by country of origin; output, capital expenditures, plant capacity utilization by user industries; market by user industries; and products with good sales potential for U.S. manufacturers. Describes the domestic market, competitive environment, major industrial users, and market access data.

3952. INDUSTRIAL PROCESS CONTROLS - SOUTH AFRICA: A FOREIGN MARKET SURVEY REPORT. *Konsulterna. U.S. Department of Commerce. International Trade Administration. NTIS# ITA 81-11-503. July 1981. 68 pg. $15.00.*

Survey of the market for industrial process controls in South Africa describes market size, suppliers, end users, and market access. Provides data on current and projected market size by product category. Gives data on imports for specific controls by country of origin. Details size and composition of the domestic industry and gives a competitive assessment of U.S. and third country suppliers and the major companies. Analyzes the sales outlook for specific U.S. products and major competing source countries, 1980-1985. Data on end user industries, by sector, includes output, capital expenditure, capacity utilization, new installations and replacements, and demand prospects for specific controls. Business practices, regulations, and technical requirements are described. Trade lists cover associations and trade events.

3953. THE MARKET FOR INDUSTRIAL PROCESS CONTROLS, ALGERIA: A FOREIGN MARKET SURVEY REPORT. *Konsulterna AB. U.S. Department of Commerce. International Trade Administration. NTIS #ITA-82-03-505. August 1981. 116 pg. $15.00.*

Survey on the market in Algeria for industrial process controls analyzes the total market (production, imports, exports), by major product categories. Provides data on imports by specific process control, by country of origin. Lists major U.S. and other firms supplying the Algerian market and gives U.S. sales potential (and major competition) for specific products. Profiles specific end user industry sectors indicating current and future needs and discusses the role of contractors / consultants / engineers. A section of the survey reports on business practices, regulations, and technical requirements. Lists associations, trade events, government agencies.

See 541, 542, 543.

Telecommunications

International

Market Research Reports

3954. COMMUNICATIONS INDUSTRIES: ELECTRONIC PUBLISHING. *FIND / SVP. December 1981. 32 pg. $250.00.*

Discusses assorted types of electronic publishing (cablenews, videotex, teletext); international electronic publishing developments; experiments in videotex and teletext in the U.S.: home shopping and banking; role of AT & T: standards and regulations; and market potential.

3955. COMMUNICATIONS / INFORMATION SYSTEMS PLANNING SERVICE. *Yankee Group. 1981. $9500.00/yr.*

Components of program include monthly series of research-oriented industry reports, with a new product opportunity section in each report; four Yankee Group seminars; an in-house seminar and consulting program; a monthly newsletter; and a call-in / call-out service. Industry reports include Advanced Local Networks, The New Generation of Data Terminals, The

AT & T Business Unregulated Utility, Auxiliary Memory Systems, Long Haul Communications: New Solutions, and others.

3956. COMMUNICATIONS / INFORMATION SYSTEMS PLANNING SERVICE: LONG HAUL COMMUNICATIONS - NEW SOLUTIONS. *Yankee Group. July 1981. 200 pg. $9500.00/yr.*

One of a series of monthly reports in Yankee Group's Communications / Information Systems Planning Service. Focuses on public and private solutions (current and upcoming) for long distance communications, with particular emphasis on satellite, microwave, radio frequency, subchannel, and fibre optics transmission media. Analyzes relative cost, performance, and functional comparisons. Discusses potential service offerings, impact of deregulation, development of new technologies, and the implementation of new applications. Survey details users' current and future requirements for long-haul transmission - speed, performance, flexibility, functionality, price sensitivity, etc.

3957. ELECTRONIC DOCUMENT DISTRIBUTION (EDD). *Institute for Graphic Communications, Inc. 1982.*

This multi-client study examines the market potential of 'electronic mail' systems. Considers technological innovations, operation costs, regulatory and standards environment, size of specific markets and extent of specific end uses, and strategies for companies entering the field.

3958. ELECTRONIC PUBLISHING. *Predicasts, Inc. July 1981. $375.00.*

Report on worldwide utilization of videotex, viewdata and teletext. Surveys and analyzes markets for services to consumers and businesses. Includes advertising expenditures and newspaper functions.

3959. THE FUTURE OF TEXT COMMUNICATIONS: THE IMPACT OF TELETEX. *Mackintosh Publications Ltd. January 1980. 88 pg. 195.00 (United Kingdom).*

Worldwide market forecast and strategic assessment of the introduction of teletex. Describes the service concept, technology, impact on other areas (e.g. word processing), market forecasts, user requirements, and tariff structure.

3960. FUTURE OF VIDEOTEXT. *Sigel, Efrem. Knowledge Industry Publications. 1982. 192 pg. $32.95.*

Covers the evolution and present status of videotext in major countries around the world; the technology of videotext, including broadcast systems (teletext), telephone line systems (viewdata), and hybrid systems using phone lines and cable or broadcast. Discusses the advantages and disadvantages of videotext compared to print media and TV. Examines the consumer market, and how consumer and business applications differ.

3961. THE GROWING IMPACT OF WIDEBAND COMMUNICATIONS. *International Resource Development, Inc. May 18, 1981. 190 pg. $985.00.*

International market analysis and forecasts through 1991 cover sales; competition between AT & T and the CATV industry; and new products, technology and services: video conferencing and electronic mail.

3962. HOME OF THE FUTURE PLANNING SERVICE: TELE-BANKING - BEYOND INTERACTIVE FINANCIAL SERVICES. *Yankee Group. August 1981. 200 pg. $8500.00/yr.*

One in a series of six reports forming one component of " "Home of the Future Planning Service.' Topics include lessons from major tests so far (Japanese banks, Banc One, UAB, Chemical, Citibank, etc.); apparent strategy of major U.S. banks; demand for home banking; format, " "friendliness' and

security issues; on-line investment transactions at home; and marketing insurance and other financial services to the home.

3963. HOME OF THE FUTURE PLANNING SERVICE: TELESHOPPING: THE NEW ELECTRONIC MARKETPLACE. *Yankee Group. June 1981. 200 pg. $8500.00/yr.*

One of six reports forming a component of " "Home of the Future Planning Service.' Topics include teleshopping's potential market size and economic impact; Viewtron, QUBE, Hi-OVIS and other experiments to date; analysis of major retailers' strategy and findings; the creative aspect of teleshopping communication; what products lend themselves to teleshopping; and logistics of orders, delivery and security.

3964. HOME OF THE FUTURE PLANNING SERVICE: THE WIRED HOME AND THE ELECTRONIC SUPERSTRUCTURE. *Yankee Group. January 1981. 200 pg. $8500.00/yr.*

One in a series of bimonthly reports from Yankee Group's Home of the Future Planning Service. Covers personal computers: configuration and marketing issues; future of phonograph and videodisc products; analysis of competing videodisc technologies; carrier channels: coax vs. paired wire vs. fiber optics; and AT & T and the regulatory environment. Examines direct broadcast satellites (DBS); digital home storage utility of the future; friendly terminal design: lessons from electronic games;and video and audio delivery systems. Includes issues in the graphical display of information, the home printer: its importance and possible configuration, and speech recognition and synthesis: technology and role in market.

3965. THE INTERNATIONAL MARKET: CURRENT STATISTICS AND FUTURE GROWTH. *Telephony Publishing Corporation. 1980. $150.00.*

Study divided into 6 geographical areas: Central and South America, Africa, Asia, Europe, Oceania and the Middle East. Provides data on total telephones by country, types of switches, and transmission methods. Gives telephones in major cities by city and number; message potential for selected countries; matrix charts of competitors in the markets, by equipment line; and construction expenditures (1975-1981) as reported by noted countries' PTT's and / or common carriers.

3966. MANAGING GLOBAL COMMUNICATIONS. *Creative Strategies International. July 1979. $1200.00.*

Defines typical corporate usage and financial requirements for both domestic and international telecommunications. Analyzes the various solutions available to management in relation to: the geographic deployment of facilities; the frequency and volume of communications; and corporate management philosophy. Evaluates the economics of alternatives as well as potential hazards of certain options.

3967. THE TELECOMMUNICATIONS INDUSTRY WORLDWIDE. *DAFSA Enterprises. French. June 1980. 165 pg. 5000.00 (France).*

Analysis of telecommunications market environments and financial analysis of 25 major corporations worldwide.

See 13, 14, 22, 546, 768, 770, 784.

Industry Statistical Reports

3968. INFORMATION ACTIVITIES, ELECTRONICS, AND TELECOMMUNICATION TECHNOLOGIES IMPACT ON EMPLOYMENT GROWTH AND TRADE. *Organization for Economic Cooperation and Development. Publications and Information Center. September 1981. 140 pg. $9.00.*

Volume investigates statistically the information sector in

OECD countries and assesses the role of information technologies in structural change in this field.

3969. SATELLITE SITUATION REPORT. *U.S. National Aeronautics and Space Administration. Bi-monthly. 1982. 47 pg.*

Provides bimonthly data on satellites in orbit, 1982. Arranged by year of launch and international designation. Gives satellite name, launching country, date of launch, and 5 types of technical data. For 20 countries and organizations provides number of objects in orbit and decayed objects.

3970. STATISTICS OF COMMUNICATIONS COMMON CARRIERS, 1980. *Federal Communications Commission. U.S. Government Printing Office. Annual. SuDoc # CC1.35:980. 1982. 216 pg.*

Contains 1980 financial and operating data taken from common carrier reports filed with the Federal Communications Commission. Gives data for specific telephone and telegraphic companies. Includes overseas operations and rates for selected large telephone and telegraph services. Provides balance sheet and income statement of communications Satellite Corporation. Shows intercorporate relations of all subject carriers and controlling companies. Indexed by company.

3971. THE WORLD'S TELEPHONES: A STATISTICAL COMPILATION AS OF JANUARY, 1979. *American Telegraph and Telephone Long Line. Annual. August 1980. 85 pg.*

International inventory of telephones, discussing volume and per capita rate of installations by world region, nation, and type of telephone (residential, business, coin box, type of automatic switching, etc.). Includes inventories for the world's largest cities, for major U.S. cities and for U.S. states. Also discusses number of telephone calls originating in each nation (by local / interurban / international), and nations most frequently called.

3972. YEARBOOK OF COMMON CARRIER TELECOMMUNICATION STATISTICS. *International Telecommunication Union. English/French/Spanish. Annual. 8th ed. 1980. 358 pg. 37.00 (Switzerland).*

International data on various branches of common carrier telecommunications cover telephones (number of stations and traffic), telegraph (number of connections and traffic), and data transmission service (number of modems). Includes demographic statistics and figures on investments and revenues.

Financial and Economic Studies

3973. ELEMENTS OF TELECOMMUNICATION ECONOMICS. *Littlechild, S.C. Institution of Electrical Engineers. 1979. 280 pg. $37.00.*

Covers production functions, cost functions, investment appraisal, pricing policy, welfare economics and marginal cost pricing. Geared for nonspecialists.

3974. VIDEOTEX REPORT SERIES 1982: VIDEOTEX TECHNOLOGY, STANDARDS AND NETWORKING. *Butler Cox and Partners Ltd. Frost and Sullivan, Inc. May 1982. $750.00.*

Reviews international developments in videotex and computer network technologies. Examines connections and competition among alternatives.

See 2145, 2146.

Forecasts

3975. FORECASTING PUBLIC UTILITIES: PROCEEDINGS OF THE INTERNATIONAL CONFERENCE HELD AT NOTTINGHAM UNIVERSITY, MARCH 25-29, 1980. *Anderson, O.D., editor. Elsevier-North Holland Publishing Company. 1980. 210 pg. $39.00.*

Contains papers presented by leading authorities from Canada, Norway, U.K. and U.S.A. Covers theoretical and practical aspects, computation and applications. Emphasizes telecommunications industry, as the public utility most rapidly expanding and developing at present.

3976. WORLD TELECOMMUNICATIONS STUDY 2: A MULTI-CLIENT STUDY: 1980-1990: VOLUME 1; OVERVIEW AND TECHNOLOGICAL TRENDS. *Arthur D. Little, Inc. 1980. 216 pg. $34000.00.*

Volume 1 of 4-volume multi-client study of telecommunications industry for over 150 countries, 1980-1990. Includes regional overview, international facilities, worldwide market summary, technological trend analysis, profiles of major manufacturers, and country-by-country chapters with projections, buyer data and competition. Study covers telephone networks, communication facilities, satellite facilities, mobile radiotelephone, international telecommunications facilities, private & government networks and CATV. Price listed is for complete set.

Directories and Yearbooks

3977. INFORMATION INDUSTRY MARKET PLACE 1982: AN INTERNATIONAL DIRECTORY OF INFORMATION PRODUCTS AND SERVICES. *R.R. Bowker. Annual. October 30, 1981. 266 pg. $37.50.*

Directory of the information industry. Includes publishers, vendors, distribution facilities, equipment manufacturers, associations, government agencies and other businesses and services. Lists business addresses, executives and lines of business. Includes bibliographic information.

3978. INFORMATION TRADE DIRECTORY 1983. *Learned Information (Europe) Ltd. Annual. October 1982. 266 pg.*

Annual contact guide to 1000 information-related businesses throughout the world includes database publishers, online vendors, information brokers, telecommunication networks, library networks and consortia, terminal manufacturers, consultants, and government agencies. Each entry includes address, contact, personnel, and product / service information. Published in the U.S. as 'Information Industry Market Place 1983.'

3979. THE INTERACTIVE CABLE TV DIRECTORY. *Walker, John, editor. Phillips Publishing Inc. Irregular. 1980. 71 pg. $39.00.*

Lists two-way interactive cable television systems and operators, and manufacturers of two-way cable equipment. Includes consultants, associations and research groups in the industry, two U.S. industry groups and four international advisory groups and their members.

3980. INTERNATIONAL CONSTRUCTION WEEK MAJOR PRODUCT REPORTS: GLOBAL REPORTS: COMMUNICATIONS. *McGraw-Hill Book Company. Annual. 1982. 8 pg. $15.00.*

One in a series of annual reports covering communications construction projects in various stages of progress (i.e., pre-investment, early planning, in process) throughout the world. Project listings include telex numbers of companies involved.

3981. JAEGER AND WALDMANN INTERNATIONAL TELEX DIRECTORY: VOLUME 3: ALPHABETICAL SECTION: AMERICA, AFRICA, ASIA, AUSTRALIA. *Belgian American Chamber of Commerce in the U.S. Inc. Irregular. 28th ed. March 1980. 2558 pg.*

Lists telex subscribers in North, South and Central America, Africa, Asia and Australia. Arranged geographically by country. Entries include full name of each business, postal address, and complete telex information.

3982. LIST OF INTERNATIONAL TELEPHONE ROUTES AS OF JANUARY 1, 1981. *International Telecommunication Union. English/French/Spanish. Annual. 1981.*

Yearly release details, for the various services, primary routes and secondary routes as of January 1 of the year of publication. Broken down by region.

3983. LIST OF TELECOMMUNICATION CHANNELS USED FOR THE TRANSMISSION OF TELEGRAMS. *International Telecommunication Union. English/French/Spanish. Annual. 3rd ed. 1980.*

Listings detail name of responsible authority; countries or geographical areas; nature of link; and nature of service.

3984. TELEPHONE ENGINEER AND MANAGEMENT DIRECTORY: THE TELEPHONE INDUSTRY DIRECTORY. *Harcourt Brace Jovanovich, Inc. July 1980. 580 pg.*

Discusses new products and their effects on telecommunications. Directory lists major telephoone industry suppliers, with brief company histories and names, addresses, telephone numbers, key executives, products and services, and branch locations. An extensive product index lists appropriate companies under each category. Gives a state-by-state listing of telephone exchanges, and other important statistics concerning executives and key employees and industrial plants and equipment. Several other directories are provided concerning telephone employees and a directory of foreign telephone systems.

3985. TELEPHONY'S DIRECTORY AND BUYERS' GUIDE: 81. *Telephony Publishing Corporation. Annual. 1981.*

Directory of U.S. and Canadian telephone utilities (Bell and independent), holding companies, regulatory associations and commissions, related carriers (satellite, telegraph, etc.), and over 2300 manufacturers / suppliers of equipment, supplies and services to the trade (which are cross-referenced by product / service group and by trade name). Entries include executives' names, products and trade names, and plant information. International section lists size of telephone system, and names of industry and government regulatory personnel, in several other nations.

3986. 1980 WORLD'S SUBMARINE TELEPHONE CABLE SYSTEMS. *Information Gatekeepers, Inc. Irregular. 1980. 350 pg. $100.00.*

Lists undersea telephone cables installed worldwide from 1950 through 1980, as well as new systems planned as of 1980, both geographically and chronologically. Includes specifications, maps, a directory of cable operators, a glossary and a bibliography.

3987. "82 VIDEOTEX DIRECTORY. *Arlen Communications. Irregular. June 1982. $150.00.*

Categories include system operators, information providers, service providers, hardware and software vendors, industry associations, government agencies, the industry in about 25 nations outside the U.S., consultants roster, and industry-related publications.

See 26, 30, 3456, 3457.

Special Issues and Journal Articles

Communication Technology Impact

3988. VIDEOTEXT FOR TECHNICAL INFORMATION SERVICES. *Technical Research Center of Finland. Elsevier Science Publishers. Irregular. v.4, no.7. October 1982. pg.8-10*

Summarizes an analysis of national videotex systems. Gives, by country, status of videotext systems, 1982. Includes service name, date, terminals, source of technology, and standard.

Infotecture

3989. INTERNATIONAL PRIVATE LEASED LINES. *Infotecture Europe. Irregular. no.14. July 26, 1982. pg.5*

Table of tariffs for M1020 voice and data lines from London to other countries. Shows annual U.K. and foreign shares and the currency.

Satellite News

3990. SATELLITE NEWS SATELLITE DIRECTORY. *Phillips Publishing Inc. Annual. 1981. $85.00.*

Annual directory of satellite end users and purchases. Includes industry analysis and historical data.

Telecommunication Journal

3991. TABLE OF ARTIFICIAL SATELLITES LAUNCHED IN 1979. *International Telecommunication Union. Annual. 1980. 12 pg.*

Inventory of artificial satellites launched during the preceding year, specifying each device's launching nation or organization, launch date and site, flight path parameters, and purpose. Communications satellites' frequencies and transmitter power are indicated where possible; on-board instruments are occasionally described. Includes a list of satellites which have 'decayed' since the publication of the previous year's edition of this table.

Telephony

3992. EVALUATING TELECOM NETS REQUIRES UPDATED INDICATORS. *Boisos, Charles. Telephony Publishing Corporation. v.203, no.14. September 27, 1982. pg.36-40*

Assessment of economic indicators for telecommunications networks shows density of units in operation, 1978, and percent of extension installations, 1960-1978.

3993. NON-U.S. TELECOM SPENDING TOPS $60 BILLION MARK. *Fargo, Dan S. Telephony Publishing Corporation. Irregular. February 23, 1981. pg.51-61*

Examines telecommunication construction expenditures (1975-1980) outside the U.S. and forecasts 1981-1982 outlook. Table details world telephone statistics, by country, including estimated population of serving area, total telephones, total telecom plant gross investment, revenues (1980) and construction expenditures (1979). Provides growth rates (1927-1980) of overseas telephone service.

3994. UNDERSEA CABLE SYSTEMS PLUNGE INTO THE WATERS OF A NEW TECHNOLOGY. *Telephony Publishing Corporation. v.203, no.8. August 23, 1982. pg.24-30*

Examines technological development in the undersea cable systems industry. Gives status of global cable network, December, 1981; growth of global cable network (including dollars invested), 1977-1981 and transoceanic cable and regional cable systems under construction, June 1982.

See 34, 35, 46, 834.

Monographs and Handbooks

3995. THE BASIS OF COMPETITION IN THIRD - WORLD TELECOMMUNICATION MARKETS. *Kamman, Alan B. Arthur D. Little, Inc. 1980.*

Forecasts to 1990 for telecommunications markets in Third World nations, government actions, local manufacturers, and production by multinationals. Includes historical data.

3996. COMING INFORMATION AGE: AN OVERVIEW OF TECHNOLOGY, ECONOMICS AND POLITICS. *DiZard, Wilson. Longman Inc. February 1982. pg.256 $24.95.*

Covers technical, economic, and political aspects of the information and communications industry.

3997. COMPENDIUM OF COMMUNICATION AND BROADCAST SATELLITES, 1958 TO 1981. *Brown, M.B. Wiley - Interscience. A Division of John Wiley and Sons. 1981. 400 pg. $32.00.*

A reference source that presents overviews and comparisons of most communication and broadcast satellite systems. Outlines major transmission parameters, physical characteristics and general information such as: number of satellites built, launch vehicle, prime contractor, and more. Gives the date of introduction into service and design life-time for each satellite, and supplies a frequency plan showing the communication and broadcast bands used.

3998. INTERNATIONAL TELECOMMUNICATIONS. *Lancaster, Kathleen Landis., ed. Lexington Books. Division of D.C. Heath and Company. March 1982.*

Investigates recent developments in communications technologies, services, and markets. Assesses international effects of these changes and evaluates new directions for corporate strategies.

3999. VIDEOTEX: THE NEW TELEVISION / TELEPHONE INFORMATION SERVICES. *Woolfe, Roger. Heyden and Son Ltd. 1980. pg.1-170*

Examines videotex information services, market prospects, costs and revenues of the services. Considers development in the Western countries, especially West Germany's Bildschirmtext.

See 839, 841, 3467.

Conference Reports

4000. COMMUNICATIONS, INFORMATION PROCESSING AND THE PRODUCTIVITY REVOLUTION: THE THIRD ANNUAL FORUM IN INTERNATIONAL TELECOMMUNICATIONS: NOVEMBER 7-9, 1979. *Kamman, Alan B. Arthur D. Little, Inc. November 7, 1979. pg.92-99 pg.100-114 pg.145-160 $750.00.*

1979 forum on international telecommunications industry. Selected three articles are on: telecommunications equipment markets (1970-1990), mainly outside U.S.; sales, revenues,

ranked leading manufacturers and markets; and business communications equipment markets (typewriters, copy machines) and products in U.S. (1980-1990). Data derived from Arthur D. Little, Inc. estimates, U.S. Dept. of Commerce, corporate reports, and AT&T.

4001. FOURTH INTERNATIONAL ONLINE INFORMATION MEETING. *Learned Information, Inc. 1981. 527 pg. $52.00.*

Sixty-five papers from 1980 conference cover such issues as the growing role of microcomputers; expansion in use of graphics equipment; and the burgeoning videotex industry.

4002. INTERNATIONAL BROADCASTING CONVENTION 1980. *Institute of Electrical and Electronics Engineers, Inc. 1980. 354 pg. $65.50.*

82 papers on broadcasting technology in the 1980's cover signal origination; digital television coding standards; teletext; transmitters and transposers; signal processing; multi-channel sound; recording and storage; signal distribution; satellites in broadcasting; and measurement techniques.

4003. NEW SYSTEMS AND SERVICES IN TELECOMMUNICATIONS. *Cantraine, G., ed. North-Holland Publishing Company. 1981. 368 pg. $69.75.150.00 (Netherlands).*

Proceedings of an international conference in Liege, Belgium, November 24-26, 1980 cover videotex and teletext systems; data broadcasting; specialized satellite telecommunication services; direct satellite broadcasting; interactive data retrieval (videotex); text communication services (teletex); data broadcasting; telematics and teleconferences; graphic display on television screen; direct satellite broadcasting; and new trends in picture and data distribution.

4004. PROCEEDINGS OF THE SIXTH ANNUAL TELECOMMUNICATIONS POLICY RESEARCH CONFERENCE. *Dordick, Herbert S., editor. Lexington Books. Division of D.C. Heath and Company. 1979. 496 pg. $21.95.*

Thirty-three conference papers concern television's social effects and the policy issues they create; computer - communications networks and the emerging information economy; spectrum allocation; economics of regulation; and economics of pricing telecommunications services.

4005. SATELLITE COMMUNICATIONS. *Online Publications, Ltd. 1980. 134 pg. 40.00 (United Kingdom).*

International 1980 conference proceedings include recent developments; SBS and the North American environment; the DARPA Atlantic packet experiment; the European satellite and TV network plan; the press and satellites; WARC-79 and satellites; and satellites and banking.

4006. TELECOMMUNICATIONS AND PRODUCTIVITY: AN INTERNATIONAL CONFERENCE JANUARY 29-30, 1980. *Moss, Mitchell L., editor. Addison-Wesley Publishing Company, Inc. 1981. 396 pg. $37.50.*

The papers deal with the potential and costs of telecommunications systems. They examine the office of the future, office automation, services to the home, selling the concept of teletext, and consumer information databases. Papers cover state and local government applications and telephone conferencing.

4007. TELECOMMUNICATIONS ENERGY CONFERENCE - INTELEC 81. *Institute of Electrical and Electronics Engineers, Inc. 1981. 371 pg. $80.00.*

65 conference papers cover microprocessors for telecommunications power; distribution and economics; batteries; energy conservation and building environment; inverters; solar and wind power systems; uninterruptible power supplies; convert-

ers; AC stand-by systems; DC power plant and rectifiers; and reliability.

4008. TELEINFORMATICS "79. *Boutmy, E.J., ed. North-Holland Publishing Company. 1979. 316 pg. $69.75.150.00 (Netherlands).*

Proceedings of June 11-13, 1979 conference in Paris cover anticipated revolution and impact of teleinformatics through the 1980's. Sections include impact on organizations; message systems; teleconferencing; computer-assisted instruction; videotex; technical issues; and case studies (man-machine with actions, political issues).

4009. VIEWDATA 81: PROCEEDINGS OF THE SECOND WORLD CONFERENCE ON VIEWDATA, VIDEOTEX AND TELETEXT. *Online Publications, Ltd. 1981. $98.00.*

Concentrating on developments in the U.K., France, U.S., Canada, Japan, and West Germany, coverage spans marketing and service providers; the impact on newspapers and publishing; transactional videotex; social and policy issues; and videotex in the home. Includes private videotex in business; systems linkages; regulations and standards; compatibility and standardization; burgeoning systems and facilities; and videotex in the public sector.

See 846, 849, 851, 854, 857.

Newsletters

4010. MONITOR. *Learned Information, Inc. Monthly. $200.00/yr.*

Monthly analytical review of events in the online and electronic publishing industries covers developments in areas such as videotext, teletext, graphics terminals, intelligent micros, consumer use of time-sharing systems, and full-text online ordering.

4011. VIDEOTEX TELETEXT NEWS. *Arlen Communications. Monthly. $180.00/yr.*

Newsletter provides international coverage of business, legal, regulatory, and technological aspects of the videotex and teletext industries.

4012. VIEWTEXT: COMPLETE COVERAGE OF THE WORLDWIDE VIEWDATA / TELETEXT MARKET. *Ferrarini, Elizabeth M. Information Gatekeepers, Inc. Monthly. v.4, no. 10. 20 pg. $125.00/yr.*

Monthly survey of service / product market, technical, and regulatory / legislative developments in the field of video data communications in North America and worldwide. Discusses activities of individual corporations.

Services

4013. THE INTERNATIONAL TELECOMMUNICATIONS MARKET. *Yankee Group. February 1980. 260 pg. $9500.00.*

Provides a market analysis for telecommunications equipment manufacturers and specialized common carriers. Includes 5-year projections on a country by country basis for specific products, peripheral equipment and technology transfer. Addresses problems of international telecommunications such as government bureaucracy, U.S. export policy and transborder data flow regulations. Statistical figures on exports 1954, 1970, government aid 1977, taxes 1977, loans 1949-1978, revenues 1951-77, markets 1979, growth rates 1960-1990, and market shares 1977. Volume 5 of Yankee Group's

Communications Inform- ation Systems Planning Service and available only to subscribers.

Indexes and Abstracts

4014. COMMUNICATIONS TECHNOLOGY. *Institute of Electrical and Electronics Engineers, Inc. Monthly. 1982. 24 pg. $80.00/yr.*

Each issue of monthly abstract service contains abstracts and bibliographic information on approximately 250 recent papers and reports published throughout the world.

4015. PREDI-BRIEFS: COMMUNICATIONS. *Predicasts, Inc. Monthly. 1981.*

One in a series of 29 industry-intensive abstract periodicals derived from Predicasts' PROMT. Abstracts 200 to 400 articles per month covering such topics as acquisitions, capacities, end uses, government regulation, market shares and statistics, and new technology.

North America

Market Research Reports

4016. THE AT & T REORGANIZATION. *Probe Research, Inc. 1980. 75 pg. $95.00.*

A market analysis reporting on the impact of AT&T's management reorganization on regulators, competitors and suppliers. Covers details of the new structure, marketing at AT&T, regulatory issues, and the strengths and weaknesses of the reorganization for those coordinating with AT&T.

4017. BELL SYSTEM SELECTED U.S. INDEPENDENT EMPLOYMENT CHANGES 1976-1980: BY OCCUPATION AND U.S. REGION. *Telephony Publishing Corporation. 1981. $35.00.*

Market report provides information on sales, advertising, and market size. Provides employment statistics by market segment.

4018. BELL TELEPHONE'S PRODUCT DEMAND AND SUPPLIERS. *National Association of Home Builders of the United States. June 1980. 448 pg.*

Describes product demand, suppliers, and the major construction projects of the Bell System. Contains statistics from 1976-1980 and forecasts to 1982. Tables describe general trade sales to the system, specific product sales, and expenditures by the Bell System.

4019. A BLURRING OF THE LINES: ADVERTISING, MARKETING AND THE NEW MEDIA TECHNOLOGIES. *Kalba, Konrad K. Kalba Bowen Associates Inc. December 10, 1979. 123 (est.) pg. $90.00.*

Examines the growth of new media technologies (cable and subscription TV, interactive cable, viewdata and teletext services, satellites and videodisc) and their feasibility as advertising and marketing devices. Discusses potential impact on consumers: benefits to consumers of media diversity; prices and other access barriers; control and monitoring; and limitations of new message formats. Includes lists of the 25 largest cable multiple system operators; the 20 largest cable TV systems; 19 pay cable services, their affiliates, subscribers and market shares; subscription TV stations; and 30 MDS pay TV services with their monthly rates, monthly revenues and number of subscribers. Reviews worldwide videotext system development. Lists 16 countries, with system and operator names. Additional services covered include multi-point distribution, superstations, telephone and standalone products (personal computers, electronic calculators and videocassettes).

4020. THE CABLE, THE PHONE OR THE STANDALONE: WHO WILL DELIVER THE ELECTRONIC MESSAGE? *Kalba, Konrad K. Kalba Bowen Associates Inc. January 1980. 16 pg. $20.00.*

Examines and forecasts to 1990 the U.S. industries which may have a major role in the new home communications market. Covers cable TV; telephone; consumer electronics; computers; publishers and programmers; and others, including broadcasters, the investment community, and the government.

4021. THE CHANGING STRUCTURE OF THE U.S. TELECOMMUNICATIONS INDUSTRY: BUSINESS INTELLIGENCE PROGRAM. *SRI International. Marketing Services Group. December 1979. 20 pg.*

A 1979 market report on the telecommunications industry with market segments analyzed, distribution, revenues, shipments and total inventories for 1979-1990 given. Also competitive leading and smaller manufacturers and vendors listed. Expenditures by users also given (1972 data). Publication is part of SRI's Business Intelligence Program series.

4022. COMMUNICATIONS IN THE U.S. *Larsen Sweeney Associates Ltd. 1981. 395.00 (United Kingdom).*

Reports on the U.S. communications industry. For 17 subsectors, gives installed base of plants and equipment by product, analysis of industry and market issues, market penetration strategies, and a survey of end users.

4023. COMMUNICATIONS INDUSTRIES: INTRODUCTION AND OVERVIEW: CONFLUENCE, CONFLICT, CONJUNCTION. *FIND / SVP. December 1981. 39 pg. $250.00.*

Analyzes size of trends in and interactions among related industries: telecommunications, common carriers, computers, consumer electronics, CATV, broadcasting, computer services, publishing, and software and hardware vendors.

4024. COMMUNICATIONS INDUSTRIES: LOCAL AREA NETWORKS. *FIND / SVP. December 1981. 18 pg. $250.00.*

Covers broadband vs. baseband modulation; media access methods; network topology; media alternatives; protocols use; gateways; alternative network offerings; presently available LANs schemes; and PBX as a local area network.

4025. COMPETITION AND STRATEGY (TELECOMMUNICATIONS). *Probe Research, Inc. Monthly. October 1980. $140.00.*

Monthly subscription publication offering news and interpretive articles on telecommunications. Company profiles, strategies and rankings; market size analysis; technological developments; and legal and regulatory developments. Publication is available as yearly subscription only.

4026. THE COMPETITIVE ENVIRONMENT: PROGRAM ON ADVANCED HOME COMMUNICATIONS. *Kalba Bowen Associates Inc. May 1981.*

One of a series of multi-client studies on advanced home communications systems. Examines the significant strengths and weaknesses of the U.S. telephone and cable TV industries, and their market penetration in the residential sector. Profiles major businesses in each industry. Discusses competition. Examines the future market role of related industries.

4027. THE CONSUMER ENVIRONMENT: PROGRAM ON ADVANCED HOME COMMUNICATIONS. *Kalba Bowen Associates Inc. September 1981.*

One of a series of reports examining the advanced home communications consumer market in the U.S. Covers such aspects as entertainment and other related consumer expenditures; societal trends and their effects on the prospects for home communications systems; and consumer attitudes toward advanced communications products and services. Some products and services examined include home shopping, pay television, video games, information retrieval, telephone banking, and home security. About 16 tables feature personal consumption expenditures and demographics of current owners of the products and services discussed.

4028. CURRENT INITIATIVES IN MASS-MARKET TRANSACTIONAL SERVICES. *Link Resources Corporation. 1981.*

Forecasts utilization and applications of, and investments in, mass-market data communications services through 1984. Discusses activities and strategies of major banks.

4029. DATA COMMUNICATIONS SERVICES MARKET. *Frost and Sullivan, Inc. October 1981. 249 pg. $1200.00.*

Contains forecasts through 1985 of expenditures by DP users for data communication line charges and hardware. Forecasts revenues to 1985 for the following: telephone carriers; record telegraph carriers; specialized common carriers; and VAN carriers. Contains user questionnaire survey results.

4030. DATA - TELECOMMUNICATIONS PROGRESS REPORT. *Business Communications Company. March 1981. 118 pg. $850.00.*

Progress report on the telecommunications industry including satellites, terminals, data and others. Includes statistics and forecasts in some 20 telecommunications areas. Discusses key participants and market penetration and segmentation.

4031. THE DEMAND FOR RESIDENTIAL TELEPHONE SERVICE. *Mahan, Gary P. Graduate School of Business Administration - Michigan State University. Michigan State University. Graduate School of BusinessAdministration. Bureau of Business & Economic Research. 1979. 210 pg. $6.00.*

Report on the demand for residential telephone service investigates the effect of such factors as price, family income, urbanization, race and social class, and calling scope on types of service for a sample of 2,000 households.

4032. DIGITAL TELECOMMUNICATIONS MARKET. *Frost and Sullivan, Inc. January 1981. 356 pg. $1100.00.*

Market analysis for digital telecommunications in the U.S. Covers major manufacturers, utilization, regulation, sales, and shipments. Forecasts to 1985.

4033. DIRECT BROADCAST SATELLITES (DBS). *Frost and Sullivan, Inc. March 1982. 30 pg. $300.00.*

Analysis by a Wall Street firm of the direct broadcast satellite industry, with forecasts, 1982-1985. Examines the corporations, market potential, proposed satellite units in operation, costs, services, technical specifications, prices, and competition.

4034. DIRECT BROADCAST SATELLITES: PRELIMINARY ASSESSMENT OF PROSPECTS AND POLICY ISSUES. *Ayvazian,Berge. Kalba Bowen Associates Inc. 1980. 83 pg. $50.00.*

Examines the direct broadcast satellite (DBS) television service in the U.S. Focuses on the technological, economic and regulatory factors that will affect its development. Discusses types of companies that could offer the service. Provides a financial model of whether a DBS service is profitable, including pro forma cash flow statements (1982-1990).

4035. ELECTRONIC CONFERENCING - IMPACT AND OPPORTUNITIES. Strategic Inc. September 1981. 115 pg. $1200.00.

Analysis of markets for electronic conferencing products, services and facilities, with forecasts to 1990. Includes audio, video, computing and transmission equipment; transmission and computing services and production; and conference facilities.

4036. ELECTRONIC MARKETING. Strauss, Lawrence. Knowledge Industry Publications. 1982. 160 pg. $34.95.

Examines the implications for marketing and advertising of new electronic media (cable TV, videotext, video discs, etc.). Describes the status of present electronic marketing experiments (Comp-U-Card, Viewtr

4037. ELECTRONIC PUBLISHING IN THE HOME OF THE 1980'S. Frost and Sullivan, Inc. December 1981. 32 pg. $350.00.

Analysis by a Wall Street firm of cable news, teletext and videotext markets. Examines costs of services, vendors, technology and regulations. Includes market surveys of user attitudes, market areas, numbers of consumers, and prices. Provides forecasts to 1985 and 1990.

4038. ELECTRONIC YELLOW PAGES. Brigish, Alan P. International Resource Development, Inc. December 1980. 178 pg. $985.00.

Analysis of the electronic yellow pages, its impacts on the industry, and competition with other forms of advertising. Covers industry structure, the role of corporations and government actions, products, services, and their utilization. Forecasts computer and other equipment units in operation, advertising expenditures, and revenues to 1990. Shows historical data on advertising 1776-1980; newsprint prices, 1965-1985; the number of newspaper businesses, and profits, 1980-1985.

4039. FIBER OPTIC TELEPHONE INDUSTRY INSTALLATIONS 1980-1984 AND MAJOR NEW DEVELOPMENTS: A FOUR-PART STUDY. Probe Research, Inc. 1980. $90.00.

This study is divided into four sections. "Fiber Optic Field Installations, 1980-1984" discusses significant features of new systems including long-wavelength 1300nm links, the 3200 km Sask Tel link, and several T3C links. "The Emergence of Long-Haul Systems" discusses economic comparisons relative to AT&T's 1983-1984 Northeast Corridor. "The Emergence Issue in Bell Fiber Optic Procurement" examines statistics relative to Bell's FOC competitiveness. "The Next Generation: An Update on Long-Wavelength R&D and Its Significance" discusses the VAD process, double mode fibers, new 1300-1700nm lasers and LEDs and detectors by the Japanese and Bell.

4040. 'GET A NEWSPAPER': VIEWDATA IN THE COMPETITIVE MARKETPLACE. Kalba, Konrad K. Kalba Bowen Associates Inc. August 1980. 7 pg. $20.00.

Report examines the emerging marketplace for viewdata in the U.S. Discusses competition from the telephone industry, consumer electronics, newspapers and the smaller cable TV industry. Examines applications of newspapers to viewdata equipment.

4041. GOVERNMENT AND MILITARY TELECOM USERS: AS COMPARED TO PRIVATE BUSINESS. Telephony Publishing Corporation. 1981. $150.00.

Market report ranks most important purchase criteria for business and government. Contains supplier information.

4042. THE HALF BILLION DOLLAR MARKET. Telephony Publishing Corporation. 1980. 72 pg. $40.00.

Surveys of 301 communications users and interconnect businesses' minimum spending, buying hierarchy and why they do, or do not look to Telcos for equipment and service advice.

4043. HOME OF THE FUTURE PLANNING SERVICE: INTERACTIVE, ONE-WAY AND STANDALONE ENTERTAINMENT. Yankee Group. April 1981. 200 pg. $8500.00/yr.

One in a series of six reports forming part of " "Home of the Future Planning Service.' Topics include role of networks and cable systems in entertainment delivery; future of subscription television (STV) and multipoint distribution services (MDS); projected future of interactive two-way cable; survey of opportunities in " "narrowcasting', downline-loaded digital music: the electronic jukebox; and ideas for electronic games and gambling- at-home.

4044. HOME TELECOMMUNICATIONS IN THE 1980'S. International Resource Development, Inc. April 1980. 196 pg. $1285.00.

Analysis of future market for residence telecommunications products and services. Includes analysis of the development of home information systems, integrated video terminals, consumer telephone interconnect, competitive MTS services, "teleshopping" and bank-at-home services. Detailed discussion of experiments in Japan, Europe, Canada and the U.S. Impact of home satellite earth stations and cellular mobile radio.

4045. INDEPENDENT TELEPHONE INDUSTRY'S EQUIPMENT FORECAST AND PLANT ANALYSES 1976-1980. Sonneville Associates. September 1979. 362 pg.

Examines the equipment expenditures and existing plants of 12 large telephone holding companies comprising 257 operating companies and 421 unaffiliated companies each having more than 3000 companies. Lists top 20 unaffiliated telephone companies. Provides general industry statistics; individual large company data; other companies having 3000 stations; touch-phone forecast; forecasts of usage of carrier equipment; and realities of the REA / RTB market.

4046. AN INTRODUCTION TO VIDEOTEX. Butler Cox and Partners Ltd. May 1980.

Overview of the world videotex scene includes a summary of videotex characteristics, end-uses, and plans in different countires.

4047. LOCAL AREA NETWORKS (LAN'S): MARKETS AND OPPORTUNITIES - VOLUME 2. Strategic Inc. November 1981. 150 pg. $1200.00.

Analysis of markets, technology, utilization and major vendors in the local area networks industry. Examines industry structure, competition with other systems and prospects for equipment.

4048. LOCAL AREA NETWORKS MARKET IN THE U.S. Frost and Sullivan, Inc. September 1981. 214 pg. $1200.00.

Contains market analysis and forecasts for local area networks. Forecasts are provided through 1990 in units and dollars for baseband and broadband networks. The major technologies and factors affecting local networks are examined. Companies competing in the network area are profiled.

4049. MARKETING BY THE BELL SYSTEM. *Probe Research, Inc. 1979. 120 pg. $240.00.*

This study provides extensive analysis of the key areas of Bell system marketing: business (team marketing, Bell's spending on marketing, advertising, salespersons and reactions of business accounts); residence (new products and sales, advertising to reach the customer, phone center stores, and openings for competitors); and directory and public service (the directory business, and coin and dial line businesses).

4050. MULTI-CLIENT STUDY: PRELIMINARY FINANCIAL MODEL AND ANALYSIS OF THE RESTRUCTURING OF AT & T UNDER THE SEPARATED ENTITY CONCEPT OF S.611. *Probe Research, Inc. September 1979. $10000.00.*

This study describes AT&T's restructure caused by the Senate's Bill S.611. The study contains estimated income statements and balance sheets, conclusions on the effectiveness of the legislation, impact on AT&T, specific strengths and weaknesses of the Bell System and FCC's development of the concept of "separation" as a tool for restructuring the industry. Probe provides six recommendations to add certainty, development, incentive, and lower costs through competition for this restructuring of AT&T.

4051. NEW OPPORTUNITIES IN SATELLITE / CABLE TELEVISION NETWORKS. *International Resource Development, Inc. 1982. 370 pg. $1285.00.*

Analysis of markets for 'blue' movies on cable and satellite TV networks. Includes advertising expenditures and leading corporations.

4052. THE OUTLOOK FOR ELECTRONIC NEWSPAPERS. *International Resource Development, Inc. 1981. 186 pg. $985.00.*

Analysis of markets for electronic newspapers. Examines new products and services, market segmentation, effects on industry structure, and outlook for vendors.

4053. THE OUTLOOK FOR VIDEOTEX: ELECTRONIC NEWS AND INFORMATION SYSTEMS. *Shapiro, Peter D. Arthur D. Little, Inc. September 30, 1980. 13 pg. $600.00.*

Discusses the progress of experimental Videotex projects, both interactive and broadcast, in the U.S., Europe and Japan. Brief descriptions of important systems being tested and some general forecasts on marketing, regulation, timing and compatibility provided.

4054. PRIVATE AND IN-HOUSE VIDEOTEXT SYSTEMS. *Butler Cox and Partners Ltd. May 1981.*

Report on private videotex systems on the market in Europe and North America. Includes purchase information and intended use. Provides 8 case studies.

4055. THE PRIVATE BUSINESS TELECOMMUNICATIONS USERS. *Telephony Publishing Corporation. January 1981. 130 pg. $250.00.*

Analysis of private business telecom user market covers U.S. businesses employing over 20 people. Examines views of 108 corporations on telcos, interconnects, manufacturers, and consultants. Matches major SCC carrier routes to 1300 corporations.

4056. PROGRAMMING AND SERVICE STRATEGIES FOR 2-WAY TV AND VIEWDATA SYSTEMS. *International Resource Development, Inc. August 1981. 228 pg. $985.00.*

Market and industry analysis with 1981-1985 forecasts for 2-way TV and Viewdata systems. Examines consumer and advertising expenditures, competition for franchises in major

market areas, regulation, consumer behavior, industry analysis, and vendors.

4057. SATELLITE TELEVISION BROADCASTING: BUSINESS OPPORTUNITIES IN SATELLITE BROADCASTING DIRECT TO THE HOME: NORTH AMERICA AND WESTERN EUROPE TO 1990. *Mackintosh Consultants, Inc. July 1981. 4700.00 (United Kingdom).*

Multi-client study exploring market opportunities through 1990 in several aspects of satellite-broadcast television. Covers provision of services, manufacture of equipment, program production, and advertising. Discusses consumer preferences, and potential costs and revenues.

4058. THE TECHNOLOGY ENVIRONMENT: PROGRAM ON ADVANCED HOME COMMUNICATIONS. *Kalba Bowen Associates Inc. June 1981.*

One of a series of multi-client studies on advanced home communications systems. Assesses U.S. telephone and cable TV technologies for application to advanced service. Examines the feasibility of delivering services to the home such as home security, information retrieval, transactions, interactive games, and institutional networks. Also assesses alternative delivery systems including microcomputer networks; fiber optics; local area networks; packet switching; and intelligent videodisc players. Discusses costs, revenues, pricing, competition and other factors.

4059. TELECOMMUNICATION ANALYSIS AND RESEARCH: BASIC REPORT 1980. *Yankee Group. Annual. 1980. 100 pg. $600.00.*

Discusses U.S. telecommunications industry (including regulatory statutes and trends and telecommunications equipment, common carriers and manufacturers) and the telecommunications equipment market 1980. Case studies, market analyses, penetration and potential, competition, new technologies and new products are analyzed for users and vendors. Statistics include: purchases 1976-1980, revenues 1976-1980, companies involved in telecommunications 1976- 1980. Accompanied by a quarterly update which profiles companies within the industry and technological developments and changes: AT&T's Dataphone Service, telephone monitoring and control equipment (TMAC), electronic message networks.

4060. TELECOMMUNICATIONS AND DATA PROCESSING. *FIND / SVP. November 1981. 15 pg. $225.00.*

Covers industry trends and the current economic and legislative environment. Assesses Dataproducts Corp.; Harris Corp.; Northern Telecom, Ltd.; M/A-Com, Inc.; Scientific-Atlanta, Inc.; Paradyne Corp.; Rolm Corp.; Cullinane Database Systems, Inc.; and Tymshare, Inc.

4061. TELECOMMUNICATIONS INDUSTRY TRENDS. *Sanford C. Bernstein and Company, Inc. Quarterly. 1981. $1000.00/yr.*

Quarterly market analysis reports. Evaluates potential of new products and services, forecasts future demand, analyzes effects of government regulations on profits, and tracks operating performance of major corporations.

4062. TELECOMMUNICATIONS MARKET OPPORTUNITIES - IMPACT OF THE AT & T SETTLEMENT. *International Resource Development, Inc. January 1982. 270 pg. $1285.00.*

Report of effects on the telecommunications industry structure from recent decisions on the regulation of AT & T. Examines new competition in equipment and services; historical data on communications legislation; projections of communications units in operation and market penetration; new lines of business for AT & T; existing vendors; forecasts of revenues to

1990; shipments of equipment, 1982-1992; and analysis of markets, 1982-1992. Lists addresses of corporations.

4063. TELEPHONE ANSWERING AND RADIO PAGING SERVICES AND ASSOCIATED EQUIPMENT MARKET. *Frost and Sullivan, Inc. September 1981. 248 pg. $1200.00.*

Contains analysis and forecasts for the telephone answering and radio paging servces and associated equipments. A 10-year market forecast is provided for TAS: by traditional equipment and manufacturers, by number of subscribers and TAS's for radio paging by wireline and radio common carriers, private users and by equipment. Gives analysis of industry changes and profiles the industry according to size, geographical breakdown and end-users.

4064. TELEPHONE COMPETITION: A PERILOUS TRANSITION. *Frost and Sullivan, Inc. January 1981. 26 pg. $300.00.*

Market report prepared by a leading Wall Street securities firm analyzes telephone industry and competition. Report includes current industry earnings estimates.

4065. THE TELEPHONE INDUSTRY: OEM MARKET. *Telephony Publishing Corporation. 1980. 19 pg. $20.00.*

Telecommunications industry design engineers describe their sources of product information and the value they place on those sources.

4066. TELEVISION AS THE HOME COMMUNICATIONS TERMINAL. *International Resource Development, Inc. 1981. 176 pg. $985.00.*

Estimates U.S. television sets produced by 1991 and their compatability with teletext systems. Projects annual spending by U.S. households on videotext services. Examines prices of videotext signal decoders and uses by manufacturers of television receivers.

4067. TWO-WAY TV AND VIEWDATA SYSTEMS. *International Resource Development, Inc. August 1981. 228 pg. $985.00.*

Analysis of market and industry structure for two-way TV and viewdata systems. Examines the technology; current research and development; market size forecasts to 1990 for various services; market penetration, 1980-1990; and vendors' marketing strategies.

4068. U.S. BUSINESS AND RESIDENTIAL PHONES. *Telephony Publishing Corporation. 1980. 72 pg. $60.00.*

Examines the distribution of Bell System and independent telephones against the background of U.S. population, income and age distribution.

4069. U.S. IMPACT OF VIDEOTEX / TELETEXT INFORMATION AND COMMUNICATIONS FOR THE HOME MARKET. *Strategic Inc. July 1981. $1200.00.*

Forecasts market for videotex / teletext services, (1985, 1990). Equipment includes interactive terminals and specially-equipped TV sets; computer and peripherals for use in CATV headends or in information databases; and printers and transaction terminals. Services include telephone and cable-TV communications facilities used in accessing specific databases provided by information providers, as well as home security, electronic banking and tele-shopping.

4070. U.S. TELEPHONE COMPANY BUYER / SPECIFIERS: HIERARCHY, SIZE AND INFORMATION SOURCES. *Telephony Publishing Corporation. 1979. 53 pg. $35.00.*

Provides information for suppliers to the telephone industry in determining individuals responsible for purchasing, and the value operating company purchase influences place on vari-

ous sources of equipment and product information. Examines questionnaires returned as a result of Fosdick Leadership Studies conducted for Telephony Magazine during 1978. Sampling base consisted of 200 Bell System companies and 100 Independents.

4071. VALUE ADDED AND SPECIALIZED COMMON CARRIERS. *Yankee Group. May 1980. 220 pg. $9500.00.*

Examines the current specialized common carrier industry (revenues, 1980-1984, markets, 1980-1984) as well as the proposed value-added networks in terms of their impact on the data communications market. Specialized common carriers are Microwave Communications Inc., Southern Pacific Communications, ITT, Western Union Telegraph Co. Value-added carriers are GTE / Telenet, Tymnet, ITT. User survey profiles usage 1980-1983 by type of carrier, interconnect device, revenues, and user interface. Volume 8 of Yankee Group's Communications Information System Planning Service and available only to subscribers.

4072. VIDEOTEX BUSINESS APPLICATIONS MARKET. *International Resource Development, Inc. April 1982. 136 pg. $1285.00.*

Analysis of market for videotex. Examines end uses of services, comparing business market size with families and households. Includes vendors; technology; market segmentation, prices, and marketing strategies; forecasts of sales, revenues, and growth rates, 1982-1992; and addresses of businesses.

4073. VIDEOTEX REPORT SERIES 1982: PRIVATE AND IN - HOUSE VIDEOTEX SYSTEMS. *Butler Cox and Partners Ltd. Frost and Sullivan, Inc. September 1982. $750.00.*

Analyzes markets for private, in-house videotex systems for corporations. Examines prices, features, utilizations, management and users' attitudes and expansion plans.

4074. VIDEOTEX REPORT SERIES 1982: VIDEOTEX APPLICATIONS AND THE MARKET. *Butler Cox and Partners Ltd. Frost and Sullivan, Inc. October 1982. $750.00.*

International analysis of videotex markets. Examines the services and facilities available to businesses and households, forecasts of market penetration, industry revenues, user attitudes and expenditures and utilization.

4075. VIDEOTEX REPORT SERIES 1982: VIDEOTEX COSTS AND ECONOMICS. *Butler Cox and Partners Ltd. Frost and Sullivan, Inc. December 1982. $750.00.*

Analyzes costs and profits for videotex vendors and the benefits for businesse and household consumers. Forecasts trends in technology costs, labor costs and disposable income. Examines prices, demand, investments, returns and industry revenues.

4076. VIDEOTEXT: THE COMING REVOLUTION IN HOME / OFFICE INFORMATION RETRIEVAL. *Sigel, Efrem, editor. Knowledge Industry Publications. 1980. $24.95.*

Directory of videotext systems such as Ceefax, Prestel, Viewdata, and two-way cable TV. Provides comparisons and analyses of products and their manufacturers.

4077. VIDEOTEXT: THE KEY ISSUES. *Butler Cox and Partners Ltd. July 1980.*

Examines the market prospects for videotext suppliers covering technological and development trends.

***See 59, 60, 73, 85, 88, 92, 93, 97, 103, 112, 117,
136, 148, 162, 163, 165, 169, 176, 177, 178, 181,
182, 189, 190, 191, 908, 912, 920, 976, 981,***

1001, 1026, 1029, 1073, 1074, 1138, 1143, 3144, 3160, 3185, 3209.

Investment Banking Reports

4078. BEAR STEARNS AND CO.: ELECTRIC, GAS AND TELEPHONE EQUITY MARKET MODEL: AUGUST UPDATE. *Mishara, Donald S. Bear Stearns and Company. Monthly. August 6, 1981. 40 pg.*

August 1981 update of electric, gas and telephone utilities financial peformance. Reviews market trends and tax legislation. Shows earnings per share, 1976-1982; dividends, 1976-1982; stock price, 1981; yield, 1981; price earnings ratio, 1981-1982; returns on capital, 1977-1981; financial requirements and sources, including 1981 expenses, short and long-term debt, preferred and common stocks; rankings of corporations by investment criteria; financial statements; 1981 trading volume; and percent of power generation by fuels, 1980, 1981, 1985.

4079. BEAR STEARNS AND CO.: ELECTRIC, GAS AND TELEPHONE EQUITY MARKET MODEL: DECEMBER UPDATE. *Mishara, Donald S. Bear Stearns and Company. December 4, 1981. 39 pg.*

Performance of electric, gas, and telephone utility investments, December 1981. Gives stock prices; dividends; yields; earnings per share, 1980-1982; return on equity, 1980; average annual growth rates, 1970-1980; rankings, 1977-1980; return on capital, 1978-1982; capitalization, 1981; and power generation by fuel consumed, 1980-1981, 1985.

4080. BEAR STEARNS AND CO.: ELECTRIC, GAS AND TELEPHONE EQUITY MARKET MODEL: SEPTEMBER UPDATE. *Mishara, Donald S. Bear Stearns and Company. September 2, 1982. 33 pg.*

Reviews electric, gas and telephone utilities' stock performance, September 1982. Includes budget for construction expenditures, 1981-1986; investment rankings of corporations, 1977-1981; recommended purchases and sales of stocks; stock prices; dividends; yield; earnings per share, 1981-1983; price earnings ratio, 1982-1983; return on equity, 1981; growth rates, 1971-1981; returns on capital, 1979-1983; valuations; and percent of power generation by type of fuel consumption, 1981, 1982, 1986.

4081. BEAR STEARNS AND CO.: ELECTRIC, GAS AND TELEPHONE UTILITY DIVIDEND INCREASE POSSIBILITIES FOR THE MONTH OF MARCH OR NEXT QUARTER (JUNE). *Mishara, Donald S. Bear Stearns and Company. February 26, 1982. pg.33-35*

Report of dividend increases for electric, gas, and telephone utility corporations. Also includes stock prices; yields; earnings per share, 1980-1982; and investment rankings.

4082. BEAR STEARNS AND CO.: RESEARCH DEPARTMENT HIGHLIGHTS: ELECTRIC, GAS AND TELEPHONE UTILITY DIVIDEND INCREASE POSSIBILITIES FOR THE MONTH OF FEBRUARY OR NEXT QUARTER (MAY). *Mishara, Donald S. Bear Stearns and Company. January 29, 1982. pg.20-23*

Report on possible dividend increases for electric, gas, and telephone utilities. Lists corporations, old and new rates, new annual rates, 1982 stock prices, yields, 1980-1982 earnings per share, and investment rankings.

4083. BEAR STEARNS AND CO.: RESEARCH DEPARTMENT HIGHLIGHTS: ELECTRIC, GAS AND TELEPHONE UTILITY DIVIDEND INCREASE POSSIBILITIES FOR THE MONTH OF NOVEMBER OR NEXT QUARTER (FEBRUARY). *Mishara, Donald S. Bear Stearns and Company. October 30, 1981. pg.29-31*

Report on electric, gas and telephone utilities with possible dividend rate increases. Shows rates for corporations with increases in October. Also shows recent 1981 stock price; dividend; yield; earnings per share, 1980-1982; quarterly payments; and investment rankings for utilities with anticipated dividend increases.

4084. BEAR STEARNS AND CO.: RESEARCH DEPARTMENT HIGHLIGHTS: ELECTRIC, GAS AND TELEPHONE UTILITY DIVIDEND INCREASE POSSIBILITIES FOR THE MONTH OF OCTOBER OR NEXT QUARTER (JANUARY). *Mishara, Donald S. Bear Stearns and Company. October 2, 1981. pg.9-12*

Analysis of electric, gas, and telephone utilities likely to raise dividends. Lists corporations with recent increases. For corporations considered likely to follow, shows 1981 stock price; current dividend and yield; earnings per share, 1980-1982; and investment ranking.

4085. BEAR STEARNS AND CO.: RESEARCH DEPARTMENT HIGHLIGHTS: ELECTRIC, GAS AND TELEPHONE UTILITY EQUITY MARKET MODEL: FEBRUARY UPDATE: SUMMARY. *Mishara, Donald S. Bear Stearns and Company. Monthly. February 5, 1982. pg.24-27*

Recommendations for investments in electric, gas and telephone utility corporations. Gives stock prices; dividends; yield; earnings per share, 1980-1982; price earnings ratio, 1981-1982; return on equity, 1980; growth rates, 1970-1980; and rankings.

4086. BEAR STEARNS AND CO.: RESEARCH DEPARTMENT HIGHLIGHTS: ELECTRIC, GAS AND TELEPHONE UTILITY EQUITY MARKET MODEL: MARCH UPDATE - SUMMARY OF REPORT DATED MARCH 5, 1982. *Mishara, Donald S. Bear Stearns and Company. March 5, 1981. pg.23-26*

Recommendations for investments in electric, gas and telephone utility corporations. Gives stock prices; dividends; yield; earnings per share, 1980-1982; price earnings ratio, 1981-1982; return on equity, 1980; growth rates, 1970-1980; and investment rankings.

4087. BEAR STEARNS AND CO.: RESEARCH DEPARTMENT HIGHLIGHTS: ELECTRIC, GAS AND TELEPHONE UTILITY EQUITY MARKET MODEL: NOVEMBER UPDATE. *Mishara, Donald S. Bear Stearns and Company. November 6, 1981. pg.30-32*

Brief summary of investment recommendations for electric, gas and telephone utilities. Gives stock price, dividend, yield, 1980-1982 earnings per share, 1981-1982 price earnings ratio. 1980 return on equity, 1970-1980 average annual growth rates, and rankings of corporations.

4088. BEAR STEARNS AND CO.: UTILITY EQUITY REVIEW: ELECTRIC, GAS AND TELEPHONE EQUITY MARKET MODEL: JUNE UPDATE. *Mishara, Donald S. Bear Stearns and Company. Monthly. June 8, 1981. 36 pg.*

Monthly financial analysis model and investment recommendations for U.S. gas, electric, and telephone utilities, June 1981. Tables include stock price index, dividends, yields, and stock prices, 1981; returns of capital, 1977-1981; sources and requirements for capital including debt, bonds, preferred stock and common stock, 1980-1981; and valuation, 1981. Examines purchase and sales recommendations; earnings per share and price earnings ratio, 1980-1982; return on equity, 1980; rankings by various investment criteria; and market shares of fuels used for power generation, 1979, 1980, 1985.

4089. BEAR STEARNS AND CO.: UTILITY EQUITY REVIEW: ELECTRIC, GAS AND TELEPHONE EQUITY MARKET MODEL: MAY UPDATE. *Mishara, Donald S. Bear Stearns and Company. Monthly. May 6, 1981. 38 pg.*

Monthly update, financial analysis, and investment recommendations for electric, gas, and telephone utility stocks, May 1981. Thirteen tables report stock performance, showing 1981 stock price, dividends, yields, earnings per share, returns of capital valuations, and price earnings ratios. Date ranges often extended for comparisons and forecasts. Tables also cover current economic indicators, including interest rates and yields on bonds, 1980-1981; financing requirements for electric utilities and sources of capital, including debt, common and preferred stocks, and construction expenditures, 1981; largest common equities in relation to trading volumes, 1980-1981; recommended purchases and sales; and some rankings of corporations.

4090. BEAR STEARNS AND CO.: UTILITY REVIEW: ELECTRIC, GAS AND TELEPHONE EQUITY MARKET MODEL: AUGUST UPDATE. *Mishara, Donald S. Bear Stearns and Company. August 6, 1982. 31 pg.*

Investment analysis model for electric, gas and telephone utilities, August 1982. Includes earnings per share, dividend, and payout ratio, 1976-1983; price earnings ratio, 1981-1983; yield; stock price; relative performance of composite indexes, 1982; expenditures for new plants and equipment, 1981-1986; investment ratings of corporations, 1977-1981; returns on capital, 1978-1982; valuation; and ranks by investment criteria, 1982.

4091. BEAR STEARNS AND CO.: UTILITY REVIEW: ELECTRIC, GAS AND TELEPHONE EQUITY MARKET MODEL: DECEMBER UPDATE. *Mishara, Donald S. Bear Stearns and Company. December 10, 1982. 36 pg.*

Provides investment rankings and stock performance valuations for electric, gas and telephone utility corporations. Includes rankings, 1977-1982; stock prices, dividends, yield, and book value; earnings per share, 1981-1983; price earnings ratio, 1982-1983; return on equity, 1981; growth rates; return on capital, 1979-1983; and composite stock price and other indexes, 1982.

4092. BEAR STEARNS AND CO.: UTILITY REVIEW: ELECTRIC, GAS AND TELEPHONE EQUITY MARKET MODEL: JANUARY UPDATE. *Mishara, Donald S. Bear Stearns and Company. Monthly. January 7, 1982. 39 pg.*

Monthly analysis of financial performance of electric, gas, and telephone utilities, including 1977-1981 rankings and investment recommendations for individual corporations. Includes dividends; returns, 1978-1982; stock price index performance, 1981; yields, 1981; earnings per share, 1980-1982; price earnings ratio, 1981-1982; return on equity, 1980; and percent of fuel types used in power generation, 1980-1981, 1985.

4093. BEAR STEARNS AND CO.: UTILITY REVIEW: ELECTRIC, GAS AND TELEPHONE EQUITY MARKET MODEL: JULY UPDATE. *Mishara, Donald S. Bear Stearns and Company. July 9, 1981. 45 pg.*

Stock performance and market analysis for electric gas, and telephone utilities. Shows stock price index, 1981; dividends and yields, 1980-1981; earnings per share, 1980-1982; price earnings ratio, 1981-1982; and stock prices, 1980-1981. Includes rankings and recommendations for purchase and sales of investments. Also includes stock-to-bond yield spreads 1978-1981; average yields 1954-1981; legislation reg-

ulating returns on equity, 1974-1981; total sales, 1978-1980; 1980 capital requirements sources, including debt and common stock, and returns on capital, 1977-1981; stock valuations; and percent of power generation by fuel type, 1980-1981, 1985.

4094. BEAR STEARNS AND CO.: UTILITY REVIEW: ELECTRIC, GAS AND TELEPHONE EQUITY MARKET MODEL: NOVEMBER UPDATE. *Mishara, Donald S. Bear Stearns and Company. Monthly. November 5, 1981. 32 pg.*

Monthly financial analysis of electric, gas and telephone utilities. Shows earnings per share, 1976-1982; dividends, 1976-1982; current stock prices; yields; price earnings ratio, 1981-1982; breakdown by corporations and lines of business; performance of indexes, 1980-1981; percent of return on capital, 1978-1982; and 1977-1980 investment rankings and recommendations.

4095. BEAR STEARNS AND CO.: UTILITY REVIEW: ELECTRIC, GAS AND TELEPHONE EQUITY MARKET MODEL: OCTOBER UPDATE. *Mishara, Donald S. Bear Stearns and Company. October 13, 1981. 42 pg.*

Financial analysis and investment recommendations for electric, gas and telephone utilities. Shows 1981 stock prices; dividends; yields; economic indicators, 1981-1982; earnings per share, 1980-1982; price earnings ratio, 1981-1982; return on equity, 1980; investment rankings, 1977-1980; and percent of power generation by fuel consumed, 1980-1985. Also comments on dividend reinvestment tax deductions.

4096. DOMINION SECURITIES AMES: INDUSTRY REVIEW: UTILITIES, PIPELINES AND RATES OF RETURN. *Coupal, C.E. Dominion Securities Ames Ltd. November 6, 1981. 24 pg.*

Financial analysis of Canadian telephone, gas, pipeline and electric utility industries, focusing on interest rates and returns. Shows stock price indexes, 1980-1981; stock price as percent of book value, 1981; book value, 1980-1981; return on average common equity, 1981; earnings per share, 1981; dividends, 1972-1980; consumer price index, 1972-1980; yields, 1972-1980; and graphs for individual corporations.

4097. DONALDSON, LUFKIN AND JENRETTE: INDUSTRY VIEWPOINT: THE COMMUNICATIONS INDUSTRIES: INTRODUCTION AND OVERVIEW: CONFLUENCE, CONFLICT, CONJUNCTION. *Wolff, Ivan L. Donaldson, Lufkin and Jenrette Inc. December 15, 1981. 39 pg.*

Analysis of communications industry and market structure, including equipment, telecommunications, broadcasting, cable and computers. Covers competition, regulation and legislation, new products and services, user attitudes, sales and growth rates, 1980-1985; costs, 1930-1980; employment; and technology.

4098. DONALDSON, LUFKIN AND JENRETTE: RESEARCH BULLETIN: AMERICAN TELEPHONE AND TELEGRAPH: THE BREAKUP OF THE BEHEMOTH. *Donaldson, Lufkin and Jenrette Inc. January 12, 1982. 6 pg.*

Report on effects on the telecommunications industry structure, competition among telephone corporations and equipment manufacturers, and results by lines of business, from the regulations requiring the breakup of AT & T. Shows AT & T's percent of revenues by lines of business, 1980.

4099. DONALDSON, LUFKIN AND JENRETTE: THE COMMUNICATIONS INDUSTRIES: DIRECT BROADCAST SATELLITES (DBS). *Leibowitz, Dennis H. Donaldson, Lufkin and Jenrette Inc. March 4, 1982. 30 pg.*

Analysis of the direct broadcast satellite industry (DBS). Includes market analysis; regulation; technology descriptions; services available; competition with other media and within the industry; applications, corporations, and construction costs; market penetration; retail prices; and revenues, expenses, depreciation and net income, 1982-1990.

4100. DONALDSON, LUFKIN AND JENRETTE: THE COMMUNICATIONS INDUSTRIES: INDUSTRY VIEWPOINT: ELECTRONIC PUBLISHING IN THE HOME OF THE 1980'S. *Bosco, Mary C. Donaldson, Lufkin and Jenrette Inc. December 15, 1981. 32 pg.*

Electronic publishing consumer and market analysis. Includes forecasts to 1990, market size in 1985, monthly personal consumption expenditures, industry structure, vendors and costs of services, activity in foreign markets, research and development projects, projections of equipment ownership, and corporations' 1981 stock prices and price earnings' ratios.

4101. DONALDSON, LUFKIN AND JENRETTE: THE COMMUNICATIONS INDUSTRIES: LOCAL AREA NETWORKS: THE NEXT GREAT GROWTH OPPORTUNITY IN OFFICE AUTOMATION. *Vitolo, Aristide J. Donaldson, Lufkin and Jenrette Inc. December 15, 1981. 18 pg.*

Analysis of industry structure, new technologies and corporations involved in local area networks fordata communications. Includes market analysis, market value forecasts, 1981-1990; competition among technologies; and corporations product lines.

4102. KIDDER, PEABODY AND CO.: TECHNOLOGY INDUSTRY PERSPECTIVES: TELECOMMUNICATIONS. *Easterbrook, William D. Kidder, Peabody and Company Inc. Annual. 1982. $350.00.*

Analysis of markets for telecommunications services and hardware products and the changing industry structure. Examines marketing strategies and operating performance of manufacturers and new technology.

4103. MORGAN STANLEY AND CO.: INVESTMENT PERSPECTIVES: TELECOMMUNICATIONS: SIGNIFICANCE OF AT & T DIVESTITURE. *Richards, Gregory P. Morgan Stanley and Company, Inc. January 11, 1982. pg.9-11*

Report on effects of regulation of AT & T on its lines of business, the telecommunications industry structure, and smaller corporations.

4104. OPPENHEIMER AND CO., INC.: INDUSTRY REPORT: THE TELEPHONE INDUSTRY: FCC AUTHORIZES HIGHER RETURN ON INTERSTATE. *Oppenheimer and Company. April 14, 1981. 2 pg. $100.00.*

Report on recent changes in FCC regulation of telephone company rates and returns on equity and the effects on profits. Shows 1981 stock price and 1980-1982 estimated earnings per share for major corporations.

4105. PAINE WEBBER MITCHELL HUTCHINS INC.: STATUS REPORT: EIGHTH ANNUAL CONFERENCE ON THE OUTLOOK FOR THE MEDIA: DECEMBER 9-12, 1980. *Noble, J. Kendrick. Paine Webber Mitchell Hutchins, Inc. August 1981. 175 (est.) pg.*

Report from a conference on print and broadcasting media and the electronic communications industry. Papers by industry and Wall Street executives cover analysis of financial performance, new products and services, recommendations for investments, revenues, costs, analysis of industry structure, markets, government regulation, and earnings and outlook for individual corporations.

4106. PAINE WEBBER MITCHELL HUTCHINS INC.: STATUS REPORT: TELECOMMUNICATIONS INDUSTRY DISCUSSION. *Peery, Bradford L. Paine Webber Mitchell Hutchins, Inc. September 25, 1981. 14 pg.*

Discussion of the telecommunications industry. Covers analysis of industry structure, regulation, competition, prices, management, individual corporations, recommendations for investments, performance of stocks, earnings, and dividends.

4107. PAINE WEBBER MITCHELL HUTCHINS INC.: STATUS REPORT: TELECOMMUNICATIONS MONTHLY. *Peery, Bradford L. Paine Webber Mitchell Hutchins, Inc. Monthly. August 6, 1981. 12 pg.*

Monthly telecommunications industry analysis, covering stock performance, regulation of rates, competition, taxes, and depreciation. Shows stock prices, 1979, 1981; total returns and growth rates, 1980-1981; revenue requirements, 1981-1986; earnings and dividends, 1980-1986; and inflation indicators, 1979-1981.

4108. PAINE WEBBER MITCHELL HUTCHINS INC.: STATUS REPORT: TELECOMMUNICATIONS MONTHLY. *Peery, Bradford L. Paine Webber Mitchell Hutchins, Inc. Monthly. July 16, 1981. 12 pg.*

Monthly analysis of telecommunications companies stock performance. Shows stock price and dividends summed to yield total returns, 1979-1981; toll message growth rates, 1974-1986; and regulation of return on equity, 1980. Includes earnings per share and dividends, 1980-1986; revenue requirements, 1981-1986; and indicators of inflation, 1979-1981.

4109. PAINE WEBBER MITCHELL HUTCHINS INC.: STATUS REPORT: TELECOMMUNICATIONS MONTHLY, APRIL 1981. *Peery, Bradford L. Paine Webber Mitchell Hutchins, Inc. Monthly. April 10, 1981. 7 pg.*

Monthly report of economic indicators for the U.S. telephone industry, 1981. Shows stock price performance, 1979-1981; utilization growth rates, 1980-1981; earnings per share forecasts, 1981-1986; dividends forecasts, 1981-1986; inflation indexes, 1979-1981; and 1981 yields and returns.

4110. PAINE WEBBER MITCHELL HUTCHINS INC.: STATUS REPORT: TELECOMMUNICATIONS MONTHLY: APRIL 1982. *Peery, Bradford L. Paine Webber Mitchell Hutchins, Inc. Monthly. April 16, 1982. 12 pg.*

Monthly news and statistics for telecommunications corporations. Shows 1982 stock prices; total returns including dividends, 1980-1982; growth rates, 1980-1982; earnings per share, 1981-1987; dividends, 1980-1987; inflation indicators, 1980-1982; and stock valuations. Also includes performance, communications legislation, regulation effects on various lines of business, and competition.

4111. PAINE WEBBER MITCHELL HUTCHINS INC.: STATUS REPORT: TELECOMMUNICATIONS MONTHLY: AUGUST 1982. *Peery, Bradford L. Paine Webber Mitchell Hutchins, Inc. Monthly. August 16, 1982. 16 pg.*

Analysis of financial performance for telephone and equipment corporations and industry news. Includes stock prices; total returns (dividends plus stock price increase), 1980-1982; yields; expected growth rates; toll message unit growth, 1980-1982; stock valuation; inflation and other economic indicators, 1980-1982; and earnings per share, 1981-1987.

4112. PAINE WEBBER MITCHELL HUTCHINS INC.: STATUS REPORT: TELECOMMUNICATIONS MONTHLY: FEBRUARY 1982. *Perry, Bradford L. Paine Webber Mitchell Hutchins, Inc. Monthly. February 11, 1982. 11 pg.*

Monthly financial analysis of telecommunications industry and corporations. Includes 1982 stock prices; total returns, 1979-

1982; dividends and growth rates, 1980-1981, 1985; performance indicators; regulations and legislation; investments and depreciation, 1980; revenue requirements, 1981-1986; earnings per share, 1981-1987; economic indicators of inflation, 1980-1982; returns; and valuation.

4113. PAINE WEBBER MITCHELL HUTCHINS INC.: STATUS REPORT: TELECOMMUNICATIONS MONTHLY: JULY 1982. *Peery, Bradford L. Paine Webber Mitchell Hutchins, Inc. Monthly. July 16, 1982. 16 pg.*

Monthly analysis of the telecommunications industry and stock performance of major corporations. Includes stock prices, 1982; total returns, 1980-1982; dividends, 1980-1987; yields; growth rates; stock valuations; toll call utilization, 1980-1982; national economic indicators, 1980-1982; news on new technology; earnings per share, 1981-1987; product lines; and operating performance and financial analysis of individual corporations.

4114. PAINE WEBBER MITCHELL HUTCHINS INC.: STATUS REPORT: TELECOMMUNICATIONS MONTHLY, JUNE 1981. *Peery, Bradford L. Paine Webber Mitchell Hutchins, Inc. Monthly. June 17, 1981. 7 pg.*

Monthly telecommunications industry analysis, June 1981. Comments on regulations, and shows stock price performance, 1980-1981; toll message growth rates, 1980-1981; earnings per share and dividends, 1980-1986. Covers inflation indicators, 1979-1981; current 1981 stock prices, dividends, yields and returns.

4115. PAINE WEBBER MITCHELL HUTCHINS INC.: STATUS REPORT: TELECOMMUNICATIONS MONTHLY: JUNE 1982. *Peery, Bradford L. Paine Webber Mitchell Hutchins, Inc. Monthly. June 15, 1982. 15 pg.*

Monthly stock performance statistics for telecommunications and equipment corporations. Includes stock prices; total returns, 1980-1982; dividends; yield; growth rates; returns; inflation indicators, 1980-1982; forecasts of earnings per share and dividends, 1981-1987; assessment of regulation effects; and financial analysis of corporations.

4116. PAINE WEBBER MITCHELL HUTCHINS INC.: STATUS REPORT: TELECOMMUNICATIONS MONTHLY, MARCH 1981. *Peery, Bradford L. Paine Webber Mitchell Hutchins, Inc. Monthly. March 26, 1981. 7 pg.*

Monthly financial performance analysis for U.S. telecommunications companies. Shows stock prices, 1979-1981; growth rates for utilization, 1980; earnings per share and dividend forecasts, 1980-1986; inflation indicators, 1979-1981; and returns and yields, 1981.

4117. PAINE WEBBER MITCHELL HUTCHINS INC.: STATUS REPORT: TELECOMMUNICATIONS MONTHLY: MARCH 1982. *Peery, Bradford L. Paine Webber Mitchell Hutchins, Inc. Monthly. March 16, 1982. 12 pg.*

Monthly report on telecommunications stocks. Includes effects of regulations on returns on equity and rates, 1980-1981; communications legislation; corporations' total returns, comprised of dividends plus changes in stock prices, 1980-1982; toll call growth rates, 1980-1982; industry structure; earnings per share, 1981-1987; inflation and economic indicators, 1980-1982; and stock valuations for 1982.

4118. PAINE WEBBER MITCHELL HUTCHINS INC.: STATUS REPORT: TELECOMMUNICATIONS MONTHLY, MAY 1981. *Peery, Bradford L. Paine Webber Mitchell Hutchins, Inc. Monthly. May 13, 1981. 7 pg.*

Financial analysis of telephone and other telecommunications industry corporations. Shows 1981 stock prices; 1980-1981 stock price performance; toll message growth rates, 1980-

1981; forecasts of earnings per share and dividends, 1980-1986; economic indicators of inflation, 1979-1981; and 1981 returns.

4119. PAINE WEBBER MITCHELL HUTCHINS INC.: STATUS REPORT: TELECOMMUNICATIONS MONTHLY: MAY 1982. *Peery, Bradford L. Paine Webber Mitchell Hutchins, Inc. Monthly. May 14, 1982. 15 pg.*

Monthly review of telecommunications stock performance data and financial analysis of corporations, May 1982. Shows stock prices, 1982; dividends; and growth rates, 1980-1982; yields; impact of rate regulations; percent of contribution to return on equity by jurisdictions, 1981-1982; forecasts of earnings, 1981-1987; and dividends, 1980-1987; inflation indicators, 1980-1982; and expected returns and relative values.

4120. PAINE WEBBER MITCHELL HUTCHINS INC.: STATUS REPORT: TELECOMMUNICATIONS MONTHLY, SEPTEMBER 1981. *Peery, Bradford L. Paine Webber Mitchell Hutchins, Inc. Monthly. September 17, 1981. 12 pg.*

Monthly financial analysis and stock performance report for the U.S. telecommunications industry. Also includes legislation, taxes, regulation, inflation, and news of major corporations. Shows total returns, including dividends and increases in stock prices, 1979-1981; toll message growth rates, 1980-1981; increases in toll rates, 1981; earnings per share, 1980-1986; dividends, 1980-1986; economic indicators, 1979-1981; and 1981 expected returns and relative value of stocks.

4121. PAINE WEBBER MITCHELL HUTCHINS INC.: STATUS REPORT: TELECOMMUNICATIONS MONTHLY: SEPTEMBER 1982. *Peery, Bradford L. Paine Webber Mitchell Hutchins, Inc. Monthly. September 22, 1982. 15 pg.*

Monthly analysis of telecommunications markets, the industry, and performance of corporations. Includes stock prices and total returns, 1980-1982; dividends, 1980-1987; yield; growth rates; economic indicators of inflation, 1980-1982; regulatory issues; and earnings per share, 1981-1987.

4122. PAINE WEBBER MITCHELL HUTCHINS INC.: STATUS REPORT: THE TELEPHONE TRANSFORMATION. *Peery, Bradford L. Paine Webber Mitchell Hutchins, Inc. December 17, 1980. 4 pg.*

Text of a speech delivered to an AT&T executive conference in December 1980. Forecasts developments in telecommunications industry structure, competition, regulation, new products, and industrial plants and equipment.

4123. PAINE WEBBER MITCHELL HUTCHINS INC.: TELEPHONE INDUSTRY: TOLL MESSAGE GROWTH: AN ECONOMIC ANALYSIS: STATUS REPORT. *Peery, Bradford L. Paine Webber Mitchell Hutchins, Inc. July 31, 1980. 5 pg.*

An economic analysis of toll message growth in the telephone industry, 1974-1985, by major company.

4124. PARKER / HUNTER INC.: RESEARCH / COMMENTS: HIGHLIGHTS OF THE C.S. LAWRENCE CONFERENCES. *Appert, Peter P. Parker / Hunter Inc. April 20, 1982. 26 pg.*

Research comments for investments in telecommunications, broadcasting, computers and consumer electronics. Includes telephone corporations, stock prices, 1978-1981; dividends and returns, 1979-1981; cable television market size and revenues, 1981, 1985; earnings per share, 1980-1982; cable TV consumers, 1982; cable corporations' operations; electronics and computer industry and market analysis; historical; and financial analysis of corporations.

4125. UNITED BUSINESS AND INVESTMENT REPORT: TELEPHONE UTILITIES' FUTURE GROWTH ON LINE. *United Business Service Company. v.73, no.45. November 9, 1981. pg.451*

Report on investments in telephone utilities, includes corporations, lines of business, and sales. Shows earnings per share, 1980-1981; stock prices, 1980-1981; price earnings ratio, 1981; and dividend and yield.

4126. WARBURG PARIBAS BECKER - A.G. BECKER: INDUSTRY REVIEW: TELETEXT AND VIDEOTEXT. *Hoffman, Anthony M. Warburg Paribas Becker - A.G. Becker, Inc. August 9, 1982. 31 pg.*

Analysis of the emerging teletext and vidotex industries and financial prospects for corporations involved. Examines the technologies and competition among them, lines of business and service by company, forecasts, and industry structure.

See 198, 1250, 1289, 1295.

Industry Statistical Reports

4127. ANNUAL TELECOMM SALARY SURVEY, 1980. *Personnel Resources International Inc. Annual. 11th ed. 1979. Free.*

Salary statistics for 16 job titles by geographic area (New York City and vicinity, elsewhere in the U.S.).

4128. CABLE ENHANCED SERVICES. *National Cable Television Association. January 28, 1982. 24 pg. Free.*

Lists services available through cable television in the areas of data transmission, electronic publishing, games, home banking and shopping, meter reading / energy management, security, videotex / information retrieval / teletext. Arrangement is by type of service, with name of company and service area, specifics of service, start-up date, number of subscribers, and cost to subscribers.

4129. CANADIAN RADIO - TELEVISION AND TELECOMMUNICATIONS COMMISSION: ANNUAL REPORT 1979-1980: CONSEIL DE LA RADIODIFFUSION ET DES TELECOMMUNICATIONS CANADIENNES: RAPPORT ANNUEL 1979-1980. *Canadian Radio - Television and Telecommunications Commission. English/French. Annual. 1980. 32 pg.*

Gives activities and performance of the Commission for the year ending March 31, 1980. Lists decisions announced by region; total number of applications for licenses; CBC radio and television coverage; number of radio and television stations in Canada; net operating revenues of telephone companies, 1972-1978; number of employees in telecommunications companies; and other statistics.

4130. CANADIAN RADIO - TELEVISION AND TELECOMMUNICATIONS COMMISSION: ANNUAL REPORT 1980-1981: CONSEIL DE LA RADIODIFFUSION ET DES TELECOMMUNICATIONS CANADIENNES: RAPPORT ANNUEL 1980-1981. *Canadian Radio - Television and Telecommunications Commission. English/French. Annual. 1981. 50 pg.*

Summarizes developments in the Canadian communications industries in 1980 and the first quarter of 1981. Covers applications for new station licenses; stations operating by medium, language, province, affiliation and ownership; new cable TV undertakings; and telephone company employment and condensed finances.

4131. COLLECTIVE BARGAINING IN THE TELEPHONE INDUSTRY. *U.S. Department of Labor. Bureau of Labor Statistics. Report 607. June 1980. 11 pg.*

Report provides summary analysis of history of collective bargaining in the telephone industry. Tables on work stoppages, 1950-1978, date of stoppage (for stoppages involving over 10,000 workers), approximate duration, establishment and location, unions involved, number of workers involved, major terms of settlements.

4132. COMMUNICATIONS SERVICE BULLETIN. *Canada. Statistics Canada. English/French. Monthly. 1981. 8 pg. 1.50 (Canada).*

Provides early release of summary information on telecommunications, including telephone industry and other telecommunications carriers; radio and television broadcasting; cable television; and miscellaneous data of general interest. Contains monthly releases on telephone statistics and radio time sales; and quarterly information on telecommunications. Data published since 1971.

4133. COMPENSATION AND BENEFITS IN THE INDEPENDENT TELEPHONE INDUSTRY - 1982. *National Telephone Cooperative Association. Annual. 14th ed. 1982. 47 pg. $25.00.*

Annual statistical summary of compensation and benefits paid to independent telephone company employees. Covers 1982 with comparative figures and percentage change from 1981. Arrangement is by 31 categories of employee. Summaries indicate averages by region, by system size / revenue, and by system size / subscribers. Includes comparison of Bell vs. non-Bell compensation.

4134. CONSTRUCTION EQUIPMENT IN THE U.S. TELEPHONE INDUSTRY. *Telephony Publishing Corporation. 1980. 40 pg. $42.00.*

Examines operating company and contractor equipment ownership, purchase versus lease approaches and the purchase decision - makers. Includes Bell System outside plant (buried, underground, aerial, coaxial and submarine cable and underground conduit and telephone poles) additions, retirements and plant totals for 1969-1979.

4135. EIS SHARE OF MARKET REPORT: TELEPHONE COMMUNICATION. *Economic Information Systems, Inc. Quarterly. 1982. $530.00.*

Shows telephone communications concentration and ownership structure. Companies in industry are listed and then ranked by production. Othe

4136. EIS SHIPMENTS REPORT: TELEPHONE COMMUNICATION. *Economic Information Systems, Inc. Quarterly. 1982. $530.00.*

Report on the telephone communications industry. Arranged by state and county, report includes every firm in the industry with annual business of $500,000 and/or 20 or more employees. Plant listings detail address, telephone, estimated annual business, and percent of market. Similar statistics for each county and state.

4137. FEDERAL COMMUNICATIONS COMMISSION, FORTY-SIXTH ANNUAL REPORT, FY 1980. *Federal Communications Commission. U.S. Government Printing Office. Annual. su. doc. no. CC1.1:980. 1982. 120 pg.*

Reviews key events in broadcasting, cable television, common carrier communications, private radio services, spectrum management and frequency allocations in U.S., 1980. Television and radio statistics cover industry financial data, nonhearing applications, and licenses granted and denied. Gives cable television financial data by state; and common carrier and private radio statistics. Provides FCC organizational chart with names; and addresses of field offices.

4138. HANDY-WHITMAN INDEX OF PUBLIC UTILITY CONSTRUCTION COSTS TO JULY 1, 1980. *Requart Whitman and Associates. October 1980. 170 pg.*

Trends of construction costs 1912-July 1980 for building construction, by electric, gas, telephone and water utilities.

4139. HOLDING COMPANY REPORT: UNITED STATES INDEPENDENT TELEPHONE ASSOCIATION, MAY 1980. *U.S. Independent Telephone Association. May 1980. 12 pg.*

Financial reports for 12 holding companies which controlled 256 operating telephone companies as of December 31, 1979. Includes information on plant investments, operating revenues, average revenues per telephone, number of common and preferred stockholders, and gross additions to plants. The report for each holding company lists executives, address, names of operating companies, and statistics on sales, employment, and revenues.

4140. INDEPENDENT PHONE FACTS "82. *U.S. Independent Telephone Association. 1982. 20 pg. Free.*

Statistics for independent telephone systems, 1982, include price index comparisons, 1960-1981; labor productivity, 1955-1981; distribution of telephone system income; installations, revenues, average revenues per unit, industrial plants and equipment, and construction costs, 1981-1982; comparisons of independents and the major corporations for units in operation, number of businesses, number of employees, and shareholders, 1981; historical data summary, 1950-1981; and rankings of largest firms.

4141. INDEPENDENT PHONEFACTS "80. *U.S. Independent Telephone Association. Annual. 1980. 24 pg.*

Contains 1980 data on U.S. independent operating telephone companies. Includes growth rates from 1950-1979, and key trends for the same period. There is a listing of the 25 largest independent telephone companies, and one of the 463 companies that grossed over $1,000,000 in 1979.

4142. INDEPENDENT PHONEFACTS "79. *U.S. Independent Telephone Association. 1979. 24 pg.*

1979 report on the independent telephone industry shows trends in patterns of growth in terms of operating revenues, construction expenditures, and industrial plants and equipment. Ranks the 25 largest companies in order of operating revenue for 1978. Lists the 422 companies which grossed more than $1 million in 1978.

4143. INDUSTRY WAGE SURVEY: COMMUNICATIONS: OCTOBER-DECEMBER 1977. *U.S. Department of Labor. Bureau of Labor Statistics. Irregular. Bulletin 2029. June 1979. 12 pg.*

Provides data on hourly pay of employees, except managers and officials, in the telegraph and telephone industries. Data is from annual industry reports filed with ICC and is tabulated separately for Bell System carriers and non-Bell carriers, by region; Western Union Telegraph Company; and international telegraph carriers. Figures are given for each occupational group.

4144. INTERNATIONAL BROADCASTING: DIRECT BROADCAST SATELLITES. *Government Information and Individual Rights Subcommittee of the Committee on Government Operations, House of Representatives, U.S. Congress. U.S. Government Printing Office. Su.Doc. # Y4.G74/7:B78. 1981. 129 pg.*

Gives proceedings of U.S. House hearing on direct broadcast satellites, October 23, 1981. Provides Voice of America's total annual budget, equipment investment, cost of equipment replacement, and budget breakdown into 5 percentages.

4145. PAYMENTS TO SUPPLIERS BY STATES AND TOWNS AND FOREIGN COUNTRIES 1979: WESTERN ELECTRIC COMPANY. *Western Electric Company, Inc. 1980.*

Summarizes Western Electric's 1979 payments to suppliers, gross payroll, expenditures for major construction, and number of employees. Contains annual report, by states, of Western Electric payments to suppliers for 1979.

4146. PHONE FACTS "82. *U.S. Independent Telephone Association. 1982. 20 pg.*

Pamphlet of statistics on independent telephone companies. Includes price index, 1960-1981; employees versus units in operation, 1955-1981; hours of labor for service, 1940-1981; distribution of telephone income; forecasts for 1982; revenues, industrial plants and equipment, construction costs, 1950-1982; units in operation and numbers of businesses, 1950-1981; shareholders; and rankings of the corporations by revenues.

4147. QUARTERLY OPERATING DATA OF TELEPHONE CARRIERS. *U.S. Federal Communications Commission. Quarterly. 1979.*

Quarterly financial review of telephone companies with annual reviews of over $1 million. Shows revenues, expenses, and operating performance.

4148. RESULTS OF THE 1979 BANK TELECOMMUNICATIONS SURVEY. *American Bankers Association. Triennial. 1979. 220 pg. $120.00.*

Reports the results of the second triennial bank telecommunications survey and is intended to indicate recent and future trends in the use of telecommunications by banks. Presents data on usage of various types of telecommunication systems, with the statistics arranged by geographic region, bank deposit size, and by unit or branch banks.

4149. STATISTICS OF COMMUNICATIONS COMMON CARRIERS, 1978. *Federal Communications Commission. U.S. Government Printing Office. Annual. SUDOC # 004-000-00371-3. 1979. 219 pg. $5.50.*

Annual review of telephone, telegraph, and communication satellites companies. Shows 1978 services, employment, rates, operating performance, and message revenues.

4150. STATUS OF COMPETITION AND DEREGULATION IN THE TELECOMMUNICATIONS INDUSTRY. *Subcommittee on Telecommunications, Consumer Protection, and Finance of the Committee on Energy and Commerce, House of Representatives, 97th U.S. Congress, 1st Session. U.S. Government Printing Office. SuDoc # Y4.En2/3:97-29. 1981. 607 pg.*

Gives proceedings of U.S. Congressional hearing, May 20, 27, and 28, 1981. Diagrams of information business show regulatory activities of 4 agencies. Includes specific companies under different circumstances. Provides following 1976, 1981, and 1986 data for 15 U.S. cities: number of cable channels and percent of homes, MDS channels, UHF and VHF channels, DBS channels, low power channels, and video cassette / disk players. Gives data on U.S. household telecommunications market.

4151. TELECOMMUNICATIONS IN TRANSITION: THE STATUS OF COMPETITION IN THE TELECOMMUNICATIONS INDUSTRY. *Subcommittee on Telecommunications, Consumer Protection, and Finance of the Committee on Energy and Commerce, U.S. House of Representatives, 97th U.S. Congress, 1st Session. U.S. Government Printing Office. Y4. En2/3:97-V. November 3, 1981. 435 pg.*

Presents report by majority staff of U.S. House subcommittee, 1981. Gives revenue data and market shares for U.S. telecommunication companies, 1979, 1980; and comparative operative statistics for one large company and all others. Supplies marketing data for all aspects of telecommunications.

4152. TELECOMMUNICATIONS STATISTICS. *Canada. Statistics Canada. English/French. Annual. 1981. 16 pg. 4.50 (Canada).*

Details telecommunications revenues, expenses, number of messages sent, mileage operated, employees, salaries and wages, by company. Data published since 1917.

4153. TELEPHONE STATISTICS. *Canada. Statistics Canada. English/French. Monthly. 1981. 4 pg. 1.50 (Canada).*

Details telephone operating revenues and expenses, salaries and wages, construction expenditures, number of toll messages, employees and telephones, by type of service. Statistics compiled from reports of 13 major Canadian telephone systems. Data published since 1977.

4154. TELEPHONE STATISTICS. *Canada. Statistics Canada. English/French. Annual. 1981. 52 pg. 6.00 (Canada).*

Details number of telephone calls; telephones, by type of service and organization; wire and pole-line mileage; employees, salaries and wages; assets, liabilities and net worth data; and revenues and expenditures, by province. Statistics issued since 1918.

4155. TELEPHONE STATISTICS: PRELIMINARY REPORT ON LARGE TELEPHONE SYSTEMS. *Canada. Statistics Canada. English/French. Annual. 1981. 4 pg. 3.00 (Canada).*

Includes statistics on telephones, by type of service; selected employment and financial statistics; and number of telephone calls for 14 of the larger telephone systems operating in Canada. Data published since 1955.

4156. U.S. INDUSTRIES TOP SAFETY SPENDER: THE COMMUNICATIONS INDUSTRY. *Telephony Publishing Corporation. 1980. 44 pg. $30.00.*

An item-by-item estimate of expenditures and turnover for new safety equipment by the Bell System, independent companies and outside contractors.

4157. VEHICLES IN THE U.S. TELEPHONE INDUSTRY: A STABLE GROWTH MARKET. *Telephony Publishing Corporation. 1980. 30 pg. $35.00.*

Gives company by company breakouts of Bell and large independents' cars, vans, light and heavy trucks and mechanics employed. Examines telephone companies' need for downsized vans, based on fleet managers' comments.

4158. 1980 ANNUAL STATISTICAL VOLUME OF THE UNITED STATES INDEPENDENT TELEPHONE ASSOCIATION. *Snyder, Courney S. U.S. Independent Telephone Association. Annual. July 1980. 189 pg. $25.00.*

Data was submitted by 755 telephone companies. Companies are ranked by revenues. Individual tables for each company list number of plants in service, number of exchanges, capitalization, revenues, expenses, net income, and dividends. Provides comparative figures for 1978 and 1979.

4159. 1980 STATISTICAL REPORT, RURAL TELEPHONE BORROWERS. *U.S. Department of Agriculture. Rural Electrification Administration. 1980. 194 pg.*

Financial analysis of rural telephone borrowers. Gives composite balance sheet data, historical data, and operating ratios, 1976-1980. Also gives state and corporation breakdown plus 1979-1980 totals for loans, miles of line, consumers, interest and principal, assets, industrial plants and equipment, new construction, reserves for depreciation, investments, liabilities, revenues, expenses, income and facilities.

See 211, 1317.

Financial and Economic Studies

4160. THE BIRTH OF ELECTRONIC PUBLISHING: LEGAL AND ECONOMIC ISSUES IN TELEPHONE, CABLE, AND OVER-THE-AIR TELETEXT AND VIDEOTEXT. *Neustadt, Richard M. Knowledge Industry Publications. July 1982. 160 pg. $32.95.*

Analysis of legislation and regulation for the electronic publishing industry. Includes economic indicators and the role of government agencies.

4161. THE CABLE / BROADBAND COMMUNICATIONS BOOK: VOLUME 2, 1980-1981. *Knowledge Industry Publications. 1981. $29.95.*

Collection of articles on the cable television industry's market and ownership structure, relevant federal and state regulations, and direct broadcast satellite facilities and availability; such issues as ascertainment of community needs and public access; and such applications as two-way programming, videotext, cable and computers, cable and libraries, and non-pay programming spinoffs via satellite.

4162. THE DECLINE OF SERVICE IN THE REGULATED INDUSTRIES. *Carron, Andrew. American Enterprise Institute for Public Policy. 1981. 73 pg.*

Contributing to the debate over regulated monopolies versus nationalization or antitrust action, this report examines recent trends in quality of service and productivity among regulated monopolies in five industries: electric utilities, natural gas distribution, telephone service, water utilities, and railroads. Statistical data, for the 1970's and often the 1960's as well, includes productivity of labor and other factors of production, price / rate increase trends, producer costs, capital investment and capacity growth, and electric utilities' capacity margins.

4163. THE ECONOMICS OF COMPETITION IN THE TELECOMMUNICATIONS INDUSTRY. *Meyer, John R. Charles River Associates. Oelgeschlager, Gunn and Hain, Inc. 1980. 341 pg.*

Coverage of the telecommunications industry (with emphasis on AT&T) includes growth and the financial impact of competition; pricing to achieve regulatory objectives; economies of scale, scope and network economies; and the interaction of market structure, regulation and technological change.

4164. THE FEDERAL SIDE OF TRADITIONAL TELECOMMUNICATIONS COST ALLOCATIONS: BASIC DATA ON THE POLITICS AND ECONOMICS OF THE INFORMATION EVOLUTION: TELECOMMUNICATION COSTS AND PRICES IN THE UNITED STATES. *Oettinger, Anthony G. Harvard University. Information Resources Policy. Order Department. 1980. 107 pg.*

Discusses the structure of the telecommunications industry and the allocation of costs. Examines state-by-state and industry segment differentials which influence cost allocation and price setting. Shows how federal, state, regional, and local interests are related and interplay with diverse consumer and supplier needs and with changing technological possibilities. Charts and tables rank and compare costs and prices.

4165. HOLDING COMPANY REPORT, MAY 1982. *U.S. Independent Telephone Association. Annual. May 1982. 24 pg. Free.*

Annual financial report for the 11 holding companies which controlled 249 operating telephone companies as of December 31, 1981. Includes information on number of stockholders, total telephones and exchanges, percentage of industry total, number of states served, total plant value, total operating revenues, number of employees, and actual and estimated construction expenditures for 1981-1983. For each holding company, lists chief personnel and names of operating companies held.

4166. MOODY'S PUBLIC UTILITY MANUAL: VOLUME 2. *Hanson, Robert P., editor. Moody's Investors Services. Annual. September 1980. pg.2201-4381 $360.00.*

Annual statistical report on all U.S. public utilities (gas, electric, telecommunications) by international investment service provides detailed 1980 stock and financial data. Alphabetical company entries include stock prices, operations overview, profits, customers, debt listings, assets, liabilities, and stockholder's letters. Maps, charts, and tables based on company research.

4167. PUBLIC UTILITIES FORTNIGHTLY STOCK SUMMARY: GAS, TELEPHONE, WATER UTILITIES. *Media General Financial Services, Inc. Public Utilities Reports. Weekly. January 15, 1981. pg.42-43*

Weekly stock report on companies which operate gas and oil pipelines, gas utilities, telephone communications, and water utilities. Data shows stock prices (high, low, close) for previous month; earnings per share for past year; price earnings ratio; dividend rate (amount, yield); and financial position (revenue, earnings, common stock equity).

4168. TECHNOLOGY AND LABOR IN FIVE INDUSTRIES: BAKERY PRODUCTS, CONCRETE, AIR TRANSPORTATION, TELEPHONE COMMUNICATION, INSURANCE. *Dept. of Labor, Bureau of Labor Statistics. U.S. Government Printing Office. Su Doc #029-001-02394-4. 1979.*

Analyzes influence of technology on five industries, 1960-1978: air transport, bakery products, concrete, insurance, and telecommunications. Covers productivity, investments, and employment. Gives forecasts to 1990.

4169. TELECOMMUNICATIONS IN CANADA: PART 1 - INTERCONNECTION. *Canada Restrictive Trade Practices Commission. Canada. Department of Supply and Services. Cat. # RG 52-17/1981E. 1981. 276 pg. Free (Canada).*

Part I of a 1977-1981 inquiry into telecommunications by Canada's Restrictive Trade Practices Commission looks into the matter of interconnection of telecommunications equipment and related monopolistic practices. The study reviews current technology and the manufacture of networks and equipment, including voice terminals; the purchase and supply of terminal equipment, with a company-by-company examination of suppliers; policies and regulations governing interconnection in Canada and in other countries; user complaints; and policy questions. The report provides conclusions and recommendations for regulation.

4170. TELECOMMUNICATIONS INDUSTRY REVIEW. *Boysen, Peter. Wall Street Transcript. November 1979. $20.00.*

U.S. telecommunications industry review and forecast (1979-1980). Toll volume growth, GNP and rate trends (1979-1980) and interest rates (1979-1982) are projected. Some topics dis-

cussed include: deregulation, competition, industry sectors, digital LSI, fiber optics and satellite transmissions. Nineteen major businesses are reviewed.

4171. U.S. COMMUNICATIONS, COPYRIGHT AND PRIVACY REGULATION. *Predicasts, Inc. January 1982. $375.00.*

Discussion of the commercial, industrial and social effects of U.S. legislation and regulations governing communications, copyrights and privacy.

4172. VIDEO IN THE "80S: EMERGING USES FOR TELEVISION IN BUSINESS, EDUCATION, MEDICINE AND GOVERNMENT. *Dranov, Paula. Knowledge Industry Publications. 1980. 180 pg. $34.95.*

Analysis of various trends affecting nonbroad cast video market. Reports on each major segment of the user market describe types of programming and recent applications. Twenty-two case studies of organizations' video operations. Data based on questionnaires completed by 11,087 organizations listed in 'The Video Register.'

See 599, 1330, 1331, 3231.

Forecasts

4173. FACSIMILE MARKETS. *International Resource Development, Inc. April 1979. 228 pg. $895.00.*

Covers developments in wideband and narrowband facsimile; current and future applications discussed. Analysis of types of equipment and markets, including 10-year projections of shipments of consumables. Discussion of Japanese and U.S. products and cross-elasticity between facsimile and electronic message transmission services. Market shares of present and future suppliers.

4174. GROWTH OF TELECOMMUNICATIONS. *Business Communications Company. April 1981. 130 pg. $850.00.*

Forecasts and statistics for about 20 fields (i.e., common carrier, voice, digital, satellites, terminals, data, interconnect) include manufacturers, equipment, and market segmentation.

4175. HOME SATELLITE TERMINAL MARKET. *Frost and Sullivan, Inc. November 1980. $1100.00.*

Forecast and analysis of home satellite terminals. History of satellite broadcasting. Technological and legal problems. Future developments of two-way satellite terminals. Marketing approaches. Competition and market share analysis. Comparison of small satellite terminals and rival hardware development.

4176. OPTICAL COMMUNICATIONS MARKETS IN THE TELEPHONE INDUSTRY, 1979-1990. *Probe Research, Inc. February 1979. 245 pg. $595.00.*

Divides the North American telephone industry into its major functional markets, including interexchange transmission, longhaul systems, subscriber loops and other markets. Analyzes optical communications and the alternative transmission technologies in detail. Contains 75 tables with critical data, such as number of trunks, circuit mileage, cross-section, length distributions, plant size, new construction. Analyzes fiber optic field installations. Detailed estimates of sales of each important fiber optic component in each major market, to 1990.

4177. PERFORMANCE AND PROSPECTS OF LEADING U.S. INDEPENDENT TELEPHONE COMPANIES. *Bender, Warren. Arthur D. Little, Inc. 1980.*

Discusses growth and profitability trends in U.S. independent telephone industry.

4178. THE REPORT ON ELECTRONIC MAIL: FACSIMILE, WORD PROCESSING AND COMPUTER-BASED MESSAGE SYSTEMS. *Yankee Group. Quarterly. 1980. 100 pg. $650.00.*

A basic study with three quarterly updates on technology, market trends and 18 month forecasts (1978-1985) in the electronic mail industry in U.S., Europe and Japan. Case studies, product analysis, information including packet switching, facsimile, mail output CRT, optical character and word processing, and comparative analysis of leading corporations given.

4179. THE RESIDENTIAL TELECOMMUNICATIONS MARKET: MAMMOTH OPPORTUNITIES, SIZEABLE CHALLENGES. *Strauss, Lawrence. Telecom Library Inc. 1980. 176 pg. $125.00.*

Statistics and forecasts on the residential telephone industry are accompanied by analyses of various marketing strategies and distribution channels.

4180. TELECOMMUNICATIONS IN THE U.S.; TRENDS AND POLICIES. *Artech House Inc. 1981. 600 pg. $45.00.*

A collection of 12 original contributions from industry experts focusing on key factors which will determine the future of telecommunications. Discusses technical developments and competition.

4181. U.S. TELCOS AND STANDBY POWER: A SURVEY OF USAGE AND PLANNED GROWTH. *Telephony Publishing Corporation. 1980. 23 pg. $35.00.*

Survey of 21,000 U.S. telephone businesses on their central office standby power systems. Forecasts planned growth over 1980-1985 period.

4182. VIDEO AGE: TELEVISION TECHNOLOGY AND APPLICATIONS IN THE 1980'S. *Knowledge Industry Publications. 1982. 225 (est.) pg. $29.95.*

Volume reviews new television applications and technologies such as home video, cable T.V., video discs, video text, non-broadcast video, corporate T.V., and video in health.

4183. WORLD TELECOMMUNICATIONS STUDY 2: A MULTI-CLIENT STUDY: 1980-1990: VOLUME 2: THE AMERICAS AND OCEANIA. *Arthur D. Little, Inc. 1980. 279 pg. $34000.00.*

Volume 2; 1980-1990 multi-client study on telecommunications industry covering telephone, satellite, mobile, radiotelephone, CATV equipment. Study covers, by country, the Americas and the Oceania regions. Charts of demographic data; financial & market structure and anaylsis; facilities; product lines; sales; revenues; and expenditures. Price listed is for complete set.

4184. A 25-YEAR FORECAST FOR COMMERCIAL COMMUNICATIONS SATELLITES AND CONGESTION OF THE GEOSTATIONARY ARC. *Information Gatekeepers, Inc. 1981. 100.00 pg.*

This report projects an increasing degree of radio transmission congestion on the geostationary arc over the next 25 years. Examines ways to stimulate world demand for U.S. - produced satellite communications equipment.

See 227, 230, 232, 237, 1338, 1370.

Directories and Yearbooks

4185. ASSOCIATED TELEPHONE ANSWERING EXCHANGES - MEMBERSHIP DIRECTORY. *Associated Telephone Answering Exchanges. Annual. 1981. 220 pg. Free.*

Free directory provides 850 geographically arranged entries detailing address and contact person.

4186. CANADIAN RADIO - TELEVISION AND TELECOMMUNICATIONS COMMISSION: ANNUAL REPORT 1979-1980. *Canadian Radio - Television and Telecommunications Commission. English/French. Annual. 1980. 32 pg.*

Numerous charts show: decisions announced; applications received, heard, and on hand; how new licenses are issued and renewed; and CBC radio and television coverage. Gives summary of broadcasting stations, originating stations, rebroadcasting stations, radio and television stations, licensed cable television undertakings; telephone and telecommunications companies absolute value, 1972-1978; and selected cable television industry statistics from 1973-1979. Also includes maps showing television and cable television coverage. Includes a summary of internal operations, including staff and structure, an organization chart, financial statement for 1978-1979, and a list of publications.

4187. THE DIRECTORY OF DATA COMMUNICATIONS NETWORKING. *Walker, John, ed. Phillips Publishing Inc. Annual. 1981. 95 pg. $67.00.*

Directory of corporations, government agencies, associations and regulatory and standard - setting bodies for the data communications network industry. Includes business addresses, executives, affiliations, and lines of business where applicable.

4188. ELECTRONIC MAIL EXECUTIVES DIRECTORY. *International Resource Development, Inc. Annual. 3rd ed. 1981. $95.00.*

Identifies some 2800 executives in charge of electronic mail and related telecommunications departments, at about 1000 leading U.S. corporations. Listings, by corporation name, also specify addresses, revenues, Fortune 500 rank, number of employees, and principal area of business.

4189. HANDBOOK OF INTERCITY TELECOMMUNICATIONS RATES AND SERVICES. *Economics and Technology, Inc. Annual. 1982. 400 (est.) pg. $225.00.*

Annual volume covers 50 types of services by Bell and other common carriers, with mileages between 30,000 city pairs. Listings include service description, availability, rates and rate formulas, including local distribution, station termination, and intercity mileage changes as well as precalculated rates for private line channels. Annual subscription includes monthly update service.

4190. HANDBOOK OF INTERCITY TELECOMMUNICATIONS RATES AND SERVICES. *Economics and Technology, Inc. Monthly. 1981. $175.00/yr.*

Explains the more than 50 different types of intercity telecommunications services available in the U.S. today. Identifies carriers operating in a particular area; includes sample costs for private line services; provides route mileage tables for over 60, 000 city pairs; and lists cities alphabetically showing services. Available with monthly or quarterly updates.

4191. THE INTERACTIVE CABLE TV HANDBOOK. *Presser, Rhonda, ed. Phillips Publishing Inc. Annual. 2nd ed. 1982. 104 pg. $49.00.*

State-by-state directory of cable TV systems which provide, are capable of providing, or plan to provide interactive two-way services. Identifies most systems' address, phone number, channel capacity, two-way capabilities, and services offered, communities served, number of subscribers, miles of cable, parent MSO (multi-system operator), and principal equipment supplier. Also indexes MSO's (with their subsidiaries identified); suppliers of hardware, software and text; research, technical and business consultants; educational institutions, associations, advocacy groups and communications law firms; and Congressional committees and their members.

4192. LOCAL AREAS NETWORKING DIRECTORY. *Phillips Publishing Inc. Irregular. 2nd ed. 1982. 150 pg. $67.00.*

Directory covers local networks, network systems and architecture including businesses, individuals, consulting firms, and government and research and development groups.

4193. MEMBERSHIP DIRECTORY, AS OF JANUARY 1, 1979: ASSOCIATED TELEPHONE ANSWERING EXCHANGES. *Associated Telephone Answering Exchanges. Annual. January 1, 1979. 275 pg.*

Associated Telephone Answering Exchanges Inc. membership directory as of January 1, 1979 covers the U.S., Canada, France, Japan and Singapore. Provides geographical breakdown of names, addresses and telephone numbers and alphabetical voting-member directory.

4194. RADIO AMATEUR CALL BOOK: 1980: UNITED STATES LISTINGS. *Radio Amateur Callback. Annual. 58th. 1980. 1085 pg. $30.70.*

Annual directory with three quarterly supplements on radio amateurs in the U.S. Directory divided into 10 call area sections listing callers names and addresses, license classes and QSL managers.

4195. SOCIETY OF TELECOMMUNICATIONS CONSULTANTS: MEMBERSHIP ROSTER. *Society of Telecommunications Consultants. Quarterly. April 20, 1982.*

Membership data for association of telecommunications consultants. Lists professionals and business addresses.

4196. TELEPHONY'S DIRECTORY AND BUYERS GUIDE: 1980-1981. *Telephony Publishing Corporation. Annual. 85th. 1980. 670 pg.*

1980 directory of the telephone industry includes products and services listed alphabetically; manufacturers and suppliers listed alphabetically, with addresses, executives and product lines; statistical data including subsidiaries, distribution of telephones, growth rates, revenues, expenditures and income statements. Agencies, organizations, commissions and operating groups in U.S. and Canada also given.

4197. THE VIDEO REGISTER: DIRECTORY OF NON-BROADCAST TELEVISION COMMUNICATIONS: 1979-1980. *Knowledge Industry Publications. Annual. 1979.*

1979 directory of non-broadcast video users, manufacturers, publishers, dealers, services and consultants for the U.S., based on responses to mail questionnaires and telephone follow-up. Listings are alphabetical by name, except for dealers, production / post-production services and consultants, which are alphabetized by state. Three indices: manufacture production index, advertisers index, general index. Provides names, addresses, telephone numbers, key personnel, product or service description.

4198. WUI'S 1979 TELEX DIRECTORY: WESTERN UNION INTERNATIONAL. *Western Union. Directory Services Department. English/German/Spanish. Annual. 1979. 366 pg.*

1979 Western Union international directory of Telex subscribers is arranged alphabetically and covers the U.S., Virgin Islands, Hawaii, Panama, Philippines, Puerto Rico. Provides names, addresses, telex numbers. An answerback listing of codes of subscribers is arranged alphabetically. There is a classified telex directory arranged by subject heading.

4199. 1980 RCA TELEX DIRECTORY: LISTING RCA SUBSCRIBERS IN THE CONTINENTAL U.S., HAWAII, DOMINICAN REPUBLIC, GUAM AND PUERTO RICO. *RCA Global Communications Inc. Annual. 1980. 272 pg.*

1980 RCA telex directory lists RCA subscribers in the continental U.S., Hawaii, Dominican Republic, Guam and Puerto Rico, alphabetically by area. Provides listing of telex subscribers by answerback codes as well. Names, addresses and codes of subscribers are given. Provides vessels equipped with marisal terminals, overseas liason offices, and overseas traffic offices.

4200. 1980 TELEX / TWX DIRECTORY AND BUYERS GUIDE. *Western Union Telegraph Company. Annual. 1979. 2080 pg.*

1980 directory of Telex / TWX subscribers provides geographical breakdown for U.S.A., Canada and Mexico. Provides name and Telex / TWX numbers. Directory also contains advertising and information pertinent to services available from Western Union Telegraph Company. There is a buyers guide arranged alphabetically by subject categories.

4202. THE 1982 SATELLITE DIRECTORY. *Phillips Publishing Inc. Annual. 4th ed. 1982. 549 pg.*

Directory and reference guide for the satellite industry contains names, addresses, telephone numbers and principal executives of satellite communication carriers, hardware manufacturers, organizations offering programming services, consulting and related services. Additional sections contain listings of trade associations, government agencies, attorneys and Congressional committees. Satellite receiving and transmitting stations are listed by state and by company. A charts and graphs section forecasts units in operation and projected demand for services to the year 2000.

4203. "82 VIDOETEX DIRECTORY. *Arlen Communications, Inc.; Institute for the Future. Institute for the Future. Annual. June 1982. $150.00.*

Details system operators (backers, participants, equipment); information providers (databases, advertising agencies); service providers (products, services, and prices of retail, financial, and customer service organizations); hardware and software vendors; professional associations; government agencies; international industry analysis; consultants directory; and publications directory.

See 248, 251, 261, 263, 266, 613, 1375, 1414, 1442, 1454, 3260.

Special Issues and Journal Articles

ABA Banking Journal

4204. HIGHLIGHTS FROM THE "79 BANK TELECOMMUNI-CATIONS SURVEY. *American Bankers Association. Triennial. February 1980. pg.92-95*

Highlights of a triennial survey of banks' usage of telecommunications: equipment, employee time spent, employment of consultants, and costs.

Adweek / East

4205. ALL YOU'LL NEED TO KNOW ABOUT THE NEW MEDIA. *A/S/M Communications. v.22, no.50. November 30, 1981. pg.CR1-CR54*

Special report on cable television and videotexts, emphasizing competition amoung developing systems, growing markets, and utilization of these new services. 15 sections are included in this report, discussing leading vendors, strategies for new ventures, and forecasting future market developments. Details potential profits, advertising expenditures and other start-up costs. Losses are noted.

American Banker

4206. IN-DEPTH: VIDEOTEX. *American Banker, Inc. v.147, no.150. August 4, 1982. pg.11-15*

Analyzes the trend toward videotex and its potential effect on the U.S. financial service industry. Includes a discussion of the following topics: the relationship of videotex to strategic, operational goals; videotex markets; and defining videotex services. Tables provide data on consumer and business expenditures for information and hypotehtical market demand for banking at home among consumer groups exposed to different prices.

Audio-Visual Communications

4207. THE SKY KINGS OF TELECONFERENCES. *United Business Publications, Inc. Irregular. v.16, no.5. May 1982. pg.24-26*

A selective listing of 64 companies offering teleconferencing products and services in at least one of five categories: audioconference producer; uplink / downlink rental center; permanent teleconference hookup site; consultant; and hardware source. Includes addresses and telephone numbers.

Business Quarterly

4208. A MULTINATIONAL COMPARISON OF INTERACTIVE COMMUNICATIONS SYSTEMS. *Ash, Stephen B. University of Western Ontario. School of Business Administration. 1981. pg.23-31 $4.95.*

Report presents a multinational comparison of four emergent systems: Prestel, in the U.K.; Teletel, in France; Telidon, in Canada; and Qube, in the U.S.

Cable Communications

4209. CABLE COMMUNICATIONS: 1981 / 82 CCTA BOARD OF DIRECTORS. *Ter-Sat Media Publications Ltd. Annual. July 1981. pg.66 $3.00.*

Lists members of the Canadian Cable Television Association (CCTA) Board of Directors, as elected in May 1981. Listings are by region; directors' companies and town locations are specified.

4210. CABLE STATISTICS CONFIRM NEED FOR NEW MARKET OPPORTUNITIES. *Ter-Sat Media Publications Ltd. v.48, no.3. March 1982. pg.11*

Compares growth and performance in the Canadian and U.S. cable industries. Statistics include number of cable systems, homes passed by cable, number of subscribers, average rates and monthly fees, pay cable and operating revenues and net income, as well as return on growth for 1976 and 1980.

4211. SUBSCRIBERS MAY PAY $78 / MONTH FOR VIDEOTEX SERVICES. *International Resource Development Inc. Ter-Sat Media Publications Ltd. v.47, no.10. October 1981. pg.32*

Summarizes a ten-year forecast of the development of U.S. home videotex services. Covers services envisioned, average expected monthly subscription fees, total revenues, competition with broadcast television, dominant systems and their owners, and emerging technical standards.

4212. 1982 ANNUAL HANDBOOK AND TRADE DIRECTORY. *Ter-Sat Media Publications Ltd. Annual. November 1981. 104 pg.*

Lists executives in the Federal Department of Communications and the Canadian Radio-Telecommunications Commission. Also lists provincial ministries responsible for communications, Canada's multi-system cable television businesses, broadcasting networks, telecommunications carriers and trade associations for cable television. Lists addresses, phone numbers, chief executives, shareholders, facilities / services as applicable, for the broadcasting and telecommunications industries. Includes the top ten multi-system cable television companies, with cable holdings outside Canada, and rankings of Canada's top 100 television systems.

4213. THE 1982 DIRECTORY / BUYER'S GUIDE: COVERING THE CABLE TELEVISION, BROADCASTING AND TELECOMMUNICATIONS INDUSTRIES. *Ter-Sat Media Publications Ltd. Annual. v.48, no.3. March 1982. pg.24-71*

Annual directory to products, services, suppliers and manufacturers for the cable television, broadcasting and telecommunications industries. Part 1 contains alphabetical listing of suppliers with addresses and telephone numbers. Part 2 contains listing of suppliers organized by products, within product group categories.

CableVision

4214. CABLEVISION PLUS: SCREENING THE FUTURE. *Rothbart, Gary. Titsch Publishing, Inc. v.7, no.25. March 8, 1982. pg.15-17*

A composite of services compares cabletext and other interactive services for three U.S. cities: Portland, Omaha, and Cincinnati. A second composite compares non-tier subscriber services for the same three cities. The projected growth of tier 4 and tier 5 subscribers to cabletext services is forecasted for 1980 to 1990.

4215. CABLEVISION: THE MARKET. *Titsch Publishing, Inc. Monthly. October 26, 1981. pg.36 $54.00/yr.*

Reports high, low and closing stock prices, during the two most recent weeks and the year to date, for about 70 cable television operators, equipment suppliers, and service / finance companies. Lists gross and net change, price / earnings ratio, and exchange where traded.

4216. CABLEVISION: 1981 WESTERN CABLE SHOW EXHIBITORS' BOOTH GUIDE. *Titsch Publishing, Inc. Irregular. November 23, 1981. pg.101-133*

Lists about 200 companies planning to exhibit their products or service operations at this December 1981 trade show. Most are producers of telecommunications equipment of interest to cable television systems; others are operators of satellite systems, program producers / telecasters, vendors of software for various on-screen services, or engineering / design firms. Products / services to be exhibited, and representatives or executives, are identified.

4217. VIDEO COMPETITION: TWO GOVERNMENT REPORTS ANALYZE THE CURRENT VIDEO MARKETS AND COME TO DIFFERENT CONCLUSIONS. *Titsch Publishing, Inc. November 23, 1981. pg.135-139*

Presents excerpts from two major government reports on competition in the telecommunications industry and its encouragement, published late in 1981: 'FCC Policy on Cable Ownership,' prepared by the FCC's Office of Plans and Policy as a rationale for altering the cable television cross-ownership rules, discusses the cable market competition and de-regulation; and 'Telecommunications in Transition,' prepared by the House Telecommunications Subcommittee to guide a rewrite of the Communications Act of 1934's common carrier provisions, examines a range of government policies which encourage competition.

Communications News

4218. COMMUNICATIONS NEWS FORECAST FOR 1982. *Wiley, Don. Harcourt Brace Jovanovich, Inc. Annual. v.19, no.1. January 1982.*

1982 annual forecast for the U.S. communications industry, covering telephone companies, satellites, radio common carriers, broadcasting, and transportation segments. Tables detail world telecom shipments (1981-1990); the top 20 telephone countries and cities; financial data for telephone companies; the top 25 independent telcos; and projected satellite earth station markets (1981, 1986, 1991). Covers cellular services revenues (1980-1989); TV and radio sets in use; cable TV growth (1973-1981); military and defense expenditures (1978-1985) on communications equipment; number and growth of 2-way radio stations, by type; and international telex, telegram and leased channels growth.

4219. COMMUNICATIONS NEWS FORECASTS FOR 1981. *Harcourt Brace Jovanovich, Inc. Annual. January 1981. pg.28-39.*

Narrative discussion by leaders in the U.S. communications industry, with forecasts (1981). Covers telephones, radio and television stations, radios manufactured, data communications terminals, satellite services, cable TV growth, and teleprocessing.

4220. REPORT SAYS 1980'S IS DECADE OF INFORMATION DISTRIBUTION AND TRANSMITTING DATA TO REMOTE AREAS WILL BE EMPHASIS. *Harcourt Brace Jovanovich, Inc. June 1982. pg.18*

Reviews a report by Ventura Development on the U.S. telecommunications industry. A graph depicts average annual growth rates (1981-1984) of PBX, Key systems, CCTV, CATV, fiber optics, CB radio, mobile radio, satellites, and facsimile equipment.

Datamation

4221. DATA COMMUNICATIONS CARRIERS. *Technical Publishing Company. Irregular. August 1980. pg.107-112*

A listing of the twelve U.S. carriers of data transmissions for 1980, with addresses and brief descriptive histories of the corporations. Includes a table ranking the companies by revenues and also listing data communications revenues as a percent of total company revenues for 1978-79.

EDP In-Depth Reports

4222. THE GROWTH OF TELECOMMUNICATIONS IN CANADA. *Evans Research Corporation. Annual. v.10, no.10. June 1981. 16 pg.*

Presents financial reports for Canada's approximately 270 common carriers, estimates of dollar value and growth rates for data communications and computer terminal population figures. Reviews revenues, new technologies, and the performance of the industry.

4223. THE GROWTH OF TELECOMMUNICATIONS IN CANADA. *Evans Research Corporation. Annual. v.9, no.10. June 1980. 16 pg.*

Financial reports for approximately 270 common carriers are listed, along with estimates of dollar value, growth rates for data communications and computer terminal population figures, industry performance, and revenues.

Electronics

4224. VIDEO CONFERENCING GROWS UP. *Lineback, J. Robert. McGraw-Hill Book Company. v.55, no.19. September 22, 1982. pg.105-106*

Analyzes the U.S. video conferencing market. Highlights companies and their products and services and reviews current technology. Forecasts annual sales, 1982-1990.

Fortune

4225. THE LOOMING BOOM IN BEEPERS AND CAR PHONES. *Morner, Aimee L. Time, Inc. v.104, no.12. December 14, 1981. pg.151-156*

Reviews the market for telecommunications services, with emphasis on stock performance and price earnings ratios. Providers of Radio Common Carrier (RCC) services are reviewed, with 1980 revenues and market value of stocks. Presents stock prices graphically for four major companies (1970-1981). Competition among industry leaders, reduction in the cost of manufacturing, new uses for RCC and paging equipment, and the impact of FCC regulations are discussed. Displays units on operation from 1976-1981, with projections to 1990 for pagers and mobile telephones.

4226. PERSONAL INVESTING: IN SEARCH OF THE MESSAGE ABOUT AT & T. *Morner, Aimee L. Time, Inc. v.105, no.5. March 8, 1982. pg.97-98*

Brief analysis of investment opportunities in the U.S. telecommunications industry provides data regarding stock price ranges in dollars for telecommunications equipment makers and telecommunications services. The 1981 stock price range is compared with those of January 7, 1982, and early February

1982. Price - earnings ratios are given based on estimates of 1982 or fiscal 1983 earnings as compiled by institutional Brokers Estimate System.

International Business Week

4227. TELECOMMUNICATIONS, EVERYBODY'S FAVORITE GROWTH BUSINESS: THE BATTLE FOR A PIECE OF THE ACTION. *McGraw-Hill Book Company. v.2760-91. October 11, 1982. pg.44-48*

Examines effect of deregulation on U.S. telecommunications industry, 1982. Lists companies participating in 6 areas of business. Shows market shares of particular telephone services for giant corporation and 4 other types of carriers, 1981, 1986.

Satellite Communications

4228. GLOBAL SATELLITE STATIONS: 1981. *Morgan, Walter L. Communications Center of Clarksburg (Maryland). Cardiff Publishing Company. June 1981. 3 pg.*

Chart showing relative locations of communications satellites in the C-band (approximately 3.4 - 4.2 GHz) and K-band (approximately 10.95 - 13.7 GHz). A narrower band width segment is specified for each satellite.

4229. SATELLITE PERFORMANCE REFERENCE WALL CHART. *Morgan, Walter. Communications Center of Clarksburg (Maryland). Cardiff Publishing Company. March 1981. 3 pg.*

Lists operational communications satellites under four categories: North American domestic satellites, international satellites in the Fixed Satellite Service (FSS), satellites capable of direct broadcasting services, and other satellites. Indicates each unit's operator and nationality, number of transponders, transmission band width, other transmission specifications, and orbit location.

4230. SATELLITE SERVICES IN NEW YORK. *Brown, Adele Q. Cardiff Publishing Company. Irregular. v.6, no.11. October 1982. pg.50, 52-53*

Provides directory information on 7 satellite services in New York that specialize in transmission / reception capabilities for video channels. Includes company name, contact personnel, telephone number, a description of services, and rates.

4231. SATELLITE SERVICES IN WASHINGTON, D.C. *Cardiff Publishing Company. Irregular. August 1982. pg.64-70*

Lists satellite services available in Washington, D.C. giving name, contact person, telephone number, description of services, and rates.

4232. SHOPPER'S GUIDE: SATELLITE SERVICES IN DENVER. *Rossie, John. Cardiff Publishing Company. Irregular. v.6, no.10. September 1982. pg.42-48*

Lists video teleconference and other satellite distribution services available in the Denver - Boulder area. Includes name of service, contact person or department, telephone number, and brief description of services.

The South

4233. THE SOUTH'S DIRECTORY OF TOP UTILITIES. *Southern Business Publishing Company, Inc. Annual. May 1980. pg.56-59*

Industry analysis of the top utilities in the Southern U.S. for 1980. Includes rankings; chief executives; and financial statements, including operating revenues, net income, earnings per

share, percent of return on equity, and assets. 1979 figures with percent of change from 1978.

Telephone Engineer and Management

4234. ANNUAL MARKETING. *Harcourt Brace Jovanovich Publications. Annual. July 15, 1981.*

Annual special issue reviews telecommunications industry markets.

4235. ANNUAL SPRING CONSTRUCTION. *Harcourt Brace Jovanovich Publications. Annual. April 15, 1981.*

Annual review of cost-effective construction techniques for use in the telecommunications industry.

4236. TWO-WAY BUSINESS COMMUNICATIONS MARKET. *Harcourt Brace Jovanovich Publications. v.86, no.17. September 1, 1982. pg.98*

Analyzes the U.S. two-way communications market. Discusses telephone interconnect, telephone answering, and communicating word processors. Presents 1980 and 1984 projected annual sale figures. Source is Frost and Sullivan, Inc.

Telephony

4237. COST MODEL PLUS FIELD TRIALS REDUCE PROFIT UNCERTAINTIES OF VIDEOTEX. *Tyler, Michael. Telephony Publishing Corporation. v.203, no.16. October 11, 1982. pg.66, 71-72, 74*

Discusses approach used to assess revenues, costs, profits, and market for videotex system. Graphs show areas of potential revenue in an average household per month. Schematic diagrams illustrate factors affecting profitability. Table compares revenue assumptions of cabletext and telephone-based system.

4238. GTE, OTHER FIRMS, REPORT EARNINGS. *Telephony Publishing Corporation. v.203, no.5. August 2, 1982. pg.18*

Reports second quarter 1982 earnings for 10 major telecommunications firms. Provides additional data on second quarter earnings per share and revenues.

4239. MONTHLY FEATURES: TELEPHONY. *Telephony Publishing Corporation. Monthly. 1981.*

Monthly statistics on telephone company stock prices, earnings, and ratios. Includes activities in foreign markets.

4240. TELCO CONSTRUCTION SPENDING TO TOP $23 BILLION IN 1981. *Anderson, Leo. Telephony Publishing Corporation. Annual. January 12, 1981. pg.28-30, 34-39*

Forecasts 1981 U.S. telephone company construction expenditures. Tables show independent and Bell construction expenditures (1961-1981); and independent and Bell company statistics including number of telephones in U.S., number of exchanges, investments in gross plants, total operating revenues and number of employees, all for 1976-1980. Plant statistics include expenditures for buildings, underground cable, aerial cable, station apparatus, underground conduit, central office equipment, buried cable, vehicles and other work equipment. Provides a breakdown, for 24 major independents, of construction spending (1979-1982).

4241. TELECOM MARKETING DURING THE RECESSION. *Fletcher, Martin W. Telephony Publishing Corporation. v.203, no.11. September 10, 1982. pg.52, 54-55*

Tables provide statistical data for total telco revenues (U.S.), including figures for revenues and real GNP (percent), 1971-1978; teleo long distance revenues and calls (U.S.), 1971-

1978; telco local revenues and calls (U.S.), 1971-1978; and telephones in service (U.S.), 1971-1978. (Sources: Federal Communications Commission; Federal Reserve Board; American Telephone and Telegraph Co.; Dataquest Inc., estimates).

4242. TELEPHONE ANSWERING SERVICES (TAS) MARKET FORECAST. *Telephony Publishing Corporation. v.203, no.19. November 1982. pg.56-61*

Examines the U.S. telephone answering service industry. Market analysis covers 1977 through 1990. Indicates shares of subscribers and revenues for telephone answering services (TAS) and telephone answering machines, 1981. Reviews and forecasts, 1972-1990, number of subscribers, number of TAS, receipts, subscribers per TAS, and receipts per TAS. Forecasts sales of automated switchboard systems, 1984-1990.

U.S. News and World Report

4243. ELECTRONIC NEWSPAPERS - WILL THEY BE HERE SOON? *Sanoff, Alvin P. U.S.News and World Report Inc. May 11, 1981.*

Analyzes newspapers' joining the electronic age as opposed to competing with it by buying into cable TV companies and broadcasting stories that readers have selected on home keyboards.

VideoPrint

4244. TELETRAVEL AND TRANSACTIONAL SERVICES. *Weissman, Steve. International Resource Development, Inc. October 22, 1981.*

Article on teletravel services available through computer - linked TV. Forecasts effect on travel agency industry structure to 1995.

See 289, 297, 302, 307, 320, 323, 328, 334, 351, 363, 375, 381, 387, 705, 1588, 1600, 1604, 1712, 1817, 3741.

Monographs and Handbooks

4245. CHANGES IN THE INFORMATION INDUSTRIES - THEIR STRATEGIC IMPLICATIONS FOR NEWSPAPERS. *Le Gates, John. Harvard University. Information Resources Policy. Order Department. I-80-5. January 1981. 20 pg. $25.00.*

Paper concerns the coming confrontations between the newspaper industry and telephone companies, cable operators, bank networks, computer data base services, and other industries capable of competing in the provision of information to the home.

4246. CTMS (COMPUTERIZED TELEPHONE MANAGEMENT SYSTEM) ADVISER. *QED Informational Services. 1980. $155.00.*

Covers computerized management tools; CTMS: equipment selection; hardware features; usage information processing; equipment survey - 16 CTMS vendor descriptions; CTMS rates and services for Bell equipment; guide and glossary for basic telephone network services.

4247. DATA COMMUNICATIONS - A USER'S HANDBOOK. *Rascal-Vadic. 1981. 76 pg.*

Adapted from a seminar series, this handbook covers such topics as data communications via telephone, telecommunications networks and services, types of modems and their roles,

modes and protocols, terminal / communications network interfaces, and diagnostics.

4248. DISTANCE INSENSITIVE UNIFORM NATIONAL TELEPHONE RATE STRUCTURE: A SPECULATIVE EXERCISE IN FEDERALISM. *Godbey, Robert Carson. Harvard University. Information Resources Policy. Order Department. P-81-4. July 1981. 83 pg. $240.00.*

Essay considers three approaches to establishing a uniform distance-insensitive national telephone rate ('postalization').

4249. THE EFFECT OF THE DEMOGRAPHICS OF INDIVIDUAL HOUSEHOLDS ON THEIR TELEPHONE USAGE. *Brandon, Belinda, editor. Ballinger Publishing Company. 1981. 403 pg.*

Covers development of customer sample; the questionnaire, calling, and billing data; data base design and implementation; box plot analysis of local and suburban calling frequencies and conservation times. Includes regression analysis of local and surburban usage; effects of time and day on local and suburban calling frequencies and conversation times. Gives the association of distance with calling frequencies and conversation times; analysis of billing data; toll usage; and other studies and suggestions for future research.

4250. ELECTRONIC FUND TRANSFERENCE IN TEXAS: THE STAGE OF ITS DEVELOPMENT AND OUTLOOK FOR THE 1980'S. *Crum, Lawrence L. University of Texas at Austin. Bureau of Business Research. March 1980. 50 pg. $5.00.*

Discusses development in electronic banking in Texas during the 1970's. Describes factors likely to affect the progress of electronic banking in the 1980's.

4251. FACTORS INFLUENCING INVESTMENT, COSTS, AND REVENUES OF SMALL, RURAL, INDEPENDENT TELEPHONE COMPANIES. *Lavey, Warren G. Harvard University. Information Resources Policy. Order Department. 1982.*

Studies the relationships of small telephone company characteristics (fewer subscribers, smaller exchanges, and lower proportions of business subscribers) to telephone plant investment, operating costs, and operating revenues. Findings are based on an econometric analysis of 1980 data from 939 telephone companies borrowing from the Rural Electrification Administration.

4252. GUIDE TO ELECTRONIC PUBLISHING. *Spigai, Fran. Knowledge Industry Publications. December 1981. 150 (est.) pg. $95.00.*

A guide to opportunities for publishers in the production of computer databases and videotext systems. Discusses types of information well-suited to these technologies, how to coordinate the publication of print and computer - based products, and how to market on-line products (including the role of the distributor). Examines systems currently in operation and identifies major telecommunications companies, common carriers, equipment suppliers, and distributors.

4253. THE HANDBOOK OF INTERACTIVE VIDEO. *Floyd, Steve, ed. Knowledge Industry Publications. Irregular. 1982. 200 pg. $34.95.*

Handbook describes the technology of interactive video and outlines available hardware. Discusses authoring systems and the design, writing and production of an interactive program. Explains differences between interactive and linear video. Appendix includes case studies. Lists manufacturers, software suppliers and other resources.

4254. IMPLICATIONS FOR THE 'COMPUNICATIONS' INDUSTRIES OF PROPOSED AMENDMENTS TO THE WEBB - POMERENE ACT. *Epperson, G. Michael. Harvard University. Information Resources Policy. Order Department. February 1982. $240.00.*

Evaluates the applicability of proposed amendments of the Webb - Pomerene Act to the computers / communications industries.

4255. LONG DISTANCE FOR LESS: HOW TO CHOOSE BETWEEN MA BELL AND THOSE OTHER CARRIERS. *Self, Robert L. Telecom Library Inc. February 1982. 160 pg. $75.00.*

Contents include the changing industry (new competition in long distance calls; entry of specialized carriers; other common carriers and resellers); Bell's long distance rates; national other common carriers (services; quality; billing; costs; basic data on ITT, MCI SPCommunication, Western Union (service descriptions; areas served; costs; options; bulk discounts); analysis of long distance calling rates and availability; basic data on each major regional reseller (services; areas; charges; directory of resellers); special long distance services for special users; and forecasts for long distance carrier industry. Appendixes cover long distance calling rates; glossary; addresses and telephone numbers of other common carriers, resellers, and Bell operating companies; and other issues.

4256. LONG DISTANCE PLEASE: THE STORY OF THE TRANSCANADA TELEPHONE SYSTEM. *Ogle, E.B. Collins Publishers. 1979. 300 pg.*

A history of Canada's long distance system. Describes the events leading up to the establishment of the TransCanada Telephone System, its growth and development. Discusses the future of the industry.

4257. THE MASTER HANDBOOK OF TELEPHONES. *Traister, Robert J. Almar Press. Irregular. 1981. 352 pg. $16.95.*

Coverage of telephone communications extends from terminology and equipment to accessories and repair. Contents include data on numerous types of telephone systems and devices.

4258. ONTARIO TELEPHONE SERVICE COMMISSION 1981 ANNUAL REPORT. *Ontario. Ministry of Transportation and Communications. Annual. 1981. 80 (est.) pg.*

Contains the activities and services of the Ontario Telephone Service Commission, government regulation, telephone service complaints / inquiries, as well as 1980 revenues and expenditures, working capital, and rate of return for the Commission. Also includes a summary of independent telephone systems as of December 31, 1981, their operating ratios, addresses, exchanges and number of phones available.

4259. PLAYERS, STAKES AND POLITICS OF REGULATED COMPETITION IN THE COMPUNICATIONS INFRASTRUCTURE OF THE INFORMATION INDUSTRY. *Oettinger, Anthony G. Harvard University. Information Resources Policy. Order Department. P-81-7. August 1981. 139 pg. $240.00.*

Analyzes pricing and costing practices principally in terms of shifting balances among diverse customers and suppliers as reflected by shifts in the conflicting pressures for competition and for costing of local telecommunications services as the residue left when politically determined (interstate and state) interexchange costs are subtracted from total costs.

4260. REVENUE AND COST ALLOCATIONS: POLICY MEANS AND ENDS IN THE RAILROAD AND TELECOMMUNICATIONS INDUSTRIES. *Godbey, Robert. Harvard University. Information Resources Policy. Order Department. I-78-8. July 1979. 33 pg. $240.00.*

Considers increasing competition within the telecommunications industry and its present system of nationwide rate averaging and subsequent revenue pooling. Describes the railroad industry's division of revenues, and discusses how problems common to both industries have been handled.

4261. STAKES IN TELECOMMUNICATIONS COSTS AND PRICES. *Oettinger, Anthony G. Harvard University. Information Resources Policy. Order Department. P.80-6. 1980. 145 pg. $240.00.*

Describes both traditional and emerging telecommunications businesses and their financial interests. Examines traditional services and facilities, prevalent labor and capital structures, and regulatory jurisdictions and price / cost relationships.

4262. TELECOMMUNICATIONS COSTS AND PRICES IN THE UNITED STATES - AN OVERVIEW. *LeGates, John C. Harvard University. Information Resources Policy. Order Department. August 1981. 27 pg.*

Discusses pricing, regulation, entry of new competitors, location of artificial intelligence, and competition of regulated and unregulated suppliers in the compunications business. Examines the long distance, local, and customer premise equipment markets. Describes the basic market structure, the presence or absence of regulation, regulatory conflict, and the major issues.

4263. TELEPHONE - LETTER MAIL COMPETITION: A FIRST LOOK. *Harvard University. Information Resources Policy. Order Department. P-81-9. December 1981. 118 pg. $240.00.*

Provides a review of telephone and letter mail use and pricing between 1950 and 1977, with special emphasis on the 1970s. Summary data for 1978 and 1979 is presented.

4264. TRADITIONAL STATE SIDE OF TELECOMMUNICATIONS COST ALLOCATIONS. *Oettinger, Anthony G. Harvard University. Information Resources Policy. Order Department. P-80-7. 1980. 177 pg. $240.00.*

Considers the current telecommunications policy debate concerning 'proper costs' and their relationships to prices, benefits, and burdens. Concentrates on pricing policy as determining the further incidence of benefits and burdens in the general absence of explicit sub-allocation of costs within the pools of costs assigned to the broad aggregates of interstate and state services.

4265. UNDERSTANDING U.S. INFORMATION POLICY: THE INFORMATION POLICY PRIMER. *Information Industry Association. 1982. $99.50.*

Offers 150 articles on political, economic, and social issues affecting computers, telecommunications, reprographics, micrographics, office automation, and other information handling technologies.

4266. UNDERSTANDING U.S. INFORMATION POLICY, VOLUME 1: RESOURCES FOR THE INFORMATION ECONOMY. *Information Industry Association. 1982. $49.50.*

Analyzes various information handling technologies (computers, telecommunications, reprographics, micrographics, office automation); how they can increase workers' productivity and quality of life; and cost effective management of information resources and assets.

4267. UNDERSTANDING U.S. INFORMATION POLICY, VOLUME 2: PARTICIPANTS IN THE INFORMATION MARKETPLACE. *Information Industry Association. 1982. $49.50.*

Covers property concepts, including intellectual property rights and obligations in such fields as copyrights, patents, trademarks, and trade secrets; government's role in the marketplace, including public and private sectors' respective roles in production and distribution of information goods and services; non-U.S. firms' threats and opportunities for U.S. trade and commerce; and interrelationships of domestic U.S. and international developments.

4268. UNDERSTANDING U.S. INFORMATION POLICY, VOLUME 3: ASSETS OF THE INFORMATION SOCIETY. *Information Industry Association. 1982. $49.50.*

Analyzes public interests in the 'Information Age' (i.e., privacy, secrecy, censorship); profiles of U.S. knowledge centers; and future technologies and potential policy conflicts.

4269. VIDEOTEX IN CANADA. *Madden, John C. Canada. Department of Communications. English/French. 1979. 32 pg.*

Based on a paper prepared for a Delta Project seminar held in Toronto 8 May 1979, report examines federal electronic system and communications policy. Explores the effects on videotex, Telidon, industrial strategy, possible markets, expenditures and estimated direct employment.

See 400, 1831, 1836, 1850, 2798, 2800.

Dissertations and Working Papers

4270. ESTIMATING THE EFFECTS OF DIFFUSION OF TECHNOLOGICAL INNOVATIONS IN TELECOMMUNICATIONS: THE PRODUCTION STRUCTURE OF BELL CANADA. *Denny, M. University of Toronto. Institute for Policy Analysis. 1979. 33 pg.*

Presents a method of estimating the cost-reducing effects of the adoption by Bell Canada of direct distance dialing. Mathematical data and statistical tables show indicators of technical change, three output cost functions, factor price elasticities, output and technical change indicator elasticities, and sources of change in production costs.

4271. THE MEASUREMENT AND INTERPRETATION OF TOTAL FACTOR PRODUCTIVITY IN REGULATED INDUSTRIES, WITH AN APPLICATION TO CANADIAN TELECOMMUNICATIONS. *Denny, M. University of Toronto. Institute for Policy Analysis. 1979. 67 pg.*

Examines total factor productivity in the Canadian telecommunications industry (Bell Canada) from 1952 to 1976. Statistical tables cover growth of outputs and inputs in current and constant dollars; average rates of growth of real outputs; average annual growth rates of real inputs; alternative output indicators; total factor productivity accounting for the growth of aggregate output; cost - output elasticities, cost - output elasticity weights and revenue share weights; sources of conventionally measured total factor productivity growth; comparison of two measures of technical change; revenues and cost; and prices and quantities of inputs and outputs.

Conference Reports

4272. CABLE MARKETING. *Quinn, Robert, ed. Hoke Communications. 1982. $15.00.*

Cassette recording of part of 1982 information utilities conference covers New Planning and Investment Approaches; What Videotex Customers Want (including home shopping and videotex-based advertising and information sciences); On-

Demand Teleshopping: The CABLESHOP Experience; and Electronic Newspapers on Cable (including new technologies' potential threat to local newspaper chains).

4273. COMMUNICATIONS: FUTURE QUESTIONS IN NETWORKING, OFFICE SYSTEMS AND REGULATION. *Johnson, Charles P. MSRA, Inc. October 1981. 131 pg. $200.00.*

Conference papers by industry executives covering networking and office systems. Includes regulation, industry structure, forecasts, new technology and products, new businesses, and competition among vendors.

4274. INFORMATION INDUSTRY: A STRATEGIC OUTLOOK. *Roseman, Fred, ed. Information Industry Association. 1982. 82 pg. $39.95.*

Proceedings of the 13th annual conference held in Boston in November 1981. Papers include Economic Outlook for the Information Industry; Telecommunications: 1990 and Beyond - A Prognosis; The Box Comes Empty; The Impact of Electronic Distribution on Industry Strategy; Deregulation Objective: Increased Competition; and Fostering Innovation in the Information Age. Covers industry restructuring, new technology, new markets, capital requirements, and need for new corporate strategies.

4275. NEW REVOLUTION IN TELECOMMUNICATIONS. *FIND / SVP. March 1981. 141 pg. $100.00.*

New York Society of Security Analysts seminar transcript considers the telecommunications revolution from industry and government standpoints. Government and industry leaders discuss communications systems utilizing new technologies; future industry needs; hardware requirements in the 1980s; optical wave guides; new switching technology; and other topics.

4276. POLITICS, POLICY AND PROFITABILITY. *Cavanagh, Michael F., ed. Hoke Communications. 1982. $27.00.*

Panel discussion from 1982 information utilities conference considers such issues as industry control, rates, and government roles (i.e., the post office in the electronic mail business). Topics include Regulating the Media; Trends and Issues Affecting Electronic Publishing Services; and Politics, Political Entities and Information Utilities.

4277. THE REAL INFORMATION UTILITY: HOME BANKING AND FINANCE. *Rockoff, Maxine, ed. Hoke Communications. 1982. $25.00.*

Cassette recording of panel discussion at 1982 information utilities conference examines operational home banking services offered in several communities, user reactions to these services, and modes of service delivery. Topics include: Potential Effects of Home Telecommunications and Personal Computers on the Financial Services Industry; Banking at the Customer's Fingertips; and Electronic Purchases from the Home: Opportunities and Obstacles. Covers general design requirements for in-home payment systems; possible role of regional EFT networks in a mature home information system; potential economic benefits to financial institutions and retailers; local versus national debit cards; and security considerations and the relationship of electronic purchase payments to bill-paying and other home banking services.

4278. SATELLITE CHERRY - PICKING: PRIVILEGE OR PIRACY. *Barlow, Portus C., ed. Hoke Communications. 1982. $10.00.*

Cassette recording from 1982 information utilities conference addresses satellite programming issues such as return on investment; risk management; range of capital opportunities; and risks of copyright infringement and Communications Act violations, with respect to unauthorized interception and use of satellite - delivered signals.

4279. TELECOMMUNICATIONS: TRENDS AND DIRECTIONS: A SEMINAR PROGRAM SPONSORED BY THE COMMUNICATIONS DIVISION, ELECTRONIC INDUSTRIES ASSOCIATION. *Electronic Industries Association. Annual. October 1981. $10.00.*

Proceedings of the sixth annual Hyannis Conference, held in May, 1981. Leaders of the telecommunications industry presented papers on the current status and future projections of the industry as a market; on the changing types of telecommunications equipment; and on the growth of alternative telecommunications systems, such as satellite data communications and electronic mail.

See 1857, 1859, 1864.

Newsletters

4280. CABLE NEWS: CABLE - PAY TV - TEXT SERVICES - MDS - STV - DBS. *Brown, Carol F., ed. Phillips Publishing Inc. Bi-weekly. 8 pg. $167.00/yr.*

Discusses marketing techniques, new technology, programming, Washington news including Senate and House bills and industry employment figures, videotext trials and ventures, satellites, business news including company earnings, revenues, executives, joint ventures and expansions, and low power TV. Cable Clips column gives brief news releases. Includes index every six months.

4281. CANADIAN COMMUNICATIONS REPORTS. *Maclean Hunter Ltd. Semi-monthly. 215.00/yr. (Canada).*

Semimonthly newsletter covers various parts of communications industry, including telecommunications, interconnection, broadcasting, cable TV, pay TV, data transmission, satellite technology, and videotex.

4282. DATA CHANNELS. *Phillips Publishing Inc. Bi-weekly. 8 pg. $147.00/yr.*

Biweekly data communications newsletter covers industry-wide and firm-specific events in North America, including new products and services and government actions. Some issues include an industry calendar.

4283. EMMS: ELECTRONIC MAIL AND MESSAGE SERVICE. *Foulger, Davis, editor. International Resource Development, Inc. Bi-weekly.*

Newsletter covering the electronic mail and message service industry. Includes new products and services, legislation, utilization trends and industry structure.

4284. F.C.C. WEEK. *Dawson-Butwick Publishers. Weekly. 8 pg. $350.00/yr.*

Covers regulation and legislation pertaining to the communications industry. Topics inlcude: national, international, broadcasting and conmon carriers. Addresses communications-related activities of the Departments of State, Commerce, Justice and Defense as well as the FCC, Congress and Executive branch.

4285. REMOTE SENSING IN CANADA. *Canada Centre for Remote Sensing. Quarterly.*

Newsletter of the Canada Centre for Remote Sensing. Intended as a vehicle for communication among members of the Canadian remote sensing community. Covers new programs and projects, research and development, symposiums

and conferences, and other news in the industry. Gives prices and specifications of satellite imagery available.

4286. TELEPHONE NEWS. *Phillips Publishing Inc. Weekly. 8 (est.) pg. $197.00/yr.*

Newsletter discusses government action; new products, activities of specific firms, (financial and personnel) and industry associations. Some issues include insider stock trade data. Concentrates on U.S. telephone industry, with some coverage of other nations and other telecommunications sectors, such as cable TV.

Services

4287. ELECTRONIC INFORMATION PROGRAM: CONTINUOUS INFORMATION SERVICE. *Link Resources Corporation. Monthly. 1981. $15000.00.*

A continuous service covering the electronic information market. Includes data on businesses involved; competition; new products; technology trends; marketing strategies; new delivery channels; and market delineation. Services include bi-monthly research reports, monthly memoranda, news briefs, online database report, and ongoing consultation.

4288. NEW ELECTRONIC MEDIA PROGRAM: CONTINUOUS INFORMATION SERVICE. *Link Resources Corporation. Monthly. 1981. $15000.00.*

Service covers the U.S. telecommunications, video, and computer industries. Focuses on competing and complementary systems (videotext, teletext, cable TV, etc.); technology trends; applications; markets; investments and returns; regulations; and businesses involved. Includes bi-monthly reports, monthly memoranda, news briefs, online database report, viewdata and videotext reports, and ongoing consultation.

See 1887.

Dictionaries

4289. THE BBDO ELECTRONIC MEDIA DICTIONARY. *BBDO International, Inc. 1981. 44 pg. $10.00.*

Defines neary 340 terms used by the electronic media industry. Also provides an appendix of broadcast and cable networks, associations, etc.

4290. DONALDSON, LUFKIN AND JENRETTE: THE COMMUNICATIONS INDUSTRY: COMMUNICATIONS INDUSTRY GLOSSARY. *Wolff, Ivan L. Donaldson, Lufkin and Jenrette Inc. April 7, 1982. 29 pg.*

Glossary of terms for the communications industry including computers, telecommunications, cable TV, and electronic publishing.

4291. TELEPHONY'S DICTIONARY. *Langley, Graham. Telephony Publishing Corporation. 1982. 315 pg. $35.00.*

Definitions of 14,500 words and terms cover such fields as computers, satellites, radio and special transmission media and techniques. Appendices cover acronyms and abbreviations, telephone signaling systems, switched data services, World Administrative Radio Conference, submarine cables, classification and designation of radio emissions, International Telecommunications Satellite Consortium, SI units (international unit system), CCITT recommentations, and logic gates and Boolean Truth Tables.

See 1897, 3374.

Central and South America

Market Research Reports

4292. TELECOMMUNICATIONS SYSTEMS AND EQUIP-MENT MARKET IN LATIN AMERICA: VOLUME 1. *Frost and Sullivan, Inc. December 1980. 250 (est.) pg. $1250.00.*

Market analysis for telecommunications systems and equipment in Latin America, with forecasts to 1990. Outlines national and regional services. Gives national surveys including economic indicators, installations, sources of capital, and competition among suppliers. First of two volumes.

4293. TELECOMMUNICATIONS SYSTEMS AND EQUIP-MENT MARKET IN LATIN AMERICA: VOLUME 2. *Frost and Sullivan, Inc. December 1980. 250 (est.) pg. $1250.00.*

Market analysis and forecasts for telecommunications systems and equipment in Latin America, 1980-1990. Gives country-by-country survey of economic indicators, system installations, capital sources, and competition among suppliers. Provides analysis of installed base and forecasts to 1990 for sales of products and services. Lists addresses of suppliers.

See 411.

Special Issues and Journal Articles

The Latin American Times

4294. BRAZILIAN SATELLITE LAUNCH LATIN AMERICA TO THE SPACE AGE. *Latin American Times Company. v.4, no.43. October 1982. pg.18-19*

Discusses Latin American satellite projects. Data provided for history of domestic Brazilian earth stations, including public and private networks.

See 416.

Western Europe

Market Research Reports

4295. ELECTRONIC DOCUMENT DELIVERY: THE ARTE-MIS CONCEPT. *Norman, Adrian. Arthur D. Little, Inc. Knowledge Industry Publications. 1981. 200 (est.) pg. $45.00.*

Examines the market potential, costs and technological requirements of a proposed pan-European satellite-linked system of teleprinters, intelligent copiers, and facsimile devices.

4296. THE EURODATA REPORTS. *Logica Ltd. 1981. 5000 pg. 6000.00 (United Kingdom).*

Eight-part market analysis report for European data communications. Forecasts demand for services and networks, installed base of equipment and utilization traffic. Examines equipment markets.

4297. EUROPEAN COMMUNICATIONS ENVIRONMENT. *Frost and Sullivan, Inc. June 1981. 216 pg. $1200.00.*

Examines structure and operating philosophies of PTT monopolies; regulatory control; and international regulatory agencies such as CCITT. Reviews European natural tariff policies and standardization. Current and future plans for telephone network development, including switching equipment and integrated services digital networks are given. National plans presented by country, vendor opportunities to supply equipment determined. PTT plans for telex, teletex, facsimile and videotex presented.

4298. LOCAL AREA NETWORK MARKETS IN EUROPE. *Frost and Sullivan, Inc. May 1981. 246 pg. $1400.00.*

Contains forecasts, 1981-1986, for non-voice local area networks and voice handling local area networks, in France, Italy, U.K., W. Germany, and the rest of Europe. Includes technology trends, survey of computer users, and market opportunities.

4299. TELECOMMUNICATIONS IN EUROPE TO 1986. *Larsen Sweeney Associates Ltd. Quarterly. 1981. 225 pg. 795.00 (United Kingdom).*

Market report on the telecommunications industry to 1986 in Western Europe. The total market is covered, as well as exchanges, switching equipment, transmission equipment, subscriber apparatus, peripherals and other components in 28 end user markets. Covers forecasts to 1986 for products, marketing, end users, distributors, production, capital investments, investments, distribution costs, profits, assets and employment and unemployment. Report updated quarterly.

4300. TELECOMMUNICATIONS IN THE U.K. TO 1985. *Larsen Sweeney Associates Ltd. Quarterly. October 1979. 58 pg. 195.00 (United Kingdom).*

Research report on telecommunications in the U.K. to 1985 (updated quarterly). Gives market analysis and forecasts to 1985 on sales, production, end users, and market value.

4301. TELECOMMUNICATIONS SECTOR WORKING PARTY: PROGRESS REPORT 1979. *National Economic Development Office. Annual. 1980. 4 pg.*

Report by the U.K. National Economic Development Council covering telecommunications gives a 1979 market analysis and market forecasts. Includes employment and unemployment statistics.

4302. TELECOMMUNICATIONS SYSTEMS AND EQUIP-MENT MARKET IN SOUTHERN EUROPEAN STATES. *Frost and Sullivan, Inc. September 1981. 402 pg. $1450.00.*

Provides analysis and forecast through 1990 for switching, transmission, subscriber terminals, and broadcasting and cable systems in Southern European states. Includes competitive assessment of suppliers' strengths and weaknesses. Gives trends and sales forecasts.

4303. TELETEXT IN THE UNITED KINGDOM: A MARKET RESEARCH STUDY. *CSP International. 1982. $2750.00.*

Survey of the U.K. teletext market examines uses by families and households, user attitudes, and competition and impacts on traditional media market structures.

4304. VIDEOTEX REPORT SERIES 1982: RELATIONSHIP OF VIDEOTEX WITH NEW MEDIA. *Butler Cox and Partners Ltd. Frost and Sullivan, Inc. July 1982. $750.00.*

Report on the relationship between videotex and other new media services and technologies. Examines effect on businesses and consumer behavior of families and households.

4305. VIDEOTEX - THE NEW TELEVISION - TELEPHONE INFORMATION SERVICES. *Woolfe, Roger. Heyden and Son Ltd. 1980. 184 pg. $7.00 (ukd).*

Monograph providing data on the videotex industry. Gives background information on services in Europe, North America and Japan and evaluates the uses and techniques involved and appraises future developments. Discusses the costs and revenues of Britain's Prestel System.

4306. WEST EUROPEAN COMMUNICATIONS NETWORKS. *Predicasts, Inc. September 1981. $375.00.*

Western European telecommunications services including satellite, telephone, telegraph, electronic mail and viewdata. Includes industry analysis, regulations, new products and markets.

4307. WESTERN EUROPEAN TELECOMMUNICATIONS MARKET. *Creative Strategies International. March 1982. $1450.00.*

Analysis of European telecommunications environment includes both voice and data transmission services. Forecasts to 1985 cover traffic flows and equipment markets. Provides profiles of major equipment vendors.

See 424, 433, 434, 436, 446, 461, 462, 463, 726, 1907, 1939.

Industry Statistical Reports

4308. BRITISH TELECOM STATISTICS 1980. *British Telecom. Annual. 1980. 100 pg.*

1980 telecommunications statistics in the U.K. produced by British Telecom. Gives full industry statistics, 1915-1980; financial statistics, 1971-1980; including profits, capital expenditure, depreciation and assets. Covers telephones, telex, equipment and hire plant, employment and unemployment and telegrams. Includes history of British Telecommunications.

4309. QUALITY OF TELEPHONE SERVICE. *British Telecom. Annual. 1981. 16 pg. Free (United Kingdom).*

Industry analysis produced annually by British Telecom gives statistics on the quality of the British Telephone Service. Includes percentage calls fouled and fault reports for the year.

4310. STATISTICAL YEARBOOK TRANSPORT, COMMUNICATIONS, TOURISM. *Publications Office of the European Communities. English/French/German/Italian. Annual. 1981. 145 pg. 400.00 (Belgium).*

Statistics on European Community nations cover transport, communications (post, telegraph, telex, telephone), and tourism, with data detailing industry infrastructure and equipment, traffic accidents, and related issues.

4311. 1981 STATISTICHES JAHRBUCH: ANNUAIRE STATISTIQUE PTT. [1981 PTT Statistical Yearbook.]. *Generaldirektion PTT. German/French. Annual. 1982. 124 pg.*

Annual of the Swiss postal services (PTT) with tables and graphs reflecting PTT and the economy, postal traffic, telecommunications traffic, installations and real estate, employees, finances and international overview.

Financial and Economic Studies

4312. ICC BUSINESS RATIO REPORT: TELECOMMUNICATIONS. *ICC Business Ratios. Annual. 3rd ed. 1982. 97.00 (United Kingdom).*

Financial analysis of 60 U.K. telecommunications and equipment corporation includes business addresses, executives, lines of business, and ownership. Company data for 1978-1981 includes fixed and intangible assets, stocks, total current assets, short-term debt, liabilities, net assets, capital employed, sales, exports, profits, interest paid, wages and salaries, and number of employees. Also includes 19 ratios of

profits, liquidity, capital utilization, gearing, productivity, exports and employees; growth rates; and performance forecasts.

4313. THE TELECOMMUNICATIONS INDUSTRY: AN INDUSTRY SECTOR ANALYSIS. *Inter Company Comparisons Ltd. 1st ed. 1981. 50 (est.) pg.*

Report giving comparisons by means of business ratios of growth of sales, assets, capital employed, employment and unemployment, profits, and stocks (1976-1979) of 60 of the leading companies in the telecommunications industry in the U.K. Also gives 1976-1979 statistics on return on capital, liquidity, exports, profit per employee, and wages paid.

See 2005.

Forecasts

4314. WORLD TELECOMMUNICATIONS STUDY 2: A MULTI-CLIENT STUDY: 1980-1990: VOLUME 3: EUROPE. *Arthur D. Little, Inc. 1980. 237 pg. $34000.00.*

Volume 3 of 1980 multi-client study on telecommunications industry for 1980-1990 throughout Europe. Equipment market by country, suppliers, equipment costs, facilities by country, leading manufacturers, growth rate, product lines and financial data & analysis by country. Equipment includes public telephone, satellite, mobile radiotelephone and CATV. Publication is part of 4 volume set. Price listed is for complete set.

See 480, 2011.

Directories and Yearbooks

4315. SWITZERLAND - YOUR PARTNER: VOLUME 1: TELECOMMUNICATIONS. *Swiss Confederation. Office for the Development of Trade. Irregular. 1979. 128 pg. 6.00 (Switzerland).*

Presents an overall analysis of the telecommunications industry in Switzerland. Includes indexes to products and services, suppliers, and a list of professional associations and groupings. Company-by-company descriptions present the range of products exported in the following categories: consultation and planning for telecommunications; telephone and telex exchanges; transmission systems; peripheral and related equipment; radio communication; and components and accessories for telecommunication systems.

Special Issues and Journal Articles

De Automatisering Gids

4316. VERVIJFROUDIGING MARKT DATA COMMUNICATIE. [Five Times the Market for Datacommunications.]. *Frost and Sullivan. Stam Tydschriften B.V. Netherlandish. v.14, no.39. September 24, 1980. pg.235*

Reviews Frost and Sullivan forecasts for the market for intelligent data communication networks in Western Europe. Examines government involvement and political aspects.

Cable Communications

4317. CABLE COMMUNICATIONS: VIDEOTEX 81: INTERNATIONAL VIDEOTEX CONFERENCE AND EXHIBITION REPORT. *Lavers, Daphne. Ter-Sat Media Publications Ltd. July 1981. pg.52-57 $3.00.*

Reports on European, North American, and Japanese subscription video text-communications systems discussed at the May 1981 Videotex conference in Toronto. Covers system capabilities and user options, types of information provided, information loads (number of 'pages'), subscription volumes, user characteristics and attitudes, subscription rates, and vendors' activities and aims, in narrative format.

Chip, Revista de Informatica

4318. EL COSTE DEL TELEPROCESO. [The Cost of Telecommunications.]. *Ediciones Arcadia, S.A. Spanish. no.06. September 1981. pg.64-65*

Cost analysis of the Spanish RETD is provided in charts and tables, showing installation and monthly costs to users. (RETD: Spanish Data Transmission Network). Lists various possibilities of use in type and services of equipment, according to user needs.

Distribution d'Aujourd'Hui

4319. UNE PREMIERE EN FRANCE: LA 'TELEMATIQUE' AU SERVICE DES COMMERCANTS ET DES CONSOMMATEURS. [A Premiere in France: 'Telematique' at the Service of Merchants and Consumers.]. *Comite Belge de la Distribution. French. August 1981. pg.47-52*

Discusses the services and usesof an experimental consumer videotext system (Teletel) operating in the town of Velizy, near Paris.

Economisch Dagblad

4320. DUITSE TELECOMMUNICATIE ZOEKT NIEUWE MARKTEN. [German Telecommunications Looking for New Markets.]. *Sijthoff Pers B.V. Netherlandish. v.42, no. 10607. February 16, 1982. pg.9*

The Bunders post, the Western German Post-Telegraph-Telephone, has started an ambitious program to find new sources of income by taking hold of new markets. Reviews the competitive position of the German telecommunications industry, research and development, sales, and profits.

Electronic Publishing Review

4321. ELECTRONIC PUBLISHING REVIEW: WHAT'S IN A NAME? *Raitt, David. Learned Information, Inc. Quarterly. March 1981. pg.16-21 $20.00.*

By name, lists European, North American, and Japanese systems for electronic transmission of printed information and graphics (including 'videotext,' 'electronic mail,' and database systems, as well as digital transmission of facsimile paper copies over telephone lines). Specifies each system's operator, base country, stage of development (startup date, number of users), and type of information transmitted.

Multichannel News

4322. BRITISH GOVERNMENT APPROVES DIRECT BROADCASTING SATELLITE. *Fallon, James. Fairchild Publications. v.3, no.11. March 22, 1982. pg.30-31*

The approval of direct broadcast satellite (DBS) development by the British government is discussed. The use of such satellites in the future for cable television and telephone channels is reviewed, and vendors and manufacturers involved in this effort are identified. Development costs are estimated, with prices for individual pieces of equipment which will be required by consumers to receive cable services.

Observateur de l'OCDE

4323. LES SYSTEMES DE VIDEOTEX INTERACTIFS ET L'INFORMATION DU CONSOMMATEUR. [Interactive Video Text Systems and Consumer Information.]. *Organisation for Economic Cooperation and Development. no.117. July 1982. pg.34-37*

Discusses services and costs of representative home data communications systems, indentifying major European, U.S., Canadian and Japanese systems and the number of terminals which they comprise. Examines such obstacles to commercialization as consumer resistance, high costs and legal questions.

Succes

4324. VIDITEL KAN GANGMAKER VOOR COMMUNICATIE-TIJDPERK WORDEN. [Viditel Might Become the Pacemaker for the Communications Age.]. *Baayens, Nico. Uitgeversmaatscuappy Success B.V. Netherlandish. v.52, no.11. November 1980. pg.87-89*

Viditel, the viewdata system with which the Dutch Post Telephone Telegraph started in summer 1980, has given the industry a chance to adapt to a growing important medium. Examines purchase costs and forecasts the outlook for the industry.

Telecommunications

4325. TELECOM CONTROL IS REGIONAL IN DENMARK: STAFF REPORT. *White, C.E. Horizon House. Irregular. v.16, no.5. May 1982. pg.25-34, 46*

Profiles Denmark's 2 major telephone companies and 8 telecommunications manufacturing firms.

Viewdata and TV User

4326. OFFICIAL PRESTEL DIRECTORY. *IPC Business Press Ltd. Irregular. July 1981. pg.49-86*

Provides an alphabetical directory, and subject - classified index, of firms and organizations providing information via, or support services to, the Prestel video data - communications system. Lists types of information provided, services, department or individual to contact, address, telephone and/or telex numbers.

See 502, 2032.

Monographs and Handbooks

4327. DIRECT BROADCASTING BY SATELLITE. *Her Majesty's Stationery Office. May 1981. 106 pg. 4450.00 (United Kingdom).*

Report of the U.K. Home Office study on direct broadcasting by satellite.

4328. ELECTRONIC DOCUMENT DELIVERY 3: TRENDS IN ELECTRONIC PUBLISHING IN EUROPE AND THE UNITED STATES. *Learned Information, Inc. May 1982. $27.00.*

Surveys the state of the publishing industry; the range of technology now confronting the industry, including videotex, online, cable TV, satellites, and videodiscs, with special attention to document delivery services; and the impact these technologies may have on industry structure and finance as well as legal, social, and economic consequences.

4329. THE IMPACT OF NEW TECHNOLOGIES ON PUBLISHING. *Commission of the European Communities. K.G. Saur Verlag KG. 1981. 195 pg. $18.00.*

Covers technological developments in the printing industry until 1990; using text processing, computer networking and satellite telecommunications to publish primary scientific and technical information; information transfer and the significance of new storage media and technolgies; teleordering in Denmark; Prestel; French policy on videodex; Euronet Diane, the European network for direct access to information; origination and editing; processing and dissemination: the wholesaler, the publisher, bookseller, librarian.

4330. LIBERALISATION OF THE USE OF THE BRITISH TELECOMMUNICATIONS NETWORK. *Beasley, M.E. Automatic Vending Machine Association of Britain. January 1981. 64 pg. 3.60 (United Kingdom).*

Report to the U.K. Secretary of State, analyzing the British Telecommunications Industry.

Conference Reports

4331. BUSINESS TELECOMS: THE NEW REGIME. *Online Inc. 1982. 197 pg. 44.00 (United Kingdom).*

Proceedings of 1981 conference in London cover 'Towards Intergrated Voice and Data Switching', 'New Electronic Message and Information Service, 'Emerging Techniques and Services for Wideband Telecoms,' 'Management of Telecoms - Practical Implications', 'Strategies for the 1980's', and 'The New Regime After the Monopoly.'

4332. VIDEOTEX AND THE PRESS. *Learned Information, Inc. May 1982. $32.00.*

Proceedings of a seminar held under the auspices of the European Commission considers how the information revolution, especially videotex, will effect newspapers and how they will adapt commercially and editorially. Issues include changing nature of coverage when the medium becomes more geared to readers' choices, and advertising on the videotex page.

4333. VIDEOTEX IN EUROPE: CONFERENCE PROCEEDINGS: (LUXEMBOURG, 19-20 JULY 1979). *Vernimb, Carlo, editor. Commission of the European Communities: Pactel (Frankfurt, B.R.D.); MWS Services (London); A.D. Little (Paris). Learned Information, Inc. 1980. 247 pg. $92.00.*

Two papers within these proceedings cover the European market outlook for videotex communications services, through 1983 and 1995. Installations are broken down by sector and by country. A third paper surveys the status, services, transmission method, terminal type, and extent and charges (where available) of individual operating and planned systems in Europe, North America and Japan.

Newsletters

See 2068.

Eastern Europe

Indexes and Abstracts

4334. REFERATIVNYI ZHURNAL: ELEKTROSVIAZ: ABSTRACT JOURNAL: ELECTROCOMMUNICATION. *Vsesoiuznyi Institut Nouchnoi Informatsii. Monthly. 1981.*

Monthly electrocommunications abstracts journal, in Russian with contents in English.

See 507, 508.

Asia

Market Research Reports

See 510, 511, 512, 513, 514, 516, 517, 529, 530, 2081.

Special Issues and Journal Articles

Economic Review

4335. TELECOMMUNICATIONS DEVELOPMENT PLAN FOR THE 80'S IN THE REPUBLIC OF CHINA. *International Commercial Bank of China. July 1981. pg.5-12*

Reviews telecommunications services in Taiwan and examines growth trends, technology, investment, and future prospects. Data includes forecasts of local telephone growth (1979-1989); toll trunk demand and construction (1979-1989); investment as percent of GDP growth (1970-1979); and estimated capital investment (1979-1989).

Oceania

Industry Statistical Reports

4336. TRANSPORT AND COMMUNICATION, NEW SOUTH WALES. *Australian Bureau of Statistics. Annual. 9101.1. 1981. 61 pg. Free.*

Provides shipping statistics relating mainly to number and net tonnages of overseas, interstate, and intrastate shipping entered and cleared at New South Wales ports; country of registration of shipping entered and type of service; quantities of overseas and interstate cargo discharged and shipped; quantities of overseas cargo classified by countries of loading and discharge by major trade areas and by type of shipping service; financial operations of Maritime Services Board of New South Wales; overseas and interstate freight rates; and ferry passenger services in Sydney and Newcastle. Railways and omnibus statistics relate primarily to traffic and financial operations of the New South Wales government railway system and omnibus services. Motor transport data mainly concerns number of licenses and vehicles of various types on register; trucks registered by capacity; and new trucks and motor vehicles registered. Other data covers finances of Dept. of Main Roads; government grants for road construction and maintenance; road traffic accidents; internal and overseas air transport operations; financial operations of the Postal Commission and Telecommunications Commission; telex services; and numbers of civil radiocommunication stations.

4337. TRANSPORT AND COMMUNICATION, SOUTH AUSTRALIA. *Australian Bureau of Statistics. Annual. 9102.4. 1981. 27 pg.*

Covers shipping (vessels entered and cleared at ports, cargo handled by trade area and type of service); civil aviation (accidents, number of aircraft, passengers, freight); railways, tram, and bus services (summary of operations); motor vehicles (licenses, fees, number on register and new registrations); communication (pst offices, telegraph and telephone services, radio and television stations); distances between Adelaide and selected cities and ports; and length of roads.

Middle East

Market Research Reports

See 535, 536, 537, 538, 540, 545.

Africa

Forecasts

4338. WORLD TELECOMMUNICATIONS STUDY 2: A MULTI-CLIENT STUDY: 1980-1990: VOLUME 4: AFRICA AND ASIA. *Arthur D. Little, Inc. 1980. 297 pg. $34000.00.*

Volume 4 of 1980 multi-client study on the telecommunications industry covering Asia and Africa for 1980-1990. Telephone, satellite, mobile radiotelephone, and CATV equipment systems discussed. Background data, financial analysis, revenues, expenditures, suppliers, market structure and areas, provided, by country. Price listed is for complete set.

Special Issues and Journal Articles

Business America

4339. KENYA GETS BANK LOAN FOR TELEPHONE SYSTEM. *U.S. Department of Commerce. International Trade Administration. v.5, no.20. October 4, 1982. pg.34*

Gives details of World Bank loan to improve telephone service in Kenya, 1982. Includes data on number of lines and subscribers.

Part 2: Section 1: Publisher Listings

A

Abbott, Langer, and Associates
P.O. Box 275
Park Forest, IL   60466
TELEPHONE: (312)756-3990

ACM
1133 Avenue of the Americas
New York, NY   10036
TELEPHONE: (212)265-6300

Acoustical Publications, Inc.
27101 E. Oviatt Rd., P.O. Box 40416
Bay Village, OH   44140
TELEPHONE: (216)835-0101

Acumen-System Three
125 New Bond St.
London, United Kingdom W1Y 9AF
TELEPHONE: (011441)4090147
TELEX: 23341
Hurwitt, Frances

Addison-Wesley Publishing Company, Inc.
Jacob Way
Reading, MA   01867
TELEPHONE: (617)944-3700

Admerca AG
Minervastrasse 46
Zurich, Switzerland CH-8032
TELEPHONE: (0114101)470044
TELEX: 57770

Administrative Management Society
Maryland Rd.
Willow Grove, PA   19090
TELEPHONE: (215)659-4300

Advanced Technology Publications
1001 Watertown
Newton, MA   02165
TELEPHONE: (617)244-2939

Advisory Group on Electronic Devices
201 Varick St.
New York, NY   10014
TELEPHONE: (212)620-3377
Weiss, M. Gerald

Agricultural Microcomputing
1119, 45 Dunfield Ave.
Toronto, Ont. Canada M4S 2H4

Air Diffusion Council
435 N. Michigan Ave.
Chicago, IL   60611
TELEPHONE: (312)527-5494

Air Movement and Control Association
30 W. University Dr.
Arlington Heights, IL   60004
TELEPHONE: (312)394-0150

Aircraft Owners and Pilots Association
7315 Wisconsin Ave.
Bethesda, MD   20014
TELEPHONE: (301)654-0500
Baker, John L.

Alberta. Department of Agriculture
9718 107 St.
Edmonton, Alta. Canada T5K 2C8
TELEPHONE: (403)427-2727

Alex. Brown and Sons
135 E. Baltimore St.
Baltimore, MD   21202
TELEPHONE: (301)727-1700

Alfred Publishing Company
15335 Morrison St.
Sherman Oaks, CA   91413
TELEPHONE: (213)995-8811

Allgemeine Treuhaud AG *see* Verlag Schweizer Automotik
Pool,Allgemeine Treuhaud AG

Alltech Publishing Company
212 Cooper Center, North Park Dr. and Browning Rd.
Pennsauken, NJ   08109
TELEPHONE: (609)662-2122

Almar Press
4105 Marietta Dr.
Binghamton, NY   13903
TELEPHONE: (607)722-6251

American Association for the Advancement of Science
1515 Massachusetts Ave., NW
Washington, DC   20005
TELEPHONE: (202)467-4400
Carey, William D., Executive Officer

American Banker, Inc.
525 W. 42nd St.
New York, NY   10036
TELEPHONE: (212)563-1900

American Bankers Association
1120 Connecticut Ave., NW
Washington, DC   20036
TELEPHONE: (202)467-4000
Alexander, Willis W.

American Bar Association
1155 E. 60th St.
Chicago, IL   60637
TELEPHONE: (312)947-4000

American Chemical Society
1155 16th St., NW
Washington, DC 20036
TELEPHONE: (202)872-4600

American Electronics Association
2600 El Camino Real
Palo Alto, CA 94306
TELEPHONE: (415)857-9300

American Enterprise Institute for Public Policy
1150 17th St., NW
Washington, DC 20036
TELEPHONE: (202)862-5800

American Institute of Aeronautics and Astronautics
555 W. 57th St.
New York, NY 10019
TELEPHONE: (212)247-6500

American Institute of Architects
1735 New York Ave., NW
Washington, DC 20006
TELEPHONE: (202)626-7300
Meeker, David Olan, Executive Vice President

American Institute of Industrial Engineers
25 Technology Park
Atlanta Norcross, GA 30092
TELEPHONE: (404)449-0460
Belden, David L., Executive Director

American Society for Information Science
1010 16th St., NW
Washington, DC 20036
TELEPHONE: (202)659-3644

American Society for Metals
Metals Park, OH 44073
TELEPHONE: (216)338-5151

American Society for Nondestructive Testing
4153 Arlingate Plaza, Caller #28518
Columbus, OH 43228
TELEPHONE: (614)274-6003

American Society of Civil Engineers
345 E. 47th St.
New York, NY 10017
TELEPHONE: (212)705-7496
Zwoyer, Eugene, Executive Director

American Telegraph and Telephone Long Line
201 Littleton Rd., Rm. 220
Morris Plains, NJ 07950
TELEPHONE: (201)631-4047
Clark, Bart, Supervisor

Amfetex, Inc.
P.O. Box 213
Arlington, MA 02174

Amivest Corporation
505 Park Ave.
New York, NY 10022
TELEPHONE: (212)688-6667
Hirsch, Michael D.

Anbar Publications Ltd.
P.O. Box 23
Wembley, Middlesex, United Kingdom HA9 8DJ
TELEPHONE: (0114401)902-4489
TELEX: 935779

Anderson Publishing Company
P.O. Box 3616
Simi Valley, CA 93063
TELEPHONE: (805)581-1184

Angelotti Editore
Via Riparmonti 115
Milan, Italy I-20141

A.P. Publications Ltd.
322 St. John St.
London, United Kingdom EC1V 4QH

Applied Computer Research
P.O. Box 9280
Phoenix, AZ 85068
TELEPHONE: (602)995-5929

Applied Management Services
Box 73
Massapequa Park, NY 11762
TELEPHONE: (516)799-3285

Architecture Technology Corporation
P.O. Box 24344
Minneapolis, MN 55424
TELEPHONE: (612)935-2035

Arlen Communications
P.O. Box 40871
Washington, DC 20016
TELEPHONE: (301)229-0909
TELEX: 440283

Artech House Inc.
610 Washington St.
Dedham, MA 02026
TELEPHONE: (617)326-8220

Arthurs Publications
5200 Dixie Rd., Suite 204
Mississauga, Ont. Canada L4W 1E4
TELEPHONE: (416)625-5277

Asian Finance Publications Ltd.
223 Gloucester Rd., Suite 9D Hyde Centre
Hong Kong, Hong Kong
TELEPHONE: (0118525)724221
Howley, Joan, Marketing Director

Aslib *see* Association for Information Management

A/S/M Communications
230 Park Ave.
New York, NY 10017
TELEPHONE: (212)661-8080

Associated Group Media Ltd.
P.O. Box 28-349,George St., Newmarket
Auckland 5, New Zealand
TELEPHONE: (011649)795-393

Associated Telephone Answering Exchanges
Bankers Square100 Pitt St.
Alexandria, VA 22314
TELEPHONE: (703)683-3770

Association for Computing Machinery
1133 Avenue of the Americas
New York, NY 10036
TELEPHONE: (212)265-6300
Weinstein, Sidney, Executive Director

Association for Information Management
3 Belgrave Square
London, United Kingdom SW1X 8PL
TELEPHONE: (0114401)235-5050

Association for Systems Management
24587 Bagley Rd.
Cleveland, OH 44138
TELEPHONE: (216)243-6900

Association Francaise des Documentalistes et des Bibliothecaires Specialises
5 Avenue Franco-Russe
Paris, France 75007
TELEPHONE: (011331)5510504

Association Nationale de la Recherche Technique
101 Avenue Raymond - Poincare
Paris, France F-75116

Association of Architectural Technologists of Ontario
see Ontario Association of Certified Engineering Technicians and Technologists

Association of Computer Users
4800 Riverbend Rd.
Boulder, CO 80301
TELEPHONE: (303)443-3600

Association of Data Processing Services Organization, Inc.
1925 N. Lynn St.
Arlington, VA 22209
TELEPHONE: (703)522-5055

Association of Industrial Chemists see Gesellschaft fur Chemiewirtschaft

Association of Public Address Engineers
47 Windsor Rd.
Slough, Berkshire, United Kingdom 5L1 2EE

Association of Sound and Communication Engineers
47 Windsor Rd.
Slough, United Kingdom SL1 2EE
TELEPHONE: (01144)39455

Auerbach Information, Inc.
6560 North Park Dr.
Pennsauken, NJ 08109
TELEPHONE: (609)662-2070
TELEX: 831464

Australian Bureau of Statistics
P.O. Box 10
Belconnen, Australia
TELEPHONE: (0116162)526636
I.P. Sharp Associates Ltd.
145 King St. West, Suite 1400
Toronto, Ont. M5H 1J8

Australian Government Publishing Service
P.O. Box 84
Canberra, Australia ACT 2000
TELEPHONE: (01161)954411

Austria Presse Agentur (APA) reg. Gen. m.b.H.
Gunoldstr. 14
Vienna, Austria A-1199
TELEPHONE: (01143222)362-550
TELEX: 074721

Austrian Institute for Economic Research see Osterreichisches Institut fur Wirtschaftsforschung (Wifo)

Automatic Information Processing Research Centre
Stadhouderskade 6
Amsterdam, Netherlands 1054 ES

Automatic Vending Machine Association of Britain
Joeden St.
Kingston-upon-Thames, Surrey, United Kingdom KT1 1EE
TELEPHONE: (0114401)549-7311

B

G.G. Baker and Associates
54 Quarry St.
Guildford, Surrey, United Kingdom GU1 3UF

Ballinger Publishing Company
54 Church St., Harvard Square
Cambridge, MA 02138
TELEPHONE: (617)492-0670

Banco de Bilbao. Service de Estudios
Paseo de la Castellano 81, Departamento Relaciones Institucionales, Planta 22
Madrid, Spain 16
TELEPHONE: (011341)455-4021
Alcaide, Julio, Director, Servicios de Estudios

Banco Hispano Americano. Servicio De Estudios
Calle Alcala 31
Madrid, Spain 14
TELEPHONE: (011341)222-4258
Gorrino, Don Vicente, Director

Bank Administration Institute
60 Gould Center
Rolling Meadows, IL 60008
TELEPHONE: (312)228-6200

Bank of Bilbao. Research Services see Banco de Bilbao. Service de Estudios

Barks Publications Inc.
400 N. Michigan Ave., Suite 1016
Chicago, IL 60611
TELEPHONE: (312)321-9440

Battelle - Frankfurt
Am Romerhof 35, P.O. Box 900 160
Frankfurt (Main) 90, West Germany D-6000

Battelle - London
15 Hanover Square
London, United Kingdom W1R 9AJ
TELEPHONE: (0114401)493-0184
TELEX: 23773

Battelle Memorial Institute
505 King Avenue
Columbus, OH 43201
TELEPHONE: (614)424-6424
TELEX: 245454
Hamilton, H. Ronald

BBDO International, Inc.
383 Madison Ave.
New York, NY 10017
TELEPHONE: (212)355-5800

Bear Stearns and Company
55 Water St.
New York, NY 10041
TELEPHONE: (212)952-5000

B.E.D. Business Books Ltd.
Restmor Way, Bridge House
Wallington, Surrey, United Kingdom SM6 7BZ
TELEPHONE: (0114401)647-1001-5

Belenes Publications Ltd.
93 Lr. Baggot St.
Dublin, Ireland

**Belgian American Chamber of Commerce in the U.S.
   Inc.**
50 Rockefeller Plaza
New York, NY 10020
TELEPHONE: (212)247-7613

Benn Publications Ltd.
25 New Street Square
London, United Kingdom EC4A 3JA
TELEPHONE: (011-44-01)353-3212

Benwill Publishing Corporation
1050 Commonwealth Ave.
Boston, MA 02215
TELEPHONE: (617)232-5470

Berkeley Enterprises, Inc.
815 Washington St.
Newtonville, MA 02160
TELEPHONE: (617)332-5453

Sanford C. Bernstein and Company, Inc.
717 Fifth Ave.
New York, NY 10022
TELEPHONE: (212)486-5810
Hollister, Fred J., Marketing Manager, Strategic Research
   Services

A.M. Best Company
Ambest Rd.
Oldwick, NJ 08858
TELEPHONE: (201)439-2200

Bill Communications, Inc.
633 Third Ave.
New York, NY 10017
TELEPHONE: (212)986-4800

Billboard Ltd.
One Astor Plaza
New York, NY 10036
TELEPHONE: (212)764-7300

Max Binkert and Company
Basterstrasse
Laufenburg, Switzerland CH-4335

BIS Marketing Research
2 Help St., P.O. Box 129
Chatswood, Sydney, Australia 2067
TELEPHONE: (011612)412-3266
TELEX: 71112

Blandford Publications
Wellesley Rd., Pembroke House
Croydon, United Kingdom CR9 2BX

Blum Publications
15 Oakridge Circle
Wilmington, MA 01887
TELEPHONE: (617)658-6876

Bobbs Merrill Educational Publishing
245 E. 24th St.
New York, NY 10010
TELEPHONE: (212)683-2396

Bohmann Druck und Verlag AG
Leberstr. 122
Vienna, Austria A-1110
TELEPHONE: (01143222)652505
Fischer, Hildegard, General Manager

Johann L. Bondi and Sons
Industriestr. 2
Perchtoldsdorf / Wien, Austria A-2380
TELEPHONE: (01143222)864-921
TELEX: 131136

Booz-Allen and Hamilton, Inc.
400 Market St.
Philadelphia, PA 19106
TELEPHONE: (215)627-8110

R.R. Bowker
1180 Avenue of the Americas
New York, NY 10036
TELEPHONE: (800)521-0600
Customer Service Department
P.O. Box 1807
Ann Arbor, MI 48107
TELEPHONE: (313)761-4700

BPA Management Services
Dene St., Concept House
Dorking, Surrey, United Kingdom RH4 2DR
TELEPHONE: (01144306)884522

**British Electrical and Allied Manufacturers Association
   Ltd.**
8 Leicester St.
London, United Kingdom WC2H 7BN
TELEX: 263536
Meigh, E.
TELEPHONE: (0114401)437-0678

British Hydromechanics Research Association
Cranfield
Bedford, United Kingdom MK43 0AJ
TELEPHONE: (011440234)750422

British Telecom
2-12 Gresham St.
London, United Kingdom EC2V 7AG

Broadband Information Services, Inc.
295 Madison Ave.
New York, NY 10017
TELEPHONE: (212)685-5320

Broadcasting Publications, Inc.
1735 DeSales St., NW
Washington, DC 20036
TELEPHONE: (202)638-1022

Irwin Broh and Associates
1001 E. Touchy Ave.
Des Plaines, IL 60018
TELEPHONE: (312)297-7515
Waitz, David

BSO Publications Ltd.
41 High St., P.O. Box 1
Wivenhoe, Colchester, United Kingdom CO79EA

Building Owners and Managers Association International
1221 Massachusetts Ave., NW
Washington, DC 20005
TELEPHONE: (202)638-2929
Abraham, A., Marketing/Sales Manager

Building Services Publications Ltd.
1-3 Pemberton Row, Red Lion Court, Fleet St.
London, United Kingdom EC4P 4HL

Bureau d'Informations et de Previsions Economiques
122 Avenue Charles De Gaulle
Neuilly-sur-Seine, Cedex, France 92522
TELEPHONE: (011331)747-1166

Business Communications Company
P.O. Box 2070
Stamford, CT 06906
TELEPHONE: (203)325-2208

Business Equipment Trade Association
109 Kingsway
London, United Kingdom WC2B 6PU
TELEPHONE: (0114401)405-6233

Business Information Services
20 Pine Mountain Rd.
Ridgefield, CT 06877
TELEPHONE: (203)748-6529

Business News Publishing Company
P.O. Box 2600
Troy, MI 48084
TELEPHONE: (313)362-3700

Business Trend Analysts, Inc.
3 E. Deer Park Rd.
Dix Hills, NY 11746
TELEPHONE: (516)462-5454

Butler Cox and Partners Ltd.
20-30 Holborn Viaduct
London, United Kingdom EC1A 2BP
TELEPHONE: (0114401)583-9381

Butterworths Publishers, Inc.
10 Tower Office Park
Woburn, MA 01801
TELEPHONE: (617)933-8260

Buyers Laboratory, Inc.
20 Railroad Ave.
Hackensack, NJ 07601
TELEPHONE: (201)488-0404

B.V. Uitgevers-Maatschappy Bonaventura
Spuistraat 110-112
Amsterdam, Netherlands 1012 VA
TELEPHONE: (01131020)244950

BYTE Publications Inc.
70 Main
Peterborough, NH 03458
TELEPHONE: (603)924-9281

C

Cahners Publishing Company
221 Columbus Ave.
Boston, MA 02116
TELEPHONE: (617)536-7780

Cahners Publishing Company
Cahners Plaza, P.O. Box 5080
Des Plaines, IL 60018
TELEPHONE: (312)635-8800

Cahners Publishing Company
270 St. Paul St.
Denver, CO 70206
TELEPHONE: (303)388-4511

Canada Centre for Remote Sensing
717 Belfast Rd.
Ottawa, Ont. Canada K1A 0Y7
TELEPHONE: (613)995-1210

Canada. Department of Communications
300 Slater St.
Ottawa, Ont. Canada
TELEPHONE: (613)995-8883

Canada. Department of External Affairs
Lester B. Pearson Bldg., 125 Sussex Dr.
Ottawa, Ont. Canada K1A 0G2
TELEPHONE: (613)593-7064

Canada. Department of Industry, Trade and Commerce
235 Queen St.
Ottawa, Ont. Canada K1A 0H5
TELEPHONE: (613)995-5771

Canada. Department of Supply and Services
11 Laurier Ave.
Hull, Que. Canada K1A 0S9
TELEPHONE: (613)994-3475

Canada. Department of Supply and Services. Publishing Centre
Ottawa, Ont. Canada K1A 0S9
TELEPHONE: (613)593-5132

Canada. Statistics Canada
Ottawa, Ont. Canada K1A OT6
TELEPHONE: (613)992-2959

Canadian Centre for Policy Alternatives
P.O. Box 4466, Station E
Ottawa, Ont. Canada K1S 5B4
TELEPHONE: (613)563-1341

Canadian Engineering Publications Ltd.
32 Front St., W., Suite 501
Toronto, Ont. Canada M5J 2H9
TELEPHONE: (416)869-1735

Canadian Information Processing Society
243 College St., W., 5th Fl.
Toronto, Ont. Canada M5T 2Y1
TELEPHONE: (416)593-4040

Canadian Radio - Television and Telecommunications Commission
Ottawa, Ont. Canada K1A 0N2
TELEPHONE: (819)997-0313

Canadian Standards Association
178 Rexdale Blvd.
Rexdale, Ont. Canada
TELEPHONE: (416)744-4000

Capital Systems Group, Inc.
11301 Rockville Pike
Rockville, MD 20852
TELEPHONE: (301)881-9400

Cardiff Publishing Company
6430 S. Yosemite
Englewood, CO 80111
TELEPHONE: (303)694-1522

Centre d'Information sur le Materiel de Bureau
4 Rue de Castellane
Paris, France 75008
TELEPHONE: (011331)265-1767

Cerberus Publishing
P.O. Box 470
Frenchtown, NJ 08825

Chambre de Commerce et d'Industrie de Reims
30 Rue Ceres
Reims, France
TELEPHONE: (0113326)881515
TELEX: CHAME 8306 42F

Channing Weinberg
950 Third Ave.
New York, NY 10022
TELEPHONE: (212)753-8922
Weinberg, Barry

Charles River Associates
200 Clarendon St.
Boston, MA 02116
TELEPHONE: (617)266-0500

Dana Chase Publications, Inc.
1000 Jorie Blvd., CS 5030
Oak Brook, IL 60521
TELEPHONE: (312)789-3484

Chase Econometrics
555 City Line Ave.
Bala-Cynwyd, PA 19004
TELEPHONE: (215)667-6000
TELEX: 831609
Nixon, David D., Vice President, Product Development

Chemical Specialties Manufacturers Association
1001 Connecticut Ave., NW
Washington, DC 20036
TELEPHONE: (202)872-8110
Engel, Ralph

Chilton Book Company
201 King of Prussia Rd.
Radnor, PA 19809
TELEPHONE: (215)964-4000

CIMAB see Centre d'Information sur le Materiel de Bureau

CISI On-Line Information Service
35, Blvd. Brune
Paris Cedex 14, France 75680
TELEPHONE: (011331)5392510

Cleworth Publishing Company, Inc.
One River Rd.
Cos Cob, CT 06807
TELEPHONE: (203)661-5000

CMP Publications, Inc.
111 E. Shore Rd.
Manhasset, NY 11030
TELEPHONE: (516)365-4600

Collins Publishers
100 Lesmill Rd.
Don Mills, Ont. Canada M3B 2R5
TELEPHONE: (416)445-8221
TWX: 06-966-673

Comite Belge de la Distribution
Marianne 34
Brussels, Belgium
TELEPHONE: (011322)3459923

Commodities Magazine, Inc.
219 Parkade
Cedar Falls, IA 50613
TELEPHONE: (319)277-6341

Communication Channels, Inc.
185 Madison Ave.
New York, NY 10016
TELEPHONE: (212)889-1850

Communication Information Management, ADP / MIS Division
GSA. 18th and F Sts. NW, Room G32
Washington, DC 20405
TELEPHONE: (202)566-1544
Holt, Mr.

Compagnie Francaise d'Edition
40, Rue Colisee
Paris, France F-75008

Compass Publications Inc.
1117 N. 19th St., Suite 1000
Arlington, VA 22209
TELEPHONE: (703)524-3136

Computer and Business Equipment Manufacturers Association
311 First St., NW, 5th Fl.
Washington, DC 20001
TELEPHONE: (202)737-8888
Henriques, Vico E., President

Computer Design Publishing Corporation
11 Goldsmith St.
Littleton, MA 01460
TELEPHONE: (617)486-8944
Shapiro, Sydney F., Managing Editor

Computer Equipment Information Bureau
1580 Massachusetts Ave.
Boston, MA 02125
TELEPHONE: (617)247-2290

Computer Graphics Software News
5857 S. Gessner, Suite 40
Houston, TX 77036

Computer Guides Ltd.
30-31 Islington Green
London, United Kingdom N1 8BJ

Computer Reference Guide
135 S. Harper Ave.
Los Angeles, CA 90048
TELEPHONE: (215)852-4886

Computer Users' Year Book
430 Holdenhurst Rd.
Bournemouth, United Kingdom BH8 9AA
TELEPHONE: (011440202)301121

Confederacion Espanola de Organizaciones Empresariales
Diego de Leon 50, Servicios Tecnicos de la CEOE
Madrid, Spain 6
TELEPHONE: (011341)262-4410
Garcia, Maximino Carpio, Assistant

Construction Products Manufacturers Council, Inc.
1717 Massachusetts Ave., NW
Washington, DC 20036
TELEPHONE: (202)667-8727

Control Data Corporation: Cybernet Services
P.O. Box 0, HQW11A, 8100 34th Ave. S.
Minneapolis, MN 55440
TELEPHONE: (612)853-8100

Copper Development Association
405 Lexington Ave.
New York, NY 10174
TELEPHONE: (212)953-7300

Crain Communications, Inc.
740 N. Rush St.
Chicago, IL 60611
TELEPHONE: (312)649-5200

Creative Strategies International
4340 Stevens Creek Blvd., Suite 275
San Jose, CA 95129
TELEPHONE: (408)249-7550

Credit Suisse. Economics, Public Relations and Marketing Division
P.O. Box 8021
Zurich, Switzerland

Croner Publications, Inc.
211-03 Jamaica Ave.
Queens, NY 11428
TELEPHONE: (212)464-0866

CSP International
90 Park Ave., 14th Fl.
New York, NY 10016
TELEPHONE: (212)599-1526
TELEX: 225115

Cuadra Associates, Inc.
2001 Wilshire Blvd., Suite 305
Santa Monica, CA 90403
TELEPHONE: (213)829-9972

Customer Satisfaction Research Institute
4901 College Blvd., Suite 107, Leawood Manor
Shawnee, KS 66211
TELEPHONE: (913)381-8209

CUYB Publications Inc.
1 Federal Bldg., Suite 401
Pottstown, PA 19464
TELEPHONE: (215)326-5188
Wilkerson, Edna, Manager

CW Communications Inc.
375 Cochituate Rd.
Framingham, MA 01701
TELEPHONE: (617)879-0700

Cybernet Services *see* Control Data Corporation: Cybernet Services

D

DAFSA Enterprises
7 Rue Begere
Paris, France 75009
TELEPHONE: (0113301)2332123
TELEX: 040472 DAFDOC

Data Consult Inc. *see* P.T. Data Consult Inc.

Data Entry Management Association
P.O. Box 3231
Stamford, CT 06905
TELEPHONE: (203)322-1166

D.A.T.A., Inc.
P.O. Box 26875
San Diego, CA 92126

Data Processing Management Association
505 Busse Highway
Park Ridge, IL   60068
TELEPHONE: (312)825-8124

Data Sources
20 Brace Rd., Suite 310
Cherry Hill, NJ   08034
TELEPHONE: (609)429-2100

Datapro Research, Inc. (McGraw-Hill)
1805 Underwood Blvd.
Delran, NJ   08075
TELEPHONE: (609)764-0100

Dataquest
19055 Pruneridge Ave.
Cupertino, CA   95014
TELEPHONE: (408)725-1200

Datek of New England
P.O. Box 68, 38 Elliot St.
Newtonville, MA   02160
TELEPHONE: (617)332-4576

Dawson-Butwick Publishers
2005 N. Trinidad St.
Arlington, VA   22213
TELEPHONE: (703)237-8967

Robert E. De La Rue Associates
P.O. Box 2370
Santa Clara, CA   95055
TELEPHONE: (408)243-2040

De Zoate and Bevan
25 Finsbury Circus
London, United Kingdom EC2M 733
TELEPHONE: (0114401)588-4141
TELEX: 888221

Dealerscope Inc.
115 2nd. Ave.
Waltham, MA   02154
TELEPHONE: (617)890-5124

Defense Market Service, Inc. *see* DMS, Inc.

DEMPA Publications, Inc.
11-15 Higashi Gotanda 1-chome, Shinagawe-ku
Tokyo, Japan 141

Derwent Publications Ltd.
Rochdale House, 128 Theobalds Rd.
London, United Kingdom WC1X 8RP
TELEPHONE: (011441)2425823
TELEX: 267487
U.S.A. Office
1735 Jefferson Davis Highway, Suite 605
Arlington, VA   22202
TELEPHONE: (703)521-7997

Developing World Industry and Technology, Inc.
915 15th St., NW, #600
Washington, DC   20005
TELEPHONE: (202)783-6060
Baranson, Jack, President

Diamond Lead Company, Ltd.
4-2 Kasumigaseki, 1-chome, Chiyoda-ku
Tokyo, Japan FP 100
TELEPHONE: (0118103)504-6796

Diamond Research Corporation
9850 Old Creek Rd.
Ventura, CA   93001
TELEPHONE: (805)649-2209
Diamond, Arthur S., Editorial Director

D. Dietrich Associates, Inc.
P.O. Box 511
Phoenixville, PA   19460
TELEPHONE: (215)935-1563

DMS, Inc.
100 Northfield St.
Greenwich, CT   06830
TELEPHONE: (203)661-7800
TELEX: 131526
Ryan, Vicki, Manager, Market Services

Documentation Francaise
29-31 Quai Voltaire
Paris, France 75340

Dominion Securities Ames Ltd.
Commerce Court S., P.O. Box 21
Toronto, Ont. Canada M5L 1A7
TELEPHONE: (416)362-5711
Marketing Department

Domus Books
400 Anthony Trail
Northbrook, IL   60062
TELEPHONE: (312)498-4000

Donaldson, Lufkin and Jenrette Inc.
140 Broadway
New York, NY   10005
TELEPHONE: (212)943-0300

Dresden Associates
P.O. Box 246
Dresden, ME   04342
TELEPHONE: (207)737-4466

Drexel Burnham Lambert Inc.
60 Broad St.
New York, NY   10004
TELEPHONE: (212)480-6000
McHale, Joseph

Dun and Bradstreet Canada Ltd. Market Services Division
365 Bloor St. E., 15th Fl.
Toronto, Ont. Canada M4W 3L4
TELEX: (416)963-6504

Dun and Bradstreet, Inc. National Credit Office Division
1290 Avenue of the Americas
New York, NY   10019
TELEPHONE: (212)957-2468

Dun and Bradstreet Publication Corporation
875 Third Ave.
New York, NY   10022
TELEPHONE: (212)605-9522

Duncker und Humblot
Dietrich-Schafer-Weg 9, Postfach 410329
Berlin 41, West Germany D-1000
TELEPHONE: (0114930)7912026

Dun-Donnelley Publishing Corporation
303 E. Wacker Dr., 3 Illinois Center Bldg.
Chicago, IL 60601
TELEPHONE: (312)938-2900

E

Eastern Economist Ltd.
Parliament St., United Commercial Bank Bldg.
New Delhi, India

Economic Information Systems, Inc.
310 Madison Ave.
New York, NY 10017
TELEPHONE: (212)697-6080

Economic Salon Ltd.
60 E. 42nd St.
New York, NY 10017
TELEPHONE: (212)986-1588

Economics and Technology, Inc.
101 Tremont St.
Boston, MA 02108
TELEPHONE: (617)423-3780

Economist Intelligence Unit Ltd.
Spence House, 27 St. James' Place
London, United Kingdom SW1A 1NT
TELEPHONE: (011441)493-6711
TELEX: 266353

Economist Newspaper Ltd.
25 St. James' Place
London, United Kingdom SW1A 1HG
TELEPHONE: (0114401)839-7000

ECRI Shared Service
5200 Butler Pike
Plymouth Meeting, PA 19462
TELEPHONE: (215)825-6700

Edi Edition Documentation Industrielle
5 Rue Jules Lefebure
Paris, France 75009
TELEPHONE: (011331)874-5370

Ediciones Arcadia, S.A.
Nunez de Balboa 49, 40
Madrid 1, Spain
TELEPHONE: (011341)2767135
Rodrigalvarez, Antonio Gonzalez, Director Gerente

Edmonton. Business Development Department
10235 101 St., Oxford Tower, Suite 2410
Edmonton, Alta. Canada T5J 3G1
TELEPHONE: (403)428-5464

EDP News Services
7620 Little River Turnpike
Annandale, VA 22003
TELEPHONE: (804)354-9400

Educational Institute of the American Hotel and Motel Association
1407 S. Harrison Rd.
East Lansing, MI 48823
TELEPHONE: (517)353-5500
Colden, Kevin, Sales Manager

EIC / Intelligence
48 W. 38th St.
New York, NY 10018
TELEPHONE: (212)944-8500

Electronic Industries Association
2001 Eye St., NW
Washington, DC 20006
TELEPHONE: (202)457-4900
McCloskey, Peter F.

Electronic Trend Publications
10080 N. Wolfe Rd.
Cupertino, CA 95014
TELEPHONE: (408)996-7401

Electronics of America
P.O. Box 2305
Chapel Hill, NC 27514
TELEPHONE: (919)929-7852

Electronics Representatives Association
233 East Erie St., Suite 1003
Chicago, IL 60611
TELEPHONE: (312)649-1333

Elsevier Science Publishers
256 Banbury Rd., Mayfield House
Oxford, United Kingdom 0X2 7DH
TELEPHONE: (0114401)511151
TELEX: 837484

Elsevier-North Holland Publishing Company
52 Vanderbilt Ave.
New York, NY 10017
TELEPHONE: (212)867-9040

Enterprise les Informations
17 Rue d'Uzes
Paris, France F-75002
TELEPHONE: (0113301)233-4435
TELEX: 680876

ERA Technology Ltd.
Cleeve Rd.
Leatherhead, Surrey, United Kingdom KT22 7SA

E.R.C. Statistics International Ltd.
P.O. Box 115, 41 Russell Square
London, United Kingdom WC1B 5DL
TELEPHONE: (011441)637-8316
TELEX: 24224

Erisa Benefit Funds Inc.
515 National Press Bldg.
Washington, DC 20045
TELEPHONE: (202)638-1914

ERL
79 Baker St.
London, United Kingdom W1M 1AJ

European Communities. Publications Office *see* Publications Office of the European Communities

Evans Research Corporation
1 Eva Rd., Suite 309
Etobicoke, Ont. Canada M9C 425
TELEPHONE: (416)621-8814

EW Communications Inc.
1170 E. Meadow Dr.
Palo Alto, CA 94303
TELEPHONE: (415)494-2800
TELEX: 9105831419

F

Fairchild Publications
7 E. 12th St.
New York, NY 10003
TELEPHONE: (212)741-4280

Financial Times Business Information Ltd.
10 Cannon St., Bracken House
London, United Kingdom EC4P 4BY
TELEPHONE: (011-44-01)623-1211
TELEX: 8814734

Financial Times Ltd.
10 Bolt Court, Fleet St.
London, United Kingdom EC4A 3HL
TELEPHONE: (0114401)248-8000

FINAT
Laan Copes van Cattenburch 79
The Hague, Netherlands 2585 EW
TELEPHONE: (011070)603837

FIND / SVP
500 Fifth Ave.
New York, NY 10110
TELEPHONE: (212)354-2424

Robert S. First, Inc.
707 Westchester Ave.
White Plains, NY 10604
TELEPHONE: (914)949-4248
TELEX: 131414

Fisher Scientific Company
711 Forbes Ave.
Pittsburgh, PA 15219
TELEPHONE: (412)562-8300

Florida Bankers Association
341 N. Mills Ave.P.O. Box 6847
Orlando, FL 32853
TELEPHONE: (305)896-6511

Focus Research Systems, Inc.
342 N. Main St.
West Hartford, CT 06117
TELEPHONE: (203)561-1047

Food Marketing Institute
1750 K. St., NW
Washington, DC 20006
TELEPHONE: (202)452-8444

Forbes, Inc.
60 Fifth Ave.
New York, NY 10011
TELEPHONE: (212)620-2200

Forecast Associates Inc.
P.O. Box 606
Ridgefield, CT 06877
TELEPHONE: (203)743-0212
Johnson, Inez H., Sales Manager

Foreign Investment Review Agency
Station D, Box 2800
Ottawa, Ont. Canada K1P 6A5

Foxlow Publications Ltd.
19 Harcourt
London, United Kingdom W1H 2AX
TELEPHONE: (0114401)724-3520

Fox-Morris Personnel Consultants
1211 Avenue of the Americas
New York, NY 10036
TELEPHONE: (212)840-6930

Freed-Crown Publishing Company
6931 Van Nuys Blvd.P.O.Box 2338
Van Nuys, CA 91405
TELEPHONE: (213)997-0644

Miller Freeman Publications Inc.
500 Howard St.
San Francisco, CA 94105
TELEPHONE: (415)397-1881

Frost and Sullivan, Inc.
106 Fulton St.
New York, NY 10038
TELEPHONE: (212)233-1080
Bonner, Rebecca, Manager, Marketing Administration

Fuel and Metallurgical Journals Ltd.
Queensway House, 2 Queensway
Redhill, United Kingdom (01144)9168611
TELEX: 948669

Fuji Bank Ltd. Research Division
1-5-5 Ote-machi, Chiyoda-ku
Tokyo, Japan

Fuji Corporation
5-29-7, Jingu-mae, Shibuya-ku, Busicen Bldg.
Tokyo, Japan

Future Computing Inc.
900 Canyon Creek Center
Richardson, TX 75080
TELEPHONE: (214)783-9375

G

G. Cam-Serveur A.
33 Avenue du Maine, Tour Maine-Montparnasse
Paris, France F-75755

Gale Research Company
Book Tower
Detroit, MI 48226
TELEPHONE: (313)961-2242
Customer Service

Gardner Publications Inc.
6600 Clough Pike
Cincinnati, OH 45244
TELEPHONE: (513)231-8020

Gas Industries
P.O. Box 558
Park Ridge, IL 60068
TELEPHONE: (312)693-3682

Generaldirektion PTT
Viktoriastr. 21
Bern, Switzerland 3013
TELEPHONE: (0114131)621111
TELEX: 32011ptt ch

Gernsback Publications, Inc.
200 Park Ave., S.
New York, NY 10003
TELEPHONE: (212)777-6400

Gesellschaft fur Chemiewirtschaft
Salesianergasse 1
Vienna, Austria 1030
TELEPHONE: (01143222)7256110

Gewiplan GmbH
Friedrich-Ebert-Analage 38
Frankfurt, West Germany D-6000
TELEPHONE: (01149611)740471

Geyer-McAllister Publications, Inc.
51 Madison Ave.
New York, NY 10010
TELEPHONE: (212)689-4411

Girard Associates Inc.
399 Howard Blvd.
Mount Arlington, NJ 07856
TELEPHONE: (201)398-5524

Globecom Publishing, Ltd.
4121 W. 83rd St., Suite 132
Prairie Village, KS 66208
TELEPHONE: (913)942-6611
Mailing Address
P.O. Box 12268
Overland Park, KS 66212

GML Corporation
594 Marrett Rd.
Lexington, MA 02173
TELEPHONE: (617)861-0515

Gnostic Concepts, Inc.
2710 Sand Hill Rd.
Menlo Park, CA 94025
TELEPHONE: (415)854-4672

O.A. Goulden and Partners
Stoke Hill, Quarry House
Stoke, Andover, Hampshire, United Kingdom

Gower Publishing Company Ltd.
Croft Rd., Gower House
Aldershot, Hampshire, United Kingdom GU11 3HR

Graft Neuhaus AG
Bachtoldstrasse 4
Zurich, Switzerland 8044
TELEPHONE: (011411)2516171

Gralla Publications
1515 Broadway
New York, NY 10036
TELEPHONE: (212)869-1300

Great Lakes Press
Maple City Postal Station
Harbor Island, MI 49664
TELEPHONE: (616)334-3651

Greene and Company
36/38 New Broad St., Bilbao House
London, United Kingdom EC2M 1NV
TELEPHONE: (0114401)628-7241

Greenwood Press
88 Post Rd., W., P.O. Box 5007
Westport, CT 06881
TELEPHONE: (203)226-3571

Groupe Tests
41 Rue de la Grange-aux-Belles
Paris, France 75483
TELEPHONE: (011331)2022910
TELEX: 230589

Guides Techniques Professionelles
13, Rue Charles-Lecocq
Paris, France F-75015

Gulf Publishing Company
P.O. Box 2608
Houston, TX 77001
TELEPHONE: (713)529-4301

H

Robert Half of New York
522 Fifth Ave.
New York, NY 10036
TELEPHONE: (212)221-6500

George D. Hall Company
20 Kilby St.
Boston, MA 02109
TELEPHONE: (617)523-3745

Halstead Press. Division of John Wiley and Sons, Inc.
605 Third Ave.
New York, NY   10016
TELEPHONE: (212)867-9800

Hanagan and Associates
999 Plaza Dr., Suite 400
Schaumberg, IL   60195
TELEPHONE: (312)843-3123

A.S. Hansen Inc.
1080 Green Bay Rd.
Lake Bluff, IL   60044
TELEPHONE: (312)234-3400

Harbor Publishing
1668 Lombard St.
San Francisco, CA   94123
TELEPHONE: (415)775-4740

Harcourt Brace Jovanovich, Inc.
757 Third Ave.
New York, NY   10017
TELEPHONE: (212)888-4444

Harcourt Brace Jovanovich Publications
124 S. First St.
Geneva, IL   60134
TELEPHONE: (312)232-1400

Harper and Row Publishers, Inc.
10 E. 53rd St.
New York, NY   10022
TELEPHONE: (800)638-3030

Louis Harris and Associates
630 Fifth Ave.
New York, NY   10020
TELEPHONE: (212)975-1600
TELEX: 148383

Harris Publishing Company
2057-2 East Aurora Rd.
Twinsburg, OH   44087
TELEPHONE: (216)425-9143

Harvard University. Graduate School of Business
Soldiers Field Rd.
Boston, MA   02163
TELEPHONE: (617)495-6000

Harvard University. Information Resources Policy. Order Department
200 Aiken St.
Cambridge, MA   02138
TELEPHONE: (617)495-4114

Harvard University. Laboratory for Computer Graphics
730 Boston Post Rd.
Sudbury, MA   01776
TELEPHONE: (617)443-4671

Hatfield Polytechnic. National Reprographic Centre for Documentation
Bayfordbury, Hereford, Hertfordshire, United Kingdom SG1 38LD

Hayden Book Company, Inc.
50 Essex St.
Rochelle Park, NJ   07662
TELEPHONE: (201)843-0550

Hayden Publishing Company, Inc. Computer Decisions Division
50 Essex St.
Rochelle Park, NJ   07662
TELEPHONE: (201)843-0550

HBJ Publications *see* Harcourt Brace Jovanovich Publications

Hearst Business Communications, UTP Division
645 Stewart Ave.
Garden City, NY   11530
TELEPHONE: (516)222-2500

Hearst Corporation
959 Eighth Ave.
New York, NY   10019
TELEPHONE: (212)262-5700

Her Majesty's Stationery Office
P.O. Box 569
London, United Kingdom SE1 9NH
TELEPHONE: (0114401)928-6977
Wilson, Raymond, Business Manager
Millbank Tower, 11th fl.
London, SW1P 4QU
TELEPHONE: (0114401)211-6112

Herman Publishing Company
45 Newbury St.
Boston, MA   02116
TELEPHONE: (617)536-5810

Heyden and Son Ltd.
Spectrum House, Hillview Gardens
London, United Kingdom NW4
TELEPHONE: (0114401)203-5171

Hirsch Organization, Inc.
6 Deer Trail
Old Tappan, NJ   07675
TELEPHONE: (201)664-3400

Hitchcock Publishing Company
Hitchcock Bldg.
Wheaton, IL   60187
TELEPHONE: (312)665-1000

Hoare and Govett Ltd.
1 King St., Atlas House
London, United Kingdom EC2
TELEPHONE: (011441)606-9800
TELEX: 885474

Hoke Communications
224 Seventh St.
Garden City, NY   11530
TELEPHONE: (516)746-6700

W.G. Holdsworth and Associates
1000 E. Northwest
Mt. Prospect, IL   60056
TELEPHONE: (312)394-2022

Hope Reports, Inc.
1600 Lyell Ave.
Rochester, NY   14606
TELEPHONE: (716)458-4250
Hope, Mabeth S.

Horizon House
610 Washington St.
Dedham, MA 02026
TELEPHONE: (617)326-8220
TELEX: 710-3480481

I

IBF Publications
121 Smallbrook QueenswayBridge House, 8th Fl.
Birmingham, United Kingdom B5 4JP
TELEPHONE: (01144021)643-4523

ICA Telecommunications Management
2175 Sheppard Ave., E., Suite 210
Willowdale, Ont. Canada M2J 1WJ
TELEPHONE: (416)494-4440

ICAP Hellos SA
64 Queen Sophia Ave.
Athens, Greece 615

ICC Business Ratios
28-42 Banner St.
London, United Kingdom EC1Y 8QE
TELEPHONE: (0114401)253-3906

I.C.C. Information Group Ltd.
81 City Rd., ICC House
London, United Kingdom EC1Y 1BD
TELEPHONE: (0114401)253-0063
TELEX: 23678
Jewitt, A.J.

IDC Europa Ltd.
2 Bath Rd.
Chiswick, London, United Kingdom W4 1LN
TELEPHONE: (0114401)995-9222
TELEX: 851-934287
Whitehead, Jane

IEEE Computer Society
10662 Los Vaqueros Circle
Los Alamitos, CA 90720
TELEPHONE: (714)821-8380

IFS (Publications) Ltd.
35-39 High St.
Kempston, Bedford, United Kingdom MK42 7BT

Illuminating Engineering Society
345 E. 47th St.
New York, NY 10017
TELEPHONE: (216)644-7926

Impact Marketing Services
10318 Globe Court
Elliot City, MD 21043
TELEPHONE: (301)465-7037

Imprint Software
420 S. Howes
Fort Collins, CO 80521
TELEPHONE: (303)482-5574

IMS America Ltd.
Butler & Maple Aves.
Ambler, PA 19002
TELEPHONE: (215)283-8500
TELEX: 846396

Inbucon Management Consultants Ltd.,Advanced Technology Group
72-80 High St.
Esher, Surrey, United Kingdom KT10 9QS
Allum, J.F.

Independent Battery Manufacturers Association, Inc.
100 Larchwood Dr.
Largo, FL 33540
TELEPHONE: (813)586-1409
Noe, Dan A., Executive Secretary

Indian Export Trade Journal
Sayajinganj
Baroda, India 390005

Industrial Communications
647 National Press Bldg.
Washington, DC 20045
TELEPHONE: (202)783-2482

Industrial Market Research Ltd.
17 Buckingham Gate
London, United Kingdom SW1
TELEPHONE: (011441)834-7814
TELEX: 917036

Industry Media Inc.
1129 E. 17th Ave., P.O. Box 18888-A
Denver, CO 80218
TELEPHONE: (303)832-1022

Infometrics, Inc.
R.R. 1, Box 194
Monticello, IL 61856
TELEPHONE: (217)333-8462

Inforesearch Institute
c/o Infodyne, Inc., 1700 W. 78th St.
Minneapolis, MN 55423

Information and Records Management Inc.
101 Crossways Park W.
Woodbury, NY 11797
TELEPHONE: (516)496-8000

Information Clearing House
500 Fifth Ave.
New York, NY 10036
TELEPHONE: (212)354-2424

Information Gatekeepers, Inc.
167 Corey Rd., Suite 111
Brookline, MA 02146
TELEPHONE: (617)739-2022

Information Industry Association
316 Pennsylvania Ave., SE, Suite 502
Washington, DC 20003
TELEPHONE: (202)544-1969
Zurkowski, Paul G., President

Information Intelligence Inc.
P.O. Box 31098
Phoenix, AZ 85046
TELEPHONE: (602)996-2283

Information Researchers, Inc.
Yoyogi-4-chome, Shibuya-kuSuite 401, No.59-3
Tokyo, Japan 151

Information Sources, Inc.
1807 Glenview Rd.
Glenview, IL 60025
TELEPHONE: (312)724-9285

Information Systems Consultants
R.R.1, Box 256 A
St. James, MO 65559

Infotecture Europe
11 Rue du March-Saint-Honore
Paris, France F-75001

Infratest-Industria GmbH
Sudliche Duffahrtsallee 75
8 Munchen 19, West Germany

INPUT
1943 Landings Dr.
Mountain View, CA 94043
TELEPHONE: (415)960-3990
Kane, Leslie

Input Two-Nine Ltd.
7 Banstead Rd.
Purley, Surrey, United Kingdom CR2 3ER
TELEPHONE: (0114401)668-5281

Insee O-E-P
Tour Gamma A 195 Rue de Bercy
Paris, Cedex 12, France 75582
TELEPHONE: (011331)345-7231

Institut fur Automation und Operations Research
Universitat Freiburg, Misericorde
Freiburg, Switzerland 1700
TELEPHONE: (011411)37219560
Kuhn, R., Professor

Institut fur Wirtschaftsforschung
Postfach 860460
Munich 86, West Germany D-8000
TELEPHONE: (01149089)9224-1
TELEX: 222691
Mueller, J., Branch Referent EDV
TELEPHONE: (01149089)9224-359

**Institut National de la Statistique et des Etudes Eco-
nomiques. Observatoire Economique de Paris** *see*
Insee O-E-P

Institute for Graphic Communications, Inc.
375 Commonwealth Ave.
Boston, MA 02115
TELEPHONE: (617)267-9425

Institute for Research on Public Policy
P.O. Box 3670
Halifax South, N.S. Canada B3J 3K6
TELEPHONE: (902)424-3801
Vagionos, Louis, Sales Manager

Institute for Research on Public Policy
2149 MacKay St.
Montreal, Que. Canada H3G 2J2
TELEPHONE: (514)879-8533

Institute for the Future
2740 Sand Hill Rd.
Menlo Park, CA 94025
TELEPHONE: (415)854-6322
Buchan, Judy, Project Secretary Program

Institute of Electrical and Electronics Engineers, Inc.
345 E. 47th St.
New York, NY 10017
TELEPHONE: (212)644-7910
Emberson, R.M., Executive Director

Institution of Electrical Engineers
Southgate House
Stevenge, United Kingdom SG1 IHG

Instrument Society of America
P.O. Box 12277, 67 Alexander Dr.
Research Triangle Park, NC 27709
TELEPHONE: (919)549-8411
Harvey, Glen F., Executive Director

Insurance Journal Inc.
5910 N. Figueroa, P.O. Box 42030
Los Angeles, CA 90042
TELEPHONE: (213)257-7591

Integrated Circuit Engineering Corporation
1522 N. 75th St.
Scottsdale, AZ 85260
TELEPHONE: (602)998-9780

Intentional Educations, Inc.
51 Spring St.
Watertown, MA 02172
TELEPHONE: (617)923-8595

Inter Company Comparisons Ltd.
81 City Rd.
London, United Kingdom EC1 1BD
TELEPHONE: (0114401)638-2946
TELEX: 23678

Interactive Data Corporation
486 Totten Pond Rd.
Waltham, MA 02154
TELEPHONE: (617)890-1234
Gibson, William

Interavia SA
86 Ave Louis Casai, CH-1216 Cointrin
Geneva, Switzerland
TELEPHONE: (01141)022-980505
TELEX: 22122

International Business Machines Corporation. IBM
Old Orchard Rd.
Armonk, NY 10504
TELEPHONE: (914)686-5680

International Commercial Bank of China
100 Chi-Lin Rd.
Taipei, China (Taiwan) 104

International Computer Programs, Inc.
9000 Keystone Crossing
Carmel, IN 46032
TELEPHONE: (317)844-7461

International Data Group
214 Third Ave.
Waltham, MA 02154
TELEPHONE: (617)890-3700

International Information / Word Processing Association
1015 N. York Rd.
Willow Gate, PA 19090
TELEPHONE: (215)657-6300

International Press Cutting Service
P.O. Box 63
Allahabad, India

International Publications Service
114 E. 32nd St.
New York, NY 10016
TELEPHONE: (212)685-9351

International Reprographic Association, Inc.
10116 Franklin Ave.
Franklin Park, IL 60131
TELEPHONE: (312)671-5356
Good, Ray A., Executive Vice President

International Resource Development, Inc.
30 High St.
Norwalk, CT 06851
TELEPHONE: (203)866-6914
TELEX: 643452
Bosomworth, Kenneth G., Sales Mgr.

International Scientific Communications Inc.
808 Kings Highway
Fairfield, CT 06430
TELEPHONE: (203)576-0500

International Self-Council Press Ltd.
306 W. 25th St.
North Vancouver, B.C. Canada V7N 2G1
TELEPHONE: (604)986-3366

International Society of Optical Engineers
P.O. Box 10
Bellingham, WA 98227
TELEPHONE: (206)676-3290

International Telecommunication Union
Place des Nations
Geneva, Switzerland CH-1211

International Television Association
136 Sherman Ave.
Berkeley Heights, NJ 07922
TELEPHONE: (201)464-6747

International Werbegesellschaft m.b.H.
Hoher Markt 12
Vienna, Austria A-1011
TELEPHONE: (0113301)758-1285
TELEX: 4067

International Word Processing Association
Maryland Rd.
Willow Grove, PA 19090
TELEPHONE: (215)657-3220
Lear, Lorraine, Director

Intertec Publishing Corporation
P.O. Box 12901
Overland Park, KS 66212
TELEPHONE: (913)888-4664

IPC Business Press Ltd.
40 Bowling Green Lane
London, United Kingdom EC1R 0NE
TELEPHONE: (0114401)837-3636
TELEX: 23839

IPC Business Press Ltd.
295 E. 42nd St.
New York, NY 10017
TELEPHONE: (212)867-2080

IPC Electrical-Electronic Press Ltd. Division of IPC Industrial Press
Stamford St., Dorset House
London, United Kingdom SE1 9LU
TELEPHONE: (0114401)261-8000
TELEX: 25137

IPC Industrial Press Ltd.
Dorset House, Stamford St.
London, United Kingdom SE1 9LU
TELEPHONE: (0114401)261-8000

IPC Transport Press Ltd. *see* IPC Industrial Press Ltd.

Iron and Steel Society of AIME
410 Commonwealth Dr., P.O. Box 411
Warrendale, PA 15086
TELEPHONE: (412)776-1535

Peter Isaacson Publications Pty. Ltd.
46-49 Porter St.
Prahan, Victoria, Australia 3101

J

Jane's Publishing Company Ltd.
238 City Rd.
London, United Kingdom EC1V 2PU
TELEPHONE: (0114401)251-9281
TELEX: 23168
U.S. Contact
730 Fifth Ave.
New York, NY 10019

Japan Electric Measuring Instruments Manufacturers Association
1-9-10 Tornomon
Minato-ku, Tokyo, Japan 105
TELEPHONE: (011813)502-0601

Japanese Semiconductor Study
P.O. Box 37000
San Francisco, CA   94137

JMA Research and Consulting
Church St. Station, P.O. Box 669
New York, NY

Johnston International Publishing Corporation
386 Park Ave. S.
New York, NY   10016
TELEPHONE: (212)689-0120

Jordan and Sons (Surveys) Ltd.
47 Brunswick Place, Jordan House
London, United Kingdom N1 6EE
TELEPHONE: (0114401)253-3030

K

Kalba Bowen Associates Inc.
12 Arrow
Cambridge, MA   02138
TELEPHONE: (617)661-2624

Kalerghi Publications
52 Welbeck St.
London, United Kingdom W1M 7HE
TELEPHONE: (0114401)935-8678

J.J. Keller Associates, Inc.
145 W. Wisconsin Ave.
Neenah, WI   54956
TELEPHONE: (414)722-2848

Keller Publishing Corporation
150 Great Neck Rd.
Great Neck, NY   11021
TELEPHONE: (516)829-9210

Kengore Corporation
9 James Ave.
Kendall Park, NJ   08824
TELEPHONE: (201)297-2526

Kerrwil Publications Ltd.
443 Mt. Pleasant Rd.
Toronto, Ont. Canada M4S 2L8
TELEPHONE: (416)482-6603

Kessler Marketing Intelligence
22 Farewell St.
Newport, RI   02840
TELEPHONE: (401)849-6771

Keynote Publications Ltd.
23 City Rd.
London, United Kingdom EC1Y 1AA
TELEPHONE: (0114401)588-2698

Kidder, Peabody and Company Inc.
10 Hanover Square
New York, NY   10005
TELEPHONE: (212)747-2000

Milton S. Kiver Publications, Inc.
222 W. Adams St.
Chicago, IL   60606
TELEPHONE: (312)263-4866

Knowledge Industry Publications
701 Westchester Ave.
White Plains, NY   10604
TELEPHONE: (914)328-9157

Kogan Page Ltd.
120 Pentonville Rd.
London, United Kingdom N1 9JN
TELEPHONE: (011441)8377851
Nichols Publishing Company
P.O. Box 96
New York, NY   10024
TELEPHONE: (212)580-8079

Kolleck and Sons Ltd.
6 Hazanovitch St., P.O. Box 7052
Jerusalem, Israel 91070
TELEPHONE: (0119722)234-13123

L

Laing and Cruickshank
The Stock Exchange
London, United Kingdom EC2N 1HA
TELEPHONE: (0114401)588-2800

Lake Publishing Corporation
700 Peterson Rd., Box 159
Libertyville, IL   60048
TELEPHONE: (312)362-8711

Wilhem Lange
Heuss - Allee 2-10
Bonn 1, West Germany 5300
TELEPHONE: (01142228)215989
TELEX: 0886420
Thorting, Kars G., Co-Publisher
Promenade 89
Bad Homburg, 6380
TELEPHONE: (011496172)29987

M. Largo-Alfonso
75 New York Stock Exchange Bldg., Apollo St.
Bombay 1, India

Larsen Sweeney Associates Ltd.
P.O. Box 36
Maidstone, Kent, United Kingdom
TELEPHONE: (011440622)678113

Latin American Times Company
660 First Ave., Suite 416
New York, NY   10016
TELEPHONE: (212)679-7241

Laurie Milbank and Company
72/73 Basinghall St., Portland House
London, United Kingdom EC2V 5DP
TELEPHONE: (0114401)606-6622

Laventhol and Horwath
1845 Walnut St.
Philadelphia, PA 19103
TELEPHONE: (215)491-1700

Cyrus J. Lawrence Inc.
115 Broadway
New York, NY 10006
TELEPHONE: (212)962-2200

Lead Industries Association, Inc.
292 Madison Ave.
New York, NY 10017
TELEPHONE: (212)578-4750

Learned Information (Europe) Ltd.
Besselsleigh Rd.
Abington, United Kingdom OX13 6EF
TELEPHONE: (011440865)730275

Learned Information, Inc.
P.O. Box 550
Marlton, NJ 08053
TELEPHONE: (609)654-6266
Hogan, Tom

Lebhar-Friedman, Inc.
425 Park Ave.
New York, NY 10022
TELEPHONE: (212)371-9400

Lesterstar Ltd.
375 Upper Richmond Rd., W.
London, United Kingdom SE14 7NX

Lexington Books. Division of D.C. Heath and Company
125 Spring St.
Lexinggton, MA 02173
TELEPHONE: (617)862-6650

Libraries Unlimited Inc.
P.O. Box 263
Littleton, CO 80160
TELEPHONE: (303)770-1220

Lighting Industry Federation
207 Balham High Rd.
London, United Kingdom SW17 7BH
TELEPHONE: (0114401)636-0766

Link Resources Corporation
215 Park Ave., S.
New York, NY 10003
TELEPHONE: (212)473-5600
TELEX: 429328

Arthur D. Little, Inc.
Acorn Park
Cambridge, MA 02140
TELEPHONE: (617)864-5770
TELEX: 921436
Marketing Department

Arthur D. Little of Canada Ltd.
120 Eglinton Ave., E.
Toronto, Ont. Canada M4P 1E2
TELEPHONE: (416)487-4114

Lloyd's of London Press Ltd.
7100 Trade Counter, Sheepen Place
Colchester, United Kingdom CO3 3LP
TELEPHONE: (0110206)69222

Logica Inc.
341 Madison Ave.
New York, NY 10017
TELEPHONE: (212)599-0828

Logica Ltd.
64 Newman St.
London, United Kingdom
TELEPHONE: (0114401)637-9111
TELEX: 27200

Longman Inc.
19 W. 44th St., Suite 1012
New York, NY 10036
TELEPHONE: (212)764-3950

William T. Lorenz and Company
311 Commonwealth Ave.
Boston, MA 02115
TELEPHONE: (617)266-1784
Lorenz, William T., Sales Manager

Luning Prak Associates, Inc.
200 Summit Ave.
Montvale, NJ 07645
TELEPHONE: (201)573-1400
Taylor, Robert L., Sales Manager

M

M and M Resources Corporation. Subsidiary of Merchants and Manufacturers Association
2300 Occidental Center,1150 S. Olive St.
Los Angeles, CA 90015
TELEPHONE: (213)748-0421

Mackintosh Consultants, Inc.
2444 Moorpark Ave.
San Jose, CA 95128
TELEPHONE: (408)998-4312

Mackintosh Publications Ltd.
Napier Rd., Mackintosh House
Luton, Bedfordshire, United Kingdom LU1 1RG
TELEPHONE: (011440582)417438
TELEX: 826818

Maclaren Publishers Ltd.
69-77 High St., P.O. Box 109
Croyden, United Kingdom CR9 1QH

Maclean-Hunter Ltd.
481 University Ave.
Toronto, Ont. Canada M5W 1A7
TELEPHONE: (416)596-5523

Maclean-Hunter Publishing Corporation
300 W. Adams St.
Chicago, IL   60606
TELEPHONE: (312)726-2802

Macro Communications, Inc.
150 E. 58th St.
New York, NY   10022
TELEPHONE: (212)826-4360

Management Information Corporation
140 Barclay Center
Cherry Hill, NJ   08034
TELEPHONE: (609)428-1020

Market and Opinion Research International
29 Queen Anne's Gate
London, United Kingdom SW1H 9DD
TELEPHONE: (0114401)222-0232

Market Studies International. Division of European Inter-Company Comparisons Ltd.
81 City Rd., ICC House
London, United Kingdom EC1Y 1BD
TELEPHONE: (0114401)250-3922
TELEX: 23678

Marketdata Enterprises
P.O. Box 436
Lynbrook, NY   11563
TELEPHONE: (516)887-1968
LaRosa, John

Marketing Development
402 Border Rd.
Concord, MA   01742
TELEPHONE: (617)369-5382
Anderson, Bud, Sales Manager

John Martin Publishing Ltd.
15 King St., Covent Garden
London, United Kingdom WC2E 8HN
TELEPHONE: (0114401)240-5024

Maruzen Company Ltd.
P.O. Box 5050, 3-10 Nihonbashi 2-chome, Chuo-ku
Tokyo, Japan 103
TELEPHONE: (0118103)2727211
TELEX: 26517

MCB (Industrial Development) Ltd.
198/200 Keighley Rd.
Bradford, West Yorkshire, United Kingdom BD9 4JQ
TELEPHONE: (01144274)499821

McFadden Business Publications
6364 Warren Dr.
Norcross, GA   30093
TELEPHONE: (404)448-1011

McGraw-Hill Book Company
1221 Avenue of the Americas
New York, NY   10020
TELEPHONE: (212)997-1221

McGraw-Hill Book Company. Electronics
1221 Avenue of the Americas
New York, NY   10020
TELEPHONE: (212)997-4597
Eyler, Janet, Books and Special Projects Manager

McGraw-Hill Information Systems Company, Sweets Division
1221 Avenue of the Americas
New York, NY   10020
TELEPHONE: (212)997-4450

McIlvaine Company
2970 Maria Ave.
Northbrook, IL   60062
TELEPHONE: (312)272-0010

McPheters, Wolfe and Jones
16704 Marquardt Ave.
Cerritos, CA   90701
TELEPHONE: (213)926-9544

Measurements and Data Corporation
2994 W. Liberty Ave.
Pittsburg, PA   15216
TELEPHONE: (412)343-9666

Meckler Publishing
520 Riverside Ave.
Westport, CT   06880
TELEPHONE: (203)226-6967

Medical Economics Company
680 Kinderkamack Rd.
Oradell, NJ   07649
TELEPHONE: (201)262-3030

Merchants and Manufacturers Association *see* M and M Resources Corporation. Subsidiary of Merchants and Manufacturers Association

Mercury House Business Publications
Waterloo Rd., Mercury House
London, United Kingdom SE1 8UL
TELEPHONE: (0114401)928-3388
TELEX: 21977

Merrill Lynch, Pierce, Fenner and Smith Inc.
One Liberty Plaza
New York, NY   10006
TELEPHONE: (212)766-1212

L. Messel and Company
100 Old Broad St., Winchester House
London, United Kingdom EC2 P2HX
TELEPHONE: (0114401)606-4411
TELEX: 883004

Metra Consulting Group Ltd.
42 Vierage Crescent, St. Mary's House
London, United Kingdom SW11 3LB

Mezhdunarodnaia Kniga
Smolenskaia Sennaia 32-39
Moscow, Union of Soviet Socialist Republics G-200

Michigan State University. Graduate School of Business Administration. Bureau of Business & Economic Research
5J Berkey Hall
East Lansing, MI   48824
TELEPHONE: (517)355-7560
Partiarche, Audrey, Sales Manager

Microcomputer Information Services
2464 El Camino Real, Suite 247
Santa Clara, CA 95051
TELEPHONE: (408)984-1097

Microform Review Inc.
P.O. Box 405, Saugatuck Station
Westport, CT 06880
TELEPHONE: (203)226-6967

Micro-Serve Inc.
P.O. Box 482
Nyack, NY
TELEPHONE: (914)358-1340

Middle East Marketing Research Bureau Ltd.
Makarios III Ave., P.O. Box 2098Mitsis Bldg.
Nicosia, Cyprus
TELEPHONE: (01135721)45413
TELEX: 2488

Mitel Publications Ltd.
20 Buckingham St.
London, United Kingdom WC2N 6EE
TELEPHONE: (0114401)839-1542

3M Canada Inc.
P.O. Box 5757
London, Ont. Canada N6A 4T1
TELEPHONE: (519)451-2500

Modern Metals Publishing Company
211 E. Chicago Ave., #800
Chicago, IL 60611
TELEPHONE: (312)337-0638

Moody's Investors Services
99 Church St.
New York, NY 10007
TELEPHONE: (212)553-0300

Morgan Stanley and Company, Inc.
1251 Avenue of the Americas
New York, NY 10020
TELEPHONE: (212)974-4000

Morgan-Grampian Book Publishing Company Ltd.
30 Calderwood St., Woolwich
London, United Kingdom SE18 6QH
TELEPHONE: (0114401)855-7777

Morgan-Grampian, Inc.
2 Park Ave.
New York, NY 10016
TELEPHONE: (212)340-9700

Morgan-Grampian Publishing Company
1050 Commonwealth Ave.
Boston, MA 02215
TELEPHONE: (617)232-5470

MSRA, Inc.
115 Broadway, Suite 200
New York, NY 10006
TELEPHONE: (201)798-1510

Multi-Science Publishing Company Ltd.
The Old Mill, Dorset Place
London, United Kingdom E15
TELEPHONE: (0114401)534-4882

Herbert Muck Ges. m.b.H.
Doblhoffg 7
Vienna, Austria A-1010
TELEPHONE: (01143222)422-574

Murphy-Ritcher Publishing Company
20 N. Wacker Dr., Suite 830
Chicago, IL 60606
TELEPHONE: (312)782-9592

MV Publishing Inc.
670 N. Batavia
Orange, CA 92660
TELEPHONE: (714)752-0271

N

National Aftermarket Audit Company
Duxbury Crossroads Bldg., P.O. Box 1509
Duxbury, MA 02332
TELEPHONE: (617)934-6577
McKendry, A.C.

National Association of Home Builders of the United States
15th and M Sts., NW
Washington, DC 20005
TELEPHONE: (202)452-0200
Stahl, David E., Executive Vice President

National Burglar and Fire Alarm Association
1133 15th St. NW
Washington, DC 20005
TELEPHONE: (202)429-9460

National Cable Television Association
918 16th St., NW
Washington, DC 20006
TELEPHONE: (202)457-6700
Schmidt, Robert L., President

National Council for U.S. - China Trade
1050 17th St., NW, Suite 350
Washington, DC 20036
TELEPHONE: (202)828-8300
Phillips, Christopher H., President

National Credit Office
1290 Avenue of the Americas
New York, NY 10019
TELEPHONE: (212)957-3800

National Economic Development Office
Millbank Tower
London, United Kingdom SW1P 4QX
TELEPHONE: (0114401)211-3000

National Electrical Contractors Association
7315 Wisconsin Ave.
Washington, DC 20014
TELEPHONE: (301)657-3110
Higgins, Robert L., Executive Vice President

National Mall Monitor, Inc.
2280 U.S. 19 N., Suite 264
Clearwater, FL   33515
TELEPHONE: (813)531-5893

National Office Products Association
301 N. Fairfax St.
Alexandria, VA   22314
TELEPHONE: (703)549-9040
Haspel, Donald P., Executive Vice President

National Outerwear and Sportswear Association
51 Madison Ave.
New York, NY   10010
TELEPHONE: (212)683-7520

National Retail Merchants Association
100 W. 31st St.
New York, NY   10001
TELEPHONE: (212)244-8780

National Safety Council
444 N. Michigan Ave.
Chicago, IL   60611
TELEPHONE: (312)527-4800

National Technical Information Service *see* U.S. Department of Commerce. National Technical Information Service

National Telephone Cooperative Association
2626 Pennsylvania Ave., NW
Washington, DC   20037
TELEPHONE: (202)342-8220

Nelson Publications
11 Elm Place
Rye, NJ   10580
TELEPHONE: (914)967-9100

New York Chamber of Commerce and Industry
65 Liberty St.
New York, NY   10005
TELEPHONE: (212)766-1300

Newsfront *see* Howard W. Sams and Company, Inc.

Newton Evans Research Company
13382 Grinstad Court
Sykesville, MD   21784
TELEPHONE: (301)465-7316

A.C. Nielsen Company
International Headquarters, Nielson Plaza
Northbrook, IL   60062
TELEPHONE: (312)498-6300

Nihon Kogyo Shimbun Publications
Sankei Bldg., 7-2, 1-chome, Ohtemachi, Chiyoda-ku
Tokyo, Japan
Nihon Kogyo Shimbun
41 E. 42nd St., Suite 518
New York, NY   10017
TELEPHONE: (212)997-6666

North American Telephone Association
1030 15th St., NW, Suite 360
Washington, DC   20005
TELEPHONE: (202)393-7444
Tobin, William J., Executive Director

Northern Business Intelligence
287 MacPherson Ave.
Toronto, Ont. Canada M4V 1A4
TELEPHONE: (416)961-1201

Northern Business Intelligence Inc.
150 Nassau St.
New York, NY   10038
TELEPHONE: (212)227-7333
White, Sean

Northern Miner Press, Ltd.
7 Labatt Ave.
Toronto, Ont. Canada M5A 3P2
TELEPHONE: (416)368-3481

North-Holland Publishing Company
Jan van Galenstraat 355, P.O. Box 211
Amsterdam, Netherlands 1000 AE
TELEPHONE: (01131020)515-9222
TELEX: 16479

Noyes Data Corporation
Mill Rd. at Grand Ave.
Park Ridge, NJ   07656
TELEPHONE: (201)391-8484

N.R.C. B.V. Afd. Tydschriften
Westblaak 180
Rotterdam, Netherlands 3012 KN
TELEPHONE: (01131010)147211

OECD Publications and Information Center see Organization for Economic Cooperation and Development. Publications and Information Center

Oelgeschlager, Gunn and Hain, Inc.
1278 Massachusetts Ave., Harvard Square
Cambridge, MA   02138
TELEPHONE: (617)876-5100
Hubert, Nancy, Marketing Director

Office Publications, Inc.
1200 Summer St., P.O. Box 1231
Stamford, CT   06904
TELEPHONE: (203)327-9670
Schulhof, William, Publisher

Official Directories of Data Processing
P.O. Box 488
Gresham, OR   97030
TELEPHONE: (503)667-4669
Knudson, R.F., Publisher

Online Inc.
11 Tannery Lane
Weston, CT   06883
TELEPHONE: (203)227-8466

Online Publications, Ltd.
Northwood Hills, Argyle House
Middlesex, United Kingdom HA6 1TS
TELEPHONE: (0114409274)28211
TELEX: 923498

Ontario Association of Certified Engineering Technicians and Technologists
400 Orchard View Blvd., Suite 253
Toronto, Ont. Canada M4R 2G1
TELEPHONE: (416)488-1175

Ontario Chamber of Commerce
2323 Yonge St.
Toronto, Ont. Canada M4P 2C9
TELEPHONE: (416)482-5222

Ontario Manpower Commission *see* Ontario. Ministry of
Labour. Ontario Manpower CommissioLabour Market
Research Group

Ontario. Ministry of Industry and Tourism
900 Bay St., Hearst Block, Queen's Park
Toronto, Ont. Canada M7A 1T2
TELEPHONE: (416)965-7075

Ontario. Ministry of Labour. Ontario Manpower Commission Labour Market Research Group
400 University Ave.
Toronto, Ont. Canada M7A 1T7
TELEPHONE: (416)965-7941

Ontario. Ministry of Transportation and Communications
1201 Wilson Ave., West Tower
Downsview, Ont. Canada M3M 1J8
TELEPHONE: (416)248-3501

Oppenheim Derrick, Inc.
P.O. Box 1783
Fort Collins, CO 80522
TELEPHONE: (303)223-3477

Oppenheimer and Company
1 New York Plaza
New York, NY 10004
TELEPHONE: (800)221-5833
Author of Report

Optical Publishing Company, Inc.
P.O. Box 1746, Berkshire Common
Pittsfield, MA 01201
TELEPHONE: (413)499-0514

Organisation for Economic Cooperation and Development
2 Rue Andre' Pascal
Paris, Cedex 16, France F75775
TELEPHONE: (011331)524-8200

Organization for Economic Cooperation and Development. Publications and Information Center
Washington Center, 1750 Pennsylvania Ave., NW
Washington, DC 20006
TELEPHONE: (202)724-1857
Ekers, Eric N., Head, U.S. Office

Organization Resource Counselors Inc.
1211 Avenue of the Americas
New York, NY 10036
TELEPHONE: (212)575-7500

Oryx Press
3930 E. Camelback Rd.
Phoenix, AZ 85018
TELEPHONE: (602)956-6233

Osterreichische Brown Boveri-Werke AG
Pernerstorfergasse 94
Vienna, Austria A-1100
TELEPHONE: (01143222)62810
TELEX: 131760

Osterreichischer Wirtschaftsverlag
Nikolsdorferg 7-11
Vienna, Austria A-1051
TELEPHONE: (01143222)555-585
TELEX: 1-11669

Osterreichisches Institut fur Wirtschaftsforschung (Wifo)
Postfach 91
Vienna, Austria A-1103
TELEPHONE: (01143222)782607
Kramer, Helmut, Editor-in-Chief
Mondweg 5/2/3
Vienna, A-1140

P

Pacific Subscription Service
FDR Station, P.O. Box 811, 245 E. 54th St.
New York, NY 10150
TELEPHONE: (212)223-0556

Pactel
33 Greycoat St.
London, United Kingdom SW1 2QF
TELEPHONE: (0114401)828-7744
TELEX: 8813082

Page Publications
380 Wellington St., W.
Toronto, Ont. Canada M5V 1E3
TELEPHONE: (416)593-0608

Paine Webber Mitchell Hutchins, Inc.
One Battery Park Plaza
New York, NY 10004
TELEPHONE: (212)623-4500

Para Publishing
P.O. Box 4232
Santa Barbara, CA 93103
TELEPHONE: (805)968-7277

Parker / Hunter Inc.
120 Broadway
New York, NY 10005
TELEPHONE: (212)732-5980

Patent Data Publications
901 N. President St.
Wheaton, IL 60187
TELEPHONE: (312)462-1222

Pedder Associates Ltd.
199 Westminister Bridge Rd.
London, United Kingdom SE1 7UT
TELEPHONE: (0114401)633-0866
TELEX: 8813024

Penguin Books Ltd.
Bath Rd.
Harmondsworth, Middlesex, United Kingdom UB7 0DA
TELEX: (0114401)759-1984

Penn Well Publishing Company
P.O. Box 1260, 1421 S. Sheridan Rd.
Tulsa, OK   74101(918)835-3161

Penton / IPC. Subsidiary of Pittway Corporation
1111 Chester Ave.
Cleveland, OH   44114
TELEPHONE: (216)696-7000

Peter Peregrinus Ltd.
Southgate House, P.O. Box 8
Stevenage, Hertfordshire, United Kingdom SG1 1H2
TELEPHONE: (011440438)3311
TELEX: 261176

Pergamon Press, Inc.
Maxwell House, Fairview Park
Elmsford, NY   10523
TELEPHONE: (914)592-7000

Personnel Resources International Inc.
342 Madison Ave., Suite 1234
New York, NY   10017
TELEPHONE: (212)682-2030

Perspective Telecommunications
15 Prospect St.
Paramus, NJ   07652
TELEPHONE: (201)845-0110

Petersen Publishing Company
8490 Sunset Blvd.
Los Angeles, CA   90069
TELEPHONE: (213)657-5100

Petroleum Publishing Company *see* Penn Well Publishing
Company

Pharmaceutical Manufacturers Association
1155 Fifteenth St. NW
Washington, DC   20005
TELEPHONE: (202)463-2000

Pharmaceutical Technology
320 N. A St., P.O. Box 50
Springfield, OR   97477
TELEPHONE: (503)726-1200

Phillips Publishing Inc.
7315 Wisconsin Ave., Suite 1200 N
Bethesda, MD   20014
TELEPHONE: (202)986-0666

Policy Studies Institute
1/2 Castle Lane
London, United Kingdom SW1E 6DR
TELEPHONE: (0114401)828-7055

James N. Porter
1224 Arbor Court
Mountain View, CA   94040
TELEPHONE: (415)961-6209

Predicasts, Inc.
11001 Cedar Ave.
Cleveland, OH   44106
TELEPHONE: (216)795-3000
TELEX: 985604

Prentice-Hall, Inc.
P.O. Box 500
Englewood Cliffs, NJ   07632
TELEPHONE: (201)592-2000

Presse - und Brancheninformationsdienst
Binsberg 28
Egmating, West Germany 8011
TELEPHONE: (01149)08095414
TELEX: 0529154

Prestwick International, Inc.
P.O. Box 205, Dept. 254
Burnt Hills, NY   12027
TELEPHONE: (518)399-6985
Roecker, Roy H.

Pritchard Publishing Company
P.O. Box 960
Durham, NH   03824
TELEPHONE: (603)868-7131

Probe Research, Inc.
P.O. Box 251
Millburn, NJ   07041
TELEPHONE: (212)732-5415
TELEX: 667812
Tannenbaum, Rita, Sales Manager

Production Publishing Company
P.O. Box 101
Bloomfield Hills, MI   48013
TELEPHONE: (313)647-8400

Professional Technical Consultants Association
1190 Lincoln Ave., Dept. RC55
San Jose, CA   95125
TELEPHONE: (408)287-8703

Progressive Grocer
708 Third Ave.
New York, NY   10017
TELEPHONE: (212)490-1000

P.T. Data Consult Inc.
P.O. Box 1081 JNG
Jakarta,
TELEPHONE: (0116221)374641
TELEX: 44328

PTN Publishing Corporation
101 Crossways Park W.
Woodbury, NY   11797
TELEPHONE: (516)496-8000

Public Press
1760 Ellice Ave.
Winnipeg, Man. Canada R3H 0B6
TELEPHONE: (204)774-1861

Public Utilities Reports
1700 N. Moore St.
Arlington, VA   22209
TELEPHONE: (703)243-7000

Publications Office of the European Communities
P.O. Box 1003
Luxembourg, Luxembourg

Putman Publishing Company
301 E. Eric St.
Chicago, IL 60611
TELEPHONE: (312)644-2020

Q

QED Informational Services
180 Linden
Wellesley, MA 02181
TELEPHONE: (617)237-5656

Quantum Science Corporation
1114 Avenue of the Americas, 28th Fl.
New York, NY 10036
TELEPHONE: (212)997-0070

R

R and B Enterprises
P.O. Box 328
Plymouth Meeting, PA 19462
TELEPHONE: (215)828-6237

Radio Amateur Callback
925 Sherwood Dr.
Lake Bluff, IL 60044
TELEPHONE: (312)234-6600

Rascal-Vadic
222 Caspian Dr.
Sunnyvale, CA
TELEPHONE: (408)744-0810
Stratton, Shelley

David Rayner
Little Waltham
Chelmsford, United Kingdom CM3 3NU
TELEPHONE: (011440245)360344

RCA Global Communications Inc.
60 Broad St.
New York, NY 10004
TELEPHONE: (212)269-9111

Thomas Reed Publications Ltd.
36-37 Cock Lane
London, United Kingdom EC1A 9BY
TELEPHONE: (0114401)248-7881
TELEX: 883526

Reifer Consultants Inc.
2733 Pacific Coast Highway, Suite 203
Torrance, CA 90505
TELEPHONE: (213)530-2274

Rentrop, Norman, Verlag *see* Verlag Norman Rentrop

Requart Whitman and Associates
1304 St. Paul St.
Baltimore, MD 21202
TELEPHONE: (301)727-3450

Resort Management, Inc.
1509 Madison Ave., P.O. Box 40169
Memphis, TN 38104
TELEPHONE: (901)276-5424

Resources for the Future, Inc.
1755 Massachusetts Ave., NW
Washington, DC 20036
TELEPHONE: (202)328-5000

Robertson and Associates, Inc.
Enterprise Mall Bldg.34 Maple St.
SUmmit, NJ 07901
TELEPHONE: (201)233-1258
Robertson, Richard B., President

Robot Institute of America
One SME Dr., P.O. Box 930
Dearborn, MI 48128
TELEPHONE: (313)271-1500

Robotics Industry Directory
P.O. Box 725
La Canada, CA 91011

Rolco Electronics
85, Rue Nollet
Paris, France F-75017
TELEPHONE: (0113301)226-0830

Royal Television Society
Tavistock Square, Tavistock House East
London, United Kingdom WC1H 9HR
TELEPHONE: (0114401)387-1970

Rydge Publications Pty. Ltd.
74 Clarence St.
Sydney, Australia NSW 2000
INTA Advertising, Inc.
1560 Broadway
New York, NY 10036
TELEPHONE: (212)275-9292

S

Verlag W. Sachon
P.O. Box 325
Mindelheim, West Germany 8948
TELEX: 539624

Sagamore Publishing Company, Inc.
1120 Old Country Rd.
Plainview, NY 11803
TELEPHONE: (516)433-6530

Howard W. Sams and Company, Inc.
4300 W. 62nd St.
Indianapolis, IN 46206
TELEPHONE: (317)291-3100

Samson Uitgeverij BV
Postbus 4
Alphen aan den Rijn, Netherlands
TELEPHONE: (0113101720)66633
TELEX: 39682

Sanford Evans Publishing Ltd.
1077 St. James St., P.O. Box 6900
Winnipeg, Man. Canada R3C 3B1
TELEPHONE: (204)775-0201

K.G. Saur Verlag KG
Possenbacherstr. 2b, Postfach 711009
Munich 70, West Germany D-8000
TELEPHONE: (0114989)798901
TELEX: 5212067
Shoe String Press, Inc.
995 Sherman Ave., P.O. Box 4327
Hamden, CT   06514
TELEPHONE: (203)248-6307

SBI, Inc. *see* Specialists in Business Information, Inc.

Schotland Business Research, Inc.
P.O. Box 511
Princeton, NJ   08540
TELEPHONE: (609)466-1400

Scrimgeour, Kemp-Gee and Company
20 Copthall Ave.
London, United Kingdom EC2R 7JS
TELEPHONE: (0114401)600-7595

Robert A. Searle
1900 W. Yale Ave.
Englewood, CO   80110

Security World. Division of Cahners Publishing Company
5670 Wilshire Blvd., Suite 2170
Los Angeles, CA   90036
TELEPHONE: (213)933-9525

Seldin Publishing, Inc.
187 W. Orangethorpe, Suite 1A
Placentia, CA   92670
TELEPHONE: (714)632-6924

Select Information Exchange
2095 Broadway
New York, NY   10023
TELEPHONE: (212)874-6408

K/S Selvig Publishing A/S
P.O. Box 9070
Vaterland, Norway

SEMA METRA
16-18 Rue Barbes, Montrouga Cedex
Paris, France 92126
TELEPHONE: (011331)6571300

Semiconductor Industry Association
20380 Town Center Lane, Suite 155
Cupertino, CA   95014
TELEPHONE: (408)255-3522

Sentinel Publishing Company
27 Centrale St.
Lasalle, Que. Canada H8R 3K2
TELEPHONE: (514)363-1484

Sentry Publishing Company
5 Kane Industrial Drive
Hudson, MA   01749
TELEPHONE: (617)562-9308

Seybold Publications Inc.
Box 644
Media, PA   19063
TELEPHONE: (215)565-2480

Siemens Aktiengesellschaft
Postfach 3240
Erlangen 2, West Germany D-8520

Sijthoff Pers B.V.
P.O. Box 19282
Den Haag, Netherlands
TELEPHONE: (01131070)190944
TELEX: 32177

Simmons-Boardman Publishing Company
1809 Capital Ave.
Omaha, NE   68102
TELEPHONE: (402)346-4300

Simon and Coates
1 London Wall Bldgs.
London, United Kingdom EC2M 5PT
TELEPHONE: (0114401)588-3644
TELEX: 885128

Martin Simpson and Company, Inc.
115 Broadway
New York, NY   10006
TELEPHONE: (212)349-7450
TELEX: 66544

Sira Institute Ltd.
South Hill
Chislehurst, Kent, United Kingdom BR7 5EH
TELEPHONE: (0114401)467-2636
TELEX: 896649

Skarbeks Computers Inc.
11990 Dorsett Rd.
St. Louis, MO   63130
TELEPHONE: (314)567-7180

Skeist Laboratories, Inc.
112 Naylon Ave.
Livingston, NJ   07039
TELEPHONE: (201)994-1050
Rocky, Joseph F., Marketing Manager

Skybird Publishing Company Inc.
508 Live Oak
Metairie, LA   70005
TELEPHONE: (504)838-8025
Mailing Address
P.O. Box 24170
New Orleans, LA   70184

Smith Barney, Harris Upham and Company
1345 Avenue of the Americas
New York, NY   10019
TELEPHONE: (212)399-6000

Stanley Smith and Company, Inc.
P.O. Box 971
Darien, CT 06820
TELEPHONE: (203)655-7664

W.R.C. Smith Publishing Company
1760 Peachtree Rd., NW
Atlanta, GA 30357
TELEPHONE: (404)874-4462

Societe Nouvelles d'Editions Publicitaires
16 Avenue Verdun
Paris, France F-75010

Society for Management Information Systems
111 E. Wacker Dr.
Chicago, IL 60601
TELEPHONE: (312)644-6610

Society of Carbide and Tool Engineers
P.O. Box 437
Bridgeville, PA 15017
TELEPHONE: (412)221-0902
Ball, Kenneth, Manager

Society of Chemical Industry
14 Belgrave Square
London, United Kingdom SW1X 8PS
TELEPHONE: (0114401)25-3681

Society of Manufacturing Engineers
P.O. Box 930
Dearborn, MI 48128
TELEPHONE: (313)271-1500

Society of Mining Engineers of AIME
Caller No. D
Littleton, CO 80123
TELEPHONE: (303)973-9550
Snedeker, Marianne, Director of Publications

Society of Telecommunications Consultants
1 Rockefeller Plaza
New York, NY 10020
TELEPHONE: (212)582-3909

Solar Vision, Inc.
Church Hill
Harrisville, NH 03450
TELEPHONE: (603)827-3347

Sonneville Associates
8604 Oakwood Dr.
Crystal Lake, IL 60014
TELEPHONE: (815)459-4358

Source EDP
Royal Bank Plaza, South Tower, Suite 2830, P.O. Box 186
Toronto, Ont. Canada M5J 2L4
TELEPHONE: (416)865-1125

Source EDP
45 William St.
Wellesley, MA 02181
TELEPHONE: (617)237-3120

Southam Business Publications Ltd. Division of Southam Communications Ltd.
310 Victoria Ave., Suite 201
Westmount, Que. Canada H32 2M9
TELEPHONE: (514)487-2302
TELEX: 055-66350

Southam Business Publications, Ltd. Division of SouthamCommunications Ltd.
1450 Don Mills Rd.
Don Mills, Ont. Canada M3B 2X7
TELEPHONE: (416)445-6641

Southern Business Publishing Company, Inc.
1621 Snow Ave.
Tampa, FL 33606
TELEPHONE: (813)251-1080

Spanish Confederation of Management Organizations
see Confederacion Espanola de Organizaciones Empresariales

Specialists in Business Information, Inc.
3375 Park Ave.
Wantagh, NY 11793
TELEPHONE: (516)781-7277

E. and F.N. Spon Ltd.
11 New Fetter Lane
London, United Kingdom EC4P 4EE
TELEPHONE: (0114401)583-9855

SRI International. Marketing Services Group
333 Ravenswood Ave.
Menlo Park, CA 94025
TELEPHONE: (415)326-6200
TELEX: 334463
Cullinan, Terrence, Sales Manager

Stam Tydschriften B.V.
Trebstraat 35
Ryswyk, Netherlands 2280AE
TELEPHONE: (0113170)991516
Frey, B.H.

Standard and Poor's Corporation
345 Hudson St.
New York, NY 10014
TELEPHONE: (212)248-2525

Steelcase, Inc.
299 Park Ave.
New York, NY 10171
TELEPHONE: (212)421-5060

Strategic Analysis, Inc.
2525 Prospect St.
Reading, PA 19606
TELEPHONE: (215)779-2800

Strategic Inc.
4320 Stevens Creek Blvd., Suite 215
San Jose, CA 95129
TELEPHONE: (408)243-8121

Strategies
15 Square de Vergennes
Paris, France 75015
TELEPHONE: (011331)828-1013

Surplus Source International
1044 S. Plymouth
Chicago, IL 60605

Survey Force Ltd.
Algarve House, 140 Borden Lane
Sittingbourne, Kent, United Kingdom ME98HR
TELEPHONE: (011440)079523778
Lainton, Keith F., Vice-President

Sutton Publishing Company, Inc.
707 Westchester Ave.
White Plains, NY 10604
TELEPHONE: (914)949-8500

Swiss Confederation. Office for the Development of Trade
Stampfenbachstrasse 85
Zurich, Switzerland CH-8035

SYBEX
2344 Sixth St.
Berkeley, CA 94710
TELEPHONE: (415)848-8233
TELEX: 336311 sybex
Weisbach, Shira, Advertising Assistant

SYMCON
14 Vanderventer Ave.
Port Washington, NY 11050
TELEPHONE: (516)883-6200

Synerjy
Grand Central Station, P.O. Box 4790
New York, NY 10017
TELEPHONE: (212)865-9595

Syracuse University Printing Services
125 College Place
Syracuse, NY 13210
TELEPHONE: (315)423-2233

Systec Consultants Ltd.
78A Grosvenor Rd.
Aldershot, Hampshire, United Kingdom GU11 3EA
TELEPHONE: (011440252)313188

T

Tab Books, Inc.
Blue Ridge Summit, PA 17214
TELEPHONE: (717)794-2191

Tactical Marketing Ltd.
73 New Bond St.
London, United Kingdom W1Y 9DB
TELEPHONE: (0114401)493-3321
Farthing, D.R., Director

Tech Tran Corporation
134 N. Washington St.
Naperville, IL 60540
TELEPHONE: (313)369-9232

Technical and Medical Studies Ltd.
29 Dorset Square
London, United Kingdom NW16QJ
TELEPHONE: (011441)7240811
TELEX: 25247

Technical Indexes Pty. Ltd.
4 Kembla St.
Cheltenham, Victoria, Australia 3192
Information Handling Services
15 Inverness Way East
Englewood, CO 80110

Technical Publications Ltd.
P.O. Box 3047
Wellington, New Zealand

Technical Publishing Company
1301 S. Grove Ave.
Barrington, IL 60010
TELEPHONE: (312)381-1840

Technical Publishing Company
666 5th Ave.
New York, NY 10019
TELEPHONE: (212)489-2200

Technical Service Council
One St. Clair Ave., E., Suite 901
Toronto, Ont. Canada M4T 2V7
TELEPHONE: (416)966-5030

Technischer Verlag Erb Ges. m.b.H.
Mariahil fer Str. F1
Vienna, Austria A-1061
TELEPHONE: (01143222)564250
TELEX: 136145
Edelbacher, Ingrid, Secretary in Charge

Technology Marketing Group Ltd.
950 Lee St.
Des Plaines, IL 60016
TELEPHONE: (312)297-1404

Technology Publishing Company
38 Commercial Wharf
Boston, 02110
TELEPHONE: (617)227-4700

Techonomics Ltd.
Manor House, Moreton
Dorchester, Dorset, United Kingdom DT2 8RG
TELEPHONE: (01144)0929-462470
Ryan, Denis M., Sales Manager

Telecom Library Inc.
205 W. 19th St.
New York, NY 10011
TELEPHONE: (212)691-8215
Newton, Harry

Telecomsept Services Inc.
10950 St. Real St.
Montreal, Que. Canada H3M 2Y4

Telephony Publishing Corporation
55 E. Jackson Blvd.
Chicago, IL 60604
TELEPHONE: (312)922-2435

Teleprocessing Products, Inc.
4565 E. Industrial St., Bldg. 7K
Simi Valley, CA 93603
TELEPHONE: (805)522-8147

Television Digest, Inc.
1836 Jefferson Place, NW
Washington, DC 20036
TELEPHONE: (202)872-9200

Television Editorial Corporation
1270 Avenue of the Americas
New York, NY 10020
TELEPHONE: (212)757-8400

C.S. Tepfer Publishing Company, Inc.
51 Sugar Hollow Rd.
Danbury, CT 06810
TELEPHONE: (203)743-2120

Ter-Sat Media Publications Ltd.
4 Smetara Dr.
Kitchener, Ont. Canada N2B 3B8

William L. Teweles and Company
777 E. Wisconsin Ave.
Milwaukee, WI 53202
TELEPHONE: (414)273-4854
TELEX: 260311

Texas A and M University. Industrial Economics Research Division see Texas Agricultural and Mechanical University. Industrial Economics Research Division

Texas Agricultural and Mechanical University. Industrial Economics Research Division
Box 83
College Station, TX 77843
TELEPHONE: (713)845-6477

Theta Technology Corporation
462 Ridge Rd.
Wethersfield, CT 06109
TELEPHONE: (203)563-9400

Thomson Publications
P.O. Box 65
Chippendale, Australia NSW

Thorne, Stevenson and Kellogg
2300 Yonge St.
Toronto, Ont. Canada M4P 1G2
TELEPHONE: (416)483-4313

Time, Inc.
Time Life Bldg., Rockefeller Ctr.
New York, NY 10020
TELEPHONE: (212)841-4800

Times Books. Division of the New York Times Company
3 Park Ave.
New York, NY 10016
TELEPHONE: (212)725-2050

Titsch Publishing, Inc.
2500 Curtis St., Suite 200; P.O. Box 5400
Denver, CO 80205
TELEPHONE: (303)573-1433

TMS
P.O. Box 2010
Orlando, FL 32802

Toshiba Corporation
1 Komukai Toshiba-cho, Saiwaiku
Kawasaka, Japan

Toyota Motor Sales
2-3-8 Kudan Minami
Chiyoda, Tokyo, Japan 102

Trade-A-Computer
P.O. Box 15842
Philadelphia, PA 19103
TELEPHONE: (215)462-4416
Van Ostrander, Philip, Publisher

Trans Data Corporation
313 High St.
Cambridge, MD 21613DK-2100
TELEPHONE: (301)338-9501
Moore, Walter W.

Transportation Research Board see U.S. Transportation Research Board. National Academy of Sciences

Tymnet Inc.
20665 Valley Green Dr.
Cupertino, CA 95014
TELEPHONE: (408)446-7000

U

Uitgeversmaatscuappy Success B.V.
Prinsevinkenpark 2
Den Haag, Netherlands 2500 CZ
TELEPHONE: (01131070)514351
TELEX: 33428

Underwriters Laboratories Inc.
1285 Walt Whitman Rd.
Melville, NY 11747
TELEPHONE: (516)271-6200
Librarian

Unipub
Murray Hill Station, Box 433
New York, NY 10016
TELEPHONE: (212)767-2791

Unisaf Publications Ltd.
32-36 Dudley St.
Tunbridge Wells, Kent, United Kingdom TN1 1LH

United Business Publications, Inc.
475 Park Ave. S.
New York, NY 10016
TELEPHONE: (212)725-2300

United Business Service Company
210 Newbury St.
Boston, MA 02116
TELEPHONE: (617)267-8855

U.K. Central Policy Review Staff
70 Whitehall
London, United Kingdom SW1A 2AS

United Kingdom. Central Statistics Office
Great George St.
London, United Kingdom SW1P 3AQ
TELEPHONE: (011441)233
I.P. Sharp Associates Ltd.
145 King Street West, Suite 1400
Toronto, Ont. M5H 1J8

U.K. Department of Industry
Victoria St., Ashdown House
London, United Kingdom
TELEPHONE: (0114401)212-3395

United Nations. Industrial Development Organization Industrial Information Section. Industrial and Technological Information Bank.
P.O. Box 300
Vienna, Austria A-1400
TELEPHONE: (01143222)26310
de Mautort, Roch J., Chief

United Nations Publications. Sales Section
801 United Nations Plaza
New York, NY 10017
TELEX: (212)754-8302

U.S. Department of Agriculture. Rural Electrification Administration
14th St. and Independence Ave., SW
Washington, DC 20250
TELEPHONE: (202)447-5606
Feragen, Robert, Administrator

U.S. Department of Commerce. Bureau of Industrial Economics. Office of Public Affairs and Publications
Washington, DC 20230
TELEPHONE: (202)377-4356

U.S. Department of Commerce. Bureau of the Census. Customer Service Branch
Washington, DC 20233
TELEPHONE: (301)449-1600

U.S. Department of Commerce. International Trade Administration
Washington, DC 20230
TELEPHONE: (202)377-2988

U.S. Department of Commerce. National Bureau of Standards
Washington, DC 20234
TELEPHONE: (301)921-1000
Ambler, Ernest, Director

U.S. Department of Commerce. National Technical Information Service
14th St. and Constitution Ave., NW Main Commerce Bldg., Rm. 1067
Washington, DC 20230
TELEPHONE: (202)377-0365

U.S. Department of Commerce. Sales Branch
14th St. between Constitution Ave. and E St., NW
Washington, DC 20230
TELEPHONE: (202)377-2000
Hartwig, Myron A., Director, Public Affairs

U.S. Department of Labor. Bureau of Labor Statistics
200 Constitution Ave., NW
Washington, DC 20210
TELEPHONE: (202)655-4000

U.S. Federal Communications Commission
1919 M St. NW
Washington, DC 20554
TELEPHONE: (202)655-4000
Ferris, Charles D., Chairman

U.S. General Services Administration
18th and F Sts., NW, General Services Bldg.
Washington, DC 20405
TELEPHONE: (202)655-4000
Freeman, Rowland G., Administrator of General Services

U.S. Government Printing Office
North Capital and H St., NW
Washington, DC 20401
TELEPHONE: (202)275-2051

U.S. Independent Telephone Association
1801 K St., NW, Suite 201
Washington, DC 20006
TELEPHONE: (202)872-1200
Pickett, George E., Executive Vice President

U.S. International Marketing Company
17057 Bellflower Ave.
Bellflower, CA 90706
TELEPHONE: (213)925-2918

U.S. International Trade Commission
701 E. St., NW
Washington, DC 20436
TELEPHONE: (202)523-0173

United States League of Savings Associations
111 E. Wacker Dr.
Chicago, IL 60601
TELEPHONE: (312)644-3100

U.S. National Aeronautics and Space Administration
400 Maryland Ave., SW
Washington, DC 20546
TELEPHONE: (202)755-2320
Frosch, Robert A., Administrator

U.S. National Science Foundation
1800 G St., NW
Washington, DC 20550
TELEPHONE: (202)655-4000
Hackerman, Norman, Chairman

U.S.News and World Report Inc.
2300 N St., NW
Washington, DC 20037
TELEPHONE: (202)861-2000

U.S. Transportation Research Board. National Academy of Sciences
2101 Constitution Ave., NW
Washington, DC 20001
TELEPHONE: (202)393-8100

University of Aston in Birmingham Technology Policy Unit
Birmingham, United Kingdom

University of Colorado. Business Research Division
Campus Box 420
Boulder, CO   80309
TELEPHONE: (303)492-8227

University of Texas at Austin. Bureau of Business Research
P.O. Box 7459, University Station
Austin, TX   78712
TELEPHONE: (512)471-1616
Glenn, Lois, Sales Manager

University of Toronto. Institute for Policy Analysis
150 George St.
Toronto, Ont. Canada N5S 1A1
TELEPHONE: (416)978-4854

University of Western Ontario. School of Business Administration
London, Ont. Canada N6A 3K7
TELEPHONE: (519)679-3222
Finnigan, Dolores, Circulation Manager

Usine Nouvelle
59 Rue du Rocher
Paris, France 75008
TELEPHONE: (011331)387-3788

V

Vance Publishing Corporation
133 E. 58th St.
New York, NY   10022
TELEPHONE: (212)755-4000

Venture Development Corporation
One Washington St.
Wellesley, MA   02181
TELEPHONE: (617)237-5080
Keith, Dorothy A., Sales Manager

Venture Magazine, Inc.
35 W. 45th St.
New York, NY   10036
TELEPHONE: (212)840-5580

Verlag Norman Rentrop
Moltkestr. 95
Bonn 2 (Bad Godesberg), West Germany 5300
TELEPHONE: (01149228)364055
TELEX: 0885651 wtagd

Verlag Schweizer Automotik Pool,Allgemeine Treuhaud AG
Bleicherweg 21, P.O. Box 5272
Zurich, Switzerland 8022
TELEPHONE: (011411)202-5950

Verlag Technik Report Ges. m.b.H.
Camillo-Sitte-g. 6
Vienna, Austria A-1150
TELEPHONE: (01143222)956-5110
Schenk, Ewald, General Manager

Verlagsges. Rudolf Muller GmbH
Stolbergerstrasse 84, Postfach 410949
Koln 41 (Braunsfeld), West Germany 5000
TELEPHONE: (011490221)5497-260
TELEX: 08881256

Vickers da Costa Ltd.
King William St., Regis House
London, United Kingdom EC4R 9AR
TELEPHONE: (0114401)623-2494
TELEX: 886004

Visual Materials Inc.
4170 Grove Ave.
Gurnee, IL   60031
TELEPHONE: (312)249-1710

Vsesoiuznyi Institut Nouchnoi Informatsii
Baltiiskaia Ul. 14
Moscow, Union of Soviet Socialist Republics A-219

W

Wakeman / Walsworth Inc.
P.O. Box 1939
Darien, CT   06820
TELEPHONE: (203)562-8518

Wall Street Transcript
120 Wall St.
New York, NY   10005
TELEPHONE: (212)747-9500

Warburg Paribas Becker - A.G. Becker, Inc.
55 Water St.
New York, NY   10805
TELEPHONE: (212)747-4400

Franklin Watts Inc.
730 Fifth Ave.
New York, NY   10019
TELEPHONE: (212)757-4050

Western Electric Company, Inc.
222 Broadway
New York, NY   10038
TELEPHONE: (212)571-2345

Western Electric Corporate Administration
P.O. Box 25000, Dept. 31GC103230
Greensboro, NC   27420
TELEPHONE: (919)697-2000

Western Union. Directory Services Department
60 Hudson St.
New York, NY   10013
TELEPHONE: (212)233-5600

Western Union Telegraph Company
One Lake St.
Upper Saddle River, NY   07458
TELEPHONE: (201)825-5000

Where to Buy, Ltd.
2 Queensway, Queensway House
Redhill, Surrey, United Kingdom RH1 1QS
TELEPHONE: (01144)68611

Whitsed Publishing Ltd.
55 Bloor West, Suite 1201
Toronto, Ont. Canada M4W 3K2
TELEPHONE: (416)967-6200

Wiley - Interscience. A Division of John Wiley and Sons
605 3rd Ave.
New York, NY   10016
TELEPHONE: (212)850-6000

John Wiley and Sons Canada Ltd.
22 Worcester St.
Rexdale, Ont. Canada M9W 1L1
TELEPHONE: (416)675-3580

John Wiley and Sons, Inc.
605 Third Ave.
New York, NY   10016
TELEPHONE: (212)850-6000

Wire Journal, Inc.
1570 Boston Post Rd.
Guilford, CT   06437
TELEPHONE: (203)453-2777

A.J. Wood Research
1405 Locust St.
Philadelphia, PA   19102
TELEPHONE: (215)546-6100

Working Women Education Fund
1224 Huron Rd.
Cleveland, OH   44115
TELEPHONE: (216)566-9308

World Wide Trade Services
P.O. Box 283
Medina, WA   98039
TELEPHONE: (206)641-8209

Worldwide Investment Research
8029 Forsyth Blvd.
St. Louis, MO   63105
TELEPHONE: (314)726-2731

Y

Yankee Group
Harvard Square, P.O. Box 43
Cambridge, MA   02138
TELEPHONE: (617)725-1100

Yugoslavia. Centre for Technical and Scientific Documentation
Sl. Penezica - Kreuna 29-31, P.O. Box 724
Belgrade, Yugoslavia 1100

Z

Zahn and Company
Salesianergasse 9
Vienna, Austria A-1037
TELEPHONE: (01143222)732126
TELEX: 01-35139

Zanny Ltd.
249 Main St.
Unionville, Ont. Canada L3R 2H3
TELEPHONE: (416)297-2922

Ziff - Davis Publishing Company
One Park Ave., Rm. 1011
New York, NY   10016
TELEPHONE: (212)725-3500

Part 2: Section 2: Publisher Index

Market Research Reports

Investment Banking Reports

Industry Statistical Reports

Financial and Economic Studies

Forecasts

Directories and Yearbooks

Special Issues and Journal Articles

Dissertations and Working Papers

Conference Reports

Newsletters

Services

Sound Recordings

Maps

Indexes and Abstracts

Bibliographic Databases

Dictionaries

Part 3: Section 1: SIC Code Index

3533   Oil field machinery 2772, 3539, 3707

3534   Elevators and moving stairways 33, 1773

3535   Conveyors and conveying equipment 33, 1773, 3705

3536   Hoists, cranes, and monorails 33, 1773, 3465, 3705, 3762, 3908

3537   Industrial trucks and tractors.....33, 688, 809, 812, 817, 828, 844, 855, 856, 860, 862, 902, 1112, 1154, 1159, 1160, 1336, 1351, 1461, 1479, 1530, 1531, 1533, 1535, 1667, 1740, 1773, 1774, 1801, 1803, 1804, 1806, 1811, 1950, 1985, 2003, 2046, 2055, 2079, 2084, 2087, 2095, 2098, 2101, 2102, 2180, 2428, 2542, 2795, 3024, 3031, 3095, 3097, 3195, 3465, 3705, 3762, 3908

354   **Metalworking machinery**
1049, 1482, 2545, 2557, 2768, 3463, 3705, 814, 833, 892, 1535, 1772, 1810, 1950, 2157, 2564, 2783, 2799

3541   Machine tools, metal cutting types 1483, 3652, 3822

3542   Machine tools, metal forming types 1483, 3652, 3822

3544   Special dies, tools, jigs & fixtures...................2126

3545   Machine tool accessories.......................522, 3800

3546   Power driven hand tools...522, 2587, 2590, 3652, 3800

3547   Rolling mill machinery 3910

3549   Metalworking machinery, nec....2126, 2129, 2585, 3073, 3076, 3744

355   **Special industry machinery**
1049, 814, 892, 1535, 3822

3551   Food products machinery.....................2701, 3910

3552   Textile machinery.....813, 1485, 3466, 1825, 1826

3554   Paper industries machinery 3563, 3567

3555   Printing trades machinery .918, 1059, 1800, 3167, 3255, 3419

3559   Special industry machinery, nec 500, 817, 828, 1160, 1802, 1806, 1811, 2079, 2102, 2126, 2287, 2323, 2552, 2557, 2642, 2974, 2999, 3045, 3663, 3664, 3671, 3673, 3705, 3749, 3751

356   **General industrial machinery**
2175, 2557, 2806, 2954, 2955, 2999, 3463, 3616, 3714, 814, 892, 1667, 3800, 3893, 3931, 3933

3561   Pumps and pumping equipment3548, 3571, 3713, 3912, 3915, 3934

3563   Air and gas compressors..................................1375

3564   Blowers and fans ..2587, 3437, 3482, 3483, 3516,. 3585, 3651, 3652, 3753, 3768, 3882, 3934

3567   Industrial furnaces and ovens 257, 3536, 3651

3568   Power transmission equipment, nec ...2622, 2623, 2777

3569   General industrial machinery, nec 71, 72, 454, 455, 456, 457, 817, 828, 1049, 1112, 1160, 1336, 1533, 1535, 1806, 1811, 1950, 2003, 2079, 2084, 2101, 2102, 2126, 2179, 2233, 2240, 2285, 2286, 2622, 2623, 2706, 2772, 2777, 3097, 3140, 3437, 3451, 3482, 3483, 3548, 3586, 3623, 3713, 3753, 3768, 3882, 3910, 3915

357   **Office and computing machines**
67, 783, 842, 875, 1049, 1103, 1126, 1238, 1241, 1242, 1290, 1302, 1308, 1327, 1364, 1425, 1426, 1573, 1597, 1723, 1724, 1778, 1812, 1819, 1877, 2042, 2119, 2324, 2383, 2384, 2385, 2388, 2463, 2471, 2516, 2518, 2571, 2959, 2964, 2976, 2980, 3120, 3135, 3136, 3138, 3142, 3143, 3145, 3146,

3159, 3162, 3164, 3165, 3166, 3173, 3174, 3175, 3177, 3178, 3179, 3183, 3184, 3186, 3187, 3188, 3189, 3191, 3198, 3199, 3211, 3213, 3219, 3220, 3224, 3229, 3231, 3232, 3233, 3234, 3237, 3238, 3240, 3244, 3251, 3253, 3254, 3255, 3259, 3263, 3265, 3273, 3275, 3290, 3293, 3294, 3295, 3321, 3325, 3331, 3332, 3333, 3334, 3335, 3338, 3359, 3361, 3366, 3370, 3372, 3374, 3378, 3382, 3383, 3385, 3386, 3391, 3392, 3396, 3405, 3406, 3407, 3409, 3410, 3411, 3412, 3415, 3416, 3421, 3429, 3430, 3436, 3446, 4000, 4273, 42, 838, 1296, 1329, 1489, 1641, 1676, 1784, 1785, 1977, 2048, 2081, 2085, 2394, 2609, 2660, 2773, 2960, 3033, 3045, 3139, 3141, 3181, 3185, 3190, 3204, 3205, 3212, 3218, 3221, 3239, 3252, 3272, 3274, 3277, 3281, 3284, 3291, 3298, 3319, 3320, 3324, 3326, 3340, 3343, 3344, 3347, 3351, 3352, 3354, 3355, 3365, 3384, 3387, 3388, 3389, 3390, 3393, 3403, 3414, 3417, 3418, 3420, 3422, 3434, 4265, 4266, 4267, 4268

3572   Typewriters 612, 815, 864, 1027, 1180, 1354, 1403, 1430, 1434, 1455, 1476, 1478, 1557, 1653, 1661, 1708, 1781, 1828, 1836, 1843, 1854, 1900, 1935, 1953, 2007, 2012, 2090, 2467, 3137, 3148, 3152, 3155, 3156, 3157, 3158, 3161, 3168, 3171, 3176, 3180, 3197, 3200, 3206, 3207, 3208, 3217, 3223, 3228, 3236, 3242, 3245, 3256, 3261, 3262, 3269, 3282, 3286, 3289, 3302, 3303, 3317, 3318, 3327, 3330, 3342, 3358, 3362, 3363, 3368, 3369, 3371, 3373, 3375, 3395, 3397, 3398, 3399, 3400, 3401, 3402, 3404, 3413, 3426, 3433, 3435

3573   Electronic computing equipment.7, 10, 31, 38, 51, 57, 65, 73, 75, 80, 86, 89, 106, 112, 117, 118, 140, 146, 151, 162, 180, 187, 194, 223, 226, 227, 235, 239, 240, 241, 243, 252, 253, 272, 330, 334, 335, 341, 351, 353, 361, 378, 380, 400, 406, 407, 410, 411, 425, 428, 437, 460, 471, 491, 508, 511, 512, 513, 514, 515, 516, 517, 524, 525, 527, 528, 529, 531, 534, 535, 536, 538, 540, 541, 542, 543, 546, 554, 556, 560, 562, 570, 575, 577, 578, 581, 595, 599, 602, 608, 612, 615, 625, 628, 629, 632, 637, 639, 645, 651, 652, 654, 658, 664, 669, 671, 678, 685, 690, 692, 698, 701, 705, 709, 722, 724, 725, 727, 728, 736, 746, 748, 750, 755, 756, 757, 758, 759, 760, 761, 762, 763, 764, 765, 766, 767, 768, 769, 770, 771, 772, 773, 774, 775, 776, 777, 778, 779, 780, 781, 782, 784, 785, 786, 787, 788, 791, 792, 793, 794, 795, 796, 797, 798, 801, 803, 804, 805, 806, 808, 809, 810, 812, 813, 814, 816, 817, 818, 819, 820, 821, 822, 823, 824, 825, 826, 827, 828, 829, 830, 831, 833, 834, 835, 836, 837, 839, 840, 841, 843, 844, 845, 846, 847, 848, 849, 850, 851, 852, 853, 854, 855, 856, 857, 858, 859, 860, 861, 862, 863, 864, 865, 866, 868, 869, 870, 871, 872, 873, 874, 876, 877, 878, 879, 881, 882, 883, 884, 885, 886, 887, 888, 889, 890, 891, 892, 893, 894, 895, 896, 897, 898, 899, 900, 902, 903, 904, 905, 906, 907, 909, 910, 911, 912, 913, 914, 915, 916, 917, 918, 919, 920, 921, 922, 923, 924, 925, 926, 927, 928, 929, 930, 931, 932, 933, 934, 935, 936, 937, 938, 939, 940, 941, 942, 943, 944, 945, 950, 951, 952, 953, 954, 955, 956, 957, 958, 959, 960, 961, 962, 963, 964, 965, 966, 967, 968, 969, 970, 971, 972, 973, 974, 975, 977, 978, 979, 980, 982, 983, 984, 985, 986, 987, 988, 989, 990, 991, 993, 994, 995, 996, 997, 998, 999, 1000, 1001, 1002, 1003, 1004, 1005, 1006, 1007, 1008, 1009, 1010, 1011, 1012, 1013, 1014, 1015, 1016, 1017, 1018, 1019, 1020, 1021, 1022, 1023, 1024, 1025, 1026, 1027, 1028, 1029, 1030, 1031, 1032, 1033, 1034, 1035, 1036, 1037, 1038, 1039, 1040, 1041, 1042, 1043, 1044, 1045, 1046, 1047, 1050, 1051,

1052, 1054, 1055, 1056, 1057, 1058, 1059, 1060,
1061, 1062, 1063, 1064, 1065, 1066, 1067, 1068,
1070, 1072, 1073, 1074, 1075, 1076, 1077, 1078,
1079, 1080, 1081, 1082, 1083, 1084, 1085, 1086,
1088, 1089, 1090, 1091, 1092, 1093, 1094, 1096,
1097, 1098, 1099, 1100, 1101, 1104, 1105, 1106,
1107, 1108, 1109, 1111, 1112, 1113, 1114, 1115,
1116, 1117, 1118, 1119, 1120, 1121, 1122, 1123,
1124, 1125, 1127, 1128, 1130, 1131, 1132, 1133,
1134, 1135, 1136, 1137, 1138, 1139, 1140, 1141,
1142, 1143, 1144, 1145, 1146, 1147, 1148, 1149,
1150, 1152, 1153, 1155, 1157, 1158, 1159, 1160,
1161, 1162, 1163, 1164, 1165, 1166, 1167, 1169,
1170, 1171, 1172, 1173, 1174, 1175, 1176, 1178,
1180, 1182, 1184, 1185, 1186, 1187, 1188, 1190,
1191, 1192, 1193, 1194, 1195, 1196, 1198, 1199,
1200, 1202, 1203, 1204, 1205, 1206, 1207, 1208,
1209, 1210, 1211, 1212, 1213, 1214, 1215, 1216,
1217, 1218, 1219, 1220, 1221, 1222, 1223, 1224,
1225, 1237, 1239, 1240, 1243, 1244, 1245, 1246,
1247, 1248, 1249, 1250, 1251, 1252, 1253, 1254,
1255, 1256, 1257, 1258, 1259, 1260, 1261, 1262,
1263, 1264, 1265, 1266, 1267, 1268, 1269, 1270,
1272, 1273, 1274, 1275, 1276, 1277, 1278, 1279,
1280, 1281, 1282, 1284, 1285, 1286, 1287, 1289,
1291, 1292, 1293, 1294, 1295, 1297, 1298, 1299,
1300, 1301, 1303, 1304, 1306, 1307, 1309, 1310,
1311, 1313, 1316, 1318, 1320, 1321, 1322, 1325,
1326, 1330, 1331, 1332, 1333, 1335, 1336, 1337,
1338, 1339, 1340, 1341, 1342, 1343, 1347, 1348,
1349, 1350, 1351, 1352, 1353, 1354, 1355, 1357,
1358, 1359, 1360, 1361, 1362, 1363, 1366, 1367,
1368, 1369, 1370, 1371, 1372, 1373, 1374, 1376,
1377, 1379, 1380, 1381, 1382, 1383, 1384, 1385,
1386, 1387, 1388, 1389, 1390, 1391, 1392, 1393,
1394, 1395, 1397, 1399, 1400, 1401, 1402, 1403,
1404, 1405, 1406, 1407, 1408, 1409, 1410, 1412,
1413, 1414, 1415, 1416, 1417, 1418, 1422, 1423,
1424, 1427, 1428, 1430, 1431, 1432, 1433, 1434,
1436, 1437, 1438, 1440, 1441, 1442, 1443, 1444,
1445, 1446, 1449, 1450, 1451, 1452, 1453, 1454,
1455, 1456, 1457, 1458, 1459, 1460, 1461, 1464,
1465, 1466, 1467, 1469, 1470, 1471, 1472, 1473,
1474, 1475, 1476, 1477, 1478, 1479, 1480, 1481,
1482, 1483, 1484, 1485, 1486, 1487, 1488, 1490,
1491, 1492, 1493, 1494, 1495, 1496, 1497, 1498,
1499, 1500, 1501, 1503, 1504, 1506, 1507, 1508,
1509, 1512, 1513, 1514, 1515, 1516, 1517, 1518,
1522, 1523, 1524, 1525, 1526, 1527, 1528, 1529,
1530, 1531, 1532, 1533, 1534, 1535, 1536, 1537,
1538, 1539, 1540, 1541, 1542, 1543, 1544, 1545,
1546, 1547, 1548, 1549, 1550, 1551, 1552, 1553,
1554, 1555, 1556, 1557, 1558, 1559, 1560, 1561,
1562, 1563, 1564, 1565, 1566, 1567, 1568, 1569,
1570, 1571, 1572, 1574, 1575, 1576, 1577, 1578,
1579, 1580, 1581, 1582, 1583, 1584, 1585, 1586,
1587, 1588, 1589, 1590, 1591, 1592, 1593, 1594,
1595, 1596, 1598, 1599, 1600, 1601, 1602, 1603,
1604, 1605, 1606, 1607, 1609, 1610, 1612, 1613,
1616, 1618, 1619, 1620, 1621, 1622, 1623, 1624,
1625, 1626, 1627, 1628, 1629, 1630, 1631, 1632,
1633, 1634, 1635, 1636, 1637, 1638, 1639, 1640,
1642, 1643, 1644, 1645, 1646, 1647, 1648, 1649,
1650, 1651, 1652, 1653, 1654, 1655, 1656, 1657,
1658, 1659, 1660, 1661, 1662, 1663, 1664, 1665,
1666, 1667, 1668, 1669, 1670, 1671, 1672, 1673,
1674, 1675, 1677, 1678, 1679, 1680, 1681, 1682,
1683, 1684, 1685, 1686, 1688, 1689, 1690, 1691,
1692, 1693, 1694, 1696, 1697, 1698, 1699, 1700,
1701, 1702, 1703, 1704, 1705, 1706, 1707, 1708,
1709, 1710, 1711, 1712, 1713, 1714, 1715, 1716,
1719, 1720, 1721, 1722, 1726, 1731, 1735, 1736,

1737, 1738, 1739, 1740, 1741, 1742, 1743, 1745,
1746, 1747, 1748, 1749, 1750, 1751, 1752, 1753,
1754, 1755, 1756, 1757, 1758, 1759, 1760, 1761,
1762, 1763, 1764, 1765, 1766, 1767, 1768, 1769,
1770, 1771, 1772, 1773, 1774, 1775, 1776, 1777,
1779, 1780, 1781, 1782, 1783, 1786, 1787, 1788,
1789, 1790, 1791, 1792, 1793, 1794, 1795, 1796,
1797, 1798, 1800, 1801, 1802, 1803, 1804, 1806,
1807, 1808, 1809, 1810, 1811, 1813, 1814, 1815,
1816, 1817, 1818, 1820, 1821, 1822, 1823, 1824,
1825, 1826, 1827, 1828, 1829, 1830, 1831, 1832,
1833, 1834, 1835, 1836, 1837, 1838, 1839, 1840,
1841, 1842, 1843, 1844, 1845, 1846, 1847, 1848,
1849, 1850, 1851, 1852, 1853, 1854, 1855, 1856,
1857, 1858, 1859, 1860, 1861, 1862, 1863, 1864,
1865, 1866, 1867, 1868, 1869, 1870, 1871, 1872,
1873, 1874, 1875, 1876, 1878, 1879, 1880, 1881,
1882, 1883, 1884, 1885, 1887, 1888, 1889, 1890,
1891, 1892, 1893, 1894, 1895, 1896, 1897, 1898,
1899, 1900, 1901, 1902, 1903, 1904, 1905, 1906,
1907, 1908, 1909, 1910, 1911, 1912, 1913, 1914,
1915, 1916, 1917, 1918, 1919, 1920, 1921, 1922,
1923, 1924, 1925, 1926, 1927, 1928, 1929, 1930,
1931, 1932, 1933, 1934, 1935, 1936, 1937, 1938,
1939, 1940, 1941, 1942, 1944, 1945, 1946, 1947,
1948, 1949, 1950, 1951, 1952, 1953, 1954, 1955,
1956, 1957, 1958, 1959, 1960, 1961, 1962, 1963,
1964, 1965, 1966, 1967, 1969, 1970, 1972, 1973,
1974, 1975, 1976, 1978, 1979, 1980, 1981, 1982,
1983, 1984, 1986, 1987, 1988, 1989, 1991, 1992,
1993, 1994, 1996, 1998, 1999, 2000, 2001, 2002,
2003, 2004, 2005, 2007, 2008, 2009, 2010, 2011,
2012, 2013, 2014, 2015, 2016, 2017, 2018, 2019,
2021, 2022, 2023, 2024, 2025, 2026, 2027, 2028,
2029, 2030, 2031, 2032, 2033, 2034, 2035, 2036,
2037, 2038, 2039, 2040, 2043, 2045, 2046, 2047,
2049, 2050, 2051, 2052, 2053, 2054, 2055, 2056,
2057, 2058, 2059, 2060, 2061, 2062, 2063, 2064,
2065, 2066, 2067, 2068, 2069, 2070, 2071, 2072,
2073, 2074, 2075, 2076, 2077, 2078, 2079, 2080,
2082, 2083, 2084, 2086, 2087, 2088, 2089, 2090,
2091, 2092, 2093, 2094, 2095, 2096, 2097, 2098,
2099, 2100, 2101, 2102, 2103, 2104, 2105, 2106,
2107, 2108, 2122, 2142, 2145, 2146, 2157, 2168,
2171, 2173, 2177, 2179, 2180, 2181, 2189, 2201,
2201, 2204, 2209, 2213, 2214, 2223, 2227, 2240,
2241, 2247, 2252, 2271, 2285, 2286, 2286, 2301,
2341, 2390, 2393, 2427, 2440, 2441, 2453, 2465,
2466, 2468, 2469, 2470, 2476, 2496, 2533, 2542,
2587, 2596, 2599, 2601, 2603, 2612, 2613, 2615,
2616, 2622, 2642, 2645, 2647, 2656, 2659, 2661,
2662, 2663, 2664, 2672, 2684, 2687, 2690, 2693,
2694, 2695, 2697, 2698, 2716, 2718, 2723, 2726,
2728, 2730, 2735, 2738, 2741, 2761, 2763, 2777,
2780, 2781, 2783, 2787, 2795, 2799, 2802, 2803,
2812, 2813, 2837, 2838, 2855, 2867, 2891, 2945,
2956, 2961, 2990, 2996, 3010, 3020, 3021, 3027,
3028, 3029, 3039, 3051, 3061, 3064, 3081, 3090,
3097, 3112, 3122, 3137, 3144, 3147, 3152, 3153,
3155, 3156, 3157, 3158, 3160, 3161, 3163, 3168,
3170, 3171, 3172, 3176, 3180, 3193, 3194, 3197,
3200, 3203, 3206, 3207, 3208, 3209, 3210, 3214,
3230, 3235, 3236, 3245, 3246, 3260, 3264, 3266,
3267, 3269, 3270, 3280, 3282, 3283, 3286, 3289,
3301, 3302, 3303, 3304, 3310, 3317, 3327, 3330,
3342, 3349, 3350, 3356, 3358, 3363, 3364, 3368,
3369, 3373, 3375, 3380, 3394, 3395, 3397, 3398,
3399, 3400, 3401, 3402, 3413, 3425, 3435, 3469,
3470, 3519, 3532, 3553, 3565, 3566, 3620, 3622,
3666, 3670, 3686, 3750, 3756, 3774, 3782, 3789,
3810, 3814, 3836, 3852, 3859, 3862, 3887, 3926,
3935, 3946, 3954, 3955, 3960, 3964, 3968, 3974,

2377, 2397, 2421, 2431, 2446, 2472, 2901, 2948, 3013, 3073, 3076, 3462

3648 Lighting equipment, nec....394, 2351, 2371, 2414, 2431, 2472, 2630

365 Radio and tv receiving equipment
3, 31, 67, 75, 106, 122, 198, 351, 390, 400, 413, 465, 482, 503, 521, 530, 539, 912, 1005, 1028, 1208, 1308, 2077, 2149, 2344, 2383, 2425, 2533, 2567, 2581, 2726, 2818, 2836, 2928, 2945, 2983, 3079, 3101, 3294, 4124, 4197, 283, 284, 337, 348, 389, 410, 425, 460, 471, 488, 493, 505, 515, 524, 525, 527, 529, 531, 534, 537, 541, 542, 1150, 1329, 1628, 1716, 1823, 1895, 2085, 2235, 2432, 2556, 2609, 2759, 2763, 2867, 2898, 2960, 3033, 3034, 3045, 3061, 3064, 3104, 3122, 3126, 4077

3651 Radio and tv receiving sets.... 35, 61, 76, 77, 110, 111, 159, 260, 276, 277, 282, 288, 313, 343, 365, 394, 397, 402, 472, 484, 533, 1026, 1619, 2080, 2213, 2227, 2243, 2249, 2300, 2301, 2343, 2388, 2516, 2539, 2551, 2555, 2558, 2569, 2615, 2735, 2760, 2761, 2775, 2794, 2852, 3390, 3852, 4027, 4035, 4066, 4067, 4076, 4182, 4213, 4215

3652 Phonograph records......151, 277, 397, 403, 1004, 2213, 2252, 2569, 2794, 2936

366 Communication equipment
3, 5, 7, 25, 26, 28, 29, 38, 45, 47, 67, 78, 79, 81, 82, 83, 96, 98, 110, 116, 122, 129, 143, 156, 160, 162, 188, 189, 192, 195, 227, 239, 240, 243, 253, 327, 329, 330, 332, 351, 353, 379, 380, 398, 399, 400, 407, 424, 430, 465, 475, 481, 499, 501, 507, 508, 510, 521, 530, 539, 545, 773, 783, 784, 912, 913, 966, 982, 985, 1025, 1063, 1064, 1074, 1308, 1310, 1330, 1389, 1907, 2042, 2100, 2119, 2181, 2230, 2231, 2232, 2247, 2328, 2344, 2383, 2425, 2441, 2477, 2516, 2581, 2663, 2687, 2766, 2818, 2836, 2896, 2928, 2945, 2983, 2991, 3060, 3079, 3145, 3156, 3159, 3162, 3191, 3209, 3361, 3771, 3772, 3877, 3955, 3964, 3966, 3976, 3984, 4009, 4019, 4020, 4035, 4038, 4058, 4097, 4172, 4173, 4176, 4183, 4225, 4234, 4279, 4290, 4292, 4314, 4334, 4338, 30, 36, 37, 42, 49, 50, 51, 53, 56, 95, 112, 194, 203, 218, 254, 268, 275, 278, 289, 296, 297, 304, 305, 315, 322, 323, 334, 337, 339, 341, 348, 355, 362, 371, 378, 382, 397, 406, 410, 416, 425, 432, 460, 471, 486, 492, 497, 506, 515, 524, 525, 527, 528, 529, 534, 537, 541, 542, 543, 669, 726, 819, 839, 967, 987, 1046, 1219, 1250, 1694, 1701, 1894, 1895, 1897, 2007, 2033, 2048, 2050, 2085, 2186, 2201, 2235, 2432, 2556, 2591, 2609, 2630, 2734, 2738, 2798, 2898, 2960, 3034, 3045, 3061, 3064, 3104, 3122, 3260, 3326, 3745, 3974, 3977, 4014, 4073, 4074, 4075, 4077, 4169, 4218, 4224, 4253, 4291, 4304, 4312, 4315

3661 Telephone and telegraph apparatus 59, 60, 63, 64, 65, 69, 73, 86, 88, 93, 97, 99, 111, 113, 117, 118, 119, 120, 121, 147, 149, 152, 153, 154, 157, 163, 168, 169, 176, 177, 178, 179, 181, 182, 183, 193, 201, 204, 206, 208, 209, 211, 212, 213, 214, 216, 220, 221, 224, 225, 229, 231, 232, 233, 234, 238, 241, 242, 248, 249, 258, 261, 262, 264, 265, 266, 302, 317, 319, 328, 333, 335, 354, 364, 367, 373, 374, 375, 383, 384, 385, 386, 408, 409, 411, 414, 426, 427, 428, 433, 434, 441, 446, 448, 449, 459, 461, 462, 463, 478, 491, 494, 502, 504, 511, 512, 513, 514, 516, 517, 522, 523, 531, 535, 536, 538, 540, 544, 676, 908, 951, 965, 976, 979, 1024, 1027, 1034, 1075, 1080, 1110, 1126, 1144, 1212, 1213, 1350, 1355, 1369, 1371, 1375, 1570, 1600, 1604, 1731, 1932, 1977, 2016, 2077, 2213, 2283, 2301, 2463, 2471, 2506, 2800, 2837, 2838,

2980, 3142, 3143, 3193, 3212, 3231, 3320, 3324, 3366, 3418, 3986, 3992, 4000, 4006, 4013, 4018, 4021, 4025, 4027, 4029, 4032, 4039, 4040, 4041, 4042, 4044, 4047, 4050, 4053, 4098, 4111, 4115, 4121, 4133, 4145, 4178, 4179, 4180, 4181, 4188, 4191, 4196, 4204, 4226, 4240, 4247, 4257, 4297, 4299, 4321, 4325

3662 Radio and tv communication equipment 1, 2, 4, 8, 9, 10, 11, 12, 14, 15, 16, 17, 18, 19, 20, 21, 22, 23, 24, 27, 31, 32, 33, 34, 35, 39, 40, 41, 43, 44, 46, 48, 52, 54, 55, 57, 58, 62, 64, 66, 68, 70, 71, 72, 73, 74, 80, 84, 87, 89, 90, 91, 94, 99, 100, 101, 102, 104, 105, 106, 107, 108, 109, 114, 115, 123, 124, 125, 126, 127, 128, 130, 131, 132, 133, 134, 135, 136, 137, 138, 139, 141, 142, 144, 145, 146, 148, 150, 155, 158, 159, 161, 164, 165, 166, 167, 170, 171, 172, 173, 174, 175, 180, 183, 184, 185, 186, 191, 196, 197, 198, 199, 200, 202, 204, 205, 207, 215, 222, 223, 225, 226, 230, 235, 236, 244, 245, 246, 247, 250, 252, 255, 256, 257, 259, 260, 263, 267, 269, 270, 271, 272, 273, 274, 276, 279, 280, 281, 283, 284, 285, 286, 287, 288, 290, 291, 292, 293, 294, 298, 299, 300, 301, 302, 303, 306, 307, 308, 310, 311, 313, 314, 316, 318, 319, 320, 321, 324, 325, 326, 331, 336, 338, 340, 342, 343, 344, 345, 346, 347, 349, 350, 352, 357, 358, 359, 360, 361, 365, 366, 368, 369, 370, 372, 375, 376, 377, 381, 387, 389, 390, 391, 392, 394, 395, 396, 401, 402, 404, 405, 411, 412, 413, 414, 415, 417, 418, 419, 420, 421, 422, 423, 429, 431, 435, 436, 437, 438, 440, 441, 442, 443, 444, 445, 447, 450, 451, 452, 453, 454, 455, 456, 457, 461, 462, 464, 466, 467, 468, 469, 470, 472, 473, 474, 476, 477, 479, 480, 483, 484, 485, 487, 488, 489, 490, 495, 496, 498, 500, 503, 509, 511, 512, 513, 514, 516, 517, 518, 519, 520, 522, 526, 532, 533, 535, 536, 538, 540, 544, 769, 814, 892, 951, 977, 1002, 1003, 1073, 1210, 1355, 1414, 1421, 1422, 1508, 1662, 1872, 1883, 1952, 1982, 1983, 2011, 2081, 2154, 2173, 2188, 2206, 2213, 2226, 2229, 2272, 2275, 2279, 2301, 2452, 2479, 2480, 2506, 2533, 2538, 2539, 2559, 2565, 2570, 2583, 2724, 2726, 2730, 2765, 2838, 2880, 2897, 2907, 3082, 3135, 3147, 3160, 3246, 3443, 3456, 3457, 3459, 3462, 3513, 3517, 3518, 3520, 3521, 3543, 3544, 3553, 3579, 3649, 3653, 3658, 3660, 3681, 3686, 3701, 3721, 3723, 3726, 3732, 3741, 3773, 3781, 3822, 3840, 3903, 3925, 3935, 3969, 3979, 3988, 3990, 3991, 3997, 4002, 4005, 4007, 4013, 4027, 4030, 4033, 4034, 4040, 4044, 4057, 4059, 4062, 4067, 4072, 4076, 4099, 4102, 4113, 4144, 4174, 4175, 4182, 4184, 4187, 4188, 4191, 4201, 4202, 4213, 4215, 4216, 4229, 4247, 4294, 4295, 4302

367 Electronic components and accessories
45, 67, 78, 82, 128, 131, 143, 156, 187, 204, 253, 321, 375, 501, 783, 784, 912, 1330, 1537, 1636, 1643, 1671, 2042, 2077, 2100, 2115, 2119, 2120, 2145, 2162, 2172, 2173, 2181, 2200, 2207, 2210, 2217, 2221, 2228, 2230, 2231, 2232, 2277, 2285, 2327, 2328, 2381, 2382, 2383, 2425, 2430, 2441, 2452, 2463, 2471, 2477, 2483, 2505, 2507, 2510, 2511, 2516, 2521, 2527, 2538, 2544, 2545, 2549, 2552, 2557, 2581, 2608, 2615, 2616, 2661, 2726, 2766, 2796, 2801, 2808, 2816, 2818, 2822, 2834, 2836, 2837, 2847, 2851, 2856, 2858, 2859, 2860, 2861, 2862, 2863, 2864, 2881, 2891, 2892, 2896, 2905, 2928, 2943, 2945, 2946, 2950, 2961, 2971, 2972, 2978, 2980, 2982, 2983, 2984, 2988, 2991, 3026, 3028, 3029, 3052, 3054, 3060, 3066, 3068, 3069, 3070, 3071, 3072, 3079, 3080, 3089, 3091, 3092, 3101, 3102, 3123, 3124, 3129, 3133, 3543, 3651, 3672, 3877, 3887, 4334, 218, 268, 315,

337, 488, 489, 529, 829, 869, 1250, 1479, 1600, 1673, 1690, 1691, 1701, 1755, 1815, 1985, 2050, 2085, 2101, 2163, 2177, 2178, 2185, 2186, 2189, 2190, 2201, 2222, 2235, 2287, 2325, 2432, 2454, 2523, 2534, 2535, 2542, 2559, 2584, 2590, 2592, 2599, 2601, 2609, 2614, 2618, 2629, 2684, 2686, 2689, 2694, 2695, 2703, 2707, 2712, 2721, 2728, 2730, 2731, 2734, 2738, 2748, 2752, 2758, 2787, 2788, 2798, 2809, 2819, 2821, 2843, 2844, 2845, 2848, 2849, 2850, 2867, 2918, 2919, 2920, 2929, 2960, 2977, 2989, 2990, 2996, 3006, 3009, 3010, 3015, 3018, 3020, 3022, 3030, 3033, 3034, 3035, 3036, 3045, 3061, 3064, 3077, 3087, 3093, 3104, 3105, 3112, 3122, 3126, 3134, 3513, 3684, 3940, 3945, 4023, 4274, 4315

3671 Electron tubes, receiving type....... 283, 284, 2249, 2362, 2405, 2566, 2833, 2853, 2900

3672 Cathode ray television picture tubes....... 283, 284, 394, 400, 1875, 2214, 2249, 2356, 2405, 2566, 2663, 2830, 2853, 4182

3673 Electron tubes, transmitting.....283, 284, 385, 400, 2239, 2249, 2363, 2405, 2622, 2785, 2938, 3456, 3457, 3763

3674 Semiconductors and related devices 98, 127, 213, 385, 400, 441, 445, 502, 578, 759, 775, 878, 979, 1065, 1099, 1215, 1218, 1466, 1484, 1545, 1595, 1647, 1687, 1709, 1912, 1913, 1914, 1949, 1969, 1970, 1986, 2047, 2063, 2080, 2097, 2109, 2110, 2122, 2123, 2124, 2126, 2127, 2138, 2139, 2140, 2141, 2143, 2144, 2146, 2147, 2150, 2156, 2159, 2166, 2167, 2169, 2176, 2187, 2198, 2209, 2213, 2216, 2220, 2224, 2236, 2238, 2240, 2242, 2243, 2247, 2255, 2258, 2259, 2261, 2264, 2265, 2269, 2270, 2274, 2275, 2278, 2280, 2281, 2283, 2286, 2288, 2289, 2290, 2292, 2293, 2295, 2301, 2308, 2309, 2311, 2312, 2318, 2319, 2320, 2322, 2323, 2326, 2329, 2330, 2331, 2332, 2333, 2334, 2335, 2340, 2342, 2343, 2346, 2347, 2375, 2388, 2389, 2390, 2395, 2424, 2427, 2433, 2436, 2440, 2442, 2443, 2444, 2445, 2447, 2451, 2457, 2462, 2464, 2465, 2466, 2467, 2469, 2470, 2473, 2475, 2478, 2485, 2486, 2487, 2488, 2489, 2490, 2491, 2492, 2493, 2494, 2495, 2496, 2497, 2499, 2500, 2501, 2502, 2503, 2504, 2526, 2530, 2532, 2548, 2554, 2565, 2567, 2568, 2594, 2598, 2600, 2603, 2604, 2606, 2611, 2612, 2617, 2619, 2644, 2667, 2668, 2673, 2674, 2676, 2678, 2679, 2680, 2682, 2683, 2687, 2693, 2696, 2697, 2702, 2709, 2716, 2717, 2725, 2732, 2736, 2742, 2743, 2751, 2753, 2765, 2779, 2786, 2789, 2790, 2802, 2803, 2815, 2823, 2826, 2831, 2839, 2853, 2868, 2870, 2873, 2874, 2875, 2876, 2877, 2879, 2884, 2885, 2893, 2894, 2897, 2899, 2902, 2903, 2906, 2907, 2909, 2910, 2913, 2915, 2916, 2917, 2923, 2924, 2933, 2934, 2935, 2942, 2975, 2979, 2998, 3007, 3008, 3012, 3037, 3059, 3074, 3075, 3078, 3081, 3084, 3098, 3103, 3107, 3109, 3115, 3116, 3117, 3118, 3119, 3456, 3457, 3515, 3516, 3550, 3562, 3685, 3694, 3697, 3741, 3800, 3838, 3842, 3853, 4007

3675 Electronic capacitors 2256, 2303, 2364, 2398, 2406, 2593, 2611, 2621, 2646, 2675, 2700, 2708, 2778, 2807, 2827, 2828, 2829, 2914, 3025, 3100

3676 Electronic resistors 2125, 2284, 2306, 2368, 2410, 2611, 2681, 2778, 2807, 2921, 2922, 3696, 3697, 4007

3677 Electronic coils and transformers 2248, 2304, 2365, 2407, 2711, 2772, 2832, 2869, 4007

3678 Electronic connectors 251, 2211, 2212, 2276,

2305, 2316, 2317, 2321, 2367, 2409, 2648, 2665, 2672, 2835

3679 Electronic components, nec 62, 85, 124, 174, 190, 214, 219, 224, 352, 381, 402, 759, 775, 785, 1180, 1186, 1571, 1839, 1930, 1931, 2109, 2121, 2159, 2196, 2213, 2237, 2248, 2250, 2251, 2253, 2254, 2260, 2262, 2263, 2271, 2298, 2309, 2311, 2366, 2408, 2427, 2498, 2506, 2548, 2597, 2598, 2612, 2617, 2619, 2620, 2658, 2666, 2667, 2670, 2671, 2688, 2704, 2705, 2710, 2724, 2744, 2765, 2778, 2784, 2794, 2803, 2823, 2826, 2869, 2870, 2873, 2877, 2882, 2883, 2885, 2890, 2895, 2897, 2906, 2911, 2912, 2913, 2917, 2921, 2925, 2934, 2936, 2937, 3008, 3011, 3012, 3078, 3082, 3084, 3099, 3108, 3689, 3696, 3741, 3852, 3853, 4007

369 **Misc. electrical equipment & supplies**
2112, 2205, 2215, 2266, 2267, 2268, 2425, 2527, 2530, 2549, 2557, 2615, 2616, 2836, 2846, 2432, 2534, 2535, 2591, 2758, 2960, 3045

3691 Storage batteries... 2162, 2261, 2313, 2348, 2376, 2386, 2397, 2420, 2428, 2429, 2438, 2449, 2460, 2462, 2536, 2797, 2866, 2927, 3044, 3047, 3096, 3132

3692 Primary batteries, dry and wet ..2113, 2161, 2162, 2261, 2313, 2314, 2315, 2348, 2373, 2397, 2417, 2428, 2429, 2438, 2449, 2462, 2536, 2560, 2561, 2562, 2563, 2595, 2866, 2927, 3015, 3040, 3044, 3047, 3096, 3132, 204, 703, 2181, 2234, 2272, 2294, 2516, 2738, 2780, 2781, 2782, 2878, 2940, 3498, 3677, 3792, 3840

3694 Engine electrical equipment 2161, 2208, 2233, 2273, 2369, 2411, 2446, 2448, 2461, 2462, 2562, 2768, 2797, 2851, 2866, 2901, 2948, 3068, 3069, 3073, 3076, 3123, 3133, 3459

3699 Electrical equipment & supplies, nec ..1716, 2115, 2223, 2227, 2294, 2361, 2404, 2567, 2630, 2735, 3038, 3095

37 **TRANSPORTATION EQUIPMENT**
2525, 2804, 2805, 2810, 2811, 3452, 486, 2142, 2202

371 **Motor vehicles and equipment**
2151, 2388, 2431, 2439, 2460, 2474, 2808, 4157, 856, 2428, 4336

3711 Motor vehicles and car bodies ..2113, 2205, 2215, 2266, 2267, 2268, 2454, 2530, 2563, 2866, 2901

3714 Motor vehicle parts and accessories ..2205, 2273, 2446, 2447, 2448, 2461, 2463, 2466, 2471, 2479, 2739, 2824, 2825, 2866, 2901, 2980, 3073, 3076, 3768

372 **Aircraft and parts**
3, 17, 34, 78, 81, 82, 143, 156, 160, 195, 196, 197, 398, 399, 405, 412, 490, 520, 2151, 2230, 2231, 2232, 2328, 2388, 2452, 2813, 271, 272, 497, 3464, 3660

3721 Aircraft 2644

3724 Aircraft engines and engine parts 2570, 3459

3728 Aircraft equipment, nec.20, 41, 2188, 2570, 3459, 3653

373 **Ship and boat building and repairing**
2151, 2388, 259, 318, 497, 2161

3731 Ship building and repairing . 78, 82, 123, 143, 412, 520

3732 Boat building and repairing 123, 2565

3743 Railroad equipment 2151

3751 Motorcycles, bicycles, and parts 2151

384 **Medical instruments and supplies**
2516, 2959, 3440, 3498, 3499, 3505, 3507, 3514, 3528, 3581, 3642, 1985, 3545, 3552, 3791, 3831

3841 Surgical and medical instruments... 703, 955, 958, 2181, 2294, 2960, 3485, 3494, 3495, 3497, 3500, 3529, 3551, 3568, 3569, 3580, 3607, 3621, 3643, 3669, 3691, 3783, 3792, 3872, 3875, 3940

3842 Surgical appliances and supplies 415, 2480, 3607

3851 Ophthalmic goods....................................3607, 3872

386 **Photographic equipment and supplies**
482, 2896

3861 Photographic equipment and supplies.... 222, 272, 274, 275, 276, 277, 282, 288, 341, 342, 394, 483, 493, 503, 998, 1119, 1199, 1385, 1388, 1736, 1800, 1839, 1930, 1931, 1983, 2080, 2516, 2782, 3150, 3163, 3167, 3170, 3230, 3243, 3257, 3258, 3263, 3268, 3281, 3287, 3292, 3300, 3305, 3306, 3308, 3309, 3311, 3312, 3321, 3323, 3333, 3345, 3346, 3353, 3360, 3372, 3387, 3388, 3393, 3408, 3419, 3428, 3430, 3480, 3633, 3750, 3826, 3828, 3832, 3861, 3875, 3977, 4197, 4265, 4266, 4267, 4268

387 **Watches, clocks, and watchcases**
2388, 2959

3873 Watches, clocks, and watchcases 2516, 2539, 2557, 2763, 3448, 3449, 3469, 3470, 3584, 3597, 3605, 3606, 3607, 3615, 3743, 3802, 3803, 3867, 3868, 3870, 3892, 3894, 3895, 3901, 3906, 3909, 3943, 3944

39 **MISCELLANEOUS MANUFACTURING INDUSTRIES**
2524

391 **Jewelry, silverware, and plated ware**
3909

394 **Toys and sporting goods**
2340

3944 Games, toys, and children's vehicles ... 198, 1150, 1628, 1716, 2224, 2343, 2763

395 **Pens, pencils, office and art supplies**
3183, 3187, 3213, 3229, 3237, 3238, 3338, 3370, 3411, 3181, 3212, 3274, 3277, 3324, 3354

3953 Marking devices 3227

3955 Carbon paper and inked ribbons 944, 3226, 3328, 3333

3991 Brooms and brushes 2762

3993 Signs and advertising displays.............2396, 2419

3999 Manufacturing industries, nec .257, 454, 455, 456, 457, 775, 1150, 1628, 1747, 1790, 2227, 2379, 2456, 2760, 2762, 2774, 2776, 2852

TRANSPORTATION AND PUBLIC UTILITIES

40 **RAILROAD TRANSPORTATION**
4162, 371, 621, 1630

401 **Railroads**
4310, 4337

4011 Railroads, line-haul operating.....1532, 4260, 4336

41 **LOCAL AND INTERURBAN PASSENGER TRANSIT**
4310, 4337

42 **TRUCKING AND WAREHOUSING**
1532, 4336

421 **Trucking, local and long distance**
4310, 4337

43 **U.S. POSTAL SERVICE**
4336

4311 U.s. postal service 1850

44 **WATER TRANSPORTATION**
1532, 1630, 4310, 4336, 4337

4411 Deep sea foreign transportation 2040

442 **Deep sea domestic transportation**
2040

4431 Great lakes transportation 2040

45 **TRANSPORTATION BY AIR**
34, 952, 1532, 4168, 1630, 3660, 4310, 4336, 4337

451 **Certificated air transportation**
3095

4511 Certificated air transportation 32

452 **Noncertificated air transportation**
3095

458 **Air transportation services**
33

46 **PIPE LINES, EXCEPT NATURAL GAS**

461 **Pipe lines, except natural gas**
4167

47 **TRANSPORTATION SERVICES**
486

4783 Packing and crating...2706

48 **COMMUNICATION**
26, 302, 461, 462, 507, 508, 510, 530, 545, 1026, 1074, 2005, 2966, 3145, 3374, 3955, 3966, 3967, 3973, 4004, 4015, 4022, 4061, 4062, 4097, 4132, 4174, 4281, 4288, 4290, 4296, 4334, 278, 323, 363, 529, 613, 849, 851, 857, 1370, 1442, 1712, 1864, 2146, 2798, 3968, 3988, 3996, 3998, 4001, 4010, 4014, 4023, 4129, 4130, 4137, 4150, 4151, 4171, 4195, 4212, 4218, 4254, 4264, 4265, 4266, 4267, 4268, 4276, 4278, 4282, 4284, 4291, 4303, 4307, 4311, 4312, 4315, 4328, 4331, 4337

481 **Telephone communication**
189, 4088, 4123, 4166, 2081, 4236

4811 Telephone communication30, 59, 66, 92, 97, 112, 176, 177, 178, 181, 182, 210, 217, 219, 261, 266, 296, 334, 351, 353, 416, 424, 463, 599, 768, 784, 834, 839, 846, 854, 920, 1001, 1250, 1289, 1330, 1375, 1414, 1454, 1588, 1600, 1817, 1831, 1836, 1850, 1887, 1897, 1907, 1939, 2032, 2068, 2145, 2800, 3160, 3209, 3209, 3260, 3956, 3958, 3959, 3961, 3964, 3965, 3970, 3971, 3972, 3974, 3975, 3977, 3978, 3982, 3984, 3985, 3986, 3989, 3992, 3993, 3994, 3995, 3999, 4003, 4008, 4011, 4012, 4016, 4017, 4018, 4019, 4020, 4021, 4024, 4025, 4026, 4027, 4028, 4029, 4030, 4031, 4032, 4035, 4036, 4037, 4038, 4040, 4041, 4042, 4045, 4047, 4048, 4049, 4052, 4055, 4056, 4058, 4060, 4063, 4064, 4065, 4066, 4068, 4070, 4073, 4074, 4075, 4078, 4079, 4080, 4081, 4082, 4083, 4084, 4085, 4086, 4087, 4089, 4090, 4091, 4092, 4093, 4094, 4095, 4096, 4098, 4101, 4102, 4103, 4104, 4105, 4106, 4107, 4108, 4109, 4110, 4111, 4112, 4113, 4114, 4115, 4116, 4117, 4118, 4119, 4120, 4121, 4122, 4124, 4125, 4126, 4127, 4131, 4133, 4134, 4135, 4136, 4138, 4139, 4140, 4141, 4142, 4143, 4145, 4146, 4147, 4148, 4149, 4152, 4153, 4154, 4155, 4156, 4157, 4158, 4159, 4160, 4162, 4163, 4164, 4165, 4167, 4168, 4169, 4170, 4177, 4180, 4181, 4186, 4187, 4188, 4189, 4190, 4192, 4194,

4196, 4199, 4200, 4203, 4205, 4208, 4213, 4219,
4224, 4226, 4227, 4233, 4234, 4235, 4237, 4238,
4239, 4240, 4241, 4245, 4246, 4247, 4248, 4249,
4251, 4253, 4255, 4256, 4258, 4259, 4260, 4261,
4262, 4263, 4269, 4270, 4271, 4273, 4274, 4275,
4279, 4280, 4283, 4286, 4287, 4292, 4293, 4295,
4297, 4298, 4299, 4300, 4301, 4302, 4304, 4305,
4306, 4308, 4309, 4310, 4313, 4316, 4318, 4319,
4320, 4323, 4324, 4325, 4329, 4330, 4335, 4336,
4339

482 Telegraph communication
4236

4821 Telegraph communication... 30, 59, 227, 328, 599,
1375, 1600, 2068, 3160, 3260, 3959, 3970, 3972,
3981, 3983, 3994, 3995, 4016, 4019, 4025, 4029,
4048, 4114, 4118, 4127, 4143, 4148, 4149, 4152,
4160, 4163, 4169, 4198, 4199, 4200, 4203, 4224,
4238, 4275, 4279, 4283, 4295, 4297, 4298, 4299,
4306, 4308, 4310, 4313, 4316, 4320, 4321, 4324,
4329, 4336

483 Radio and television broadcasting
247, 294, 300, 303, 483, 3956, 4105, 4184, 4186,
4219, 4289, 50, 246, 283, 284, 286, 292, 295,
314, 397, 726, 839, 2081, 4002

4832 Radio broadcasting.......................... 228, 293, 4144

4833 Television broadcasting . 2, 8, 12, 23, 27, 35, 107,
145, 198, 250, 263, 277, 293, 296, 298, 299, 301,
306, 307, 309, 310, 311, 313, 320, 369, 390, 391,
403, 404, 465, 472, 487, 532, 533, 1211, 1249,
1818, 3160, 3954, 3979, 4012, 4019, 4026, 4028,
4036, 4037, 4043, 4051, 4056, 4057, 4066, 4067,
4069, 4124, 4126, 4128, 4160, 4161, 4172, 4182,
4191, 4205, 4208, 4209, 4210, 4211, 4213, 4215,
4216, 4217, 4220, 4222, 4223, 4244, 4272, 4280,
4286, 4317, 4319, 4322, 4323, 4326

489 Communication services, nec
295, 2081, 4236

4899 Communication services, nec .. 2, 7, 8, 12, 13, 22,
23, 27, 30, 35, 46, 50, 85, 88, 97, 107, 109, 136,
145, 148, 155, 165, 171, 181, 190, 191, 194, 198,
227, 246, 250, 263, 293, 294, 296, 299, 300, 301,
306, 307, 308, 309, 310, 312, 314, 320, 334, 351,
353, 369, 375, 388, 391, 403, 404, 434, 436, 472,
533, 546, 599, 705, 726, 768, 770, 784, 841, 846,
854, 920, 976, 1022, 1029, 1073, 1117, 1138,
1143, 1249, 1295, 1317, 1330, 1331, 1375, 1414,
1454, 1588, 1600, 1817, 1831, 1836, 1850, 1857,
1859, 1883, 1887, 1897, 1907, 1939, 2011, 2032,
2068, 2145, 2235, 2898, 3160, 3185, 3209, 3253,
3254, 3260, 3380, 3954, 3956, 3957, 3958, 3959,
3960, 3961, 3962, 3963, 3964, 3966, 3969, 3974,
3977, 3978, 3979, 3987, 3990, 3992, 3995, 3997,
3999, 4002, 4003, 4005, 4007, 4008, 4009, 4011,
4012, 4013, 4019, 4020, 4021, 4025, 4026, 4027,
4028, 4029, 4030, 4033, 4034, 4035, 4036, 4037,
4038, 4040, 4043, 4044, 4049, 4050, 4051, 4052,
4056, 4057, 4058, 4059, 4066, 4067, 4069, 4071,
4072, 4073, 4099, 4100, 4101, 4105, 4112, 4114,
4118, 4124, 4126, 4127, 4128, 4144, 4148, 4149,
4160, 4161, 4163, 4168, 4169, 4170, 4172, 4175,
4178, 4180, 4182, 4184, 4186, 4187, 4188, 4191,
4192, 4197, 4198, 4200, 4201, 4202, 4203, 4205,
4206, 4207, 4208, 4209, 4210, 4211, 4213, 4214,
4215, 4216, 4217, 4219, 4220, 4221, 4222, 4223,
4224, 4228, 4229, 4230, 4231, 4232, 4237, 4243,
4244, 4247, 4250, 4252, 4253, 4269, 4272, 4273,
4277, 4279, 4280, 4283, 4285, 4286, 4287, 4289,
4292, 4293, 4294, 4295, 4297, 4299, 4305, 4306,

4308, 4313, 4316, 4317, 4318, 4319, 4320, 4322,
4323, 4324, 4325, 4326, 4327, 4329, 4332, 4333

49 ELECTRIC, GAS, AND SANITARY SERVICES
4138, 3055, 3746

491 Electric services
4166, 2137, 3017

4911 Electric services 2345, 3473, 3508, 3975, 4078,
4079, 4080, 4081, 4082, 4083, 4084, 4085, 4086,
4087, 4088, 4089, 4090, 4091, 4092, 4093, 4094,
4095, 4096, 4162, 4233

492 Gas production and distribution
3975, 4078, 4079, 4081, 4082, 4083, 4084, 4085,
4086, 4087, 4088, 4089, 4092, 4093, 4094, 4095,
4096, 4162, 4166, 4080, 4090, 4091

4922 Natural gas transmission 4167

493 Combination utility services
4081, 4083, 4084, 4088, 4089, 4095, 4096, 4166,
4091

4931 Electric and other services combined............ 4233

4932 Gas and other services combined 4233

4941 Water supply.. 843, 3451, 3893, 3975, 4162, 4167

495 Sanitary services
4162, 3893

4953 Refuse systems......................................3548, 3623

4971 Irrigation systems... 843

WHOLESALE TRADE

50 WHOLESALE TRADE-DURABLE GOODS
1189

5013 Automotive parts and supplies........................2536

5021 Furniture 3276, 3325, 3338, 3414

5043 Photographic equipment and supplies............. 277

506 Electrical goods
407, 2143, 2624, 2632, 2653, 2747, 2871, 2943,
2983, 2985, 2987, 3000, 3001, 3085, 3089

5063 Electrical apparatus and equipment....71, 72, 100,
101, 102, 267, 277, 281, 304, 461, 462, 478, 479,
1075, 2024, 2162, 2438, 2990, 3083, 3088, 3666,
3702, 4216

5064 Electrical appliances, tv and radios................2555

5065 Electronic parts and equipment.... 254, 304, 1575,
2153, 2162, 2210, 2222, 2280, 2282, 2293, 2424,
2483, 2541, 2546, 2807, 2843, 2844, 2845, 2846,
2971, 2972, 2989, 2990, 3093

5081 Commercial machines and equipment.... 623, 654,
792, 840, 975, 978, 1021, 1030, 1054, 1072,
1075, 1088, 1106, 1134, 1135, 1199, 1431, 1456,
1480, 1538, 1540, 1547, 1551, 1590, 1731, 1795,
1885, 3146, 3161, 3170, 3195, 3222, 3245, 3276,
3277, 3296, 3309, 3325, 3335, 3336, 3337, 3338,
3411, 3414, 3415, 3425, 4247

5083 Farm machinery and equipment 2555

5084 Industrial machinery and equipment 3531, 3534

5086 Professional equipment and supplies .3345, 3491,
3503, 3542, 3554, 3558, 3580, 3638, 3643, 3646,
3650, 3655, 3656, 3662, 3791, 3888, 3938, 3947

5094 Jewelry, watches, & precious stones...3453, 3584

51 WHOLESALE TRADE-NONDURABLE GOODS

511 Paper and paper products
3222

5112 Stationery supplies......................3277, 3338, 3411

5122 Drugs, proprietaries, and sundries..................1631

668, 671, 672, 673, 676, 678, 683, 686, 690, 698, 701, 702, 704, 706, 707, 709, 710, 711, 712, 714, 720, 724, 725, 726, 728, 734, 737, 738, 744, 746, 748, 749, 750, 751, 753, 767, 819, 820, 830, 837, 839, 843, 845, 848, 850, 858, 859, 868, 871, 872, 873, 885, 905, 987, 1055, 1084, 1111, 1136, 1154, 1192, 1220, 1251, 1284, 1325, 1326, 1331, 1335, 1402, 1404, 1414, 1434, 1437, 1473, 1474, 1477, 1480, 1486, 1522, 1528, 1536, 1576, 1581, 1591, 1600, 1648, 1659, 1677, 1691, 1695, 1700, 1701, 1709, 1719, 1736, 1751, 1765, 1817, 1821, 1841, 1860, 1870, 1888, 1889, 1890, 1896, 1897, 1905, 1963, 1976, 1999, 2013, 2023, 2033, 2035, 2050, 2065, 2068, 2104, 2171, 2235, 2440, 2802, 3252, 3324, 3414, 4010, 4023, 4046, 4054, 4203, 4245, 4259, 4262

7372 Computer programming and software....... 10, 227, 406, 548, 550, 551, 553, 555, 570, 574, 579, 580, 584, 586, 592, 593, 594, 595, 601, 602, 604, 613, 614, 616, 617, 619, 620, 623, 624, 628, 629, 630, 632, 633, 636, 638, 639, 643, 646, 647, 648, 649, 650, 653, 658, 660, 670, 674, 675, 677, 679, 680, 681, 682, 684, 685, 688, 691, 692, 695, 699, 703, 705, 708, 713, 715, 716, 717, 718, 721, 723, 727, 730, 732, 733, 735, 736, 739, 743, 747, 754, 758, 792, 802, 815, 900, 902, 923, 924, 928, 931, 934, 936, 937, 949, 950, 966, 967, 972, 986, 988, 989, 991, 1004, 1011, 1019, 1026, 1031, 1036, 1037, 1041, 1043, 1048, 1062, 1071, 1081, 1083, 1085, 1086, 1094, 1095, 1102, 1105, 1118, 1123, 1129, 1134, 1137, 1168, 1170, 1173, 1177, 1179, 1201, 1324, 1344, 1345, 1356, 1396, 1399, 1405, 1415, 1416, 1417, 1418, 1419, 1420, 1421, 1429, 1435, 1442, 1445, 1446, 1447, 1449, 1450, 1459, 1463, 1465, 1469, 1470, 1488, 1491, 1497, 1498, 1499, 1505, 1510, 1511, 1514, 1515, 1517, 1518, 1519, 1520, 1521, 1524, 1526, 1542, 1545, 1556, 1560, 1562, 1563, 1568, 1572, 1575, 1595, 1598, 1601, 1611, 1612, 1614, 1615, 1616, 1617, 1630, 1676, 1705, 1708, 1711, 1718, 1720, 1725, 1726, 1728, 1729, 1732, 1733, 1734, 1741, 1743, 1745, 1746, 1757, 1771, 1772, 1774, 1782, 1799, 1814, 1818, 1829, 1831, 1832, 1838, 1840, 1842, 1843, 1844, 1852, 1855, 1867, 1868, 1869, 1873, 1874, 1878, 1884, 1885, 1906, 1909, 1910, 1925, 1929, 1936, 1942, 1943, 1944, 1961, 1964, 1965, 1966, 1968, 2020, 2069, 2082, 2089, 2168, 2213, 2258, 2465, 2613, 2687, 2716, 2761, 2780, 2781, 2837, 2838, 3246, 3266, 3365, 3373, 4012, 4252, 4317, 4323

7374 Data processing services 121, 548, 577, 588, 590, 591, 595, 600, 607, 612, 614, 615, 629, 630, 633, 636, 651, 657, 663, 665, 667, 669, 696, 697, 719, 745, 755, 766, 776, 800, 827, 867, 874, 880, 901, 933, 946, 947, 948, 976, 991, 1031, 1034, 1050, 1062, 1087, 1151, 1158, 1181, 1189, 1200, 1209, 1211, 1300, 1324, 1344, 1345, 1356, 1362, 1398, 1400, 1418, 1429, 1445, 1446, 1448, 1449, 1450, 1494, 1504, 1505, 1511, 1514, 1515, 1517, 1518, 1519, 1520, 1521, 1524, 1526, 1556, 1563, 1579, 1597, 1599, 1705, 1727, 1728, 1730, 1814, 1829, 1831, 1840, 1852, 1853, 2018, 2041, 2044, 2458, 3231, 3235, 4012, 4060, 4252, 4317, 4323

7379 Computer related services, nec 564, 566, 572, 587, 595, 611, 614, 629, 630, 633, 745, 880, 888, 1078, 1079, 1080, 1087, 1110, 1152, 1158, 1197, 1324, 1328, 1356, 1372, 1431, 1439, 1448, 1459, 1505, 1511, 1514, 1515, 1517, 1519, 1520, 1521, 1524, 1526, 1575, 1579, 1735, 1829, 3184, 3194, 4187

739 **Miscellaneous business services** 1323

7391 Research & development laboratories 3004, 3654, 3657, 3672, 3742, 4191

7392 Management and public relations 611, 901, 1209, 1211, 4191

7393 Detective and protective services 71, 72, 185, 235, 470

7394 Equipment rental and leasing................ 277, 3184

7397 Commercial testing laboratories 2557, 3527, 3672, 3740

7399 Business services, nec.......... 277, 463, 700, 3140, 3309, 4063, 4185, 4193, 4225, 4242

76 **MISCELLANEOUS REPAIR SERVICES**

7622 Radio and television repair..................... 277, 2964

7631 Watch, clock, and jewelry repair................... 3909

7692 Welding repair 2585

78 **MOTION PICTURES**
107

781 **Motion picture production & services**
277, 294, 505

7813 Motion picture production, except tv.....31, 75, 76, 77, 200, 245, 1005, 1574, 4197

7814 Motion picture production for tv...... 246, 304, 393, 483, 487, 393

782 **Motion picture distribution & services**
277

7823 Motion picture film exchanges 4197

7824 Film or tape distribution for tv..294, 304, 483, 487

7829 Motion picture distribution services 4197

79 **AMUSEMENT & RECREATION SERVICES, NEC**
952

7929 Entertainers & entertainment groups 304

799 **Misc. amusement, recreational services**
4310

7993 Coin-operated amusement devices................ 2774

80 **HEALTH SERVICES**
557, 1019, 1847, 658, 709, 1192, 1984

8011 Offices of physicians 1528

806 **Hospitals**
693, 694, 988, 850

8071 Medical laboratories 3530

82 **EDUCATIONAL SERVICES**
1084

822 **Colleges and universities**
1374

8231 Libraries and information centers .. 548, 711, 1848

86 **MEMBERSHIP ORGANIZATIONS**

8611 Business associations 1440, 2514, 2649

8621 Professional organizations.......... 1517, 1518, 2649

89 **MISCELLANEOUS SERVICES**
1323

8911 Engineering & architectural services 481, 600, 628, 667, 697, 700, 701, 1486, 1546, 2387, 2534, 2605, 2719, 2720, 2820, 2994, 3004, 3005, 3062, 3121, 3129, 3345, 3472, 3650, 3672, 3742, 3768, 3947

387

Part 3: Section 2: Subject Index

Part 4: Title Index

A

B

C

D

E

G

I

J

K

L

M

N

O

Q

R

S

T

U

V

W

X

Y

Z

Part 4: Title Index